35th UIT (Italian Union of Thermo-Fluid-Dynamics) Heat Transfer Conference (UIT2017)

Journal of Physics: Conference Series Volume 923

Ancona, Italy
26 - 28 June 2017

Editors:

Renato Ricci
Valerio D'Alessandro

ISBN: 978-1-5108-5271-6
ISSN: 1742-6588

Printed from e-media with permission by:

Curran Associates, Inc.
57 Morehouse Lane
Red Hook, NY 12571

Some format issues inherent in the e-media version may also appear in this print version.

Printed by Curran Associates, Inc. (2017)

For permission requests, please contact the Institute of Physics at the address below.

Institute of Physics
Dirac House, Temple Back
Bristol BS1 6BE UK

Phone: 44 1 17 929 7481
Fax: 44 1 17 920 0979

techtracking@iop.org

Additional copies of this publication are available from:

Curran Associates, Inc.
57 Morehouse Lane
Red Hook, NY 12571 USA
Phone: 845-758-0400
Fax: 845-758-2633
Email: curran@proceedings.com
Web: www.proceedings.com

Table of contents

Volume 923

35th UIT Heat Transfer Conference (UIT2017)26–28 June 2017, Ancona, Italy

Accepted papers received: 1 November 2017

Published online: 20 November 2017

Heat transfer in nuclear plants

Multi-phase fluid dynamics, heat transfer and interface phenomena

Computational fluid dynamics and heat transfer

Heat transfer and efficiency in energy systems, environmental technologies and buildings

Natural, forced and mixed convection

Conduction

Heat transfer in Fire Engineering

Other

IOP Conf. Series: Journal of Physics: Conf. Series **923** (2017) 011001 doi:10.1088/1742-6596/923/1/011001

35th UIT (Italian Union of Thermo-fluid dynamics) Heat Transfer Conference – Ancona (Italy) 26-28 June 2017

Andrea Agostini, **In volo gioioso**

Preface

The 35th UIT (Italian Union of Thermp-Fluid Dynamics) Heat Transfer Conference was organized by Dept. of Industrial Engineering and Mathematical Sciences, Marche Polytechnic University (Itay) and was held at Faculty of Engineering of the same University, Ancona, June 26-28, 2017.
The annual UIT Conference, which has grown over time, comes back in Ancona after 27 years.
The scope of the conference covers a range of major topics in theoretical, numerical and experimental heat transfer and related areas, ranging from energy efficiency to nuclear plants. The Organizing Committee was also honored to continue the initiative about the new section devoted to Technical Seminars introduced 2015 within UIT conferences. In particular a seminar regarding the adoption of Unmanned Aerial Vehicles for Thermo-Fluid Dynamics research was held. Moreover a photo exhibition supported by "Aeronanutica Militare Italiana" was also offered to the participants.
In this conference edition, 73 papers were contributed by about 300 authors on eight different topics: micro and nano scale thermo-fluid dynamics; multi-phase fluid dynamics, heat transfer and interface phenomena; computational fluid dynamics and heat transfer; heat transfer in nuclear plants; heat transfer and efficiency in energy systems, environmental technologies and buildings; natural, forced and mixed convection; conduction; heat transfer in fire engineering.
The conference program scheduled plenary, oral and poster sessions. The two invited plenary Keynote Lectures were given by Prof. Giovanni Latini (Marche Polytechnic University, Italy) and Prof. Peter Stephan (Technische Universitat Darmstadt, Germany). A Technical Seminar was given by Prof. Renato Ricci (Marche Polytechnic University, Italy). There were also 9 oral sessions and a poster sessions.
This special issue collects a paper presented in a keynote lecture plus 56 papers selected from those presented and discussed during the congress.

IOP Conf. Series: Journal of Physics: Conf. Series **923** (2017) 011001 doi:10.1088/1742-6596/923/1/011001

The UIT 2017 conference was a useful occasion to stimulate discussion, further the understanding of heat transfer and related phenomena, present the state-of-the-art of some topics, discuss emerging trends and promote collaborations. We hope this issue will maintain and extend some of these features.
A special thank you is due to the Organizing and Scientific Committees, to the sponsors and to all the participants.

Renato Ricci and Valerio D'Alessandro, editors.

IOP Conf. Series: Journal of Physics: Conf. Series **923** (2017) 011001 doi:10.1088/1742-6596/923/1/011001

35th UIT Conference Ancona 2017 – Scientific Committee

Renato Ricci (chair)	Università Politecnica delle Marche, Italy
Dario Ambrosini	Università degli Studi dell'Aquila, Italy
Giovanni Sebastiano Barozzi	Università degli Studi di Modena e Reggio Emilia, Italy
Gian Paolo Beretta	Università degli Studi di Brescia, Italy
Gian Piero Celata	Centro Ricerche ENEA Casaccia, Roma, Italy
Gianni Cesini	Università Politecnica delle Marche, Italy
Michele Ciofalo	Università degli Studi di Palermo, Italy
Davide Del Col	Università degli Studi di Padova, Italy
Wilson K. S. Chiu	University of Connecticut, USA
Paolo Di Marco	Università di Pisa, Italy
Oronzio Manca	Università degli Studi della Campania Luigi Vanvitelli, Italy
Rita Mastrullo	Università degli Studi di Napoli Federico II, Italy
Gian Luca Morini	Università degli Studi di Bologna, Italy
Vincenzo Naso	Università degli Studi di Napoli Federico II, Italy
Alfonso Niro	Politecnico di Milano, Italy
Enrico Nobile	Università degli Studi di Trieste, Italy
Carlo Nonino	Università degli Studi di Udine, Italy
Stefano Piva	Università degli Studi di Ferrara, Italy
Sara Rainieri	Università degli Studi di Parma, Italy
Luisa Rossetto	Università degli Studi di Padova, Italy
Peter Christian Stephan	Technische Universität Darmstadt, Germany
Giovanni Tanda	Università degli Studi di Genova, Italy
John Richard Thome	École Polytechnique Fédérale de Lausanne (EPFL), Switzerland

35th UIT Conference Ancona 2017 – Organizing Committee

Renato Ricci (Chair)	Università Politecnica delle Marche, Italy
Valerio D'Alessandro (Co-chair)	Università Politecnica delle Marche, Italy
Fabio Polonara	Università Politecnica delle Marche, Italy
Gaetano Borrelli	Università Politecnica delle Marche, Italy
Maurizio Spegni	Università Politecnica delle Marche, Italy
Michela Mengaroni	Università Politecnica delle Marche, Italy
Antonella Rossi	Università Politecnica delle Marche, Italy
Erika Costanzo	Università Politecnica delle Marche, Italy
Francesco Garbuglia	Università Politecnica delle Marche, Italy
Giacomo Clementi	Università Politecnica delle Marche, Italy
Luca Giammichele	Università Politecnica delle Marche, Italy

IOP Conf. Series: Journal of Physics: Conf. Series **923** (2017) 011001 doi:10.1088/1742-6596/923/1/011001

UIT Sterring Committee

Alfonso Niro (President)	Politecnico di Milano, Italy
Michele Ciofalo (Vice-President)	Università degli Studi di Palermo, Italy
Davide Del Col	Università degli Studi di Padova, Italy
Dario Ambrosini	Università degli Studi dell'Aquila, Italy
Pietro Asinari	Politecnico di Torino, Italy
Vincenzo Naso	Università degli Studi di Napoli Federico II, Italy
Sara Rainieri	Università degli Studi di Parma, Italy

UIT Accounting Auditors

Mario De Salve	Politecnico di Torino, Italy
Marco Spiga	Università degli Studi di Parma, Italy
Luca Tagliafico	Università degli Studi di Genova, Italy

UIT Conferences

I	1983, Trieste	XIX	2001, Modena
II	1984, Bologna	XX	2002, Maratea (PZ)
III	1985, Palermo	XXI	2003, Udine
IV	1986, Genova	XXII	2004, Genova
V	1987, Torino	XXIII	2005, Parma
VI	1988, Bari	XXIV	2006, Napoli
VII	1989, Firenze	XXV	2007, Trieste
VIII	1990, Ancona	XXVI	2008, Palermo
IX	1991, Pisa	XXVII	2009, Reggio Emilia
X	1992, Genova	XXVIII	2010, Brescia
XI	1993, Milano	XXIX	2011, Torino
XII	1994, L'Aquila	XXX	2012, Bologna
XIII	1995, Bologna	XXXI	2013, Moltrasio (CO)
XIV	1996, Roma	XXXII	2014, Pisa
XV	1997, Torino	XXXIII	2015, L'Aquila
XVI	1998, Siena	XXXIV	2016, Ferrara
XVII	1999, Ferrara	**XXXV**	**2017, Ancona**
XVIII	2000, Cernobbio (CO)		

Peer review statement

All papers published in this volume of *Journal of Physics: Conference Series* have been peer reviewed through processes administered by the proceedings Editors. Reviews were conducted by expert referees to the professional and scientific standards expected of a proceedings journal published by IOP Publishing.

IOP Conf. Series: Journal of Physics: Conf. Series **923** (2017) 012001 doi:10.1088/1742-6596/923/1/012001

Thermophysical properties of fluids: dynamic viscosity and thermal conductivity

G Latini
Professor Emeritus, Dipartimento di Ingegneria Industriale e Scienze Matematiche
Università Politecnica delle Marche, Via Brecce Bianche n.10, 60131 Ancona, Italy

E-mail: giovanni.latini@univpm.it

Abstract. Thermophysical properties of fluids strongly depend upon atomic and molecular structure, complex systems governed by physics laws providing the time evolution. Theoretically the knowledge of the initial position and velocity of each atom, of the interaction forces and of the boundary conditions, leads to the solution; actually this approach contains too many variables and it is generally impossible to obtain an acceptable solution. In many cases it is only possible to calculate or to measure some macroscopic properties of fluids (pressure, temperature, molar volume, heat capacities…). The ideal gas "law," PV = nRT, was one of the first important correlations of properties and the deviations from this law for real gases were usefully proposed. Moreover the statistical mechanics leads for example to the "hard-sphere" model providing the link between the transport properties and the molecular size and speed of the molecules. Further approximations take into account the intermolecular interactions (the potential functions) which can be used to describe attractions and repulsions. In any case thermodynamics reduces experimental or theoretical efforts by relating one physical property to another: the Clausius-Clapeyron equation provides a classical example of this method and the PVT function must be known accurately. However, in spite of the useful developments in molecular theory and computers technology, often it is usual to search for physical properties when the existing theories are not reliable and experimental data are not available: the required value of the physical or thermophysical property must be estimated or predicted (very often estimation and prediction are improperly used as synonymous). In some cases empirical correlations are useful, if it is clearly defined the range of conditions on which they are based. This work is concerned with dynamic viscosity μ and thermal conductivity λ and is based on clear and important rules to be respected when a prediction or estimation method is proposed.

1. Introduction

An acceptable estimation method of a given thermophysical property must be characterized as follows:

- the values of the thermophysical property must be reliable for pure substances and for mixtures at any given temperature, pressure and composition;
- the phase (solid, liquid, or gas) must be clearly indicated;
- the number of input data has to be minimized;
- the ranges of the probable errors must be clearly indicated;
- the computation time must be minimized.

Obviously very few of the methods satisfy these conditions, and among the different molecular theories useful for data correlation, a theory is particularly helpful: *the law of corresponding states* or *the corresponding-states principle*, originally presented by macroscopic arguments, but successively grounded on a molecular basis. The law of corresponding states (van der Waals [1]) is based on the hypothesis that equilibrium properties are linked in a universal way with the critical properties, and allows the development of correlations and estimation methods: two-parameter corresponding states

IOP Conf. Series: Journal of Physics: Conf. Series **923** (2017) 012001 doi:10.1088/1742-6596/923/1/012001

theory asserts that if pressure, volume, and temperature are divided by the corresponding critical properties, the function relating reduced pressure to reduced volume and reduced temperature is the same for all substances. Of course small molecules with spherical symmetry are satisfactorily covered by a two-constant law of corresponding states, but complex, non-spherical and weakly polar molecules do not follow the law as well, and a third parameter, for example the acentric factor ω has to be added. Summarizing, good correlations exist for nonpolar fluids, but analogous correlations are often not available for polar fluids or show very low reliability.

Another useful method has to be presented: a macroscopic property clearly depends upon the molecular structure and can be estimated by the *group contributions method* (generally by an algebraic operation that sums the contributions from the molecule's parts).

In any case reliable experimental data are preferred with respect to results obtained by estimation methods and several data banks are available containing experimental and calculated data. Some estimation methods are based on theory with empirical corrections, others are based on empirical generalizations and totally empirical correlations are also useful when theory or general principles are not acceptable.

Several estimation methods require parameters that are characteristic of the single pure compound or of the constituents of a mixture and classical examples are critical temperature T_c, normal freezing point T_f and the normal boiling point T_b. Pure component properties include for example the vapor-liquid critical properties, atmospheric boiling and freezing temperatures, dipole moments, liquid molar volume, heat capacities, acentric factor and so on. These properties are calculated by several methods and often can be compared with the corresponding experimental values. The properties of a molecule are usually calculated by the contributions of its elements (atom contributions, bond contributions, group interaction contributions…). Some examples are presented in this work.

1.1. Critical Prtoperties: Method of Joback [2,3]

The relations for the critical properties are as follows:

$$T_c(K) = T_b [\![0.584 + 0.965\{\textstyle\sum_k N_k(tck)\} - \{\sum_k N_k(tck)\}^2]\!]^{-1} \quad (1)$$

$$P_c(bar) = [\![0.113 + 0.0032 N_{atoms} - \textstyle\sum_k N_k(pck)]\!]^{-2} \quad (2)$$

$$V_c(cm^3 mol^{-1}) = 17.5 + \textstyle\sum_k N_k(vck) \quad (3)$$

where tck, pck and vck are referred to the contributions. The group identities and Joback's values for contributions to the critical properties are tabulated in [4]. T_c needs the value of the normal boiling point T_b. The Joback method for critical properties is reliable for Tc (errors usually less than 5%), but there is a significant increase in error when estimated (and not experimental) values of T_b are used. For the calculated values of Pc the reliability is less. For Vc, the average error is generally greather than 5%.

1.2. Critical Properties: Method of Constantinou and Gani [5]

The method of Constantinou and Gani for critical properties is based on an advanced group contribution method ("second order"level):

$$T_c(K) = 181.128 \ln [\![\textstyle\sum_k N_k(tc1k) + W \sum_j M_j(tc2j)]\!] \quad (4)$$

$$P_c(bar) = [\![\textstyle\sum_k N_k(pc1k) + W \sum_j M_j(pc2j) + 0.10022]\!]^{-2} + 1.3705 \quad (5)$$

$$V_c(cm^3 mol^{-1}) = -0.00435 + [\![\textstyle\sum_k N_k(vc1k) + W \sum_j M_j(vc2j)]\!] \quad (6)$$

In this case Tc does not depend on the value ofr Tb. Nk is the number of first-order groups of type k in the molecule; the labels 1 and 2 are referred to the first-order and second-order group; Mj is the number of second-order groups of type j in the molecule; the value of W is 0.0 and 1.0 respectively for first-order and second-order calculations.

The first-order contributions are are usually sufficient. The second- order contributions are useful, but in some cases they are very small and can worsen the results. Several other estimation methods for critical properties exist in the literature and can be divided into two classes: the first class is based on so-called factor analysis (correlations built from data of measurable, macroscopic properties such as densities, molecular weight, boiling temperature, etc), the second class is based on so-called molecular descriptors which are not normally measurable.

1.3. Normal boiling point T_b and normal freezing point T_f

Some estimation methods need T_b as input information for the calculation of T_c and obviously the errors can increase (greater than 10%).

1.3.1.-Method of Joback [3, 4]

Joback, using a group contribution scheme with several new functional groups, proposes for T_f and T_b the following relations:

$$T_f = 122 + \sum_k N_k(tfk) \tag{7}$$

$$T_b = 198 + \sum_k N_k(tbk) \tag{8}$$

The labels **tfk** and **tbk** are for the contributions to T_f and T_b respectively. The errors are usually greater than 10%.

2. Viscosity of gases and liquids

Dynamic viscosity μ is a measure of the internal fluid friction; molecular theories for the gases were satisfactorily developed by non equlibrium statistical mechanics and kinetic theory, but the corresponding theories for the liquids are far away from the desired accuracy in the results. In this paper the discussion is concerned with newtonian fluids, that is with fluids in which the viscosity does not depend upon either the magnitude of the shearing stress or velocity gradient.

Estimation methods are presented able to:
- correlate viscosities with temperature and pressure;
- estimate viscosities when no experimental data are available;
- estimate the viscosities of mixtures.

2.1. Low-pressure gas viscosity, simple theoretical approach and estimation methods

The different gas viscosity estimation methods are based on the Chapman-Enskog theory [6] or on the law of corresponding states. In a very simple way a gas can be modeled to obtain general relations among viscosity, temperature, pressure, and molecular size. The Chapman-Enskog model shows:

$$\mu = \frac{(26.29)(MT)^{1/2}}{\sigma^2 \Omega_v} \tag{9}$$

where μ is in μP, M is the molecular weight in g/mol, T the temperature in K, σ the so-called hard-sphere diameter in Å and Ω_v a particular collision integral, different for each given intermolecular potential model.

2.1.1-Method of Chung et al [7,8]

Chung et al start from eq (9) and obtain:

$$\mu = 40.785 \cdot \frac{F_C (MT)^{1/2}}{V_C^{2/3} \Omega_V} \tag{10}$$

where μ is in μP, M is the molecular weight in g/mol, T is the temperature in K, Vc is the critical volume in cm^3/mol, Ω_V the collision integral and Fc a factor necessary to take into account the molecular shapes and polarities of dilute gases. The average absolute error appears to be about 1.9%.

2.1.2. Corresponding states methods
Starting from eq(9) Lucas [9] obtains:

$$\mu\xi = \begin{bmatrix} 0.807T_r^{0.618} - 0.357\exp(-0.449T_r) + \\ +0.340\exp(-0.458T_r) + 0.018 \end{bmatrix} F_P^0 F_Q^0 \tag{11}$$

where μ is in μP, T_r is the reduced temperature T/T_c, $\xi = 0.176\left(\frac{T_c}{M^3 P_c^4}\right)^{1/6}$, M is the molecular weigth in kg/kmol, P_c is the critical pressure in N/m^2. F_P^0 and F_Q^0 are correction factors introduced to take into account polarity and quantum effects.

Reichenberg [10,11] has suggested an other corresponding states relation for low-pressure gas viscosity of organic compounds which appears as follows:

$$\mu = \frac{M^{1/2} T}{a^* \left[1 + (4/T_c)\right] \left[1 + 0.36 T_r (T_r - 1)\right]^{1/6}} \frac{T_r(1 + 270\mu_r^4)}{T_r + 270\mu_r^4} \tag{12}$$

μ is in μP, T is the temperature in K, T_r is the reduced temperature T/T_c, , M is the molecular weigth in kg/kmol, μ_r is a particular reduced dipole moment and a* depends upon the groups of the different compounds. The errors are usually in the range 0.5% to 1.5% for nonpolar compounds and 2% to 4% for polar compounds.

2.2. The pressure dependence of the viscosity of pure gases
Usually, at reduced temperatures $T_r > 1.5$ and reduced pressures $P_r < 2$, the viscosity shows a small dependence on the pressure; on the contrary if $1 < T_r < 1.5$ and $P > P_c$, pressure has a a very large effect on viscosity. Near the critical point T_c , the viscosity diverges and its value is larger than expected, even if this effect (the so-called critical enhancement) is smaller than the analogous enhancement for thermal conductivity. The thermal conductivity usually increases by a factor of two near the critical point, while the increase in viscosity is about 1%.

At high reduced pressures the viscosity behavior appears to simulate a liquid state, and an increase in temperature results in a decrease in viscosity. Finally, at very high-reduced temperatures, again pressure has little effect and viscosities increase with temperature. This behavior strongly suggests that density is a variable to take into account for a correct descrition of viscosity at high pressures.

Very few theoretical studies were developed to estimate the dependence of the viscosity on the pressure; Chapman and Cowling [6] , following the theory due to Enskog, faced the problem for dense gases assuming that the gas consists of dense hard spheres and behaves as a low-density hard-sphere system when the events occur at a faster rate owing to the higher rates of collision.

2.3. Viscosities of gas mixtures at low pressures
The kinetic theory developed by Chapman and Enskog [6] can be extensively used to calculate the viscosity of low-pressure multicomponent mixtures The formulas are very complicated and are rarely used without strong simplifications. Some methods are presented in [4] and usually they are interpolative (the viscosities of the pure compounds must be available).

Lucas [9] and Chung et al. [7,8] proposed estimation methods over the entire range of composition. The errors by these two methods are higher than those of the interpolative techniques, but usually less than ±5%. Such errors could be reduced if pure component viscosity data are available and employed by a simple linear correction scheme.

2.4. Viscosities of gas mixtures at high pressures
The methods due to Lucas [9] and Chung et al [7,8] show smaller accuracies when applied to gas mixtures at high pressures. The accuracies are very poor in the critical region or at densities close to those of a liquid at the same temperature. The lack of accurate high-pressure gas mixtures experimental viscosity data does not allow accurate tests, but reasonable absolute average deviations of 8% to 9% for both polar and nonpolar dense gas mixtures can be supposed. In any case it is possible, when using the above cited methods, to program the same relations for low-pressure pure gases, low-pressure gas mixtures, and high-pressure pure gases. One program can cover all those cases and high-pressure gas mixtures.

2.5. Liquid viscosity
In many cases it is necessary to determine the liquid dynamic viscosity from experimental data when such data exist. The viscosities of liquids are much larger than those of gases at the same temperature. For example, the viscosity of liquid benzene near the normal boiling point (353.4 K) is more than 30 times the vapor viscosity, and at lower temperatures, this ratio increases. Moreover, below T_c, low-pressure gas viscosities vary in a nearly linear way with the temperature; on the contrary low-pressure liquid viscosities show a strong curvature which can be eliminated if the logarithm of the viscosity is plotted as a function of the reciprocal absolute temperature. Usually the normal boiling point for organic compounds would be at a value of $T_r^{-1} \approx 1.5$. For temperatures below the normal boiling point ($T_r^{-1} > 1.5$), the logarithm of the viscosity varies linearly with T_r^{-1}, but above the normal boiling point, this is not verified. In the nonlinear region, several corresponding states estimation methods have been suggested. In the linear region most of the corresponding states methods are not reliable, and generally the estimation methods are used to emphasize the effects of the chemical structure. At comparable reduced temperatures, viscosities of polar fluids are higher than those of nonpolar liquids. For engineering purposes it is preferable to use the ratio of the dynamic viscosity to the density, that is the kinematic viscosity, normally expressed in m^2/s or in stokes. The kinematic viscosity $\mathbf{v}$ decreases with increasing temperature so that that $\ln\nu$ is nearly linear in temperature for both the saturated liquid and vapor.

Summarizing, pure liquid viscosities at high reduced temperatures are normally correlated with some extension of the law of corresponding states. At lower temperatures, several methods are empirical in nature and based on a group contribution approach. Liquid mixture correlations are essentially mixing rules relating pure component viscosities to composition.

2.5.1. Liquid viscosity at high pressure
Lucas [12] has suggested the following equation:

$$\frac{\mu}{\mu_{sL}} = \frac{1+D\left(\frac{\Delta P_r}{2.118}\right)^A}{1+C\omega\Delta P_r} \quad (13)$$

μ = viscosity of the liquid at pressure P

μ_{sL} = viscosity of the saturated liquid at P_{vp}

$\Delta P_r = (P - P_{vp})/P_c$

ω = acentric factor

$A = 0.9991-[4.674\cdot10^{-4}/(1.0523T^{-0.03877} - 1.0513]$

$D = [0.3257/(1.0039 - T_r^{2.573})^{0.2906}] - 0.2086$

$C = -0.07921+ 2.1616T_r - 13.4040T_r^2 + 44.1706T_r^3 - 84.8291T_r^4 +$
$+96.1209T_r^5 - 59.8127T_r^6 + 15.6719T_r^7$

Lucas tested 55 polar and non polar liquids and claimed errors less than 10%.

2.5.2. Liquid viscosity as function of temperature
The viscosities of liquids decrease with the increasing temperature in isobaric conditions and along the saturation line. From the freezing point to close the normal boiling point, it is a good approximation to assume a linear relation between ln μ and the reciprocal of the absolute temperature; i.e.:

$$\ln\mu = A + \frac{B}{T} \tag{14}$$

This simple equation is usually referred to as the Andrade equation [13,14]. Several modifications of Eq. (14) have been proposed to improve its accuracy by including some function of the liquid molar volume or introducing the use of a third constant. In conclusion, from the freezing point to close the normal boiling point, Eq. (14) is a reliable function correlating the liquid viscosity with the temperature even if two datum points are required. At low temperature, in order to estimate the liquid viscosity, structural-sensitive parameters are often employed which are valid only for certain homologous series or are found from group contributions. These methods often use some extension of Eq. (14) and are acceptable for reduced temperatures less than about 0.75. In literature are also described techniques emploing corresponding states methods. These methods are not particularly reliable.

2.5.3. Liquid viscosity estimation at high temperatures
Low-temperature viscosity correlations usually assume that lnμ is a linear function of reciprocal absolute temperature. Because above T_r near to 0.7 this relation is not accurate, from about Tr $\approx$ 0.7 to near the critical point, some estimation methods based upon corresponding states principle are proposed which are similar to those used for gases. In particular Sastri [4] proposed:

$$\ln\mu = \left[\!\left[\frac{\ln\mu_b}{\ln(\alpha\mu_b)}\right]\!\right]^{\phi} \ln(\alpha\mu_b) \tag{15}$$

where μ is in nPa·s, μ_b is the viscosity at T_b, the value of α is 0.1175 for alcohols and 0.248 for other compounds and $\phi = (1 - T_r)/(1 - T_{br})$.

Sastri claimed average deviations of 10% for $T_r > 0.9$ and 6% for $T_{br} < T_r < 0.9$. The problem of joining high-temperature and low-temperature estimation methods for liquid viscosities should be faced.

2.5.4. Viscosity of liquid mixtures
All the correlations for liquid mixture viscosity are generally accurate at reduced temperatures (of the pure components) less than 0.7, when pure liquid and liquid mixture viscosities strongly depend upon the structure of the constituent molecules. The trend of the curve of viscosity as a function of composition can be nearly linear for so-called ideal mixtures, but mixtures containg alcohols and/or

water sometimes show a maximum and/or a minimum. Substantially the methods to estimate liquid mixture viscosities need the knowledge of the values of the viscosities of the pure components and so the methods are interpolative. A simple example of a correlation for the low-temperature liquid mixture viscosity is given by the correlation of Grunberg and Nissan [15]:

$$\ln\mu_m = \sum_i x_i \ln\mu_i + \frac{1}{2}\sum_{i=1}^{n}\sum_{j=1}^{n} x_i x_j G_{ij} \qquad (16)$$

where x represents le liquid mole fractions and Gij is an interaction parameter depending on the components and on the temperature.

3. Thermal conductivity of gases and liquids

Some estimation techniques for thermal conductivities **λ** of gases and liquids are covered in this section with the effects of temperature and pressure. Estimation techniques for gas and liquid mixtures are also discussed. The units used for thermal conductivity are W/(m K).

3.1. Theory of thermal conductivity of monoatomic and poliatomic gases

A detailed treatment for monatomic gases, without rotational or vibrational degrees of freedom, yields:

$$\lambda = 2.63\cdot 10^{-23}\frac{\left(\frac{T}{M}\right)^{1/2}}{\sigma^2\Omega_v} \qquad (17)$$

where:
T : the temperature in K

k : the Boltzmann constant = $1.3805\cdot 10^{-23}$ J/K

M : the molecular weight in kg/mol

σ : the characteristic dimension of the molecule in m

Ω_v : the collision integral, dimensionless

Eucken [4] proposed some modifications in Eq. (17) to take into account the polyatomic gases by introducing translational and internal degrees of freedom. Generally the methods useful for estimating the pure gas thermal conductivity at atmospheric pressure use a particular so-called Eucken [4] factor **(λM)/(μCv)** which usually depends upon T_r, and ω; in any case independent values of the gas viscosity and heat capacity are necessary. Some methods require only critical temperature and pressure and are based on the group contribution method in order to take into account the internal degrees of freedom. Summarizing, the Eucken factor should decrease with temperature, but, from experimental measures, in several cases the factor appears to increase. Generally all the proposed methods show errors near 10%.

3.2. The temperature dependence of thermal conductivities of gases at low-pressure

Thermal conductivities of low-pressure gases increase with temperature at low-pressure. The exact function λ(T) is difficult to propose in analitic form: some other temperature-dependent parameters are necessary in the equations. The range of dλ/dT is from 4×10^{-5} to 1.5×10^{-4} W/(m K^2): the complex and polar molecules show the larger values. Some power laws relating λ with T have been proposed without acceptable results.

3.3. The thermal conductivities of pure gases as function of pressure
The thermal conductivities of all gases increase with pressure, and three regions can be usefully described [4]:

3.3.1. Very Low Pressure
Below about 103 bar (the mean free path of the molecules is comparatively larger with respect to the dimensions of a measuring cell), the gas thermal conductivity can be treated as proportional to the pressure. The term zero-pressure is usually used.

3.3.2. Low Pressure
From about 103 to 10 bar the gas thermal conductivity increases about 1% or less per bar and the 1-bar value or the "zero-pressure"extrapolated value can be used.

3.3.3.High Pressure
The gases thermal conductivity increases with pressure, and close the critical point the effect of pressure is very high. When the temperature increases at low pressures the gas thermal conductivity becomes larger, but at high pressure the opposite effect is shown. Similar behaviour is evident for the region below T_c, where λ for liquids decreases with temperature, while for gases there is an increase of λ with T. Substantially pressure effects are small below T_c, exception doing for the behavior of λ near the critical point; in this region λ can increase by a factor of six (the so-called enhancement in the critical region). This enhancement is bigger for the thermal conductivity with respect to the analogous one evidenced for viscosity.

3.4. Thermal conductivities of gas mixtures
When between the molecules of the mixture there is a great difference in polarity, usually the mixture thermal conductivity is larger with respect to the thermal conductivity predicted from a mole fraction average; for nonpolar molecules, the opposite trend is evident and is very high if the difference in molecular weights or sizes of the constituents is high. It is very difficult to modify monatomic mixture correlations to cover polyatomic molecules. The proposed techniques proposed are essentially empirical and the corresponding states methods for low-pressure were proposed with poor results. Very few experimental data [4] for gas mixtures at high pressures are presented in literature and generally the researches are concerned with simple gases and light hydrocarbons: nitrogen-carbon dioxide, ethylene mixtures with nitrogen and carbon dioxide, rare gases and binaries containing carbon dioxide, nitrogen, and ethane, methane-carbon dioxide and methane-carbon tetra-fluoride binaries. The proposed estimation methods are substantially modifications of procedures developed for low- and high-pressure pure gas thermal conductivities: they are tested by the available experimental data and average errors of about 7% are found.

4. Thermal conductivities of liquids
The thermal conductivities of organic compounds in the liquid phase are usually about 10 to 100 times larger than those of the low-pressure gases at the same temperature. The dependence on pressure is generally small (at pressures less than about 50 bar) and the increase of the temperature produces a decrease of the thermal conductivity. The values of λ for the common organic liquids vary in the range 0.100 to 0.200 W/(m· K) at temperatures below the normal boiling point, exception doing for some compounds as for example water, ammonia and other highly polar molecules.

4.1- Thermal conductivityof pure liquids as function of temperature
Almost all estimation methods for pure liquids thermal conductivity are empirical, even if they appear reliable below the normal boiling point. Two methods in particular are selected and tested in [4]:

4.1.1. Latini, et al. Method [4,16]
The method of Latini et al is very simple and the suggested correlation is as follows:

$$\lambda = A\frac{(1-T_r)^{0.38}}{{T_r}^{1/6}} \quad (18)$$

where:

λ = liquid thermal conductivity in W/(m K)

$$A = \frac{A^*T_b^\alpha}{M^\beta T_c^\gamma} \quad (19)$$

T_b = normal boiling temperature (at 1 atm), K

T_c = critical temperature, K

M = molecular weight, g/mol

$T_r = T/T_c$

The parameters A*, α, β and γ are proposed for several organic families (Saturated Hydrocarbons, Olefins, Cycloparaffins, Aromatics, Alcohols, Organic Acids, Ketones, Esters, Ethers, Refrigerants); errors are usually less than 10%.

4.1.2. Sastri Method [4]
The Sastri method is represented by the following correlation:

$$\lambda=\lambda_b a^m \quad (20)$$

where:

$$m = 1 - (\frac{1-T_r}{1-T_{br}})^n \quad (21)$$

λ_b is the thermal conductivity at the normal boiling point and is calculated by the group contribution values and corrections; T_r and T_{rb} are the reduced temperature and the reduced normal boiling point; for alcohols and phenols a=0.856 and n=1.23, for other compounds a=0.16 and n=0.2. Sastri claims an absolute average deviation of 8%.

4.2. Thermal conductivities of pure liquids as function of pressure
At pressures up to 50 to 60 bar, the dependence of the liquid thermal conductivity on the pressure is negligible (exception doing for the range close to the critical point, where the so-called critical enhancement occurs). At lower temperatures λ increases with pressure.

A useful way (Lenoir [17]) of estimating the effect of pressure on λ is given by the relation:

$$\frac{\lambda_2}{\lambda_1} = \frac{L_2}{L_1} \quad (22)$$

IOP Conf. Series: Journal of Physics: Conf. Series **923** (2017) 012001 doi:10.1088/1742-6596/923/1/012001

λ_2 and λ_1 refer to liquid thermal conductivities at T and pressures P_2 and P_1; λ_2 and λ_1 are functions of the reduced temperature and pressure, as shown in particular graphs. A test with experimental data for 12 liquids, both polar and nonpolar, shows errors of only 2% to 4%.

Missenard [18] and Latini and Baroncini [19] proposed other simple correlations extending the calculations to very high pressures.

4.3. Thermal conductivities of liquid mixtures

Several organic liquids mixtures show thermal conductivities generally less than those predicted by a mole or weight fraction average. The methods present in the literature are usually proposed for temperatures near 20°C.

4.3.1. Filippov correlation [20,21]

Filippov proposed the following equation:

$$\lambda_m = w_1\lambda_1 + w_2\lambda_2 - 0.72 w_1 w_2(\lambda_2 - \lambda_1) \tag{23}$$

where w_1 and w_2 are the weight fractions of components 1 and 2; λ_1 and λ_2 are the thermal conductivities of the pure components. Obviously $\lambda_2 > \lambda_1$. The constant 0.72 may be adjustable if binary mixture data are available. The correlation is not suitable for multicomponent mixtures.

4.3.2 Jamieson et al. Correlation [22]

Jamieson et al proposed the follozing equation:

$$\lambda_m = w_1\lambda_1 + w_2\lambda_2 - \alpha(\lambda_2 - \lambda_1)[\![1 - (w_2)^{1/2}]\!]w_2 \tag{24}$$

where w_1 and w_2 are weight fractions and $\lambda_2 > \lambda_1$. The parameter α is equal to unity if mixture data are unavailable. The authors claim an average error within about 7% for all types of binary mixtures. Also in this case the correlation is not suitable for multicomponent mixtures.

4.3.3 Baroncini et al. [23,24,25] correlation

The correlation of Baroncini et al is an extension of eq(18) to binary liquid mixtures and is given by the following equation:

$$\lambda_m = [x_1^2 A_1 + x_2^2 A_2 + 2.2\left(\frac{A_1^3}{A_2}\right)^{\frac{1}{2}} x_1 x_2]\frac{(1-T_{rm})^{0.38}}{T_{rm}^{1/6}} \tag{25}$$

where x_1 and x_2 are the mole fractions of the components 1 and 2. The parameters A_1 and A_2 can be estimated from Eq.(19) or can be calculated from pure component thermal conductivities. The mixture reduced temperature is $T_{rm} = T/T_{cm}$, where:

$$T_{cm} = x_1 T_{c1} + x_2 T_{c2} \tag{26}$$

with T_{c1} and T_{c2} the pure component critical temperatures. The choice between components 1 and 2 is made with the criterion $A_1 \leq A_2$. The test of the correlation with over 600 datum points for 50 binary systems including those with highly polar components produced an average error of about 3%. The method is not suitable for multicomponent mixtures.

The methods suggested in this section for estimating λ_m have been tested by binary mixture data, they require the thermal conductivities of the pure components and so they are interpolative in nature.

5. An up-to-date review

In the scientific and technical literature are available a lot of experimental and estimated dynamic viscosity and thermal conductivity data for organic gases and liquids at the atmospheric pressure in the temperature range from the normal freezing point to the normal boiling point (and over) at saturation pressures. A complete and updated source of data can be found in the DIPPR 801 database [26]. For data elaboration a large collection of data is also due to Vargaftik et al [27]. Both the collections are presented with the accuracies for the experimental data; when experimental data are not available, suggested accuracies are presented and often they are greater than 25%. Useful reviews of the researches on thermophysical properties are due to Assael et al [28], Millat et al [29] and Horvath [30].

A last estimation method for liquid thermal conductivity can be considered and is due to Gharagheizi et al [31] : the method is based on a group contribution method and concerns 1635 pure chemical compounds with about 19,000 liquid thermal conductivity data extracted from the DIPPR 801 database at different temperatures. The parameters of the model cover 49 chemical substructures plus temperature. This method is cited for its applicability in terms of number of compounds and large temperature ranges. The authors claim an average absolute relative deviation less than 7.1%.

6. Conclusions

Several techniques to estimate dynamic viscosity and thermal conductivity of fluids [28, 29, 30] have been published including the theoretical and semi-theoretical investigations concerned with the atomic and molecular structure of the different compounds in the gaseous or the liquid phase. In this paper separate sections are presented to describe the more important methods able to estimate with acceptable accuracy dynamic viscosity and thermal conductivity of gases and liquids. These methods, very useful for engineering purposes, are essentially based on group contributions, or on the corresponding stetes principle, or on empirical calculations obtained from the experimental available data. Sometimes it is difficult to realize the difference among these categories, and the distinction is not clear (many methods present both corresponding-states and empirical features). Equations based on group contributions and empirical correlations often lead to big tabulations of constants for individual compounds. The methods adopted for mixtures are often valid at ambient temperature and the lack of experimental data make difficult to known their degree of accuracy. Summarizing, empirical estimation techniques find their greatest applications for low-pressure gas and gaseous mixtures. The dense fluid region is generally described by corresponding-states methods.

7. Nomenclature

ω	acentric factor
T	absolute temperature, K
T_c	critical temperature, K
T_b	normal boiling point, K
T_f	Normal freezing point, K
P_c	critical pressure, bar
V_c	critical volume, $cm^3 mol^{-1}$
T_r	reduced temperature = T/T_c
M	molecular weight, g/gmol or kg/kmol
μ	dynamic viscosity, poise (P)
ν	Kinematic viscosity, stoke (St)

8. References

[1] van der Waals J D 1873 Over de continuiteit van den Gas – en Vloeistoftoestand University of Leiden
[2] Joback K G 1984"A Unified Approach to Physical Property Estimation Using Multivariate Statistical Techniques," S.M. Thesis, Department of Chemical Engineering, Massachusetts Institute of Technology, Cambridge, MA
[3] Joback K G and Reid RC 1987 *Chem. Eng. Comm.* **57**: 233

IOP Conf. Series: Journal of Physics: Conf. Series **923** (2017) 012001 doi:10.1088/1742-6596/923/1/012001

[4] Poling BE, Prausnitz JM and O'Connell JP 2001 *"The Properties of Gases and Liquids"*, 5th ed McGrawHill
[5] Constantinou L and Gani R 1994 *AIChE J.* **40**: 1697
[6] Chapman S and Cowling TG 1939 *The Mathematical Theory of Nonuniform Gases*, Cambridge, New York
[7] Chung T, Ajlan M, Lee LL and Starling KE 1988 *Ind. Eng. Chem. Res.* **27**: 671
[8] Chung TH, Lee LL and Starling KE 1984 *Ind. Eng. Chem. Fundam.* **23**: 8
[9] Lucas K, VDI-Warmeatlas, Abschnitt DA, 1984 ''Berechnungsmethoden für Stoffeigenschaften,*''Verein Deutscher Ingenieure*, Düsseldorf,
[10] Reichenberg D, August 1971 *DCS report 11*, National Physical Laboratory, Teddington, England
[11] Reichenberg D, 1979 *Symp. Transp. Prop. Fluids and Fluid Mixtures* Natl. Eng. Lab., East Kilbride, Glasgow, Scotland
[12] Lucas K 1981 *Chem. Ing. Tech.* **53**: 959
[13] Andrade E N da C, 1930 *Nature* **125**: 309
[14] Andrade E. N. da C 1934 *Phil. Mag.* **17**: 497, 698
[15] Grunberg L and Nissan AH 1949 *Nature* **164**: 799
[16] Baroncini C, Di Filippo P, Latini G and Pacetti M 1981 *Intern. J. Thermophys.* 2(1): 21
[17] Lenoir J M 1957 *Petrol. Refiner* **36(8)**: 162
[18] Missenard A 1970 *Rev. Gen. Thermodyn.* **101(5)**: 649
[19] Latini G and Baroncini C 1983 *High Temp.-High Press.* **15**: 407
[20] Filippov L 1955 *Vest. Mosk. Univ., Ser. Fiz. Mat. Estestv. Nauk* **(8)10(5)**: 67 - 69
[21] Filippov L P and Novoselova NS 1955 *Vestn. Mosk. Univ., Ser. Fiz. Mat. Estestv* Nauk (3)10(2):37 - 40
[22] Jamieson D T, Irving J B and Tudhope JS 1975 *Liquid Thermal Conductivity. A Data Survey to 1973*, H. M. Stationary Office, Edinburgh
[23] Baroncini C, Di Filippo P, Latini G and Pacetti M : 1981 (pub. 1983) *Thermal Cond. 17th* Plenum Pub. Co., p. 285
[24] Baroncini C, Di Filippo P and Latini G March 1983 *Workshop on Thermal Conductivity Measurement*, IMEKO, Budapest
[25] Baroncini C, Latini G and Pierpaoli P 1984 *Intern. J. Thermophys.* **5(4)**: 387
[26] DIPPR-801, Project 801 2006 *Evaluated Process Design Data*, Public Release Documentation, Design Institute for Physical Properties (DIPPR),American Institute of Chemical Engineers, AIChE
[27] Vargaftik NB, Filippov LP, Tarzimanov AA and Totskii EE 1994 *"Handbook of Thermal Conductivity of Liquids and Gases"*, Boca Raton: CRC Press
[28] Assael MJ ,Trusler JPM and Tsolakis TF 1996 *"Thermophysical Properties of Fluids"*, Imperial College Press
[29] Millat J, Dymond JH and Nieto de Castro CA 1996 *"Transport of Fluids"*, Cambridge University Press
[30] Horvath AL 1992 *"Molecular Design, Chemical Structure Generation from the Properties of Pure Organic Compounds"*, Elsevier
[31] Gharagheizi F, Ilani-Kashkouli P, Sattari M, Mohammadi AH, Ramjugernath D and Richon D 2013 AIChE J. **59**:1702

IOP Conf. Series: Journal of Physics: Conf. Series **923** (2017) 012002 doi:10.1088/1742-6596/923/1/012002

Electro-osmotic flow in rectangular microchannels: geometry optimization

M Lorenzini

Università di Bologna - DIN Dipartimento di Ingegneria Industriale Campus di Forlì - Via Fontanelle 40, I-47121 Forlì, ITALY

E-mail: marco.lorenzini@unibo.it

Abstract. Micro-flow devices have turned over the years from the subject of fundamental research to fully-fledged industrial applications. Although for some cases the transport phenomena in micro-devices can be handled satisfactorily by using the same approach as for their larger-size counterparts, micro-effects such as electro-osmotic flow (EOF) are typical of small scales and can be employed to circulate coolant through heat sinks by means of electro-osmotic pumping. In order to obey the constraints that different engineering applications impose, design criteria must be employed to optimize the performance of the devices according to the criteria chosen. In this work optimization of the cross-sectional area of a microchannel where EOF and heat transfer at the walls occur is employed to demonstrate how approaches based on the first or second law of thermodynamics may yield opposite results, which are strongly dependent on the peculiarities of EOF, i.e. the ratio of Joule heating to heat transfer at the walls.

1. Introduction

Starting from the early years of this century microchannels have gradually evolved from the subjects of fundamental research to constitutive elements of so-called micro-flow devices (MFDs), as can be seen by comparing older reviews on the subject to more recent ones [1; 2]. In spite of some research still being addressed to fundamental subjects [3–9], MFDs have now progressed beyond the level of prototypes and find applications in several fields [10]; in particular, micro heat exchangers (MHXs) are employed in air conditioning systems [11] and heat pumping equipment [12]. Only at small scales, however, can flow devices exploit micro-effects such as electro-osmosis, which allows the motion of a liquid relative to a charged surface (e.g. due to chemical equilibrium between the surface and an electrolyte), induced by an applied external potential gradient across a microchannel, which acts on the layer of mobile ions, known as Electric Double Layer (EDL) [13]. Electro-osmotic flow in microchannels have received considerable theoretical attention, both for traditional geometries, such as circular ducts and parallel plates[14; 15], but also for more complex geometries such as rectangular, elliptical and triangular ducts [16; 17] and the use of electro-osmotic pumps (EOPs) has been investigated both theoretically and experimentally [18; 19] to circulate a polar fluid through the channels of heat sinks. As in several other engineering fields [20; 21] also MHXs are liable to optimization in order to enhance their performance or to reduce manufacturing and pumping costs. This can be achieved e.g. through changes of the basic configuration, treating the problem from the perspective of the first law of thermodynamics, which has led to the introduction of performance evaluation criteria, (PECs) [22; 23]. PECs adopt the view of the first law of thermodynamics, but the works of Bejan

[24; 25] through the entropy generation minimton (EGM) stress the importance of the second law. More recently, researchers have dealt with the optimization of micro heat exchangers [26–28] combining both approaches, with new and interesting results obtained when also conjugate effects are considered.
Among the many shapes of the ducts of heat exchangers which have been investigated, square cross-sections with smoothed corners have recently been studied in two works [29; 30], for laminar flow and H1 and H2 boundary conditions, whilst Lorenzini and Morini [31] and Lorenzini [32] extended the analysis to the case of rectangular and trapezoidal cross-sections, with and without viscous dissipation. So far, to the best of the Author's knowledge, no attempt has been made to optimize microchannels where an electro-osmotic flow occurs combining first- and second-law approaches through PECs and EGM. Hence the motivation for this work, where the effect of the change in the shape of the cross-section on the flow and heat transfer performance of an electro-osmotic flow is investigated in terms of maximum heat transfer and minimum entropy generation.

2. Mathematical Model

The non-dimensional concentration of the ions is obtained by solving the incompressible Nernst-Planck equation for negligible Peclet numbers, which yields the so-called Boltzmann distribution

$$c_i^* = \exp\left(\mp\frac{z_i e_0 \psi}{\kappa_B T}\right) \tag{1}$$

where c_i^* is the concentration of the i-th species normalized by the average concentration of the same species c_0, z_i its valence, e_0 the unit electric charge, ψ the potential of the electric field, κ_B is the Boltzmann constant, and T the thermodynamic reference temperature of the fluid. The $\mp$ sign depends on whether co-ions or counter-ions are involved. Knowledge of the concentration of co-ions and counter-ions allows the calculation of the density of charge in the fluid, which can be used in the constitutive equation for the electric field to obtain the potential distribution function in non-dimensional form $\psi^* = \psi\frac{z_i e}{\kappa_B T}$ by solving the Poisson-Boltzmann equation:

$$\nabla^2\psi^* = (kD_h)^2 \sinh(\psi^*) \tag{2}$$

where $k = \sqrt{2e^2 c_0 z_i^2/\varepsilon\kappa_B T}$ is the Debye-Hckel parameter, reciprocal to the Debye length, c_0 is the average concentration of the species, ε the electrical permittivity of the medium. To solve the equation, boundary conditions on the potential at the channel wall ψ_0 must be given: neglecting the compact layer, this equals the value of the potential at the Stern plane ζ, so that

$$\psi_0^* = \zeta^* = \frac{z_i e}{\kappa_B T}\zeta \tag{3}$$

The dependence of ψ^* on geometry, ζ^* and kD_h is thus apparent.
The non-dimensional velocity profile in the channel is obtained by solving the suitable form of the Navier-Stokes equation. The assumptions made in this work are steady, fully-developed laminar flow of a Newtonian fluid with thermophysical properties independent of temperature inside a microchannel of constant cross-section with rigid, non-porous walls. Further, no pressure gradients act along the flow direction and the only body force present is an external electric field along the channel's axis, z, $\overline{E}_{ext} = E_z\hat{i}_z$. As a consequence, only the velocity component along z is non-zero, and its magnitude at each point of the cross-section depends on the other two coordinates: $\vec{u} = u_z(x,y)\,\hat{i}_z$. The momentum transport equation in non-dimensional form becomes

$$\nabla^{2^*} u^* = ME_z^* \sinh(\psi^*) \tag{4}$$

IOP Conf. Series: Journal of Physics: Conf. Series **923** (2017) 012002 doi:10.1088/1742-6596/923/1/012002

where the non-dimensional quantities are $E_z^* = \frac{E_z L}{\zeta}$ (electric field), $u^* = u/U$ (velocity), and $M = \frac{2n_0 ze\zeta D_h^2}{\rho\mu U}$. The normalizing velocity U is chosen such that $M = 1$, so

$$\nabla^{2^*} u^* = E_z^* \sinh(\psi^*) \tag{5}$$

The boundary conditions for the problem are given by the no-slip velocity condition at the walls:

$$u_{wall} = u_0 = 0 \Rightarrow u_0^* = 0 \tag{6}$$

Given the depndence of the velocity problem on the shear layer potential and on the applied electric field, one can write

$$u^* = u^* (geometry, \zeta^*, \kappa D_h, E_z^*) \tag{7}$$

The energy equation is then solved to obtain the temperature distribution. Under the additional assumptions of thermally fully developed flow with negligible viscous dissipation effects, heat generation in the fluid due to Joule heating, uniform cross-section area A_c and $H1$ thermal boundary conditions (uniform heat flux along the channel's axis, q', and uniform temperature over the perimeter of the cross-section), the non-dimensional form becomes

$$\frac{u^* U}{u_m}\left(\frac{1}{A_c^*} + M_z\right) = \nabla^{2^*} T^* + M_z \tag{8}$$

where $A_c^* = A_c/D_h^2$ and u_m the average cross-sectional velocity defined as

$$u_m = \frac{U}{A_c^*}\int_{A_c^*} u^* dA_c^* \tag{9}$$

The quantity M_z relates the thermal input due to Joule heating to the linear heat flux q':

$$M_z = \frac{E_z^2 \lambda_0 D_h^2}{q'} \tag{10}$$

where λ_0 is the electrical conductivity of the fluid. Imposing $H1$ boudnary conditions results in

$$T^*\Big|_{P_h^*} = 0 \tag{11}$$

The line heat flux q', is identified given by specifying the value of M_z.
In view of the above discussion, the temperature field is dependent on the flow field and on the heat flux over the cross-section, so that

$$T^* = T^* (geometry, \zeta^*, \kappa D_h, M_z) \tag{12}$$

It is to be remarked that the dependence on the non-dimensional external electric field E_z^* has disappeared; this is due to the presence of the u^*/u_m ratio: in the momentum transport equation the term E_z^* introduces but a scaling factor in the solution, which appears both in u^* and in u_m. Solving the momentum and energy transport equations yields the velocity and temperature profiles over a cross-section, from which the Poiseuille and Nusselt numbers can be obtained:

$$Po = f Re = -\frac{2}{P^*}\int_P^* \frac{\partial u^*}{\partial \hat{n}} \tag{13}$$

which is also independent of E_z^*, and

$$Nu = -\frac{1}{P_h^*}\frac{\int_{A_c^*} u^* dA_c^*}{\int_{A_c}^* T u^* dA_c^*} \tag{14}$$

3. Performance Evaluation Criteria and Objective Functions

Microchannels are also employed in the manufacture of heat sinks: indeed, when conjugate effects are not relevant, the performance of the heat sink is determined by the performance of the microchannel. In order to obtain acceptable values of the heat transfer surface, pumping power, thermal power exchanged, every engineering application imposes specific constraints. These are often fewer than the degrees of freedom of the problem, which is thus amenable to optimization.

PECs, [23], define the kind of optimization carried out by specifying which quantities are left unchanged with respect to a reference configuration and the objective functions: in this work only the geometry of the microchannel shall be modified. Starting from a rectangular cross-section with sharp corners, they are rounded progressively up to a radius of curvature corresponding to half the length of the shorter side. The rectangular geometry is uniquely defined by two non-dimensional parameters, the aspect ratio, β, and the radius of curvature, R_c, which are defined with reference to Fig.1 as

$$\beta = \frac{2a}{2b} \quad R_c = \frac{r_c}{a} \tag{15}$$

The cross-sectional area, A_c and perimeter P, have the following expressions

$$A_c = a^2 \left[\frac{4}{\beta} - R_c^2 \left(4 - \pi\right) \right] \tag{16}$$

$$P = 4a \left[1 + \frac{1}{\beta} - 2R_c \left(1 - \frac{\pi}{4}\right) \right] \tag{17}$$

which are made non dimensional using the hydraulic diameter, D_h:

$$D_h = \frac{4A_c}{P} = a \frac{\frac{4}{\beta} - R_c^2 \left(4 - \pi\right)}{1 + \frac{1}{\beta} - 2R_c \left(1 - \frac{\pi}{4}\right)} \tag{18}$$

so that $A_c^* = A_c/D_h^2$, $P^* = P/D_h$; likewise $a^* = a/D_h$, $b^* = b/D_h$.

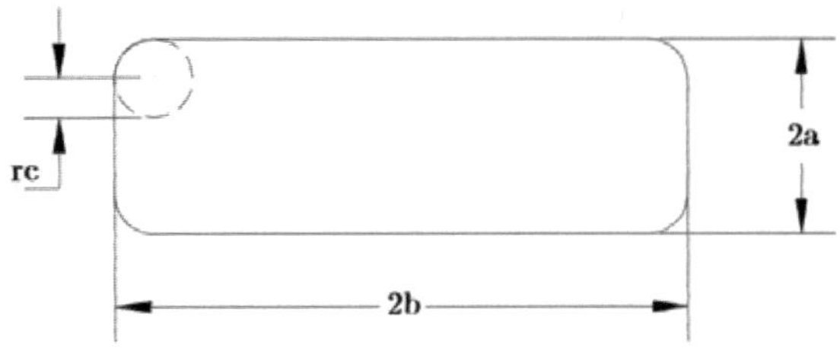

Figure 1. Geometry investigated.

For the applications of PECs the modified geometry (i.e. with rounded corners) is compared to the reference one, and three new quantities are defined:

$$S' = \left(\frac{a}{a_{ref}}\right)^2 \left[1 - \frac{\beta}{4} R_c^2 (4 - \pi)\right] \tag{19}$$

$$P_h' = \frac{a}{a_{ref}} \frac{1 + 1/\beta - 2R_c(1 - \pi/4)}{1 + 1/\beta} \tag{20}$$

$$D_h' = \frac{a}{a_{ref}} \frac{4/\beta - R_c^2(4 - \pi)}{1 + 1/\beta - 2R_c(1 - \pi/4)} \frac{1 + 1/\beta}{4/\beta} \tag{21}$$

where the superscript $'$ indicates the pertaining ratio between the actual and the reference values. The ratio a/a_{ref} appears in all the above equations, further geometrical constraints need to be introduced to obtain expressions that only depend on β and R_c. In this work, $D'_h = 1$, and Eq.(21) is used to obtain a suitable expression for a/a_{ref}.
The objective functions are the heat duty $\dot{Q}$, the pumping power W, the length of the microchannel, L, the mass flow rate $\dot{m}$ and the temperature difference between wall and bulk fluid $\Delta T = T_w - T_b$. They can be related by applying a mechanical and thermal energy balance to the single channel, and are expressed as the ratio of their values in the modified configuration to those for the reference ($R_c = 0$) geometry:

$$\dot{Q}' = \frac{\dot{Q}}{\dot{Q}_{ref}} = \frac{Nu' P'_h L' \Delta T'}{D'_h} \tag{22}$$

$$W' = \frac{W}{W_{ref}} = \frac{\dot{m}'^2 (fRe)' L'}{S' D'^2_h} \tag{23}$$

Equations(22) and (23) contain all five objective functions, and these are valid for a fixed geometry; the problems has thus four degrees of freedom, which means that each PEC is defined by holding the three other quantities constant.

4. Entropy balance and entropy generation number

Optimization through PECs is based ultimately on first-law considerations: in order to obtain information as to how efficiently available energy is transformed, entropy generation rate, $\dot{S}_{gen}$, must be calculated, and it can be split into two contributions, one due to heat transfer, the other to pressure drop, and can be derived through the combined use of the energy and entropy balance equations.:

$$\dot{S}_{gen} = \dot{S}_{gen,\Delta T} + \dot{S}_{gen,\Delta p} \tag{24}$$

Details for the case of pressure-driven flows with viscous dissipation details can be found in [28]. Under the assumptions of negligible fluid temperature difference between wall and bulk fluid at the inlet as compared to the bulk temperature and of fluid temperature increase as compared to the inlet temperature($(T_i - T_m)/T_m << 1$, $(T_e - T_i)/T_i << 1$), in the case of electro-osmotic flow, the first term of Eq.(24) becomes

$$\dot{S}_{gen,\Delta T} = \frac{\dot{Q}\Delta T}{T_i^2 \left(\frac{T_e}{T_i}\right)} \left[1 + M_z A_c \left(1 + \frac{T_e}{\Delta T}\right)\right] \tag{25}$$

And the second term has the same form as for the case of pressure-driven flow

$$\dot{S}_{gen,\Delta p} = \frac{2\mu \dot{m}^2 fReL}{\rho^2 A_c D_h^2 T_i} \tag{26}$$

Equations (25), (26) represent the entropy generation rate due to heat transfer and to frictional losses respectively; they can be related by introducing the augmentation irreversibility ratio,

$$\phi = \frac{\dot{S}_{gen,\Delta p}}{\dot{S}_{gen,\Delta T}} \tag{27}$$

With a view to comparing the influence of the radius of curvature of the corners on the constrained performance of microchannels of rectangular cross-section, it is expedient to employ the augmentation entropy generation number N_S as introduced in [24]

$$N_S = \frac{\left(\dot{S}_{gen}\right)_{opt}}{\left(\dot{S}_{gen}\right)_{ref}} = \frac{\frac{\left(\dot{S}_{gen,\Delta T}\right)_{opt}}{\left(\dot{S}_{gen,\Delta T}\right)_{ref}} + \frac{\left(\dot{S}_{gen,\Delta p}\right)_{ref}}{\left(\dot{S}_{gen,\Delta T}\right)_{ref}} \frac{\left(\dot{S}_{gen,\Delta p}\right)_{opt}}{\left(\dot{S}_{gen,\Delta p}\right)_{ref}}}{1 + \frac{\left(\dot{S}_{gen,\Delta p}\right)_{ref}}{\left(\dot{S}_{gen,\Delta T}\right)_{ref}}} = \frac{N_T + \phi_{ref} N_P}{1 + \phi_{ref}} \tag{28}$$

In the case of N_T, $T_i/T_{ref}T_i' = 1$ and

$$N_T = \frac{\dot{Q}'\Delta T'}{T_e'} + \frac{1 + M_z A_c^* \left(1 + \frac{T_e'}{\Delta T'}\frac{T_{e,ref}}{\Delta T_{ref}}\right)}{1 + M_{z,ref} A_{c,ref}^* \left(1 + \frac{T_{e,ref}}{\Delta T_{ref}}\right)} \tag{29}$$

the normalized exit temperature T_e' is obtained from an energy balance

$$T_e' = \frac{T_i}{T_{e,ref}} + \left(1 - \frac{T_i}{T_{e,ref}}\right) \frac{\dot{Q}'}{\dot{m}'} \frac{1 + M_z A_c^*}{1 + M_{z,ref} A_{c,ref}^*} \tag{30}$$

For N_P

$$N_P = \frac{\dot{m}'^2 (f Re)' L'}{A_c' D_h'^2} = W' \tag{31}$$

The entropy generation number is thus expressed in terms of the objective functions, Eqs.(22) and (23), shown above. Some quantities must be known to compute $\dot{S}_{gen}$, and for this work their value has been chosen as follows: $T_i/T_{ref} = 0.95$, $\Phi_{ref} = 0.1$, $(T_e/\Delta T)_{ref} = 20$ and $M_{z,ref} = 0.001 - 0.01 - 0.1$.

In Eqs. (29) and (30) no direct or indirect (i.e. through objective functions) of the quantity M_z on the geometrical parameters β and R_c has been shown yet. This is a point deserving attention: when Nu is computed, a certain value of M_z is chosen, which binds E_z to q'. A similar problem is encountered when imposing E_z^*, which relates L and E_z, but fRe is independent of E_z^*, so using the Poiseuille number is sufficient to overcome the difficulty. In the case of M_z, E_z is decoupled from q' if M_z is written in terms of the objective functions. It can be demonstrated that this leads to the following expression

$$M_z = \frac{(f Re')^2 \dot{m}'^2 L'}{D_h^2 A_c'^2 \dot{Q}'} M_{z,ref} \tag{32}$$

Equation (32) is not explicit, as $\dot{Q}'$ depends on M_z through Nu, yet a dependence $M_z = M_z(R_c, \beta)$ can be established once the PEC has been defined.

5. Results

5.1. Velocity and temperature distribution

The Poiseuille and Nusselt numbers must be calculated in order to carry out the optimization procedure. To this extent, the velocity and temperature distributions for different radii of curvature were computed by solving Eqs. (2), (5), (7) considering $D_h = 24\,\mu m$ and a fully heated perimeter. For all simulations, an externally imposed electric field $E_z = 100\,kV \cdot m^{-1}$ was chosen, while the microchannel length was $L = 0.01\,m$. The fluid considered was de-ionized ultra-filtered water (DIUF), with $n_0 = 10^{-6}\,mol \cdot m^{-3}$ at a reference temperature was $T = 293\,K$ and with $\zeta = 0.2\,V$. The corresponding non dimensional quantities appearing in the equations become $\zeta^* = 8$, $E_z^* = 5000$, $\kappa D_h = 9.85$, 78.84. For each geometry four different aspect ratios

were considered ($\beta = 0.1,\ 0.25,\ 0.5,\ 1.0$), radii of curvature from $R_c = 0.0$ to $R_c = 1.0$ in steps of 0.1 and $M_z = 10^{-3},\ 10^{-2}$ and M_z from 0.1 to 1.0 in steps of 0.1.
The Poiseuille numbers for a given hydraulic diameter are independent of M_z and can be expressed as a function of the non-dimensional radius of curvature and of the aspect ratio of the channel in the form

$$fRe\left(\beta, R_c\right) = \sum_{i=0}^{3} d_i R_c^i \tag{33}$$

where $d_0 = fRe_0\left(\beta\right)$ is the value of the Poiseuille number for the reference cross-section.The Nusselt number is likewise expressed as

$$Nu\left(\beta,\ R_c,\ M_z\right) = Nu\left(\beta,\ R_c,\ M_{z,ref}\right) - \left|a\left(\beta,\ R_c\right)\right| M_z \tag{34}$$

$$Nu\left(\beta,\ R_c,\ M_{z,ref}\right) = \sum_{i=0}^{3} c_i R_c^i \tag{35}$$

where $c_0 = Nu_0\left(\beta,\ M_{z,ref}\right)$ is the value of the Nusselt number for the reference cross-section. Once the functional forms of the Poiseuille and Nusselt numbers are known, they can be used in the expressions of $\dot{Q}'$ and P' to apply the PECs and to compute N_S.

5.2. PEC and entropy analysis:FG1a $\left(L', \dot{m}, \Delta T' = 1;\ \dot{Q}'\right)$

In the following the results of the application of the FG1a PEC (fixed-geometry, with $\dot{Q}'$ as objective function) are analysed and compared to the conclusion of a second-law analysis. The results are presented as two- and three-dimensional plots with quantities normalized by dividing them for a reference configuration with $R_c = 0$ and $\beta = 1$. The two-dimensional plots are useful for engineering problems where β is assigned, whereas thei three-dimensional counterparts can be employed when the aspect ratio may be chosen freely. The length, mass flow ratc and temperature difference remain unchanged. Equation (22) becomes

$$\dot{Q}' = Nu' P_h' \tag{36}$$

and M_z takes a form which only depends on the geometrical parameters and which must be solved numerically

$$M_z = \frac{\left(fRe'\right)^2}{A_c'^2 Nu' P_h'} M_{z,ref} \tag{37}$$

Figure 2 shows the dependence of M_z on the radius of curvature at a reference value of $M_{z,ref} = 10^{-3}$; it is to be remarked that increasing $M_{z,ref}$ only scales the results accordingly, without any changes in the shape or relative position of the curves, and therefore no other such plots are shown. This behaviour is readily explained by recognizing that in Eq.(37) $M_{z,ref}$ only appears as a proportionality factor. The same holds true for all PECs. It is clear that M_z is always increasing with the radius of curvature, and is higher for larger values of the aspect ratio. Considering the constraints imposed by the FG1a criterion, this implies that the line heat flux at the walls q' decreases with increasing radius of curvature, and this trend is more pronounced the higher the aspect ratio: all the others quantities being the same, M_z is inversely proportional to its denominator, q'. This is confirmed by the trend of $\dot{Q}'$, see Fig. 3, which is in fact unaffected by the value of $M_{z,ref}$: this can be understood if one considers how Nu' can be expressed through Eq.(37), that is as a function of geometrical parameters only.

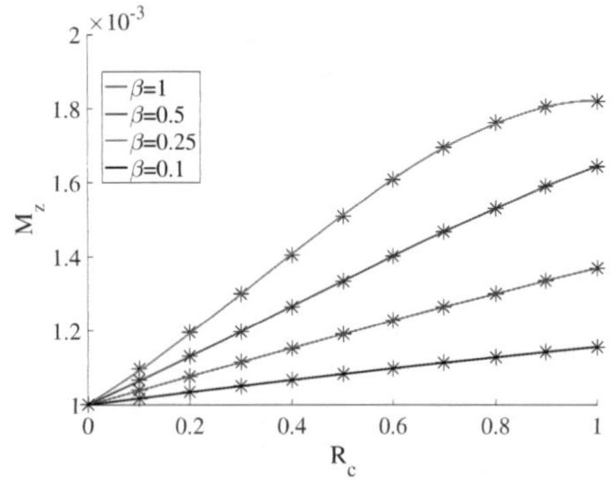

Figure 2. M_z versus R_c, $M_{z,ref} = 10^{-3}$.

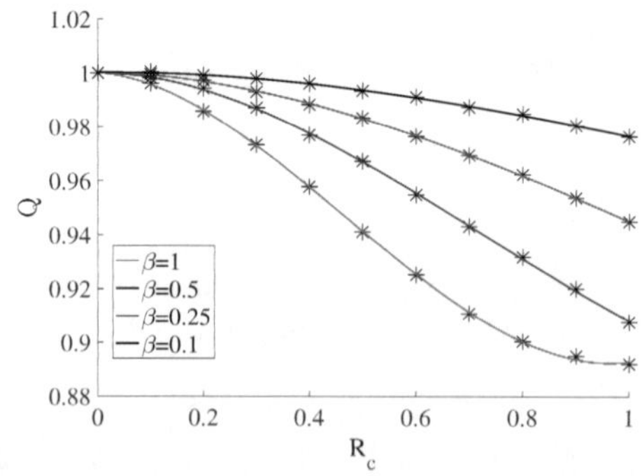

Figure 3. $\dot{Q}'$ versus R_c.

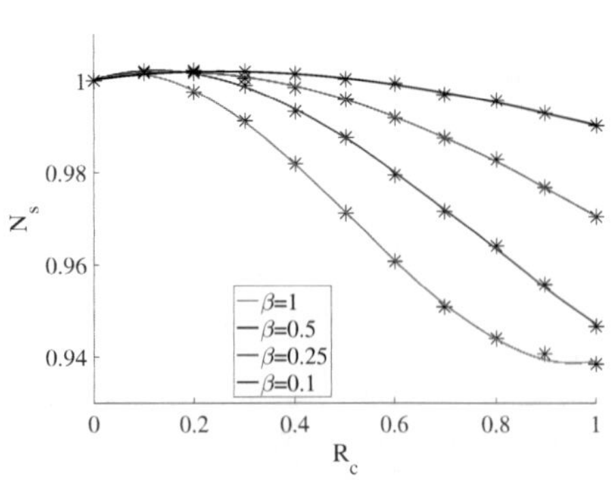

Figure 4. N_S versus R_c, $M_{z,ref} = 10^{-3}$.

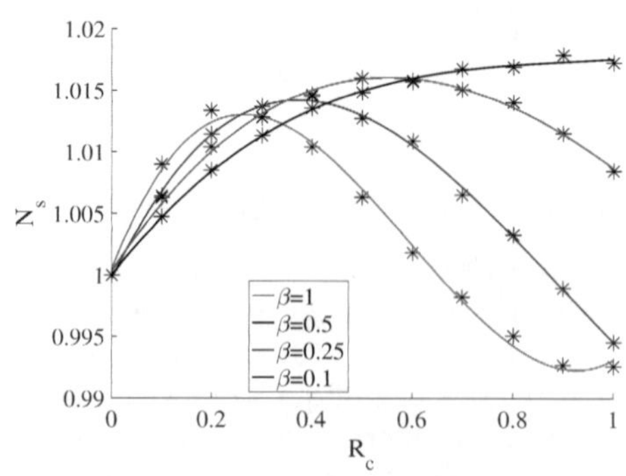

Figure 5. N_S versus R_c, $M_{z,ref} = 10^{-2}$.

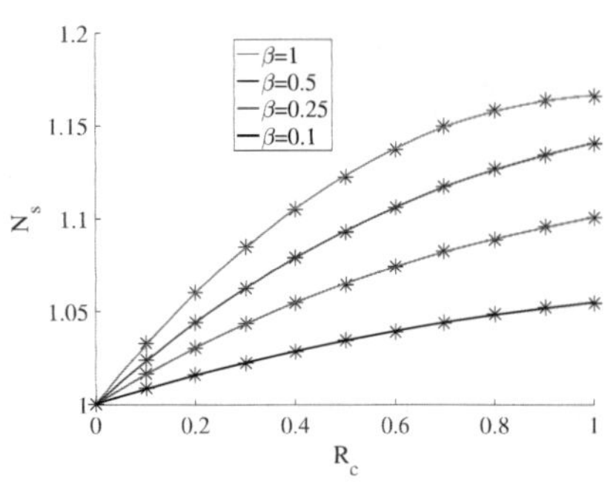

Figure 6. N_S versus R_c, $M_{z,ref} = 10^{-1}$.

Without considering a second-law approach, it can be concluded that under FG1a constraints the maximum heat transfer is achieved by a cross-section with sharp corners. Also, a geometry with high aspect ratio is less sensitive to the variation of R_c and could be therefore preferable, as it can compensate possible manufacturing imprecisions more efficaciously. If the entropy generation number is also considered, the value of $M_{z,ref}$ does play a significant role, as shown in the Figs. 4-6, whose trends are easily explained if one considers the forms of Eqs. (29)-(31). When $M_{z,ref}$ is low, so is M_z, and N_T depends almost exclusively on $\dot{Q}'$, and for this reason the trends of the quantities is almost the same: entropy generation is caused for the most part by actual heat transfer, also thanks to the modest relative contribution of N_P ($\phi_{ref} = 0.1$). When $M_{z,ref}$ increases, Fig. 5, the trends are inverted up to about $R_c = 0.2$, then are partially restored; for large enough values of $M_{z,ref}$, Fig. 6, the trends are fully inverted, increasing with R_c and decreasing with β. This means that Joule heating gives an ever increasing contribution to entropy generation, which is opposite in trend to that of $\dot{Q}$. Hence N_s exhibits a maximum at $R_c \approx 0.2$: below this value, heat transfer from the walls dominates, above it Joule heating prevails. As $M_{z,ref}$ increases further, Joule heating becomes the only significant contributor to N_S. The factor that guarantees the maximum heat transfer is that with the lowest aspect

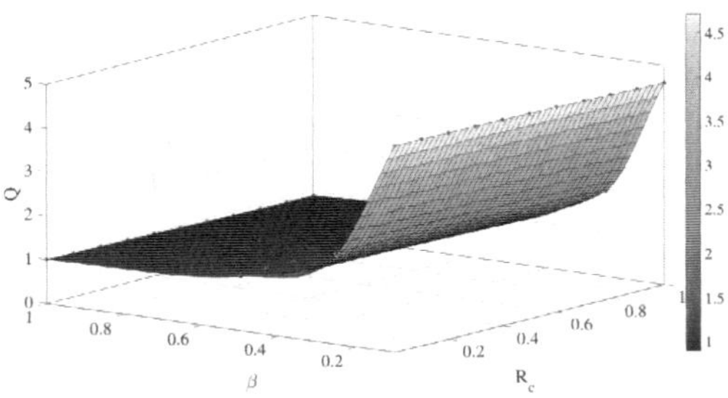

Figure 7. Three-dimensional plot of $\dot{Q}'(\beta, R_c)$.

ratio ($\beta = 0.1$), R_c giving a much smaller contribution. The entropy generation number has a dependence on the aspect ratio which is very similar to that of $\dot{Q}'$, Figs. 8 and 9, which was to be expected, since such a high increase with β of the thermal power transferred could not leave entropy generation unaffected. Concerning the influence of R_c, as $M_{z,ref}$ varies, the same considerations made for the bi-dimensional plots can be made, and the inversion in the trend of N_S can be seen from comparison of Fig. 8 and Fig. 9. Also, the increase in the contribution of Joule heating to N_S makes the variation of this latter quantity with β and R_c smaller as $M_{z,ref}$ increases. Finally, it should be noticed that the minimum value of N_S for $M_{z,ref} = 10^{-3}$ is obtained at low β and $R_c = 1$, whereas for $M_{z,ref} = 10^{-1}$ the geometry that minimizes entropy generation is a square cross section with sharp corners. In such cases optimization becomes multi-objective.

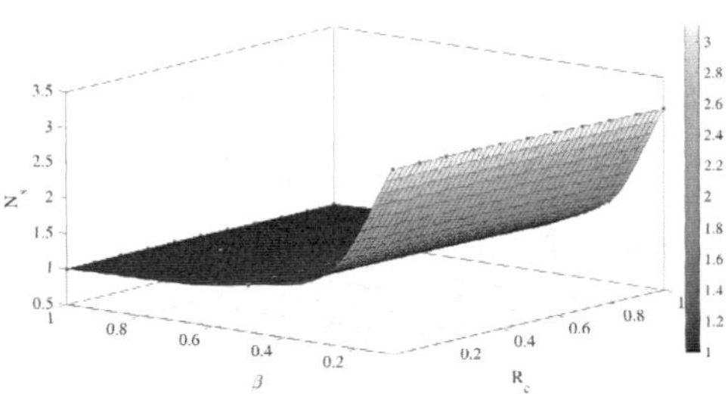

Figure 8. Entropy generation number, $M_{z,ref} = 10^{-3}$.

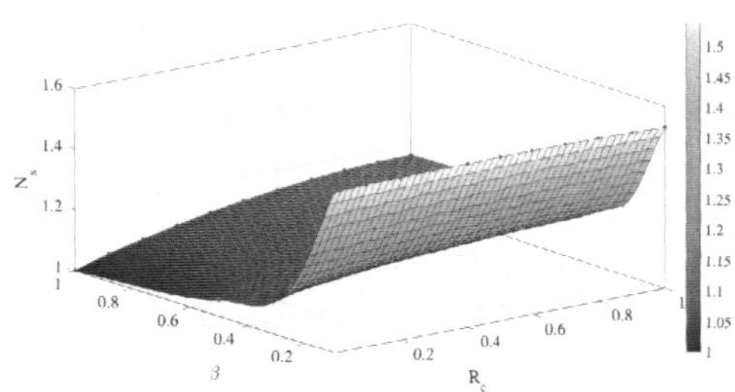

Figure 9. Entropy generation number, $M_{z,ref} = 10^{-1}$.

6. Conclusions

In this work the influence of the smoothing of the corners of rectangular microchannels on the transport phenomena associated to electro-osmotic flow has been studied using an approach that combines techniques based on the first law (PECs) and the second law (EGM) of thermodynamics. Although sometimes the outcomes are concordant, it has been demonstrated that for cases such the FG1a criterion (maximize $\dot{Q}$ when $\dot{m}$, L and ΔT remain constant) for fixed D_h the configuration that maximizes $\dot{Q}$ is also that which corresponds to maximum entropy generation, in particular when the value of $M_{z,ref}$ are high enough, i.e. when Joule heating within the fluid gains of importance with respect to heat transfer. When the aspect ratio has not been dictated by some constraints, wall heat transfer is maximized at low aspect ratios, whereas the combination of β and R_c for minimum entropy depends on $M_{z,ref}$. In such cases the problems becomes one of multi-objective optimization.

References

[1] Morini G 2004 *Int. J. Th. Sc.* **43** 631
[2] Kandlikar S, Colin S, Peles Y, Garimella S, Pease R, Brandner J and Tuckerman D 2013 *J. of Heat Tr.* **135** 091017
[3] Lorenzini M, Daprá I and Scarpi G 2017 *App.Th. Eng.* **122** 118
[4] Vocale P, Morini G L and Spiga M 2014 *Int. J. Heat Mass Tr.* **71** 376
[5] Morini G, Lorenzini M, Salvigni S and Spiga M 2009 *Micr. Nanof.* **7** 181
[6] Henning T, Brandner J, Schubert K, Lorenzini M and Morini G 2007 *H. Tr. En.* **28** 834
[7] Morini G, Lorenzini M, Colin S and Geoffroy S 2006 *Pr. ICNMM2006* (NY: ASME) p 411
[8] Morini G, Yang Y and Lorenzini M 2012 *Exp. H. Tr.* **25** 151
[9] Vocale P, Morini G and Spiga M 2016 *J. of Heat Tr.* **138** 044502
[10] Ohadi M, Choo K, Dessiatoun S and Cetegen E 2013 *Next Generation Microchannel Heat Exchangers* (NY: Springer)
[11] Han Y, Liu Y, Li M and Huang J 2012 *Energ. Proc.* **14** 148
[12] Kew P and Reay D 2011 *App. Th. Eng.* **31** 594
[13] Bruus H 2010 *Theoretical microfluidics* (Oxford: OUP)
[14] Rice C and Whitehead R 1965 *J. of Ph. Ch.* **69** 4017
[15] Mala G, Li D, Werner C, Jacobasch H J and Ning Y 1997 *Int. J.. Heat Fluid F.* **18** 489
[16] Vocale P, Geri M, Cattani L, Morini G and Spiga M 2013 *Int. J. Th. Sc.* **72** 92
[17] Vocale P, Geri M, Morini G L and Spiga M 2014 *JOP: Conf Ser* **501** 012026
[18] Morini G, Lorenzini M, Salvigni S and Spiga M 2006 *Int. J. Th. Sc.* **45** 955
[19] Geri M, Lorenzini M and Morini G 2012 *Int. J. Th. Sc.* **55** 114
[20] Renzi C 2016 *Int. J. Adv. Man. Tech.* **86** 3365
[21] Renzi C and Leali F 2016 *Journal of Multi-Criteria Decision Analysis* **23** 87
[22] Webb R and Eckert E 1972 *Int. J. Heat Mass Tr.* **15** 1647
[23] Bergles A, Blumenkrantz A and Taborek J 1974 *Pr. 5th Int. H. Tr. Con.* (Tokyo) p 239
[24] Bejan A 1982 *Entropy generation through heat and fluid flow* (New York: Wiley)
[25] Bejan A 2001 *Exergy* **1** 269
[26] Saenen T and Baelmans M 2012 *Int. J. Th. Sc.* **59** 214
[27] Stevens T and Baelmans M 2008 *App. Th. Eng.* **28** 2353
[28] Lorenzini M and Suzzi N 2016 *H. Tr. En.* **37** 1096
[29] Ray S and Misra D 2010 *Int. J. Th. Sc.* **49** 1763
[30] Chakraborty S and Ray S 2011 *Int. J. Th. Sc.* **50** 2522
[31] Lorenzini M and Morini G 2011 *H. Tr. En.* **32** 1108
[32] Lorenzini M 2013 *Houille Bl.* 64

A multiscale approach to interpret and predict the apparent slip velocity at liquid-liquid interfaces

P Poesio[1], **A Damone**[1] **and O K Matar**[2]

[1] Universitá degli Studi di Brescia, Department of Mechanical and Industrial Engineering, Via Branze 38, 25123 Brescia, IT

[2] Imperial College London, Department of Chemical Engineering, South Kensington Campus, London SW7 2AZ, UK

E-mail: pietro.poesio@unibs.it

Abstract. The classical boundary conditions at flat liquid-liquid interfaces are continuity of the velocity and of the tangential component of stress; for curved interfaces, one also demands that the jump in the normal stress at the interface is balanced by the product of the interfacial tension and curvature. While these conditions are widely accepted, and are often used at the macro scale, the recent interest in micro and nano-fluidics challenges their validity. At molten polymer-polymer interfaces, for instance, it has been consistently shown by direct and indirect measurements that, apparent, velocity jumps exist and can be modelled effectively via a Navier slip condition (NSC). Here, we discuss that if a viscosity, which accounts for the density and mole fraction distributions, is included in the Navier-Stokes equations, we can describe, naturally, and without recourse to *ad-hoc* models such as the NSC, the velocity profile in the interfacial region separating two fluids. This approach is supported by the observation that there is a relation between apparent slip and density distribution across the interface.

1. Introduction

Liquid-Liquid Interfaces (LLI) characterise and influence the behaviours of several macro- and micro-systems. For instance, in biology, interfaces between two immiscible liquid electrolyte solutions are of great importance as they occur in tissues and cells of all living organisms. In oil and gas industry, the balance between break-up and coalescence (both interfacial phenomena) determine the occurrence of phase inversion, a process that can lead to the blockage of the entire pipeline. Furthermore, in recent lab-on-a-chip technology, liquid droplets, moving through an immiscible liquid, are used as chemical (and biological) reactors, where reactions are carried out while the droplet is transported along micro-channels. The proper understanding and modelling of liquid-liquid interface dynamics is not only a challenging problem from a fundamental viewpoint, but it is of great significance to many industries. Experiments at liquid-liquid interfaces are difficult to perform due to the presence of thermal and hydrodynamic fluctuations and direct visualisations are carried out only at very particular conditions. One of the problems when dealing with flows in micro- and nano-scales is the proper formulation of boundary conditions. The classical boundary conditions at liquid-liquid interfaces are continuity of the velocity and of the tangential component of stress and jump in the normal stress at the interface is balanced by the product of the interfacial tension and curvature [1]. However, the recent interest in micro-[2] and nano-fluidics [3] challenges their validity. At molten polymer-polymer interfaces, for instance, it has been shown by direct and indirect measurements [4] that, apparent, velocity jumps exist and can be modelled effectively via a Navier slip condition. In this paper, we show that if a viscosity profile is included in the Navier-Stokes equations, we can describe, the velocity profile in the interfacial region separating two fluids. This approach is supported by the observation that there is a relation between apparent slip and the viscosity profile across the interface [5]. For simple fluids, only few recent MD simulations pinpoint the presence of an apparent slip velocity, i.e. a region, whose thickness is of the molecular size, close to the interface where the velocity changes very rapidly. Similarly to solid-liquid interfaces [6], this rapid change is replaced by a velocity jump, Δu, and described by a NSC [5], $\Delta u = \alpha\tau$, where Δu is proportional to the shear stress, τ, through a slip length, α. Another possibility is the introduction of an interfacial viscosity [7] within a layer (whose thickness must be determined by previous MD simulations) across the interface. While both strategies prove effective in describing the velocity profile far from the interface, they require case-specific information: a MD simulation must be carried out to extract, for instance, the slip length, which is, then, used as an interfacial condition for the solution of the Navier-Stokes equations.

These approaches lack generality and cannot describe the very rapid velocity transition across the interface, which is simply replaced by an unphysical discontinuity in the velocity profile. These difficulties can in principle be alleviated by full MD simulations, which are, however, computationally impossible in large scale geometries.

The aim of this work is to further investigate this aspect and to relate the velocity discontinuity at liquid-liquid interfaces to the degree of mixing of the liquid. In particular we want to question the validity of simple potentials – such as Lennard-Jones interaction – on the few results on the slip velocity obtained so far. Different mixtures are used to show that if the effect of viscosity profile is accounted for, then the Navier-Stokes equations can predict the full velocity profile, including, crucially, the rapid change in velocity across the interface, which is often interpreted as, apparent, slip.

Our method represents a mulitscale approach where pieces of information collected from MD simulation are enforced into Navier-Stokes continuum theory. In this work, we only limit ourselves to the investigation of Lennard-Jones fluids, while we leave to Ref. [8] the problem for molecular liquids.

2. Molecular dynamics simulations details

As a test geometry, we consider a simple steady Couette flow [1] between two planes at $y = \pm$ L with the streamwise component of the velocity u $(y = \pm \text{ L}) = \pm$ V, of two

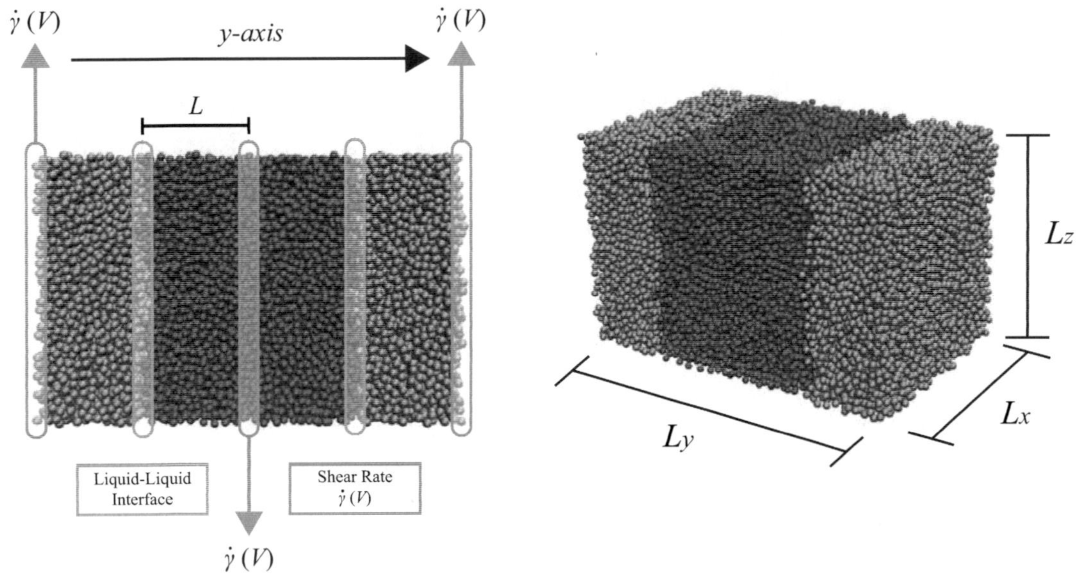

Figure 1: Numerical set-up

liquids, with the interface at $y = 0$; here, y represents the wall-normal direction, and L and V are constants as shown in Fig. [1].

All the Molecular Dynamics simulations presented in the paper are carried out by the code LAMMPS [9]. In the following, details are provided. The simulations are performed using the properties of liquid Argon and a modified Lennard Jones potential is implemented as follow:

$$U_{ij} = 4\varepsilon \left[\left(\frac{\sigma}{r_{ij}} \right)^6 - \beta_{ij} \left(\frac{\sigma}{r_{ij}} \right)^{12} \right]$$

where the parameter β_{ij} allows to change the miscibility between the two fluids. $\beta_{ij} = 1$ if $i = j$; if $i \neq j$ the parameter β_{ij} is used to change the miscibility of the two liquids. $\beta_{ij} = 0$ corresponds to immiscibility, $\beta_{ij} = 1$ corresponds to full miscibility; $0 < \beta_{ij} < 1$ corresponds the partial miscibility. The values of ε and σ are $\varepsilon/k_B = 120$ K and $\sigma = 3.4$ Å, where k_B is the Boltzmann constant. The mass is taken equal to M = 39.948 Kg/kmol. All the simulations are NVT thermostatted at T = 130 K using a Nosé-Hoover thermostat. A fully period box ($L_y = 160.82$ Å, $L_x = L_z = 102.34$ Å) is used in all the simulations. Each box is divided into 700 bins ($\delta_y = 0.23$ Å, $\delta_x = \delta_z = L_x = L_z$). The resulting constitutive relations are provided in Figs. [2]

2.1. *Calculation of the viscosity*

After stabilizing the simulation at equilibrium conditions, the viscosity is computed using the Muller-Plathe method – see Figs. [2] where the constitutive relation between the shear viscosity μ and the density ρ and mole fraction χ is provided, *i.e.* $\mu = \mu(\rho, \chi)$.

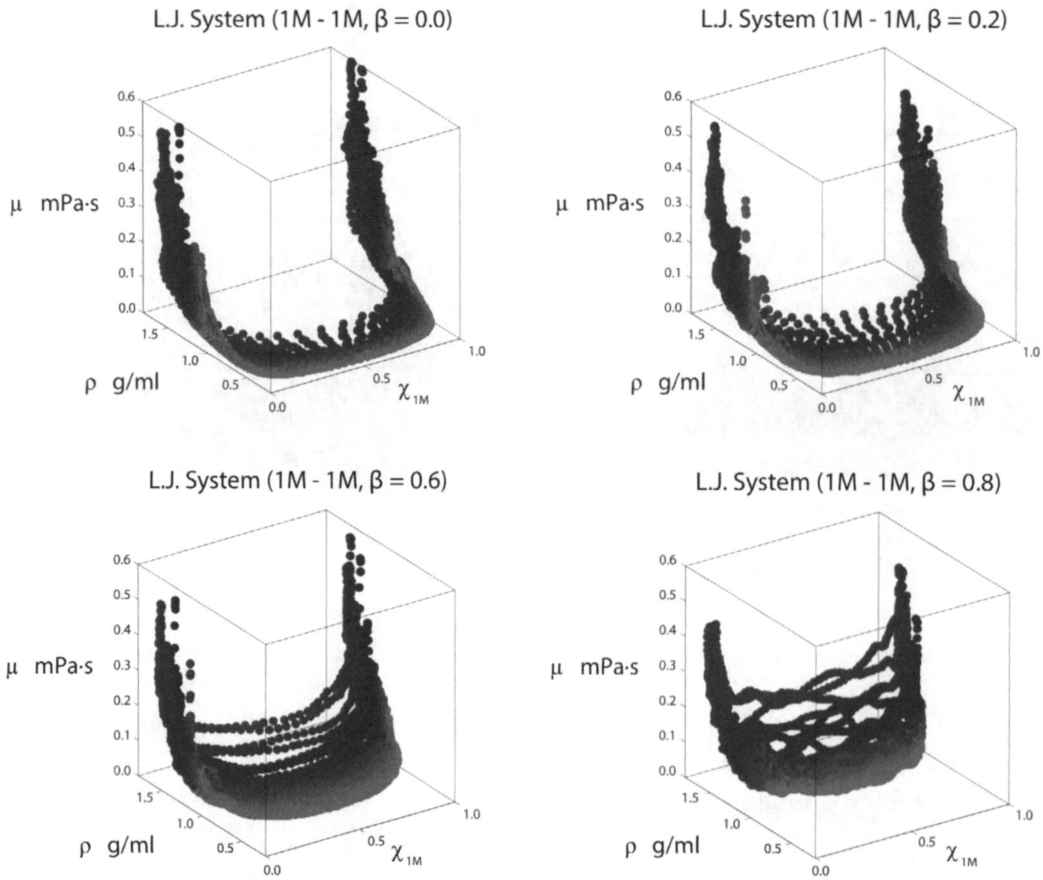

Figure 2: Constitutive relations $\mu = \mu(\rho, \chi)$ for the Lennard-Jones fluids at different values of β. The two fluids have the same mass M.

The shear viscosity μ is computed using the well-known Muller-Plathe method by imposing a shear rate $\dot{\gamma}$ that induces a velocity gradient dv/dy and by computing the virial stress σ as

$$\mu = -\frac{\sigma_{yz}}{dv/dy}$$

$$\sigma_{yz} = \frac{1}{V}\sum_{i=1}^{N}\left\langle S_{yz}^{\,i}\right\rangle$$

$$S_{yz}^{\,i} = -m_i v_{iz} + \frac{1}{2}\sum_{j\neq i}^{N}\frac{\partial U(r_{ij})}{\partial r_{ij}}\frac{r_{ijz}r_{ijy}}{r_{ij}}$$

where N is the total number of particles and V is the volume, v_{iz} and v_{iy} are the z and y component of velocity vector $\mathbf{v}_i$, r_{ijz} and r_{ijy} are the z and y component of vector $\mathbf{r}_i$ and $\mathbf{r}_j$; m_i is the mass of the ith particle.

IOP Conf. Series: Journal of Physics: Conf. Series **923** (2017) 012003 doi:10.1088/1742-6596/923/1/012003

2.2. *Simulation with imposed shear*

A Couette flow, with a given shear rate $\dot{\gamma}$ is obtained by applying opposing velocities $\pm V$ along the y-axis. The velocity V is enforced by applying an additional body force F_V in the y direction of each atom within the two external bins and within the central bin. Further details can be found in Ref. [10]. The application of the force F_V is based on the following (leap-frog) scheme. For a system on n particles, the center of mass velocity along one dimension is given as

$$v_{com}^{t-\frac{1}{2}} = \frac{1}{n}\sum_{i=1}^{n} v_i^{t-\frac{1}{2}};$$

if the total force at step t is $F_{com}^t = \sum_{i=1}^n F_i^t$, then we look for an additional body force F_V^t to be uniformly applied to each atom so to drive the flow with a velocity u_c. By setting $v_{com}^{t-\frac{1}{2}} = V$, we get

$$V = v_{com}^{t-\frac{1}{2}} + \frac{\delta t}{n\,M} F_{com}^t + \frac{\delta t}{n\,M}\sum_{i=1}^{n} F_V^t$$

where $n\,M$ is the total mass of the system and δt is the imposed time step. Solving for the total mass force, we obtain :

$$\sum_{i=1}^{n} F_V^t = \left[V - v_{com}^{t-\frac{1}{2}}\right]\frac{\delta t}{n\,M} - F_{com}^t$$

that translates into a force to be applied to each particle

$$F_V^t = \left[V - v_{com}^{t-\frac{1}{2}}\right]\frac{\delta t}{M} - \frac{F_{com}^t}{n}.$$

3. Navier-Stokes – continuum scale – solution

A plane Couette flow, with non uniform viscosity, is described by the following equation

$$\frac{d}{dy}\left[\mu(\rho,\chi)\left(\frac{du}{dy}\right)\right] = 0. \qquad (1)$$

A constitutive relation to describe how viscosity changes as function of density and composition, $\mu = \mu(\rho,\chi)$, is needed to solve this equation. Here we assume that the conditions are laminar, and therefore $Re << 1$. Such a relation is obtained, in our case, by MD simulations $\mu = \mu(\rho,\chi)$.

4. Results

The first example is given by two identical immiscible atomic fluids. This simplified case allows us to compute the density and mole fraction profiles across the interface using DFT (see Fig. [3]); the viscosity distribution is then computed using a theoretical approach developed for a strongly inhomogeneous Lennard-Jones fluid [11]. Substitution

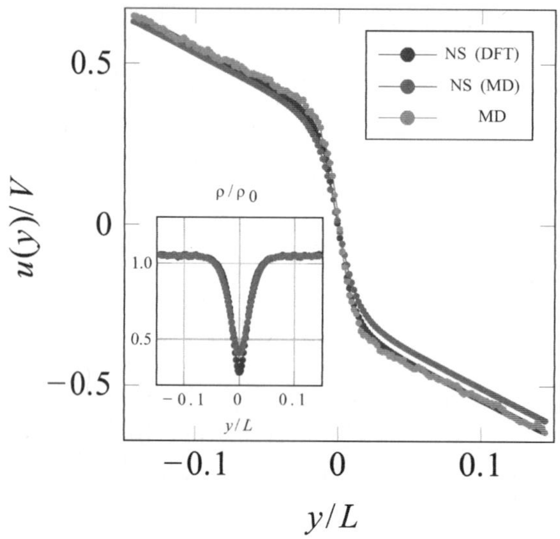

Figure 3: Close-up of the normalized streamwise velocity component in the interfacial region, obtained via MD simulations (orange line), and solution of the Navier-Stokes equations with $\mu(\rho)$ calculated using DFT (blue line) and MD (red line) for the case of two immiscible, symmetrical Lennard-Jones fluids ($V = 25$ m/s, $L = 46.65$ Å, $L_x = L_z = 108.34$ Å). The density (see insert, $\rho_0 = 1.1705$ g/ml) used to solve the NS equations was obtained from DFT [12] (blue line) and MD simulations (red line). The MD simulation uses 17248 particles for each fluid.

of the relation $\mu = \mu(\rho, \chi)$ into the Navier-Stokes equations, and their subsequent numerical solution yields the results shown in Fig. [3]. As can be seen, the agreement with the predictions [12] of the full MD simulations is very good.

The viscosity distribution $\mu = \mu(\rho, \chi)$ is obtained using the distribution of $\rho(y)$ and $\chi(y)$ at equilibrium conditions and, therefore, does not depend upon the specific state. This is in contrast with the other approaches where the non-equilibrium properties, such as slip length [13] and interfacial viscosity [14], depend upon the specific state, characterised, for instance, by the imposed shear rate $\dot{\gamma}$. As the fluid structure and the intermolecular interactions become more complex, however, the required information can not be obtained via a purely theoretical approach and MD becomes the only feasible option. If we extract from the MD simulations the relation $\mu = \mu(\rho, \chi)$ for the case we have just described, we can see from Fig. [3] that the results compare very well in this case as well; here, the constitutive relation $\mu = \mu(\rho, \chi)$ is computed through equilibrium MD simulations. The constitutive relation here is characteristic for this type of fluids, while is other fluid will be characterised by other relations. The density distribution across the interface is computed by equilibrium simulation, namely without any shear velocity imposed at the boundary.

In Fig. [4], we show the same results for two identical Lennard-Jones fluids with different levels of miscibility, induced by changing the interaction parameter β.

Changing the interaction (between the two fluids) results in different density profiles at the interface; when we move from complete immiscibility ($\beta = 0$) to full miscibility ($\beta = 1$), the density profile changes and the dip (molecular depletion region) gradually reduces to zero at full miscibility. Due to molecular depletion, the shear viscosity at the interfaces is lower than in the bulk region showing that the transport of momentum in the former region is less efficient than in the bulk; to balance the viscosity decrease, the velocity gradient increases, as shown in Fig. [4], leading to apparent slip. As the density contrast disappears, the associated variation in the velocity profile flattens and the apparent slip is not present anymore. If we introduce the corresponding density variation into the Navier-Stokes equations, numerical solutions of the latter are in excellent agreement with MD predictions.

In Fig. 5, we show that our approach is valid for different imposed shear rate ($\dot{\gamma}$), without requiring new, or *ad-hoc* conditions, as opposite to to approaches that rely on the use of slip models [15]. The density distribution does not depend on $\dot{\gamma}$, supporting the choice of computing at equilibrium conditions.

5. Conclusions

We have shown that the apparent slip velocity results from a molecular depletion at the interface, which is a direct consequence of the molecular interactions and, therefore, a property of the fluids not of shear rate $\dot{\gamma}$. When the two fluids are well-mixed at the interface, no apparent slip is present. This conclusion is further supported by the observation that the density distribution, which, in turn, determines the apparent slip, does not depend on the shear rate (as long as the fluids behave as Newtonian), but only on the intermolecular interactions. Furthermore, we have shown that if we account for the viscosity dependence on the density variation across the interface, the Navier-Stokes equations can be used to predict the velocity distribution throughout the domain, without the need for *ad-hoc* boundary conditions such as slip models, but all the properties can be independently measured.

With such a multi-scale approach, we can enforce some microscopic features into Navier-Stokes equations; the final result is therefore a better and physically sound description of liquid-liquid interactions.

Acknowledgements

All authors would like to acknowledge the support of the support of the NANO-BRIDGE project, namely "Heat and mass transport in NANO-structures by molecular dynamics, systematic model reduction, and non-equilibrium thermodynamics: An international collaborative effort to develop authentically multi-scale engineering recipes to BRIDGE the gap among scales and to validate them for carbon binders in thermal storage and carbon-based nano-filters" (PRIN 2012 call, code 2012LHPSJC, duration 36 months), involving Politecnico di Torino and Universitá di Brescia. PP and OKM thank

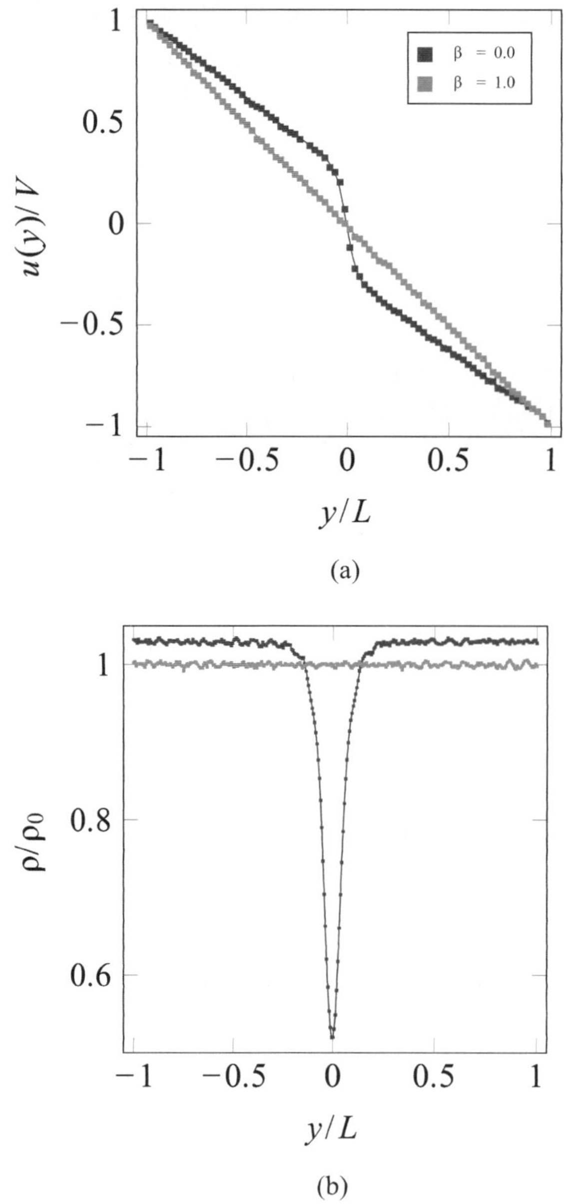

Figure 4: Normalized streamwise velocity component (V = 50 m/s, L = 40.205 Å, L_x = L_z = 108.34 Å) across the interface (a), obtained for different density distributions ($\rho(\dot{\gamma} = 0) \equiv \rho_0 = 1.36$ g/ml), (b), calculated by varying the parameter β_{ij} in the intermolecular potential used in the MD simulations. The Navier-Stokes solutions that use these distributions and MD velocity predictions are shown in (a) by the solid lines and symbols, respectively. The MD simulations use 17248 particles for each fluid.

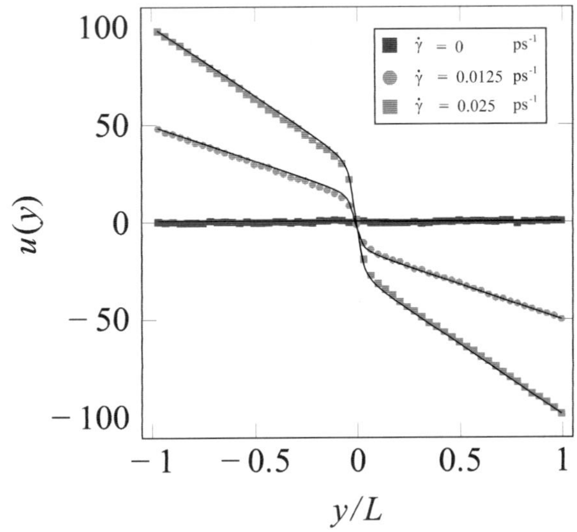

Figure 5: Streamwise velocity component calculated from the Navier-Stokes equations (solid lines) and MD simulations (symbols), obtained for different shear rates (L = 40.205 Å, $L_x = L_z$ = 108.34 Å), $\dot{\gamma}$ with $\beta_{ij} = 0$. The MD simulations use 17248 particles for each fluid.

acknowledge the support of the Royal Society through their International Exchanges Scheme (2014/R3), and the Engineering and Physical Sciences Research Council, UK, through the MEMPHIS programme grant (EP/K003976/1).

References

[1] G K Batchelor 1967 *An Introduction to Fluid Dynamics* (Cambridge University Press)
[2] J Atencia and D Beebe 2005 *Nature* **437** 648–655
[3] S Pleasants 2014 *Nature Photonics* **8**
[4] W Sparreboom, A van der Berg, J C T Eijkel 2009 *Nature Nanothechnology* **4** 713–720
[5] R Zhao and C W Macosko 2002 *J. Rheol.* **46** 145–167
[6] R Mauri 2013 *Non-Equilibrium Thermodynamics in Multiphase Flows* (Springer)
[7] S Senapati and ML Berkowitz 2001 *Phys. Rev. Lett.* **87** 176101
[8] P Poesio, A Damone, and OK Matar 2017 *Phys. Rev. Fluids* **2** 044004
[9] Sandia National Laboratories Temple University URL `http://lammps.sandia.gov/`
[10] K Issa and P Poesio 2014 *Phys. Rev. E* **89** 043307
[11] H Hoang and G Galliero 2013 *J. Phys. Condens. Matter* **25** 485001
[12] P Geysermans, N Elyeznasni, and V Russier 2005 *J. Phys. Chem* **123** 204711
[13] PA Thompson and S Troian 1997 *Nature* **389** 360–362
[14] G Galliero 2013 *Phys. Rev. E* **81** 056306
[15] S Razavi, J Koplik and I Kretzschmar 2014 *Langmuir* **30** 11272–11283

Nanoparticle Deposition During Cu-Water Nanofluid Pool Boiling

L Doretti[1], G A Longo[2], S Mancin[2], G Righetti[2] and J A Weibel[3]

[1]Dept. of Civil, Architectural and Environmental Engineering, University of Padova, Via Venezia 1, 35131, Padova, IT
[2]Dept. of Management and Engineering, University of Padova, Str.lla S. Nicola 3, 36100, Vicenza, IT
[3]School of Mechanical Engineering, Purdue University, 585 Purdue Mall, 47907-2088, West Lafayette, IN

E-mail: tony@gest.unipd.it

Abstract. The present research activity aims to rigorously investigate nanofluid pool boiling in order to definitively assess this as a technique for controlled nanoparticle coating of surfaces, which can enhance the nucleate boiling performance. This paper presents preliminary nanoparticle deposition results obtained during Cu-water (0.13 wt%) nanofluid pool boiling on a smooth copper surface. The tests were run in an experimental setup designed expressly to study water and nanofluid pool boiling. The square test sample block (27.2 mm × 27.2 mm) is equipped with a rake of four calibrated T-type thermocouples each located in a 13.6-mm deep holes drilled every 5 mm from 1 mm below the top surface. The imposed heat flux and wall superheat can be estimated from measurement of the temperature gradient along the four thermocouples. The samples are characterized by scanning electron microscopy (SEM) to analyse the morphological characteristics of the obtained thin, Cu nanoparticle coating.

1. Introduction

Pool boiling is widely used in many different engineering systems: chemical and nuclear reactors, refrigerating and air conditioning equipment, and thermal management of electronic devices. These applications have a shared limitation on the maximum heat flux that can be rejected by the cooling systems under safe, reliable, and efficient operation.

The study of the pool boiling has been a topic of the worldwide research since Nukiyama [1] conducted experiments to develop a boiling curve in stagnant liquid. It is well known that various surface treatment approaches can effectively enhance boiling heat transfer. In particular, microparticle coatings have been experimentally demonstrated to have promising capabilities for the enhancement of nucleate boiling heat transfer coefficients and critical heat flux (CHF).

Recent work has led to new concepts for surface modification at the nanoscale. Over the last decade, nanostructured materials (*e.g.*, nanowire coatings, nanoporous layers, carbon nanotube arrays, *etc.*) have been also shown to enhance nucleate boiling (NB) by several authors [4-6]. A comprehensive review can be found in Ref. [7].

An alternative strategy to enhance boiling heat transfer is by using fluid additives, for example by seeding the fluid with a small concentration of nanoscale particles to produce a nanofluid. The heat transfer behaviour of nanofluids has been extensively studied by many researchers since the late 1990s. For application of nanofluids under pool boiling conditions, some significant, though quite scattered, enhancements from 10% to 400% of the pool boiling CHF have been reported [8-12]; however, there are also contrasting reports of large deterioration [13-16]. Results in the literature are inconsistent even for the same nanoparticle size/type under similar experimental conditions.

There are a few possible parameters affecting boiling heat transfer with nanofluids, which includes morphological and thermophysical properties of nanoparticles and nanofluids, the stability of nanofluids, the content of nanofluids such as the presence of surfactants and ions, and the deposition and interaction of nanoparticles with the heating surface. As boiling heat transfer is very sensitive to surface characteristics, especially the number and shape of potential nucleation sites, any change in the surface would probably result in different boiling behaviours. There is a general scientific agreement on the fact that the enhancement or the deterioration observed during nanofluid boiling can be attributed to modification of the surface via nanoparticle deposition [17].

The present research activity aims to rigorously investigate nanofluid pool boiling in order to clearly asses the potential of this technique for obtaining either higher CHF values directly or to produce coated surfaces, which can enhance the nucleated boiling performance. This paper presents preliminary nanoparticle deposition results obtained during Cu-water (0.13 wt%) nanofluid pool boiling on a smooth copper surface.

2. Experimental setup and data reduction

As shown in Figure 1, tests are performed in an experimental setup designed and built to study water and nanofluid pool boiling on smooth or enhanced surfaces. In order to avoid any contamination of the components by the nanoparticles contained in the nanofluid, two identical setups were used, one for the pure fluid tests and one for the nanofluid tests.

The heater assembly consists of two heater block sections, both made of copper. The first 19 mm-tall test sample block has the top surface exposed to the working fluid; the second heater block section contains nine electrical cartridge heaters (200 W/240 V) controlled by a motorized Variac for heating of the test sample from below.

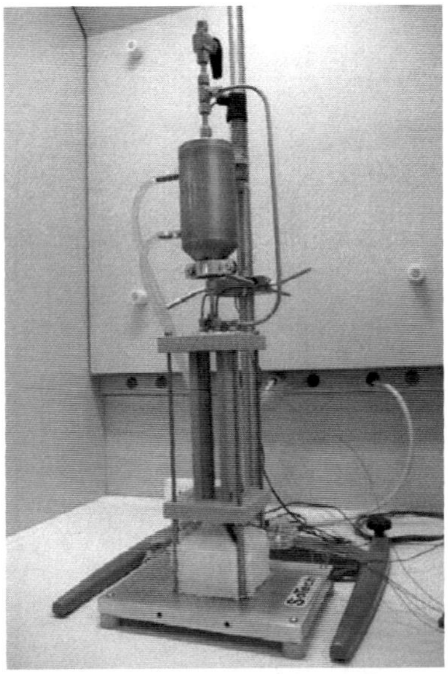

Figure 1. Photograph of the experimental setup. The test chamber for confined boiling experiments is mounted to the top of the heater block assembly.

The test sample block is designed to have a characteristic heater size equal to the Rayleigh-Taylor wavelength, being 27.2 mm in the case of water at ambient conditions. Each test sample block is glued in a polyether ether ketone (PEEK) plate (thermal conductivity, $\lambda = 0.028$ W m^{-1} K^{-1}) and exposes a smooth or enhanced top surface area (27.2 mm × 27.2 mm) to the working fluid (water or nanofluid). Two different boiling chambers were designed to analyse the boiling behaviour in confined and unconfined conditions (*i.e.*, whether the chamber is equal to or larger than the boiling surface, respectively). Both 200 mm-tall chambers consist of two glass and two PEEK walls; sealing is accomplished by means of a silicone o-rings. One chamber has exactly the same cross sectional size of the heating block to study confined pool boiling, while the other has a larger cross section (54.2 mm × 54.2 mm) to study the unconfined pool boiling behaviour. Nanoparticle deposition tests were always run using the confined chamber in order to avoid the contamination of the PEEK plate during the nanofluid boiling, whereas the other two chambers were used for the pure water tests.

Each boiling chamber is closed by a top PEEK plate and the vapour space is directly connected to the condenser, which is located on the top and fed with tap water. There are four threaded holes in the top PEEK plate to allow direct access to the boiling chamber. One hole is used to insert a temperature probe equipped with a T-type thermocouple (uncertainty ±0.05 K, for coverage factor, k = 2) to monitor the fluid temperature 20 mm above the heated surface while another hole is connected to an absolute pressure transducer (uncertainty ±0.065% f.s., f.s. = 20 bar, k = 2) to monitor the saturation pressure. The third hole is used to convey the vapour to the top of the condenser, while the last hole is used to charge the boiling chamber with fluid.

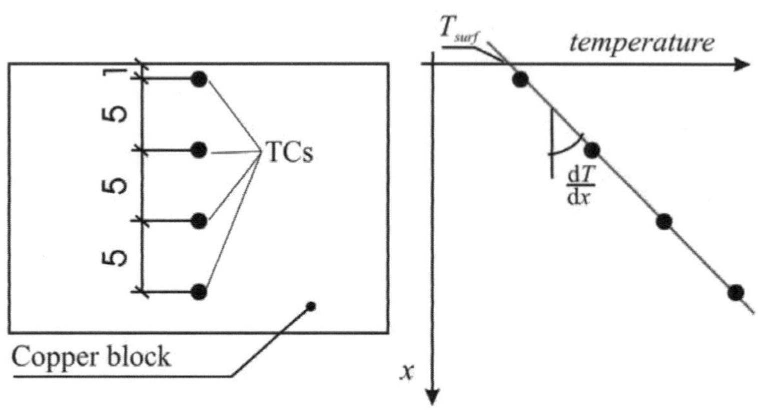

Figure 2. Schematic drawing of the test sample block with thermocouple locations indicated (dimensions in mm), and representative plot the temperature profile along the block height.

As illustrated in Figure 2, the copper test sample block is equipped with four calibrated T-type thermocouples located in as many 13.6 mm-deep holes drilled every 5 mm from 1 mm below the top surface. The imposed heat flux and wall superheat can be estimated from analysis of the temperature gradient measured by the thermocouples. The tests were run by increasing the heat flux up to the point of CHF. The data logging frequency was set at 1 Hz and, at steady state conditions, the experimental data collected over 100 s were time-averaged for subsequent post-processing of the actual surface heat flux and corresponding surface temperature.

In fact, assuming one-dimensional conduction, the temperature along the sample can be approximated by the following expression:

$$T(x) = T_{surf} + \left|\frac{\mathrm{dT}}{\mathrm{dx}}\right| \cdot x \tag{1}$$

where T_{surf} is the surface temperature at the position $x = 0$, and dT/dx is the temperature gradient estimated by a simple linear regression of the recorded temperatures. Thus, the heat flux q is given by:

$$|q| = \lambda \cdot \left|\frac{dT}{dx}\right| \tag{2}$$

where λ is the thermal conductivity of the sample block (387 W m^{-1} K^{-1} for 99.99% pure copper). The wall heat transfer coefficient h can be calculated by dividing the heat flux by the wall superheat, as:

$$h = \frac{q}{T_{surf} - T_{sat}} \tag{3}$$

Four additional calibrated T-type thermocouples were located in the lower heater block section to verify the estimated heat flux. A high-speed video camera is used to visualize the boiling phenomenon through the glass walls.

As described before, the heat flux and the wall temperature are obtained from a linear fit to the temperature data; the uncertainties in the linear fit by a least-squares regression are obtained as described by Brown *et al.* [18]. The surface temperature and heat flux uncertainties are approximately ±0.06 °C and ±4.0 kW/m^2 over the range of heat fluxes investigated.

3. Setup calibration

The experimental setup and procedure were validated through a set of calibration boiling curves obtained for water on a smooth copper sample, using both the confined and unconfined chambers. The copper sample topography was characterized by means of a 3D optical profilometer; Table 1 lists main surface roughness parameters of the sample.

Table 1. Surface roughness parameters of the smooth copper calibration sample according to ISO 4287 [19].

Roughness parameter	Calibration Sample
Ra (arithmetical mean deviation of the roughness profile)	0.070 μm
Rz (maximum height of the roughness profile)	0.695 μm
Rt (total height of the roughness profile)	1.83 μm
RSm (mean width of the roughness profile elements)	82.4 μm

The main results are plotted in Figure 3, which shows a water boiling curve obtained for the smooth copper sample in unconfined conditions. As described before, the tests were run by increasing the heat flux up to the point of CHF (yellow dots in Figure 3), which was found to be 1265 kW m^{-2} for this sample. Figure 3 also reports the value of the CHF as calculated using the well-known Zuber [20] correlation as reported by Lienhard and Dhir [21]:

$$CHF = 1.14 \cdot \frac{\pi}{24} \cdot \sqrt{\rho_V} \cdot h_{LV} \cdot [\sigma \cdot g \cdot (\rho_L - \rho_V)]^{0.25} \tag{4}$$

where the subscripts L and V refer to saturated liquid and vapour, respectively, ρ is the density, h_{LV} is the latent heat of vaporization, σ is the surface tension, and g is the gravitational acceleration.

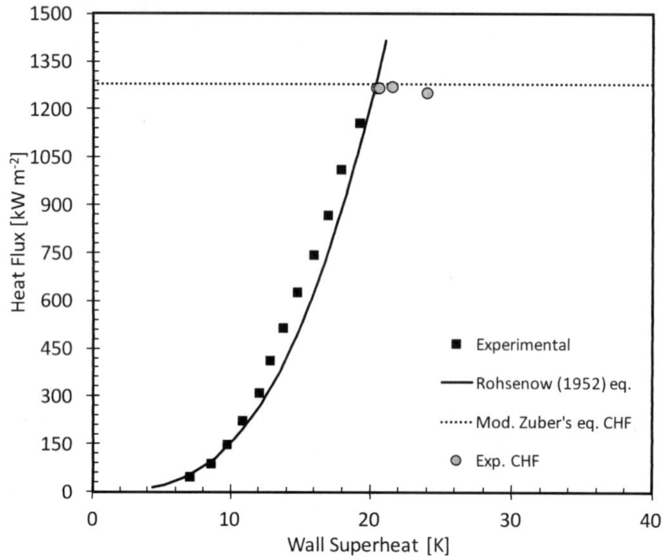

Figure 3. Boiling curve for water on the smooth copper surface in the unconfined test chamber.

When applied to the current experimental test conditions, the calculated CHF is equal to 1283 kW m^{-2}, which is in excellent agreement with the experimental value.
Furthermore, the diagram also reports the values of the heat flux as a function of the wall superheat calculated with the Rohsenow [22] equation:

$$q_{Ro} = \mu_L \cdot h_{LV} \cdot \left[\frac{g\cdot(\rho_L-\rho_V)}{\sigma}\right]^{0.5} \cdot \left(\frac{c_{p,L}\cdot(T_{surf}-T_{sat})}{C_{sf}\cdot h_{LV}\cdot Pr_L}\right)^3 \quad (4)$$

where μ_L, $c_{p,L}$, and Pr_L are the liquid dynamic viscosity, specific heat capacity, and Prandlt number, respectively. C_{sf} is the surface-liquid coefficient, which in this case (i.e. polished copper-water) was taken equal to 0.0128. The results are in fair agreement with the experimental values, showing relative and absolute deviations equal to -27.8% and 35.5%, respectively.

4. Nanoparticle deposition

The nanoparticle deposition was obtained via vigorous boiling of a Cu-water nanofluid obtained by seeding 25 nm copper nanoparticles into distilled pure water. Specifically, 25 g of copper nanoparticles were dispersed in 2 liters of distilled water, obtaining a 1.2 wt% (*i.e.*, 0.13 vol%) Cu-water nanofluid, which was then stirred for 4 hr and sonicated for 3 hr to ensure uniform dispersion.

The surface topography of the smooth copper sample used for the nanoparticle deposition test was also characterized; Table 2 lists the values of the measured roughness parameters. Moreover, the morphology of the surface was further analyzed using scanning electron microscope (SEM) imaging (FEG-ESEM, Quanta 250); a few selected SEM images are shown in Figure 4. Finally, the contact angle measured for a 5 μL water droplet deposited on the smooth copper surface was 88.2 deg ± 2.0 deg.

The experimental campaign consisted of two different tests: the first test investigated pure water boiling performance on the smooth copper surface while the second test was performed with the Cu-water nanofluid to obtain the nanoparticle deposition on the same copper block.

Table 2 Main surface roughness parameters of the smooth copper deposition sample according to ISO 4287 [19].

Roughness parameter	Deposition Sample
Ra	0.054 μm
Rz	0.473 μm
Rt	1.02 μm
RSm	55.6 μm

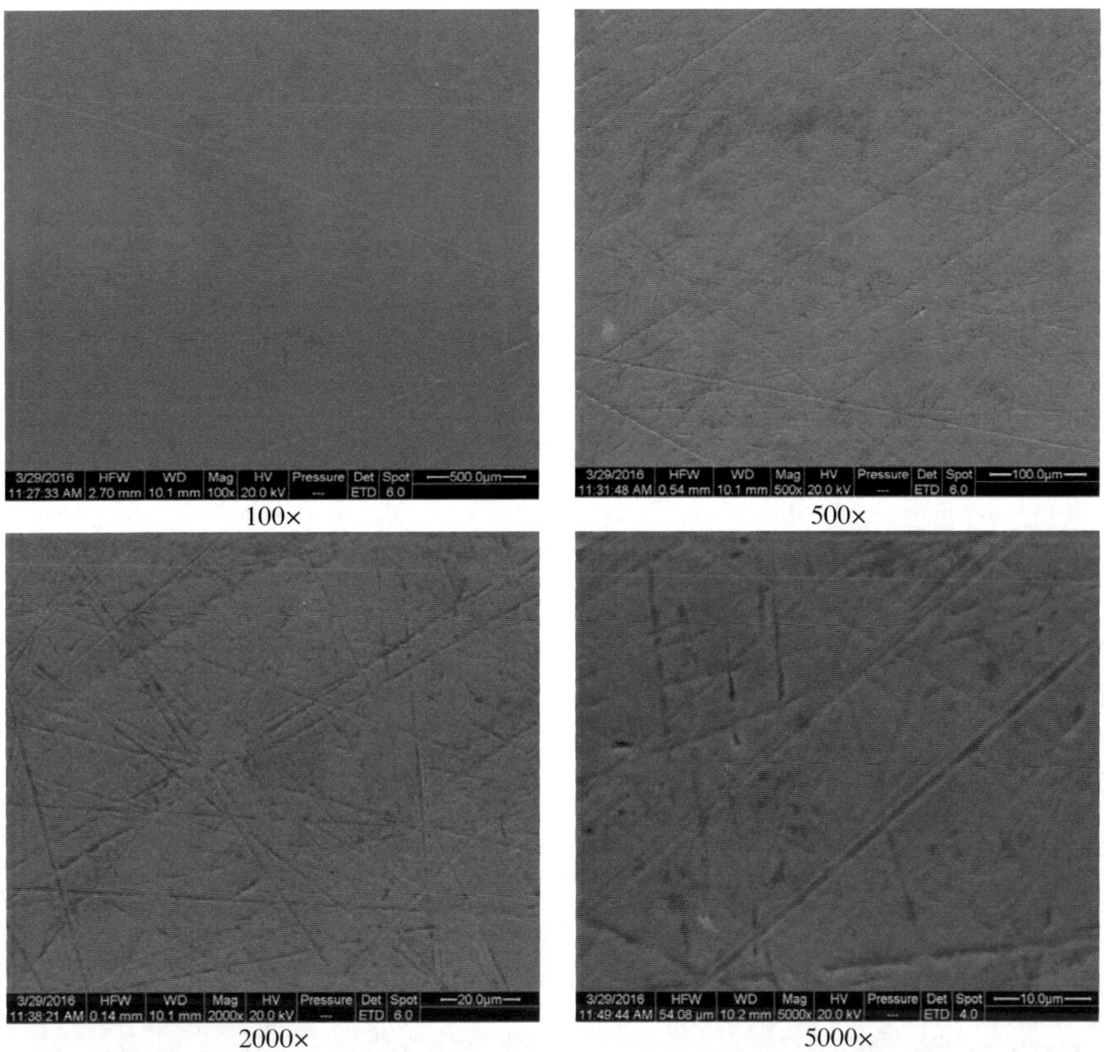

Figure 4. SEM images of the smooth copper surface before nanoparticle deposition.

Figures 5 and 6 report the boiling curves and values of the heat transfer coefficient as a function of heat flux that were measured during the pure water and Cu-water nanofluid pool boiling experiments. As shown in Figure 5, for both tests the heat flux was limited to around 30% of the CHF observed for the smooth surface to avoid this extreme event. The deposition was performance at approximately 300 kW m^{-2} and 400 kW m^{-2}; at both these heat fluxes, boiling was sustained for 30 min at steady state.

Analysing the results plotted in Figures 5 and 6, it appears that the boiling performance of the Cu-water nanofluid is better as compared to that measured during pure water boiling. In particular, at all the investigated heat fluxes, the associated wall superheats are lower in the case of nanofluid boiling, and thus the heat transfer coefficients are from 15% to 34% greater than those measured during pure water boiling.

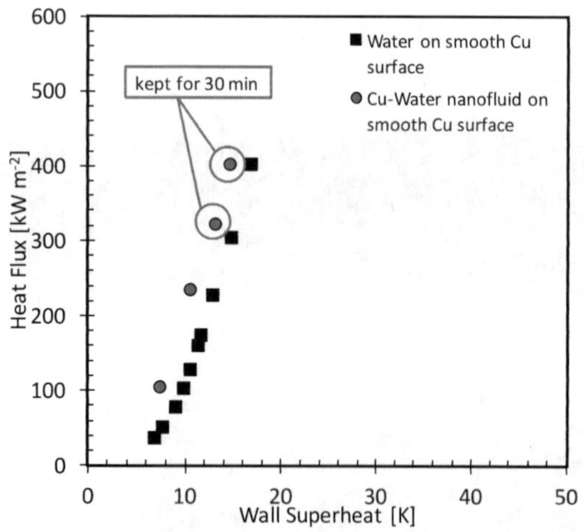

Figure 5. Boiling curve (*i.e.*, heat flux versus wall superheat) for pure water and Cu-water nanofluid boiling.

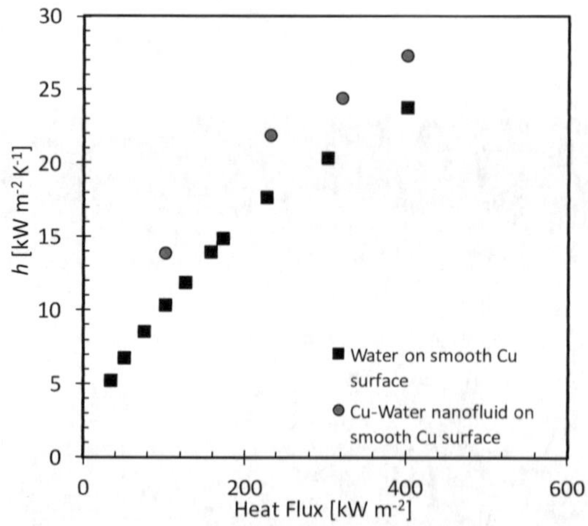

Figure 6. Heat transfer coefficient versus wall superheat for pure water and Cu-water nanofluid boiling.

Figure 7 presents two SEM images of the test surface after Cu-water nanofluid boiling taken at 90× and 100× magnification. The images reveal a thin layer of nanoparticles deposited onto the smooth surface, which is flaking up from the surface in several locations.

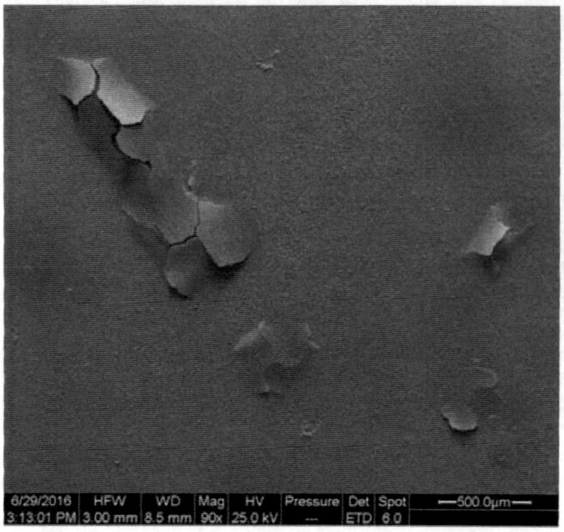

90×

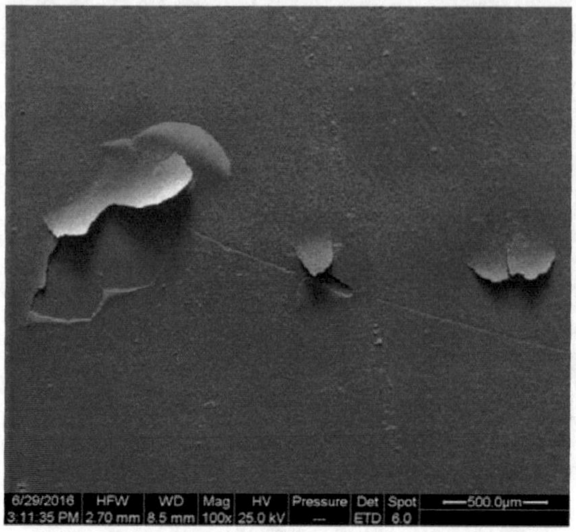

100×

Figure 7. SEM images of the coated copper surface after nanoparticle deposition by boiling of the Cu-water nanofluid.

The observed flaking process might be caused by cleaning of the sample before the SEM imaging procedure. The coating seems to be uniform and smooth, and it follows the topography of the uncoated surface. Figure 8 reports four more SEM images taken at higher magnifications from 1000× to 10000× in order to further inspect the coating.

Considering the first image taken at 1000×, the coating seems to consist of a sublayer of nanoparticles that is relatively uniform and dense; on top of which there is a secondary layer of additional nanoparticles that are more sparsely deposited. This layered deposition structure is more easily recognized in the image taken at higher magnification (5000×). When further zooming in to 10000× magnification, the additional nanoparticles, sparsely deposited, are more evident.

The contact angle measurement was repeated by depositing a 5 μL water droplet on the coated surface. It was found to be 107.9 deg, which is 19.7 deg higher than that measured for the uncoated surface before nanofluid boiling.

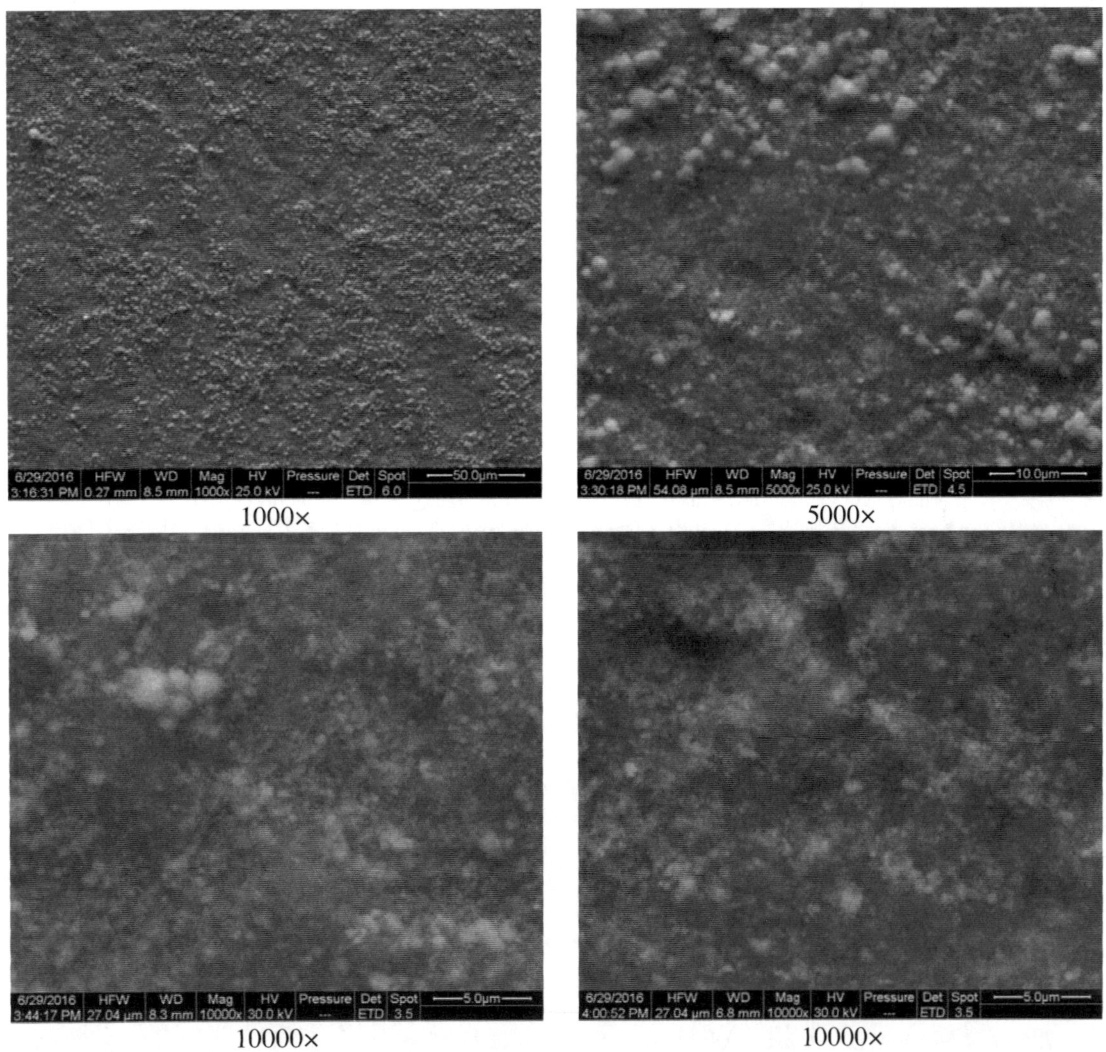

Figure 8. High-magnification SEM images of the coated copper surface after nanoparticle deposition by boiling of the Cu-water nanofluid.

5. Conclusions

This paper presents preliminary nanoparticle deposition results obtained during Cu-water (0.13 wt%) nanofluid pool boiling on a smooth copper surface. The results showed that the Cu-water nanofluid exhibited better boiling heat transfer performance, showing higher heat transfer coefficients as compared to boiling with pure water, on the smooth copper sample. After boiling with the nanofluid, a thin coating was formed on the surface by deposition of the nanoparticles; this coating seems to be uniform, with only a few defects found that might have been caused by post-test cleaning. Scanning electron microscopy images taken of the sample after nanofluid boiling also showed that the coating consists of a dense, uniform bottom layer onto which a secondary nanoparticle deposition occurred, creating another thin more sparsely porous layer. Finally, a contact angle measurement made on both the uncoated and coated surfaces highlighted the effect of the nanoparticle deposition on wetting: in fact, the contact angle increased from 88.2° to 107.9°. These preliminary results show the interesting possibilities offered by the nanofluid boiling as possible new particle deposition technique to realize thin coatings for nucleate boiling enhancement. Further experimental analyses are surely needed to understand the underlying deposition mechanisms in order to develop a rigorous and repeatable deposition protocol.

6. Acknowledgment

This research project was funded by University of Padova through the project CPDA149720: "An Innovative Nanoparticles Deposition Method for Nucleate Boiling Enhancement". The Authors would thank Eng. Roberto Meneghello and Dr. Enrico Della Rovere for their fundamental contribution to the surface characterization.

7. References

[1] Nukiyama S, 1934 J. *of the Japan Society of Mech. Eng.* **37** (206) 367-374.
[2] Hwang G S and Kaviany M 2006 *Int. J. Heat Mass Transf.* **49** 844-849
[3] Kim J H, Rainey K N, You S M and Pak J Y 2002 *J. Heat Transf*er **124** 500-506.
[4] Lu Y W and Kandlikar S 2011 *Heat Transf. Eng.* **32** (10) 827-842.
[5] Ujereh S, Fisher T, and Mudawar I 2007 *Int. J. Heat Mass Transf.* **50** 4023-4038.
[6] Li S, Furberg R, Toprak M S, Palm B and Muhammed M 2008 *Adv. Funct. Mater.* **18** 2215-2220.
[7] El-Genk M S and Ali A F 2010 *Int. J. Multiphase Flow* **36** 780-792.
[8] Kedzierski 2009 *J. Heat Transfer* **131** 043205.
[9] Park K-J and Jung D 2007 *Int. J. Heat Mass Transf.* **50** 4499-4502
[10] Liu Z, Xiong J and Bao R 2007 *Int. J. Multiphase Flow* **33** 1284-1295.
[11] Park K-J, Jung D and Shim SE 2009 *Int. J. Multiphase Flow* **35** 525-532.
[12] Kathiravan R, Kumar R, Gupta A, and Chandra R, 2009 *J. Heat Transfer* **131** 081902
[13] Trisaksri V and Wongwises S 2009 *Int. J. Heat Mass Transf.* **52** 1582-1588.
[14] Kim H D and Kim MH 2007 *App. Phys. Lett.* **91** 014104.
[15] Bang I C and Chan S H, 2005 *Int. J. Heat Mass Transf.* **48** 2407-2419
[16] Das S K, Putra N, Roetzel W 2003 *Int. J. Heat Mass Transf.* **46** 851-862.
[17] Wen 2012 *Applied Therm. Eng.* **41** 2-9.
[18] Brown K, Coleman H and Steele W 1995 In: *Proceedings of the 33rd Aerospace Science Meeting and Exhibit, Reno, NV, USA.*
[19] International Organization for Standardization, 1997 ISO4287
[20] Zuber N 1952 *Physics and Mathematics* **AECU 4439**.
[21] Lienhard JH and Dhir VK 1973 *National Aeronautics and Space Administration Report* NASA CR-2270
[22] Rohsenow WM 1952 *Transactions of the ASME*, **74**, 969.

IOP Conf. Series: Journal of Physics: Conf. Series **923** (2017) 012005 doi:10.1088/1742-6596/923/1/012005

Pre-test analysis of protected loss of primary pump transients in CIRCE-HERO facility

V Narcisi [1], F Giannetti [1], A Del Nevo [2], M Tarantino [2] and G Caruso [1]

[1] DIAEE - Nuclear Section, "Sapienza" University of Rome, Corso Vittorio Emanuele II, 244, 00186, Roma, Italy

[2] Italian National Agency for New Technologies, Energy and Sustainable Economic Development, C.R. ENEA Brasimone, Italy

E-mail: vincenzo.narcisi@uniroma1.it

Abstract. In the frame of LEADER project (Lead-cooled European Advanced Demonstration Reactor), a new configuration of the steam generator for ALFRED (Advanced Lead Fast Reactor European Demonstrator) was proposed. The new concept is a super-heated steam generator, double wall bayonet tube type with leakage monitoring [1]. In order to support the new steam generator concept, in the framework of Horizon 2020 SESAME project (thermal hydraulics Simulations and Experiments for the Safety Assessment of MEtal cooled reactors), the ENEA CIRCE pool facility will be refurbished to host the HERO (Heavy liquid mEtal pRessurized water cOoled tubes) test section to investigate a bundle of seven full scale bayonet tubes in ALFRED-like thermal hydraulics conditions. The aim of this work is to verify thermo-fluid dynamic performance of HERO during the transition from nominal to natural circulation condition. The simulations have been performed with RELAP5-3D© by using the validated geometrical model of the previous CIRCE-ICE test section [2], in which the preceding heat exchanger has been replaced by the new bayonet bundle model. Several calculations have been carried out to identify thermal hydraulics performance in different steady state conditions. The previous calculations represent the starting points of transient tests aimed at investigating the operation in natural circulation. The transient tests consist of the protected loss of primary pump, obtained by reducing feed-water mass flow to simulate the activation of DHR (Decay Heat Removal) system, and of the loss of DHR function in hot conditions, where feed-water mass flow rate is absent. According to simulations, in nominal conditions, HERO bayonet bundle offers excellent thermal hydraulic behavior and, moreover, it allows the operation in natural circulation.

1. Introduction

The Horizon 2020 SESAME project supports the development of the heavy liquid metal (HLM) system. The project aims to develop and validate advanced numerical code for design and safety evaluation of innovative HLM cooled reactors. Intense experimental campaigns will be performed in order to achieve a suitable database for the verification and validation of numerical codes.

In this frame, CIRCE (CIRColazione Eutettico) pool facility, at ENEA Brasimone Research Center, will be refurbished to host a test section aimed at investigating the primary system of a fast liquid metal cooled reactor. The test section will be equipped with an innovative heat exchanger (HX), named HERO. The objective of the experimental campaign will be to simulate the transition from forced to natural circulation in a protected loss of flow accident scenario, in order to provide an experimental database to promote the realization of a validation benchmark.

The objective of this work is to carry out pre-test simulations aiming to provide the preliminary test-matrix of the experimental campaign, which will be performed at the end of this year. Simulations are carry out with RELAP5-3D© (R5-3D) code upgrading the nodalization scheme validated during post-tests analysis on the CIRCE-ICE test facility [2].

The results of R5-3D simulations are presented comparing different accident scenario.

2. CIRCE-HERO test facility

CIRCE is a multipurpose pool facility designed to study innovative heavy liquid metal (HLM) cooled nuclear system. It consists of a main cylindrical vessel, which aims to contain different test sections, and two auxiliary tanks to store liquid metal during the maintenance phases and to host LBE during the transfer phases [3].

Previous test section installed in CIRCE, named ICE (Integral Circulation Experiment), has been upgraded in order to host the innovative HERO heat exchanger (HX) which consists of seven double wall bayonet tubes representing, as much as possible, ALFRED Steam Generator conditions. Figure 1 shows longitudinal section of the bayonet tubes, composed of four concentric tubes with an active length of 6 m (1:1 to ALFRED SG tubes). Feed-water enters the unit at the top edge of the slave tube; it flows downward, slightly increasing the temperature, and it is collected into the lower plenum. Then water flows upward through the volume between inner and second tubes, reaching superheated conditions. Finally steam is collected into the steam chamber, at the top of the unit. The main feature of HERO steam generator bayonet tubes (SGBTs) is the leakage monitoring system. It is obtained by filling the volume between second and third tubes (which offer a double physical separation between primary and secondary fluids) with pressurized helium and high thermal conductivity powder in order to detect any leakages, monitoring helium pressure, and while maintaining excellent heat exchange capability thanks to the metallic powder. Moreover the gap between slave and inner tubes is filled by an insulator layer in order to prevent the steam condensation on the outer surface of the inner tube [1].

HERO SGBTs are arranged into the hexagonal wrapper with a pitch-to-diameter ratio of 1.42 and the relative position between tubes and wrapper are fixed by five hexagonal spacer grids. The heat exchanger is contained into a double-wall cylindrical shell with air gap, in order to reduce heat losses towards the cold pool [1].

The test section is completed with a Venturi-nozzle flow meter, installed upstream the fuel pin simulator (FPS), which represents the core of the facility. The heat source (HS) consists of a bundle of 37 electrical pins, arranged in a wrapped hexagonal lattice with *p/d* of 1.8, characterized by the nominal thermal power of 800 kW and an active length of 1000 mm. The argon injection system is installed downstream the FPS, at the inlet section of the riser. At the top of the test section, the separator allows separation of hot LBE, flowing downward into the HX, and the argon, which flows upward to the gas plenum of the facility [2].

The test section is not equipped with a decay heat removal system (DHR) but this function will be performed by HERO heat exchanger, reducing the mass flow rate of the feed-water.

The main flow path of the test facility is depicted in Figure 2.

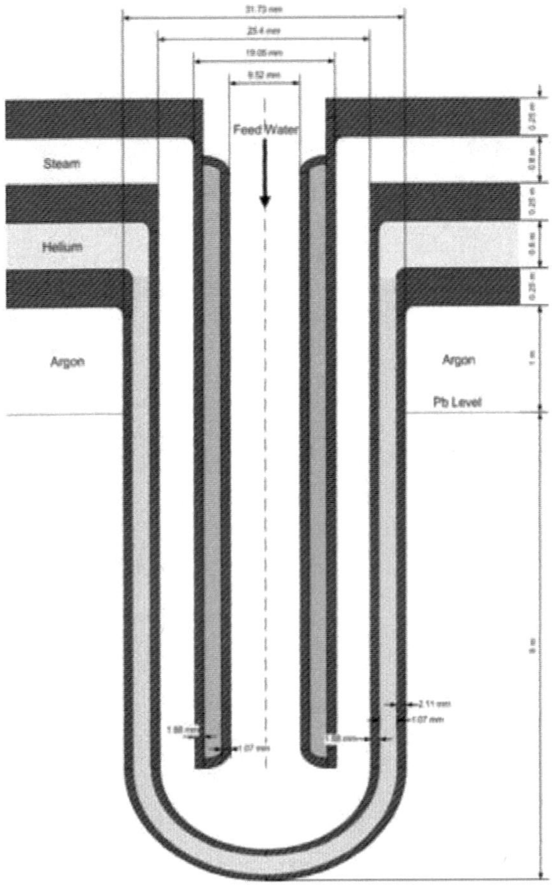

Figure 1. HERO steam generator bayonet tube

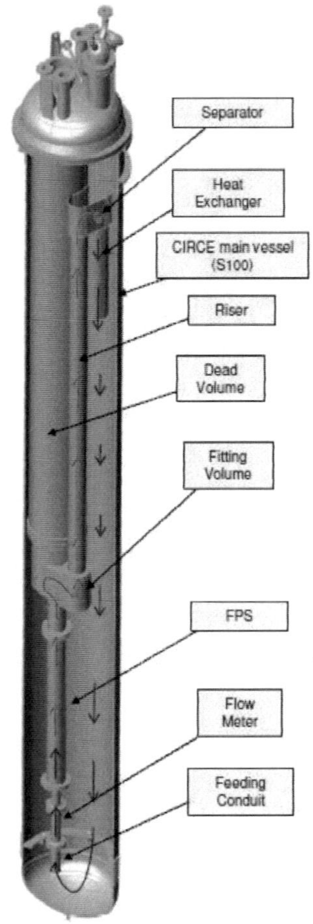

Figure 2. HERO test section: primary system main flow path caption.

3. RELAP5-3D© modelling

RELAP5-3D© is the last version of R5 series, which was developed to simulate accidental thermal-hydraulic transient in light water cooled reactors. The two main improvement of R5-3D from previous versions are the capability to simulate multi-dimensional component and the addition of new working fluids, including HLM [4].

The validated R5-3D nodalization scheme of CIRCE-ICE test facility [2] have been upgraded in order to carry out pre-test analysis of HERO experimental campaign. Three are the main differences between two test sections: HERO SGBTs substitutes for previous HX, the holes on the cylindrical shell of HS are closed, in order to limit the heat losses between FPS and cold pool, and the DHR system is removed.

R5-3D nodalization scheme consists of two macro regions, coupled to simulate the global facility: a mono-dimensional model is conceived to simulate thermal-hydraulic behaviour of the primary main flow path (see Figure 3) and three-dimensional component, shown in Figure 4 where the internals are depicted only to display the positioning, investigates thermal stratification and mixing convection phenomena into CIRCE main pool.

IOP Conf. Series: Journal of Physics: Conf. Series **923** (2017) 012005 doi:10.1088/1742-6596/923/1/012005

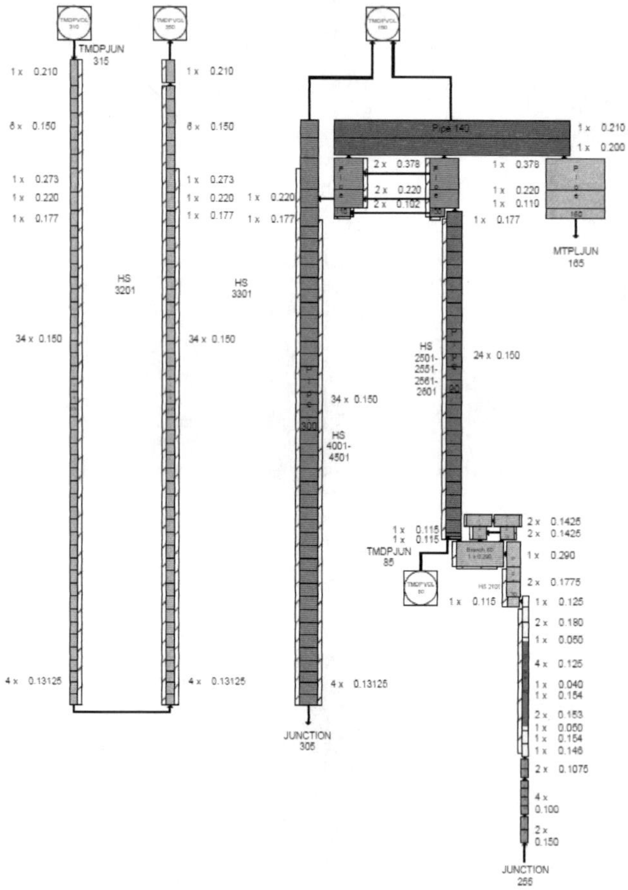

Figure 3. Region #1: mono-dimensional scheme

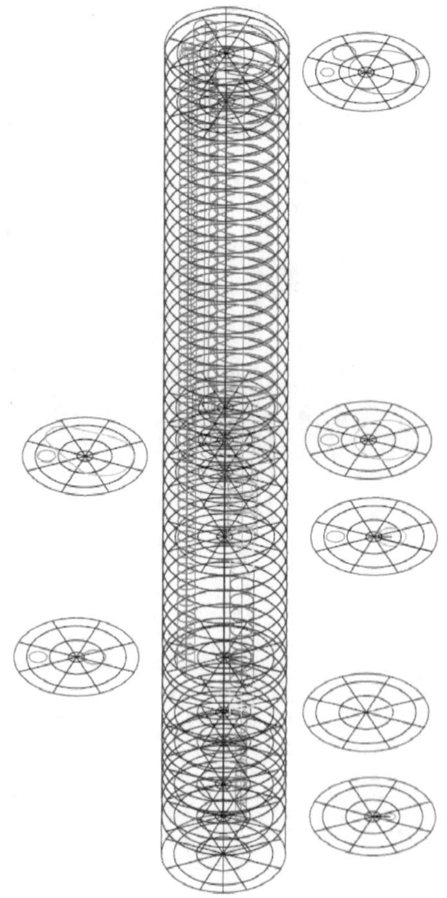

Figure 4. Region #2: multi-dimensional scheme

The 1-D nodalization scheme are hydraulically coupled with multi-dimensional component by three junctions (J). J-255 achieves connection between the feeding conduit and CIRCE lower plenum. The HS is simulated by a single equivalent pipe and one heat structure, which models the thermal power supplied by 37 electrical pins. The artificial fouling factor of 0.86, calibrated during post-test analysis on CIRCE-ICE campaign [2], is applied in order to correct the *p/d* value (the maximum allowed by R5-3D is 1.6) and to improve the heat transfer coefficient (HTC) according to Ushakov correlation [5], which better evaluates Nusselt number than the R5-3D default HTC correlation (Todreas & Kazimi correlation [6]), as showed in [7] for HLM system characterized by pitch-to-diameter ratio greater than 1.3.

At the inlet section of the riser, the argon injection system is simulated by boundary conditions, defining pressure, temperature and mass flow rate of the gas.

HERO LBE side is simulated by a single pipe composed of 43 volumes and coupled with the separator, at the top, and the downcomer of the 3-D component at the bottom. Seven bayonet tubes are concentrated into a single equivalent tube which includes feed-water inlet (boundary conditions define temperature, pressure and mass flow rate), descending tube, water/steam annular riser, steam chamber and steam outlet. Eight heat structures achieve thermal behavior of the unit; in particular, heat exchange between the water/steam and the LBE sides is increased by the calibrated fouling factor equal to 1.01, according to Ushakov correlation.

The pressure drop along the main flow path was evaluated during ICE post-tests and the same calibrated coefficients, depended on the flow conditions, are used on HERO model [2].

Multi-dimensional component consists of 51 axial levels, 4 radial meshes and 8 azimuthal intervals. The whole facility is modelled by 1877 control volumes, 4797 junctions and 15235 heat transfer nodes.

4. Pre-Test Calculations

4.1. Identification of full power steady state conditions

The experiment consists of the transition from forced to natural circulation in a protected loss of flow accidental scenario. In order to identify the initial conditions of the transient test, several calculations are performed.

Case 1 is set to achieve a constant temperature drop across the FPS equal to 80 K in the range of about 673-753 K, representative of the temperature drop across ALFRED core. The thermal power supplied by FPS is set to the nominal value of 450 kW, temperature pool is initialized to the average value of 690 K and argon and feedwater mass flow rate are respectively to 1.28 Nl/s and 0.3308 kg/s.

Cases from 2 to 7 consist of the same boundary conditions of case 1, except pool initial temperature, which is respectively set to: 670, 675, 665, 662, 660 and 655 K.

Case 8 aims to achieve LBE mass flow rate across the HX equal to 44.7 kg/s, representative of the scaled down SG of ALFRED. Boundary conditions are the same of case 3, except argon injection, set to the value of 2.354 Nl/s. Cases from 9 to 12 also use same conditions of case 3, except gas flow rate, respectively 2.242, 2.13, 1.85 and 1.79 Nl/s.

All simulations show that the trend of the main variables result stable up to 2500 s. The main results are summarized in Table 1. Cases from 1 to 7 show the temperature drop across the HS equal to 80 K but only cases from 4 to 7 comply with the range of temperature previous established.

Table 1. Preliminary calculations: main results

	Case 1	Case 2	Case 3	Case 4	Case 5	Case 6	Case 7	Case 8	Case 9	Case 10	Case 11	Case 12
FPS inlet T [K]	673.4	668.4	669.3	667.2	662.1	665.5	663.7	672.4	672.2	672.0	671.4	671.1
FPS outlet T [K]	754.2	749.7	750.9	748.4	747.4	746.8	745.1	740.8	741.3	741.9	743.7	744.2
LBE - HERO inlet T [K]	745.1	738.9	740.6	737.0	736.2	735.6	732.8	732.2	733.1	733.5	734.6	734.8
LBE - HERO outlet T [K]	671.8	668.3	669.4	667.6	666.8	666.4	665.1	672.4	672.1	672.1	671.4	671.0
LBE mass flow HERO inlet [kg/s]	37.6	37.5	37.5	37.5	37.4	37.4	37.3	44.5	43.9	43.4	42.1	41.7
Power removed by HERO [kW]	421.0	405.0	409.7	400.2	398.1	395.0	389.0	410.0	409.1	409.1	407.9	410.0
Steam max temp T [K]	663.4	657.8	659.0	682.8	654.6	656.6	652.0	659.1	659.0	659.0	659.0	658.7

The full power steady state, identified as initial conditions of the transient testes, is case 4, characterized by a temperature drop across the FPS of about 80 K and the average temperature into the pool in the range 665 to 680 K. Figure 5 shows LBE mass flow rate at the inlet section of the heat exchanger. The Ar injection starts at 50 s and it reaches the nominal value after 5 s. LBE mass flow rate immediately assumes the value of 35 kg/s, due to the enhanced circulation only. At 220 s from the beginning of the simulations, HS starts to supply thermal power up to the nominal value at 390 s. LBE temperature at the outlet section of the FPS reaches the maximum value of 750 K and temperature at inlet and outlet section of the HX starts to increase with a delay time (see Figure 6); meanwhile LBE mass flow rate reaches the maximum value of 38.5 kg/s. After 488 s from the beginning of the test, feedwater system is activated, instantly reaching the nominal mass flow rate value. Temperature at the outlet section of the HX decreases to the minimum value of about 658 K and then increases to the steady state value. After a delay time, LBE temperature at the inlet section of the FPS reaches the value of 748 K and the primary mass flow rate decrease to the nominal value of 37.5 kg/s.

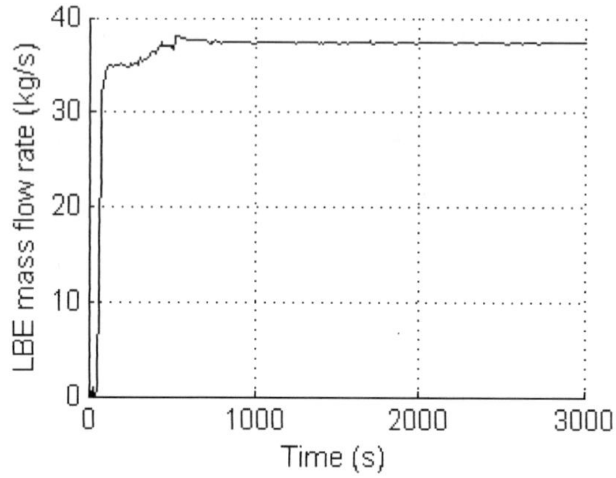

Figure 5. Case 4: LBE mass flow rate

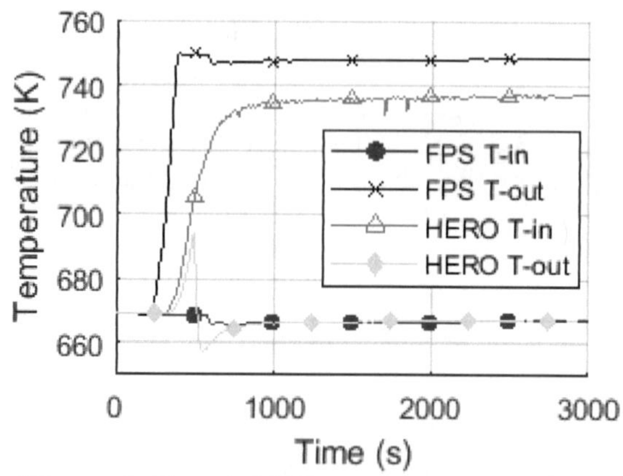

Figure 6. Case 4: LBE temperature at FPS and HX inlet/outlet

Previous analysis was carried out using thermophysical properties correlations implemented in R5-3D. The same simulations are repeated using the most recent correlations derived from [9], which have been implemented in Relap5-3D [10]. Figure 7 and Figure 8 show the comparison between R5-3D and NEA properties correlations on case 4. Figure 7 compares LBE mass flow rate at the inlet section of the HX. At the beginning of the test (from 50 to 220 s) the calculation with NEA properties underestimates mass flow rate following the overestimation of the concentrated and wall friction pressure drop. After the activation of HS and feedwater pump, temperature drop across FPS and HX increases of about 5% with NEA correlations (Figure 8) and it causes an increase in the driving force which compensates the overestimation of the total pressure drop.

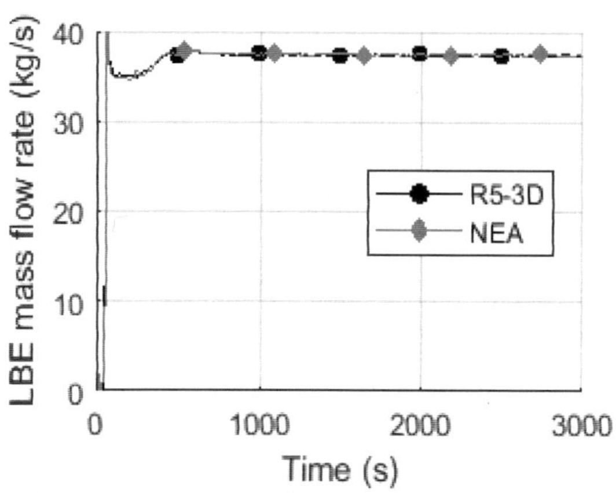

Figure 7. LBE mass flow rate: R5-3D and NEA properties comparison

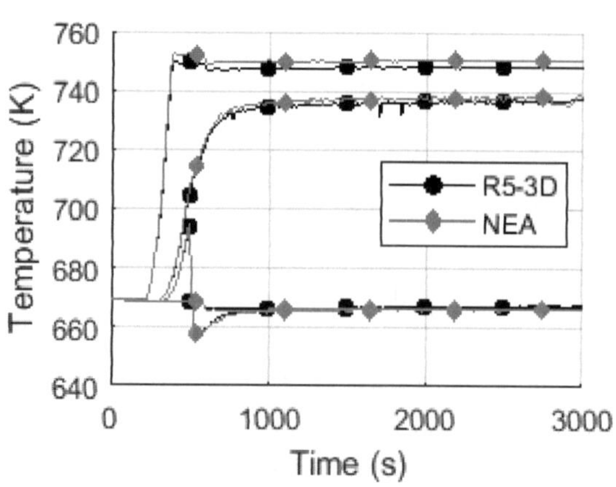

Figure 8. LBE temperature: R5-3D and NEA properties comparison

4.2. Transient analysis

The starting point for the transient analysis is considered case 4. Seven transient tests are considered in order to individuate the reference accidental conditions.

Test 1 consists of a protected loss of LBE pump, decreasing to 0 the argon injection with a curve simulating the presence of a pump flywheel. The FPS power simulates a decay heat curve, scaled down by a typical curve for fast reactor and compensated with heat losses through main vessel, and feedwater mass flow rate decreases to 10% of the nominal value to simulate the activation of the DHR system.

From test 2 to 6, boundary conditions are the same of test 1, except feedwater mass flow rate, set to 15%, 20%, 5%, 2% and 1% of the nominal value.

The last test 7 consists of a loss of LBE pump plus loss of DHR function in hot conditions: boundary conditions are the same of test 1, except feedwater mass flow rate which decreases to 0.

Figure 9 and Figure 10 depicts the evolution of LBE mass flow rate and LBE temperature for 1500 s from the transient test 1 start up. Immediately after the beginning of the accident, LBE temperature at the inlet section of the heat exchanger start to increase, reaching the maximum value of 740 K after 24 seconds. At the same time, thermal power has nearly reached the value of the compensated decay power and LBE mass flow rate is close to 50% of the nominal value. It results in the minimum value of temperature drop across the FPS and, with a delay time, across the HX; LBE reaches the temperature of 685 K at the FPS outlet and 705 K at the end of HERO. Figure 11 depicts the trend of the temperature into the pool, considering the main representative longitudinal section, which includes the inlet of the feeding conduit and the outlet of the steam generator, and also highlights the hot-spot downstream the HX at 85 seconds. Then, LBE mass flow rate reaches the minimum value and temperature at the outlet section of FPS and HX assumes the peak value at about 350 seconds; after mass flow tends to the nominal natural circulation value and the average temperature of the system starts to decrease. At the end of the transient test, the thermal stratification level, which occurs at 2.5 m above the bottom of the main vessel at the beginning of the transient test, is shifted downward of about half meter (see Figure 11).

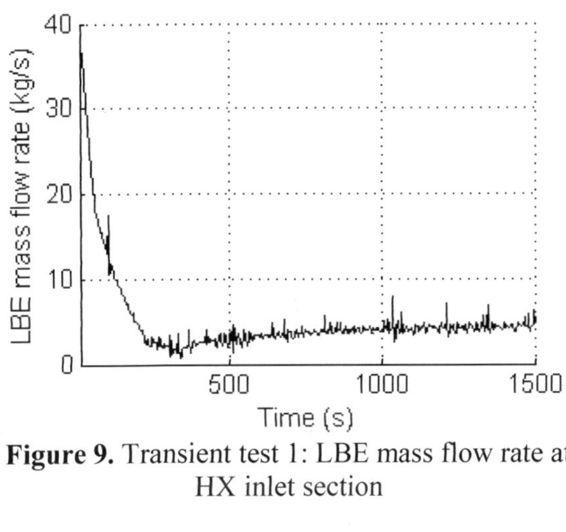

Figure 9. Transient test 1: LBE mass flow rate at HX inlet section

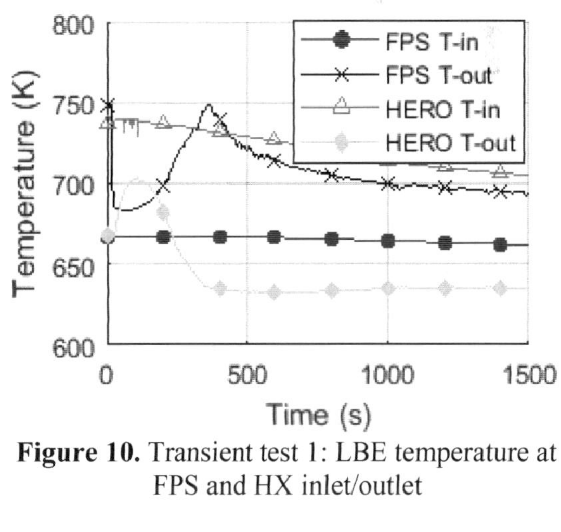

Figure 10. Transient test 1: LBE temperature at FPS and HX inlet/outlet

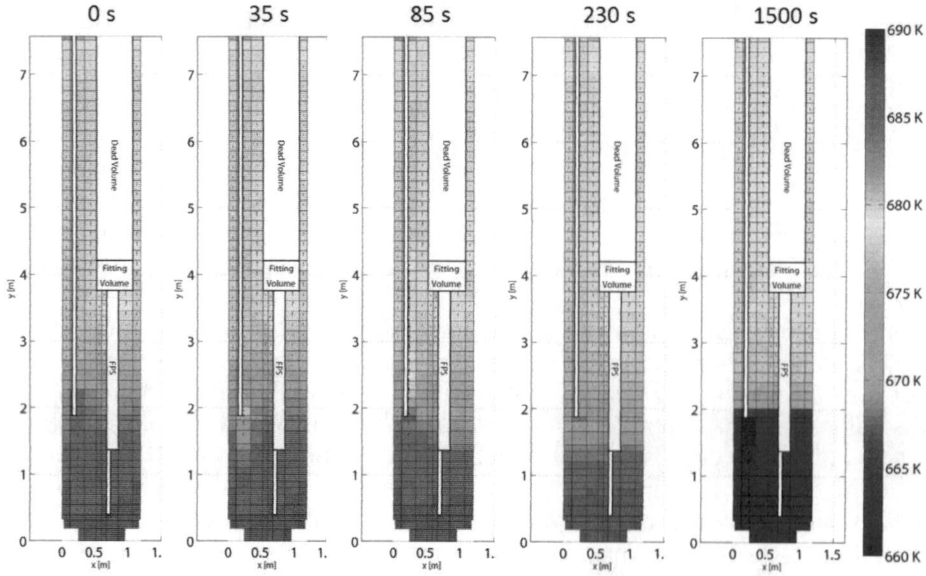

Figure 11. Transient test 1: pool temperature

Following figures highlight the comparison of test 3 and 7 with the reference transient test 1. Figure 12 shows the LBE mass flow rate at the inlet section of HERO. The driving force is deeply influenced by thermal power removed by the HX and LBE mass flow rate, during test 3, is constantly greater than test 1 and 7. In particular, during test 7, the driving force depends only on heat losses, since feedwater mass flow rate is stopped. Between 250 and 350 s the LBE reverse flow occurs caused by the hot fluid exiting the bottom edge of HX, which, meeting cold LBE into the pool (see Figure 14), starts to flow upward provoking the reverse flow into the heat exchanger.

Figure 13 depicts temperature at inlet and outlet of the FPS and the HX. Test 3 follows the same qualitative trend of test 1, while assuming lower temperature than the reference test. Significant differences occur during test 7; in the first 200 s of transient simulation, test 7 follows the same trend of the reference one, maintaining higher temperatures. After the establishment of the reverse flow, temperature at the inlet section of the HS reaches the pick value of about 740 K and the same effect occur at the outlet section of HERO, where LBE assumes the minimum value of 680 K. After that, the LBE mass flow rate reaches the nominal value of about 2 kg/s and temperature at the outlet section of the core assumes and maintains the maximum value of about 830 K.

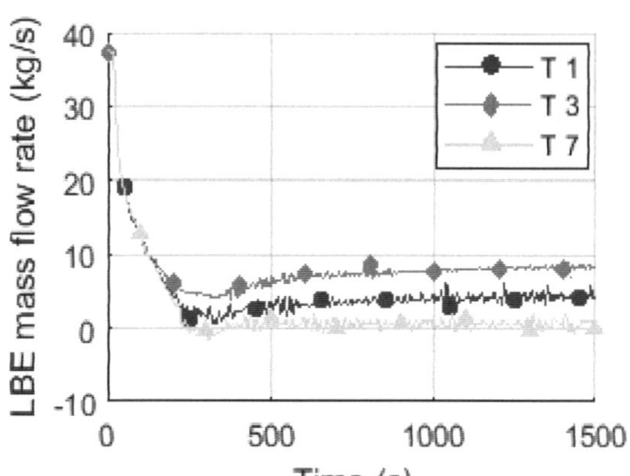

Figure 12. Transient test comparison: LBE mass flow rate

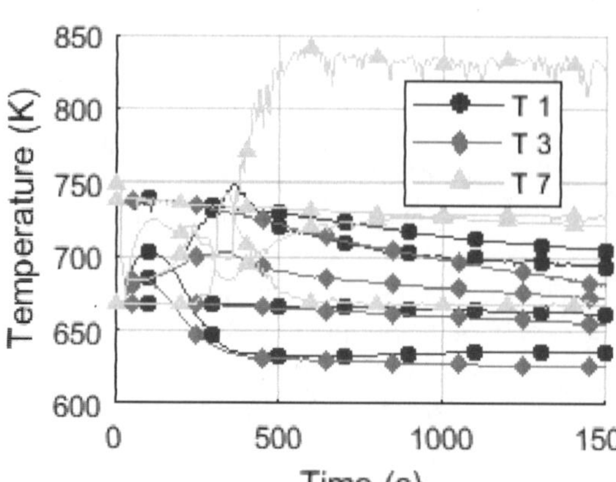

Figure 13. Transient test comparison: T at FPS and HX inlet/outlet

Figure 14 depicts the temperature evolution into the main vessel, showing, at the end of the simulation, the thermal stratification level is shifted downward of about one meter, bove the outlet of the heat exchanger.

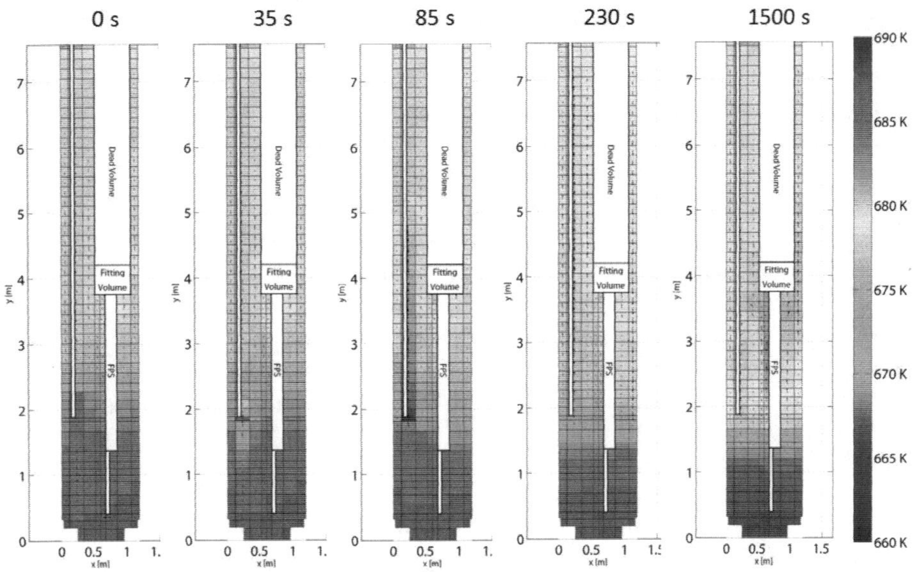

Figure 14. Transient test 7: pool temperature

The reference test is also performed using NEA recommended properties and comparison with R5-3D default properties is depicted in following figures which highlight same considerations of the steady state comparison.

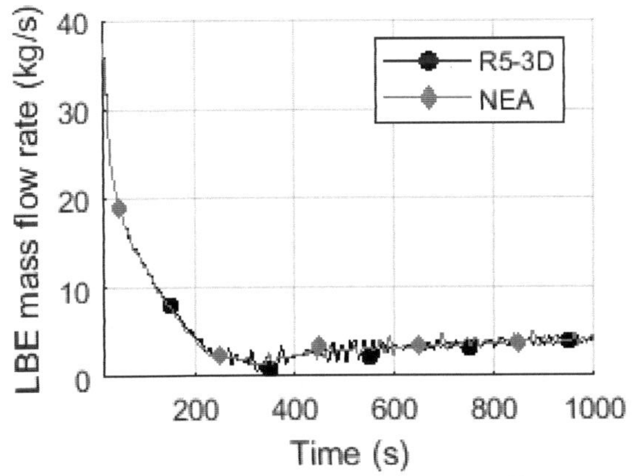

Figure 15. Transient test 1: LBE mass flow rate - R5-3D/NEA properties

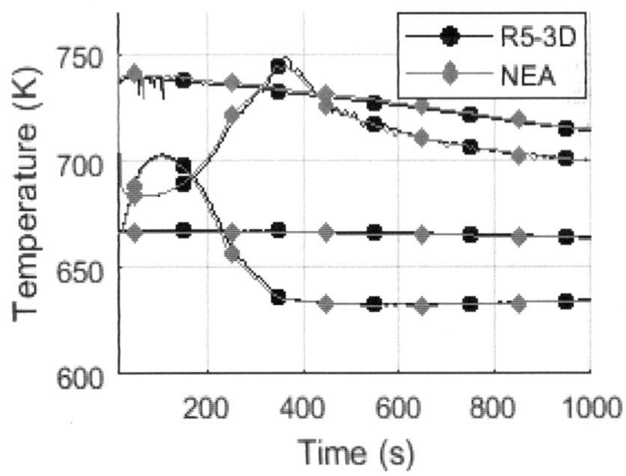

Figure 16. Transient test 1: temperatures - R5-3D/NEA properties

5. Conclusions

The main objective of this work is to present pre-test analyses on CIRCE-HERO test facility, to identify the test matrix for the experimental campaign. The simulations are performed with Relap5-3D© code, upgrading the validated model of CIRCE-ICE facility.

The analysis aims to investigate the transition from forced to natural circulation in a protected loss of flow accident scenario. At first a sensitivity analysis has been performed in order to identify steady state conditions which represents the initial conditions for the transient tests. After that, several

IOP Conf. Series: Journal of Physics: Conf. Series **923** (2017) 012005 doi:10.1088/1742-6596/923/1/012005

transient calculations have been carried out, reducing the feedwater mass flow rate from 20% to 0% of the nominal value. The simulation results highlight that HERO test section does not permit excellent natural circulation conditions. In particular, when feedwater mass flow rate decreases under 2% of the nominal value, code predicts the reverse flow of the primary coolant. In this frame, further investigations are necessary in order to confirm this phenomenon.

Transient test 1 is identifies as the reference scenario; further researches will verify the physical feasibility of the compensated power shutdown curve, with control system of CIRCE facility, and the possibility to reproduce a pump flywheel curve with the argon injection system.

References

[1] Narcisi V, Giannetti F, Caruso G, Del Nevo A, Di Piazza I and Tarantino M 2016 Installazione della Sezione di Prova HERO nella Facility CIRCE: Sviluppo Nodalizzazione e Progettazione della Campagna Sperimentale *ENEA Research Center Brasimone Report*
[2] Narcisi V, Giannetti F, Tarantino M, Martelli D and Caruso G 2017 *J. Phys. Conf. Ser. (this volume)*
[3] Tarantino M, Martelli D, Barone G, Di Piazza I and Forgione N 2015 *Nucl. Eng. Des.* **286** 261–77
[4] Team T R C D 1995 RELAP5-3D© Code Manual Vol.1 Code Structure, System Models and Solution Methods NUREG/CR55
[5] Ushakov P A, Zhukov A V and Matyukhin N M 1977 *High Temp.* **15**
[6] Todreas N E, .Kazimi M S, Schlumberger, Maersk, IMO, Giordano N, Brogan R J, Ahlgren F, Mondejar M E, Genrup M and Thern M 2015 *Taylor Fr. Publ.* **138** 1–10
[7] Giannetti F, Vitale Di Maio D, Naviglio A and Caruso G 2016 *Nucl. Eng. Des.* **305** 168–78
[8] Schikorr M, Bubelis E, Mansani L and Litfin K 2010 *Nucl. Eng. Des.* **240** 1830–42
[9] NEA 2015 Handbook on Lead-bismuth Eutectic Alloy and Lead Properties, Materials Compatibility, Thermal- hydraulics and Technologies 647–730
[10] Balestra P, Giannetti F, Caruso G and Alfonsi A 2016 *Sci. Technol. Nucl. Install.* **2016**

IOP Conf. Series: Journal of Physics: Conf. Series **923** (2017) 012006 doi:10.1088/1742-6596/923/1/012006

Pool temperature stratification analysis in CIRCE-ICE facility with RELAP5-3D© model and comparison with experimental tests

V Narcisi [1], F Giannetti [1], M Tarantino [2], D Martelli [3] and G Caruso [1]

[1] DIAEE - Nuclear Section, "Sapienza" University of Rome, Corso Vittorio Emanuele II, 244, 00186, Roma, Italy

[2] Italian National Agency for New Technologies, Energy and Sustainable Economic Development, C.R. ENEA Brasimone, Italy

[3] Dipartimento di Ingegneria Civile e Industriale, University of Pisa, Largo Lucio Lazzarino, Pisa, Italy

E-mail: vincenzo.narcisi@uniroma1.it

Abstract. In the frame of heavy liquid metal (HLM) technology development, CIRCE pool facility at ENEA/Brasimone Research Center was updated by installing ICE (Integral Circulation Experiments) test section which simulates the thermal behavior of a primary system in a HLM cooled pool reactor. The experimental campaign led to the characterization of mixed convection and thermal stratification in a HLM pool in safety relevant conditions and to the distribution of experimental data for the validation of CFD and system codes. For this purpose, several thermocouples were installed into the pool using 4 vertical supports in different circumferential position for a total of 119 thermocouples [1][2].The aim of this work is to investigate the capability of the system code RELAP5-3D© to simulate mixed convection and thermal stratification phenomena in a HLM pool in steady state conditions by comparing code results with experimental data. The pool has been simulated by a 3D component divided into 1728 volumes, 119 of which are centered in the exact position of the thermocouples. Three dimensional model of the pool is completed with a mono-dimensional nodalization of the primary main flow path. The results obtained by code simulations are compared with a steady state condition carried out in the experimental campaign. Results of axial, radial and azimuthal temperature profile into the pool are in agreement with the available experimental data Furthermore the code is able to well simulate operating conditions into the main flow path of the test section.

1. Introduction

The Fukushima Daiichi nuclear power plant (NPP) accident, happened on March 11, 2011, highlighted the need of NPP capability to assure residual heat removal for long periods and to limit significant off-site releases after the occurrence of a severe accident.

The lesson was acknowledged by Generation IV International Forum (GIF) and it established the requirement of highest level of safety for innovative nuclear systems [3].

Lead-cooled fast reactor (LFR) belongs to the six concepts selected by GIF as Generation IV systems and includes lead and lead-bismuth eutectic alloy (LBE) technologies; both coolants are chemically inert and they offer other attractive characteristics in terms of interaction with structural materials and thermodynamic features. LFR systems also well respond to lesson of Fukushima accident allowing natural circulation both in nominal and accident conditions. This feature offers considerable grace time in order to cope with unprotected loss of flow transient and it permits to introduce fully passive decay heat removal system (DHR), assuring very high safety features over long periods without need for operator actions, combined with active systems [3].

Considerable research and development (R&D) activities were carried out in Europe. In this frame, the CIRCE pool facility (CIRColazione Eutettico), at ENEA Brasimone research center, was refurbished in order to host the test section ICE (Integral Circulation Experiment) which aims to simulate the thermal-hydraulic behavior of the primary system in a HLM cooled pool reactor. The test section is equipped with a fuel pin simulator (FPS), which electrically simulates the core of the facility, the heat exchanger (HX) and the DHR system, immersed in the upper part of the pool. The LBE circulation into the main flow path is enhanced by the injection of argon at the inlet section of the riser [5].

The experimental campaign performed on CIRCE-ICE facility aims to investigate mixing and thermal stratification phenomena in a pool type reactor, which should induce thermo-mechanical stress on the structures [2]. Moreover, the validation of TH (Thermal Hydraulic) codes against experimental data is a fundamental step in order to justify their use in the design phase for improving safety aspects.

In this work the experimental campaign was reproduced using the system code RELAP5-3D© in order to investigate the capability to simulate thermal phenomena which occur into the pool of innovative HLM reactor. The RELAP5-3D© model has been set up by coupling mono-dimensional model of the main flow path, HX and DHR secondary side included, with the 3D model of the pool.

2. Experimental campaign

2.1. CIRCE-ICE test facility

CIRCE is a multipurpose pool facility designed to host different test sections welded to and hung from bolted vessel heads for the investigation of thermal-hydraulic aspects related to the HLM pool system.

The facility consists of a main vessel, earmarked for containing test section and filled with about 70 tons of molten LBE, two auxiliary tanks, dedicated to store LBE during maintenance phases and to transfer liquid metal during loading and drainage phases, and data acquisition system [6].

The ICE test section, placed inside the main vessel, aims to simulate the primary system of HLM pool-type reactor. Main components and their positioning are showed in Figure 1 while the main flow path is highlighted in Figure 2. The feeding conduit is the inlet pipe of the test section which allows connection between the lower plenum and the fuel pin simulator. A Venturi-nozzle flow meter is installed into the feeding conduit in order to measure the primary flow rate through the "fuel" bundle. Molten LBE flows upward and it increases temperature passing through the FPS, which consists of an electrical pin bundle with a nominal thermal power of 800 kW and an active length of 1000 mm. It is composed of 37 electrically heated pins arranged in a wrapped hexagonal lattice and characterized by a pitch to diameter ratio equal to 1.8 (Figure 3). Each pin has an outer diameter of 8.2 mm, a thermal power of 25 kW and a heat flux at the pin wall of 1 MW/m^2. The hot fluid, exiting the heat source (HS), is introduced into the fitting volume and then it flows upward into the riser which is a double wall pipe connecting the fitting volume with the separator. At the inlet section of the riser, a nozzle is installed allowing the injection of argon. The mixture flows upward and it collects in the separator where the separation of LBE and Ar occurs (LBE enters the HX while Ar flows upward into the gas plenum through the free surface). The heat exchanger is made of 91 bayonet tubes with an active length of 3462 mm; Figure 4 shows a sketch of the bayonet element which consists of three concentric tubes. The feed-water flows downward into the inner tube and then upward into the annular region between inner and middle tube, where the change of phase take place; outside the tubes, the LBE flows downward decreasing the temperature. The volume between middle and outer tube is filled by pressurized helium to detect any leakage. Exiting the HX, primary coolant flows through the downcomer reaching the lower plenum [6].

The positioning of the DHR system is shown in Figure 1. It consists of only one bayonet tube and the decay power is removed by forced circulation of air. The bayonet is inside a double wall shell with a thin air insulation gap to thermally decouple the DHR from the external LBE pool. Hot LBE enters the DHR by the upper inlet section, it flows downward decreasing the temperature and it exits the component in the downcomer [1].

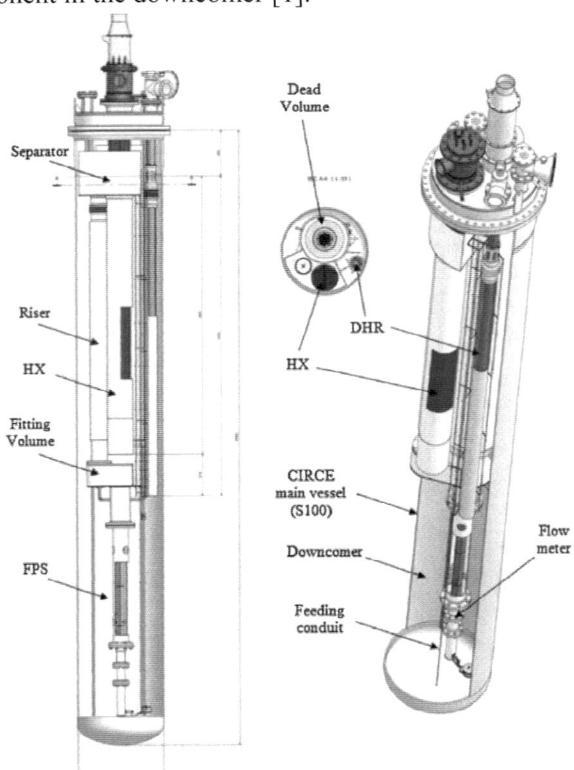

Figure 1. ICE test section[1]

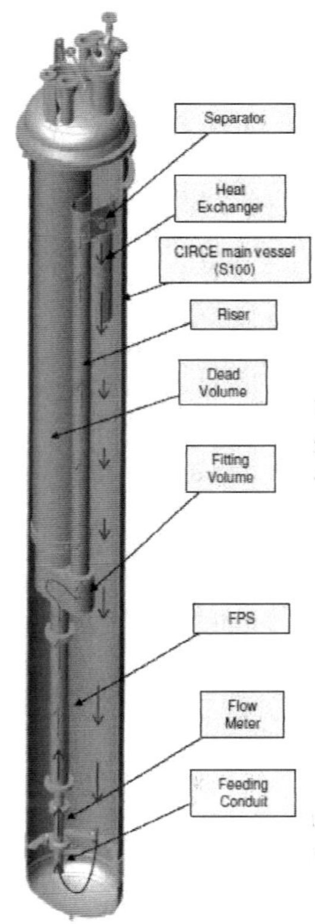

Figure 2. ICE test section: primary system main flow path

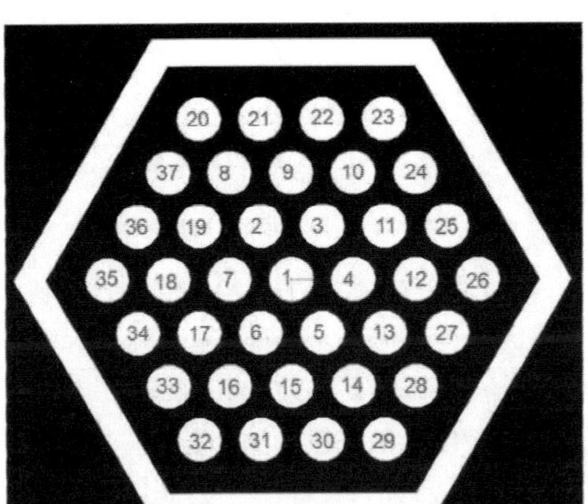

Figure 3. FPS cross section

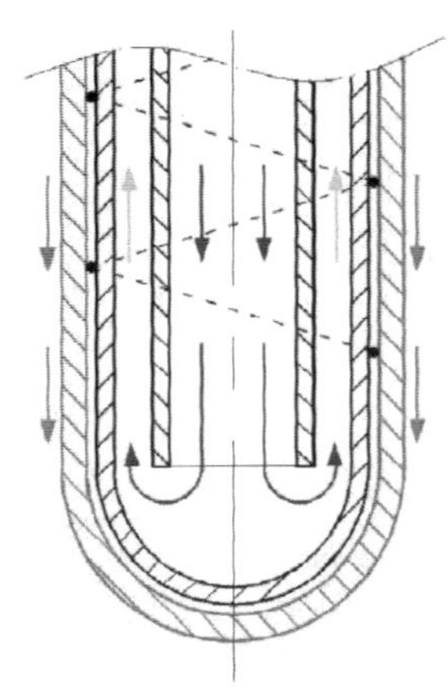

Figure 4. HX bayonet tube[6]

2.2. Test matrix

In this work, the steady state condition reached during the full power operation is analysed. The FPS supplies a power of about 720 kW and 0.65 kg/s of feed-water flows into the main HX; during the test, the DHR system is not activated.

The test considered is TEST I [1]; the main parameters are showed in Table 1.

Table 1. Initial and boundary conditions of the test

	Duration (h)	Electrical power supplied(kW)	Initial LBE temperature (K)	LBE mass flow rate (kg/s)	Feed-water mass flow rate (kg/s)
TEST I	7	720	600	55	0.65

3. RELAP5-3D© modelling

Post-test simulations are carried out using RELAP5-3D© v. 4.3.4 (R5-3D). RELAP5 code was developed to allow transient simulation of light water reactor during postulated accidents. R5-3D is the last version of the series of RELAP5 code and it contains several improvement; the two most enhancements from the previous versions are the multi-dimensional thermal-hydraulic capability and the addition of new working fluids, including heavy liquid metals [7].

The nodalization scheme of the CIRCE-ICE facility is divided into two macro-regions. The first one is depicted in Figure 5 and represents the mono-dimensional model of the ICE test section.

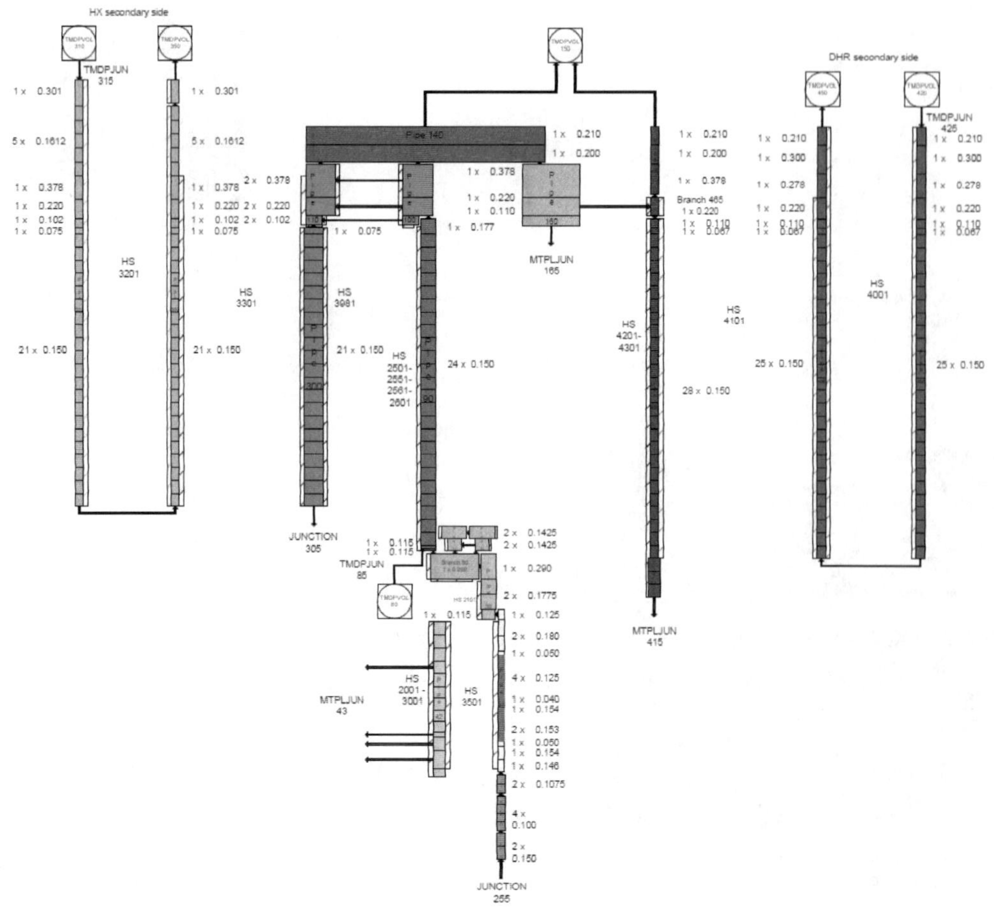

Figure 5. Region #1: mono-dimensional scheme of the ICE test section, DHR and HX

IOP Conf. Series: Journal of Physics: Conf. Series **923** (2017) 012006 doi:10.1088/1742-6596/923/1/012006

The 1-D scheme reproduces all components described in previous section. The HS is simulated by a single equivalent pipe and one heat structure, which reproduces the 37 electrical pins. In R5-3D, for the evaluation of the heat transfer coefficient (HTC) on heavy liquid metals, Todreas & Kazimi correlation [8] is implemented. Previous simulations of HLM system showed that this correlation underestimates the Nusselt number for pitch-to-diameter ratio greater than 1.3 [9]. Additionally, R5-3D does not permit a pitch-to-diameter ratio of 1.8 and the *p/d* of the pin bundle is set to the maximum allowed value of 1.6. In order to improve the HTC according to Ushakov correlation [10] (more accurate in this case) and to correct the heat exchange to experimental *p/d* value, an artificial fouling factor of 0.86 is applied to the heat transfer coefficient.

Upstream the FPS, the pressure drop of the Venturi nozzle is simulated by a concentrated pressure loss coefficient *K*, dependent on the flow conditions [11], according to equation (0.1).

$$K_{Venturi} = 10.5\,\mathrm{Re}^{-0.014} \tag{0.1}$$

The argon injection at the inlet section of the riser is simulated by boundary conditions: the time dependent volume sets gas inlet conditions and the time dependent junction adjusts the mass flow rate injection. An additional time dependent volume sets the pressure of the gas plenum of the facility.

The HX primary side is simulated by a single equivalent pipe and one heat structure, which thermally couples LBE and water side. A calibrated fouling factor of 1.02 is applied on the LBE side to increase the HTC, according to Ushakov correlation. The bayonet tubes are modelled by only two pipes, in order to simulate the descending and ascending side of water/steam tubes, and one heat structure, to model heat dispersion between the two pipes.

The pressure losses due to grids installed into FPS and heat exchanger are calculated by the Rheme correlation [12]:

$$\Delta p_{grid} = C_v \cdot \varepsilon^2 \cdot 0.5 \cdot \rho \cdot v^2 \tag{0.2}$$

where ρ and v are respectively the density and the velocity of the fluid while ε represents the blockage factor of the grids, calculated as:

$$\varepsilon = \frac{A_{grid}}{A_{flow}} \tag{0.3}$$

The C_v parameter is a modified drag coefficient and it is calculated as [12]:

$$C_v = MIN\left[3.5 + \frac{73.14}{\mathrm{Re}^{0.264}} + \frac{2.79 \cdot 10^{10}}{\mathrm{Re}^{2.79}}, \frac{2.6}{\varepsilon^2}\right] \tag{0.4}$$

The bayonet tube of the DHR system is also simulated and it is composed of one pipe for the LBE channel and two pipes to model the descending and ascending air side.

The second region, depicted in Figure 6, represents the 3D model of the main vessel, composed of 51 axial levels, 4 radial meshes and 8 azimuthal intervals. The internals depicted in Figure 6 are only representative of the positioning into the multi-dimensional component.

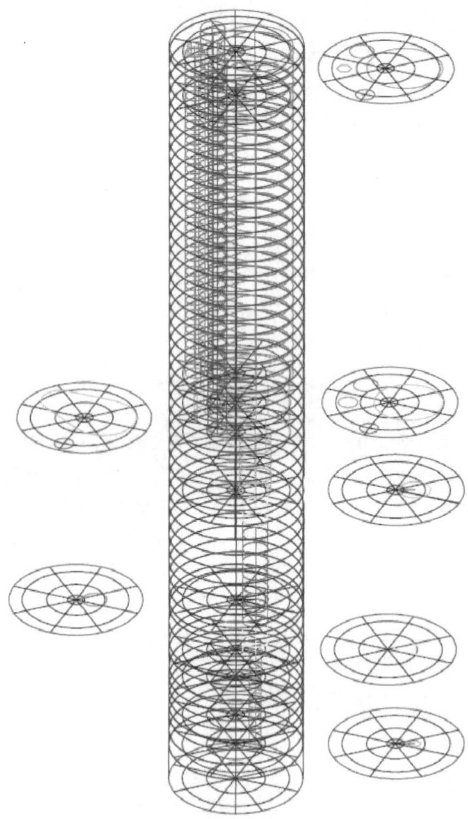

Figure 6. Region #2: multi-dimensional scheme of the CIRCE pool

Mono-dimensional and multi-dimensional components are coupled by junctions, showed in Figure 5. Several heat structures model the heat losses between the LBE cold pool and the hot internal components, which influence the LBE temperature at the inlet section of HX. Additionally, main vessel heat losses are simulated by calibrated heat transfer coefficients.

CIRCE-ICE model is completely composed of 1929 control volumes, 4856 junctions and 15353 heat transfer nodes.

4. Results analysis

The simulation analysis are conducted using default thermophysical properties correlations of R5-3D and repeated using the most recent correlations for LBE [13], implemented in R5-3D as described in [14].

Table 1 summarize the boundary conditions of the experimental tests. TEST I simulations are carried out reducing by 5% the electrical power supplied to the FPS, to take into account the power dissipated by joule effect in the cables and connectors, which does not contribute to the thermal power supplied to the LBE by the HS.

The LBE mass flow rate, measured by the Venturi-nozzle, is represented in Figure 7. The strong oscillations, during first phase of the experimental test, are due to the volumetric blowers used to inject argon into the riser and, after 13000 seconds, they are dumped by installing a check valve into the gas injection system [1]. The experimental data are compared to the simulated values, calculated with default correlations of R5-3D and Nuclear Energy Agency (NEA) recommended correlations. Unless the fluctuations, Figure 7 highlights that both calculated values well reproduce the experimental data, reaching a value of about 55 kg/s. According to [14], the simulation with default correlations underestimates the LBE mass flow of about 1%, because the smaller value of the natural driving force, depending on temperature drop across the core and the heat exchanger. The experimental trend of LBE temperature at the inlet and outlet sections of the HS, both represented by the average value of three thermocouples, is compared in Figure 9 with the calculated values. Obtained numerical results well reproduce the experimental trend. In particular the simulation with NEA correlations highlights a higher temperature drop of about 3% (mainly due to the different specific heat capacity) and, as a consequence, a slight overestimation of the experimental data.

Figure 10 depicts the comparison of the LBE temperature at the heat exchanger inlet and outlet sections, obtained averaging respectively the values of three and six thermocouples. The temperature at inlet section is lower than the value at the outlet of the FPS, due to the heat losses through the hot leg (towards the pool), which are well predicted by the simulations. The difference between the experimental and calculated temperature at the outlet of the HX depends on the relative positioning between the thermocouples and the volume control, arranged 45 mm upstream. The use of the NEA correlations results an increment of about 4% in the temperature drop, and a better agreement with the experimental data.

Figure 8 shows the comparison of the thermal power removed by the he13at exchanger. The calculated values are essentially the same and they both slightly overestimate the experimental data, probably due to the fouling which is not considered during the simulations.

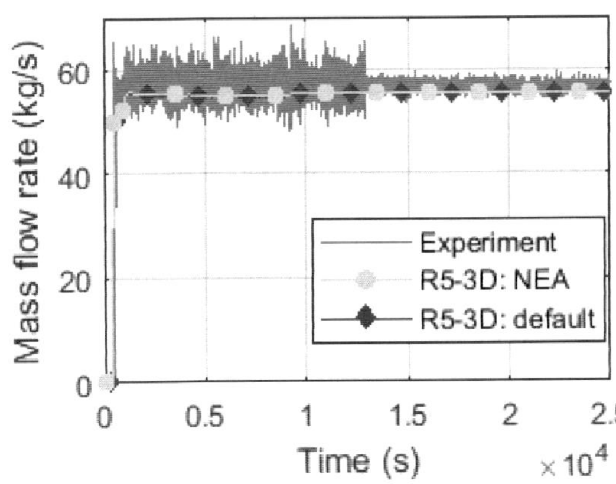

Figure 7. TEST I: LBE mass flow rate

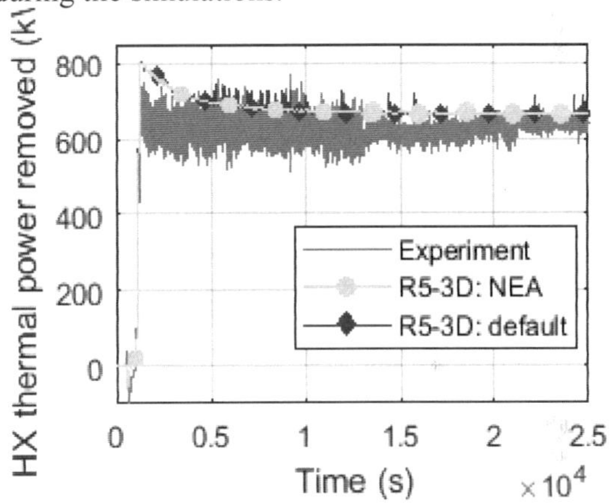

Figure 8. TEST I: HX removed power

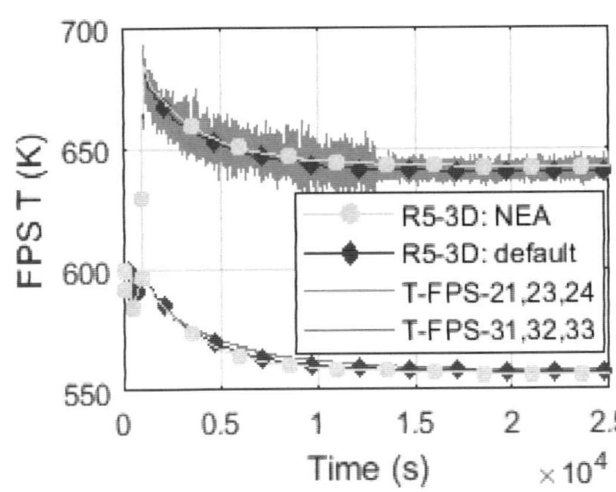

Figure 9. TEST I: temperature at FPS inlet/outlet

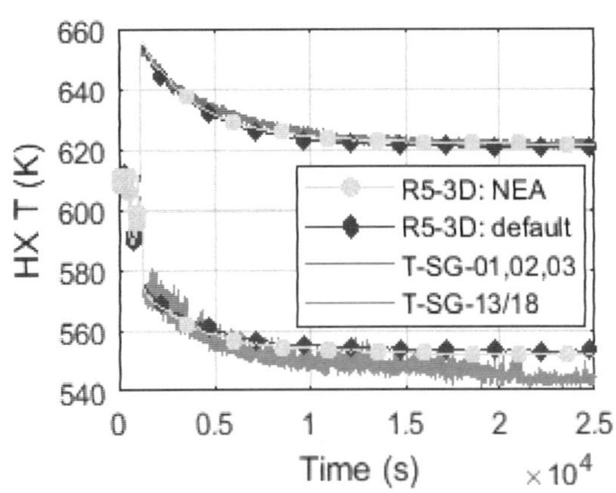

Figure 10. TEST I: temperature at HX inlet/outlet

In order to investigate thermal stratification and mixing convection phenomena into the pool, CIRCE is equipped with several thermocouples (TCs), arranged in different vertical and

IOP Conf. Series: Journal of Physics: Conf. Series **923** (2017) 012006 doi:10.1088/1742-6596/923/1/012006

circumferential positions, as showed in Figure 11 [1]. The multidimensional component, on R5-3D model, is conceived to compare LBE temperature in the exact position of the TCs.

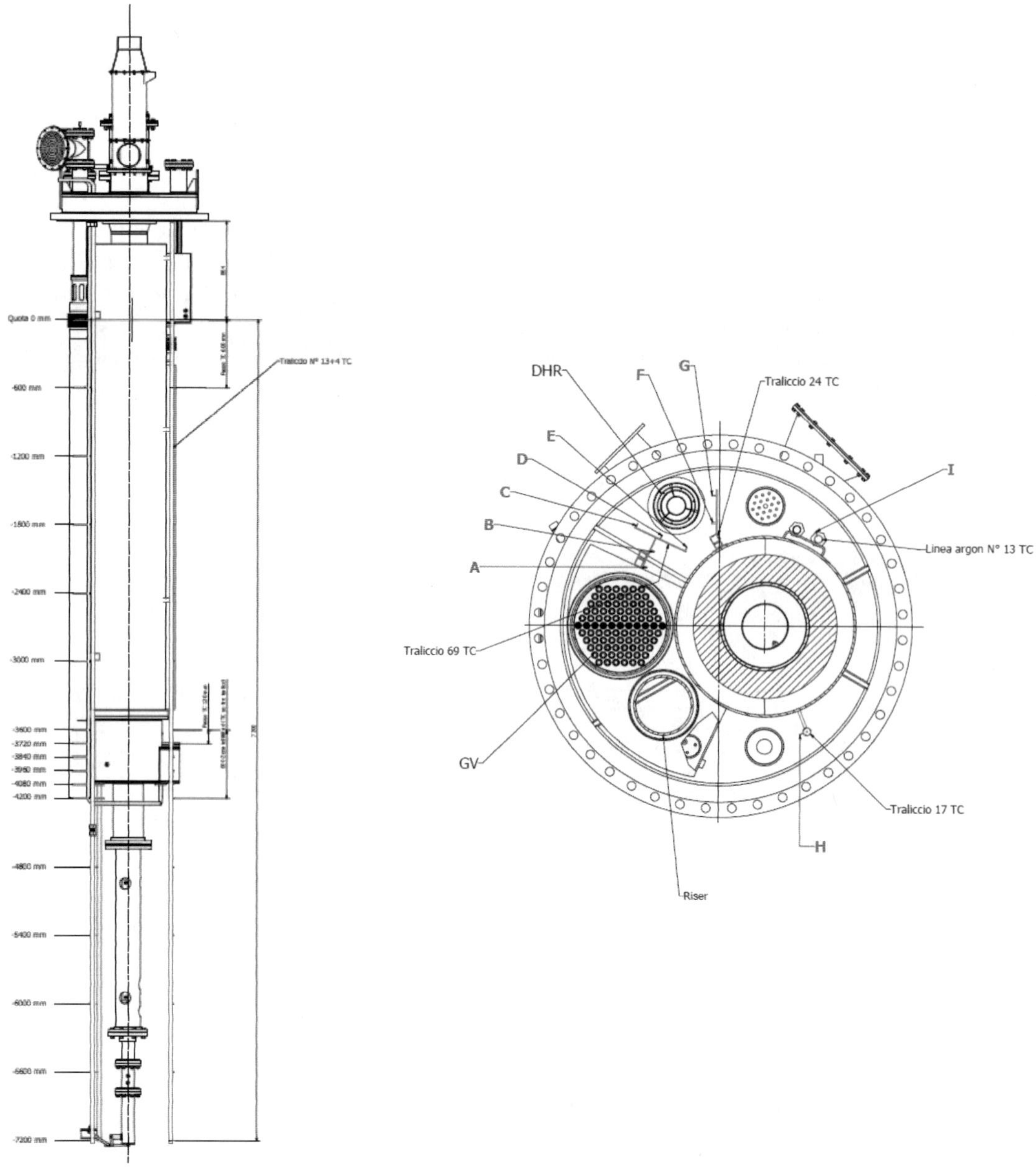

(a) (b)

Figure 11. Thermocouples positioning

Figure 12 shows a comparison between experimental and calculated axial temperature profile in different circumferential positions, at the end of the test. In Figure 12a and b, the experimental value is obtained as the average temperature measured respectively by six and two TCs. Temperature profiles show a similar (but not equal) trend in the different positions. Both simulations rather well reproduce the experimental data, except in the region between -1 and -4 m, where pool temperature is underestimated by R5-3D. The differences in the temperature profile are probably due to the the absence of the axial liquid conductivity correlation into the code.

IOP Conf. Series: Journal of Physics: Conf. Series **923** (2017) 012006 doi:10.1088/1742-6596/923/1/012006

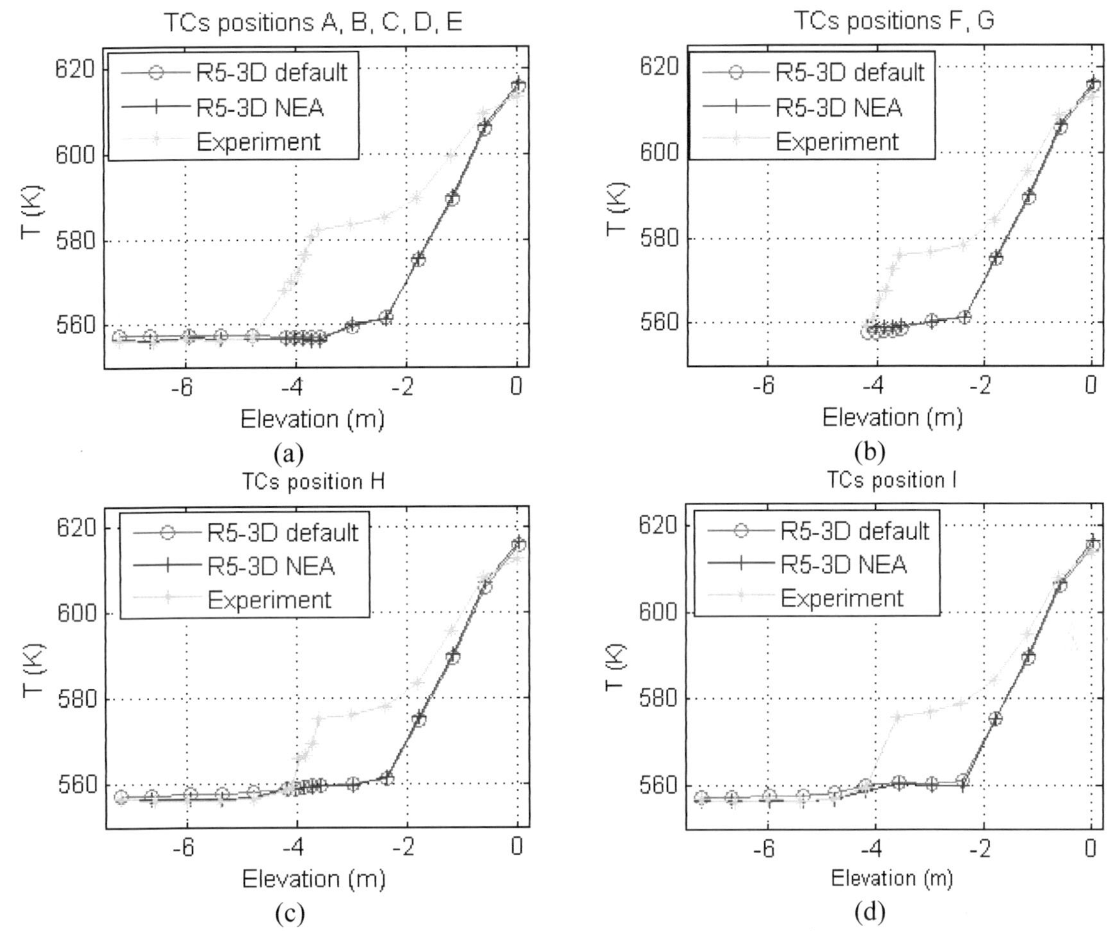

Figure 12. Axial temperature profile in the LBE pool at 25200 s

5. Conclusions

The experimental campaign conducted on CIRCE-ICE facility offers additional data for the validation of R5-3D code in the frame of HLM pool-type reactors.

The geometrical nodalization scheme consists of a mono-dimensional model, which reproduces the primary main flow path and the secondary side of the HX and DHR system, coupled with the three-dimensional model of the pool.

The thermal-hydraulic behavior of the primary flow path is well reproduced by the code. In particular the implementation of new thermophysical properties correlations, recommended by NEA, allows a better estimation of the Nusselt number and, consequently, a better reproduction of LBE temperature variation along the flow path and the heat losses through the internal structures.

The main objective of the 3D component, used in this work, was to investigate thermal stratification and mixing convection phenomena into the pool. The LBE temperature is well reproduced by the code, except between -1 and -4 m, where the simulations underestimate the experimental value. Further investigation for the thermal stratification profile in this zone is needed, with a heat losses detailed characterization which could be carried out with CIRCE-HERO experimental data contribution.

IOP Conf. Series: Journal of Physics: Conf. Series **923** (2017) 012006 doi:10.1088/1742-6596/923/1/012006

References

[1] Tarantino M, Martelli D, Barone G, Di Piazza I and Forgione N 2015 *Nucl. Eng. Des.* **286** 261–77
[2] Martelli D, Tarantino M and Piazza I 2016 *ASME Proceeding* 1–8
[3] OECD Nuclear Energy Agency 2014 Technology Roadmap Update for Generation IV Nuclear Energy Systems 1–66
[4] ESNII Task Force 2010 Demonstration Programme for Fast Neutron Reactors
[5] Bandini G, Piazza I Di, Gaggini P, Nevo A Del, Tarantino M, Tarantino M, Ciampichetti A and Agostini P 2011 CIRCE experimental set-up design and test matrix definition *ENEA Research Center Brasimone Report*
[6] Tarantino M, Agostini P, Benamati G, Coccoluto G, Gaggini P, Labanti V, Venturi G, Class A, Liftin K, Forgione N and Moreau V 2011 *J. Nucl. Mater.* **415** 433–48
[7] Team T R C D 1995 RELAP5/MOD3 Code Manual Vol.1 Code Structure, System Models and Solution Methods NUREG/CR55
[8] Todreas N E, .Kazimi M S, Schlumberger, Maersk, IMO, Giordano N, Brogan R J, Ahlgren F, Mondejar M E, Genrup M and Thern M 2015 *Taylor Fr. Publ.* **138** 1–10
[9] Giannetti F, Vitale Di Maio D, Naviglio A and Caruso G 2016 *Nucl. Eng. Des.* **305** 168–78
[10] Ushakov P A, Zhukov A V and Matyukhin N M 1977 *High Temp.* **15**
[11] Team T R C D 1997 Volume II , Appendices II 1–277
[12] Schikorr M, Bubelis E, Mansani L and Litfin K 2010 *Nucl. Eng. Des.* **240** 1830–42
[13] Anon 2015 Handbook on Lead-bismuth Eutectic Alloy and Lead Properties, Materials Compatibility, Thermal- hydraulics and Technologies 647–730
[14] Balestra P, Giannetti F, Caruso G and Alfonsi A 2016 *Sci. Technol. Nucl. Install.* **2016**

IOP Conf. Series: Journal of Physics: Conf. Series **923** (2017) 012007 doi:10.1088/1742-6596/923/1/012007

Analysis of a full-scale integral test in PERSEO facility by using TRACE code

S D'Amico[1], C Lombardo[2], I Moscato[1], M Polidori[2] and G Vella[1]
[1] University of Palermo, Dipartimento di Energia, Ingegneria dell'Informazione e Modelli Matematici, Building 6, Viale delle Scienze, 90128 Palermo, Italy
[2] ENEA-FSN/SICNUC/SIN, Via Martiri di Monte Sole 4, 40129 Bologna, Italy

E-mail: giuseppe.vella22@unipa.it

Abstract. Over the last decades a lot of experimental researches have been done to increase the reliability of passive decay heat removal systems implementing in-pool immersed heat exchanger. In this framework, a domestic research program on innovative safety systems was carried out leading the design and the development of the PERSEO facility at the SIET laboratories. The configuration of the system consists of an heat exchanger contained in a small pool which is connected both at the bottom and at the top to a large water reservoir pool.
Within the frame of a national research program funded by the Italian minister of economic development, the DEIM department of the University of Palermo in cooperation with ENEA has developed a computational model of the PERSEO facility in order to simulate its behaviour during an integrated test. The analysis here presented has been performed by using the best-estimate TRACE code and - in order to highlight the capabilities and limits of the TRACE model in reproducing qualitatively and quantitatively the experimental trends - the main results have been compared with the experimental data. The comparison shows that the model is able to predict the overall behaviour of the plant during the meaningful phases of the transient analysed. Nevertheless, some improvements in the modelling of certain components in which take place complex three-dimensional phenomena are suggested in order to reduce some discrepancies observed between code results and test measurements.

1. Introduction

In the framework of the research program devoted to improve the reliability of residual heat removal systems characterized by in-pool heat exchangers as heat sink and requiring the actuation of a triggering valve to start their operation, ENEA and CEA have developed an innovative device called *thermal valve* [1]. The thermal valve, which was conceived for advanced light water nuclear reactors, rely on gravity injection and natural circulation and consists of a diving bell surrounding an in-pool heat exchanger with a pilot valve located on the top. During an accidental scenario, the pilot valve is opened allowing the steam to condensate in the pool through an effective natural circulation which assures a passive removal of the decay heat. Because of some operating and construction problems, mainly due to the need of a pilot valve with a large flow area, ENEA and SIET have proposed a new configuration, the so-called PERSEO (In-Pool Energy Removal System for Emergency Operation), which consists of two pools and an heat exchanger. In stand-by conditions the pool containing the heat exchanger is empty while the pool representing the water reservoir is full of cold water. The pools are connected to each other at the bottom and the top to allow the natural circulation and to guarantee a long term energy removal during the accidental scenarios. A triggering valve, which determines the

device operation is installed on the pool bottom connecting pipe (liquid line). The valve opening causes the flooding of the heat exchanger allowing the heat transfer from the primary side to the pool side. An injector is installed at the exit of the steam duct located at the top of the pools (steam line) in order to improve the pool water mixing and to avoid thermal stratification phenomena. The system has been tested at the SIET laboratories in Piacenza by modifying the existing PANTHERS IC-PCC facility (Performance Analysis and Testing of Heat Removal System Isolation Condenser - Passive Containment Condenser), which was utilized in the past for testing a full scale module of SBWR in-pool heat exchanger. Two types of tests were performed on PERSEO test facility: *integral tests*, aimed at verifying the behaviour and performance of the device following a request of operation and *stability tests*, finalised to study some critical problems occurring in case of sudden condensation at the steam-water interface or in case of triggering valve re-opening. The experimental tests have allowed a complete evaluation of the efficiency and reliability of the PERSEO innovative decay heat removal system and they have confirmed its effectiveness [2], [3], [4].
The analysis reported in the paper have been performed by using the best-estimate TRACE code with the purpose of evaluating the prediction capability of the computational model, developed by the DEIM thermal-hydraulic research unit, in analysing an integral test devoted mainly to investigate the flow regime variation and the trend of power as a function of the water level in the pools.
The main results have been compared with the experimental data in order to highlight the capabilities and limits of the TRACE model in reproducing qualitatively and quantitatively the experimental trends. The comparison shows that the model is able to predict the overall behaviour of the plant during the transient, but some improvements in modelling specific components in which take place complex three-dimensional phenomena have been suggested, in order to reduce the discrepancies observed between code results and test measurements.

2. PERSEO facility description

The PERSEO facility, whose scheme is shown in Fig. 1, was built at the SIET laboratories by modifying the existing PANTHERS IC-PCC facility that was utilised for testing a full scale module of the GE-SBWR in-pool heat exchanger [2].

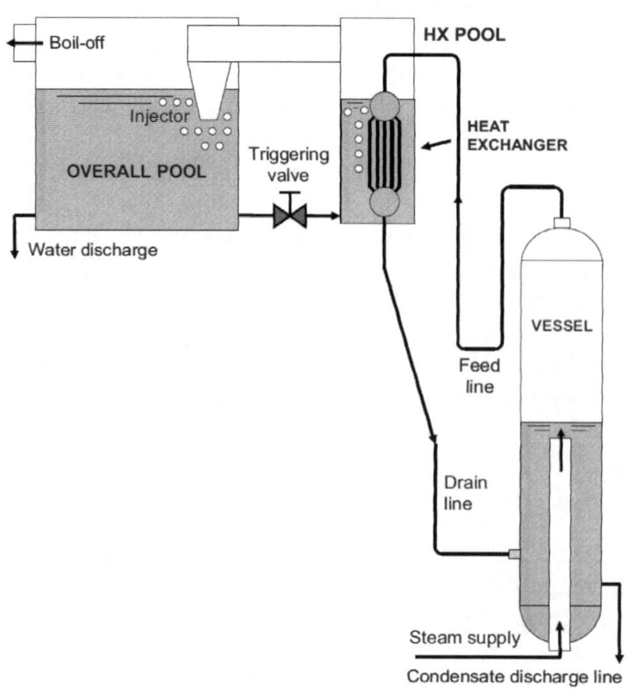

Figure 1. PERSEO facility layout [5].

The PERSEO design was an evolution of the thermal valve concept and they differ mainly for the location of the triggering valve which is installed liquid side in PERSEO, namely on the line connecting two pools at the bottom. The main features of the PERSEO design are reported in Table 1.

Table 1. PERSEO facility design features [2].

Quantity	Value
Power	20 [MW]
Vessel pressure	10 [MPa]
Vessel temperature	310 [°C]
Heat exchanger pressure	8.62 [MPa]
Heat exchanger temperature	302 [°C]
Super-heated steam flowrate	12 [kg/s]
De-superheating water flow rate	3 [kg/s]
Pool side pressure	0.15 [MPa]
HX Pool temperature	300 [°C]
Overall Pool temperature	130 [°C]
Pool side water flowrate	25 [kg/s]

As already said, the triggering valve is closed during normal operation and the pool containing the heat exchanger (HX pool) is empty, while the other pool (Overall pool) is full of cold water. During emergency conditions the valve is opened which leads to the heat exchanger (HX) flooding and the consequent heat transfer from the primary side to the pool side. The pools are connected at the top by means of a pipe ending with an injector which promotes the steam circulation and the homogenization of water temperature inside the Overall pool. The primary side of the facility is maintained at the saturation conditions by supplying the steam coming from the nearby power station: the pressure is kept constant by controlling the steam supply and discharge valves.
Before starting the experiment, the primary side reaches the saturation conditions specified for the test and the proper water level. At the same time the feed line, the heat exchanger and the drain line are being filled of saturated steam. The HX pool is full of air or steam according to the test matrix, while the Overall pool is full of cold water at the nominal level.
Once reached the initial test conditions, the triggering valve is opened and the HX pool is flooded by cold water. Steam condensation at the primary side of heat exchanger occurs soon after the HX pool flooding and the heat is transferred from the primary side to the pool side. Steam produced in the HX pool is driven to the Overall pool through the steam duct. The only connection between the pool side and the environment occurs throughout the Boil-off pipe located at the top of the Overall pool.
In such kind of system, the heat transfer rate decreases according to the water reserve.

3. TRACE code brief description

Best-estimate thermal-hydraulic system codes are complex tools developed to simulate both normal and transient conditions in nuclear installations (power plants, research reactors, spent fuel pools and experimental facilities). Among them, the TRACE (TRAC/RELAP Advanced Computational Engine) developed by the USNRC is an advanced, two-fluid, thermal-hydraulic computer code designed for use in realistic studies of light water reactors. It consolidates the capabilities of four codes (TRAC-P, TRAC-B, RELAP 5 and RAMONA) into one. In fact, system thermal-hydraulic analysis addresses the simulation of complex thermal-hydraulics phenomena by a best estimate approach in order to get an evaluation of the plant behaviour which will be as realistic as possible. The TRACE code is capable of analysing of LOCAs, operational transients, accident scenarios and phenomena in nuclear power plants and related experimental facilities. TRACE uses different types of components which could be further subdivided into various numbers of physical volumes (or cells), to model each physical piece of equipment comprised in the reactor system. Over each finite volume cells, the fluid, conduction and kinetics equations are averaged. The hydraulic components in TRACE include BREAKs, CHANs

IOP Conf. Series: Journal of Physics: Conf. Series **923** (2017) 012007 doi:10.1088/1742-6596/923/1/012007

(BWR fuel channels), VESSELs, PIPEs, FILLs, PRIZERs (pressurizers), HEATRs (feed-water heaters), PLENUMs, PUMPs, SEPDs (separators), TEEs, TURBs (turbines), JETs (jet pumps) and VALVEs and so on. Models supplied in TRACE include multidimensional two-phase flow, such as critical flow model, offtake model and counter current flow model, non-equilibrium thermo-dynamics, re-flood, level tracking and generalized heat transfer. In addition, it contains not only a point kinetics model but also the capability to be coupled with external 3D neutronic codes-and the possibility to be managed through the graphical user interface SNAP [6], [7].

4. TRACE model

The PERSEO TRACE nodalization has been made by using TRACE V5.840 patch 4 and it models in detail the experimental facility [8]. The total number of thermal-hydraulic components presented in TRACE model are 113, being 33 PIPEs, 5 VESSELs, 9 VALVEs, 7 BREAKs, 1 FILLs, 3 TEEs, 2 PUMPs and 53 HEAT STRUCTUREs. In addition, 34 TRIPs, 89 CONTROL BLOCKs and 104 SIGNAL VARIABLEs complete the model. In particular, where the 1D approach is not sufficient (i.e. secondary flows are not negligible) as in the pressure vessel, in the Overall pool and in the HX pool, the 3D component "VESSEL" has been used.

4.1. Principal physical models adopted

In order to better simulate the behavior of the various systems during an integral test, several physical models made available by the code have been used [9].

The "level tracking" model have been adopted in all vertical components to localize, as well as possible, the liquid-gas interfaces. Models of counter-current flow limitation have been used at the cell junctions where this event was expected during flooding and reflux condensation phases. The "offtake" model has been applied at the side junctions of ducts where a horizontal stratified two-phase flow could have occurred to predict the correct offtake flow quality that would have come out from the main tube. At cell edges that simulate the break and the ADS orifices the "critical flow" model has been used.

4.2. Primary side model description

The primary circuit has been reproduced from the vessel to the heat exchanger, while the steam supply has been provided by means of a proper control volume. The pressure vessel and its riser, Fig. 2, have been described using the VESSEL component: its nodalization consists of 21 axial levels, 2 radial rings and 8 azimuthal sectors. The azimuthal nodalization has been chosen in order to better simulate the circumferential position of the nozzles and with the purpose of reproducing as best as possible the pressure drops. The feed-line and the by-pass line have been modelled by means of 9 PIPEs and 2 TEEs, while the drain line has been described by using 2 PIPEs and 2 VALVEs. The heat exchanger has been nodalized by means of 2 cartesian VESSELs and 15 PIPEs: the component VESSEL has been used to simulate the two header and their nodalization consists of 2 axial levels, 17 X-sections and only 1 Y-sectors; each PIPE simulates 8 tubes of the tube bundle. Fig. 3 shows the nodalization of the heat exchanger. The pressure and level controls systems have been also modelled by using 4 VALVEs, 1 TEE and the needed SIGNAL VARIABLEs and CONTROL BLOCKs.

4.3. Pool side model description

The Overall pool nodalization has been done through the application of a cartesian VESSEL that is divided in 14 axial levels, 2 X-sections and 2 Y-sectors and by using a PIPE for the upper part; The boil-off has been simulated by means of another PIPE component. The discharge line has been simulated by means of a PUMP and a BREAK allowing to simulate a constant flow rate leaving the water reservoir. The HX pool has been described by using another cartesian VESSEL: its nodalization is characterized by 73 axial levels, 1 X-section and 1 Y-sectors. The Steam-duct and the cone-shaped injector have been simulated by using a PIPE, while the vacuum breaker has been simulated through a BREAK and a VALVE component. Fig. 4 shows the nodalization of the pool side.

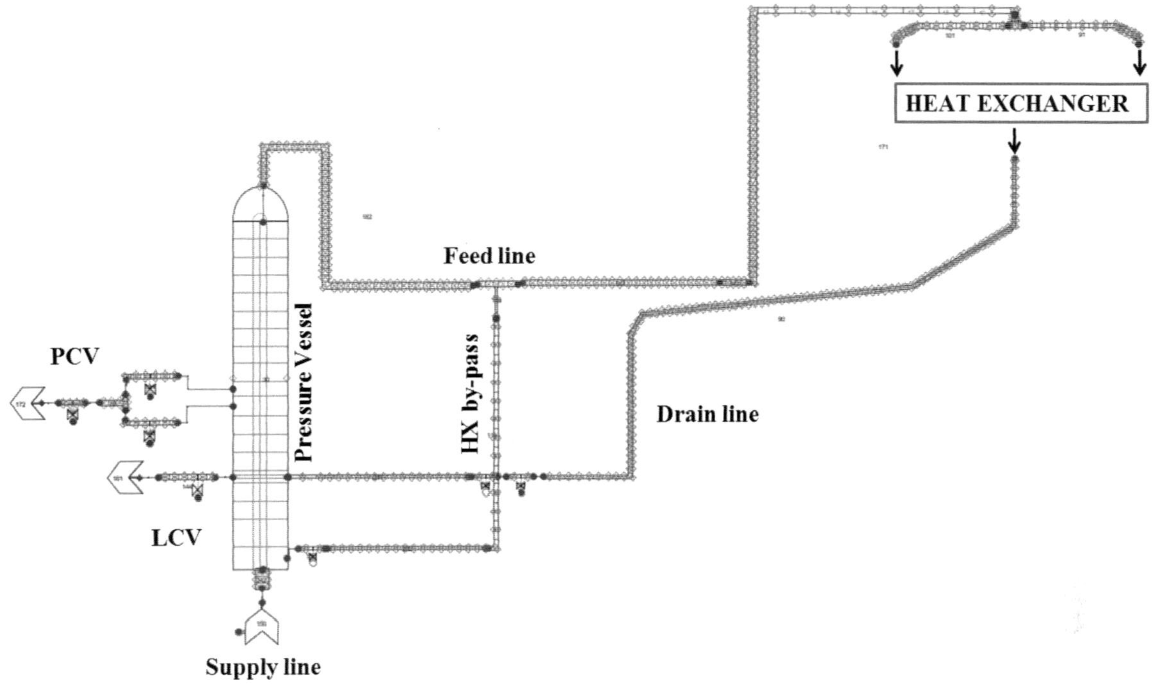

Figure 2. Primary side nodalization .

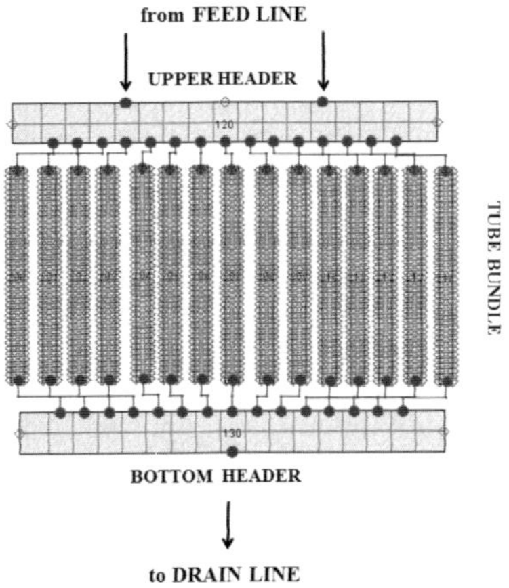

Figure 3. HX nodalization.

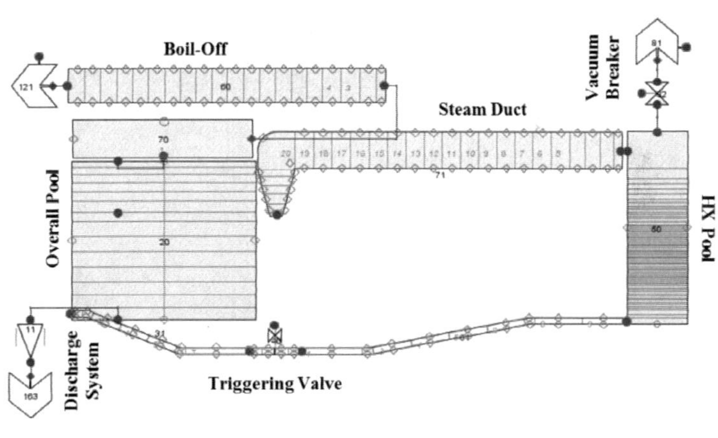

Figure 4. Pool side nodalization.

5. PERSEO facility test program objectives

The PERSEO facility test program was aimed at verifying the new system operation, steadiness and performance. The performance concerned the heat transfer start-up after the triggering valve opening, its steadiness with the two pools full of water and the exchanged heat trend according to the water level in the pools. The PERSEO facility set-up was conducted performing a series of shake down tests suitable to verify the correct plant operation and to characterise the main parameters of the facility such as the water pouring-off from the Overall pool to the HX pool.

The PERSEO experimental tests can be distinguished in: *integral* and *stability* tests; both of them suitable for the system concept demonstration. The integral tests were aimed at demonstrating the behaviour and performance of the system following an accidental scenarios during all its main phases.

The stability tests were aimed at studying some critical problems happening in case of sudden condensation at the steam-water interface or in case of a re-opening of the triggering valve.
The main conditions and a short description of the integral and stability tests has been reported in Table 2, together with a short description of tests [2].

Table 2. PERSEO facility test matrix [2].

Test	Test conditions	Description
6	Primary side pressure: 7 [MPa]	Integral test interrupted at the beginning of the pool level decreasing
7	Primary side pressure: 7 [MPa]	Stability and integral test: partial and subsequent HX Pool filling with reaching of boiling conditions and level decreasing
8	Primary side pressure: 7 [MPa]	Stability test: partial HX Pool filling with reaching of boiling conditions
9	Primary side pressure: 4 [MPa]	Integral test

The tests confirmed the effectiveness of the PERSEO innovative system: the heat transfer from the primary side to the pool side is soon actuated after the triggering valve opening and it is stable, decreasing as the HX pool level [2], [3]. The steam generated in the HX pool accelerated into the Overall pool by means of the injector promotes the water circulation and avoids the thermal stratification in the Overall poll. Instabilities due to the sudden steam condensation, pointed out after an early interruption of the heat transfer and during the HX pool re-flooding were dumped very soon by using the vacuum breaker installed at the HX pool top or on the Steam-duct according to the test [4], [5].

6. PERSEO test data analysis

The test 6 has been selected by the authors in order to carry out the comparison between the experimental data and the TRACE numerical results. In fact, the test has been considered the most representative from the performance and stability point of view and for thermal-hydraulic system code assessment.

6.1. Experimental test description

The experimental test analysed is an integral test with the primary side pressure of 7 MPa; the main steps have been reported in the following:

- pressurization of the primary circuit to the pressure of 7 MPa,
- opening of the triggering valve,
- reaching of saturation conditions in HX pool and Overall pool,
- level decrease in the pools until the injector outlet uncovering,
- level decrease accelerated by water discharge from the Overall pool,
- triggering valve closure and HX pool boil-off,
- depressurization of the primary circuit.

The aim of the test was devoted to investigate the items below:

- the system actuation and reaching of the thermal regime in both the pools,
- the effectiveness of the injector in mixing the Overall Pool water,
- the power variation after the Overall Pool level decreasing below the injector outlet,
- the trend of power as a function of the water level in the pools, decreasing for the loss of mass through the boil-off.

The test starts at time 1100 s by opening the triggering valve which determines the HX water level increasing through the cold water flowrate between the two pools; once reached the bottom header of the heat exchanger, the heat begins to be exchanged from the primary to the pool side. The primary

circuit pressure decrease rapidly due to the heat removal and it is manually increased. After about 250 s, the pressure average value returns to 7 MPa until 4967 s when the primary circuit is quickly depressurized and the test concluded due to a break on a vessel discharge line. The HX pool level begins to increase at time 1114 s and reaches the maximum value at 1550 s.
After the Overall pool level decreasing for the HX pool fill-up, its average level remains quite constant until about 4400 s when the valve of the discharge line is opened. Regarding the removed heat from the primary circuit, it is possible to observe an increase according the HX pool level trend up to stabilise at about 17.7 MW when in the HX pool are reached the boiling conditions. Before the water injection, the temperature of the air contained in the HX pool is around 240 °C due to the HX radiation, then the temperature decreases for the incoming of cold water. The steam, produced in the HX pool and accelerated into the Overall pool by means of the injector, promotes the water circulation and mixing avoiding the thermal stratification in the Overall pool. Unfortunately, for problems on the pressure vessel control system, a fast discharge of hot water from the vessel is operated and causes a break of the line towards the condenser leading the end of the test [2].

6.2. Comparison between experimental data and numerical results
The PERSEO test analysis here presented has been performed by the thermal-hydraulic research groups of the DEIM department of the University of Palermo. In order to improve the prediction capability of the PERSEO TRACE model, several nodalizations of the experimental plant have been studied; they have been used to perform simulations of different tests comparing the results with the experimental ones [8]. Unfortunately, a pressure boundary condition at the HX pool top has been maintained active for 150 s during the start-up to avoid large instabilities calculated by the code; these instabilities are generated by the simple models adopted to treat non-condensable gases. A great importance has been given to the proper simulation of the connection between the injector outlet and the Overall pool, which affects the water circulation and mixing. The simulation of the test by using TRACE code has been performed up to the phase in which the discharge of the water from the Overall pool takes place since the main phenomena have been investigated. The TRACE numerical results show a good agreement with experimental data. The calculated primary pressure well reproduces the facility behaviour during the main phases of the test though some mismatch is present in timing and value when the pressure is manually increased operating through the supply valve, Fig. 5. The Overall pool level predicted by TRACE code is quite similar to experimental data in trend and values, as well as the calculated HX pool level, Fig. 6. The Fig. 7 reports the steam flow rate supplied to the heat exchanger and the condensate flowrate discharged in the pressure vessel (the condensate flowrate experimental measure is not valid before time 1000 s). As can be seen, the flow rates predicted by the code are in good agreement with the experimental data even though the oscillations due likely to the pressure control system operation.
An important limit of the model must be underlined, that is the difficulty of the code of reproducing the experimental exchanged power without using a proper fouling material in the HEAT STRUCTUREs of the heat exchanger, Fig. 8. This difficulty depends on the heat transfer regime selection logic for the pre-CHF (Critical Heat Flux) and condensation regimes [9] which means an overrate of the control volumes involved in the subcooled nucleate boiling regime (pool side): the overestimation strongly depends on the discretization used to simulate the facility components.
The steam produced in the HX pool is driven and accelerated into the Overall pool through the injector which promotes water mixing in the pool reducing the thermal stratification that is quite well predicted by the code, Fig. 9.
Fig. 10 shows the HX top pressure: it can be observed a quite good agreement in curve trends, but a general overestimation of the calculated variable for 1000 s after the transient starting.
Anyway the overall behaviour of plant is well predicted by TRACE code providing a confirmation of the model capability to simulate the thermal hydraulic phenomena that occur in this advanced safety system.

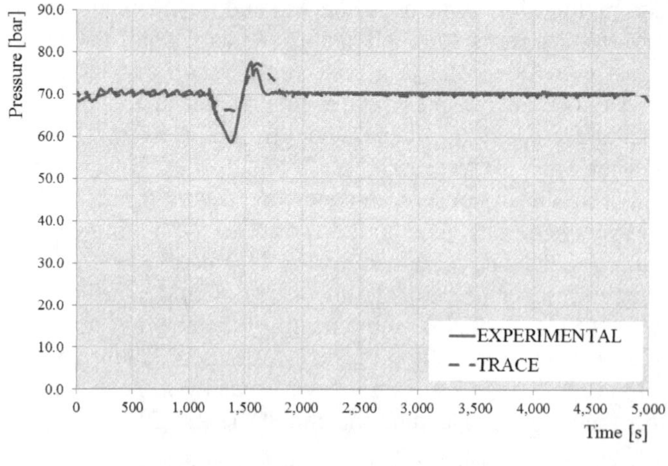

Figure 5. Primary side pressure.

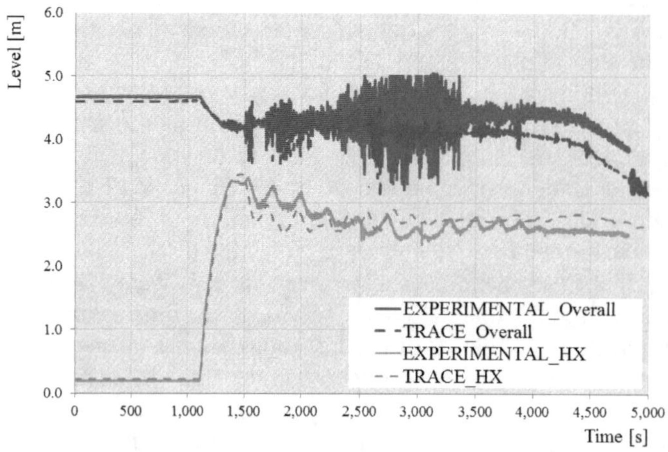

Figure 6. Overall and HX pools water level.

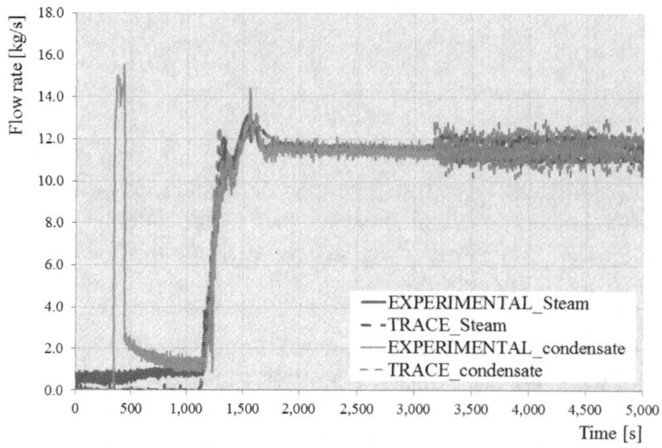

Figure 7. Steam and condensate flowrate.

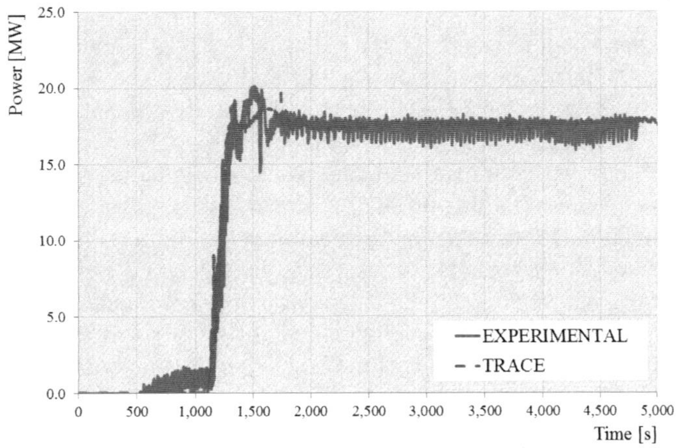

Figure 8. HX exchanged power.

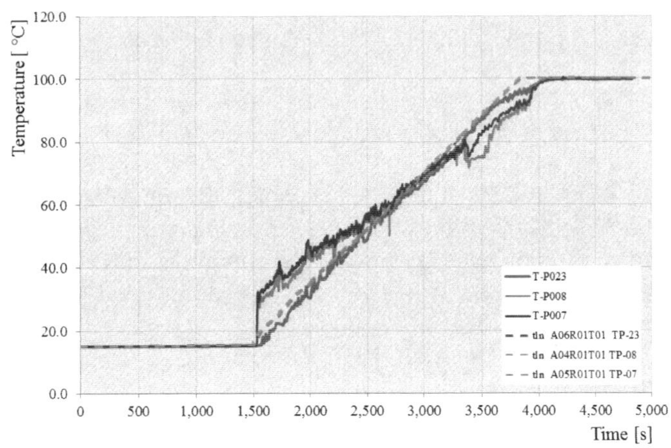

Figure 9. Overall pool temperature.

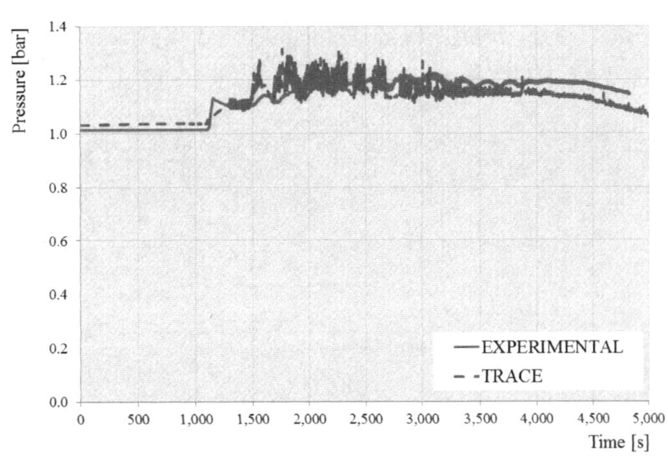

Figure 10. HX Top pressure.

IOP Conf. Series: Journal of Physics: Conf. Series **923** (2017) 012007 doi:10.1088/1742-6596/923/1/012007

7. Conclusions

The present work concerns a comparison between experimental and numerical results obtained for the integral test 6 of the test matrix prepared for the PERSEO facility. The numerical analysis here presented has been performed by using TRACE thermal-hydraulic code with the purpose of evaluating its prediction capability. The main results show that the TRACE model well predict the overall behaviour of the plant during the transient, even if some limits of the code were identified in simulating particular phenomena and quantities. As already highlighted, the code limit to reproduce the experimental exchanged power depends on the heat transfer regime selection logic for the pre-CHF and condensation regimes. Moreover, others discrepancies observed between code results and test measurements can be attributed to complex three-dimensional phenomena suggesting improvements in the modelling of some components. As recommended in [4] a wider comparison with other best estimate codes is necessary in order to set-up a definitive model suitable to simulate such innovative decay heat removal system. The activity has been funded by Italian Minister of Economic Development.

8. References

[1] Bianchi F Meloni P Pignatel J.F. Gautier G.M 1997 Thermal valve system for LWR applications *Proc. Post-SMIRT14 Seminar* Pisa.

[2] Ferri R Achilli A Gandolfi S 2002 *PERSEO Project Experimental Data Report* 01014RP02 Piacenza.

[3] Ferri R Achilli A 2003 In-Pool Energy Removal System for Emergency Operation: Experiment and Analytical Assessment *Proc. Of Nuclear Energy for New Europe 2003 International Conference* Portoroz.

[4] Ferri R Achilli A Cattadori G Bianchi F Meloni P Design, experiments and Relap5 code calculations for the PERSEO facility *Nuclear Engineering and Design* **235** (2005) 1201-1214.

[5] Bandini G Meloni P Polidori M Lombardo C 2011Validation of CATHARE V2.5 thermal-hydraulic code against full-scale PERSEO tests for decay heat removal in LWRs *Nuclear Engineering and Design* **241** (2011) 4662-4671.

[6] U.S. Nuclear Regulatory Commission 2010 *TRACE V5.0 User's Manual Volume 1: Input Specification* Washington.

[7] Applied Programming Technology Inc. 2012 *SNAP 2.2.1 User's manual* Bloomsburg.

[8] Chiovaro P D'Amico S Lombardo C Moscato I Polidori M Vella G 2016 *Miglioramento dei modelli dei codici integrali CATHARE e TRACE per la simulazione di specifici aspetti della facility integrale SPES2.* (Bologna: CIRTEN-ENEA report).

[9] U.S. Nuclear Regulatory Commission 2010 *TRACE V5.0 Theory Manual: Field Equations, Solution Methods and Physical Models* Washington.

Verification of RELAP5-3D code in natural circulation loop as function of the initial water inventory

C Bertani[1], N Falcone[1], A Bersano[1], M Caramello[1], T Matsushita[2], M De Salve[1] and B Panella[1]

[1] Energy Department
Politecnico di Torino
Corso Duca degli Abruzzi 24, 10129 Torino (TO), Italy
[2] Graduate School of Energy Science
Kyoto University
2-1010 Asasiro-nishi, Kumatori-cho, Sennan-gun, Osaka 590-0494, Japan

E-mail: andrea.bersano@polito.it

Abstract. High safety and reliability of advanced nuclear reactors, Generation IV and Small Modular Reactors (SMR), have a crucial role in the acceptance of these new plants design. Among all the possible safety systems, particular efforts are dedicated to the study of passive systems because they rely on simple physical principles like natural circulation, without the need of external energy source to operate. Taking inspiration from the second Decay Heat Removal system (DHR2) of ALFRED, the European Generation IV demonstrator of the fast lead cooled reactor, an experimental facility has been built at the Energy Department of Politecnico di Torino (PROPHET facility) to study single and two-phase flow natural circulation. The facility behavior is simulated using the thermal-hydraulic system code RELAP5-3D, which is widely used in nuclear applications. In this paper, the effect of the initial water inventory on natural circulation is analyzed. The experimental time behaviors of temperatures and pressures are analyzed. The experimental matrix ranges between 69 % and 93%; the influence of the opposite effects related to the increase of the volume available for the expansion and the pressure raise due to phase change is discussed. Simulations of the experimental tests are carried out by using a 1D model at constant heat power and fixed liquid and air mass; the code predictions are compared with experimental results. Two typical responses are observed: subcooled or two phase saturated circulation. The steady state pressure is a strong function of liquid and air mass inventory. The numerical results show that, at low initial liquid mass inventory, the natural circulation is not stable but pulsated.

1. Introduction

The increasing request for energy supply is a well-recognized problem worldwide. In particular, the request of electricity is expected to grow considerably in the next years mainly due to the development of emerging countries. This fact poses huge challenges for the future also taking into account the efforts that many countries are making to reduce the greenhouse gases emissions. A possible source of

IOP Conf. Series: Journal of Physics: Conf. Series **923** (2017) 012008 doi:10.1088/1742-6596/923/1/012008

abundant and clean electricity is nuclear energy. Even if the risks related to nuclear are considerably lower than other energy sources, the public opinion is generally against this possibility. This is one of the reasons that have pushed the civil nuclear industry to improve continuously the safety of commercial nuclear power plants. In this framework, the international community is devoting many efforts to develop Generation IV reactors. This new family of nuclear reactors is characterized by improved safety, higher economic benefits and a stronger resistance to nuclear proliferation [1]. This generation includes a wide use of passive systems in particular for the operation of the plant in case of an accident. Passive systems rely on physical principles to operate, without the need of external energy to operate or human intervention [2]. This characteristic offers great advantages particularly concerning safety system that should operate in every condition.

Among the six proposed technologies for Generation IV reactors, Italy is particularly involved in the design of the lead-cooled fast reactor (LFR). The main current project is the design of a European demonstrator of the LFR technology called Advanced Lead Fast Reactor European Demonstrator (ALFRED), which is carried out by FALCON (Fostering ALfred CONsortium) consortium.

The use of lead as coolant offers several advantages, such as good heat transfer properties, neutron and gamma shielding and decay heat removal by natural circulation. On the other hand, some drawbacks should be carefully considered; one of them is the freezing of liquid lead below 600.65 K (327.5°C). This means that inside the reactor this temperature should always been guaranteed, which is relatively easy during the normal operation of the plant but not so obvious in accidental conditions. In case of an accident the reactor is shut down and the emergency cooling is activated. Under this condition the cooling rate needs to be controlled not to reach the lead freezing temperature. The emergency cooling of the reactor is achieved by a Decay Heat Removal system (DHR) that uses water in natural circulation to operate. These systems are constituted by a bundle of bayonet heat exchangers that remove power from the lead transferring it into a water pool where an Isolation Condenser (IC) is submerged.

In the first proposed configuration, the heat removal was so effective that lead freezing emerged as a major issue. Due to this reason Ansaldo Nucleare (one member of the FALCON consortium along with ENEA and Nuclear Research Institute “ICN”) proposed an “anti-freezing” system to be added to DHR (Italian Patent MI2013A001778) [3]. This anti-freezing system consists in the addition of non-condensable gases inside the DHR; based on the operating temperature and pressure of the system, the quantity of non-condensable gases is automatically controlled and the excess gas is collected in a dedicated tank. When the temperature of the system decreases during the operation the pressure is reduced and non-condensable gases can enter in the Isolation Condenser. The effect is a deterioration of the heat transfer capacity and a consequent lower reduction rate of the lead temperature [4].

To verify the effectiveness of this system it is important to study the effect of non-condensable gases on natural circulation. Many researches focus on the reduction of the heat transfer coefficient due to the presence of air [5]. This paper focuses on the effect of air on the global system behavior. In addition to the reduction of the heat transfer capability, the position of air in the system is very important because it can create plugs and therefore decrease or stop the water recirculation.

This problem has been studied both experimentally and with computer simulations. The experiments have been carried out on PROPHET facility (PROtotype Passive Heat Exchanger sysTem, whose design has been presented in the 34th UIT Heat Transfer Conference [6]) at Dipartimento Energia of Politecnico di Torino; to perform the numerical simulation the system code RELAP5-3D [7,8,9] has been used. RELAP5-3D is a thermal-hydraulic code developed by Idaho National Laboratories (INL) widely used in the nuclear field for transient analysis of complex systems. The effect of the presence of air in the system has been observed in terms of operative pressure reached at steady state conditions. Pressure and temperature time behaviors have been measured during the transient and compared to the numerical results for several initial water inventory values.

2. Experimental facility and methodology

The experimental facility is a loop consisting of a bayonet heat exchanger (heat source), a condenser (heat sink) and the connection piping [6,10]. All the components are made of stainless steel AISI 304. The heat source consists in two electric heating tapes wrapped around the bayonet heat exchanger for a length of 1.07 m; the total electric nominal power is 1.566 kW at 240 V. In Figure 1 the scheme of the facility is shown, highlighting the position of the instrumentation nozzles and geometrical and physical data of the main components of PROPHET are listed in Table 1.

K-type thermocouples measure the fluid temperature at eight locations (T1-T8, red circles), and the bayonet outer wall temperature in three points (TA, TB and TC, orange circles); two absolute pressure measurements (p1 at the bayonet bottom, and p2 upstream the condenser, green circles) and three differential pressure measurements (Δp1, Δp2 and Δp3, purple circles) are also carried out by Rosemount transducers. The bayonet wall temperatures are measured merely to avoid the electric resistances breakage; in fact the measured values are not reliable because of local distortion of the thermal field at the measurement points.

A thermographic camera was also used to measure the temperature distribution all over the experimental facility.

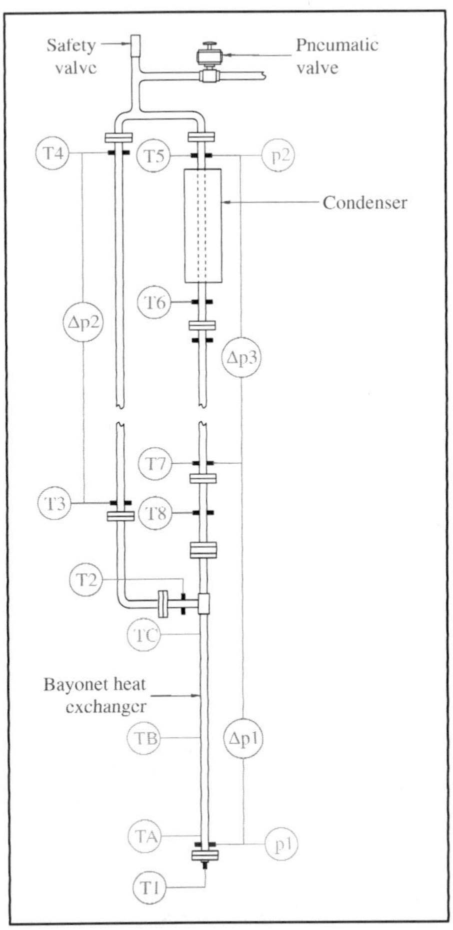

Figure 1: Schematic of PROPHET facility with numbered instrumentation nozzles

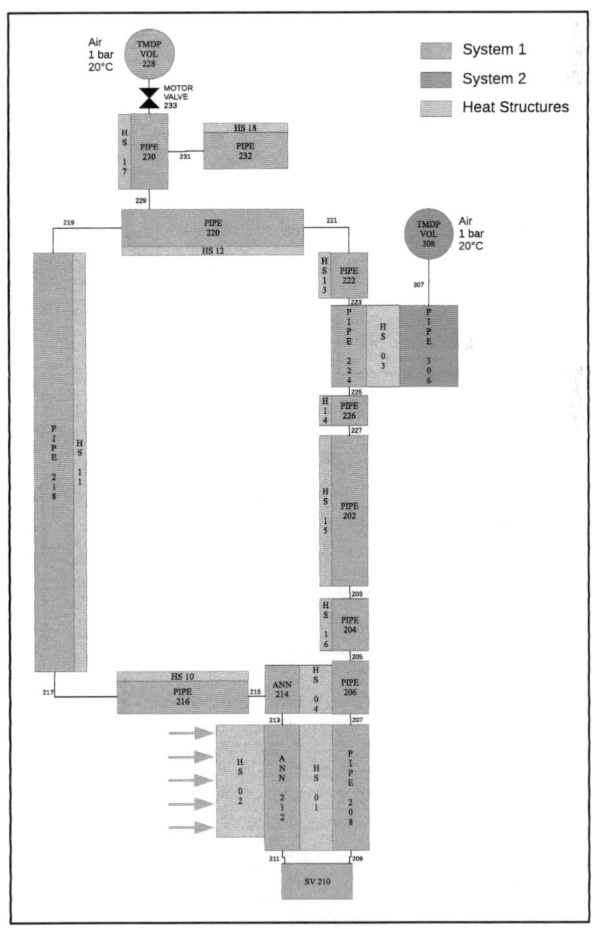

Figure 2: RELAP5-3D nodalization of the facility

IOP Conf. Series: Journal of Physics: Conf. Series **923** (2017) 012008 doi:10.1088/1742-6596/923/1/012008

Table 1: Geometrical and physical data of PROPHET main components.

Total Volume [dm^3]= 3.20	**Volume [dm^3]**	**% of total**	**Length [m]**	**Pipe Heat Capacity [kJ/K]**
Bayonet Downcomer (Downward)	0.07	2	1.36	0.16
Bayonet Annulus (Upward)	0.35	11	1.35	3.00
Lower Horizontal Pipe	0.08	3	0.26	0.59
Hot Leg(Vertical Upward)	1.24	39	3.74	8.31
Upper Horizontal Pipe	0.11	3	0.36	0.81
Hot Leg (Vertical Downward)	0.07	2	0.21	0.46
Condenser	0.17	5	0.50	1.12
Cold Leg (Vertical Downward)	0.91	28	2.74	6.10
Connections to Safety/Pneumatic Valves	0.20	6	0.60	1.34
Total	3.20	100	11.12	21.88

The fluid inside the loop is demineralized water. In this experimental campaign, the system was operated in both single-phase and two-phase flow natural circulation with presence of a non-condensable gas (air). Since no insulation has been installed, the actual configuration is characterized by distributed heat sink (along a length of 7.81 m); being the thermal power relatively low, the thermal losses are sufficient to compensate the power provided to the bayonet and therefore the shell-side of the condenser is filled with air. In this configuration the power removed in the condenser is almost negligible with respect to heat losses; moreover, due to heat losses in the hot leg, the average temperature in the pipe is lower, thus the average density is higher and the natural circulation is expected to be reduced.

Seven sets of tests have been carried out filling the circuit with different amounts of water, as shown in Table 2. Because of the geometric configuration of the facility, some air remains trapped in the upper part of the bayonet annulus and in the pipes connecting the cooling loop to the safety valve and to the pneumatic valve. The facility is initially filled completely with water and then a controlled amount of water (0, 100, 200, 300, 400, 600 and 800 g) is discharged. The nominal volume of the loop is 3.20 dm^3 and the maximum water inventory has been of 2.98 dm^3. Every set of heat-up experiments is composed of five tests with the same initial conditions. The power source is turned on after 1 minute from the beginning of the experiment and maintained constant at 100% for the rest of the transient duration (5 hours in total). The input electric power is 1.75 kW; the heat losses of the heating tapes have been measured to be around 30%, so the effective input thermal power is 1.225 kW. At the end of each experiment the remaining mass in the loop has been measured. The standard deviation has been evaluated for both the mass in the loop and the steady state operating pressure. The data acquisition is performed with a dedicated LabVIEW program, which has been specifically designed for the facility using a timestep of 0.1 s

Table 2: Reference test matrix.

Run	**1**	**2**	**3**	**4**	**5**	**6**	**7**
Liquid mass [kg]	2.98	2.88	2.78	2.68	2.58	2.38	2.18
Filling ratio [%]	93	90	87	84	81	75	69
Initial void fraction [%]	7	10	13	16	19	25	31
Estimated initial air volume [dm^3]	0.22	0.32	0.42	0.52	0.62	0.82	1.02
Air mass (p=1.01325 bar, T=20 °C) [g]	0.313	0.434	0.554	0.674	0.795	1.036	1.277

IOP Conf. Series: Journal of Physics: Conf. Series **923** (2017) 012008 doi:10.1088/1742-6596/923/1/012008

3. RELAP5-3D Model

The experimental transients have been simulated by RELAP5-3D, which is a widely-used code to simulate thermal hydraulic transients in nuclear reactors. The adopted nodalization is shown in Figure 2. System 1 represents the main loop in PROPHET facility, and system 2 the condensation pool.

The bayonet heat exchanger is composed by a downcomer region simulated by pipes 206 and 208, an inversion chamber (single volume 210), and an annular riser represented by annulus 212 and 214. Heat structures 01 and 04 represent the inner pipe wall that separates the downcomer from the annulus, through which regenerative heat transfer occurs. Heat structure 02 represents the external wall of the bayonet and a constant heat flux boundary condition is imposed on its outer surface.

The condenser consists of pipe 224 which is concentric with an annular pool simulated by pipe 306. In the condenser heat is transferred from the fluid of pipe 224 to the one in pipe 306 by the heat structure 03. Pipe 306 is connected to time dependent volume 308 to simulate the ambient boundary conditions (air at 20°C and 1 bar). All the network piping except for the bayonet and the condenser (which in this case is full of air) are connected to heat structures to simulate the heat losses to the environment. Time dependent volume 228 and motor valve 233 model the safety valve connected to the experimental facility. The valve opens if the pressure in the loop exceeds 7 bar and closes when it returns below 6 bar. In the upper part of the facility, pipe 230 and pipe 232 model respectively the vertical connection to the safety valve and the horizontal connection to the pneumatic valve used to discharge water and air during the filling procedure of the facility.

Convective boundary conditions have been imposed on heat structures from 10 to 18. The heat transfer coefficient on the surface of heat structures was determined as the sum of the contributions of free external convection and radiative heat transfer. Its value has been tabulated as a function of the wall temperature and provided as input to RELAP5-3D.

Regarding the initial conditions, both water and air are initially present in the experimental facility loop, while the shell-side of the condenser (pipe 306) contains only air. The initial water inventory in the loop is adjustable by selecting the initial amount of water that fills each component's volumes. The total facility volume computed by RELAP5-3D differs by 1% from the measured one (effect of a discrete nodalization and of the curves shape). For this reason, the measured mass inventory and the numerical one differ by 1% for comparable percentages of initial fillings. Non-equilibrium model is adopted everywhere.

4. Results and discussion

Experimental heat-up transients have been carried out with seven initial masses ranging from 2.19 kg to 2.98 kg, which corresponds to the maximum possible filling of the loop.

The experimental results showed that, at high percentage on initial filling, single-phase natural circulation occurs and that at the end of the heat-up transient steady state values of pressures and temperatures are reached. On the other hand, at lower percentages of initial filling, two-phase flow natural circulation is reached and instability phenomena related to this specific flow regime occurs, which induces oscillations in the measured valued of pressure and temperatures along the loop.

This chapter will first of all analyse the experimental results of two tests, one of high percentage of filling (~ 91%) and the other at low percentage (~ 69%), and compare them with the predictions of RELAP5-3D. In the condition of initial percentage of filling equal to 91% the water inventory is equal to 2.96 kg and the air-water interface is located in pipe 220. With an initial percentage of filling of 69% the water inventory is equal to 2.25 kg and the air-water interface is located between pipe 226 and pipe 202. For every experimental test, the average values of pressure during the last 500 s of the transient have been evaluated. The range of oscillation of each parameter around its average value is negligible in the test of high filing percentage, while it is more relevant at low filling percentage because of the occurrence of two-phase natural circulation instabilities. Similarity, also the average values of the parameters evaluated by RELAP5-3D have been calculated. In the final part of this chapter, experimental and predicted steady-state average values of pressure are compared.

4.1. Pressure and temperature history

Figure 3, Figure 4 and Figure 5 present the history of the absolute pressure in the bayonet inversion chamber, as well as of the bayonet inlet and outlet temperatures for two water inventories: a high one (about 91% of the filling) that corresponds to liquid single phase flow, and a relatively low one (about 69% of the filling) at which there is two-phase flow. Each experimental result is compared with two different simulations by RELAP5-3D.

The first simulation (Simulation A) is performed with an input thermal power to the bayonet of 1225 W. This value is the result of several single-phase tests performed in forced circulation by inserting a pump in the loop. The measurement of the mass flow rate and of the temperature difference in the bayonet allows the computation of the thermal power received by the water in the bayonet. The obtained value is lower than the electric power provided to the heater, since a part of it is dissipated to the environment. From Figure 3 it can be seen that even if the shapes of predicted and experimental pressures are very similar, the agreement at the end of the transient is not satisfying.

The analysis of the fluid temperatures predicted by RELAP5-3D shows that temperatures in pipes 230 and 232 were much lower than the temperature of the pipes outer surface revealed by the thermographic camera. This is due to the fact that no net flow occurs through pipes 230 and 232 and, using a one-dimensional modeling, RELAP5-3D is not able to describe the recirculation phenomena actually occurring inside these pipes. The underestimation of the temperature in such volumes results in an underestimation of the pressure.

In order to improve the prediction, a second simulation set (Simulation B) has been performed: a lower power to the bayonet is used (1150 W) and the critical components (pipe 230 and 232) have been considered adiabatic by removing the heat structures connected to them; in this way, the predicted temperatures reach the value comparable with the upper part of the loop. The outer surface of these pipes corresponds to approximately 6% of the networking piping and condenser surface; in order to compensate for the lower heat transfer surface to the environment, the thermal power to the outer surface of the bayonet has been reduced of 6%, which lead to a power of 1150W.

The boundary conditions of Simulation B improve the overall prediction, but the final pressure is slightly overestimated.

The absolute pressure in the inversion chamber in both simulations are reported together with the experimental data in Figures 3a and 3b respectively for high and low percentages of filling.

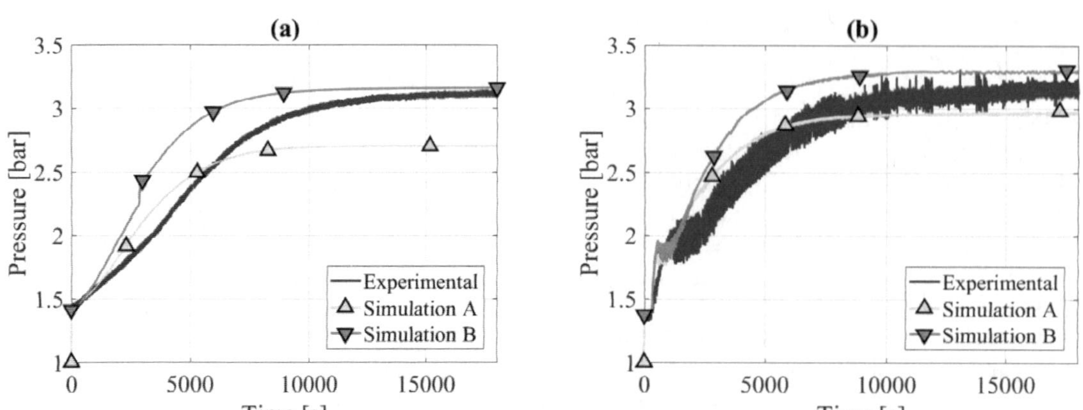

Figure 3: Bayonet inversion chamber pressure for high (~91%) and low (~69%) percentage filling.

It can be seen that, except for the first part of the transient, the experimental results are in between the two numerical prediction for both high and low percentage of filling, but the predictions by Simulation B are in general the best ones as regards the final steady state values of the thermal-hydraulic parameters. Therefore, only the prediction obtained by Simulation B will be compared with the experimental results from now on.

Figures 3a and 3b shows that the pressure curve for high water inventory is smooth and neat; on the other hand, at lower initial filling the pressure presents relevant fluctuations due to the occurrence of the pulsated two-phase flow in the loop. Conditions in the bayonet are very similar to pool boiling with intermittent flow and high pressure and pressure drop oscillations, which also affect the fluid temperature history.

The experimental and predicted temperature time behaviors at the bayonet inlet and outlet at high percentage of filling, together with the saturation temperature, are plotted respectively in Figures 4a and 4b; there is a good matching between experimental and predicted values in the first part of the transient, while the code slightly underestimates the temperatures (approximately 10°C) once the steady state conditions are reached. Since the temperature at both the bayonet inlet and outlet always remains below the saturation value, natural circulation of subcooled liquid occurs in the loop at initial high percentage of filling.

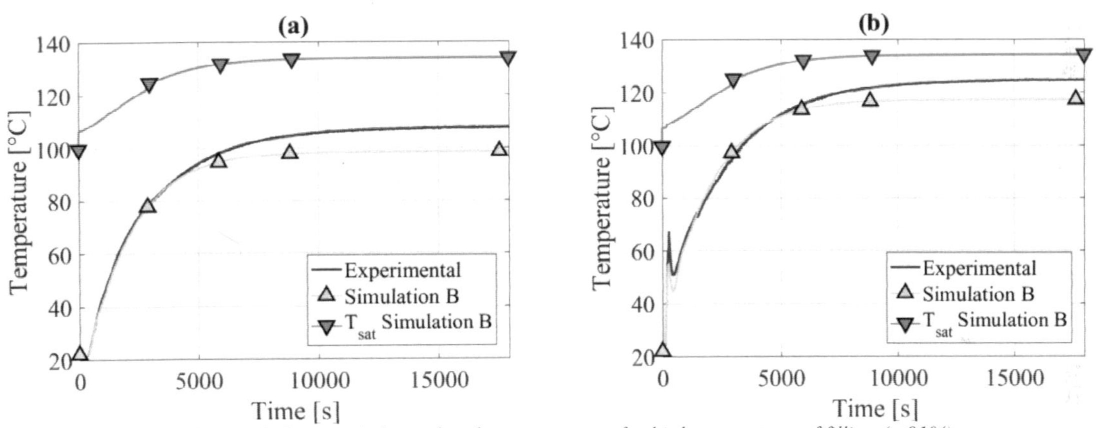

Figure 4: Bayonet inlet and outlet temperature for high percentage of filling (~ 91%)

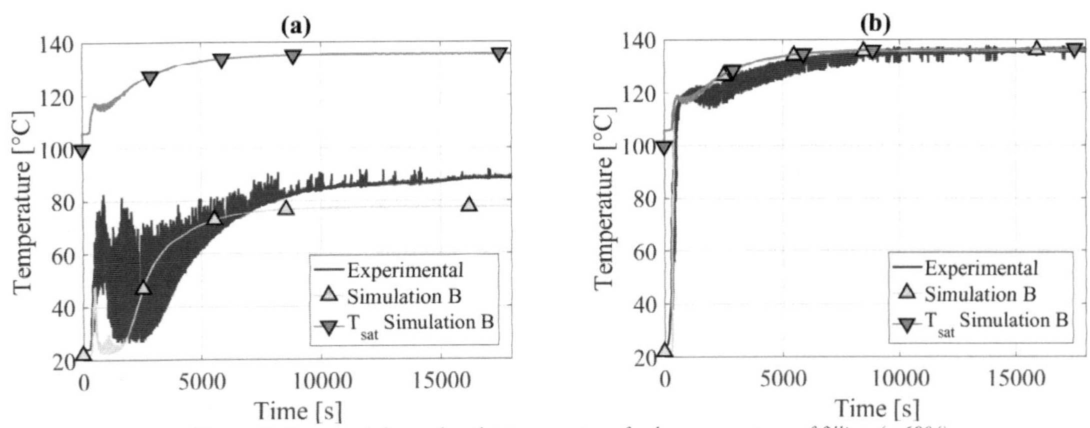

Figure 5: Bayonet inlet and outlet temperature for low percentage of filling (~ 69%).

The history of experimental and predicted temperatures at the inlet and at the outlet of the bayonet at low percentage of filling is plotted in Figure 5a and 5b. Unlike the previous case, wide oscillations occurred during the experiments, especially during the first half of the transient; during this period of time, higher discrepancies are also found between experimental temperature at the bayonet inlet and the values predicted by Simulation B.

Since saturation temperature is reached at the bayonet outlet (Figure 5b), this test is characterized by two-phase natural circulation, which shows oscillations due to pulsated flow. Moreover, the simulation results show that air tends to be collected at the top and in the coldest part of the loop (condenser and cold leg). The experimental bayonet outlet temperature trends are in good agreement with the predicted results, even though the oscillations are not predicted by the code. The strong temperature oscillations at the bayonet inlet (with peaks larger than 40°C) are related to the pulsated flow in the loop due to the mixing with cold water that intermittently flows in the cold leg.

Comparing the two test conditions, the experimental temperature at the bayonet inlet is considerably higher at high water inventory (20°C higher than at low mass inventory). The temperature at the bayonet outlet, instead, is slightly higher in the low water inventory case. Looking at the plots it can be seen that, assuming the upper part of the loop is adiabatic, good results are found for the pressure history, whereas the temperature behavior is not well reproduced. The discrepancies between experimental and predicted temperatures might be due to the reduction of the power for the adiabatic assumption.

4.2. Steady state operating pressure and mass flow rate

After having analyzed two characteristic cases representative for high and low water inventory, the steady state absolute pressure and the predicted mass flow rate have been plotted as a function of the percentage of filling.

Figure 6 presents the final steady state pressure in the bayonet inversion chamber (the lowest part of the loop), by comparing the experimental results with the ones predicted by RELAP5-3D in both Simulations A and B. The experimental results show a minimum for the pressure (at approximately 87% of filling), which represents a boundary between two regions:

- a region of higher initial inventory (greater than 87%), where a stable single-phase natural circulation occurs; the pressure decreases rapidly as the initial filling decreases because of the increase of the available expansion volume.
- a region of relatively low initial filling (lower than 87%), where boiling occurs in the bayonet; the lower the initial filling is, the higher is the amount of liquid that undergoes the phase change and therefore the higher is the final pressure. In this condition, the instability of the two-phase flow increases as the water mass decreases and the oscillations become larger.

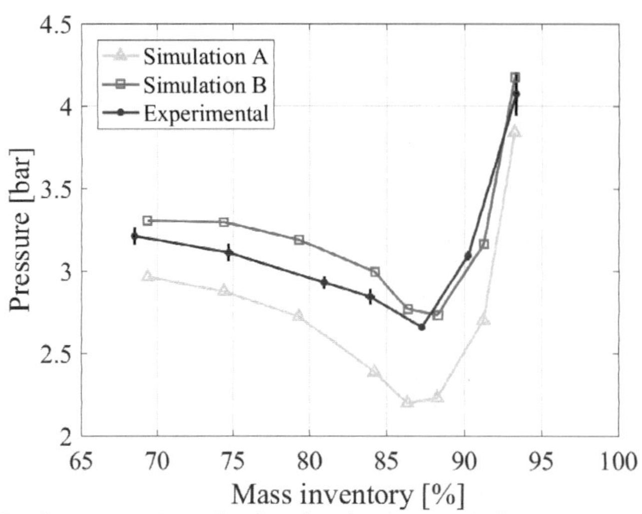

Figure 6: Steady state experimental and predicted pressure in the bayonet inversion chamber

As mentioned in the previous section, the numerical Simulation A is widely underestimating (~ 0.5 bar) the absolute pressure with respect to the experimental results. The Simulation B is slightly overestimating it, but it better represents the experimental trend.

Figure 7 shows the predicted values of the steady state mass flow rate through the loop (evaluated at the inlet of the bayonet heat exchanger) in the two simulations. The vertical bars reported in the figure represent the amplitude of the oscillations at the end of the heat-up transient, computed as standard deviations of the data.

The predicted value are particularly low and, at present, a comparison with measured values is not possible. Nevertheless, it is reasonable to suppose that, similarly to the pressure curves, the actual value of the flow rate will be within the range delimited by the results of simulations A and B.

The difference between the values predicted by the two simulations is very low; both curves show a constant mass flow rate value for high water inventory (>87% of filling). Below this limit value, the flow rate tends to decrease for decreasing initial fillings, since the natural driving force is reduced by the increasing amount of air present in the loop.

Anyway the experimental results show a strong increase of oscillations for decreasing initial water inventories. This is an indication of the instability of the flow that becomes pulsated as the air in the loop increases.

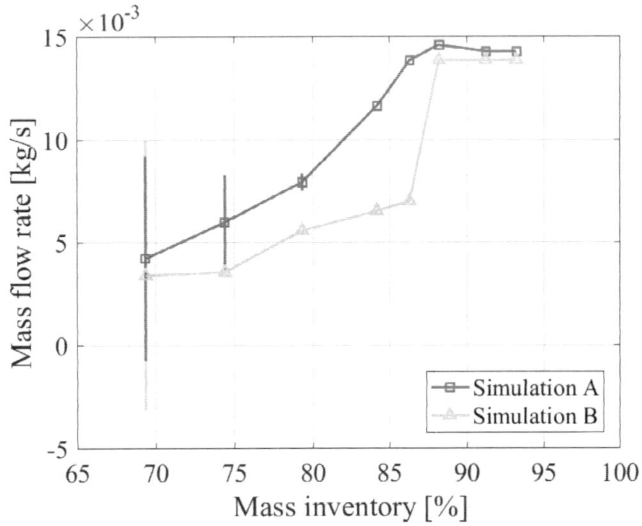

Figure 7: Steady state predicted mass flow rate at bayonet inlet

5. Conclusions and future work

All the proposed Generation IV reactors adopt several passive systems to enhance their safety. One possible safety system is the removal of the decay heat through natural circulation. To study natural circulation in decay heat removal system, the PROPHET facility has been built and tested at Politecnico di Torino.

In this experimental campaign, the effect of the presence of non-condensable gases on natural circulation has been studied both experimentally and with RELAP5-3D simulations. The aim is not focused on the heat transfer coefficient reduction due to the presence of air but on the effect of air on the overall system behavior. The most significant experimental result is the beginning of a pulsated circulation for low water inventories. If the air quantity is significant and the water circulation path is interrupted, there is the beginning of oscillations that become stronger as the air in the system increases.

Simulations performed using the RELAP5-3D code show generally a good agreement with experimental results. Steady state operative pressure is correctly reproduced for different water inventory values. Non-negligible discrepancies have been noticed in correspondence of the connection between the system and the safety and pneumatic valves, where the air temperature is strongly underestimated by the 1-D simulation with RELAP5-3D. These differences are probably due to the presence of two pipes filled with air, through which the net flow is null. The adopted nodalization seems unable to describe local phenomena of fluid recirculation. The present paper shows that some improvements can be obtained by considering adiabatic these pipes. Nevertheless, in future works, the multidimensional component available in the newest version of RELAP5-3D will be used in order to further improve the predictions.

The facility configuration is, at present, characterized by a heat sink distributed along the networking piping. The installation of pipe insulation is planned, in order to rise the temperature level in the system and to enhance the natural circulation. By reducing the thermal losses to the environment and by increasing the heat removal in the condenser, the accuracy of the experimental measurements will increase and the flow instabilities will likely be reduced.

Acknowledgments

The present research has been supported by ENEA and by the Italian Ministry of Economic Development (PAR 2016). Thanks to Rocco Costantino for the technical support in the laboratory.

References

[1] Nuclear Energy Agency (NEA) on behalf of the Generation IV International Forum (GIF) 2014 *OECD Nuclear Energy Agency*

[2] International Atomic Energy Agency (IAEA) 1991 *IAEA-TECDOC-626*

[3] Alemberti A 2015 *Consultants' Meeting: Education Training Seminar on Fast Reactors Science and Technology (ITESM Campus Santa Fe, Mexico City, 29 June – 03 July 2015).*

[4] Caramello M, Gregorini M, Bertani C, De Salve M, Alemberti A and Panella B 2017 *Prog. Nucl. Energy* **99** 127-139

[5] Huang J, Zhang J and Wang L 2015 *Appl. Therm. Eng.* **89** 469-484

[6] Bertani C, De Salve M, Caramello M, Falcone N, Bersano A and Panella B 2017 *J. Phys. Conf. Ser.* **796** Number 1 1-10

[7] The RELAP5-3D© Code Development Team 2015 *Idaho National Laboratory INL/MIS-15-36723*

[8] The RELAP5-3D© Code Development Team 2015 *Idaho National Laboratory INL/MIS-15-36723*

[9] The RELAP5-3D© Code Development Team 2015 *Idaho National Laboratory INL/MIS-15-36723*

[10] Caramello M, Gregorini M, De Salve M, Bertani C, Alemberti A and Panella B 2015 *Proc. ICAPP – Int. Cong. on Advances in Nuclear Power Plants (Nice 3-6 May 2015)* pp 41-48

Analysis of unmitigated large break loss of coolant accidents using MELCOR code

M Pescarini[1], F Mascari[2], D Mostacci[1], F De Rosa[2], C Lombardo[2] and F Giannetti[3]

[1] Alma Mater Studiorum - Università di Bologna, Via Zamboni n. 33 – 40126 Bologna, Italia
[2] ENEA, Via Martiri di Monte Sole n. 4 - 40129 Bologna
[3] Sapienza, University of Rome, DIAEE, Nuclear Section - C.so Vittorio Emanuele II, 244 – 00186 Rome, Italy

E-mail: marco.pescarini@gmail.com

Abstract. In the framework of severe accident research activity developed by ENEA, a MELCOR nodalization of a generic Pressurized Water Reactor of 900 MWe has been developed. The aim of this paper is to present the analysis of MELCOR code calculations concerning two independent unmitigated large break loss of coolant accident transients, occurring in the cited type of reactor. In particular, the analysis and comparison between the transients initiated by an unmitigated double-ended cold leg rupture and an unmitigated double-ended hot leg rupture in the loop 1 of the primary cooling system is presented herein. This activity has been performed focusing specifically on the in-vessel phenomenology that characterizes this kind of accidents. The analysis of the thermal-hydraulic transient phenomena and the core degradation phenomena is therefore here presented. The analysis of the calculated data shows the capability of the code to reproduce the phenomena typical of these transients and permits their phenomenological study. A first sequence of main events is here presented and shows that the cold leg break transient results faster than the hot leg break transient because of the position of the break. Further analyses are in progress to quantitatively assess the results of the code nodalization for accident management strategy definition and fission product source term evaluation.

1. Introduction

Following the Fukushima accident events, a particular attention on unmitigated Nuclear Power Plant (NPP) accidents and their mitigation has been addressed by the International Scientific Community [1-3]. This interest has induced the study of different unmitigated accidents in order to understand their progression and, starting from a thermal hydraulic analysis of the transient, to evaluate their consequences in terms of core damage and fission products release. Nuclear reactors are designed to maintain the fuel damage and radioactive release within authorized limits during selected postulated accidents - Design Basis Accident (DBA) -. A severe accident is a Beyond Design Basis Accident (BDBA) where significant core degradation takes place [4]. Considering the duration of these transients and the several interacting phenomena taking place, integral codes, based on the "lumped - parameters" approach, are used and permit a transient simulation, in a reasonable time, reproducing the different complex coupled phenomena. In the framework of severe accident research activity developed by ENEA, funded by the Italian Ministry of Economic Development (ENEA-MISE agreement), a MELCOR nodalization of a generic Pressurized Water Reactor (PWR) of 900 MWe [1,3,5] has been developed. MELCOR is a fully integrated severe accident code, based on the control volume approach, developed by the Sandia National Laboratories (SNL) for the US Nuclear Regulatory Commission (USNRC). With respect to the activities dedicated to the unmitigated transient analysis in Light Water Reactors (LWR) (e.g. [6,7,8]), the aim of this paper is to present the results of the analysis of MELCOR 2.1 calculations concerning two independent unmitigated Large Break Loss of Coolant Accident (LBLOCA) transients, occurring in the cited type of reactor. In particular, the analysis and comparison between the transients initiated by an unmitigated double-ended Cold Leg (CL) rupture and an unmitigated double-ended Hot Leg (HL) rupture in the loop 1 of the Primary Cooling System (PCS) is presented herein. This activity has been performed focusing specifically on the in-vessel phenomenology that characterizes this kind of accidents. The thermal-hydraulic phenomena analysis, deeply affecting the evolution of both of the transients, and the core degradation processes, caused due to the unmitigated postulated accidents, are here presented.

2. Plant and core modelling using MELCOR

2.1. MELCOR code description

MELCOR [9] is a fully integrated severe accident code developed by SNL for the USNRC. The code is characterized by a modular structure based on packages that simulate the different phenomenologies involved in a severe accident occurring in a LWR. MELCOR is able to simulate all the main phenomena characterizing the Reactor Pressure Vessel (RPV), the reactor cavity, the containment and the confinement building during an accident. The estimation of the source term is obtained by the MELCOR code as well. In particular, the code uses a control volumes approach and simulates thermal-hydraulic phenomena in steady and transient conditions, as well as core degradation phenomena occurring during a severe accident. The Control Volume Hydrodynamics (CVH) package and the Flow Path (FP) package simulate the mass and energy transfer between control volumes, the Heat Structure (HS) package simulates the thermal response of the active and passive heat structures. The CORe (COR) package simulates the behavior of the fuel and the structures contained in the core and lower plenum region during the degradation process. MELCOR can be used through Symbolic Nuclear Analysis Package (SNAP) [10] graphic interface in order to develop nodalizations and for data post-processing, taking advantage of its animation capabilities.

2.2. MELCOR nodalization of the plant

The MELCOR thermal-hydraulic nodalization of the reference reactor has been developed by ENEA with SNAP and has been realized with the aim of minimizing its components, and therefore the calculation time, while maintaining the most realistic and accurate prediction of the phenomena involved during the transients [1,3,11]. The thermal-hydraulic nodalization of the plant (figure 1) consists in 55 control volumes, modeled by the CVH package, interconnected by 70 flow paths, modeled by the FL package. The heat structures, realized to simulate the thermal response of structures characterizing the plant, are 62 and are modeled by the HS package. The three loops of the PCS have been modeled separately. In particular, each loop comprises the HL, the Steam Generator (SG) inlet chamber, two equivalent control volumes representing ascending side and descending side of the U tubes of the SG, the SG outlet chamber, the loop seal and the CL. In each loop is also modeled the Main Coolant Pump (MCP) through the FL package. The ascending and descending side of the U tubes are thermally coupled with the riser of the SG. Two separate control volumes, representing respectively the riser and the downcomer, constitute the secondary side of each SG. The Pressurizer (PRZ) and the surge line have been represented separately and are connected to the loop 1 of the PCS. An accumulator, modeled by a control volume and a second one representing the connection pipe, is connected to the CL of each loop of the PCS. The RPV has been modeled using six separate control volumes representing its main regions, namely: downcomer, lower plenum, core, bypass, upper plenum and upper head. The reactor containment has been modeled with two separate control volumes representing respectively the reactor cavity and all of the other regions of the containment building. The Safety Relief Valves (SRV) and the Pilot-Operated Relief Valve (PORVs) were modeled as well.

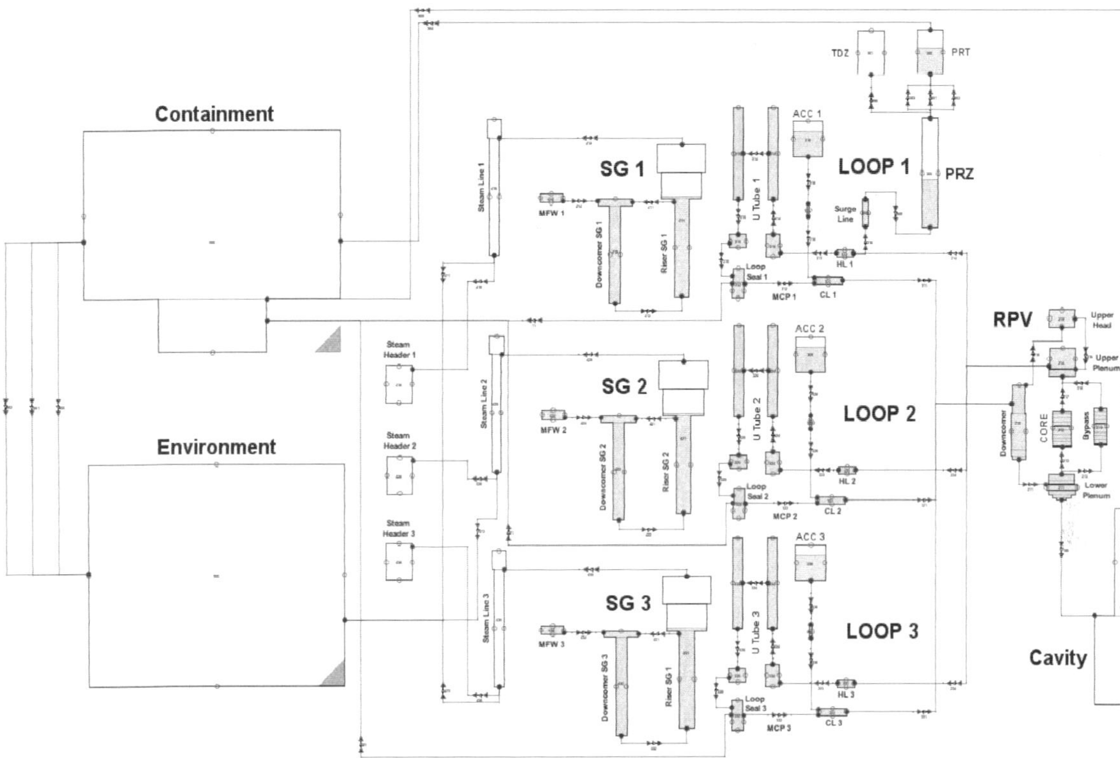

Figure 1. MELCOR nodalization of the reference reactor.

The core and lower plenum regions were further modeled using COR package in order simulate the degradation phenomena/processes and to distribute masses in those regions. The nodalization developed consists in 6 radial rings and 17 axial levels. The axial levels 9÷16 model the active core regions. Moreover, just five radial rings have been used to describe the core region, while the sixth one simulates the portion of that region which extends beyond the radial perimeter of the core in the lower plenum region. Figure 2 shows the 3D SNAP visualization of the COR package nodalization. All of the masses (fuel, cladding, steel etc.) were distributed in the cells of the COR package nodalization, as reported in [3,11], to model in-vessel core degradation phenomena [14]. The lower head nodalization is composed of 15 rings with 20 lower head nodes. The parameters and models, related to the degradation phenomena, have been implemented following the NUREG/CR-7008 [12].

3. MELCOR calculated results

3.1. MELCOR steady state calculation results

MELCOR steady state calculation of the generic PWR of 900 MWe was performed in order to check the operational point predicted by MELCOR against the reference generic plant data. In particular, 1000 s of steady state operation were simulated. Table 1 shows the comparison between the results of steady state simulation, obtained using MELCOR code, and the reference data [1,5].

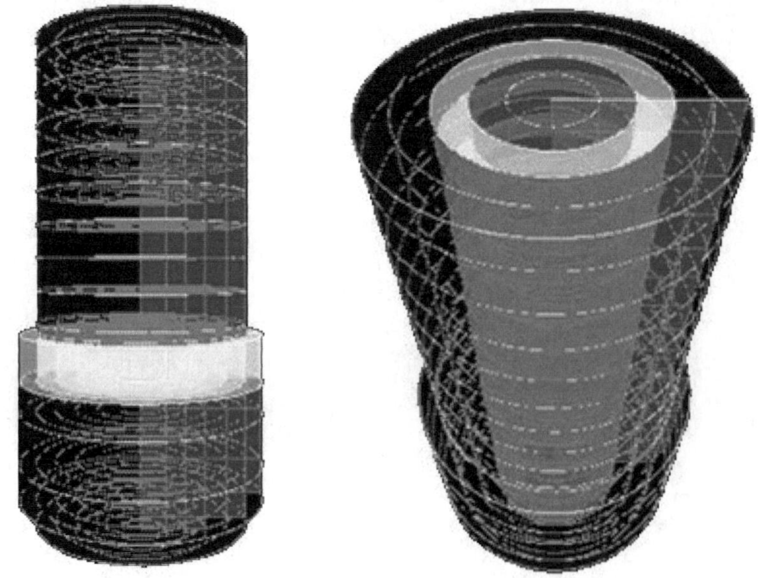

Figure 2. 3D SNAP visualization of COR package nodalization.

Table 1. Main steady state parameters obtained with MELCOR.

Parameter	MELCOR 2.1	Reference Data [1,5]	Relative Error
Reactor thermal power	2775 MWth	2775 MWth	0%
PCS pressure	15.51 MPa	15.51 MPa	0%
Total mass flow rate	13968.9 kg/s	13734 kg/s	1.7%
Core inlet temperature	564.49 K	564.85 K	0.06 %
Core outlet temperature	598.53 K	601.75 K	0.53 %

3.2. Analysis of Double-ended CL and HL LBLOCA transients

For the purpose of this study, two unmitigated Loss Of Coolant Accident (LOCA) scenarios were simulated and analysed. Specifically, a LBLOCA [13] consists in a significant loss of primary coolant from the PCS due to a large size rupture in the hydraulic circuit. Therefore this kind of accident causes a deterioration of core cooling capabilities due to the loss of coolant. In particular, phenomenology and timings characterizing a LOCA transient depend strongly on the size of the rupture and on its position in the PCS. The transients selected for the present analysis are initiated respectively by:

- Double-ended rupture of CL 1 in loop 1 (CL1 accident);
- Double-ended rupture of HL 1 in loop 1 (HL1 accident).

As a matter of fact, the transients initiated by these two events are critical for the reactor since they cause a drastic or even total deterioration of core cooling capabilities if the Emergency Core Cooling System (ECCS) is unavailable. This last condition causes the transients to be "unmitigated" and thus becoming BDBA, which may lead to the degradation of the core, becoming severe accidents.

After 1000s of steady state, the initiator event and plant conditions implemented in MELCOR nodalization for both analyzed scenarios are:

- Break occurrence at t = 0;
- ECCS unavailable except for accumulators injections.

3.3. Analysis of the progression of the transients

The timings of the main events characterizing the two scenarios selected for the analysis are presented in table 2.

Table 2. Chronology of the main events characterizing the selected transients.

Event	CL 1 Break	HL 1 Break
Steady state reactor operation	< 0 s	< 0 s
Break opening	0 s	0 s
Reactor SCRAM	~100 ms	~100 ms
Top of active fuel uncovery	~200 ms	~300 ms
MFWs trip	~1 s	~1 s
Turbine isolation	~1 s	~1 s
MCPs trip	~2 s	~3 s
Start of ACC1 discharge	~2 s	~7.5 s
Start of ACC2 discharge	~5 s	~7.5 s
Start of ACC3 discharge	~5 s	~7.5 s
Bottom of active fuel uncovery	8 s[a]	2784 s
Isolation of ACC1	34 s	40 s
Isolation of ACC2	36 s	40 s
Isolation of ACC3	36 s	40 s
Oxidation onset	875 s[b]	1270 s
Start of core materials melting process	1160 s	1542 s
Oxidation rate peak temperature achievement	1260 s	1650 s
First fuel rod failure	2003 s	2452 s
First lower core plate failure	4426 s	5110 s
First core support plate failure (Slumping)	5339 s	5862 s
Lower head failure	6223 s	9756 s

[a] This determines a temporary core uncovery.
[b] During CL 1 break scenario a first temporary oxidation is predicted at about 2 s from the Start Of the Transient (SOT).

At the SOT the blowdown phase is initiated in both transients by the high pressure difference between the PCS and the containment, inducing reactor SCRAM, turbine isolation, MCPs trip and MFWs trip due to the logics implemented in the nodalization. The results of the calculated data, as expected, show that the CL1 accident is characterized by a mass flow rate way higher than the one that is predicted in the HL1 accident [11]. The reason of this different behavior resides in the fact that in the CL1 case the rupture is modeled right after the MCP 1, which is the position characterized by the highest pressure in the hydraulic circuit constituting loop 1. Figure 3 shows the comparison between CL1 and HL1 accidents regarding the trend of PCS pressure function of time. In general, it is notable how the CL1 accident is the fastest between the two. As expected, in both of the scenarios MELCOR predicts a first subcooled blowdown phase, characterized by a higher depressurization rate, followed by a two-phase blowdown phase, characterized by a lower depressurization rate. This second phase starts when the coolant in the PCS reaches the local saturation conditions, starting the flashing process from the hottest regions, namely the core and the upper head [11].

Figure 4 shows the collapsed coolant level evolution in the core region during the two transients analyzed. MELCOR predicts in both scenarios a first drastic drop of the collapsed coolant level at the beginning of the transient, induced by the blowdown phase. During the CL1 accident, this decrease of the collapsed coolant level is such as to cause the total temporary uncovery of the core active region (bottom of active fuel uncovery). Following the PCS pressure drop, the reflooding phase is started in both of the transients by the accumulators injection, restoring partially the collapsed coolant level in the core (more significantly in the HL1 case). Since the pressure inside the PCS during the accumulators injection is still way higher than the containment one, a significant amount of emergency coolant is lost through the break. At the end of the accumulators injection, the collapsed coolant level inside the core starts to decrease again due to the unmitigation of the transients.

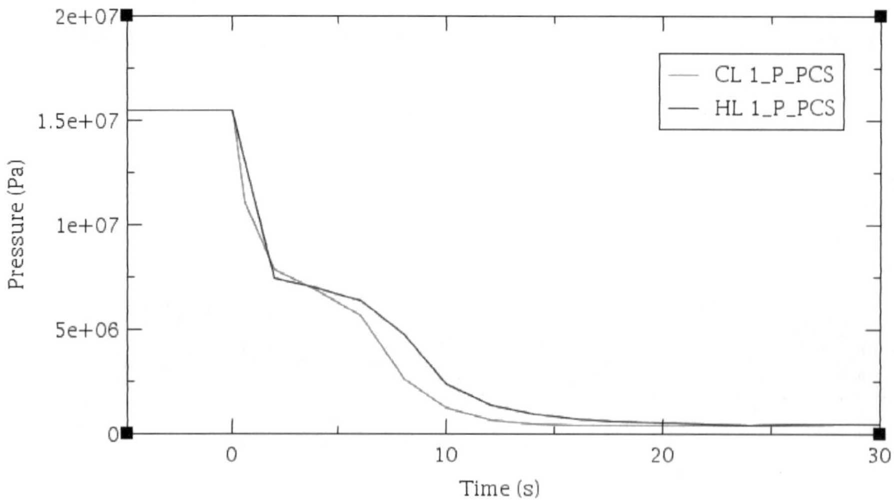

Figure 3. PCS pressure function of time.

This time the decrease is slower, compared to the one occurring during the blowdown phase, because of the lower pressure characterizing the PCS. The progressive core uncovery starts the core heat-up process in the transients analyzed since the core cooling capabilities, at this stage of both of the scenarios, are completely deteriorated. Figure 5 shows, on the other hand, the collapsed coolant level trend in the lower plenum region during the two transients studied. Following complete core uncovery, the level of the collapsed coolant undergoes a relative stabilization in the lower core region. During this phase of both transients the core heat-up process leads to the oxidation of metallic materials inside the core region (especially the Zircaloy cladding enveloping the fuel rods) with the steam produced inside the RPV. The progressive heating-up of the core leads eventually to the failure and collapse of the fuel, relocating onto the lower core plate. The code predicts the failure of the lower core plate 4426 s after the SOT in the CL1 accident and 5110 s after the SOT in the HL1 accident, causing the corium relocation upon the core support plate. The failure of this latter plate is predicted 5339 s after the SOT in the CL1 case and 5862 s after the SOT in the HL1 case. Following the core support plate failure, the corium relocates into the lower plenum (slumping), interacting with the residual water inside of it and leading eventually to the complete debris dryout in both of the transients. The failure of the two plates causes the collapsed coolant level drops in the lower plenum region shown in figure 5. Figure 4 and figure 5 also underline how the CL1 transient is in general predicted by MELCOR as the fastest, leading to the complete debris dryout sooner than in the HL1 transient.

Figure 6 shows the trend of the cladding temperature function of time for transients analyzed in the core upper central region, selected as representative. The cladding temperature sharply increases during the first instants of both transients due to the rapid core uncovery, induced by the blowdown phase. In the CL1 accident, this first peak of the cladding temperature reached is above 1100 K, thus inducing a temporary oxidation of the Zircaloy with steam. In the HL1 accident, the first peak of the cladding temperature predicted is significantly lower than the other transient's one. The reason of this different behavior is that the break of a HL pipe causes a dragging of the PCS fluid toward the rupture passing "naturally" through the core region and thus generating a core cooling contribution. If a CL pipe breaks, on the contrary, the fluid inside the RPV is dragged backwards towards the rupture, causing the complete core uncovery and a higher cladding temperature peak. Successively MELCOR predicts a sharp decrease of the cladding temperature in both transients due to the reflooding phase. After the accumulators isolation the fuel rod cladding temperature starts to rise again in both scenarios because of the unavailability of the ECCS, starting the core heat-up process. The first phase of this process is governed by fission products decay heat, with an average heat-up rate lower than 1 K/s, according with literature [14].

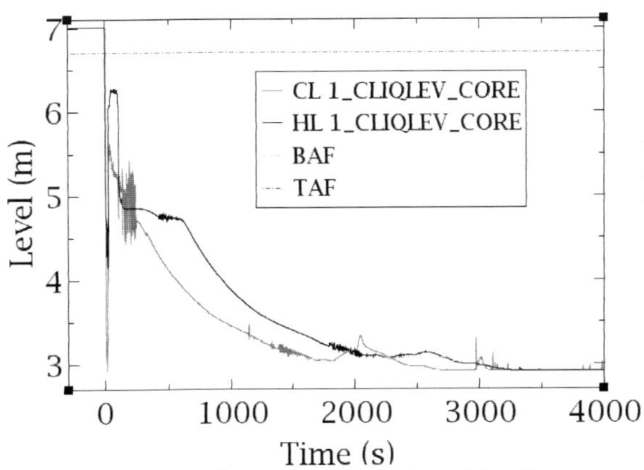

Figure 4. Collapsed coolant level in the core region function of time.

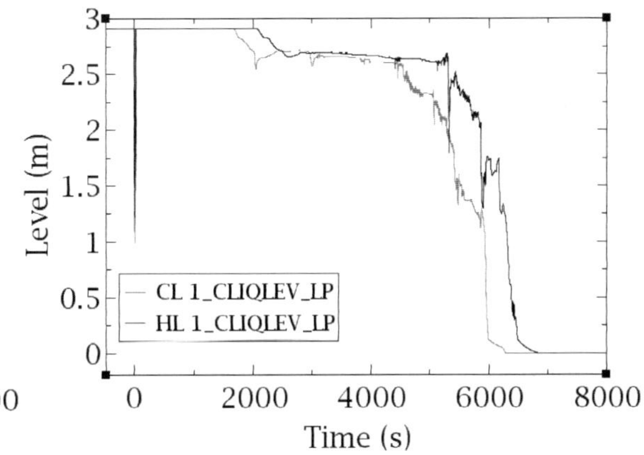

Figure 5. Collapsed coolant level in the lower plenum region function of time.

When the cladding temperature reaches 1100 K, the Zircaloy oxidation process with steam starts. Once the temperature of about 1850 K is reached, the Zircaloy oxidation reaction rate rises abruptly. This condition, reached 1260 s after the SOT during the CL1 transient and 1650 s after the SOT during the HL1 transient, causes, in turn, a rapid increase of the cladding temperature. This temperature reaches the Zircaloy melting point (2098K) in both transients, starting the candling process and the relocation of core molten material into lower core regions. These regions are characterized by lower temperatures because of the residual coolant inside RPV, therefore inducing the solidification of the relocated materials, possibly creating blockages obstructing coolant flow. Following the progression of the core heat-up process, MELCOR predicts the fuel rods failure starting from the upper central core region. The first fuel rod failure is predicted 2003 s after the SOT in the CL1 accident and 2452 s after the SOT in the HL1 accident. Figure 7 shows the cumulative hydrogen production by oxidation for the two transients function of time. This process involves the oxidation of metallic materials inside the RPV (Zircaloy and stainless steel) with steam; therefore it depends from the flow blockage phenomena and available area for oxidation. During the first instants of the CL1 transient, MELCOR predicts a first temporary hydrogen production. This is due to the oxidation process, mentioned earlier, taking place during the blowdown phase. Successively, the reflooding phase reduces the cladding temperature stopping the oxidation process. In both transients the significant oxidation process starts when the cladding temperature increases above 1100 K during the core heat-up process. When the temperature of about 1850 K is reached, the hydrogen production increases abruptly in both transients because of the oxidation reaction rate rise. The code finally predicts another significant hydrogen production in both transients following the lower core plate failure and the core support plate failure. These events induce further steam creation, available to the oxidation process, due to the interactions between the relocating corium and the residual coolant inside the RPV. MELCOR predicts the production of about 224 kg of hydrogen during the CL1 transient and about 227 kg during the HL1 transient. In both scenarios, about 205 kg of hydrogen are produced by Zircaloy oxidation, while the rest is produced by stainless steel oxidation.

Figure 8 shows the main degradation process phases of the CL1 accident while figure 9 regards HL1 accident. Figure 8 and figure 9 show that the degradation phases predicted by MELCOR are basically the same in both transients except for timings. This underlines how the main differences between the two scenarios analyzed are predicted during the first thermal-hydraulic phase of the transients rather than during the degradation process.

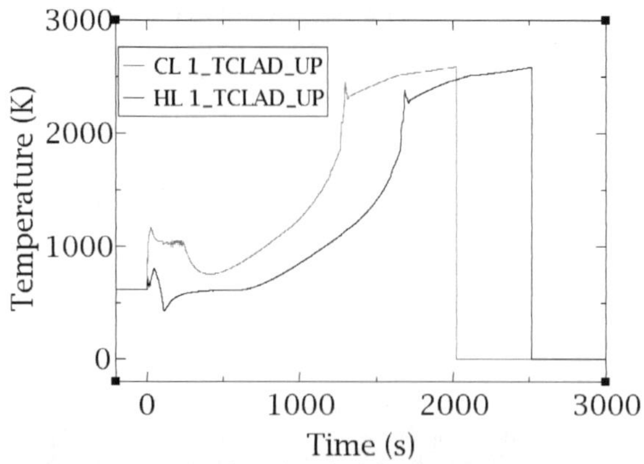

Figure 6. Fuel rods cladding temperature in the core upper central region function of time.

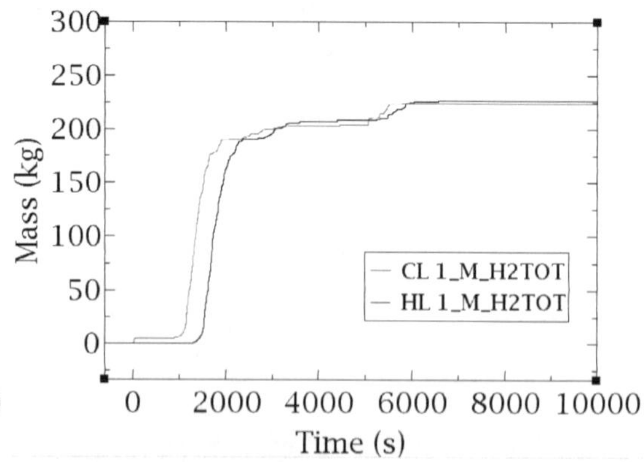

Figure 7. In-vessel hydrogen production by oxidation function of time.

Successively the fuel rods collapse takes place in both transients starting from the upper central core region and follows in to the relocation of most of the core materials upon the core lower plate. Considering the degradation parameter implemented by the code user, the majority of the relocated material is in the form of particulate debris, but the core predicts also the formation of a stratified molten pool (a lighter metallic molten pool upon a heavier oxide molten pool) in both transients. After the slumping of the corium into the lower head the quenching process starts, causing the solidification of most of the molten material. Prior to the lower head failure, most of the material inside the RPV is in facts in the form of particulate debris in both transients. During the CL1 transient the failure of the lower head is predicted 6223 s after the SOT. On the other hand, during the HL1 transient, the failure of lower head is predicted after 9756 s from the SOT.

4. Conclusions

A first MELCOR 2.1 analysis of the two unmitigated LBLOCA transients selected shows the code capability to predict the expected thermal hydraulic and core degradation phenomena/processes typical of these transients. The main differences between the two transients are predicted by the code during their first phases, governed by thermal-hydraulic phenomena. Because of the position of the break in the hydraulic circuit and the consequent faster blowdown, the CL1 accident is predicted to be characterized by a complete temporary core uncovery during the blowdown phase and a lower collapsed coolant level increase during the reflooding phase. During the first collapsed coolant level decrease, due to the blowdown phase, both transients are characterized by a first cladding temperature peak. The temperature peak predicted during the HL1 transient is significantly lower than the one predicted in the CL1 transient and is cooled faster during the reflooding phase. The unmitigation of both transients causes the start of the core heat-up process which ultimately leads to the core degradation process. The two transients are characterized by similar phenomenological degradation phases: the oxidation of metallic materials (especially Zircaloy) with steam, generating hydrogen and rapidly rising the cladding temperature, the fuel rods failure and relocation in the lower core regions, the lower core plate and core support plate failure and, finally, the lower head failure. The lower head failure takes place in the CL1 transient before of the HL1 transient. The unmitigated CL1 transient is therefore confirmed as the fastest and so the most severe for the reactor, leading sooner to the lower head failure. Further analyses are in progress to quantitatively assess the results of the current MELCOR nodalization. In particular, sensitivity analyses are in progress to characterize the break flow through the double-ended ruptures, determining the blowdown velocity and the consequent sequence of main events.

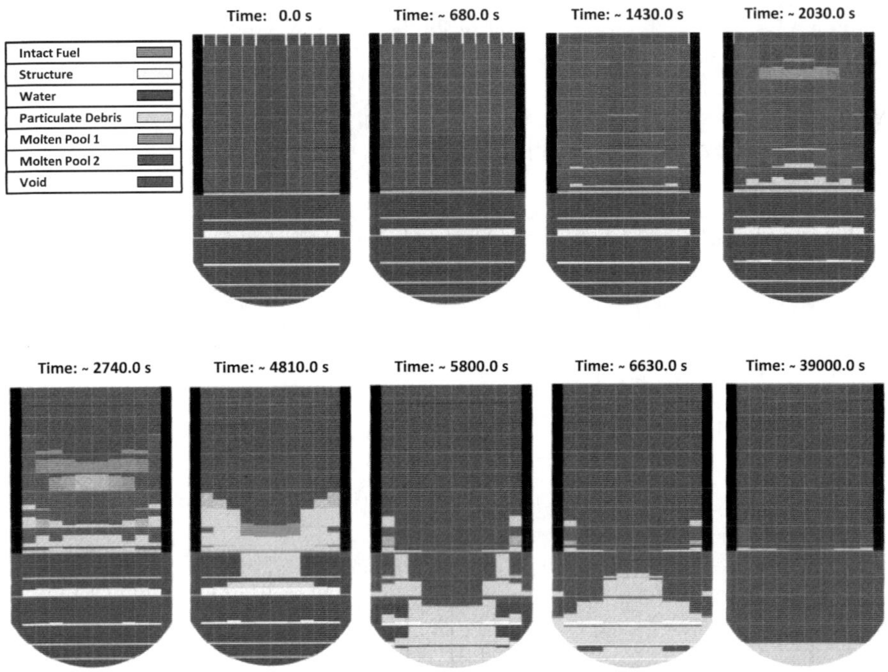

Figure 8. SNAP visualization of the main core degradation phases (CL 1 break).

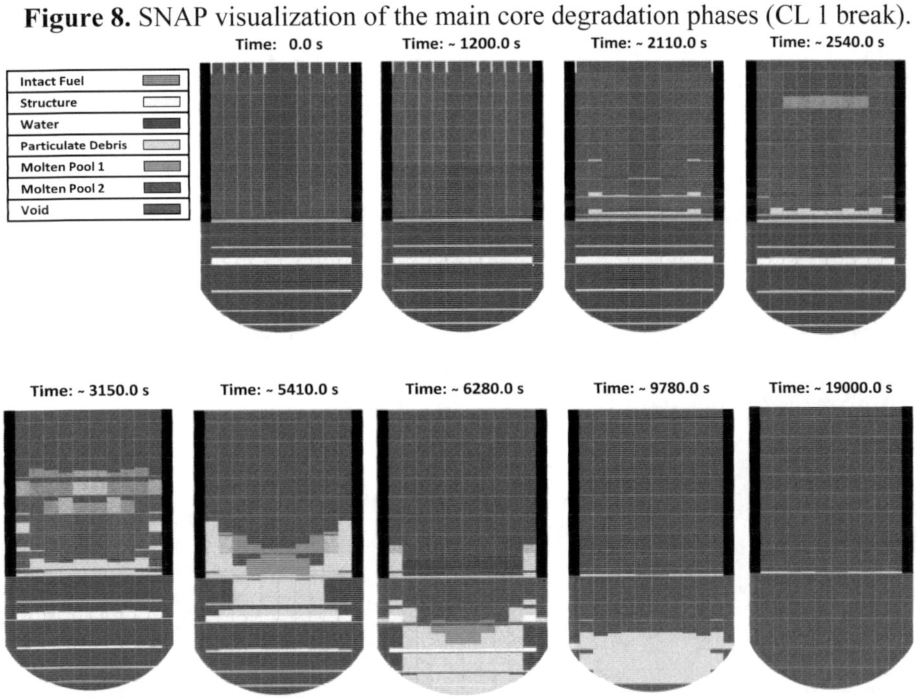

Figure 9. SNAP visualization of the main core degradation phases (HL 1 break).

Further sensitivity studies are in progress in order to study the hydrogen generation, during the core degradation phase, and the lower head behaviour and failure, and the effects of a potential ex-vessel cooling. The quantitative assessment of the calculated data is crucial to define possible accident management strategy, for plant consequences reduction, and for fission product source term evaluation. This source term evaluation, as radiological release, could be the input for atmospheric dispersion code for emergency preparedness strategy.

Acknowledgments
The activity has been funded by Italian Minister of Economic Development.

References

[1] Lombardo C, Mascari F, Buffa P, Castiglia F, Giardina M and Palermo G 2014 *Nodalizzazione MELCOR per lo Studio Integrale di Sequenze Incidentali su Reattori PWR da 900 MWe e valutazioni preliminari d'impatto a breve e medio raggio*, ADPFISS-LP1-026 (Bologna)
[2] Mascari F, Giannetti F, Balestra P, Zoino A and Caruso G 2015 *Integral study of accident sequences with reference to NPPs next to the Italian borders*, ENEA Internal Technical Report ADPFISS-LP1-026 (Bologna)
[3] Mascari F, Pescarini M, Giannetti F, Luciani I, Caruso G, 2016, *Integral calculations of severe accident Scenarios in PWR and BWRs*, ADPFISS-LP1-075
[4] *Severe Accident Management Programmes for Nuclear Power Plants*, IAEA Safety Standards, Safety Guide No. NS-G-2.15, 2009
[5] *Progetto Unificato Nucleare, Direzione delle Costruzioni DIR-PUN*, ENEL, I Quaderni dell'Energia No. 12, 1986
[6] Jin Y, Xu W, Liu X and Cheng X 2015 *Ann. of Nuc. En.* **80** 322-37
[7] Li G, Liu M, Wang J, Chong D and Yan J 2016 *Ann. of Nuc. En.* **90** 195-204
[8] Hashim M, Ming Y and Ahmed A S 2013 *Research J. of Appl. Sci., Eng. and Technol.* **5**(12) 3320-35
[9] *MELCOR Computer Code Manuals*: Vol. 1, SAND 2015-6691 R; Vol. 2, 2015-6692 R; Vol.3 SAND 2015-6693 R, 2015, Sandia National Laboratories
[10] *Symbolic Nuclear Analysis Package (SNAP)* User Manual, 2012, Applied Programming Technology, Inc., Bloomsburg, PA.
[11] Pescarini M 2016 *Analisi di un incidente non mitigato di tipo loca in un reattore pwr mediante il codice MELCOR 2.1* Master Degreee Thesis, Alma Mater Studiorum - Università di Bologna (Bologna)
[12] Ross K, Phillips J, Gauntt R O and Wagner K C 2014 *MELCOR Best Practices as Applied in the State-of-the-Art Reactor Consequence Analyses (SOARCA) Project* NUREG/CR-7008
[13] Jones O C 1981 *Nuclear Reactor Safety Heat Transfer* (Hemisphere Publishing Corporation)
[14] 2001 *In-Vessel Core Degradation Code Validation Matrix Update 1996-1999* NEA/CSNI/R(2000)21 (OECD Nuclear Energy Agency)

IOP Conf. Series: Journal of Physics: Conf. Series **923** (2017) 012010 doi:10.1088/1742-6596/923/1/012010

Study of steam condensation at sub-atmospheric pressure: setting a basic research using MELCOR code

A Manfredini and M Mazzini

ACTA Srl Via di Torretta snc, 56122 – Pisa

E-mail: mazzini@ing.unipi.it

Abstract. One of the most serious accidents that can occur in the experimental nuclear fusion reactor ITER is the break of one of the headers of the refrigeration system of the first wall of the Tokamak. This results in water-steam mixture discharge in vacuum vessel (VV), with consequent pressurization of this container. To prevent the pressure in the VV exceeds 150 KPa absolute, a system discharges the steam inside a suppression pool, at an absolute pressure of 4.2 kPa. The computer codes used to analyze such incident (eg. RELAP 5 or MELCOR) are not validated experimentally for such conditions. Therefore, we planned a basic research, in order to have experimental data useful to validate the heat transfer correlations used in these codes. After a thorough literature search on this topic, ACTA, in collaboration with the staff of ITER, defined the experimental matrix and performed the design of the experimental apparatus. For the thermal-hydraulic design of the experiments, we executed a series of calculations by MELCOR. This code, however, was used in an unconventional mode, with the development of models suited respectively to low and high steam flow-rate tests. The article concludes with a discussion of the placement of experimental data within the map featuring the phenomenon characteristics, showing the importance of the new knowledge acquired, particularly in the case of chugging.

1. Introduction

One of the most serious accidents that can occur in the experimental nuclear fusion reactor ITER is the break of one of the headers of the first wall cooling system of the Tokamak. This results in water-steam mixture discharge in vacuum vessel (VV), with consequent pressurization of this container.

To prevent the pressure of the VV exceeding 150 KPa absolute, a venting system discharges the steam inside a Suppression Pool (SP). The design foresaw initially the use of 24 spargers, each with about thousand holes of 10 mm diameter, venting the steam under a water head between 1 m and 2 m. (It has to be noted that at present the solution adopted by ITER Organization for VVPSS is different, but the range of steam flowrate per unit area of the sparger remains about the same).

Ultimately it is an accident similar to that of LB-LOCA in a BWR, with the important difference that the suppression pool in ITER is initially at a sub-atmospheric pressure of 4.2 KPa.

The computer codes used to analyze such accident (e.g. RELAP 5 or MELCOR) are not validated experimentally for such conditions. Therefore, during a meeting with the staff of ITER, the Director of the DIMNP of the University of Pisa and ACTA Srl, M. Mazzini proposed to run a basic experimental research on the phenomenology now mentioned, in order to have experimental data useful to validate the heat transfer correlations used in these codes.

The main objectives of the research program were:

- the clear understanding of VVPSS prototypical thermal-hydraulic (TH) conditions impact on direct contact condensation: bubble formation, bubble condensation, bubble size distribution, bubble rise velocity, potential for bubble coalescence, etc.

- to acquire data for thermal hydraulic code qualification.

As already stated, the phenomenological basis of VVPSS behavior cannot be considered completely covered by previous experience, due to the fact that the VVPSS thermal-hydraulic conditions are rather unique:

- initial pressure of SP free space: 4.2 kPa
- steam pressure at VVPSS inlet: 106-150 kPa
- SP initial water temperature: 280-300 K
- steam flow-rate into the suppression pool: 0.3-5 g/s per sparger hole (corresponding to a total flow-rate in the range 5-100 kg/s)
- water head above the exit hole: 1-2 m

The experimental steam flow-rate range was established taking into account the results of the MELCOR calculations of the reference accident (Fig. 1) and the possibility of a not uniform distribution of the steam among the 24 spargers.

After a thorough literature search on this topic, ACTA, in collaboration with the staff of ITER, defined the experimental matrix and performed the design of the experimental apparatus.

The experimental equipment is extremely simple: a container of adequate size, initially at a pressure of 4.2 KPa absolute, in which a pipe discharges steam at specified conditions of pressure and temperature, through one or more holes of 10 mm diameter. Of course, the equipment includes the required systems for steam flow generation, realization of vacuum, water temperature and level adjustment in the condensation tank, in addition to the instrumentation and control systems.

For the thermal-hydraulic of the experiments, a series of calculations was executed by the MELCOR code.

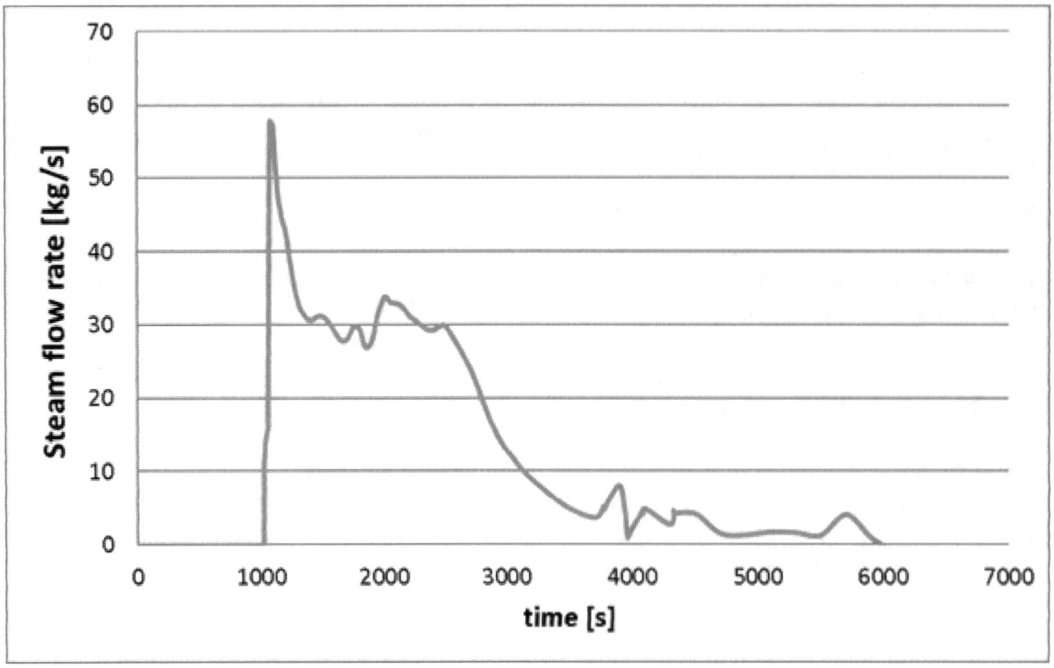

Figure 1: Trends of steam flow-rate in VVPSS in case of DBA in ITER plant

2. Bibliographic research

The bibliographic research aimed at the understanding the fundamental characteristics of a condensation jet and the resulting thermal mixing in a subcooled water pool.

To analyse the thermal mixing induced by a steam discharge into a pool, it is essential to understand the phenomena occurring in a condensing jet. Both the thermal aspects and the hydraulic ones must be studied.
When the steam jet penetrates into a water pool, it entrains liquid from the pool water and induces a turbulent jet, the velocity of which decays slowly along its axial distance. The pool water entrained into the vapour core forms a two-phase mixture jet region. Therefore it is important to consider the interfacial transfer of mass and energy.
When the pool temperature increases or the steam mass flux decreases, the circulation flow pattern induced by the horizontal turbulent jet diminishes and a buoyant plume appears distinctively near the sparger due to the increased effect of buoyancy and/or the decrease of condensation efficiency.
In general, the behaviour of a steam condensing jet immersed in a pool of water is characterized by three distinct regions:

1) The steam core or cavity of vapour,
2) The mixing region, with "vapour entrapment" and "liquid entrainment" phenomena,
3) The region of turbulent jet generated by the condensation of the jet.

These can be described by introducing some characteristic parameters that include the form of the steam jet, the jet expansion rate or the length of penetration, the distribution of temperature within the vapour cavity, the features of the turbulent jet, etc..
In particular, high condensation rates are achievable thanks to the high temperature difference between the two phases.
The condensation process can directly generate instability at the interface of condensation, which is inherently unstable in a macroscopic sense and may be more unstable in some conditions.
Actually, in practical applications in nuclear technologies at atmospheric pressure, the behaviour of a vapour jet in a condensation pool is generally governed by two factors: the steam mass flow rate for unit area at the exit hole and the temperature of the pool water Tp. The experimental observation shows also that, at atmospheric pressure, the shape of the vapour cavity is rather random and condensation instabilities of various type occur for mass flow-rates lower than 270 $kg/s.m^2$ in the entire temperature range tested. With a mass flow-rate higher than 300 $kg/s.m^2$, instead, the steam jet becomes stable. In this case, the shape of the cavity of the steam jet is conical or ellipsoidal, depending on the mass flow and the temperature of the pool. Also the temperature of pool water has an important effect on the thermal-hydraulic behaviour of the condensing steam. These two factors are used for the construction of stability maps, that effectively represent a reference for the results of many studies (References (1), (2), (3), (4), (5), (6)). In particular, Cho et al. proposed a map covering a wide range of mass flows of steam and temperature of the pool (Fig. 2). The condensation regimes are classified into the following types: chugging (C), transitional chugging (TC), condensation oscillation (CO), bubbling (BCO), stable condensation (SC) and interfacial condensation oscillation (IOC). The researchers recorded the highest noise and higher vibrations in unstable condensation regimes, such as chugging and condensation oscillation.
Typically, the single phase turbulent jet becomes a nearly flat, turbulent front and a mixing layer forms at the tip of the nozzle exit. In the region close to the nozzle, the turbulence penetrates inward toward the axis of the jet and forms a wedge-like region, with the potential core surrounded by a mixing layer. In the fully developed region, the turbulence penetrates to the axis and the potential core disappears. Down-stream of the potential core, the flow begins to develop into the Gaussian shape of a velocity profile, which reaches and maintains a self-preserving shape. This asymptotic self-similarity of the free turbulent jet flow is normalized reasonably well with its maximum values and characteristic width. The static pressure in the jet core region is below the surrounding fluid pressure and the fluid is entrained considerably as the jet travels forward, and then the mass flow of the jet increases downstream.

This introduces another point of view, which underlines that the dynamic pressure characteristics determine the stability of the jet of steam. The dynamic pressure shows a trend that is strictly dependent on the mass flow-rate of steam and the temperature of the pool water.

The pressure load increases gradually with the increase of the steam mass flow-rate in the CO scheme, but decreases rapidly in case of higher undercooling (Tp = 30°C or 40° C); lower pressure loads indicate that the steam jet is condensed with efficiency and stability. On the contrary, with a very low sub-cooling (Tp higher than 90° C), at atmospheric pressure, the condensation of steam is inefficient, and this causes a small change of the pressure loads. Experimental data show that the interference of the steam with water during the condensation process is stationary until the temperature is low and the steam mass flow-rate is high. At high temperature of the pool water, the rate of condensation at water-steam turbulent interface is reduced below that necessary to condense the steam.

Besides, one can observe that the effect of non-condensable gases, either on steam condensation or on the resulting dynamic pressure, is very pronounced.

The rate of condensation is limited by the effectiveness of the thermal diffusion process from the interface to the mass of the liquid; in turn, it depends on the intensity of turbulence on the liquid side and therefore on the distribution of water temperature in the pool. The local temperature distribution is an effect of the turbulent mixing due to the jet, of the pool water circulation and of stratification due to gravity.

Predicting the thermodynamic behaviour of the steam discharged in a pool requires, in general, three-dimensional analysis capabilities.

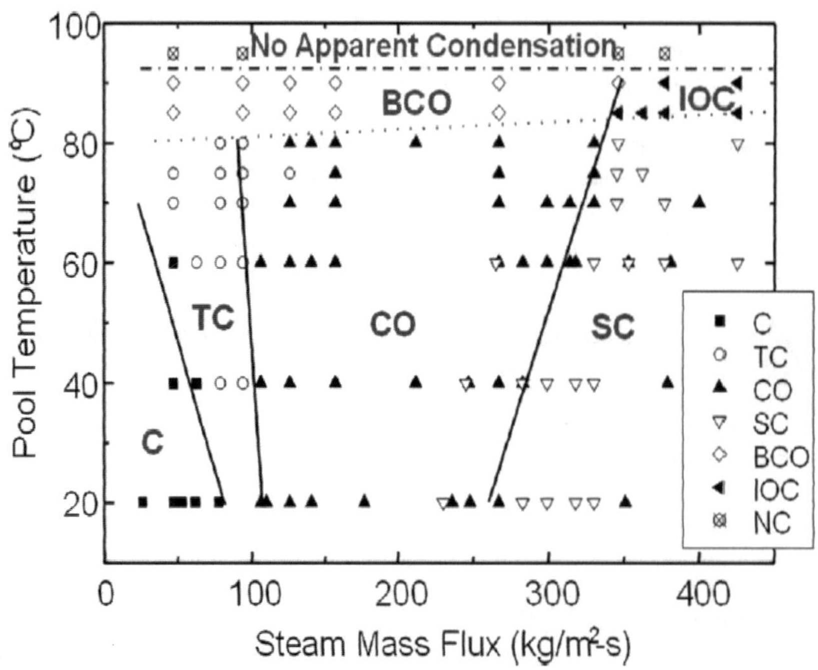

Figure 2: Map of the various regimes of steam condensation as a function of pool temperature and steam specific mass flow-rate

3. Planned experimental matrix

IOP Conf. Series: Journal of Physics: Conf. Series **923** (2017) 012010 doi:10.1088/1742-6596/923/1/012010

With respect to the experimental ranges indicated in the introduction, due to the limited funding available, the minimum test flow-rate has been established to 0.5 g/s (for characterizing the VVPSS behavior during the last part of VV depressurization transient - Fig. 1). This low value of steam flow-rate is applied only to experiments with the lower head of water above the sparger hole: 1.3 or 1.6 m (we did not consider tests with water head above the exit hole of 1 m). The value of 2.5 g/s corresponds to the mean flow-rate foreseen during the first 500 s of this transient, while the value of 5 g/s overcomes the maximum steam flow-rate, also considering possible not uniform distribution of steam among the various spargers.

The test matrix (Table 1) comprises 3 series of test sequences:

A. 1 Hole of 10 mm diameter
B. 3 Vertical Holes of 10 mm diameter and pitch equal to actual sparger
C. 9 Holes of 10 mm diameter and pitch equal to actual sparger.

The results of the first series of test sequences will constitute the data basis needed for the validation of the computer codes used for simulating VVPSS behavior. Besides, they will give important information on related phenomena (bubble formation, bubble condensation, bubble size distribution, bubble rise velocity, potential for bubble coalescence, etc.), as well as on possible dynamic loads originated by the steam condensation. Apart from the above-mentioned limitations about the hydraulic head, they cover the whole spectrum of foreseen TH conditions.

The following series B and C are aimed at investigating possible interaction between one hole and the neighboring holes, firstly in one dimension (series B) and finally in two dimensions (series C). In these series, the test TH conditions are limited to those more prototypical of VVPSS behavior.

Actually, each experiment is a sequence of 4 or 5 tests (depending from the water temperature foreseen for the beginning of the sequence). Each test sequence will be executed with the following procedure.

The condensation tank will be filled with normal water at the temperature foreseen for the sequence (10 or 20 °C), up to the specified level (according with the values foreseen in the test matrix). Thereafter the vacuum system will establish the required pressure in the SP free volume, the test will be started injecting steam at the specified value of steam flow-rate in the condensation tank. After 15 minutes of stationary conditions, the test will be finished, isolating the condensation tank. The pool water temperature will be increased of 10 °C with respect to the previous test value, achieving the new test initial conditions. Immediately after, the new test will start with the same modalities. This last part of the procedure will be repeated two or three times, completing the test sequence.

4. Thermal-hydraulic design of tests

A sketch of the model, initially set up for the MELCOR code to simulate the VVPSS experimental apparatus, is rather conventional, as shown in Fig. 3. However, this model results in a prompt mixing of all the pool water, with very limited increases of the temperature anywhere.

Thereafter, a preliminary effort was done, studying both experimentally and analytically, by the use of a CFD code, the discharge of air in a water pool. These runs showed that:

- at low flow-rate (0.3-1 g/s), small bubbles of air form at the hole exit, moving immediately upward along the vent pipe;
- at high flow-rate (2.5-5 g/s), a jet forms pushing away the water in front of the exit hole.

This confirms the conclusions drawn from the bibliographic research: it is very important to accurately simulate the two different steam condensation modalities, "forced jet" and "buoyant plume", which directly affect the mass transfer from the steam cavity to the surrounding fluid.

On the basis of these observations, a new nodalization was set-up for the condensation pool; Figures 4 and 5 show respectively its vertical and horizontal sections in correspondence to the steam injection volume (named CV105). In this second nodalization we split the condensation pool into 114 control volumes interconnected by appropriate junctions (a total of 342), such as to simulate the steam jet

IOP Conf. Series: Journal of Physics: Conf. Series **923** (2017) 012010 doi:10.1088/1742-6596/923/1/012010

dispersion along the directions more relevant. In particular, as shown in Fig. 4, we split the pool volume in 6 slabs with the same number and layout of control

Table 1. Initial conditions of the various series of basic experiments

Series	Test no.	Sparger inlet		Condensation tank	
		Pressure	Steam flow-rate	Water temp.	Water level
A) 1 Hole 10 mm	1	< 150 kPa, s.h.	0.3 g/s	10 °C	1.3 m
	2	< 150 kPa, s.h.	2.5 g/s	10 °C	1.3 m
	3	< 150 kPa, s.h.	2.5 g/s	20 °C	1.3 m
	4	< 150 kPa, s.h.	5.0 g/s	10 °C	1.3 m
	5	< 150 kPa, s.h.	0.3 g/s	10 °C	1.6 m
	6	< 150 kPa, s.h.	2.5 g/s	10 °C	1.6 m
	7	< 150 kPa, s.h.	5.0 g/s	10 °C	1.6 m
	8	< 150 kPa, s.h.	2.5 g/s	10 °C	2.0 m
	9	< 150 kPa, s.h.	5.0 g/s	10 °C	2.0 m
	10	< 150 kPa, sat.	2.5 g/s	10 °C	1.3 m
	11	< 150 kPa, sat.	2.5 g/s	10 °C	1.6 m
	12	< 150 kPa, sat.	2.5 g/s	10 °C	2.0 m

Series	Test no.	Sparger inlet		Condensation tank	
		Pressure	Steam flow-rate	Water Temp.	Water level
B) 3 Holes 10 mm	13	< 150 kPa, s.h.	0.3 g/s	10 °C	1.3 m
	14	< 150 kPa, s.h.	2.5 g/s	10 °C	1.3 m
	15	< 150 kPa, s.h.	5.0 g/s	10 °C	1.3 m
	16	< 150 kPa, s.h.	2.5 g/s	10 °C	2.0 m
	17	< 150 kPa, s.h.	5.0 g/s	10 °C	2.0 m

Series	Test no.	Sparger inlet		Condensation tank	
		Pressure	Steam flow-rate	Water Temp.	Water level
C) 9 Holes 10 mm	18	150 kPa, s.h.	0.3 g/s	10 °C	1.3 m
	19	150 kPa, s.h.	2.5 g/s	10 °C	1.3 m
	20	150 kPa, s.h.	5.0 g/s	10 °C	1.3 m
	21	150 kPa, s.h.	2.5 g/s	10 °C	2.0 m
	22	150 kPa, s.h.	5.0 g/s	10 °C	2.0 m

Notice: the steam flow-rate is that of each sparger hole. The water level is that above the simulated sparger hole centerline.

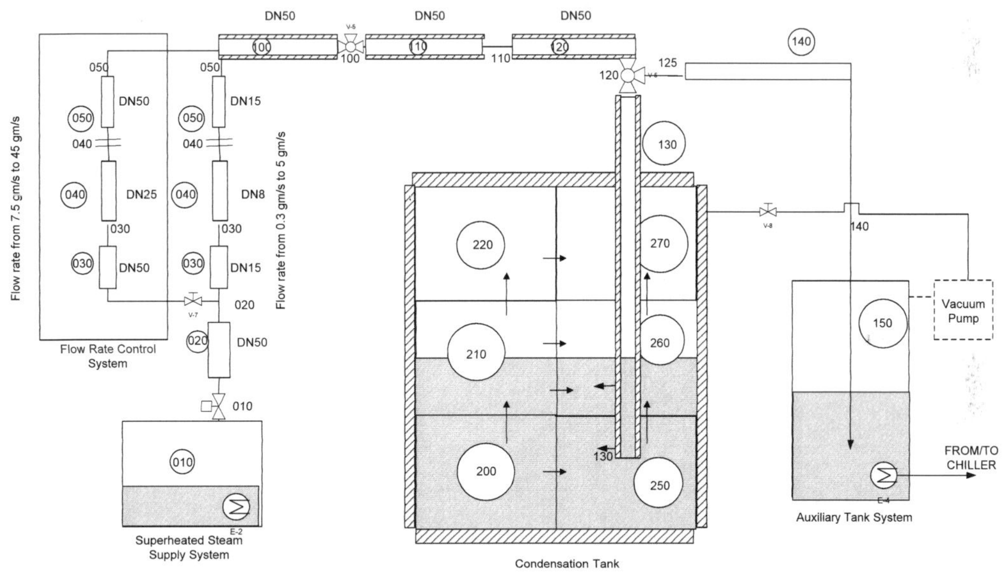

Figure 3: First nodalisation of the condensation pool for MELCOR code

volumes, but with different heights: the first slab goes from 0 m to 0.2 m; the second from 0.2 m to 0.4 m; the third from 0.4 m to 0.6 m; the fourth from 0.6 m to 1 m; the fifth one from 1 m to 2 m; the sixth from 2 m to 3.2 m. In this way the zone closest to the injection hole/s, simulated by smaller control volumes, is described in a greater detail. The same principle was applied to the control volumes subdivision in the horizontal plane, as shown in Fig. 5. The volumes immediately adjacent and nearest to the injection hole/s were shaped with a rectangular basis of 0.2 x 0.22 m^2 up to the opposite wall of the tank. A total of 15 of these small volumes are present for each plan. Excess space was modeled with much broader control volumes, taking also into account the cylindrical form of the tank.

Applying to the junctions appropriate values of the local, direct k_d and reverse k_r pressure drop coefficients (i.e., k values up to three orders of magnitude higher than those for the junctions on the

IOP Conf. Series: Journal of Physics: Conf. Series **923** (2017) 012010 doi:10.1088/1742-6596/923/1/012010

sides of the control volume that correspond to the jet direction), we can effectively simulate the jet formation, the natural circulation induced by steam condensation and the energy released from the latter.
Due to the intrinsic characteristics of MELCOR code, the steam condensation is complete in all the simulation, but the effects of the steam jet for the higher flow-rates seem simulated adequately.

CV502	CV505	CV506	CV507	CV508	CV509
CV402	CV405	CV406	CV407	CV408	CV409
CV302	CV305	CV306	CV307	CV308	CV309
CV202	CV205	CV206	CV207	CV208	CV209
CV102	CV105	CV106	CV107	CV108	CV109
CV002	CV005	CV006	CV007	CV008	CV009

Figure 4: Vertical section of the condensation pool in the second nodalization

<table>
<tr><td rowspan="3">CV102</td><td colspan="5">CV104</td></tr>
<tr><td>CV125</td><td>CV126</td><td>CV127</td><td>CV128</td><td>CV129</td></tr>
<tr><td>CV105</td><td>CV106</td><td>CV107</td><td>CV108</td><td>CV109</td></tr>
<tr><td rowspan="2">CV101</td><td>CV115</td><td>CV116</td><td>CV117</td><td>CV118</td><td>Cv119</td></tr>
<tr><td colspan="5">CV103</td></tr>
</table>

Figure 5: Horizontal section of the condensation pool in the second nodalization

Some calculation results are reported in Figure 6 for a high flow-rate test. The calculation is done with 600 s of stationary initial conditions; at this moment the test starts, with the injection of steam in the CV105 volume. The effects of the steam induced turbulence and of natural circulation correspond to a quick water temperature increase in the steam discharge volume (CV105) and in the two forward volumes (CV106 and CV107), while the rest of the pool has slower and modest increases of temperature.
Analogously, we tried to simulate the buoyant plume, which develops for small steam flow-rates, as shown in Fig. 7. A small peak in water temperature occurs initially in the discharge volume CV105, but the heating of all other volumes is very low.

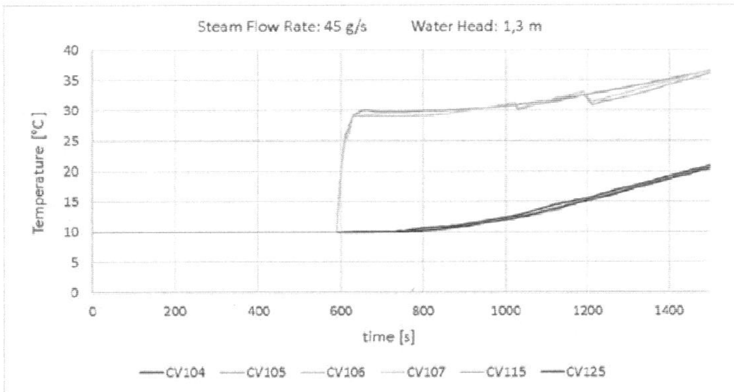

Figure 6: Temperature in the horizontal section at the discharge height in a high flow-rate test

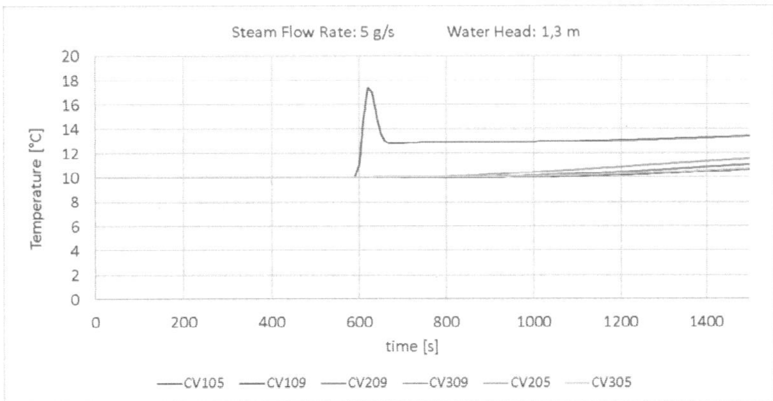

Figure 7: Temperature in the vertical section at the discharge height in a low flow-rate test

5. Concluding Remarks

In the attempt of generalize the results of the research, we tried to derive from Figure 2 a more general map, considering instead of pool temperature the “nominal” degree of sub-cooling ΔT. This quantity is calculated assuming as reference saturation temperature 100°C, because all the experiments of Fig. 2 were performed at ambient pressure. The new map is shown in Fig.8, where also the points corresponding to the tests foreseen in this research are indicated.

The experiments carried-out at the lower flow-rate are all concentrated in the chugging zone, while the others at intermediate or greater flow-rates lie mostly in the transitional chugging zone. Only a few tests, carried out with the condensing pool at a temperature of 50 °C (and therefore with the lower sub-cooling) should enter in the bubbling zone.

The experimental results will extend the map in a considerable way with respect to the previous situation toward the lower flow-rates. This is particularly interesting, because it corresponds to the occurrence of chugging and transitional chugging.

It has to be noted that the experimental team, together with the new representatives of ITER Organization, presented at ICONE 24 the results of the tests actually carried out (Ref. (7)). These showed a behaviour sometimes different from that presented in Fig. 2: at sub-atmospheric pressure, the region of stable condensation is largely extended (a factor of about 5) toward the lower steam flow-rates, while the areas of chugging, transitional chugging and condensation oscillations are highly reduced. Good news, from the safety point of view.

Chugging can generate noticeable, negative pressure peaks, which shall be taken in due account in the design of the condensing apparatus: the VVPSS of the ITER plant. Obviously, MELCOR cannot reproduce these pressure spikes, due to its nature of lumped parameters thermal equilibrium code.
The steam condensation should occur in all the tests. The mean temperature increase in the condensation pool in all test is limited to 1 or a few °C, due to the modest energy injection. The related experimental results will be particularly useful for the validation (or the modification) of the correlations used in the codes of thermal-hydraulic design of the VVPSS of the ITER plant.
The last observation concerns the possibility of simulating by MELCOR the natural circulation patterns, which develop in the condensing pool at low and high steam flow-rates, both in clock wise and in counter-clock wise sense. These effects can be reproduced also by a lumped parameter code, if properly used.

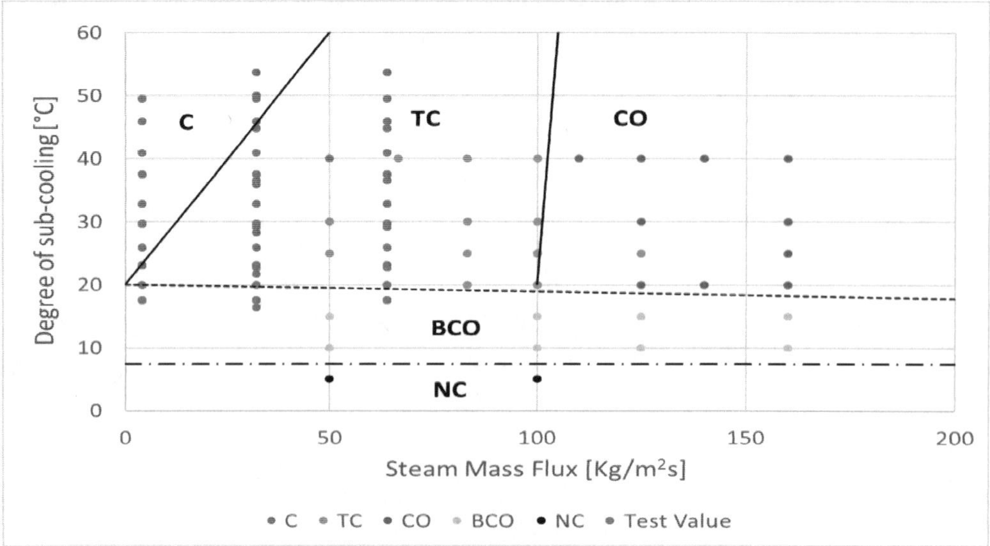

Figure 8: Map of various regimes of steam condensation as a function of pool sub-cooling and steam specific mass flow-rate

6. Acknowledgements
The authors acknowledge ITER Direction, and particularly Ing. Sergio Orlandi, for making possible this publication and for the useful suggestions for its final drawing.
Thanks are due also to Mr. Mike Meekins, at the time ITER responsible of the VVPSS research for his contribution to the set-up of the experimental matrix.
Finally, we thank very much Mr. Marco Galassi, who executed the MELCOR calculations inside his degree thesis in nuclear engineering, contributing also to the issue of this article.

7. Disclaimer
The views and opinions expressed herein do not necessarily reflect those of the ITER Organization. The results should be considered preliminary until they are accepted by the ITER licensing agency.

References

[1] Fukuda S 1982 *J. Nucl. Sci. Tech.* **24** pp 466–474
[2] Narai H and Aya I 1986 *J. Nucl. Eng. Des.* **95** pp 35–45
[3] Chun M H, Kim Y S and Park J W 1996 *Int. Con. Heat Mass Transfer* **23** pp 947–958
[4] Petrovic A, Calay R K and De With G *Int. J. Heat Mass Transfer* **50** pp 1762–1770
[5] Wu X Z, Yan J J, Shao S F, Cao Y and Liu J P 2007 *Int. J. Multiphase Flow*, **33**(12) pp 1296–1307
[6] Wu X Z, Yan J J, Li W J, Pan D D and Liu G Y 2010 *Exp. Therm. Fluid Sci.* **34** pp 10–19
[7] Mazed D, Lo Frano R, Aquaro D, Del Serra D, Sekachev I and Orlandi F 2016 *Proc. of the 24th Int. Conf. on Nuclear Engineering* ICONE, Vol **2**

Novel gas holdup and regime transition correlation for two-phase bubble columns

G Besagni[1] and F Inzoli[1]

[1] Politecnico di Milano, Department of Energy, Via Lambruschini 4a, 20156, Milano

E-mail: giorgio.besagni@polimi.it

Abstract. The gas holdup is dimensionless parameter of fundamental and practical importance in the operation, design and scale-up of bubble columns. Unfortunately, the many relationships between the bubble column fluid dynamic parameters and the various variables characterizing the system make it difficult to find general correlations for the precise estimation of the gas holdup. Wilkinson et al. (1992), in their pioneering paper, proposed a correlation to predict the gas holdup in industrial-scale bubble columns, based on the physical properties of the phases and the operating conditions. However, this correlation lacks in generality, as it does not take in account the bubble column design. In this paper, we propose a generalization of the Wilkinson et al. (1992) gas holdup correlation to take into also the bubble column design parameters. Starting from considerations concerning the flow regime transition, corrective parameters are included to account for the effects introduced by the gas sparger openings, the bubble column aspect ratio and the bubble column diameter. The proposed correlation has been found to predict fairly well previously published gas holdup and flow regime transition data.

1. Introduction

Two-phase bubble columns are multiphase reactors where a gas phase is dispersed into a liquid phase in the form of dispersed bubbles or of "*coalescence-induced*" bubbles. The simplest bubble column configuration consists in a vertical cylinder, in which the gas enters at the bottom—through a gas sparger—and the liquid phase is supplied in batch mode or it may be led in either co-currently or counter-currently to the upward gas stream. Despite the simple bubble column design, complex fluid dynamics interactions and coupling between the phases—manifesting in the prevailing flow regimes [1]—exist. Therefore, the correct design and operation of these reactors rely on the proper prediction of the "*reactor-scale*" and the "*bubble-scale*" fluid dynamic properties. Among the many fluid dynamic properties, the gas holdup (ε_G)—a dimensionless parameter defined as the volume of the gas phase divided by the total volume of the system—is a "*reactor-scale*" fluid dynamic property of fundamental and practical importance. The gas holdup determines the mean residence time of the dispersed phase and, in combination with the size distribution of the dispersed phase, the interfacial area for the rate of interfacial heat and mass transfer. The many relations between the bubble column fluid-dynamic parameters and the variables characterizing the system make it difficult to find general correlations for the precise estimation of the gas holdup curve, which is the analytical relation between the gas superficial velocity U_G (viz. the ratio between the volumetric flow rate and the cross-sectional area) and ε_G ($\varepsilon_G = f(U_G)$). Numerous correlations are available in literature, although due to the fluid dynamic complexity of the problem none of them can be either considered of general validity or applied to a wide range of geometrical parameters and operating conditions. Wilkinson et al. [2], in their pioneering paper, proposed a correlation to predict the gas holdup in industrial bubble columns, based on the physical properties of the phases and the operating conditions (i.e., high pressure operation). The main idea is that the total gas holdup is the sum of the contribution of the gas holdup before and after the flow regime transition between the homogeneous and the transition flow regime (see ref. [3] for a discussion concerning the flow regime transitions). The homogeneous flow regime—associated with small superficial gas velocities—is referred as the flow regime where only "*non-coalescence-induced*" bubbles exist (e.g. as detected by the gas disengagement technique); the transition flow regime, is identified by the appearance of the "*coalescence-induced*" bubbles [4] and is characterized by large flow macro-structures with large eddies and a widened bubble size distribution. Thus, accordingly with Wilkinson et al. [2], the gas holdup can be estimated based on:

$$\varepsilon_G = \begin{cases} \dfrac{U_G}{U_{\text{"non-coalescence induced" bubbles}}} & U_G < U_{G,trans} \\[2ex] \dfrac{U_{G,trans}}{U_{\text{"non-coalescence induced" bubbles}}} + \dfrac{U_G - U_{G,trans}}{U_{\text{"coalescence induced" bubbles}}} & U_G > U_{G,trans} \end{cases} \tag{1}$$

Where $U_{\text{"non-coalescence induced" bubbles}}$ and $U_{\text{"coalescence induced" bubbles}}$ refer to the rising velocity of the "*non-coalescence induced*" bubble and the "*coalescence induced*" bubbles, respectively. In particular, the terms in Eq. (1) have been evaluated by Wilkinson et al. [2], by using a non-linear regression of the experimental data obtained in two bubble columns having different inner diameters ($d_c = 0.15$ and $d_c = 0.23$ m), operated at different operating pressures and filled by different liquid phases (n-heptane, monoethylene glycol, and water):

$$U_{G,trans} = 0.5 U_{\text{"non-coalescence induced" bubbles}} \exp\left(-193 \rho_G^{-0.61} \mu_L^{0.5} \sigma^{0.11}\right) \tag{2}$$

$$U_{\text{"non-coalescence induced" bubbles}} = 2.25 \frac{\sigma}{\mu_L} \left(\frac{\sigma^3 \rho_L}{g \mu_L^4}\right)^{-0.273} \left(\frac{\rho_L}{\rho_G}\right)^{0.03} \tag{3}$$

$$U_{\text{"coalescence induced" bubbles}} = \frac{\sigma}{\mu_l} \left\{ \frac{\mu_l U_{\text{"non-coalescence induced" bubbles}}}{\sigma} + 2.4 \left[\frac{\mu_l \left(U_G - U_{G,trans}\right)}{\sigma}\right]^{0.757} \left(\frac{\sigma^3 \rho_l}{g \mu_l^4}\right)^{-0.077} \cdot \left(\frac{\rho_l}{\rho_g}\right)^{0.077} \right\} \tag{4}$$

In Eqs. (2-4) μ_L is the liquid phase viscosity, ρ_L is the liquid phase density and σ is the surface tension. Unfortunately, the above correlation, despite is considered the state-of-the-art in bubble columns, it lacks in generality, as it does not take in account the bubble column design parameters (i.e., sparger openings, aspect ratio and bubble column inner diameter). In this respect, the reader should refer to our recent paper concerning the importance of bubble column design parameters in gas holdup correlations as well as scaling up criteria [4].

Therefore, in this paper, we propose a generalization of the Wilkinson et al. [2] gas holdup correlation to take in account both the physical properties and the operating conditions, as well as the bubble column design. The proposed correlation aims to provide an engineering practical tool able to estimate the global bubble column parameters. Starting from considerations concerning the flow regime transition, corrective parameters are included in the gas holdup correlation to account for the effects introduced by the gas sparger openings, the bubble column aspect ratio and the bubble column diameter.

2. The proposed scheme of correlation

A general and precise correlation for the gas holdup depends on the many parameters characterizing the system: (a) the pressure and the temperature (p_c and T_c); (b) the gas and liquid superficial velocities (U_G and U_L); (c) the physical and the interface properties; (d) the bubble column design parameters (i.e., bubble column inner diameter, d_c, bubble column height, H_c, initial liquid height, H_0, and diameter of the gas sparger openings, d_o, see Figure 1).

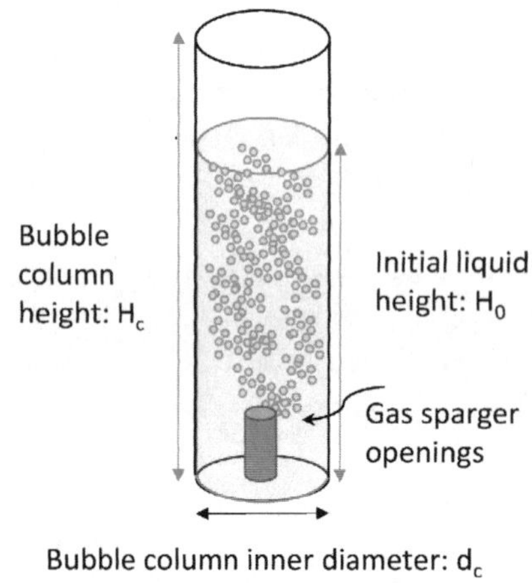

Figure 1. Bubble column design parameters.

The "*reactor-scale*" and the "*bubble-scale*" fluid dynamic properties are also related to the prevailing flow regime. Therefore, a gas holdup correlation should take into account the flow regime transition point between the homogeneous flow regime and the transition flow regime ($U_{G,trans}$). This concept has the underlying assumption that systems having similar flow regime transition points show similar "*reactor-scale*" behaviour. Hence, a correlation for the gas holdup would be as follows:

$$\varepsilon_G = f\left(U_G, U_L, U_{G,trans}, p_c, T_c, \mu_L, \mu_G, \rho_L, \rho_G, n, n^*, \sigma, d_c, H_0, d_o, g, \ldots\right) \quad (5)$$

In Eq. (5), n and n^* are the actual and the critical concentration of active compounds [4]. Indeed, the fluid dynamics in bubble columns filled with binary liquid phases cannot be entirely modelled by the bulk physical properties and the properties of the interface should be considered [5, 6]. This issue, as well as the influence of U_L, has been addressed in ref. [4] and, to keep the discussion as simple as possible, their influence has been neglected in the following (viz. pure liquid phase in the batch mode). Thus, Eq. (5) simplifies to Eq. (6):

$$\varepsilon_G = f\left(U_G, U_{G,trans}, p_c, T_c, \mu_L, \mu_G, \rho_L, \rho_G, \sigma, d_c, H_0, d_o, g\right) \quad (6)$$

Now we assume that (a) the physical properties and the operating conditions and (b) the bubble column design parameters provide two different effects. Thus, from Eq. (6) follows Eq. (7):

$$\varepsilon_G = f\left(U_G, U_{G,trans}, p_c, T_c, \mu_L, \mu_G, \rho_L, \rho_G, \sigma, g, \ldots\right) \cdot f\left(d_c, H_0, d_o\right) \quad (7)$$

In the right side of Eq. (7) the geometrical parameters are the same as the ones listed in the three Wilkinson et al. [2] scale-up criteria. At this point, a scheme of correlation is selected: the dependence on the physical properties have been accounted by using the Wilkinson et al. [2] correlation:

$$\varepsilon_G = f_{Wilkison\ et\ al.(1992)}\left(U_G, p_c, T_c, \mu_L, \mu_G, \rho_L, \rho_G, \sigma, g, \ldots\right) \cdot f\left(d_c, H_0, d_o\right) \quad (8)$$

To generalise Eq. (8), the geometrical parameters are modelled by using the corresponding non-dimensional groups, which are related to the fluid dynamics in the bubble column:

$$\varepsilon_G = f_{Wilkison\ et\ al.(1992)}\left(U_G, p_c, T_c, \mu_L, \mu_G, \rho_L, \rho_G, \sigma, g, \ldots\right) \cdot f\left(D_H^*, d_o^*, AR\right) \quad (9)$$

The non-dimensional groups in Eq. (9) are defined as follows:

- **Dimensionless bubble column height (*AR, aspect ratio*).** The aspect ratio is defined as the ratio between the initial liquid height, H_0, and the diameter of the column, d_c (see ref. [4]):

$$AR = \frac{H_0}{d_c} \tag{10}$$

  In the systems where the bubble sizes are not at their maximum equilibrium size (and where coalescence may occur), the liquid height will influence the extent of the coalescence. Consequently, the gas holdup would decrease with H_0, because the higher the column the longer the time the bubbles have to coalesce and, thus, the lower the mean residence time of the gas phase. Furthermore, in shorter bubble columns, the liquid circulation patterns (that tends to decrease the gas holdup) are not fully developed and the end-effects are more evident.
- **Dimensionless sparger opening (d^*_o).** Depending on the gas sparger opening, a bubble size (d_b) is imposed as boundary condition at the lower level of the bubble column. Hence, the diameters of the sparger openings can be related to the bubble size produced by the bubble nucleation at the sparger, estimated by the well-known Gaddis and Vogelpohl correlation [7]:

$$d_o^* = \frac{d_b}{d_o} = \left(\frac{6 d_o \sigma}{\rho_L g}\right)^{1/3} / d_o \tag{11}$$

- **Dimensionless diameter (D^*_H).** The dimensionless diameter represents the quantification of the Rayleigh–Taylor instabilities [8] at the "*reactor-scale*". In particular, the dimensionless diameter is related to the Eötvös number of the slug bubbles as follows:

$$D_H^* = \frac{D_H}{\sqrt{\sigma / g(\rho_L - \rho_G)}} \xrightarrow{Circular\,bubble\,column} \frac{d_c}{\sqrt{\sigma / g(\rho_L - \rho_G)}} = \frac{1}{\sqrt{\sigma / d_c^2 g(\rho_L - \rho_G)}} = \frac{1}{\sqrt{1 / Eo_c}} = \sqrt{Eo_c} = \sqrt{Eo_{slug-bubble}} \tag{12}$$

  Where D_H is the hydraulic diameter, d_c is the bubble column inner diameter, σ is the surface tension, g is the acceleration due to gravity, ρ_L - ρ_G is the density difference between the two phases, and $Eo_c = Eo_{slug\text{-}bubble}$ is the Eötvös number computed using the bubble column diameter, which is also the characteristic length of the slug bubbles. Bubble columns with D^*_H greater than the critical value $D^*_{H,cr} = 52$ [9] are considered to be large-diameter bubble columns. When the $d_c > D^*_{H,cr}$, the stabilizing effect of the channel wall on the interface of the Taylor bubbles decreases, and slug flow cannot be sustained anymore because of the Rayleigh-Taylor instabilities.

Thus, the terms in Wilkinson et al. [2] correlation (Eq. (1)) have been modified as follows:

$$U_{G,trans} = A\, U_{G,trans,Wilkison\,et\,al.(1992)} D_H^{*\,m_1} AR^{m_2} d_o^{*\,m_3} \tag{13}$$

$$U_{"non-coalescence\,induced"bubbles} = B\, U_{"non-coalescence\,induced"bubbles,Wilkison\,et\,al.(1992)} D_H^{*\,n_1} AR^{n_2} d_o^{*\,n_3} \tag{14}$$

It worth noting that the coalescence-induced term has not been modified, as the influence of the bubble column design in the heterogeneous flow regime is lower compared with the homogeneous flow regime. In Eqs. (13-14), A, B, (m_1, m_2, m_3) and (n_1, n_2, n_3) are calibration parameters; Eqs. (13-14) are related to the Wilkinson et al. [2] correlation, by the following terms:

$$U_{G,trans,Wilkison\,et\,al.(1992)} = 0.5\,U_{"non-coalescence\,induced"bubbles,Wilkison\,et\,al.(1992)} \exp\left(-193 \rho_G^{-0.61} \mu_L^{0.5} \sigma^{0.11}\right) \tag{15}$$

$$U_{"non-coalescence\,induced"bubbles,Wilkison\,et\,al.(1992)} = 2.25 \frac{\sigma}{\mu_L} \left(\frac{\sigma^3 \rho_L}{g \mu_L^4}\right)^{-0.273} \left(\frac{\rho_L}{\rho_G}\right)^{0.03} \tag{16}$$

3. The experimental dataset and methods

To calibrate and validate the proposed scheme of correlation, gas holdup and flow regime transition point data are needed. To this end, a comprehensive experimental dataset of gas holdup data has been taken from the literature. These gas holdup data have been post-processed, by using a systematic approach, to estimate the regime transition point. In the following, firstly, the method to estimate the regime transition point has been discussed (Section 3.1) and, secondly, the whole dataset has been presented (Section 3.2).

3.1. Flow regime transition analysis

Although the flow transition from the homogeneous to the transition flow regime does not happen instantaneously [36], the definition of an approximate transition point is helpful for modelling the hydrodynamic of bubble columns. In this study, we employ two methods to invetsigate the flow regime transition (as discussed in ref. [4]): (a) the swarm velocity and (b) the drift flux method. In particular, the flow regime transition point is computed as the average of the values obtained by the two methods:

$$U_{G,trans} = \frac{U_{G,trans,swarm} + U_{G,trans,Wallis}}{2} \tag{17}$$

$$\varepsilon_{G,trans} = \frac{\varepsilon_{G,trans,swarm} + \varepsilon_{G,trans,Wallis}}{2} \tag{18}$$

3.1.1. Swarm velocity method

This method, proposed by Zuber and Findlay [37], is based on the evaluation of the swarm velocity:

$$U_{swarm} = U_G / \varepsilon_G \tag{19}$$

In this method, the swarm velocity is plotted against the superficial gas velocity: U_{swarm} is constant in the homogeneous flow regime and starts to increase as the system enters the transition/heterogeneous flow regime ($U_{G,trans}$). The appearance of the first large bubble is responsible for the increase in the swarm velocity and is an indication of flow regime transition.

3.1.2. Wallis plot method

The drift-flux method was proposed by Wallis [38]. This method is based on the drift flux (that represents the gas flux through a surface moving with the speed of the two-phase mixture) and it is experimentally obtained as follows:

$$J_T = U_G (1 - \varepsilon_G) \tag{20}$$

Theoretically, the drift flux is written in terms of a parameter, U_b, whose dependence upon ε_G varies with the prevailing flow regime:

$$J_E = U_b (1 - \varepsilon_G) \tag{21}$$

The idea is to employ a model for U_b valid for the homogeneous flow regime and to plot J_E and J_T in the same graph as a function of ε_G. In the homogeneous flow regime J_E is equal to J_T and the transition point is, thus, defined when:

$$J_E = U_b (1 - \varepsilon_G) \tag{22}$$

The evaluation of U_b is a matter of discussion in the literature, as different models were proposed and applied. In this study, we follow the approach of Krishna et al. [39], which is based on the empirical model of Richardson and Zaki [40]:

$$U_b = u_\infty (1 - \varepsilon_G)^{n-1} \tag{23}$$

where n is fluid-dependent ($n \cong 2$ for water) and u_∞—the terminal velocity of an isolated bubble—should be fitted with the aid of the experimental data in the determination of the flow regime transition point. Combining Eq. (21) and Eq. (23) results:

$$J_E = u_\infty \varepsilon_G (1 - \varepsilon_G)^n \quad (24)$$

3.2. Dataset employed

Table 1 displays the dataset employed as well as the corresponding transition points, compued by the methods described in Section 3.1. It is worth noting that AR has been computed by using H_0 and, when this parameters was not available, by an estimation based on H_c. In data post-processing, the mid-point conditions have been used to compute the physical properties (see ref [4]).

Table 1. Dataset employed for calibration and validation of the correlation with corresponding transition point (using the methods in Section 3.1)

Reference	d_c [m]	d_o [mm]	Free area [%]	H_C [m]	H_0 [m]	$U_{G,trans}$ [m/s]	$\varepsilon_{G,trans}$ [-]
Chaumant et al. [10]	0.200	1.0	0.80	1.60	-	0.0391	0.1224
Chilekar [11]	0.290	0.5	0.46	1.60	-	0.0439	0.1796
Han and Al-Dahhan [12]	0.162	0.5	0.16	-	1.81	0.0640	0.2193
Jin et al. [13]	0.162	1.0	0.21	2.50	0.45	0.0408	0.1411
Kemoun et al. [14]	0.162	0.4	0.04	2.50	1.80	0.0498	0.1882
Krishina et al. [15]	0.160	10.0	7.42	1.20	-	0.0569	0.1263
Krishina and Ellenberger [16]	0.630	2.5	0.10	4.00	-	0.0976	0.1583
Krishina and Ellenberger [16]	0.380	0.5	-	4.00	1.70	0.0694	0.1843
Krishina et al. [17]	0.150	1.0	2.78	-	1.00	0.0285	0.1289
Kumar et al. [18]	0.154	1.0	0.51	2.72	1.15	0.0339	0.1275
Letzel et al. [19]	0.150	0.5	0.22	1.20	-	0.0524	0.1814
Ohki and Inoue [20]	0.160	3.0	1.30	3.00	0.40	0.0594	0.1299
Veera and Jochi [21]	0.385	1.0	0.42	3.20	-	0.0500	0.1365
Reilly et al. [22]	0.300	1.5	0.75	5.00	-	0.0421	0.1754
Sal et al. [23]	0.330	2.0	1.08	3.00	0.66	0.0736	0.1128
Sal et al. [23]	0.330	1.0	0.98	3.00	0.66	0.0616	0.1066
Deckwer et al. [24]	0.200	1.0	0.14	4.44	-	0.0418	0.2078
Vandu and Krishna [25]	0.380	0.5	0.48	-	1.60	0.0443	0.1941
Akita and Yoshida [26]	0.152	5.0	0.11	4.00	-	0.0313	0.0777
Patil et al. [27]	0.380	3.7	-	-	0.684	0.0730	0.1068
Patil et al. [27]	0.380	3.7	-	-	1.026	0.0744	0.1052
Patil et al. [27]	0.380	3.7	-	-	1.406	0.0764	0.1019
Godbole et al [28]	0.305	1.66	--	2.44	0.60	0.0590	0.1261
Gopal and Sharma [29]	0.200	6.0	0.09	-	0.80	0.0449	0.0909
Gopal and Sharma [29]	0.600	2.0	0.07	-	3.60	0.0592	0.1411
Schumpe and Grund [30]	0.300	1.0	-	-	3.60	0.0400	0.2416
Schumpe and Grund [30]	0.300	1.0	-	-	3.60	0.0600	0.1237
Wilkinson et al. [2]	0.150	2.0	-	-	1.50	0.0461	0.0918
Wilkinson et al. [2]	0.230	7.0	-	-	1.20	0.0518	0.0961
Thorat et al. [31]	0.385	1.5	0.11	-	1.54	0.0612	0.1817
Sasaki et al. [32]	0.200	1.4	0.18	2.00	0.80	0.0500	0.1500
Passos et al. [33]	0.090	0.4	-	-	2.00	0.0180	0.1129
Pjontek et al. [34]	0.102	3.18	2.24	1.80	-	0.0300	0.1616
Ruzicka et al. [35]	0.140	0.5	0.20	-	0.40	0.0360	0.2753

4. Novel correlation: calibration and validation procedures

4.1. Correlation calibration

Based on the experimental dataset employed (viz. Table 1 data and flow regime transition points), the coefficients A, B, (m_1, m_2, m_3) and (n_1, n_2, n_3) in Eqs. (13-14) have been obtained. These coefficients have been obtained by mean of regression analysis and the results are as follows:

$$U_{G,trans} = 2.1 U_{G,trans,Wilkison\ et\ al.(1992)} D_H^{*0.5} AR^{-0.31} d_o^{*0.01} \tag{25}$$

$$U_{\text{"non-coalescence induced" bubbles}} = 0.5 U_{\text{"non-coalescence induced" bubbles},Wilkison\ et\ al.(1992)} D_H^{*0.4} AR^{-0.3} d_o^{*-0.39} \tag{26}$$

4.2. Correlation validation

In this section, the proposed scheme of correlation has been compared with the experimental dataset obtained (Table 1). Figure 2 compares the experimental and the predicted flow regime transition points (in terms of the gas superficial velocity and gas holdup). The proposed correlation has been able to predict the transition points, for the different experimental setups considered, within the boundaries of ±20% of error. This result is an advancement compared with the previous correlation to estimate the flow regime transition point in bubble columns. Indeed, at present, the Wilkinson et al [2] correlation was not able to take into account the bubble column design (as observed in our previous studies, see for example ref. [4]).

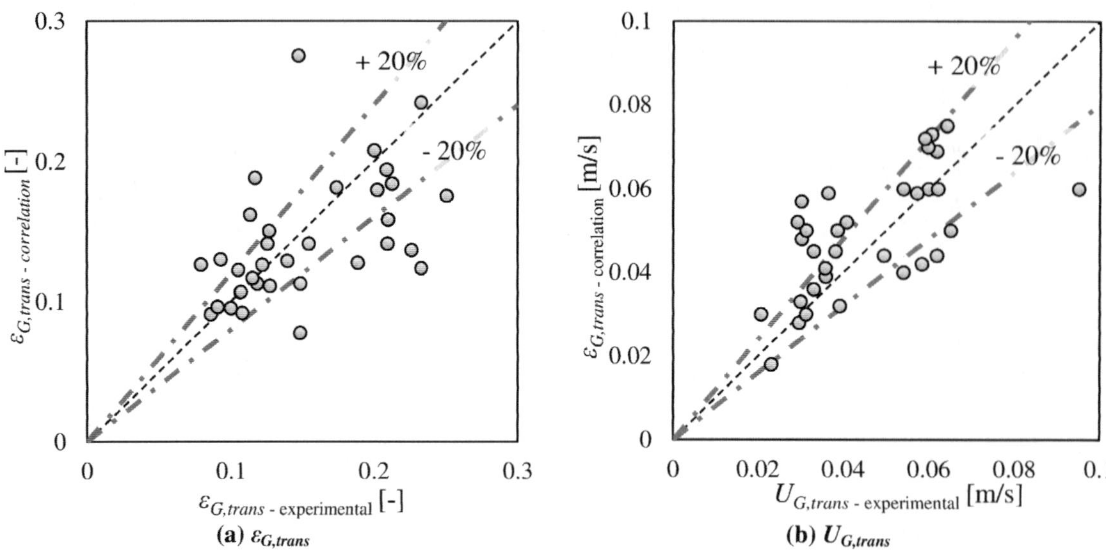

Figure 2. Comparison between predicted and experimental transition points.

Figure 3 compares the experimental and the predicted rising velocity of the "*non-coalescence induced*" bubbles. According to Wilkinson et al. [2], the rising velocity of the "*non-coalescence induced*" bubbles is related to the rising velocity of bubbles in the homogeneous regime, so it can be easily computed by using the definition of the swarm velocity. The proposed correlation is able to predict the experimental data within the boundaries of ±10% of error, for the different datasets considered. Finally, Figure 4 and Figure 5 compare the experimental and the predicted gas holdup curves for some of the studies listed in Table 1. It is shown that the proposed correlation has been found to predict fairly well previously published gas holdup data. Therefore, the proposed scheme of correlation can be used as a tool to estimate the main bubble column global parameters in the design of bubble columns as a function of the operating conditions and design parameters.

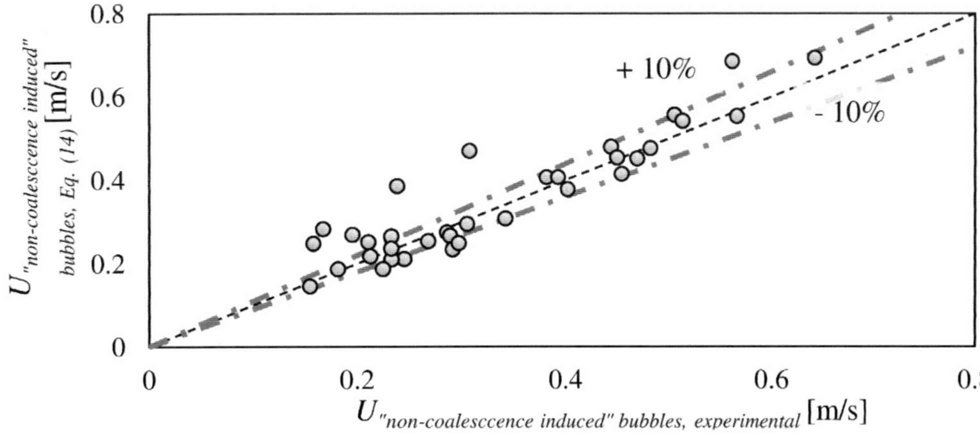

Figure 3. Comparison between predicted and experimental rising velocity of the "*non-coalescence induced*" bubbles

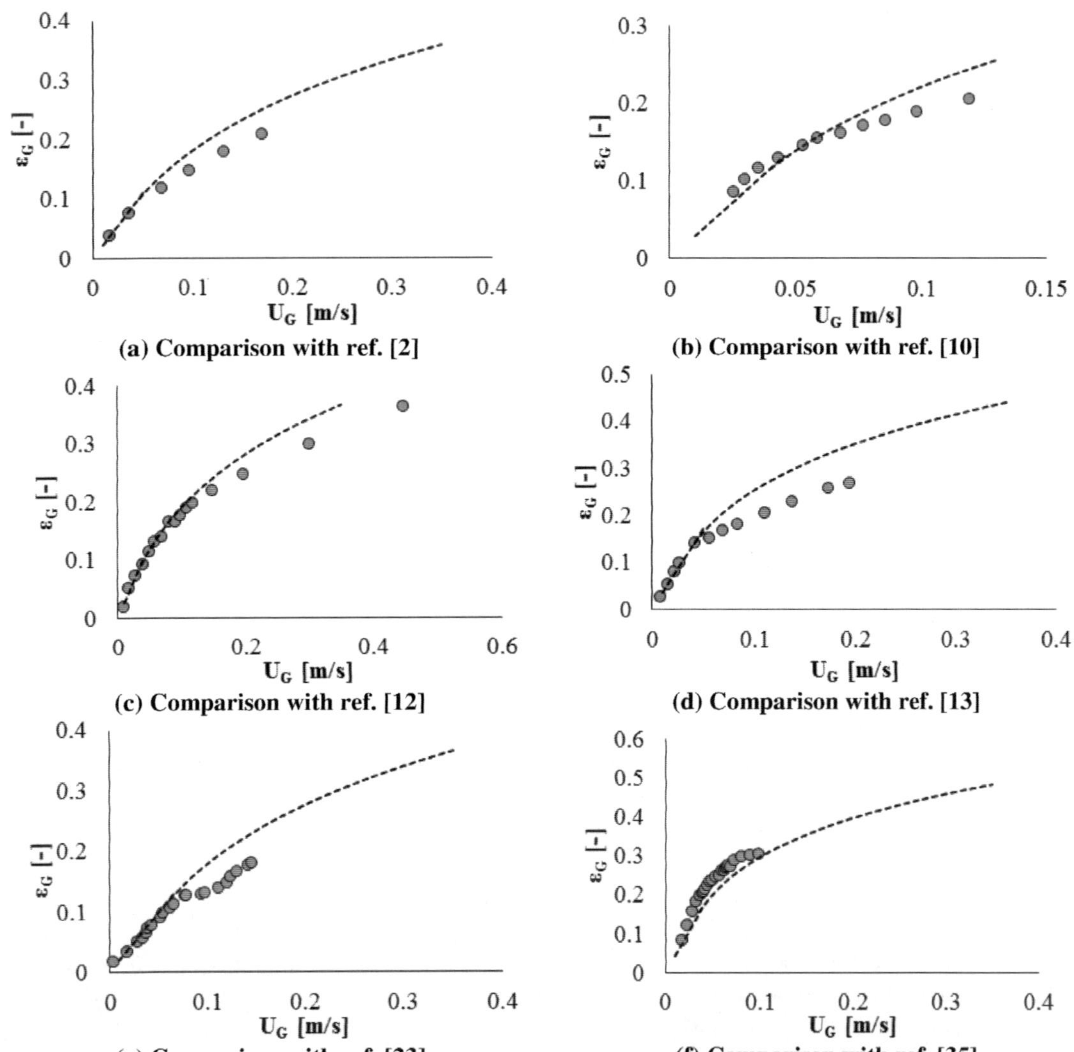

Figure 4. Comparison between predicted and experimental gas holdup curves.

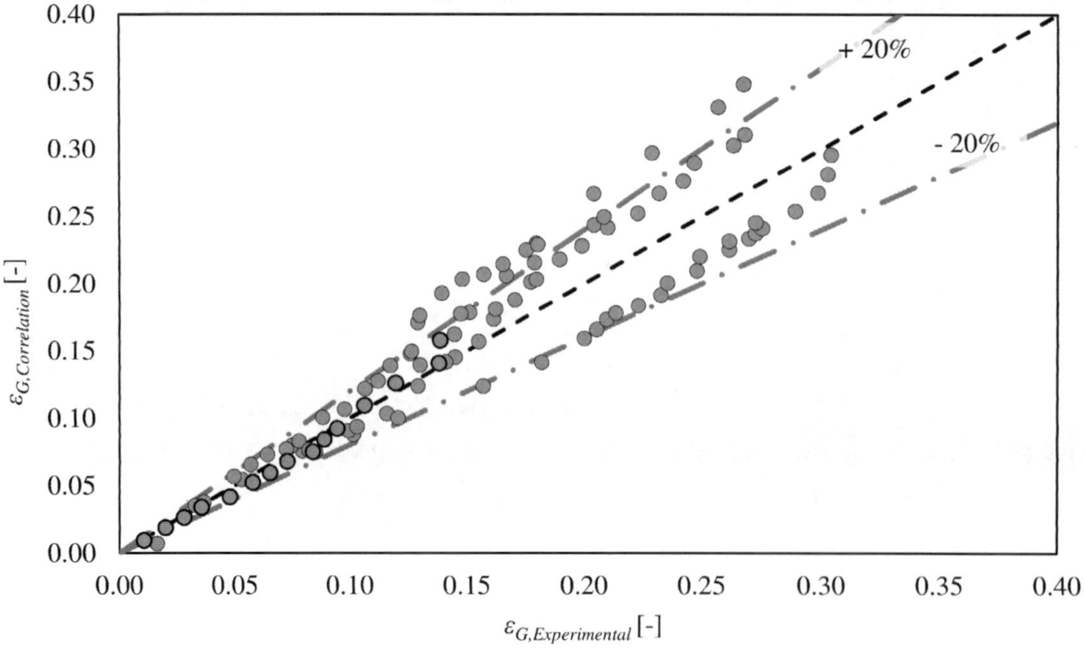

Figure 5. Comparison between predicted and experimental gas holdup curves: parity plot.

5. Conclusions

We have proposed a generalised version of the well-known Wilkinson et al. [2] gas holdup correlation. Starting from considerations concerning the flow regime transition, corrective parameters are included in the gas holdup correlation to account for the effect of the changes introduced by the sparger openings, the aspect ratio and the diameter. The proposed correlation is able to predict the gas holdup and the flow regime transitions, by taking into account the physical properties of the phases, the bubble column operating conditions, as well as the bubble column design. The proposed scheme of correlation has been calibrated and validated over a wide range of literature studies. The main outcomes of this study are as follows:

- the proposed correlation has been found to predict the rising velocity of the "*non-coalescence induced*" bubbles within the boundaries of ±10% of error;
- the proposed correlation has been found to predict previously published gas holdup and flow regime transition data within the boundaries of ±20% of error;
- the proposed scheme of correlation can be used as a tool to estimate the main bubble column global parameters in the design of bubble columns as a function of the operating conditions and design parameters.

On the other hand, it should be noted that the proposed correlation is valid only for two-phase bubble columns in the batch mode without the contribution of active agents. Future studies should extend the proposed scheme of correlation by taking into account: (a) the influence of the solid load is not studied; (b) co-current and counter-current bubble column; (c) the influence of active agents. In this respect, the proposed scheme of correlation can be coupled with another gas holdup correlation recently proposed by the authors in ref. [4] (which takes into account the influence of the liquid superficial velocity and the concentration of the active agents).

References

[1] Majumder S K 2016 *Hydrodynamics and Transport Processes of Inverse Bubbly Flow* (Amsterdam) Elsevier
[2] Wilkinson P M, Spek A P and van Dierendonck L L 1992 *AIChE J.* **38** 544-54
[3] Nedeltchev S 2015 *Chem. Eng. Sci.* **137** 436-44
[4] Besagni G, Di Pasquali A, Gallazzini L, Gottardi E, Colombo L P M and Inzoli F 2017 *Int. J. Multiphas. Flow* **94** 53–78
[5] Guo K, Wang T, Yang G and Wang J. 2017 *J. Chem. Technol Biot* **92**(2) 432-441.
[6] Shah Y T, Joseph S, Smith D N and Ruether J A 1985 *Ind. J. Chem. Process Des. Dev.* **24** 1140-48
[7] Gaddis E S and Vogelpohl A *Chem. Eng. Sci.* **41** 97-105
[8] Kitscha J and Kocamustafaogullari G 1989 *Int. J. Multiphas. Flow* **15** 573-88
[9] Brooks C S, Paranjape S S, Ozar B, Hibiki T and Ishii M. 2012 *Int. J. Heat. Fluid Fl* **37** 196-208.
[10] Chaumat H, Billet-Duquenne A M, Augier F, Mathieu C and Delmas H 2005 *Chem. Eng. Sci.* **60** 5930-5936
[11] Chilekat V P 2007 *Hydrodynamics and mass transfer in slurry bubble columns: Scale and pressure* (Ph.D Thesis)
[12] Han L and Al-Dahhan M H 2007 *Chem. Eng. Sci.* **62** 131-139
[13] Jin H, Yang S, He G, Wang M and Williams R A 2010 *J. Chem. Technol. Biot.* **85** 1278–83
[14] Kemoun A, Ong B C, Gupta P and J and Al-Dahhan M H 2001 *Int. J. Multiphas. Flow* **27** 929-46
[15] Krishna R, Wilkinson P M and Van Dierendonck L L 1991 *Chem. Eng. Sci.* **46** 2491-6
[16] Krishna R and Ellenberger J 1996 *AICHE J.* **42** 2627-34
[17] Krishna R, Baten J M and Urseanu M I 2001 *Chem. Eng. Technol.* **24** 451-8
[18] Kumar S, Munshi P and Khanna A *Procedia Eng.* **42** 842-853
[19] Letzel H M, Schouten J C, van den Bleek C M and Krishna R 1997 *Chem. Eng. Sci.* **52** 3733-9
[20] Ohki Y and Inoue H 1970 *Chem. Eng. Sci.* **46** 1-16
[21] Veera U P and Jochi J B 2000 *Chem. Eng. Res. Des.* **46** 425-34
[22] Reilly I G, Scott D S, De Bruijn T, Jain A and Piskorz J 1986 *Can. J. Chem. Eng.* **64** 705-17
[23] Sal S, Gul O F and Ozdemir M 2013 *Chem. Eng. Process.* **64** 259-66
[24] Deckwer W D, BurckhartR and Zoll G 1974 *Chem. Eng. Sci.* **29** 2177-88
[25] Vandu C O and Krishna R 2004 *Chem. Eng. Process.* **43** 575-79
[26] Akita K and Yoshida F 1973 *Ind. Eng. Chem. Process Des. Dev.* **12** 76-80
[27] Patil V K, Jochi J B and Sharma M M 1984 *Can. J. Chem. Eng.* **24** 228-32
[28] Godbole S P, Honath M F and Shah Y T 1984 *Chem. Eng. Comm.* **16** 1-6
[29] Gopal J S and Sharma M M 1983 *Can. J. Chem. Eng.* **61** 517-26
[30] Schumpe A and Grund G 1986 *Can. J. Chem. Eng.* **64** 891-6
[31] Thorat B N, Shevade A V, Bhilegaonkar K N, Aglawe R H Parasu Veera R, Thakre S S, Pandit A B, Sawant S B and Jochi J B 1998 *Chem. Eng. Res. Des.* **76** 823-34
[32] Sasaki S, Hayashi K and Tomiyama A 2016 *Exp. Therm. Fluid. Sci.* **72** 67-74
[33] Passos A D, Voulgaropoulos V P, Paras S V and Mouza A A 2015 *Chem. Eng. Res. Des.* **95** 93-104
[34] Pjontek D, Parisien V and Macchi A 2014 *Chem. Eng. Sci.* **111** 153-69
[35] Ruzicka M C, Vecer M M, Orvalho S and Drahoš J 2008 *Chem. Eng. Sci.* **63** 951-67
[36] Nedeltchev S 2015 *Chem. Eng. Sci.* **137** 436-44
[37] Zuber N and Findlay J A 1965 *J. Heat Transf* **87** 453-68
[38] Wallis G B 1969 *One-dimensional two-phase flow* (New York) McGraw-Huill
[39] Krishna R, Ellenberger J and Maretto C 1999 *Int. Commun. Heat Mass* **26** 467-75
[40] Richardson J F and Zaki W N 1997 *Chem. Eng. Res. Des.***75**, Supplement S82-S100

IOP Conf. Series: Journal of Physics: Conf. Series **923** (2017) 012012 doi:10.1088/1742-6596/923/1/012012

Water holdup estimation from pressure drop measurements in oil-water two-phase flows by means of the two-fluid model

L P M Colombo, M Guilizzoni, G Sotgia, P Babakhani Dehkordi and A Lucchini

Politecnico di Milano, Department of Energy, Via Lambruschini 4, 20156 Milano, Italy

E-mail: manfredo.guilizzoni@polimi.it

Abstract. The Two-Fluid Model (TFM) has been applied to determine water holdup from pressure drop measurements for core-annular flows in horizontal pipes. The fluids are Milpar 220 oil (ρ_o=890 kg/m^3, μ_o=0.832 Pa·s at 20 °C) and tap water (μ_w=1.026x10^{-3} Pa·s at 20 °C). The investigated volume flow rates range from 2 to 6 m^3/h, for water, and from 1 to 3.5 m^3/h, for oil, respectively. The results are in very good agreement with available experimental data from the literature and a simple correlation between water holdup and water input fraction has been benchmarked to the overall data set. Eventually, the TFM endowed with the holdup correlation has been adopted to predict the pressure drop with quite satisfactory results: 98% of data fall within a percentage error of ±10%, 99% of the data fall within ±15%, and all the data are predicted within ±20%. On the other hand, the mean absolute relative error for the pressure drop reduction factor is 5.5%.

1. Introduction

The flow of oil-water mixtures in pipes is widely studied, mainly due to the importance of long-distance transportation of oil products. Growing interest is devoted, in particular, to heavy-oils because of the progressive depletion of on-shore fields and light-oil reserves. However, the increased pressure drop resulting from the higher viscosity raises both extraction and transportation energy costs, which can represent up to one third of the overall operational expense [1]. A technique to reduce the pressure drop is water injection aimed at creating the so-called core annular flow (CAF), a flow regime characterised by the presence of an oil core enveloped in a water annulus wetting the pipe wall, so that the apparent viscosity of the mixture is considerably reduced. Fundamental literature on this subject, focusing on heavy-oil flows in horizontal tubes, may be summarized as follows. Clark and Shapiro [2] got the earliest patent of an injector able to reduce the pressure drop up to 30%. Charles et al. [3] considered three oils and concluded that flow patterns were largely independent of the oil viscosity. They also found that the pressure gradient was reduced to a minimum by the addition of water in case of laminar flow of the oil core. In addition, the maximum pressure gradient reduction (10 times) was achieved with the most viscous oil. Charles and Redberger [4] proposed a numerical procedure to solve the Navier-Stokes equations for the general case, in which the liquids are stratified as a result of the different densities of oil and water. The calculated pressure gradient reduction resulted considerably lower than experimental values. The authors concluded that wave motion and mixing at the oil-water interface might act as mechanisms of further reduction. Ooms et al. [5] developed a model based on the hydrodynamic lubrication theory that highlighted the importance of interface phenomena for CAF. They showed that the ripples on the interface, moving with respect to the pipe wall, can generate pressure variations in the annular layer. These result in a force acting normal to the core, which can counterbalance the buoyancy effect. Oliemans et al. [6] added the effect of turbulence in the water film surrounding the oil core, improving the predicting ability, provided that actual wave amplitudes and wavelengths observed during experimental tests are used as input data. On the other hand, Brauner [7] developed an analytical model based on the integral forms of the momentum equation for each phase (Two-fluid Model – TFM) regardless of the flow regime. The wall and the interfacial shear stresses are expressed in terms of the phase actual average velocities and the corresponding friction factors, hence empirical closure equations are needed. For the case of horizontal laminar core, explicit solutions for the holdup and the pressure gradient reduction factor are presented. Ullman and Brauner [8] have more recently provided further refinement of closure equations. Grassi et al. [9] assessed the effectiveness of such an approach in the prediction of pressure drop for two-phase liquid-liquid flows with high-viscosity ratio in horizontal and slightly inclined pipes. Differently, Arney et al. [10] reported on holdup and pressure drop measurements for waxy crude oil and No. 6 fuel oil (ASTM Classification). The comparison with previous sources [3, 11-13] led to a correlation formula for the water holdup as a function of the input fraction. Accordingly, they defined a modified Reynolds number to extend the single-phase friction factor definition to two-phase flows. In this paper, the TFM is used to relate the holdup and the pressure drop so that the former can be determined from the usual measurement of the latter. This is convenient as, generally, it is simpler to measure pressure drop rather than holdup, and in many practical situations, it is impossible to directly measure holdup. The simple scheme of correlation provided by Arney et al. [10] is then adopted to provide a correlation between water input fraction and holdup to be used in the TFM instead of a correlation for the interfacial shear stress. The results are then discussed and compared with the other available approaches.

2. Theoretical background

The Two-Fluid Model (TFM) for horizontal pipes, assuming fully developed flow, writes

$$\begin{cases} -A_o \dfrac{dp}{dx} - \tau_i S_i = 0 \\ -A_w \dfrac{dp}{dx} - \tau_w S_w + \tau_i S_i = 0 \end{cases} \tag{1}$$

where "o" denotes the core phase (oil) and "w" the phase in contact with the wall (water), S the wetted perimeter and A the cross-sectional area of the single phase, as indicated in Fig. 1. Eliminating the interfacial shear stress, and denoting the overall cross-sectional area $A = A_o + A_w$, it follows

$$-A\frac{dp}{dx} - \tau_w S_w = 0 \tag{2}$$

The wall shear stress can be replaced making use of the friction factor as

$$\tau_w = f_w \frac{\rho_w U_w^2}{2} \tag{3}$$

where the friction factor depends on the water Reynolds number. Introducing the water holdup as the ratio of the superficial velocity J_w to the actual one U_w

$$H_w = \frac{J_w}{U_w} \tag{4}$$

and the hydraulic diameter for the water phase (adjoining the pipe wall, see Fig. 1)

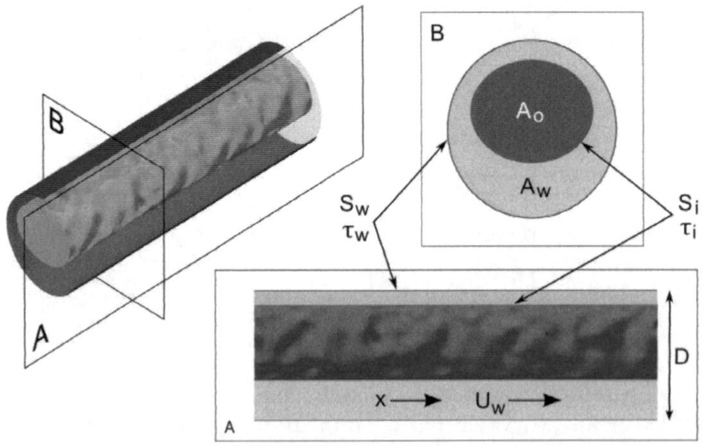

Figure 1. Sketch of an oil-water annular flow.

$$D_w = \frac{4A_w}{S_w} = H_w D \tag{5}$$

the Reynolds number is then

$$\mathrm{Re}_w = \frac{D_w U_w}{\nu_w} = \frac{D J_w}{\nu_w} \tag{6}$$

so that

$$f_w = C_w \mathrm{Re}_w^{-n_w} \tag{7}$$

For the laminar flow regime $C_w = 16$ and $n_w = 1$, whereas for developed turbulent flows, the Blasius formulation is often used; accordingly, $C_w = 0.079$ and $n_w = 0.25$ for Re < 50000. $C_w = 0.046$ and $n_w = 0.2$ for Re > 50000. Replacing (4) to (7) in (2), the water holdup as a function of the superficial velocity and the measured pressure drop per unit length results

$$H_w = \frac{C_w \left(\frac{D J_w}{\nu_w}\right)^{-n_w} \rho_w J_w^2}{\left(-\frac{dp}{dx}\right)\frac{D}{2}} \tag{8}$$

3. Experimental setup and data acquisition

The liquid-liquid flow facility is shown in Fig. 2. Oil (Milpar 220, ρ_o=890 kg/m^3, μ_o=0.832 Pa·s at 20 °C) and water (tap water, μ_w=1.026×10^{-3} Pa ·s at 20 °C) are pumped from their respective storage tanks. A magnetic flow meter (uncertainty ±0.5 % of the reading) and a calibrated metering pump (uncertainty ±2 % of the set point) are used for measuring water and oil flow rates, respectively. The range of investigated volume flow rates is 2 to 6 m^3/h for water and 1 to 3.5 m^3/h, for oil, respectively. The two fluids are pumped into the test section after going through a coaxial mixer, where oil flows parallel to the pipe axis while water is injected through an annulus into the oil stream. This procedure ensures the onset of stable annular flows provided that the superficial velocities of the phases lie in the range of existence of this flow regime, as reported in flow pattern maps available in [14]. The test section consists of a 12 m long circular pipe made of Plexiglas® that can be composed by segments with different diameters to realize sudden variations in the cross-sectional area. Pressure gradients along the pipe are measured by connecting 15 pressure taps (500 mm apart from each other) to

IOP Conf. Series: Journal of Physics: Conf. Series **923** (2017) 012012 doi:10.1088/1742-6596/923/1/012012

differential piezoresistive pressure transducers (Kulite IPTE, 0-350 mbar, uncertainty ±2% full scale, and SETRA 0-1 PSI, uncertainty ±1.5% full scale). The temperature dependence of oil viscosity has been measured in the range 20 °C ≤ T ≤ 40 °C where it is well described as $\mu_o = 3.444 \cdot \exp(-0.071 \cdot T)$. Flow temperature during the experiments is detected by a K-type thermocouple (uncertainty 0.2 °C) located 50 mm before the first pressure tap to avoid flow disturbance. All sensor signals are collected by means of National Instruments acquisition boards and processed by *ad hoc* software. Operating conditions are defined by varying the superficial velocities of the phases within the range of existence of annular flow. In particular, water is supplied starting from the maximum value of the superficial velocity $J_{w,max}$. Then, oil is added at the selected superficial velocity J_o. At each run, J_w is decreased until its minimum value is reached. The value of J_o is then changed and the sequence is repeated.

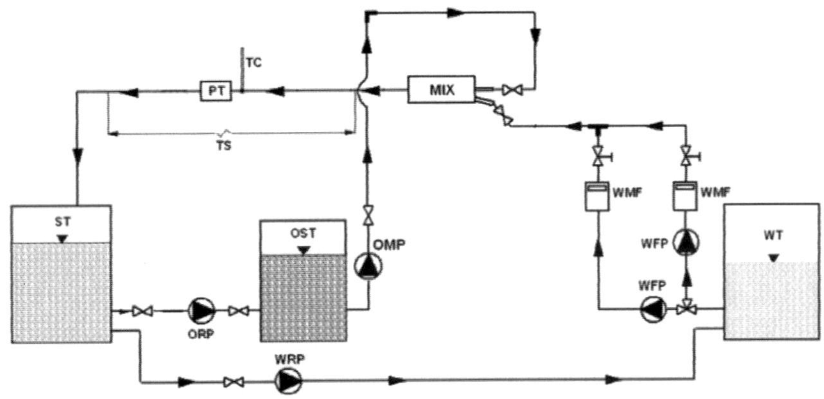

Figure 2. Schematic representation of the oil-water loop. MIX phase inlet mixer, OMP oil metering pump, ORP oil recovering pump, OST oil supply tank (0.5 m^3), PT pressure transducer, ST phase collector/separator tank (1.0 m^3), TC thermocouple (K type), TS test section, WFP water feeding pump, WMF water magnetic flow meter, WRP water recovering pump, WT water supply tank (5 m^3).

4. Results and discussion

4.1. Holdup estimation

Different experimental campaigns were run to analyse test section configurations including pipes with uniform diameter or upstream/downstream of sudden variations in the cross-sectional area. In all the cases, the Reynolds number of water, Eq. (6), resulted within about 17000 and 102000 (turbulent range) and the qualitative behavior of the pressure drop data showed quite similar characteristics. Thus, in the following, reference is made to pipes with D = 30 mm, for which data can be compared in all the tested configurations. Starting with the straight tube, the pressure gradient is reported as a function of the water input fraction, ε_w, for the different oil superficial velocities, J_o, in Fig. 3. As expected, the pressure gradient increases with J_o, at constant ε_w. On the other hand, the data points at constant J_o show that the pressure gradient increases with the water content. This confirms that for core-annular flows the pressure drop lowers by reducing the water content and hence thinning the water annulus [4]. Incidentally, parabolic fitting seems to reproduce very well the behavior with a regression coefficient always higher than 0.99. Moreover, it has been observed that interface instability may determine a sudden transition to stratified-wavy flow, if the mixture velocity, J_{mix}, lowers below a critical value [14]. This would cause an abrupt increase in the pressure drop. It is customary to define the pressure reduction factor, R, as the inverse of the ratio between the two-phase pressure drop and the pressure drop of the oil-only flow with the same superficial velocity. Figure 4 shows the pressure reduction factor as a function of the water input fraction, ε_w, for the different oil superficial velocities, J_o. The monotonic behavior shows that all the operating conditions correspond to stable core-annular flow regimes. A straight line fits properly the data and the slope seems to lower by increasing J_o. It is worth noting that the same pressure reduction factor is achieved at a reduced water input fraction as the oil superficial velocity increases. Quite similar characteristics have been found for the flow upstream and downstream of an abrupt change in the cross-sectional area. Figure 5 compares the pressure gradient as a function of the water input fraction, ε_w, for the different oil superficial velocities, J_o, for all the configurations. Evidently, the plot reproduces the behavior of Fig. 3 with a slightly higher dispersion of data points. This seems to arise from the perturbation caused by the sudden contraction: though the flow regime remains annular, it has been observed that the oil-water interface becomes more irregular and shows a tendency to form small drops [15,16].

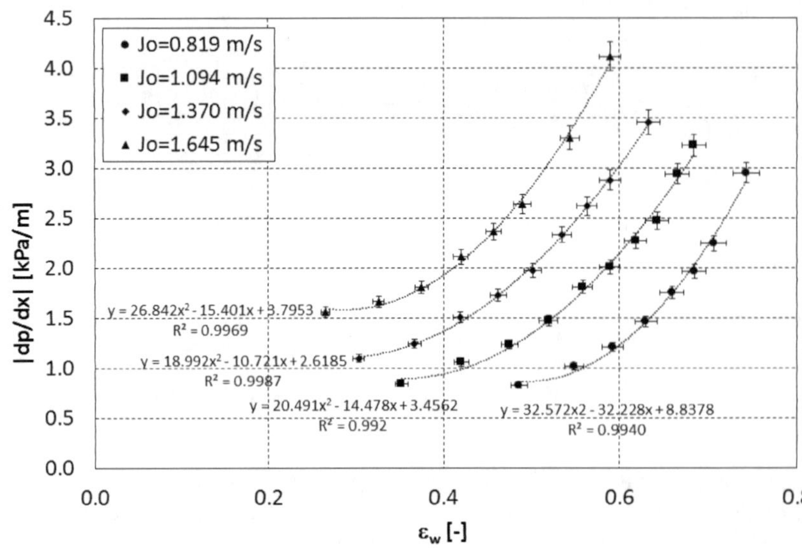

Figure 3. Pressure gradient versus water input fraction at constant oil superficial velocity (straight tube, D = 30 mm).

The water holdup is calculated from the measured pressure gradient according to Eq. (8). It is reported in Fig. 6(a) together with the experimental data collected by means of the quick closing valves technique for pipes of 30 mm and 40 mm i.d., respectively [16]. Substantial agreement is found between the measured and the calculated values. According to Arney et al. [10], the water holdup as a function of the water input fraction can be expressed by a parabolic fitting

$$H_w = \varepsilon_w\left[1 + C\left(1 - \varepsilon_w\right)\right] \tag{9}$$

Least square fitting gives $C = 0.356$ with regression parameter $R^2 = 0.98$ for the calculated holdup, $C = 0.358$ with regression parameter $R^2 = 0.95$ for the measured holdup, and $C = 0.357$ with regression parameter $R^2 = 0.95$ for both calculated and measured holdup. Hence, a unique value of 0.36 is assumed without significant differences (solid line in Fig. 6(a)). Mean Percentage Error (MPE) and Mean Absolute Percentage Error (MAPE) are defined as:

$$MPE = \frac{100\%}{N}\sum_{i=1}^{N}\frac{a_i - f_i}{a_i} \qquad MAPE = \frac{100\%}{N}\sum_{i=1}^{N}\left|\frac{a_i - f_i}{a_i}\right| \tag{10),(11}$$

where a_i is the actual value of the quantity being forecast, f_i is the forecast, and N is the population of the sample. Accordingly, Eq. (9) predicts the holdup data with $MPE = 0.1\%$ and $MAPE = 2.9\%$.

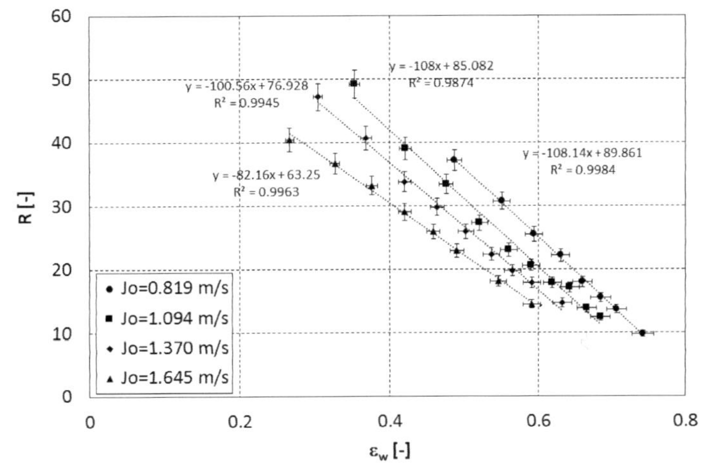

Figure 4. Pressure reduction factor versus water input fraction at constant oil superficial velocity (straight tube, D = 30 mm).

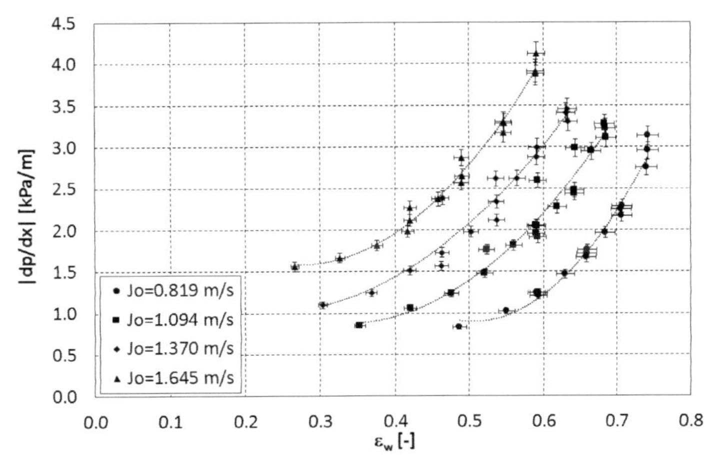

Figure 5. Pressure gradient versus water input fraction at constant oil superficial velocity (all data, D = 30 mm).

These findings are in very good agreement with Arney et al. [10] that deals with holdup measurements for core-annular flows of waxy crude oil and No. 6 fuel oil, leading to $C = 0.35$ (see also Table 1). Figure 6(b) reports the comparison of Eq. (9) and all the available experimental data from [3,11-13]. *MPE* is 0.2% and *MAPE* is 3.8%. *MAPE* values indicate a very satisfactory accuracy of the prediction, whereas the very small *MPE* indicates that the forecast by Eq. (9) is not significantly biased. On the other hand, Fig. 6(c) reports the comparison between Eq. (9) and the other models available in the literature for core-annular flow of liquid-liquid mixtures, listed in Table 1 (Arney et al. correlation is not reported since, being in the same form as Eq. (9), with $C = 0.35$ instead of 0.36, it cannot be distinguished). The model by Oliemans [11] always underestimates the water holdup with $MPE = -13\%$ and $MAPE = 16\%$. Ullmann and Brauner [8] provided an analytical solution of the Two-Fluid Model introducing as a closure equation a suitable correlation for the interfacial shear stress. According to Grassi et al. [9] the parameter c_i^0, reported in Table 1, has been set to 1.17. This approach, which generalizes a former result from Brauner [7], results slightly underestimating with *MPE* = –0.5% and *MAPE* = 8.4%.

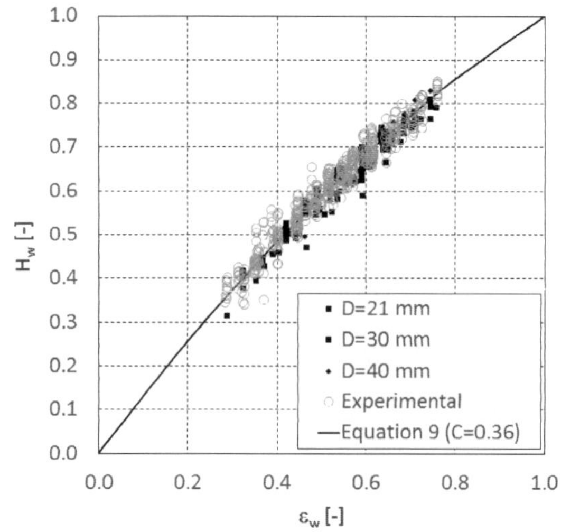

Figure 6(a). Water holdup versus water input fraction. Comparison between quick-closing valves data and TFM prediction.

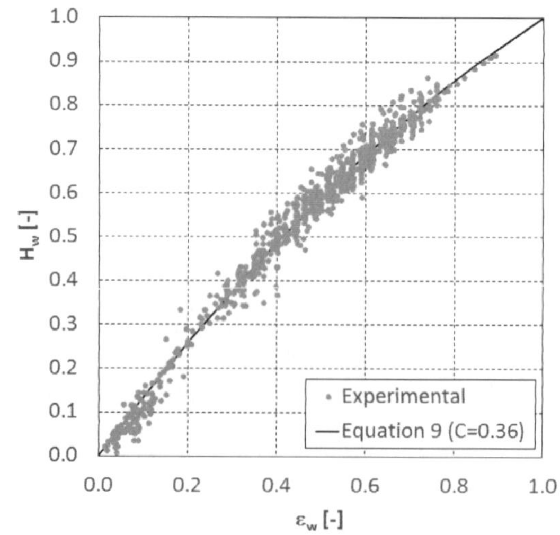

Figure 6(b). Water holdup versus water input fraction. Comparison between the proposed correlation and all the available data from the literature.

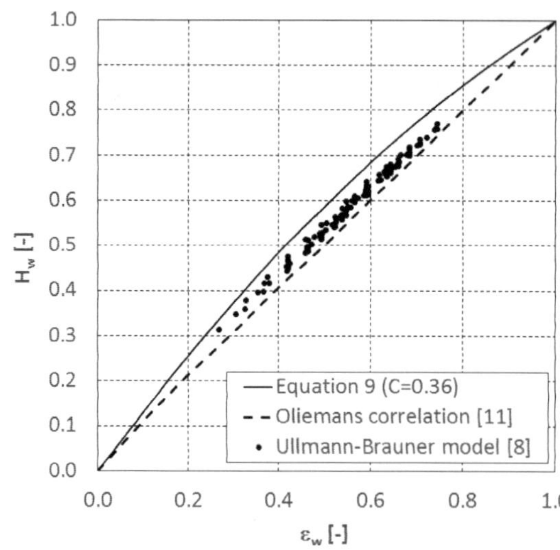

Figure 6(c). Water holdup versus water input fraction. Comparison between the proposed correlation and available models from the literature.

4.2. Pressure gradient estimation

The semi-empirical expression of the water holdup, Eq. (9), can be used in the Two-Fluid Model to predict the pressure drop, simply rearranging Eq. (8):

$$-\frac{dp}{dx} = 2C_w\left(\frac{DJ_w}{\nu_w}\right)^{-n_w}\frac{\rho_w J_w^2}{DH_w} \tag{12}$$

IOP Conf. Series: Journal of Physics: Conf. Series **923** (2017) 012012 doi:10.1088/1742-6596/923/1/012012

As shown in the parity plot of Fig. 7, the result is satisfactory. In particular, the model is able to predict 98% of the data within a percentage error of ±10%, 99% of the data fall within ±15%, and all the data are predicted within ±20%. Globally, *MPE* = 0.5% and *MAPE* = 4.4%. Similarly, for the pressure reduction factor 96% of the data are predicted within a percentage error of ±10%, 99% of the data fall within ±20%, and all the data are predicted within ±25%. Globally, *MPE* = 1.6% and *MAPE* = 5.5%. Figure 8 shows the comparison with the pressure reduction factor according to the Ullmann-Brauner model [8]. With the latter, *MPE* = –18.0% and *MAPE* = 18.1% hence the pressure reduction factor is generally underestimated. Nevertheless, such a model allows acceptable results in a broad range of operating conditions.

Table 1. Holdup models for oil-water flows from the literature.

Ref.	Model	Additional information
Arney [10]	$H_w = \varepsilon_w\left[1+0.35(1-\varepsilon_w)\right]$	-
Oliemans [11]	$H_w = \varepsilon_w\left[1+0.2(1-\varepsilon_w)^5\right]$	-
Ullmann and Brauner [8]	$H_w = \dfrac{c_i^0/2-\chi^2\phi/F_i+\dfrac{c_i^0}{2}\left[1+4\chi^2\left(\dfrac{\phi}{c_i^0}\right)^2\Big/F_i\right]^{1/2}}{c_i^0+\phi-\chi^2\phi/F_i}$ $c_i^0 = 1.15 \div 1.2$ $F_i = 1$	$\phi = J_o/J_w$ $\chi^2 = \dfrac{0.046}{16}\left(\dfrac{\mu_w}{\mu_o}\right)^{0.2}\left(\dfrac{\rho_w}{\rho_o}\right)^{0.8}\dfrac{1}{\phi^{1.8}}\mathrm{Re}_o^{0.8}$ (laminar oil core – turbulent water annulus) $\mathrm{Re}_o = \rho_o J_o D/\mu_o$

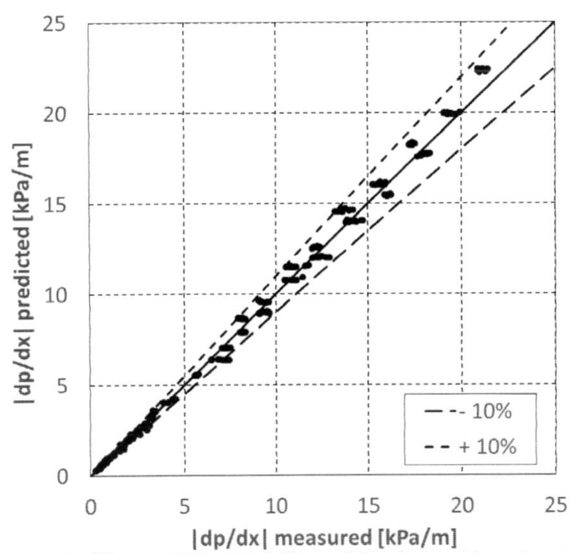

Figure 7. Predicted vs. measured pressure gradient.

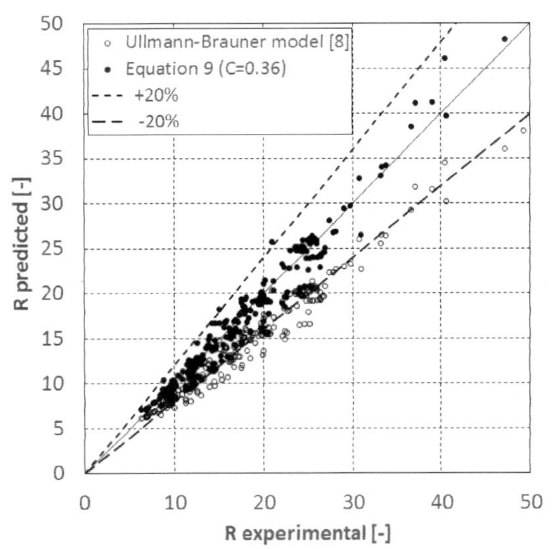

Figure 8. Comparison between predictions of the pressure drop reduction factor.

5. Conclusions

The water holdup has been estimated from the measurements of the pressure drop by means of the Two-Fluid Model. This is convenient as in general it is simpler to measure pressure drop than holdup. Moreover, holdup estimation is thus possible also when holdup cannot be directly measured. Hence, a semi-empirical formulation of the holdup as a function of the input water fraction has been derived, which generalizes on a broader data set an approach presented in the literature [10]. The expression for the holdup is then used in the Two-Fluid Model to predict the pressure drop and, in particular, the pressure reduction factor. This approach leads to better results compared with the use of a correlation for the interfacial shear stress in the Two-Fluid Model.

6. Nomenclature

A	cross-sectional area (m^2)
C	constant (-)
D	pipe diameter (m)
f	Fanning friction factor (-)
H	holdup (-)
J	superficial velocity (m/s)
MPE	mean percentage error (%)
MAPE	mean absolute percentage error (%)
p	pressure (Pa)
Re	Reynolds number (-)
S	wetted perimeter (m)
TFM	Two-Fluid Model
T	temperature (°C)
U	average velocity (m/s)

Greek Symbols

ε	volume ratio (-)
μ	dynamic viscosity (kg/m-s)
ρ	density (kg/m^3)
τ	shear stress (N/m^2)

Subscripts and Superscripts

n	power-law exponent
o	oil
w	water

References

[1] Appert O and Favennec J P 2008 *Encyclopaedia of Hydrocarbons* **IV** ch.2.3
[2] Clark A F and Shapiro A 1949 U.S. patent No. 2 533 878
[3] Charles M E, Govier G W and Hodgson G W 1961 *Can. J. Chem. Eng.* **39**(1) 27–36
[4] Charles M E and Redberger P J 1962 *Can. J. Chem. Eng.* **40**(2) 70–75
[5] Ooms G, Segal A, Van der Wees A J, Meerhoff R and Oliemans R V A 1983 *Int. J. Multiphas. Flow* **10**(1) 41–60
[6] Oliemans R V A, Ooms G, Wu H L and Duijvestijn A 1987 *Int. J. Multiphas. Flow* **13**(1) 23–31
[7] Brauner N 1991 *Int. J. Multiphas. Flow* **17**(1) 59–76
[8] Ullmann A and Brauner N 2004 *Multiphase Science and Technology* **16**(4) 355–87
[9] Grassi B, Strazza D and Poesio P 2008 *Int. J. Multiphas. Flow* **34** 950–65
[10] Arney M S, Bai R, Guevara E, Joseph D D and Liu K 1993 *Int. J. Multiphas. Flow* **19**(6) 1061–76
[11] Oliemans R V A 1986 *The Lubricating-film Model for Core-annular Flow* Ph.D. Thesis, Technishe Hogeschool Delft, Delft University Press, The Nederlands
[12] Sinclair A R 1970 *J. Petrol. Technol.* **6**(22) 711–19
[13] Bai R, Chen K and Joseph D D 1992 *J. Fluid Mech.* **240** 97–142
[14] Colombo L P M, Guilizzoni M and Sotgia G M 2012 *Exp. Fluids* **53** 1617–25
[15] Balakhrisna T, Ghosh S, Das G and Das P K 2010 *Int. J. Multiphas. Flow* **36** 13–24
[16] Colombo L P M, Guilizzoni M, Sotgia G M and Marzorati D 2015 *Int. J. Heat Fluid Fl.* **53** 91–97

Characterization of highly hydrophobic textiles by means of X-ray microtomography, wettability analysis and drop impact

M Santini[1], M Guilizzoni[2], S Fest-Santini[3], M Lorenzi[4]

[1] Department of Engineering and Applied Sciences, University of Bergamo, viale Marconi 5, 24044 Dalmine, Italy
[2] Politecnico di Milano, Department of Energy, Via Lambruschini 4, 20156 Milano, Italy
[3] Department of Management, Information and Production Engineering, University of Bergamo, viale Marconi 5, 24044 Dalmine, Italy
[4] School of Engineering and Mathematical Sciences, City University London, Northampton Square, London EC1V 0HB, United Kingdom

maurizio.santini@unibg.it

Abstract. Highly hydrophobic surfaces have been intensively investigated in the last years because their properties may lead to very promising technological spillovers encompassing both everyday use and high-tech fields. Focusing on textiles, hydrophobic fabrics are of major interest for applications ranging from clothes to architecture to environment protection and energy conversion. Gas diffusion media – made by a gas diffusion layer (GDL) and a micro-porous layer (MPL) – for fuel cells are a good benchmark to develop techniques aimed at characterizing the wetting performances of engineered textiles. An experimental investigation was carried out about carbon-based, PTFE-treated GDLs with and without MPLs. Two samples (woven and woven-non-woven) were analysed before and after coating with a MPL. Their three-dimensional structure was reconstructed and analysed by computer-aided X-ray microtomography (μCT). Static and dynamic wettability analyses were then carried out using a modified axisymmetric drop shape analysis technique. All the surfaces exhibited very high hydrophobicity, three of them near to a super-hydrophobic behavior. Water drop impacts were performed, evidencing different bouncing, sticking and fragmentation outcomes for which critical values of the Weber number were identified. Finally, a μCT scan of a drop on a GDL was performed, confirming the Cassie-Baxter wetting state on such surface.

1. Introduction

Engineered surfaces have been gaining more and more importance since the last years, as the ability to finely tune the surface properties, at least in part independently from the bulk material properties, may lead to very promising technological applications. Among the different surface properties, hydrophobicity and super-hydrophobicity have been intensively investigated because water repellency, low adhesion, low friction, high restitution coefficient may offer potential benefits in a wide range of fields from everyday use to high-tech applications (e.g. self-cleaning coatings, water repulsion and prevention of ice formation, reduced friction and directional flow, capillary devices in micro-electromechanical systems and microfluidic devices). An important sub-topic is the study of highly hydrophobic textiles (cloths, felts, papers), that is of major interest for applications ranging from manufacturing of consumer goods to Architecture and Civil Engineering, to Design, environment protection and energy conversion (clothes, umbrellas, tents, tensile structures and coverings, innovative surfaces with peculiar properties – e.g. conductive or antibacterial – membranes and diffusion layers) [1–4]. Within the latter field, gas diffusion media (GDMs) – made by gas diffusion layers (GDLs) and micro-porous layers (MPLs) – for fuel cells [5,6] are a very good benchmark to develop techniques aimed at characterizing the wetting performances of engineered textiles.

Hydrophobicity obviously depends on the intrinsic wettability of the material, related to its surface energies in contact with water and with air (or more in general with the wetting liquid and the surrounding ambient – gas, vapor or another immiscible liquid). Polytetrafluoroethylene (PTFE) and other fluoropolymers may grant contact angles with water higher than 110° [7], but unfortunately the known materials cannot at present reach the so-called super-hydrophobicity (contact angles higher than 150° and contact angle hysteresis under 10°) for smooth surfaces [8]. Hence, the actual strategy for obtaining super-hydrophobicity must also include a proper design of the morphology of the surface, aimed at increasing the specific surface of the material and at getting a suitable roughness (possibly a two-scale hierarchical or fractal roughness, mimicking natural super-hydrophobic surfaces [9–11]). It is necessary to reach the so-called Cassie-Baxter wetting state [12], a composite wetting regime in which the drop does not fill the grooves in the surface texture. For textiles, another aspect to be considered is the fact that the drop may be lift by the fibers protruding from the surface, without truly touching the same. It thus experiences a sort of extreme Cassie-Baxter effect which is not, however, really due to the surface in itself [13]. Regrettably, the Cassie-Baxter wetting state – also known as the ''fakir'' state [14] – is unstable and for impacting drops over a certain impact velocity a transition occurs towards the Wenzel wetting behavior [15], in which the drop completely fills the grooves (and super-hydrophobicity cannot be reached). For some applications, e.g. GDMs, the behavior in contact with sessile drops is the most representative of real conditions. Even if some authors [16] questions the effective validity of the "external" contact angles (static, maximum advancing and minimum receding) for the prediction of in-pore phenomena, the contact angle and contact angle hysteresis are still the most used "lumped parameters" to provide valuable information about the liquid-surface interaction. For other applications, e.g. for clothes, umbrellas and architectural surfaces, the response to impacting drops is the most representative. Studies about the latter are available in the literature, but almost only concerning smooth surfaces or surfaces where the super-hydrophobic behavior is obtained by regular patterns of micro- and nano- pillars [17] or square posts [18]. On these surfaces four different outcomes of the impact are reported: deposition, sticking, rebound and fragmentation [17,18] and critical velocities separating these outcomes are given. The same outcomes are reported, but at different impact velocities, for surfaces that are made hydrophobic by suitable coatings: on porous super-hydrophobic polymer surface, transitions limits for rebound and fragmentation is determined for significantly lower Weber ($We = \rho V^2 D / \sigma$, where ρ is the density, V is the impact velocity and σ is the surface tension) numbers: 60 [19] in confront to about 400 [20], even if the static contact angle values and hysteresis were similar to the uncoated, micro-structured surfaces. There is on the contrary a lack of information about the correlation between the static and dynamic wetting behavior for surfaces characterized by a complex, three-dimensional structure, as textiles are. The aim of the present study is to partially fill such gap by performing an analysis that combines information about the surface structure, topology and roughness (by means of computer-aided X-ray microtomography, μCT), static and dynamic wettability analyses (by means of a modified axisymmetric drop shape analysis technique) and water drop impacts (by means of high speed imaging), on pseudo-isotropic non regular rough substrates having different morphologies.

2. Investigated surfaces

As representative examples of engineered textiles, gas diffusion media (GDMs) for proton exchange membrane fuel cells (PEMFC) and direct methanol fuel cells (DMFC) were selected. Such layers are used in the fuel cells to reduce the problems of water flooding and methanol cross-over. They consist of a gas diffusion layer (GDL) – usually made of porous carbon paper or carbon cloth, wet-proofed with PTFE or other fluoropolymers – additionally coated (to increase the performances) with a thin microporous layer (MPL) – made by a mixture of carbon powder, PTFE and other additives [21,22].

Four surfaces were considered in this work: two commercial substrates (SEAL SCCG5N carbon cloth, named CC in the following, and SIGRACET 10CA carbon woven–non-woven, named WNW in the following) analyzed as they are and after coating with a carbon and PTFE microporous layers (coated surfaces will be named CC40 and WNW40 in the following). Full details about the surfaces and the MPL preparation can be found in [5]. All of them have a pseudo-isotropic surface structure, but with different characteristics. Figure 1 shows some examples of the investigated surfaces, including some common issues that may occur in their preparation.

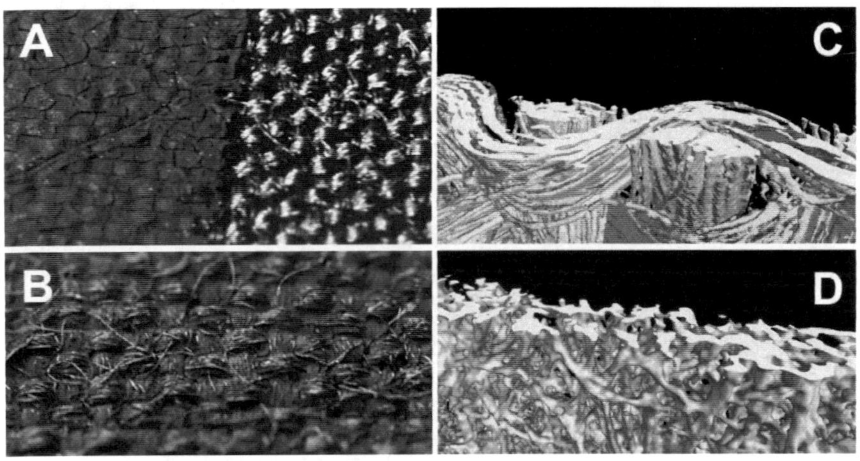

Figure 1. Examples of the investigated surfaces: (A, right part) uncoated CC; (A, left part): CC coated with a thick MPL showing diffuse cracking; (B) CC surface coated with a very thin MPL that is not able to completely cover the CC warp-and-weft structure and protruding fibers; (C) μCT reconstruction of uncoated CC; (D) μCT reconstruction of uncoated WNW.

3. Surface analysis

The four investigated layers were preliminarily observed using a SEM microscope [5], obtaining images characterized by a very high resolution and quality; the disadvantage is that such visualization is only two-dimensional and cannot be used to reconstruct the three-dimensional structure. The latter can on the contrary be obtained by computer-aided X-ray microtomography (μCT) [23-25], that was therefore used in this work to describe the surfaces both qualitatively and by means of suitable quantitative parameters, namely Wenzel roughness ratio, lift and complexity. The first one is defined as the ratio between the actual surface area and the top-view surface area; for the Wenzel wetting state, it also corresponds to the ratio between the real and apparent wetted areas. For each sample, a 3D digital volume was reconstructed by μCT with a resolution between 4 and 8 μm, from which the surface was extracted using a threshold calculated on the basis of the whole 3D volume intensities and saved in stereolitography format (STL) as a triangle mesh. Figure 2 shows two examples of the acquired layers: the woven CC (Fig. 2A) and the woven-non-woven WNW (Fig. 2B), also evidencing the top surface. The extracted surfaces were then sliced, and a contour following technique was implemented to follow the surface profile also within the caves that have some hidden parts when seen from above, but that in the Wenzel assumption are considered as filled with liquid. This approach was carried out on the binary version of each of the vertical, 1-voxel thick, slices of the reconstructed volume. Using the same slice-based approach, the two other surface parameters – lift and complexity – were also calculated. In this case, 40 horizontal lines were analysed for each slice, at depths uniformly spaced between the minimum and maximum height of each slice. More specifically, the surface lift along each investigated line is the ratio between the number of pixels belonging to the layer and the total length in pixels of the layer along that line. Concerning the surface complexity, it is calculated as the number of transitions between pixels belonging to the layer and pixels belonging to cavities (or vice versa) over the total length in pixels of the layer along that line. It thus counts the number of layer/air or air/slayer interfaces along the chosen line, with an approach completely analogous to the one used for two-phase flow [26,27].

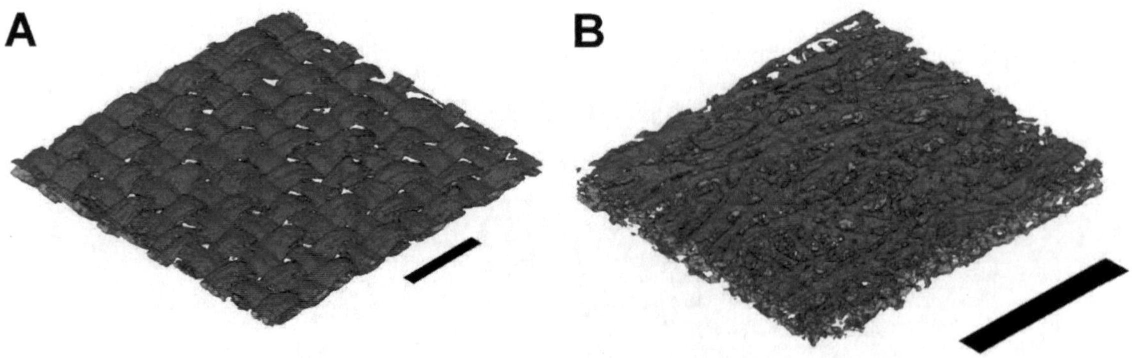

Figure 2. Two clippings from the CC (A) and WNW (B) reconstructed samples, showing in red the facets that are visible in a top view. Black bars represent 1 mm.

Thus lift and complexity profiles were obtained as functions of the height, both as single-slice profiles and as profiles averaged on all the slices. The globally averaged values of lift and complexity on the whole surface can also be calculated. Their values are summarized in Table 1. The roughness of the surface was not included among the descriptive parameters, as it would not add much information with respect to the Wenzel roughness ratio.

Table 1. Surface parameters (all of them are dimensionless) for the investigated layers (average values and standard deviation).

Surface	Wenzel ratio (-)	Lift (-)	Complexity (-)
CC	1.22±0.11	0.458± 0.008	0.024± 0.128
CC40	1.18±0.09	0.474± 0.106	0.028± 0.009
WNW	2.39±0.26	0.255± 0.059	0.058± 0.012
WNW40	2.32±0.17	0.324± 0.046	0.140± 0.021

The Wenzel roughness ratio is higher for the woven-non-woven samples, due to the fiber irregular arrangement and compression. The variability in the measurements is higher for the WNW cases too, both due to the same aspects and to the greater complexity of the μCT acquisition for such samples. The difference in the Wenzel roughness ratio between the uncoated (GDL) and coated (GDL+MPL) surfaces is very slight, contrarily to what could be expected. The cause may be that even if the MPL acts as a film which levels the unevenness of the substrate, it is not able to cover it completely (see Figure 1B). Figure 3A reports the results in terms of averaged (on all the tomographic slices for a layer) surface lift profiles with depth. The analysis of the lift (see Table 1 for the globally averaged values) evidence how the woven surfaces (CC and CC40) offer a higher lift, due to their more compact structure. In their case, the addition of the MPL does not change significantly the lift profile. It just moves the maximum lift region towards the highest lines, where the compact MPL layer is deposed. On the contrary, for the woven-non-woven surfaces (WNW and WNW40) the MPL increase slightly the lift, moreover it enlarges the region of maximum lift towards the lowest lines. Due to the sparser structure of the base substrate, the MPL soaks it more deeply. This reflects also on the average lift value for WNW40. Surface complexity profiles for CC and WNW are shown in Fig. 3B. WNW shows a much more complex structure than CC. This is consistent with the fact that from both the SEM and μCT analysis, the woven-non-woven surface appears to be similar to a "mangrove forest", with a lot of unordered fibers that creates a tangle and a maze of tunnels between them. On the contrary, the complexity of the WNW40 surface seems odd: it is much higher than all the others, in contrast with the physical appearance of the coated surface. The extraction of the iso-lift contour lines for the CC and WNW surfaces also reproduces the warp-and-weft structure of the woven CC, which is not present in the WNW.

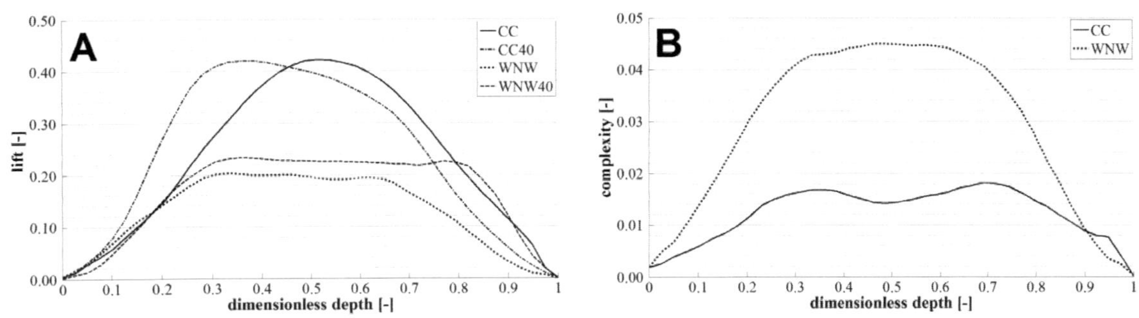

Figure 3. Averaged lift (A) and complexity (B) profiles for the investigated surfaces.

4. Wettability analysis

Micro-tomography can also be used from the quantitative point of view, calculating contact angles on slices of the reconstructed volume, as reported in [28], but for this work wettability analyses were performed using a conventional optical approach, in terms of static "as placed" [29], advancing and receding contact angles. Static "as placed" contact angles and advancing contact angles were measured by means of a modified version of the axisymmetric drop shape analysis technique [30,31], while receding contact angles are estimated from the previous two using a slightly modified (to adapt it to non-wetting drops) version of the model by Tadmor [29]. Figure 4 shows two examples of drops on the WNW and WNW40 surfaces, with the theoretical Laplace-Young contour superposed to the experimental side-views. Figure 5 shows the results of the experimental tests using box plots for the static and advancing contact angles and single markers for the receding contact angle. The whiskers ends represent minimum and maximum values, while the box body covers the interquartile range of the data. Table 2 summarizes the main statistics of the results. The data spreading is quite high but this could be expected due to the rough and not completely homogeneous nature of the investigated surfaces and to the very high values of the contact angles on the same, which makes the measurement extremely difficult [32]. In terms of median or average contact angles, all the investigated surfaces can be regarded as super-hydrophobic. Solely, the woven-non-woven texture WNW goes below the lower limit of super-hydrophobicity if the estimated standard derivation is deducted from the median contact angle. In terms of contact angle hysteresis, the Cassie-Baxter regime is characterized by low contact angle hysteresis and the surfaces CC, CC40 and WNW40 are, therefore, expected to present liquid-solid interaction in this wetting regime. The low receding angle of WNW should instead indicate that the surface exhibits Wenzel behavior and also raises a doubt on the super-hydrophobic behavior of this surface. It is also interesting to note how for the CC surfaces the coating does not significantly enhance the super-hydrophobicity, while for WNW structured surfaces it seems to modify the wetting regime. An explanation of such difference may be found in the different level of coverage that the coating reaches with respect to the asperities of the base substrate: both μCT and SEM analyses of the surfaces [5] show that for woven textiles the coating still follows the warp and weft of the original GDL structure, while on WNW fabrics the coating completely covers the texture of the rough material.

Figure 4. Drop side views on WNW and WNW40, with superposition of the fitted Laplace-Young profile (enlarged to 3 pixels for visualization purposes).

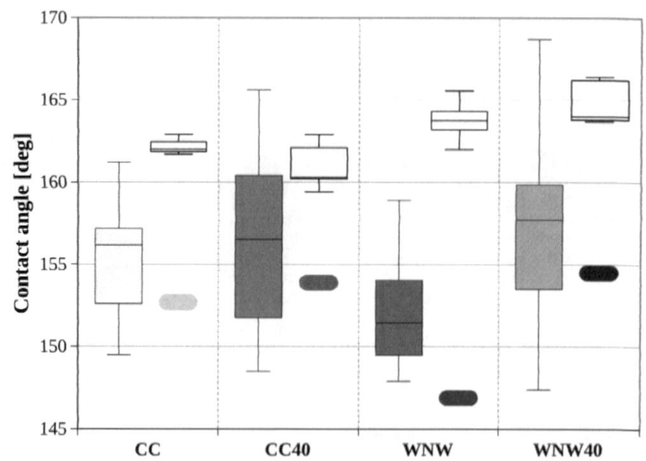

Figure 5. Contact angles of water drops in air on the investigated surfaces: static "as placed" (box plots with coloured background), advancing (box plots with white background) and receding (superellipses markers).

Table 2. Statistical summary of the measured contact angles.

Surface	Median [deg]	Avg. [deg]	Std. dev. [deg]
"As placed" contact angles			
CC	156	155	4.4
CC40	156	156	6.4
WNW	152	152	5.0
WNW40	158	157	5.2
Advancing contact angles			
CC	162	161	5.9
CC40	160	161	1.5
WNW	164	166	4.7
WNW40	164	165	1.4
Receding contact angles (by Tadmor model)			
CC	CC40	WNW	WNW40
153	154	147	154

In previous studies, computer-aided X-ray microtomography (μCT) was also used with very promising results to investigate the real shape of the contact surface and the contact angle profile along the contact line for water drops on smooth and rough surfaces [28,33,34]. Thus, in this study μCT was also used to directly scan a drop on one of the investigate surfaces, to verify the wetting state by a three-dimensional acquisition of the real system, with micrometric resolution. The non-intrusive, fully 3D nature of microCT allows visualizing the contact region of the drop-surface system. Figure 6 shows the result, after extraction of the drop and CC surfaces as isosurfaces at the proper level of X-ray attenuation. Panel A shows a clipping of the drop and surface to evidence the complexity of the contact between water and GDL, that is further underlined by the extraction of the drop bottom surface in Panel B.

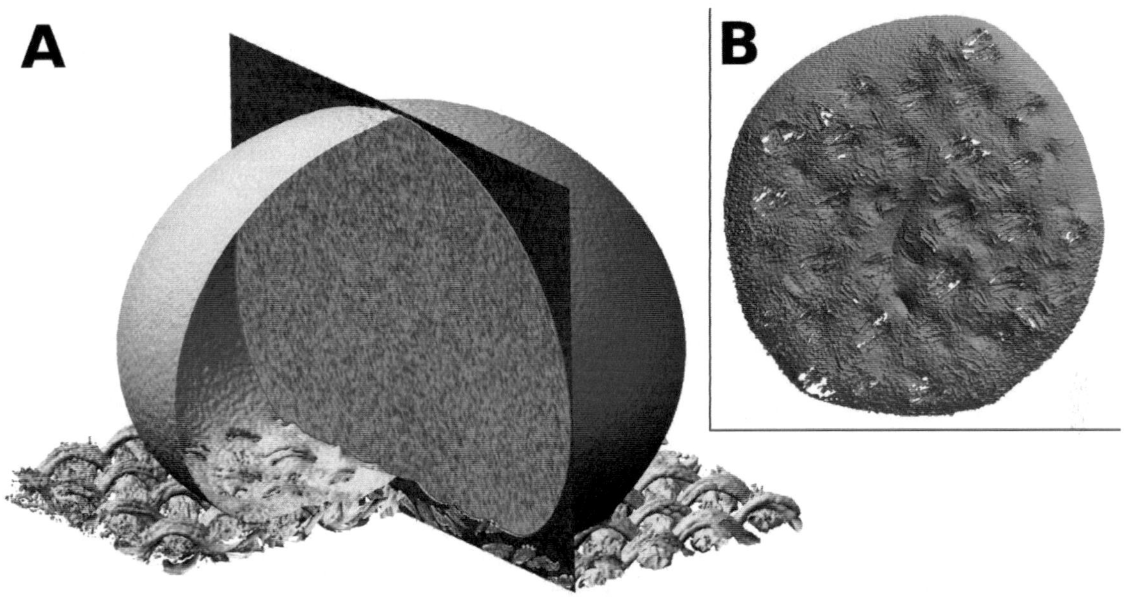

Figure 6. Rendering of drop and CC surfaces extracted as isosurfaces of X-ray attenuation from μCT: A) clipping of the drop and CC surface, B) extraction of the drop bottom surface.

5. Drop impact

Drop impact experiments were carried out using high speed imaging, following a procedure which is described in details in [35,36]. As discussed in the Introduction, four different outcomes of drop impact may be observed, whereby rebound and fragmentation is related to the Cassie-Baxter regime, sticking to the Wenzel regime and deposition could occur for both wetting states [19]. To observe the transition between the different outcomes, the impact velocity was systematically varied by changing the impact height, while the droplet diameter was kept constant to 3.5 mm. The corresponding Weber number was analyzed as a function of a mean contact angle, calculated using the mean advancing and the receding contact angle $(\theta_A + \theta_R)/2$ summarized in Table 1. As an example, two impact events at same impact conditions but on different surfaces are shown in Fig. 7. Figure 8 reports the outcomes of the impact events for all the investigated impact conditions, in terms of contact angle and Weber numbers (open symbols: bouncing/sticking, filled symbols: fragmentation). Experimental data for bouncing and fragmentation from the work of Rioboo et al. [19] and the corresponding contact angle measurements from Rioboo et al. [37] (black triangles) are also reported for comparison. For the surfaces CC, CC40 and WNW40, which behave following the Cassie-Baxter wetting model, transition between bouncing and fragmentation is observed. The minimal required Weber number to observe fragmentation (which will be called "critical Weber number" in the following) and the number of fragmented drops differ for these three surfaces. All the droplets above the critical Weber number are rejected from the surface. The uncoated carbon cloth structure CC and the coated woven-non-woven surface WNW40 exhibit the same critical Weber number of about 50 despite they have different mean contact angles and surface topologies.

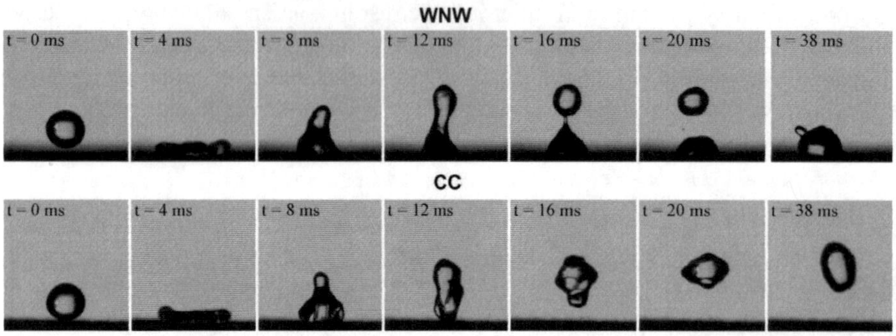

Figure 7. Impact sequence at We = 37.3 on WNW (1st row, with pinning and partial bouncing) and CC (2nd row, with bouncing).

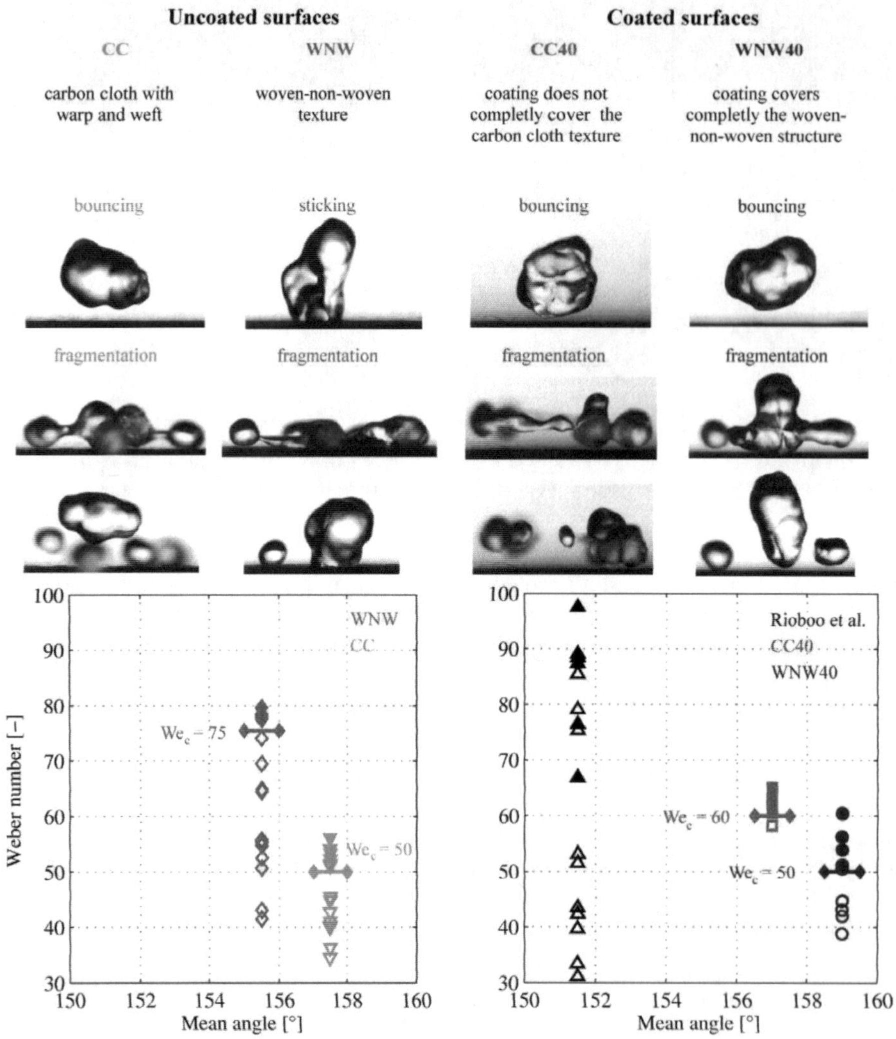

Figure 8. Outcomes of the impact events for different We (open markers: bouncing / sticking, filled markers: fragmentation). Experimental data from the work of Rioboo et al. [19,37] (black triangles) are also reported for comparison.

The CC texture seems to feature the highest lift, whereby the coated woven-non-woven surface WNW40 features the highest observed mean contact angle. However, the number of fragmented drops is higher for the CC substrate. Comparing the two coated structures CC40 and WNW40, transition between bouncing and fragmentation is shifted to higher critical Weber numbers with decreasing mean contact angle. This tendency is confirmed by the experimental results of Rioboo et al. [19,37], which are also shown in Figure 8 for comparison. The critical Weber number is found to be about 60 for the coated carbon cloth structure CC40. It is worth noting that the carbon cloth structures CC and CC40 feature different critical Weber numbers despite they have similar contact angles. Thereby, it is assumed that the observed effects are caused by the topology modifications, which affect the impact behavior more than the wettability. For the woven-non-woven surface WNW, whose low receding contact angle indicates Wenzel wetting characteristics, sticking is observed for Weber numbers below the critical one of about 75. Above this limit, fragmentation is found. Nevertheless, contrarily to what happens with the other investigated surfaces, the fragmented drops still stick on the surface and thereby seem to behave following Wenzel wetting model, as they do below the critical Weber number. The cause of such behavior has probably to be searched in the peculiar morphology of the WNW surface: its uncoated tangle of unordered and in part protruding fibers entangles the drop, thus preventing its rebound.

Conclusions

The wettability and interactions with sessile and impacting drops of carbon-based, PTFE-treated, surfaces used as GDL and GDM for fuel cells were investigated by joint use of optical techniques for contact angle measurement and drop impact analysis, and X-ray microtomography for surface characterization and direct visualization of the drop-surface couples. Given the rising importance of super-hydrophobic surfaces and engineered textiles, obtaining thorough and reliable information about the surface structure and wetting behavior is of high interest in a wide range of fields, from improvement of consumer goods to architectural and industrial applications. A woven and a woven-non-woven GDLs, both as they are and after coating with a MPL, were analyzed. Even if a deeper investigation is still needed to fully understand some of the outcomes of drop impacts on these kinds of surfaces and the spreading of the measured contact angles is significant, the combined use of such information gave generally consistent results (also in good agreement with literature data). More specifically, the woven-type surfaces, both uncoated and coated, and the coated woven-non-woven surface appear to feature a Cassie-Baxter super-hydrophobic behavior, while the uncoated woven-non-woven surface showed a very high static contact angle, but also a large contact angle hysteresis, more typical of a Wenzel wetting state. The woven-non-woven sample showed also that a surface coating of sufficient thickness, which hides the texture of the base substrate, seems to grant a significant enhancement of the contact angles, probably also changing the governing wetting regime. For coated super-hydrophobic textures, the transition between bouncing and fragmentation is shifted to lower Weber numbers with increasing contact angles. Finally, a microtomographic scan of a drop on a woven surface was acquired, to directly verify the real wetting state. The results confirmed the hypotheses, as the drop on such surface is in a Cassie-Baxter state, with very evident air pockets under its bottom.

Funding sources and acknowledgements

S. Fest-Santini acknowledges funding by *Pro Universitate Bergomensi* within their internationalization projects.
The investigated MPLs were prepared and coated on the GDLs by the Materials for Energy and Environment ($MAT4EN^2$) group, Dipartimento di Chimica, Materiali e Ingegneria Chimica "G. Natta", Politecnico di Milano. Their contribution is gratefully acknowledged.

References

[1] Shuhui Li, Jianying Huang, Zhong Chen, Guoqiang Chen and Yuekun Lai 2017 *J. Mater. Chem. A* **5** 31–55
[2] Sohyun Park, Jooyoun Kim and Chung Hee Park 2015 *J. Eng. Fiber. Fabr.* **10**(4) 1-18
[3] Latthe S S, Gurav A B, Maruti C S and Vhatkar R S 2012 *J. Surface Engineered Materials and Advanced Technology* **2** 76-94
[4] Xue C-H, Chen J, Yin W, Jia S-T and Ma J-Z 2012 *Appl. Surf. Sci.* **258**(7) 2468–72

[5] Gallo Stampino P, Cristiani C, Dotelli G, Omati L, Zampori L, Pelosato R and Guilizzoni M 2009 *Catal. Today* **147S** S30–S35
[6] Gallo Stampino P, Balzarotti R, Cristiani C, Dotelli G, Guilizzoni M and Latorrata S 2013 *Chemical Engineering Transactions* **32** 1603–08
[7] Lee S, Park J-S and Lee T R 2008 *Langmuir* **24** 4817–26
[8] Arkles B, Pan Y and Kim Y M 2009 The Role of Polarity in the Structure of Silanes Employed in Surface Modification, in Mittal K L, Silanes and Other Coupling Agents, **5**(1) Brill K NV, Leiden, Netherlands 51-64
[9] Guo Z, Liu W and Su B-L 2011 *J. Colloid Interf. Sci.* **353**(2) 335–355
[10] Kang M, Jung R, Kim H-S and Jin H-J 2008 *Colloid Surface A* **313–314** 411–414.
[11] Shibuichi S, Onda T, Satoh N and Tsujii K 1996 *J. Phys. Chem.* **100** 19512–17
[12] Cassie A B D and Baxter S 1944 *T. Faraday Soc.* **40** 546–551
[13] Wang T, Hu X and Dong S 2007 *Chem. Commun.* **18** 1849-51.
[14] Moulinet S and Bartolo D 2007 *Eur. Phys. J. E* **24** 251–260
[15] Wenzel R N 1936 *Ind. Eng. Chem.* **28**(8) 988–994
[16] Gurau V, Bluemle M J, De Castro E S, Tsou Y-M, Mann J A Jr and Zawodzinski T A Jr 2006 *J. Power Sources* **160** 1156–62
[17] Bartolo D, Bouamrirene F, Verneuil É, Buguin A, Silberzan P and Moulinet S 2006 *Europhys. Lett.* **74**(2) 299–305
[18] Smyth K, Paxon A, Kwon H-M, Deng T and Varanasi K K 2010 Dynamic wetting on superhydrophobic surfaces: droplet impact and wetting hysteresis Thermal and Thermomechanical Phenomena in Electronic Systems (ITherm), 12th IEEE Intersociety Conference 2–5 June 2010
[19] Rioboo R, Voué M, Vaillant A and De Coninck J 2008 Drop Impact on porous superhydrophobic polymer surfaces *Langmuir* **24** 14074–77
[20] Reyssat M, Pépin A, Marty F, Chen Y and Quéré D 2006 Bouncing transition on microtextured materials *Europhys. Lett.* **74**(2) 306–312
[21] Guilizzoni M, Gallo Stampino P, Cristiani C, Dotelli G and Latorrata S 2013 *Chemical Engineering Transactions* **32** 1657–62
[22] Latorrata S, Gallo Stampino P, Cristiani C and Dotelli G 2014 *Int. J. Hydrogen. Energ.* **39**(10) 5350 – 57
[23] James J P, Choi H-W and Pharoah J G 2012 *Int. J. Hydrogen. Energ.* **37** 18216–30
[24] Sasabe T, Inoue G, Tsushima S, Hirai S, Tokumasu T and Pasaogullari U 2012 *ECS Transactions* **50**(2) 735–744
[25] Santini M, Guilizzoni M, Lorenzi M, Atanassov P, Marsili E, Fest-Santini S, Cristiani P and Santoro C 2015 *Biointerphases* **10**(3) 031009 1-9
[26] Guilizzoni M *Int. J. Multiphas. Flow* **51** 1–10
[27] Arosio S and Guilizzoni M 2006 *J. Visual.-Japan* **9** 275–282
[28] Santini M, Guilizzoni M and Fest-Santini S 2013 *J. Colloid Interf. Sci.* **409** 204–210
[29] Tadmor R 2004 *Langmuir* **20** 7659–64
[30] del Rio O I, Neumann A W 1997 *J. Colloid Interf. Sci.* **196**(2) 136–147
[31] Guilizzoni M 2011 *J. Colloid Interf. Sci.* **364** 230–236
[32] Extrand C W and Moon S I 2010 *Langmuir* **26**(22) 17090–99
[33] Santini M. and Guilizzoni M 2014 *Colloids and Interface Science Communications* **1** 14–17
[34] Santini M, Guilizzoni M, Fest-Santini S and Lorenzi M 2015 *Rev. Sci. Instrum.* **86** 023708
[35] Santini M, Fest-Santini S and Cossali G E 2013 *Exp. Fluids* **54** 1593–07
[36] Santini M, Fest-Santini S and Cossali G E 2017 *Eur. J. Mech. B-Fluid* **62** 21–31
[37] Rioboo R, Voué M, Vaillant A, Seveno D, Conti J, Bondar A I, Ivanov D A and De Coninck J 2008 *Langmuir* **24**(17) 9508–14

IOP Conf. Series: Journal of Physics: Conf. Series **923** (2017) 012014 doi:10.1088/1742-6596/923/1/012014

Bubble aspect ratio in dense bubbly flows: experimental studies in low Morton-number systems

G Besagni[1], F Inzoli[1], T Ziegenhein[2], H Hessenkemper[2] and D Lucas[2]

[1] Politecnico di Milano, Department of Energy, Via Lambruschini 4a, 20156, Milano
[2] Helmholtz-Zentrum Dresden-Rossendorf e. V. Institute of Fluid Dynamics, 01314 Dresden, Germany

E-mail: giorgio.besagni@polimi.it

Abstract. Almost every modelling approach of bubbly flows includes assumptions concerning the bubble shape. Such assumptions are usually made based on single bubble experiments in quiescent flows, which is far away from the flow field observed in large-scale multiphase facilities. Considering low Morton-numbers and the highly deformable interface at medium and large Eötvös-numbers, the evaluation of the bubble shape in such systems under real flow conditions is highly desirable. In this study, we experimentally evaluate the bubble shape (in terms of aspect ratio), at low Morton-numbers, in different bubble column setups and a pipe flow setup under different operating conditions. The bubble shape in the bubble column experiments were obtained with cameras at Politecnico di Milano and Helmholtz-Zentrum Dresden Rossendorf (HZDR) whereas the shapes in the pipe flows were measured by the ultrafast electron beam X-ray tomography system (ROFEX) at HZDR. In the bubble column experiments almost the same shape is observed; conversely, the shape in the pipe flows distinctly depends on the flow conditions. In conclusion, in bubble columns the assumption of a constant shape regardless of the flow conditions is valid whereas in pipe flows the turbulence and shear rates can be strong enough to deform distinctly the bubbles.

1. Introduction

Bubbly flows are observed in different industrial applications and in multi-phase reactors. The correct design and scale-up of these reactors rely on the precise knowledge of the fluid dynamics at the different scales: maily, the "*reactor-scale*" and the "*bubble-scale*". In particular, at the "*bubble-scale*", the knowledge of the interface size and shape is crucial to characterize the heat and mass transfer phenomena and to model multiphase flows properly. Generally, the bubble shape depends on the fluid dynamic phenomena and the phase properties (viz, the physical and the interfacial properties, as discussed in ref. [1]). On the practical point of view, the bubble shape can be estimated by the aspect ratio (E, the ratio between the minor and the major axes), as discussed by Ellingsen and Risso [2]. Of course, the bubble shape can be related, by the dimensional analysis to non-dimensional groups; for example, the well-known Clift diagram [3] relates the bubble shape to the Morton, the Eötvös and the Reynolds numbers. Another common strategy is to study the influence of the flow phenomena and the phase properties on single bubbles and to apply the obtained models to more complex flows. In this respect, the reader may refer to the literature surveys in refs. [4, 5] and to the discussions in some of the very recent studies [6, 7]. Despite its importance, a comprehensive evaluation of the aspect ratio under realistic flow conditions (viz. swarm of bubbles) is missing and limited to single bubbles. This study contributes to the existing discussion and it investigates the integral dependencies (i.e., the gas and liquid flow rates, the column design and the sparger setup) on the bubble shape, in low Morton-numbers systems (air-water). To this end, bubble shape data were obtained with cameras at Politecnico di Milano (PoliMI) and Helmholtz-Zentrum Dresden Rossendorf (HZDR); in addition, bubble shapes were measured in pipe flows by using the ultrafast electron beam X-ray tomography system at HZDR.

2. The experimental setups and experimental methods

In this section, the experimental setups and methods are described. It is worth noting that, when analyzing the bubble shape data obtained in the different experimental setups, following the integral approach of the present work, we evaluated the bubbles over the complete flow cross-sections.

2.1. Politecnico di Milano (PoliMI) bubble columns

The experimental facility (Figure 1) is a non-pressurized "*large-diameter*" and "*large-scale*" circular bubble column (inner diameter 0.24 m, height 5.3 m); plese note that the discussion concerning the "*large-diameter*" and "*large-scale*" effects has been proposed in ref. [1]. The bubble column has been tested in the open tube configuration with a spider sparger (sparger openings: d_o = 2-4 mm) and in the annular gap configuration with a pipe sparger (d_o = 3 mm). The gas sparger openings classify the distributor as a "*coarse*" type [1]. Filtered air and deionizer water have been used and their temperatures were maintained constant at room temperature (295 ± 1 K). Experiments have been conducted both in the batch mode and in the counter-current mode (Table 1 summarizes the tested conditions). The photos have been taken using a NIKON D5000 camera (general settings as follow: Nikon 10-24mm lenses, f/3.5; 1/1600s; ISO400; 4288 x 2848 pixels and a spatial resolution of 11.8 pixel/mm). A 500W LED halogen lamp has been used as light source. The measurement location was located at approximately 2.4 m from the gas sparger, which corresponds to the developed region of the two-phase flow. In dense bubbly flows, the three dimensional shape is not accessible with photographic techniques; however, taking into account the discussion in ref. [2], each bubble is approximated and reconstructed by ellipses. The projected bubble size area has been used to calculate the major and minor axes as well as the bubble size (computed as the spherical equivalent diameter of the rotational volume of the projected area). In order to sample the ellipsoidal bubbles, the approach described by Besagni and Inzoli [1] has been applied. It is worth noting that ref. [1] collects, summarizes and extend all our previous studies concerning bubble column flows.

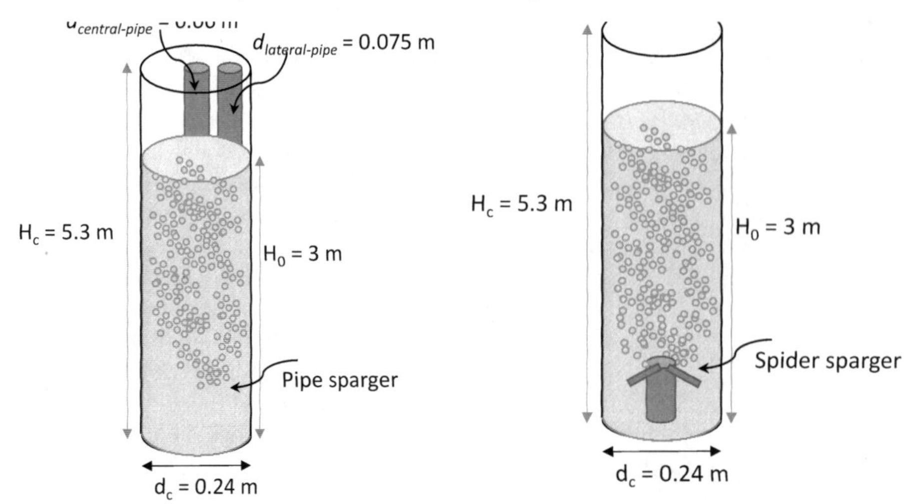

(a) Annular gap bubble column ***(b) Open tube bubble column***

Figure 1. PoliMI bubble column experimental setups.

Table 1. Experimental matrix (PoliMi bubble columns): gas flow rates (U_G) vs. liquid flow rates (U_L).

U_L [m/s] \ U_G [m/s]	0.0037	0.0074	0.0087	0.0111	0.0149	0.0188	0.0220	0.0313
0	*OT*	*OT*	*AG*	*OT*	*OT*	*OT*	*AG*	*AG*
-0.040			*AG*					
-0.066	*OT*	*OT*		*OT*	*OT*	*OT*		

Open tube bubble column (OT), annular gap bubble column (AG)

2.2. *Helmholtz-Zentrum Dresden Rossendorf (HZDR) bubble columns*

The experimental facilities (Figure 2) are non-pressurized "*small-scale*" and "*small-diameter*" rectangular-section bubble column. Different configurations were tested [8-11]: (a) two "*bubble plume*" experiments (Figure 2a), (b) an airlift reactor (the width of the downcomer is $W = 0.06$ m; the sparger size varies in the range $S = 0.085$ - 0.035 m, Figure 2b); (c) a homogenously aerated bubble column (Figure 1c); (d) single needle configuration (the bubble plume setup with one needle in the center); (e) single bubble in different shear flows configuration (single bubbles have been generated in the left side whereas at the right wall many large bubbles rise generating a vortex in which the areas with a linear shear field). The height of the bubble plume and the single needle experiments ranges of $H = 0.7$ - 0.9 m (Figure 2a); the sparger size varies in the range $S = 0.085$ - 0.035 m (Figure 2). In Table 2, the inner diameter of the used needles in the sparger (d_o) are shown with the corresponding flow rates. For some experiments, plastic "*caps*" on the needles have been used to generate large bubbles at low flow rates. It is worth noting that the sparger openings classifies the gas spargers ranging from "*fine*" to "*coarse*" type [1]. Filtered air and purified water have been used and their temperatures were maintained constant at room temperature (293 ± 1 K). All experiments have been conducted in batch mode.

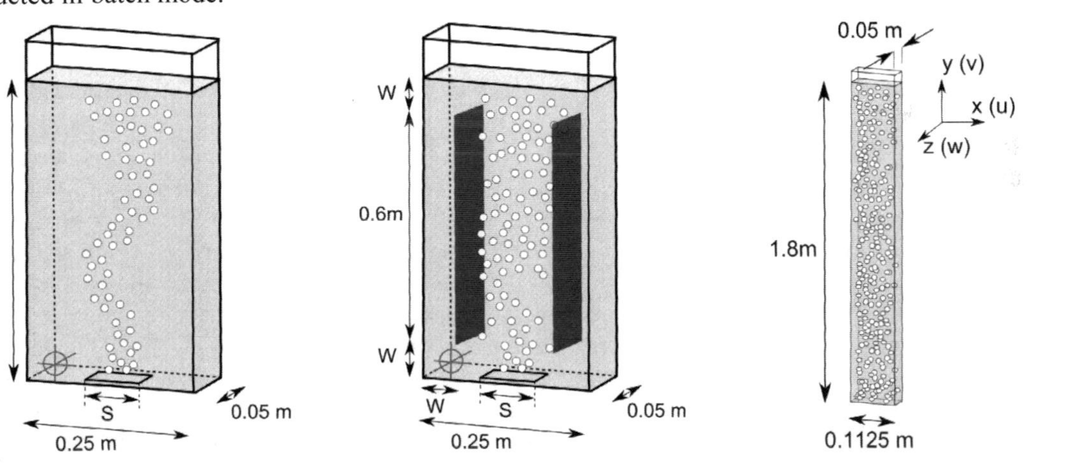

(a) Bubble plume and single needle experiments ***(b) Airlift*** ***(c) Homogenously aerated bubble column***

Figure 2. HZDR bubble column experimental setups.

Bursts of 10 Frames with 400 frames per second have been taken every second by using a Redlake Motion Pro high speed camera (general settings as follow: Samyang 135mm lens; f/5.8; 1/10000 s; 1280x1024 pixels and a spatial resolution of 10 pixel/mm). A 200W LED backlight has been used as light source. The same considerations as reported above (Section 2.1) for the limitations in the photography technique also apply here. For the single bubble experiments the photographs have been evaluated automatically and for all other experiments by hand with support of automated image analyzes: both techniques are described in refs. [8, 11]. The bubble sizes have been evaluated at different levels in the bubble columns, which have been all sufficiently away from the sparger. Since we have not observed any height dependency of the bubble shape, the results of the different levels were summed up for the single experiments. It is worth noting that, liquid velocity data are available for some of these experimental setup and were discussed by Zieghenein et al. [10, 11]

Table 2. Experimental matrix (HZDR bubble columns): gas flow rates (U_G) vs. inner diameter of the needles (d_o).

<table>
<tr><th></th><th></th><th colspan="11">U_G [m/s]</th></tr>
<tr><th></th><th></th><th>0</th><th>0 - 0.0009</th><th>0.0008</th><th>0.0024</th><th>0.0030</th><th>0.0040</th><th>0.0053</th><th>0.0060</th><th>0.0064</th><th>0.0071</th><th>0.0080</th></tr>
<tr><td rowspan="10">d_o [mm]</td><td>0.12</td><td rowspan="9">Single bubble in shear flow</td><td rowspan="8">Single needle</td><td></td><td></td><td>Hom</td><td></td><td></td><td></td><td></td><td></td><td></td></tr>
<tr><td>0.2</td><td>BP1</td><td>BP1</td><td></td><td>BP1</td><td></td><td></td><td></td><td>Hom</td><td></td></tr>
<tr><td>0.3</td><td></td><td>BP1</td><td>Hom</td><td></td><td></td><td></td><td>BP1</td><td></td><td></td></tr>
<tr><td>0.5</td><td></td><td></td><td></td><td></td><td></td><td></td><td></td><td></td><td></td></tr>
<tr><td>0.6</td><td></td><td></td><td>Hom</td><td>AL | BP2</td><td>BP2</td><td>AL</td><td></td><td></td><td>AL</td></tr>
<tr><td>0.7</td><td>BP1</td><td></td><td></td><td>BP1</td><td></td><td></td><td>BP1</td><td></td><td>BP1</td></tr>
<tr><td>0.9</td><td></td><td></td><td>Hom</td><td></td><td></td><td></td><td></td><td></td><td></td></tr>
<tr><td>1.5</td><td></td><td>BP1</td><td></td><td>BP1|BP2</td><td>BP2</td><td></td><td>BP1</td><td>Hom</td><td>BP1</td></tr>
<tr><td>Caps</td><td></td><td></td><td></td><td>Hom</td><td></td><td></td><td></td><td></td><td></td><td></td></tr>
<tr><td>1.5/0.6</td><td></td><td></td><td></td><td></td><td></td><td>BP2</td><td></td><td></td><td>BP2</td><td></td><td></td></tr>
</table>

Bubble plume 1 (BP1), bubble plume 2 (BP2), homogenously aerated bubble column (Hom), air lift reactor (AL)

2.3. Vertical pipe experimental setup (HZDR)

The experimental facility is a pressurized "*large-diameter*" and "*large-scale*" vertical pipe (Figure 3a, inner diameter 0.548 m, height 4.95 m). The gas is injected in the lower part of the pipe and flows co-currently with the liquid phase (see ref. [12] for more details). Filtered air and deionizer water have been used; their temperatures and pressures were maintained constant at 303 ± 1 K and 40 bar, respectively. Bubble shape data have been obtained in two flow regimes: the homogeneous flow regime (d_o = 0.8 mm - two conditions tested: (i) U_G = 0.00835 m/s, U_L = 0.405 m/s; (ii) U_G = 0.00835 m/s, U_L = 1.017 m/s) and the slug flow regime (d_o = 1.2 mm - U_G = 0.219 m/s, U_L = 0.405 m/s). Bubble shape information have been obtained, at 3.270 m above the gas sparger (which is the fully developed two-phase flow) by using the ROFEX (*Rossendorf Fast Electron beam X-ray tomograph*) technique. This technique, based on a scanned electron beam approach, was further developed towards a fully operable scanner at HZDR. An electron gun provides an electron beam of 150 kV acceleration voltage and up to 10 kW power. The beam is focused onto a circular X-ray target approximately 1 m away from the gun and is circularly scanned around this target by means of a special electron beam deflection system. The scanner is further complemented by a static circular X-ray detector arranged in the opening inside the scanner head. Fast beam deflection (< 8 kHz), small detector pixels (~1.3 mm pitch) and high read-out frequency (1 MHz) give a high scanning frame rate. Its maximum is 8 kHz but since at that rate spatial resolution deteriorates considerably, scanning speeds up to 2 kHz are used. The measurements were performed in two layers with a distance of 0.102 m to evaluate interfacial velocities. The raw date were processed by a tomographic reconstruction resulting in two stacks of cross-sectional images of the gas volume distribution according to the two measuring layers. A binarisation procedure described by Banowski et al. [13] is applied and single bubbles are identified. By using these data the bubble size distributions as well as bubble shapes are determined. Figure 3b show three-dimensional reconstructed pictures of the flow from ultrafast X-ray tomography measurements, when the time-resolved 2D-Images of the measurement cross-sectional plane get stacked on top of each other.

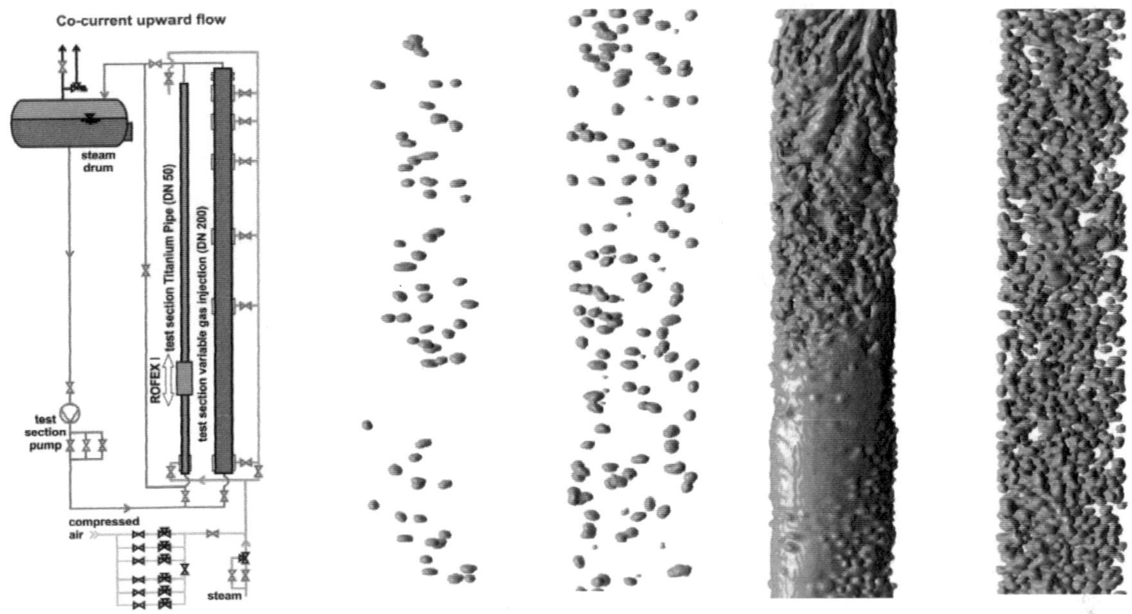

(a) Experimental setup [13]   *(b) Reconstructed 3D-Images of the homogeneous flow regime*   *(c) Reconstructed 3D-Images of the slug flow regime*

Figure 3. HZDR pipe flow: experimental setup (a) and post-processing methods (b and c).

3. Experimental results

Herein the experimental results are discussed. First, the influence of the gas and liquid superficial velocities on bubble shape is discussed. Second, the influence of the experimental setup on bubble shape is discussed. Finally, the data are compared with literature correlations. It is worth noting that, the aspect ratio data are generally scattered broadly; therefore, the data are grouped into classes of equivalent diameters. Each class has been represented by the average aspect ratio of the bubbles belonging to that class.

3.1. Influence of the gas and liquid superficial velocity

As discussed by Zieghenein and Lucas [8], the study of the relationships between the aspect ratio E and the gas and liquid superficial velocities provides information concerning if and in which extent the bubble shape is affected when the flow conditions are changed. Nevertheless, with this approach it cannot be determined to which extent local flow values like the shear rate influence the bubble shape. In the worst case, it is possible that different effects are compensated by just increasing the flow rate.

3.1.1. Bubble column experiments

Figure 4 and Figure 5 display the relationships between the aspect ratio E and the gas and liquid superficial velocities for the PoliMi and HZDR (airlift and BP1) bubble columns. The general trend of almost spherical small bubbles and ellipsoidal large bubbles has been observed for all the flow conditions in all the experimental setups. Indeed, in all the experimental setups, a general relationship between the Eötvös number (Eo, related to the bubble size) and E seems to exist: small bubbles have a high aspect ratio, whereas larger bubbles seem to be characterized by lower aspect ratios. Generally, bubbles with equivalent diameters less than 1 mm have an aspect ratio greater than 0.7. This means that the small bubbles tend to be spherical; conversely, bubbles with higher equivalent diameters are characterized by lower aspect ratios (E in the range of 0.4 - 0.7), which reveals the trend of larger bubbles to be flatter. It is also observed that, in all the different experimental setups considered, changing the gas and liquid superficial velocities does not (generally) influence the bubble shape significantly overall.

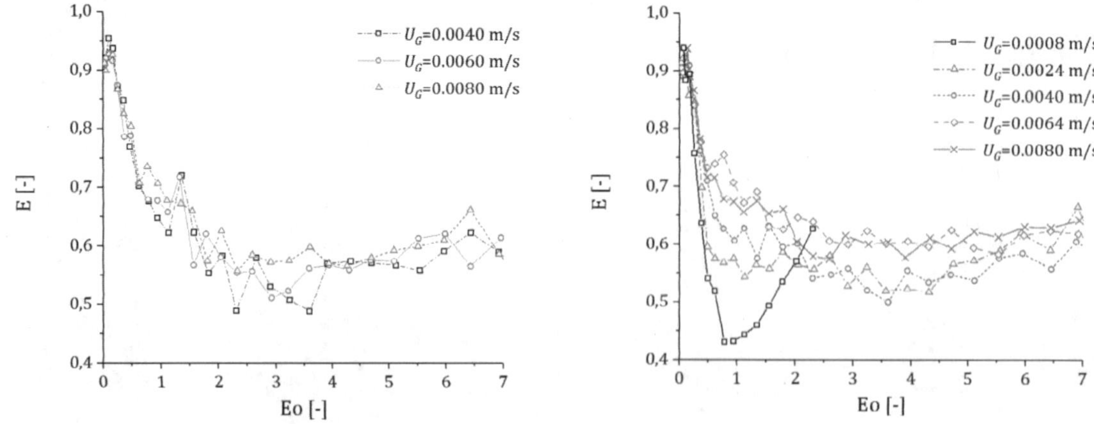

(a) Airlift configuration ***(b) Bubble plume experiment BP1***

Figure 4. Influence of the gas superficial velocity (HZDR bubble columns) on E - batch mode.

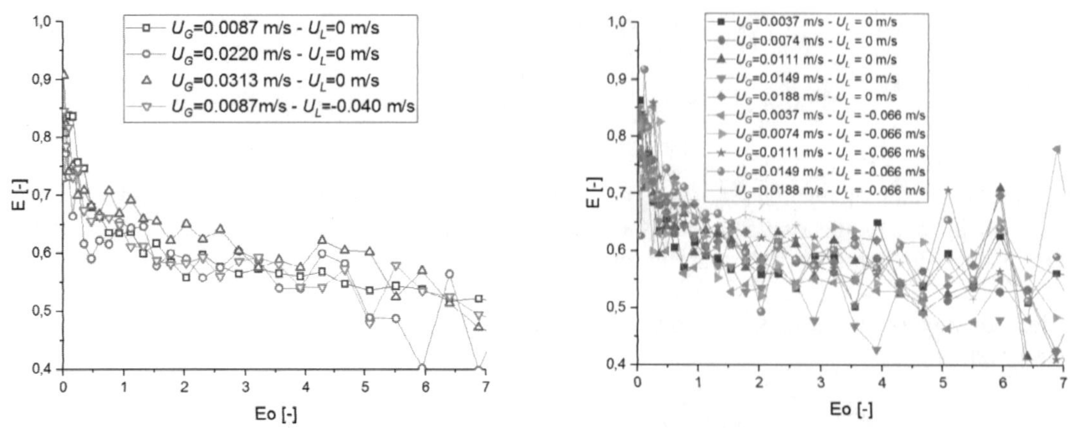

(a) Annular gap bubble column ***(b) Open tube bubble column***

Figure 5. Influence of the gas and liquid superficial velocities on E (PoliMi bubble columns).

However, looking closer to the different experimental setups, some additional considerations can be drawn. Considering the airlift configuration (Figure 4a), since the bubble shape is not changing with higher gas superficial velocities, it might be concluded that the bubble shape depends rather on the bubble size distribution and the prevailing flow regime rather than the void fraction (in this respect, see the data and the discussions in ref. [8]). Conversely, the BP1 experiment (Figure 4b) shows a different picture: the aspect ratios increase while increasing the flow rates, and they approach each other for larger bubble sizes. To explain this behaviour, Zieghenein and Lucas [11] compared the flow field of the bubble plume with the airlift reactor; they observed a distinct backflow at the walls with a slightly steeper averaged velocity profile in the airlift configuration; conversely, normal Reynolds stress tensor components ($v'v'$) are similar. The $v'v'$ profiles in the BP1 experiment originate predominantly from the large-scale transient swinging motion of the bubble plume, especially at low gas flow rates. In addition, the bubble size distribution in the BP1 experiment is changing with the flow rates in contrast to a constant bubble size distribution in the airlift reactor for all the flow rates. Since the bubble shape is not changing with higher flow rates in the airlift reactor, it might be concluded that the bubble shape depends rather on the bubble size distribution and present flow regime than the void fraction for this case (see the discussion in ref. [11]). However, due to the larger bubbles in a wide bubble size distribution, the bubbly flow can be considered "*instable*" (see the lift force discussion proposed in ref. [1]) so that higher local velocity gradients, possible more bubble-bubble interactions, and higher turbulence can be expected. Similar conclusions can be applied to the PoliMI bubble columns in both the batch and in the counter-current mode. It is worth noting that the study of the counter-current mode is interesting, as it produce an additional contribution to the liquid velocity; indeed, in the batch mode the liquid velocity field is completely determined by the recirculation induced by buoyancy counter-current mode. The PoliMI bubble columns are characterised by poly-dispersed bubble size distributions (see, for example, ref. [1], summarizing all our previous studies) that shift towards larger diameters when increasing the gas and liquid superficial velocities (see our previous papers). In the PoliMi bubble columns the bubble shape is not changing with higher gas and liquid superficial velocities, even if the open tube data are more scattered compared with the annular gap data. This observation can be explained considering the higher liquid superficial velocities in the open tube configuration (higher turbulence) and the larger spider sparger openings (resulting in a more poly-dispersed bubble size distribution). Taking into account our other papers focused on the influence of the liquid phase properties [1], we may speculate that the relationship between bubble sizes and shapes (in the homogeneous flow regime) mainly depends upon the system considered (i.e., the liquid phase properties) and not on the flow conditions. This statement has been verified for the PoliMI configuration and its validity to other systems should be checked and verified by means of future studies.

3.1.2. Pipe flow experiments

Figure 6 displays the relationships between the aspect ratio E and the gas and liquid superficial velocities for the pipe flow experimental setup. In this case, a larger dependency of the E on the flow rates has been observed, if compared with the previously reported experimental results (Section 3.1.1). When considering the data in Figure 6, it is worth noting that, when computing the Eötvös number, beside the bubble size information, physical properties of the air-water system at 303 ± 1 K and 40 bar are needed. It is known that the surface tension is influenced more by temperature than by pressure and, for this reason, it has been estimated based on the data provided in ref. [14].

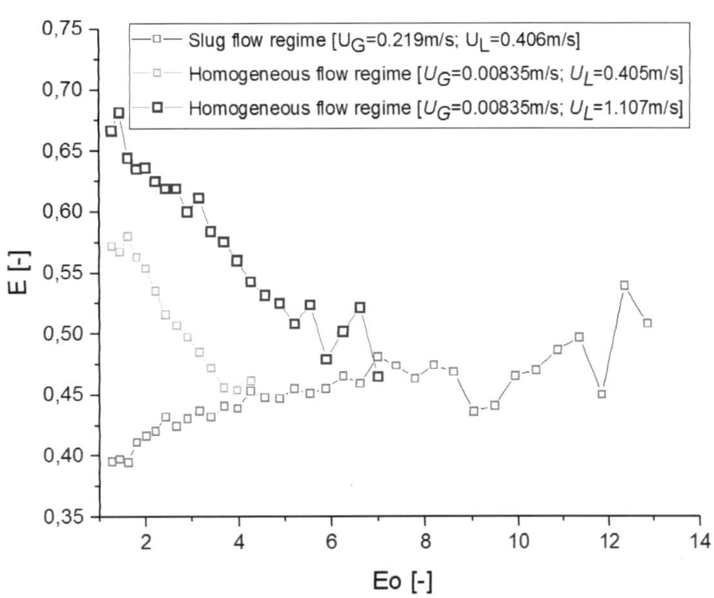

Figure 6. Relationship between *Eo* and E in pipe flows.

When considering the homogeneous flow regime, we observed that (a) E decreases while increasing Eo (larger bubbles tends to be flatter) and (b) E decreases while increasing U_L. It is worth noting that the relationship between E and Eo follows nearly the same trend between the two liquid superficial velocities: increasing U_L, increases the bubble deformation, but the trend does not change. Conversely, the slug flow experiments show a complete different trend: (a) small bubbles are characterised by lower E (b) increasing the bubble size, E increases. These experiments are characterized by a high gas holdup (up to 50 %). Therefore, bubble interactions may have larger influence on the bubble shape as well as wake effects and induced turbulence of larger bubbles. Interestingly, the shape of the smaller bubbles is slightly more influenced by this effect compared to larger bubbles. It is worth noting that pipe flow experiments have bene performed at high U_L; the influence of U_L on the fluid dynamics is probably due to the comparable order of magnitude of the liquid and gas velocities: it is known that if U_L is low compared with the bubble rise velocities, no impact of U_L on the fluid dynamics is expected because the acceleration of the bubbles will be negligible [15, 16].

3.2. Influence of the experimental setup

The conclusion that may be drawn from the previous sections is that in bubble columns the assumption of a constant shape regardless of the flow conditions is valid whereas in pipe flows the turbulence and shear rates can be strong enough to deform the bubble distinctly. In this section, we focus on the influence of the experimental setup on the relationship between E and Eo in bubble column flows. Indeed, besides the influence of the gas flow rate, also the experimental setup determines the flow conditions. In the homogenously aerated column large-scale fluctuations are not dominant, whereas such fluctuations dominate the bubble plume experiments as well as in the PoliMi bubble columns (where further experimental investigation on the liquid local velocities are planned). Inside the riser of airlift reactors, a distinct background flow is observed so that large scales (bubble plume) and small scales are present. The different single-phase background flows, on the other hand, completely determine the single bubble experiments. The single needle experiments can be interpreted as bubble plumes at very low gas flow rates; the bubbles do not ascent in a steady state but might be still less influenced by the background flow. Comparing E of the different experiments, which are averaged over all gas flow rates and needle setups, relatively small but distinct differences can be observed (Figure 7): generally, different aspect ratios are observed in different experimental setups.

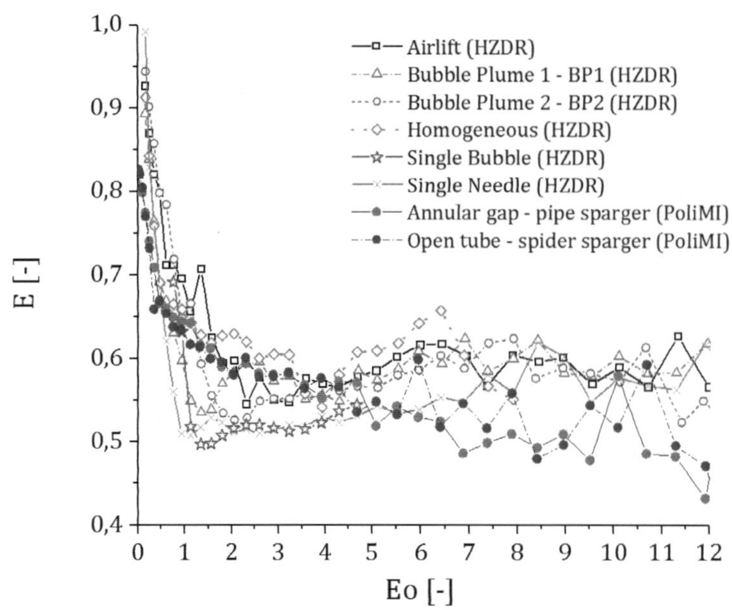

Figure 7. Relationship between Eo and E for the different bubble columns experiments.

The PoliMi bubble column exhibit the smallest aspect ratios for $Eo > 5$ (possibly owing to the very poly-dispersed homogeneous flow regime). Subsequently, the single needle and single bubble experiments exhibit slightly higher aspect ratios compared with PoliMi bubble columns. In addition, the single needle experiment and the single bubble experiment with background flow, however, are almost equal. Interestingly, the aspect ratios measured in the homogenously aerated bubble column are comparable to the aspect ratios observed in these two experiments. In addition, the reproducibility of the experiments is indicated by the almost equal ratios obtained for the two different bubble plume experiments. As would be expected, the aspect ratios observed in the airlift reactor, in which the highest turbulence was measured, are mostly the highest. Generally, distinct differences are observed at smaller bubbles, whereas the aspect ratios approach each other with increasing bubble size.

3.3. Comparison with the literature

Several attempts have been made in the literature to relate the aspect ratio to dimensionless parameters. In this respect, many correlations to relate bubble sizes and shapes were proposed in the last decades; however, most of the studies dealt with single rising bubbles by experimental and numerical approaches and reliable correlations for swarm of bubbles are still not available (see the literature survey in ref. [17] and the analysis of Zieghenein and Lucas [8]). However, when considering dense bubbly flows, these correlations may not be suitable (as deeply discussed by Besagni and Inzoli [1, 17]). The goal of this section is to compare experimental data previously presented and some literature correlations. Detailed studies about the bubble shape from steady state rising single bubbles in quiescent flow can be found in the literature. To compare the results from the present study with these studies, the aspect ratios obtained from the single experimental setups are averaged, thus providing a very large dataset: (a) approximately 7000 and 5000 bubbles in the Polimi open tube and annular gap configurations, respectively; (b) concerning the HZDR experiments, over one million bubbles in the single bubble experiments and around twenty thousand bubbles measured for the airlift experiments. The experimental data have been compared with (a) the fit of Besagni and Inzoli [17] (which has been obtained for dense bubbly flows in annular gap bubble column), (b) the well-known correlation of Wellek et al. [18] and (c) the correlation for super-purified water measured by Sanada et al. [19]. Figure 8 compares the different correlations and the experimental datasets.

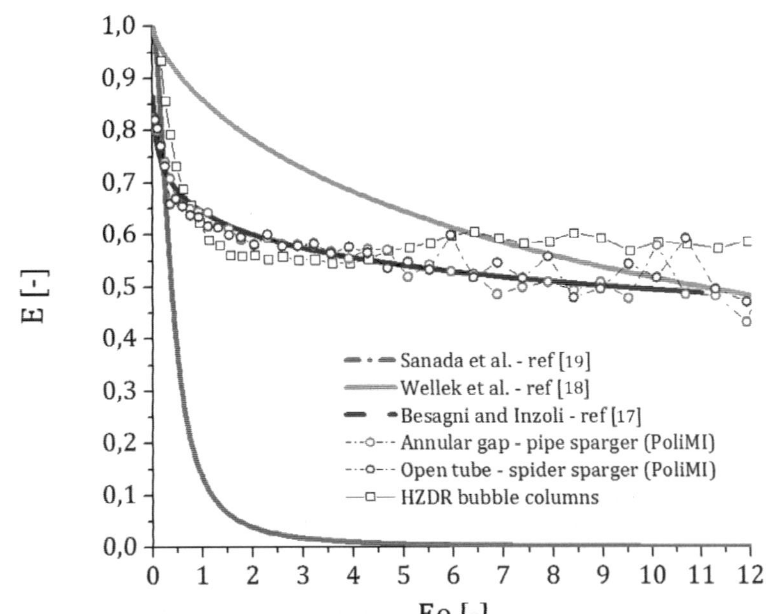

Figure 8. Comparison between literature correlations and aspect ratio experimental data.

The Wellek et al. [18] correlation overestimates the aspect ratio, the Sanada et al. [19] correlation is unable to correlate the aspect ratio; conversely, the correlation proposed by Besagni and Inzoli [17] (obtained for dense bubbly flows) fits the data fairly well. This suggests that correlations taken from the literature should be carefully evaluated before being applied to dense bubbly flows. In future studies, the present dataset should be coupled all together to further improve the scheme of correlation which is under development by some of the authors: indeed, the correlation of Besagni and Inzoli [17], firstly developed for swarm of bubbles in an annular gap bubble column (Figure 1a), has been later modified in a more complex analytical form and has been applied to binary liquid phases to consider active compound systems and viscous liquid phases (see the discussion in Section 4.3.3 of ref. [1]).

4. Conclusions

This paper contributes to the existing discussion concerning the relationships between the bubble shape, size and flow conditions. In particular, we have experimentally evaluated the bubble shapes (in terms of the aspect ratio), at low Morton-numbers, in different bubble column setups and in a pipe flow setup under different operating conditions (representative of real dense bubbly flows). In the bubble column experiments almost the same shape is observed; conversely, the shape in the pipe flows distinctly depends on the flow conditions. Therefore, we may conclude that in bubble columns the assumption of a constant shape regardless of the flow conditions is valid whereas in pipe flows the turbulence and shear rates can be strong enough to deform the bubbles distinctly. In addition, we have demonstrated the importance of developing bubble shape correlations based on data obtained in dense bubbly flows. Future studies will be devoted to extend the present dataset with bubble shape data obtained in binary system and, finally, to propose a general correlation to estimate the bubble shape in real flow conditions.

Acknowledgement

Giorgio Besagni is deeply indebted to Prof. Paolo Di Marco for all the advices and discussions.

References

[1] Besagni G and Inzoli F 2017 *Chem. Eng. Sci.* **170** 270-296
[2] Ellingsen K and Risso F 2001 *J. Fluid Mech.* **440** 235–68
[3] Clift R, Grace J R and Weber M E 1978 *Bubbles, drops, and particles* (New York)
[4] Loth E 2008 *Int. J. Multiphas. Flow* **34** 523-46
[5] Liu L, Yan H and Zhao G 2015 *Exp. Therm. Fluid. Sci.* **62** 109-21
[6] Aoyama S, Hayashi K, Hosokawa S and Tomiyama A 2015 *Exp. Therm. Fluid. Sci.* **79** 23-30
[7] Aoyama S, Hayashi K, Hosokawa S, Lucas D and Tomiyama A 2017 *Int. J. Multiphas. Flow.* **96** 113-122
[8] Ziegenhein T and Lucas D 2017 *Exp. Therm. Fluid. Sci.* **85** 248-256
[9] Ziegenhein T, Zalucky J, Rzehak R and Lucas D 2016 *Chem. Eng. Sci.* **150** 54-65
[10] Ziegenhein T and Lucas D 2016 *Flow. Meas. Instrum.* **48** 36-41
[11] Ziegenhein T, Garcon M and Lucas D 2016 *Chem. Eng. Sci.* **153** 155-64
[12] Banowski M, Beyer M, Lucas D, Hoppe D and Barthel F 2016 *Experiments on vertical gas-liquid pipe flows using ultrafast x-ray tomography* (Technical report HZDR-075, Helmholtz-Zentrum Dresden – Rossendorf)
[13] Banowski M, Lucas D and Szalinski L 2015 *Int. J. Therm. Sci.* **90** 311-22
[14] Vargaftik N B, Volkov B N and Voljak L D 1983 *J. Phys. Chem. Ref. Data* **12** 817-20
[15] Hills J H 1976 *Chem. Eng. J.* **12** 89-99
[16] Rollbusch P, Becker M, Ludwig M, Bieberle A, Grünewald M, Hampel U and Franke R 2015 *Int. J. Multiphas. Flow* **75** 88-106
[17] Besagni G and Inzoli F 2016 *Exp. Therm. Fluid. Sci.* **74** 27-48
[18] Wellek R M, Agrawal A K and Skelland A H P 1966 *AlChE J.* **12** 854–862
[19] Sanada T, Sugihara K, Shirota M and Watanabe M 2008 *Fluid Dyn. Res.* **40** 524-545

IOP Conf. Series: Journal of Physics: Conf. Series **923** (2017) 012015 doi:10.1088/1742-6596/923/1/012015

Flow boiling of R32 in a horizontal smooth tube of 6.0 mm internal diameter: heat transfer coefficient and pressure drop

R MASTRULLO, A W MAURO and L VISCITO

Department of Industrial Engineering, Federico II University of Naples, P.le Tecchio 80, 80125, Naples (Italy)

E-mail: alfonsowilliam.mauro@unina.it

Abstract. This paper presents a series of experiments on flow boiling heat transfer coefficient and pressure drop using refrigerant R32. Tests were performed in a smooth horizontal stainless steel tube with an internal diameter of 6.00 mm, with the heat directly applied on the test section tube through DC current. Trends of the heat transfer coefficient and pressure drop data are displayed for different operating conditions. Specifically, the saturation temperature was fixed to 25 and 35 °C, the heat flux varied from 5 to 50 kW/m^2 and the mass velocity was fixed to the low values of 150 and 250 kg/m^2 s. The effects of the operating parameters on the heat transfer results are also shown.

1. Introduction

The research on flow boiling of refrigerants is of primary importance in several fields, such as air conditioning, refrigeration, nuclear and cooling electronic systems. Therefore, the capability to determine the two-phase heat transfer and pressure drop behavior is important for the correct design of evaporators and heat spreaders systems. The European Union has already planned a gradual removal of high-GWP substances in the next years [1] and R32 is one of the candidates for the substitution of R410A in the air-conditioning sector, due to its relatively low GWP index of 675 when used as pure fluid. Moreover, the peculiar thermodynamic characteristics of R32 make its employment in domestic refrigerators and roof-top systems particularly appealing [2]-[3], due to a reduction of the fluid mass charge, a higher compactness of heat exchangers and an overall increased efficiency.

Different studies are accessible in open literature coping with flow boiling of refrigerants in macro and mini tubes of different geometries and configurations, but relatively few data are currently available for refrigerant R32 and most of them present flow boiling data for R32 included in HFC mixtures.

Shin et al. [4] studied convective boiling heat transfer of different pure refrigerants and refrigerants mixture, including R32, in a horizontal seamless stainless steel tube with an inner diameter of 7.7 mm. The saturation temperature was fixed to 12 °C, whereas the heat flux and mass flux varied from 10 to 30 kW/m^2 and from 424 to 742 kg/m^2 s, respectively. The authors found the heat transfer coefficient strongly dependent on heat flux in the low-quality region and almost constant at higher vapor qualities.

Li et al. [5] presented flow boiling data for R1234yf/R32 mixtures in a smooth horizontal tube with an inner diameter of 2 mm. The experiments were conducted for two mass concentrations (80/20 and 50/50), by varying the heat flux from 6 to 24 kW/m^2 and the mass flux from 100 to 400 kg/m^2 s. The evaporation temperature was fixed to 15 °C. The authors found that the heat transfer coefficients of the mixtures were about 20-50% lower than those of pure R32.

Kondou et al. [6] studied flow boiling heat transfer coefficient and pressure drop of R1234ze/R32 mixtures in a microfin tube of 5.21 mm internal diameter at a saturation temperature of 10 °C. They varied the mass flux from 150 to 400 kg/m^2 s, for two different heat fluxes of 10 and 15 kW/m^2. In case of pure R32, the authors found a small influence of both vapor quality and mass velocity.

Azzolin et al. [7] presented flow boiling data of a zetropic mixture of R1234ze/R32 (50/50) in a single microchannel of 0.96 mm internal diameter. Tests were conducted at a saturation temperature of 26.2 °C, mass velocities from 300 to 600 kg/m^2 s and heat fluxes from 27 to 165 kW/m^2. The authors found a strong influence of the imposed heat flux above all other parameters, especially in the low-quality region.

Jige et al. [8] studied flow boiling of pure R32 in smooth tubes of different diameters (1.0, 2.2 and 3.5 mm), examining the influence of mass velocity (from 50 to 600 kg/m^2 s) and of heat flux (from 5 to 40 kW/m^2) on the heat transfer coefficient and the pressure drop. They found that the heat transfer coefficient increased with reduced tube diameter and that the pressure drop was enhanced for higher mass velocities and vapor qualities and lower tube diameters.

This work presents two-phase flow boiling heat transfer coefficient and pressure drop data for refrigerant R32 in a smooth horizontal stainless steel (SS316) tube with an internal diameter of 6.00 mm. Heat transfer coefficient data are taken at both top and bottom surface of the heated tube. Different trends were shown at several operating conditions, by changing the saturation temperature (25 and 35 °C), heat flux (from 5 to 50 kW/m^2) and mass velocity (150 and 250 kg/m^2 s), to focus the attention on the effect of the heat flux and of the asymmetric annular flow structure on the local heat transfer coefficients.

2. Experimental apparatus

The test facility has been assembled in the Refrigeration Laboratory, at the Industrial Engineering Department, Università degli Studi di Napoli Federico II. The experimental test rig consists of a main refrigerant loop, in which the test section inlet conditions in terms of saturation temperature, mass flux and vapor quality are set and monitored, and a secondary water close loop, whose temperature is controlled by a thermostatic bath. Both loops are schematized in Fig. 1.

2.1. Main and secondary loops description

The black line in Fig. 1 portrays the main refrigerant loop, that is made up of a magnetic gear pump, a preheater, a diabatic test section, a brazed plate condenser, a tube-in-tube sub-cooler, a liquid receiver and a micrometric throttling valve.

The sub-cooled refrigerant passes through the magnetic gear pump, whose velocity can be set from 1650 to 3400 rpm by means of an inverter coupled with the electrical motor, able to provide volumetric flow rates within a range of 1.3-2.5 dm^3/min. A throttling valve on the liquid line is employed during the experiments to adjust the system pressure and the mass flow rate to the desired values. The liquid then enters the preheater section, in which the heat is supplied by two fiberglass heating tapes (each of them with a nominal heat power of approximately 900 W at 240 V and 25 °C, as indicated by the manufacturer). The fluid then passes through an adiabatic, smooth, horizontal part, whose length is able to obtain a fully developed flow at the inlet of the tube section. A diabatic test section of 192.5 ±0.47 mm allows the heat transfer coefficient and pressure drop measurements. Another micrometric throttling valve at the test section outlet is used for the single-phase tests to adjust the system pressure and mass flow rate. The liquid/vapor refrigerant mixture is condensed with a plate heat exchanger and then flows into a liquid receiver. The working fluid is then sub-cooled by means of a double pipe heat exchanger before the pump suction head which closes the loop. A by-pass circuit that recirculates part of the refrigerant flow into the liquid receiver allows the investigation of very low mass fluxes.

The secondary fluid in the heat exchangers is demineralized water, whose temperature is controlled by a thermostatic bath. Its circuit is displayed with a blue line in Fig. 1. Both the tube-in-tube and the brazed plate heat exchangers may be excluded from the water flow by means of two ball-cock valves. The secondary fluid specific volume variations are restricted with the use of an expansion vessel.

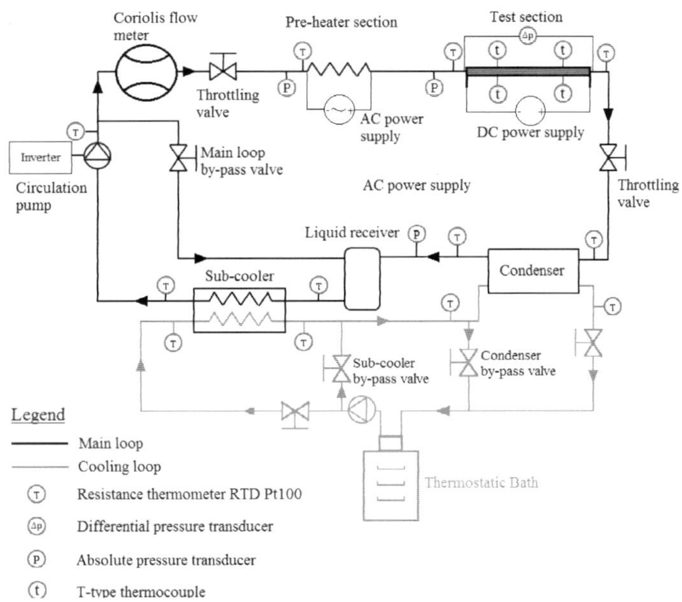

Figure 1. Schematic view of the experimental plan

2.2. Test section

The test section employed in this work is a smooth, horizontal, circular stainless steel (type 316) tube with an internal diameter of 6.00 ±0.05 mm and an outer diameter of 8.00 ±0.05 mm. Fig. 2 exposes a picture and a schematic view of the test section with its geometric characteristics. The heat is applied to the fluid by Joule effect, by means of a DC power supply unit and two copper electrodes welded on the external tube surface (see points A and F in Fig. 2), placed at a distance of 192.5 ±0.47 mm one another. At approximately the same distance, two ports are employed for the pressure drop measurements. Two different measurement positions are provided along the tube, which are respectively 77.2 ±0.41 mm and 154.5 ±0.61 mm far from the diabatic inlet section (see C and D in Fig. 2). At each point, two T-type thermocouples are placed on the top and bottom surface of the tube for the measurement of the outer wall temperature. A high temperature epoxy resin and a Kapton adhesive layer guarantee the sensors fastening and their electrical insulation from the heated tube. The measurement points for the DC voltage is not located on the copper electrodes, since they might suffer of locally concentrated tension drop. Two measurement wires were instead clamped at a certain distance (see points B and E in Fig. 2) to guarantee the heat flux uniformity. The SS316 test was supplied with DC current (in the range 0-8 V and 0-300 A), and its electrical resistance is estimated to be 5.1 ±0.084 mΩ. A suitable amount of synthetic rubber (λ = 0.040 W/m K at 40 °C) covers the test section and the whole experimental facility in order to minimize the heat losses. The preheater section, due to the higher thermal loads handled, was firstly covered by a high-temperature insulation wool (λ = 0.050 W/m K at 200 °C).

IOP Conf. Series: Journal of Physics: Conf. Series **923** (2017) 012015 doi:10.1088/1742-6596/923/1/012015

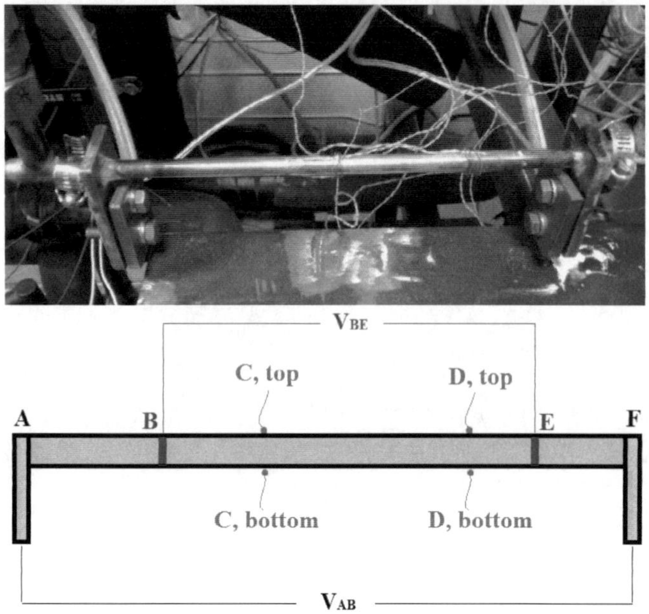

Geometrical features	
A, F	copper electrodes
B, E	voltage measurement points
C, D	temperature measurement points
AF	192.5 ±0.47 mm
BE	121.0 ±0.39 mm
AC	77.2 ±0.41 mm
AD	154.5 ±0.61 mm

Figure 2. Geometrical details of the test section

2.3. Measurement instrumentation
With the exception of the test section, all the temperature measurements in the test rig were obtained with high-accuracy flat ceramic resistance thermometers, carrying an overall uncertainty of ±0.180 °C.

The absolute pressure measured at the inlet of the test section, preheater and liquid receiver were obtained with absolute pressure transducers within the range 0-50 bar. Their overall uncertainty, taking into account the non-linearity and repeatability effects is ±0.5% of the read value.

Four T-type thermocouples were employed for the outer wall test tube temperature measurements. All sensors were calibrated in the range 5-55 °C by using a high-accuracy resistance thermometer. Their B-type uncertainty is assumed to be ±0.1 °C.

The mass flow rate was measured by means of a Coriolis flow meter, calibrated up to 2% of the full scale, with a maximum uncertainty of 1% of the measurement.

Two differential pressure transducers were employed for the measurement of the test section pressure drop. The transducers were calibrated with two different full scale of 30 kPa and 50 kPa, respectively, and with an overall uncertainty of ±0.05 kPa.

A digital wattmeter provided the heat load applied to the preheater section, by separately measuring the voltage (100 mV-500 V) and current (1 mA-16 mA). Its uncertainty is 1% of the reading, as indicated by the manufacturer.

The DC voltage applied to the test section was evaluated with an electrical voltage transducer of 0-5 V carrying an uncertainty of ±0.03% of the reading, while the DC current was directly measured by the DC power supply unit (0-300 A). The manufacturer guarantees an overall uncertainty of 1% of the measurement.

2.4. Experimental procedure and data acquisition
All two-phase test were recorded in steady state conditions. The saturation temperature, mass flux, heat flux and vapor quality were remotely set by Labview software and Arduino One controller.

Specifically, the saturation temperature was imposed by setting the water temperature in the thermostatic bath; the mass velocity was instead controlled by changing the inverter frequency of the motor coupled with the magnetic gear pump. Small adjustments of the mass flow rate and system pressure (i.e. saturation temperature) were possible by manipulating the main circuit by-pass valve and

IOP Conf. Series: Journal of Physics: Conf. Series **923** (2017) 012015 doi:10.1088/1742-6596/923/1/012015

the micrometric throttling valve. The heat flux applied to the test section was imposed by remotely changing the voltage from the DC power supply unit. Precisely, the low electrical resistance of the steel tube required very low voltages (up to 1.0 V) and high currents (more than 100 A). The fiberglass heating tape of the preheater section was instead supplied with AC power. By varying the applied voltage by means of a TRIAC electronic unit (up to 240 V), different thermal loads were set to obtain a desired vapor quality at the inlet of the test section.

Data from sensors were recorded with an acquisition frequency of 1.0 Hz and the arithmetic average over 5 minutes was taken as nominal value of the sample. The system was considered stabilized when the calculated deviation of the main parameters was sufficiently low (±0.2 °C for the wall outer temperature obtained by the T-type thermocouples, ±0.1 bar for the inlet saturation pressure, ±10 kg/m^2 s for the mass velocity, ±1.0 kW/m^2 for the imposed heat flux).

3. Method

3.1. Data reduction and uncertainty analysis

The local heat transfer coefficient is evaluated with the Newton equation:

$$h_i = \frac{\dot{q}}{T_{wall,i} - T_{fluid,i}} \tag{1}$$

where the subscript i refers to the specific measurement location (point C or point D in Fig. 2). The heat flux is evaluated as:

$$\dot{q} = \frac{V_{BE} \cdot I}{\pi d \overline{BE}} \tag{2}$$

in which V_{BE} is the voltage applied in the section $\overline{BE}$ and I is the DC current occurring in the test tube.

The wall temperature $T_{wall,i}$ is evaluated from the measured wall outer temperature $T_{th.i}$ by considering 1-D heat transfer and uniform generation in the metal tube:

$$T_{wall,i} = T_{th.i} + \frac{V_{BE} \cdot I}{4\pi \lambda_{tube} \overline{BE}} \cdot \frac{\left(\frac{D}{d}\right)^2 \cdot \left(1 - \log\left(\left(\frac{D}{d}\right)^2\right)\right) - 1}{\left(\frac{D}{d}\right)^2 - 1} \tag{3}$$

In the above equation, d and D represent the inner and outer diameter of the tube, respectively. λ_{tube} is the tube thermal conductivity, considered equal to 16.26 W/(m K) for all the experiments performed. The fluid saturation temperature $T_{fluid,i}$ at the measurement points is evaluated by considering a linear pressure drop from the tube inlet.

The local vapour quality at the measurement points is:

$$x_i = \frac{i_i - i_{L,i}}{i_{LV,i}} \tag{4}$$

in which $i_{L,i}$ and $i_{LV,i}$ are the local saturated liquid and vapor enthalpies and i_i is the local enthalpy at the measurement point, which is computable as:

$$i_i = i_{in} + \frac{\dot{Q}_i}{\dot{m}} \tag{5}$$

where $\dot{Q}_i$ is the heat load applied up to the measurement section, obtained as a linear interpolation of the total heat applied, $\dot{m}$ is the refrigerant mass flow rate and i_{in} is the test section inlet enthalpy, whose value is obtained from an energy balance applied to the preheater section.

$$i_{in} = i_{in,preh} + \frac{\dot{Q}_{preh}}{\dot{m}} \tag{6}$$

In the above equation, $i_{in,preh}$ is the preheater inlet enthalpy (obtained with measured temperature and pressure) and $\dot{Q}_{preh}$ is the preheater load directly measured by the digital wattmeter.

All refrigerant thermodynamic properties are evaluated with the software REFPROP 9.0, developed by NIST [9]. The whole data reduction is carried out with MATLAB software [10].

The uncertainty analysis of the results was also carried out in this paper. The instrumental B-type uncertainties were composed to the standard deviation evaluated in the recording time for all the measured parameters. For the derived quantities, the law of propagation of uncertainties was implemented. The final results provided a maximum heat transfer coefficient uncertainty equal to ±35% and an average uncertainty of ±15% for all the operating conditions far from the occurrence of dry-out. In this condition, substantial fluctuations of the wall temperatures were detected, increasing the A-type contribution and therefore the overall sample uncertainty. The maximum uncertainty recorded for pressure drop is ±45% in case of low vapor qualities and low pressure gradients, while the average uncertainty in the high vapor quality region was ±8%.

3.2. Test section validation

Preliminary tests were performed in liquid single-phase with refrigerant R134a in order to verify the energy balance in the preheater and in the test section. Moreover, liquid single-phase heat transfer coefficients were obtained and compared with the Dittus-Boelter correlation [11].

Specifically, 16 tests were conducted to verify the correct insulation of the preheater, using mass flow rates from 30 to 40 g/s, heat loads from 100 to 600 W and temperatures from 32 to 38 °C. The difference between the tube and the ambient temperature is approximately the same encountered in the two-phase tests performed for this paper. Fig. 3a compares the electrical heat imposed versus the effective heat absorbed by the fluid. The heat dispersed, calculated by the energy balance, is approximately 7% of the total heat imposed for all the single-phase tests and therefore the real heat considered in the data reduction has been corrected accordingly.

As regards the test section, 29 liquid single-phase tests with refrigerant R134a were carried out, using mass flow rates from 4 to 31 g/s, heat fluxes from 3 to 42 kW/m^2 and inlet temperatures from 18 to 36 °C. The energy balance led to an agreement always better than 5% and therefore the heat losses were neglected. On the other hand, the liquid heat transfer coefficients compared with the Dittus-Boelter, as portrayed in Fig. 3b, showed deviations lower than 15%.

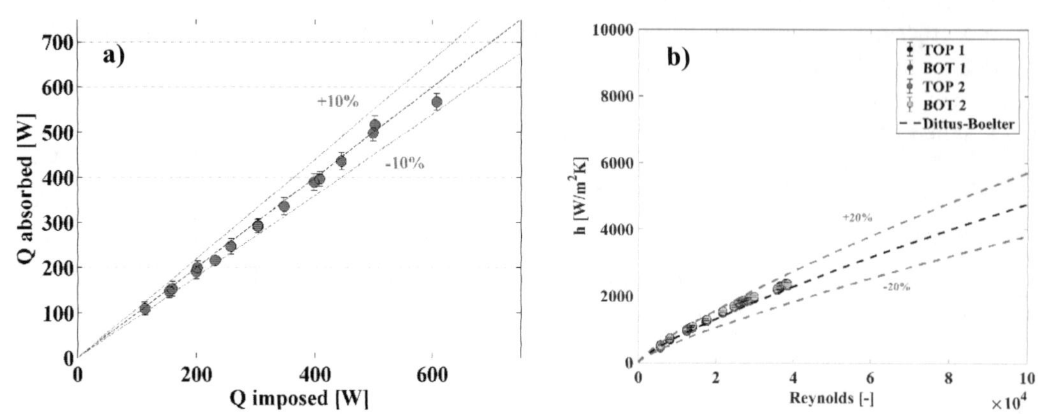

Figure 3. R134a liquid single-phase tests. (a) Energy balance on the preheater section. (b) Experimental and predicted liquid single-phase heat transfer coefficient on test section, for the two measurement points, top and bottom data.

4. Results

4.1. Local heat transfer coefficients

Six different operating conditions were considered for the two-phase heat transfer experiments, by using two different inlet saturation temperatures T_{sat} (25 and 35 °C) and two mass velocities G (150 and 250 kg/m^2 s), by varying the heat flux q from 5 to 50 kW/m^2. Fig. 4 shows the local heat transfer coefficients with their uncertainty for an inlet saturation temperature of 35.1 °C, a mass velocity of 152 kg/m^2 s and an imposed heat flux of 5.1 kW/m^2. The results are shown for measurement points C and D (see Fig.2): as it can be seen, except for the small differences due to vapor quality variations, the trends are in line as a double check of the results from different thermocouples and locations. This behavior was also recorded for other operating conditions and, for this reason, all the remaining diagrams of this section will refer only to the first measurement point C. The top heat transfer coefficients h_{top} tend to initially decrease up to a vapor quality of 0.4, then a mild increase is detected before the final drop, that in this case occurs for x=0.80. The bottom heat transfer coefficients h_{bot} are instead always increasing with vapor quality up to the dry-out condition, which occurs for higher qualities (x=0.92 in this case). Both h_{bot} and h_{top} curves almost blend one another, with the small differences included in the uncertainty range.

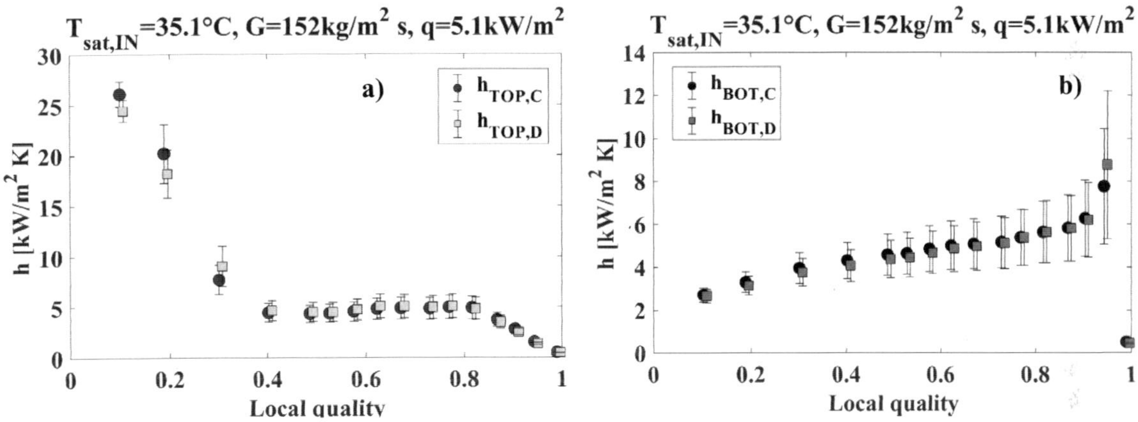

Figure 4. R32 local heat transfer coefficient evaluated at T_{sat} = 35.1°C, G = 152 kg/m^2 s and q = 5.1 kW/m^2 for measurement points C and D. a) top surface. b) bottom surface

The effect of the heat flux on the top and bottom heat transfer coefficient is shown in Fig. 5a, in which the saturation temperature and the mass velocity are fixed to 25.2 and 150 kg/m^2 s, respectively. When the heat flux changes from 5 to 20 kW/m^2, h_{bot} increases (+50% at low vapor qualities and +25% at x=0.87), whereas the heat flux has a negligible effect on h_{top} and the recorded differences between the values are included in the error band, suggesting a poor contribution of the nucleate boiling mechanism on the upper surface of the tube, for this range of values. Remarkably, the top and bottom contributions have different trends: in particular, while that at the bottom the heat transfer coefficient increases monotonically with the vapor quality, the one at the top has a strong dependence on the flow pattern (film thickness and slug frequencies in the slug region; asymmetry and fluid velocity into the asymmetric annular flow: these results are consistent with previous similar studies, Grauso et al. [12]). To highlight the contribution of the heat flux on the heat transfer at the bottom, Fig. 5b displays instead h_{bot} at the same conditions for heat fluxes up to 50 kW/m^2. The enhancement of the heat transfer coefficient is significant, while the vapor quality at the dry-out incipience remains substantially the same (≈0.92) up to 20 kW/m^2, whereas it drops to 0.78 at q=50kW/m^2.

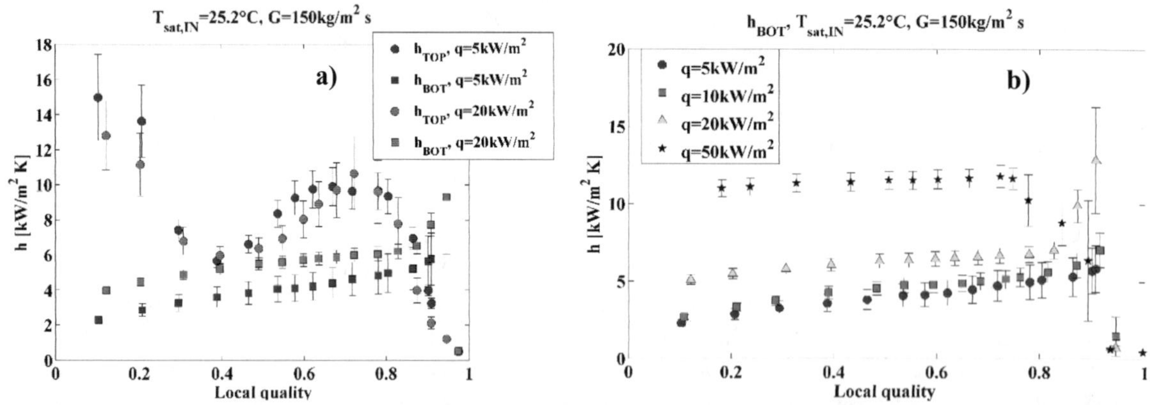

Figure 5. R32 local heat transfer coefficient evaluated at T_{sat} = 25.2°C, G = 150 kg/m^2 s for different heat fluxes. a) Top and bottom heat transfer coefficients for heat fluxes of 5 and 20 kW/m^2. b) Bottom heat transfer coefficients for heat fluxes from 5 to 50 kW/m^2

The effect of the saturation temperature (equal to 25 and 35 °C) on h_{bot} and h_{top} is displayed in Fig. 6, for a heat flux of 20.3 kW/m^2 and an average mass flux of 152 kg/m^2 s. While the bottom heat transfer coefficient (see Fig. 6a) is always enhanced (about +20%) with increasing saturation temperature, the effect on the top heat transfer coefficient (see Fig. 6b) changes according to the vapor quality. For $x < 0.3$, the saturation temperature has a positive effect on h_{top}, while the trend is reversed for higher vapor qualities. As suggested by Grauso et al. [12], a higher reduced pressure causes a higher liquid density and therefore a lower fluid velocity, resulting in a lower heat transfer coefficient during annular flow. Both curves reach the same heat transfer coefficient when dry-out occurs (at approximately $x = 0.90$).

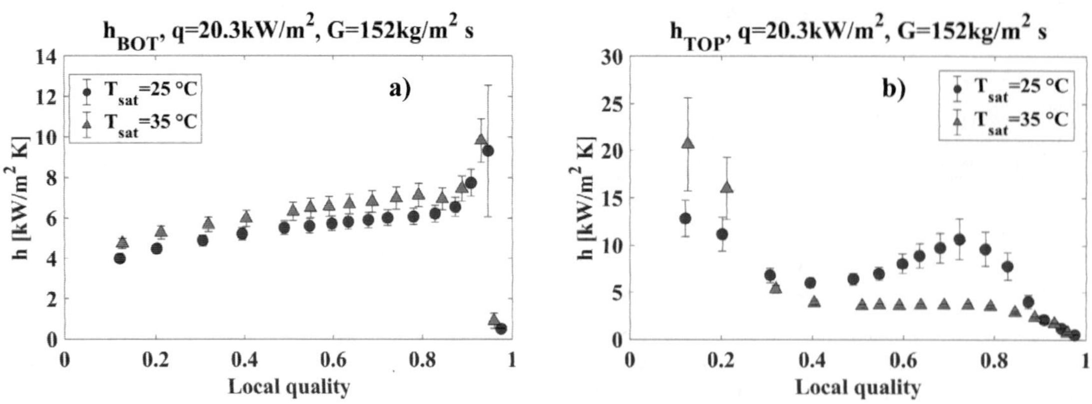

Figure 6. R32 local heat transfer coefficient evaluated at q = 20.3 kW/m^2, G = 152 kg/m^2 s for two different saturation temperatures. a) Bottom heat transfer coefficients for T_{sat} of 25 and 35 °C. b) Top heat transfer coefficients for T_{sat} of 25 and 35 °C.

The effect of mass velocity on bottom and top heat transfer coefficient is portrayed in Fig. 7a and Fig. 7b, respectively. In these tests, the saturation temperature is fixed at 35 °C and the heat flux at 5.1 kW/m^2. As regards the bottom section (see Fig. 7a), the mass flux increases the heat transfer efficiency and its effect is more evident for higher vapor qualities. The top heat transfer coefficient (see Fig. 7b) is instead significantly improved (more than +100%) with increasing mass flux in case of vapor qualities included between $x = 0.25$ and $x = 0.90$, whereas h_{top} is penalized for lower vapor qualities.

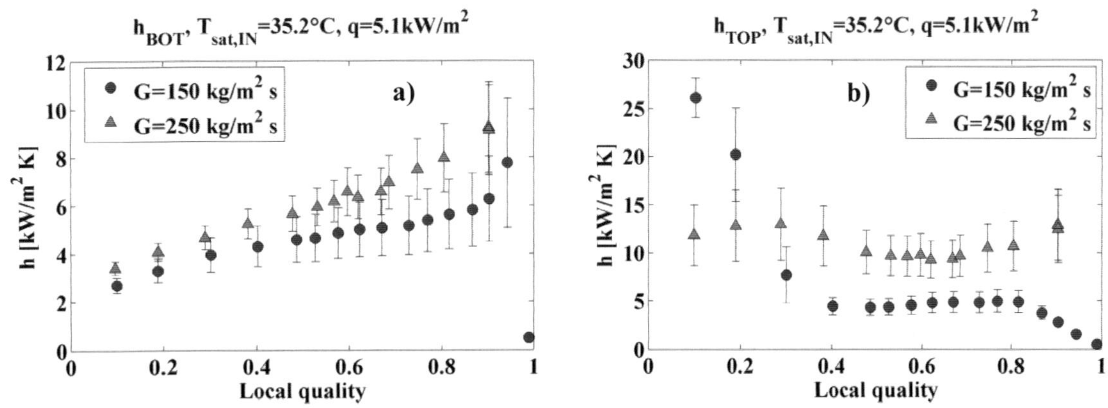

Figure 7. R32 local heat transfer coefficient evaluated at T_{sat} = 35.2°C, q = 5.1 kW/m² for different mass fluxes. .a) Bottom heat transfer coefficients for G equal to 150 and 250 kg/m² s. b) Top heat transfer coefficients for G equal to 150 and 250 kg/m² s.

4.2. Adiabatic pressure gradients

Under the same operating conditions for the heat transfer, data were also used for the evaluation of the adiabatic pressure gradient along the test section. The following diagram refers to the total pressure gradient measured by the pressure transducer with the lower full scale of 30 kPa and divided over the lube length. Due to the very small variation of the vapor quality between the inlet and outlet section of the tube under adiabatic conditions, the momentum contribution was neglected.

Fig. 8 exposes the total pressure gradient as a function of the local vapor quality, for a saturation temperature of 35 °C. The blue and red markers refer to mass velocities of 150 and 250 kg/m² s, respectively. As expected from previous studies [13], the pressure drop increases with ongoing evaporation for both mass fluxes, even if the influence of vapor quality is greater for the higher mass flux of 250 kg/m² s. Pressure drop is also significantly increased with mass velocity: at x = 0.6, the pressure gradient for G = 250 kg/m² s is approximately 180% higher than that obtained for G = 150 kg/m² s. Finally, the maximum was found for lower a lower vapor quality (0.78 instead of 0.82) for G = 250 kg/m² s.

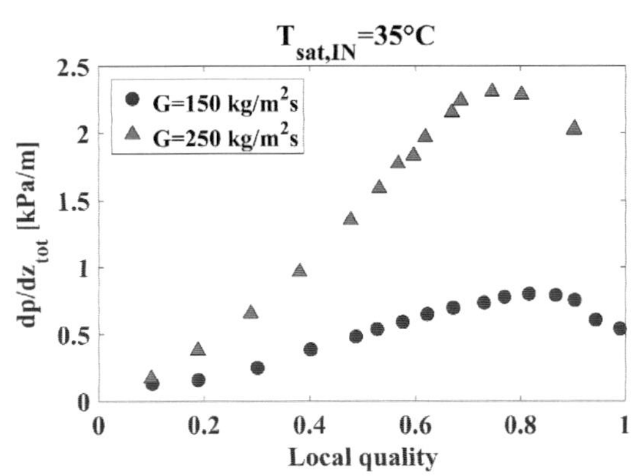

Figure 8. R32 total pressure gradient vs local vapor quality, for T_{sat} = 35 °C. The mass velocity is fixed to 150 and 250 kg/m² s.

5. Conclusions

New experimental data on local two-phase heat transfer coefficients and adiabatic pressure gradients were collected in this work, for a horizontal tube of 6.00 mm internal diameter. The main conclusions can be summarized as follows:

- The heat flux enhances the bottom heat transfer coefficient and has instead a poor effect on h_{top}, which does not seem to be significantly affected by the nucleate boiling contribution. Consistently, also a variation from low to higher reduced pressures lead to higher bottom heat transfer coefficients, whereas h_{top} is conversely reduced (at least in the annular flow region) due to the reduction of the liquid velocity caused by the increased liquid density. Also, the trend of h_{top}, being this heat transfer contribution related only to the convection, is strongly dependent on the flow regime.
- The bottom heat transfer coefficient is increased with higher mass flux, with greater differences for higher vapor qualities, whereas h_{top} is enhanced only in the annular flow region.
- The total pressure gradient is significantly dependent on vapor quality, showing a maximum for vapor qualities approximately equal to 0.80. The mass velocity (from 150 to 250 kg/m^2 s) also increases the pressure drop, which passes from 0.5 to 1.6 kPa when the vapor quality is fixed to 0.60.

References

[1] Regulation (EU) No 517/2014 of the European Parliament and the Council of 16 April 2014 *Off. J. Union.*
[2] Shuxue X, Guoyuan M, Qi L and Zhongliang L 2013 *Int. J. Th. Sc.* **68** 103
[3] Han X H, Wang Q, Zhu Z W and Chen G M 2007 *App. Th. Eng.* **27** 2559
[4] Shin J Y, Kim M S and Ro S T 1997 *Int. J. Ref.* **20** 267-275
[5] Li M, Dang C and Hihara E 2012 *Int. J. Heat Mass Tr.* **55** 3437
[6] Kondou C, Baba D, Mishima F and Koyama S 2013 *Int. J. Ref.* **36** 2366
[7] Azzolin M, Bortolin S and Del Col D 2016 *Int. J. Th. Sc.* **110** 83
[8] Jige D, Sagawa K and Inoue N 2017 *Int. J. Ref.* **76** 206
[9] Lemmon E W, Mc Linden M O and Huber M L 2009 *Standard Reference Database* **23**
[10] MATLAB release. *Natick, Massachussets, United States, The MathWorks, Inc.*
[11] Dittus F W and Boelter L M K 1930 *Univ. Calif. Publ. Engin.* **2** 443
[12] Grauso S, Mastrullo R, Mauro A W and Vanoli G P 2013 *Int. J. Heat Mass Tr.* **56** 107
[13] Grauso S, Mastrullo R, Mauro A W and Vanoli G P 2014 *Exp. Th. Fl. Sc.* **52** 79

R245fa Flow Boiling inside a 4.2 mm ID Microfin Tube

G A Longo, S Mancin, G Righetti and C Zilio

Dept. of Management and Engineering, University of Padova, Str.lla S. Nicola 3, 36100, Vicenza, IT

E-mail: tony@gest.unipd.it

Abstract. This paper presents the R245fa flow boiling heat transfer and pressure drop measurements inside a mini microfin tube with internal diameter at the fin tip of 4.2 mm, having 40 fins, 0.15 mm high with a helix angle of 18°. The tube was brazed inside a copper plate and electrically heated from the bottom. Sixteen T-type thermocouples are located in the copper plate to monitor the wall temperature. The experimental measurements were carried out at constant mean saturation temperature of 30 °C, by varying the refrigerant mass velocity between 100 kg m^{-2} s^{-1} and 300 kg m^{-2} s^{-1}, the vapour quality from 0.15 to 0.95, at two different heat fluxes: 30 and 60 kW m^{-2}. The experimental results are presented in terms of two-phase heat transfer coefficient, onset dryout vapour quality, and frictional pressure drop. Moreover, the experimental measurements are compared against the most updated models for boiling heat transfer coefficient and frictional pressure drop estimations available in the open literature for microfin tubes.

1. Introduction

Microfin tubes have been deeply investigated and used in many technical applications (e.g., air conditioning and refrigeration systems) since their first introduction in 1977 by Fujie et al. [1]. In fact, they potentially have many advantages with respect to smooth tubes, mainly when applied during refrigerant phase change. Despite a pressure drop increase, they provide higher heat transfer coefficients and a delayed onset of dryout during the boiling process. Furthermore, the fins realized along the tube circumference should contribute to an easier and quicker transition to annular regime with a consequent increase of the heat transfer performance [2-3]. Beside, in the last years microfin tubes with relatively lower diameters have been investigated. A reduction in diameter leads to a reduction of the refrigerant charge, so these mini microfin tubes can provide more compact heat exchangers with a minimization of the whole system refrigerant hold-up, while maintaining high efficiencies. This latter feature could result attractive also to cope with the new and even more stringent national and international environmental regulations that limit the refrigerant charge inventory and its maximum GWP, e.g. the new European F-gas regulation (Regulation (EC) No 517/2014). The scientific literature is strongly motivated in continuously studying microfin tubes in terms of heat transfer performance to provide an updated experimental database including different geometries and refrigerants, and to identify and/or develop reliable heat transfer and pressure drop models. Despite that, in the last decade or so, just few research groups have been focusing their attention microfin tubes smaller than D=6 mm.

Baba et al. [4] presented an experimental work on R1234ze(E), R32, and a zeotropic R1234ze(E)/R32 (50:50) by mass% mixture flow boiling heat transfer inside a 4.86 mm ID at the fin tip microfin tube. The mass velocity ranged from 150 kg m^{-2} s^{-1} to 400 kg m^{-2} s^{-1} at a saturation temperature of 10°C.

Kondou et al. [5] tested several refrigerants during flow boiling inside a water heated microfin tube having 4.94 mm ID at the fin tip: R32, R1234ze(E), a R32/R1234ze(E) (20:80) by mass% mixture, and a R32/R1234ze(E) (50:50) by mass% mixture. The saturation temperature was fixed at 10°C, the heat flux was set at 10 kW m^{-2} and 15 kW m^{-2}, and mass velocity ranged from 150 kg m^{-2} s^{-1} to 400 kg m^{-2} s^{-1}.

Furthermore, Kondou et al. [6] analysed the heat transfer performance in terms of heat transfer and pressure drop during both the condensation and the vaporization processes inside the same horizontal microfin tube (4.94 mm ID at the fin tip) of four refrigerant mixtures: two R744/R32/R1234ze(E) mixtures with compositions: (4:43:53) and (9:29:62) by mass% and two R32/R1234ze(E) mixtures with compositions: (30:70) and (40:60) by mass%. The authors carried out the flow boiling experimental tests at a saturation temperature of 10°C, a heat flux equal to 10 kW m^{-2}, and mass velocity ranging from 150 kg m^{-2} s^{-1} to 600 kg m^{-2} s^{-1}. In addition, the same 4.94 mm ID at the fin tip microfin tube was investigated by Kondou et al. [7] during R1234ze(E), R1234ze(Z), and R134a condensation and vaporization. During flow boiling tests, the heat flux was fixed at 10 kW m^{-2}, the saturation temperature ranged from 0°C to 30°C, and the mass velocity was set at 150 kg m^{-2} s^{-1}, 200 kg m^{-2} s^{-1}, and 300 kg m^{-2} s^{-1}.

In a different laboratory, a 3.4 mm ID at the fin tip electrically heated microfin tube was investigated during flow boiling of several refrigerants at 30°C of saturation temperature, heat fluxes ranging from 10 kW m^{-2} to 50 kW m^{-2} and mass velocities from 190 kg m^{-2} $s^{-1}$to 940 kg m^{-2} s^{-1}. Mancin et al. [8] tested R134a, Diani et al. [9] presented R1234ze(E) data, and Diani et al. [10] measured R1234yf heat transfer performance. The same approach was used to test a 2.4 mm ID at the fin tip electrically heated microfin tube during flow boiling of R1234ze(E) and R134a [11], and R1234yf [12] at 30°C of saturation temperature, with mass velocities ranging from 375 kg m^{-2} s^{-1} to 940 kg m^{-2} s^{-1}, and heat fluxes from 10 kW m^{-2} to 50 kW m^{-2}.

Wu et al. [13] investigated R22 and R410A flow boiling inside one smooth tube and five microfin tubes having the same outer diameter (5 mm) and different number of fins and helix angles. The investigated mass velocities ranged from 100 to 620 kg m^{-2} s^{-1}, at a saturation temperature of 6 °C. They found that microfin tubes are more efficient at low mass velocities, since the heat transfer coefficient per unit pressure drop decreases with mass velocity.

He et al. [14] investigated R410A and a near azeotropic R290/R32 (68:32) by mass% mixture during flow boiling inside three different microfin tubes, having 4.3 mm, 6.1 mm, and 8.48 mm diameter at the fin tip, respectively. The saturation temperature was kept equal to 7 °C, 9 °C, and 11 °C, the mass velocity ranged from 50 kg m^{-2} s^{-1} to 250 kg m^{-2} s^{-1}, and the heat flux from 10 kW m^{-2} to 30 kW m^{-2}. The authors measured heat transfer coefficients also inside smooth tubes with 4.0 and 6.0 mm respectively. The HTC measured during flow boiling inside the microfin tubes having similar fin tip diameter were consistently higher than the ones inside smooth tubes at the same working conditions.

Despite the few recent works listed above, the available experimental data set in different microfin geometries having small diameter is still rather limited. So new sets of data obtained on different microfin tubes are indeed very helpful to increase the knowledge on the subject and to properly assess the classical correlations proposed during the years.

This paper focuses on R245fa flow boiling inside a new geometry microfin tube, having a fin tip diameter of 4.2 mm. Experimental measurements were collected at several mass velocities, from 100 kg m^{-2} s^{-1} to 300 kg m^{-2} s^{-1} at two different heat fluxes 30 kW m^{-2} and 60 kW m^{-2}, and by keeping the mean saturation temperature equal to 30°C. The new data sets permit to investigate the vapor quality, heat flux, and mass velocity effects on heat transfer coefficient and pressure drop. Furthermore, several literature correlations were assessed against the experimental database to test the suitability of the most common models also in this microfin geometry.

2. Experimental setup and data reduction

As shown in Figure 1, the experimental test facility consists of three loops: refrigerant, cooling water, and hot water loops. The rig was designed for heat transfer and pressure drop measurements and flow visualization during either vaporization or condensation of pure refrigerants and refrigerants mixtures inside structured micro- and nano-geometries. The refrigerant is pumped through the circuit by means of a variable speed volumetric gear pump, then it is vaporized in a Brazed Plate Heat Exchanger (BPHE) fed with hot water to achieve the desired value of vapour quality. The hot water is supplied by a thermostatic bath; both water flow rate and water temperature can be independently set. The heat flow rate exchanged at the BPHE evaporator is accurately measured by means of a magnetic flow meter and a calibrated T-type thermopile; furthermore, preliminary tests were run to verify the heat balance between refrigerant and water sides, the results showed a misbalance always less than 2%. The refrigerant enters the microfin test tube at a known mass velocity and vapour quality and then it is vaporized by means of a calibrated Ni-Cr wire resistance.

The electrical power supplied to the sample is indirectly measured by means of a calibrated reference resistance (shunt) and by the measurement of the effective electrical difference potential of the resistance wire inserted in the copper heater. The current can be calculated from the Ohm's law. The fluid leaves the test section and enters in a post-condenser, a brazed plate heat exchanger fed with tap water, where it is fully condensed and subcooled. A damper connected to the compressed air line operates as pressure regulator to control the saturation conditions in the refrigerant loop.

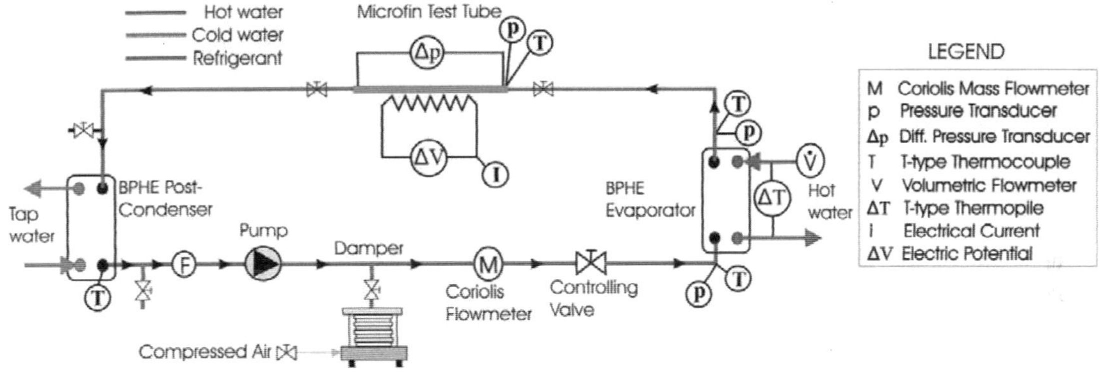

Figure 1. Experimental setup.

Table 1. Instruments uncertainty.

Transducer	Uncertainty (k=2)
T-type thermocouples	± 0.1 K
T-type thermopiles	± 0.05 K
Electric power	± 0.26% of the reading
Coriolis mass flowmeter (refrigerant loop)	± 0.10% of the reading
Magnetic volumetric flowmeter (hot water loop)	± 0.2% of FS= 0.33 10^{-3} m^3 s^{-1}
Differential pressure transducer (test section)	± 0.075% of 0.3 MPa
Absolute pressure transducers	± 0.065% of FS= 4 MPa

As shown in Figure 1, the refrigerant pressure and temperature are measured at several locations throughout the circuit to know the refrigerant properties at the inlet and outlet of each heat exchanger. The refrigerant mass flow rate can be independently controlled by the gear pump and it is measured by means of a Coriolis effect flowmeter. No oil circulates in the refrigerant loop. Table 1 lists the values of uncertainty (coverage factor, k=2) of the instruments used in the experimental facility. The mini microfin tube was brazed inside a guide milled on the top surface of a copper plate, which is 200 mm long, 10 mm wide, and 20 mm high. 16 holes were drilled just 1 mm below the microfin tube, in order to locate as many T-type thermocouples to monitor the wall temperature distribution. Another guide was milled on the bottom side of the copper plate, to host a Nickel-Chrome wire resistance connected to a DC current generator, which supplies the heat flow rate needed to vaporize the refrigerant flowing inside the tube. In order to avoid the abrupt pressure drops due to flow contraction and expansion, a suitable smooth connection to the refrigerant circuit having the same fin tip diameter (D=4.2 mm) was designed and realized to join the test tube with inlet and outlet pipes. Pressure ports are located about 25 mm downstream and upstream of the copper plate, thus the length for pressure drop measurements is 250 mm. The test section is located inside an aluminum housing filled with 15 mm thick ceramic fiber blanket, to minimize the heat losses due to conduction to the ambient. Figure 2 and Table 2 summarise the main geometrical characteristics of the tested tube while Figure 3 shows a cross section of the tested microfin tube. Given the reported dimensions, the area enhancement with reference to the smooth tube having the same fin tip diameter is equal to 1.62.

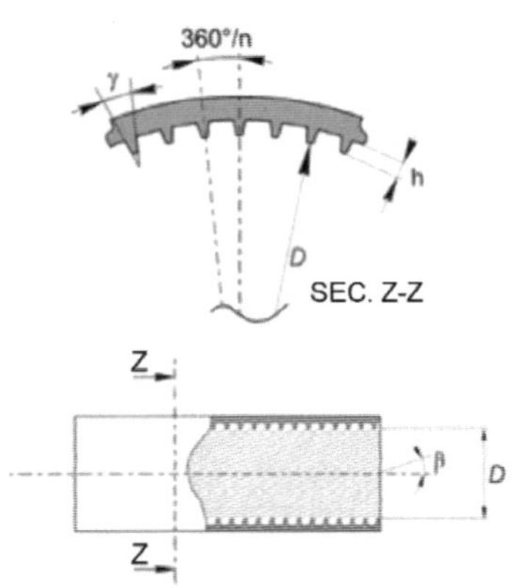

Figure 2. Microfin tube drawing.

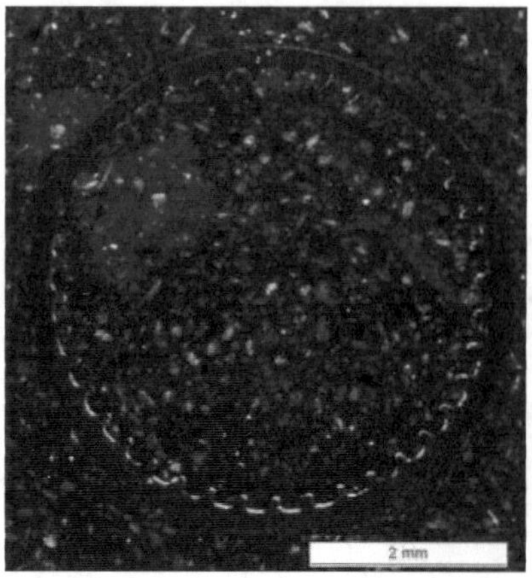

Figure 3. Cross section of the tested Microfin tube.

Table 2. Microfin tube geometry

Parameter	**Value**
Outer diameter	5 mm
Fin tip diameter, D	4.2 mm
Number of fins, n	40
Fin height, h	0.15 mm
Apex angle, γ	42°
Helix angle, β	18°
Tube thickness, t	0.25 mm

3. Data reduction

As described in the previous section, the subcooled liquid pumped by the volumetric gear pump is vaporized into a BPHE fed with hot water. Thus, the vapour quality at the inlet of the test section can be calculated from a thermal balance at the evaporator, as:

$$q_{evap} = \dot{m}_w \cdot c_{p,w} \cdot (t_{w,in} - t_{w,out}) = \dot{m}_r \cdot (J_{in,TS} - J_{L,sub}) \quad (1)$$

where $\dot{m}_w$ is the water mass flow rate, $c_{p,w}$ is the specific heat capacity of the water, $t_{w,in}$ and $t_{w,out}$ are the inlet and outlet water temperatures. The right-hand side term of eq. (1) reports the refrigerant side heat flow rate where $\dot{m}_r$ is the refrigerant mass flow rate while $J_{in,TS}$ and $J_{L,sub}$ are the unknown specific enthalpy at the inlet of the test section and the specific enthalpy of the subcooled liquid entering the BPHE, respectively. Once calculated $J_{in,TS}$, the vapour quality at the inlet of the test section can be estimated by:

$$x_{in,TS} = \frac{J_{in,TS} - J_L}{J_V - J_L} \quad (2)$$

where J_L and J_V are the specific enthalpies of the saturated liquid and vapour, respectively, evaluated at the saturation pressure of the refrigerant measured at the inlet of the test section. As already described, the electrical power supplied to the sample is indirectly measured by means of a calibrated reference resistance (shunt) and by the measurement of the effective electrical difference potential of the resistance wire inserted in the copper heater. Preliminary heat transfer measurements permitted to estimate the heat loss (q_{loss}) due to conduction through the test section as a function of the mean wall temperature. The tests were run under vacuum conditions on the refrigerant channel by supplying the power needed to maintain the mean wall temperature at a set value. The measurements were carried out by varying the mean wall temperature from 27 °C to 64 °C. The results show that the heat loss increases linearly as the mean wall temperature increases (R>0.99). In the tested range of wall temperature, the heat loss by conduction through the test section can be estimated by:

$$|q_{loss}| = 0.1965 \cdot \bar{t}_{wall}\ [°C] - 4.3574 \quad [W] \quad (3)$$

thus, the actual heat flow rate q_{TS} supplied to the sample is given by:

$$q_{TS} = P_{EL} - |q_{loss}| = \Delta V \cdot I - |q_{loss}| \quad (4)$$

It is worth underlying that the q_{loss} varied from 1.0% to 5.0% of the electrical power supplied. The mean vapour quality, x_{mean} is the average value between the inlet and outlet ones. The two-phase heat transfer coefficient *HTC*, referred to the nominal area A, can be defined as:

$$HTC = \frac{q_{TS}}{A \cdot (\bar{t}_{wall} - \bar{t}_{sat})} = \frac{q_{TS}}{\pi \cdot D \cdot L \cdot (\bar{t}_{wall} - \bar{t}_{sat})} \quad (5)$$

where $\bar{t}_{wall}$ and $\bar{t}_{sat}$ are defined by the followings:

$$\bar{t}_{wall} = \frac{1}{16}\sum_{i=1}^{16} t_{wall,i} \text{ and } \bar{t}_{sat} = \frac{t_{sat,in}(p_{sat,in})+t_{sat,out}(p_{sat,out})}{2} \quad (6)$$

The hydraulic performance of the microfin tube is given in terms of frictional pressure drop, which was calculated from the measured total pressure drop by subtracting the momentum and gravitational pressure gradients, as:

$$\Delta p_f = \Delta p_t - \Delta p_c - \Delta p_a \quad (7)$$

The momentum pressure drops are estimated by the homogeneous model for two-phase flow as follows:

$$\Delta p_a = G^2(v_V - v_L)\,|\Delta x| \quad (8)$$

where G is the refrigerant mass flux, v_L and v_V are the specific volume of liquid and vapour phase, $|\Delta x|$ is the absolute value of the vapour quality change through the whole test section. The gravitational contribution Δp_c is equal to 0 Pa because the microfin tube is horizontally located. Thermodynamic and transport properties are estimated from RefProp v9.1 [15]. A detailed error analysis was performed in accordance with Kline and McClintock [16] using the values of the uncertainty of the instruments listed in Table 1; it was estimated that the uncertainty (k=2) on the two-phase heat transfer coefficient showed a maximum value of ±8.4%, while the uncertainty on the vapour quality was ±0.03. The pressure drops showed a mean uncertainty of around 9%.

4. Experimental results

This section presents the experimental results collected during vaporization of the R245fa inside the mini microfin tube. The data were collected at constant, mean saturation temperature of 30 °C. Table 3 lists the main thermo-physical properties of the R245fa at this reference saturation temperature. This refrigerant can be classified as a low pressure fluid; its critical pressure is 36.51 bar and thus, at this operating conditions, it has a relatively low reduced pressure, around 0.049.

Table 3. Thermo-physical properties of the R245fa at t_{sat}= 30°C. (Lemmon et al. [15])

Properties		**Value**
Saturation pressure, p_{sat} [bar]		1.78
Density, ρ [kg m^{-3}]	Liquid	1324.9
	Vapour	10.2
Thermal Conductivity, λ [mW m^{-1} K^{-1}]	Liquid	86.53
	Vapour	13.27
Dynamic viscosity, μ [μPa s]	Liquid	382.3
	Vapour	10.4
Surface tension [mN m^{-1}]		13.40
Latent heat of vaporization [kJ kg^{-1}]		187.33

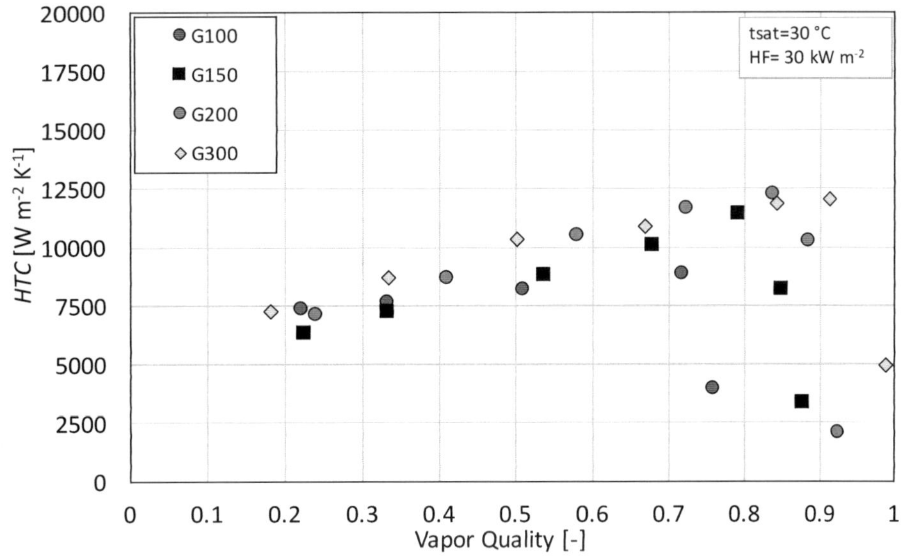

Figure 4. Flow boiling heat transfer coefficient vs. vapour quality as a function of the mass velocity at constant heat flux of 30 kW m^{-2}. G expressed in [kg m^{-2} s^{-1}]

Furthermore, it has a relevant difference between the liquid and vapour densities (ρ_L/ρ_V=132), and a relatively high surface tension (13.4 mN m^{-1}), if compared against others HFCs and HFOs refrigerants. It is well known that actual flow boiling heat transfer inside a tube is the result of the combination of the two main heat transfer mechanisms: nucleate boiling and forced convective boiling. Different scenarios can occur as a function of the actual operating test conditions: mass velocity, vapour quality, heat flux, and thermo-physical properties, and reduced pressure.

In particular, mass velocity, vapour density, and vapour quality mainly affect the forced convective contribution while heat flux and reduced pressure mainly affect the nucleate boiling. Considering the thermo-physical properties reported in Table 3, this fluid exhibits low vapour density and high surface tension, which might promote the forced convective boiling rather than the nucleate boiling which is also limited by the very low reduced pressure. These considerations are confirmed by the experimental results plotted in Figure 4 and 5 where the effects of the mass velocity at the two investigated heat fluxes of (HF) 30 kW m^{-2} and 60 kW m^{-2}, respectively, are showed.

At HF=30 kW m^{-2} (Figure 4), it clearly appears that at all the investigated mass velocities, the heat transfer coefficient increases almost linearly with the vapour quality, meaning that the two-phase forced convection is mainly affecting in the phase-change process. Furthermore, the slope of the measured profiles increases when passing from 100 kg m^{-2} s^{-1} to 150 kg m^{-2} s^{-1} and then to 200 kg m^{-2} s^{-1}. However, when further increasing mass velocity to 300 kg m^{-2} s^{-1}, the enhancement decreases and, at high vapour qualities, i.e. x_{mean}>0.6, the values of the heat transfer coefficient are similar, even slightly lower, to those measured at 200 kg m^{-2} s^{-1}. This behaviour might be due to the saturation temperature drop related to the high two-phase pressure drop, which increases with both mass velocity and vapour quality. The vapour quality at the onset of the dryout phenomenon occurs earlier at low mass velocity; in fact, it was observed at around x_{mean}=0.72, x_{mean}=0.78, x_{mean}=0.86, and x_{mean}=0.92 at G=100 kg m^{-2} s^{-1}, G=150 kg m^{-2} s^{-1}, G=200 kg m^{-2} s^{-1}, and G=300 kg m^{-2} s^{-1}, respectively.

As reported in Figure 5, when increasing the heat flux to HF=60 kW m^{-2} slightly different results are found. The heat transfer coefficients measured at the three investigated mass velocities seem to be almost similar and they exhibit similar behaviour up to the respective vapour qualities at the onset of dryout.

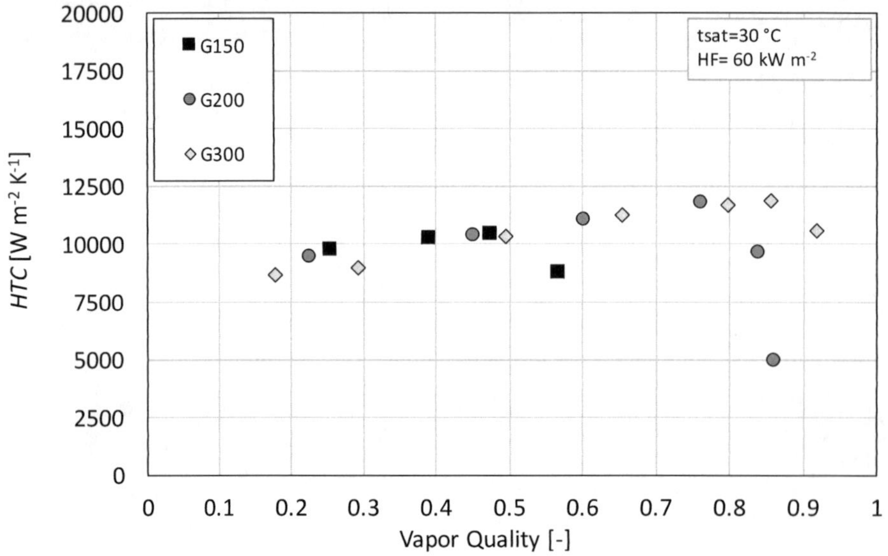

Figure 5. Flow boiling heat transfer coefficient vs. vapour quality as a function of the mass velocity at constant heat flux of 60 kW m^{-2}. G expressed in [kg m^{-2} s^{-1}]

In fact, there is not any noticeable effect of the mass velocity while there still remains a weak effect of the vapour quality since the heat transfer coefficient increases as the vapour quality increases but, for a given mass velocity, the slope of the experimental profile is lower as compared to that reported in Figure 4 at 30 kW m^{-2}. At these operating conditions, the nucleate boiling seems to be the most effective heat transfer mechanism. At all the investigated mass velocities, the onset of the dryout phenomenon occurs earlier as compared to the data collected at 30 kW m^{-2}; in fact, it was observed at x_{mean}=0.50, x_{mean}=0.82, and x_{mean}=0.85 at G=150 kg m^{-2} s^{-1}, G=200 kg m^{-2} s^{-1}, and G=300 kg m^{-2} s^{-1}, respectively.

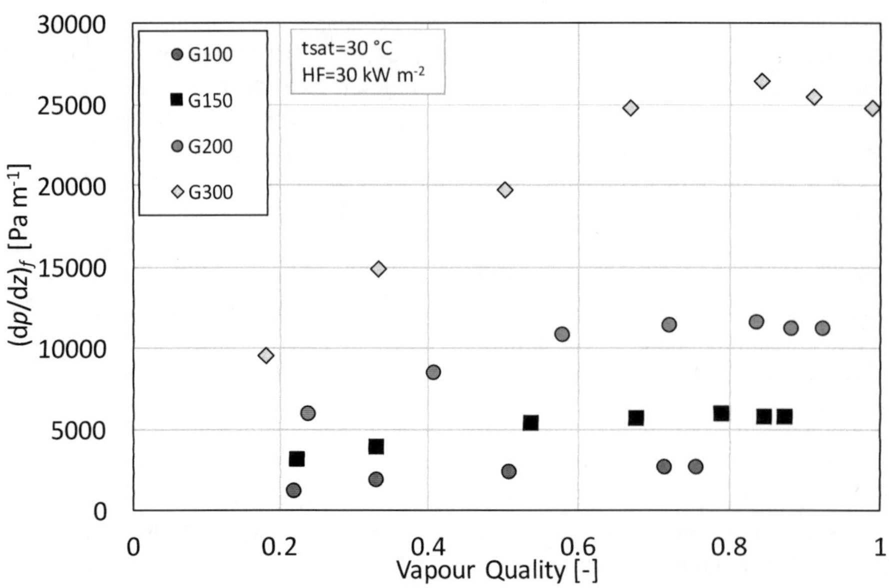

Figure 6. Frictional pressure gradient vs. vapour quality as a function of the mass velocity at constant heat flux of 30 kW m^{-2}. G expressed in [kg m^{-2} s^{-1}]

The experimental frictional pressure gradients are plotted in Figure 6 as a function of the mean vapour quality; for the sake of clarity, only data points relative to an imposed heat flux of 30 kW m^{-2} are reported. The results show that the frictional pressure gradient increases with both the mass velocity and vapour quality.

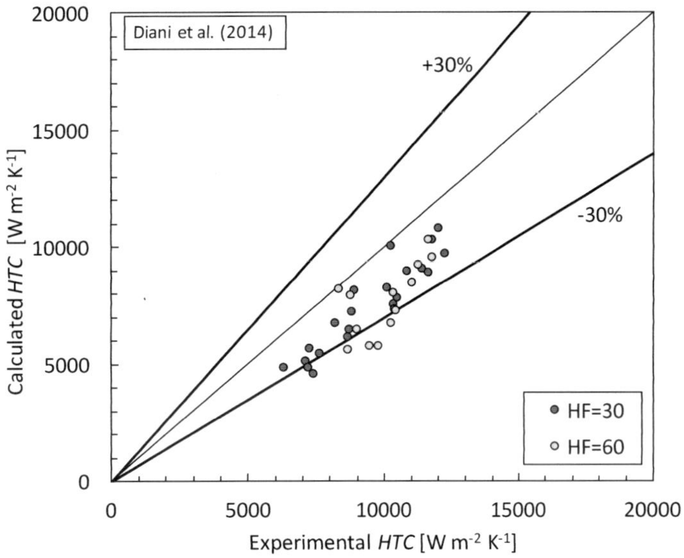

Figure 7. Calculated vs. experimental heat transfer coefficients. HF expressed in [kW m^{-2}].

Figure 7 reports the model proposed by Diani et al. [9], which even if it slightly underestimated the experimental data, it still showed the acceptable prediction capabilities with relative and absolute deviations equal to -23.1% and 23.1%, respectively. Figure 8 presents the results obtained by applying the model proposed by Diani et al. [9], which presented a relative deviation of 26.7% and an absolute deviation of 26.7%.

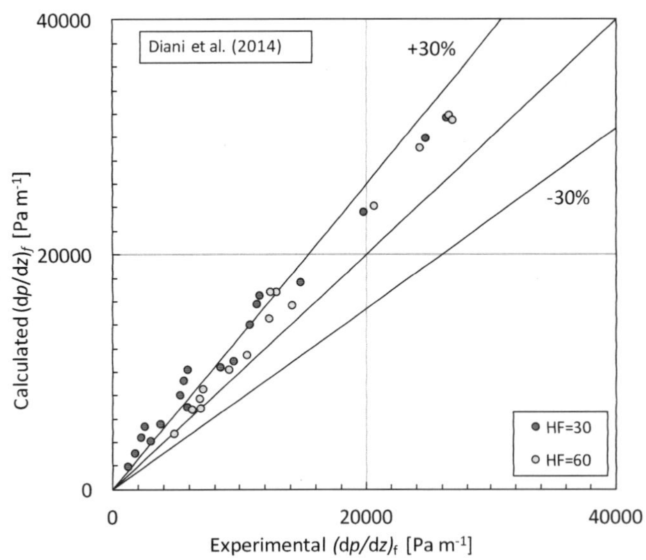

Figure 8. Calculated vs. experimental frictional pressure gradients. HF expressed in [kW m^{-2}].

5. Conclusions

The paper presents experimental heat transfer coefficients and pressure drops measured during R245fa flow boiling inside a mini microfin tube with an inner diameter at the fin tip of 4.2 mm. Tests were run at a constant mean saturation temperature of 30 °C, by varying the vapour quality from 0.15 to 0.95, the mass velocity from 100 kg m^{-2} s^{-1} to 300 kg m^{-2} s^{-1}, at two imposed heat fluxes, 30 kW m^{-2} and 60 kW m^{-2}. The results confirm that the heat transfer process is controlled by the two main mechanisms that govern the flow boiling phenomenon, i.e. nucleate boiling and two-phase forced convection, and that the prevailing one depends upon the actual operating test conditions. The two-phase frictional pressure drops were also measured. They increase with both mass velocity and vapour quality. Finally, the results were compared against the models for flow boiling heat transfer coefficient and frictional pressure drop estimations proposed by Diani et al. [9], which demonstrated to be satisfactory consistent with the present data.

6. Acknowledgment

Wieland-Werke AG, and Generalmeccanica Snc are gratefully acknowledged.

7. References

[1] Fujie K, Itoh N, Kimura H, Nakayama and Yanugi T 1977 Heat transfer pipe, US Patent 4044797, assigned to Hitachi.
[2] Han D and Lee KJ 2005 *Int. J. Heat Mass Transf.* **48** 3804-3816.
[3] Doretti L, Mancin S, Zilio C and Cavallini A 2013. *Int. J. of Refrigeration* **36** 567-587.
[4] Baba D, Nakagawa T and Koyama S 2012 In: *Proceedings of the International Refrigeration and Air-Conditioning Conference at Purdue*, West Lafayette, IN.
[5] Kondou C, Baba D, Mishima F and Koyama S 2013 *Int. J. of Refrigeration* **36**, 2236–2378.
[6] Kondou C, Mishima F, Liu JF and Koyama S 2014 In: *Proceedings of the International Refrigeration and Air Conditioning Conference at Purdue*, West Lafayette, IN, USA, Paper 1448.
[7] Kondou C, Mishima F, Liu JF and Koyama S 2014 In: *Proceedings of the International Refrigeration and Air Conditioning Conference at Purdue*, West Lafayette, IN, USA, Paper 1446.
[8] Mancin S, Diani A and Rossetto L 2015 *Heat Transfer Engineering*, **36**(14-15) 1218-1229.
[9] Diani A, Mancin S and Rossetto L 2014 *Int J. Refrigeration*, **47**, 105-119.
[10] Diani A, Mancin S and Rossetto L 2015 *Exp. Therm. and Fluid Science* **66** 127–136.
[11] Diani A, Mancin S, Cavallini A and Rossetto L 2016 *Int J. Refrigeration* **69** 272–284.
[12] Diani A and Rossetto L 2015 *Journal of Physics: Conference Series* **655** 012032.
[13] Wu Z, Wu Y, Sunden B and Li W 2013 *Exp. Therm. and Fluid Science* **44** 298-408.
[14] He G, Liu F, Cai D and Jiang J 2016 *Int. J. Heat Mass Transf.* **102** 561–573.
[15] Lemmon EW, Huber ML and McLinden MO 2013 NIST Standard Reference Database 23: Reference Fluid Thermodynamic and Transport Properties- REFPROP, Version 9.1, National Institute of Standards and Technology, Standard Reference Data Program, Gaithersburg.
[16] Kline SJ and McClintock FA 1953 *Mech. Eng.* **75** 3–8.

Heat transfer degradation during condensation of non-azeotropic mixtures

M Azzolin, A Berto, S Bortolin, D Del Col

Department of Industrial Engineering, University of Padova, Via Venezia 1, 35131 Padova, Italy

Corresponding author: Davide Del Col; e-mail: davide.delcol@unipd.it;
tel.: +39-049-8276891, fax: +39-049-8276896

Abstract. International organizations call for a reduction of the HFCs production and utilizations in the next years.
Binary or ternary blends of hydroflourocarbons (HFCs) and hydrofluoroolefins (HFOs) are emerging as possible substitutes for high Global Warming Potential (GWP) fluids currently employed in some refrigeration and air-conditioning applications.
In some cases, these mixtures are non-azeotropic and thus, during phase-change at constant pressure, they present a temperature glide that, for some blends, can be higher than 10 K. Such temperature variation during phase change could lead to a better matching between the refrigerant and the water temperature profiles in a condenser, thus reducing the exergy losses associated with the heat transfer process. Nevertheless, the additional mass transfer resistance which occurs during the phase change of zeotropic mixtures leads to a heat transfer degradation. Therefore, the design of a condenser working with a zeotropic mixture poses the problem of how to extend the correlations developed for pure fluids to the case of condensation of mixtures. Experimental data taken are very helpful in the assessment of design procedures.
In the present paper, heat transfer coefficients have been measured during condensation of zeotropic mixtures of HFC and HFO fluids. Tests have been carried out in the test rig available at the Two Phase Heat Transfer Lab of University of Padova. During the condensation tests, the heat is subtracted from the mixture by using cold water and the heat transfer coefficient is obtained from the measurement of the heat flux on the water side, the direct measurements of the wall temperature and saturation temperature. Tests have been performed at 40°C mean saturation temperature. The present experimental database is used to assess predictive correlations for condensation of mixtures, providing valuable information on the applicability of available models.

1. Introduction

In the last three decades, growing attention to global warming and environmental issue has led to a reconsideration of the fluids traditionally employed in refrigeration and air-conditioning applications. Starting from 1987, the international environmental agreement called Montreal Protocol, signed by all UN and EU members, established the phase-out of ozone depleting CFCs (chlorofluorocarbons), such as R12 and R502. With the Copenhagen Amendment to the Montreal Protocol (1992) also HCFCs (hydrochlorofluorocarbons), such as R22, have been affected by the regulation. Then the Kyoto Protocol (1997) fixed emission reduction targets for the six main greenhouse gases, including HFCs (hydrofluorocarbons) such as R134a and R32. The EU Regulation No. 517/2014 disposed the phase-out of most refrigerants commonly used in refrigeration and air conditioning systems, in particular R404A, R410A and R134a, due to their high GWP (Global Warming Potential). The aim is to reach a two-thirds reduction of EU F-gas emissions by 2030 compared to 2014 levels. In 2016 the Kigali Amendment to the Montreal Protocol was adopted to significantly reduce the consumption and production of HFCs on a global scale.

As HFCs need to be replaced to meet European and international requirements, considerable research has been fostered on possible substitutes, both natural and synthetic. Recently, low GWP hydrofluoroolefins (HFOs) such as R1234yf ($GWP_{100\text{-}years}$=4) and R1234ze(E) ($GWP_{100\text{-}years}$<1) have been studied as possible candidates for the replacement of R134a. Currently, no pure fluids competitive in terms of thermodynamic performance, low GWP and non-flammability are available as alternatives to R410A in chiller and air-conditioning applications and to R404A in commercial and transport refrigeration. For these applications a possible solution should be represented by refrigerant mixtures of HFOs and other HFCs (i.e. R32, R125). These mixtures present a temperature variation during phase change at constant pressure, due to the temperature glide between the dew and the bubble point, which can lead to a better matching between the refrigerant and the water temperature profiles. On the other hand, the additional mass transfer resistance which occurs during the phase change of zeotropic mixtures causes a heat transfer degradation.

In the literature, few studies are available on condensation and evaporation of HFOs/HFCs mixtures. Hossain et al. [1] investigated the flow boiling heat transfer characteristics of the mixture R1234ze(E)/R32 (mass fraction 55/45%) in a horizontal tube with inner diameter of 4.35 mm, comparing them with those of R410A and R32 at the same operating conditions. The local heat transfer coefficient of the mixture was found to be lower than that of R32 and R410A in the whole quality region. With regard to R1234ze(E), the heat transfer coefficient of the zeotropic blend is lower only in the high quality region. Kondou et al. [2] investigated the condensation and evaporation heat transfer coefficients of the mixtures R744/R32/R1234ze(E) and R32/R1234ze(E) at different composition in a horizontal microfin tube (5.35 mm equivalent inner diameter). The authors found that both condensation and in particular evaporation heat transfer coefficients of the mixtures are penalized with respect to those of single components and they tried to quantify the contribution of the additional mass transfer resistance. Wang et al. [3] studied the condensation heat transfer characteristics of R32/R1234yf non-azeotropic mixtures (mass fraction 52/48% and 77/23%) inside a horizontal smooth tube with inner diameter equal to 4 mm. They developed also a prediction model for the forced convective condensation of zeotropic mixtures in reasonable agreement with experimental data.

In this paper, two mixtures of R1234ze(E) and R32 at different mass compositions (23/77% and 46/54% by mass) have been investigated during condensation inside a single circular cross section microchannel with inner diameter equal to 0.96 mm. Tests have been carried out on the experimental apparatus available at the Two-Phase Heat Transfer Lab of the University of Padova. The new experimental data are compared to those of pure components R1234ze(E) and R32 in order to analyze the heat transfer penalization due to the mass transfer resistance of these zeotropic mixtures and to assess predictive models.

2. Test facility

The heat transfer test section is located in the experimental apparatus schematized in Figure 1. In the primary loop, the refrigerant is subcooled in a post-condenser and then dried up before entering an independently controlled gear pump, magnetically coupled to a variable-speed electric motor. The mass flow rate is measured by a Coriolis effect mass flow meter. Hence, the fluid passes through a mechanical filter and a tube-in-tube heat exchanger where the primary fluid can be vaporized and superheated using hot water flowing in a secondary loop with PID-controlled electrical heaters. The refrigerant is finally sent through the test section for heat transfer measurements. The test section is characterized by a pre-conditioning sector, used to reach the desired thermodynamic conditions of the refrigerant at the inlet of the measuring sector. The refrigerant pressure at the inlet of each sector is measured by means of relative pressure transducers, whereas a differential pressure transducer is employed to measure pressure drop along the measuring sector. All the temperatures are detected using T-type thermocouples which have been calibrated following the technique in Del Col et al. [4]. Two thermal baths are employed on the secondary fluid side: the first serves brine at 5 °C to the post-condenser while the second sets the temperature of the distilled water entering the measuring and pre-conditioning sectors. In every test run, when the apparatus is working in steady state conditions, measurements of thermo-fluid-dynamic parameters are recorded for 50 s with a time step of 1 s and then averaged.

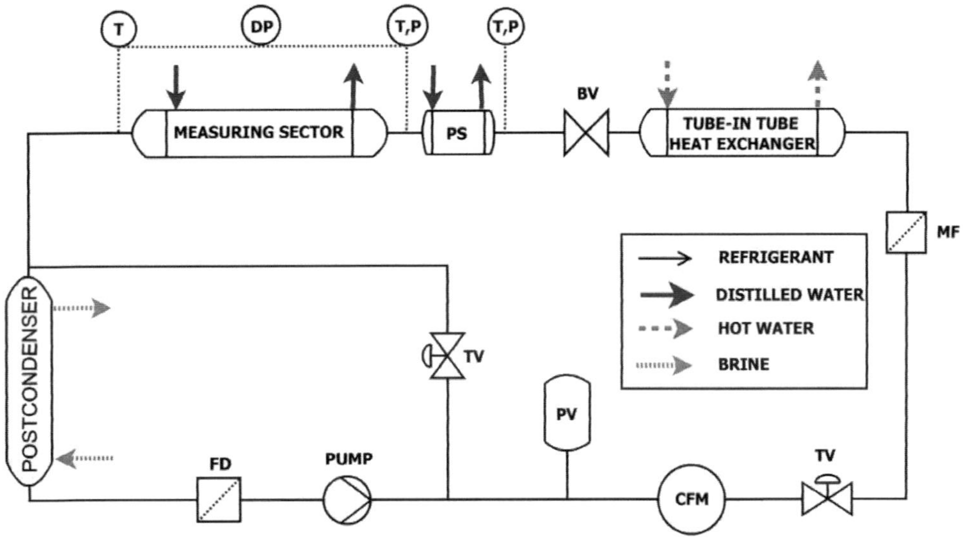

Figure 1. Experimental test rig: PS pre-conditioning sector; MF mechanical filter; FD filter dryer; PV pressure vessel; CFM Coriolis-effect mass flow meter; TV valve; P pressure transducer; DP differential pressure transducer; T thermocouple.

2.1. Heat transfer coefficient test section

The test section is composed by two parts: a 50 mm long pre-conditioning sector and a 230 mm long measuring sector. Two independent distilled water loops serve the the pre-conditioning sector and the measuring sector and each of them is provided by a flow regulating valve to set the flow rate measured by a dedicated Coriolis effect mass flow meter. When necessary, the water entering the pre-conditioning sector is additionally heated up.

Both the pre-conditioning and measuring sectors work as counter-current heat exchangers and have been obtained from a 8 mm external diameter copper rod having a 0.96 mm internal bore and an inner surface roughness *Ra* equal to 1.3 μm. The external surface of the copper rod is machined to embed the secondary fluid channel (Figure 2), which is externally closed by plastic sheath. The tortuous path geometry on the water side enables good water mixing for precise local coolant temperature and meanwhile it reduces the external heat transfer resistance in favour to low uncertainty measurements. The two sectors are connected between them and with the experimental apparatus using stainless steel adiabatic tubes to allow thermal separation and the insertion of two pressure ports.

On the coolant side, the inlet and outlet water temperatures are measured by T-type thermocouples at the ends of each sector, while the water temperature differences across both sectors are measured by multi-junction copper-constantan thermopiles. In the measuring sector, the local water temperature profile is determined from the measurements of 15 thermocouples placed in the water channel while the local wall temperatures are gauged by 13 thermocouples embedded in the wall. More details on the test section are given in Matkovic et al. [5].

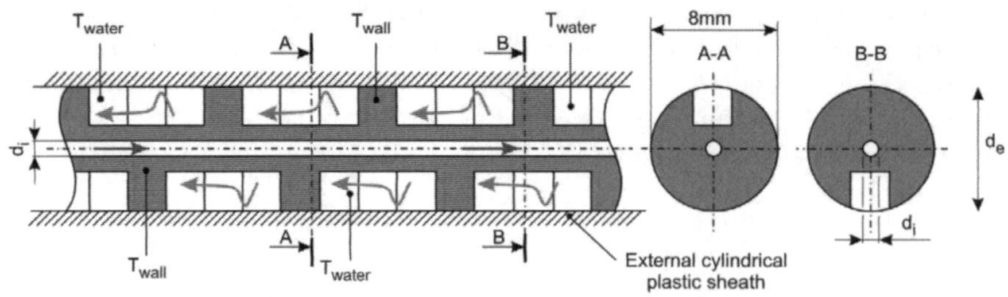

Figure 2. Sketch of the water coolant path in the microchannel.

3. Mixture composition

In this work, condensation heat transfer characteristics of two R32/R1234ze(E) mixtures with different compositions (0.23/0.77 and 0.46/0.54 by mass) have been investigated. In Table 1, some thermophysical and transport properties of the refrigerant blends at 40 °C saturation temperature are reported, compared with those of pure fluid components, R32 and R1234ze(E). Properties are computed using NIST Refprop 9.1 (Lemmon et al., [6]) with the Akasaka model [7], specifically developed for the evaluation of thermodynamic properties of R32/R1234ze(E) mixtures.

From Table 1 it can be noticed that for R32/R1234ze(E) mixture at 23/77% composition the temperature glide is equal to 11 K, while at 46/54% it is equal to 8.2 K. These mixtures present a $GWP_{100\text{-years}}$ value that is significantly lower than those of traditionally used refrigerants, such as R404A ($GWP_{100\text{-years}}$=3922), R410A ($GWP_{100\text{-years}}$=2088) and R32 ($GWP_{100\text{-years}}$=675).

Table 1. Properties of pure R32 and R1234ze(E) and R32/R1234ze(E) mixtures at 40 °C saturation temperature (data from NIST Refprop 9.1, Lemmon et al. [6]).

Fluid	p_{sat} [bar]	ρ_l [kg m^{-3}]	ρ_g [kg m^{-3}]	μ_l [μPa s]	λ_l [W m^{-1} K^{-1}]	t_{dew} [°C]	t_{bub} [°C]	*Glide* [K]	*GWP* [-]
R32	24.8	893.0	73.3	95.0	0.115	/	/	/	675
R1234ze(E)	7.7	1111.3	40.7	167.0	0.069	/	/	/	0.97
R32/R1234ze(E) 23/77%	13.1	1075.3	56.3	143.8	0.083	45.6	34.6	11	155
R32/R1234ze(E) 46/54%	17.4	1017.9	64.8	122.0	0.094	44.4	36.2	8.2	309

4. Data reduction

The local heat flux is determined from the slope of the water temperature profile along the test section (obtained from the thermocouples inserted along the coolant path):

$$q' = \frac{\dot{m}_{water}\, c_{water}}{\pi\, d} \frac{\mathrm{d}T_{water}(z)}{\mathrm{d}z} \tag{1}$$

where z is the axial coordinate along the tube and $\mathrm{d}T_{water}(z)/\mathrm{d}z$ is the derivative of the polynomial equation interpolating the measured water temperature. Thus a proper interpolating function of the measured water temperatures is needed (complete description of the whole procedure is reported in Del Col et al. [8]).

The heat flow rate transferred to the secondary fluid up to a certain position z is obtained integrating the local heat flux from the refrigerant inlet to the position z:

$$q(z)=\pi\, d_i \int_0^z q'(z)\, dz \tag{2}$$

Finally, the local heat transfer coefficient inside the microchannel is obtained as:

$$\alpha(z)=\frac{q'(z)}{T_{SAT}(z)-T_w(z)} \tag{3}$$

where T_w is the wall temperature, measured with the thermocouples embedded in the wall while T_{SAT} is the saturation temperature calculated from the pressure, enthalpy and mixture composition X_{mix}, assuming an equilibrium state:

$$T_{SAT}(z) = f\left(p(z), h(z), X_{mix}\right) \tag{4}$$

Along the measuring sector, the refrigerant enthalpy at the position z can be calculated as follows:

$$h(z)=h_{in,MS}-\frac{q(z)}{\dot{m}_{ref}} \tag{5}$$

During a condensation test, the refrigerant enters the pre-conditioning sector as superheated vapour and is cooled down by the secondary fluid. The enthalpy of the entering vapour in the pre-conditioning sector is determined from the measured temperature and pressure, while the refrigerant enthalpy at the inlet of the measuring sector $h_{in,MS}$ is determined from a thermal balance on the water side.

The pressure of the mixture is measured at the inlet and outlet of the measuring sector. The local pressure could be obtained from a linear interpolation of inlet and outlet measured values but in the present case a more precise iterative procedure is needed. In fact, the pressure profile is not linear along the microchannel and the interpolation between inlet and outlet pressure should account for the frictional pressure drop in the stainless steel tubes and in the copper microchannel, the pressure variations due to changes in the geometry and the pressure recovery due to momentum variation during condensation. A detailed description of the models applied to determine these local and distributed contributions is available in Del Col et al. [9].

As for the refrigerant temperature, the vapor quality along the test section is evaluated using the local pressure, the local enthalpy and the mixture composition:

$$x(z) = f\left(p(z), h(z), X_{mix}\right) \tag{6}$$

Experimental uncertainties of the measured parameters (Type B) are reported in Table 2.

Table 2. Type B uncertainty of measured parameters.

Temperature	± 0.05°C
Temperature difference (with thermopile)	± 0.03°C
Water flow rate in pre-conditioning sector	± 0.2 % at 10 kg h^{-1}
Water flow rate in measuring sector	± 0.14 % at 10 kg h^{-1}
Refrigerant flow rate	± 0.2 % at 2 kg h^{-1}
Absolute pressure	± 5 kPa
Pressure difference (higher than 1 kPa)	± 0.12 kPa
Pressure difference (lower than 1 kPa)	± 0.1 %

5. Experimental results and comparison to pure components

Experimental results obtained during condensation of R32/R1234ze(E) mixtures at 23/77% and 46/54% composition in a 0.96 mm diameter microchannel are reported in Figure 3. Condensation tests have been performed at a pressure of 13.7 bar and 17.7 bar respectively for the 23/77% and the 46/54% mixture. The local heat transfer coefficients for the two mixtures are plotted against local vapour qualities for mass velocities ranging from 200 to 800 kg $m^{-2}s^{-1}$.

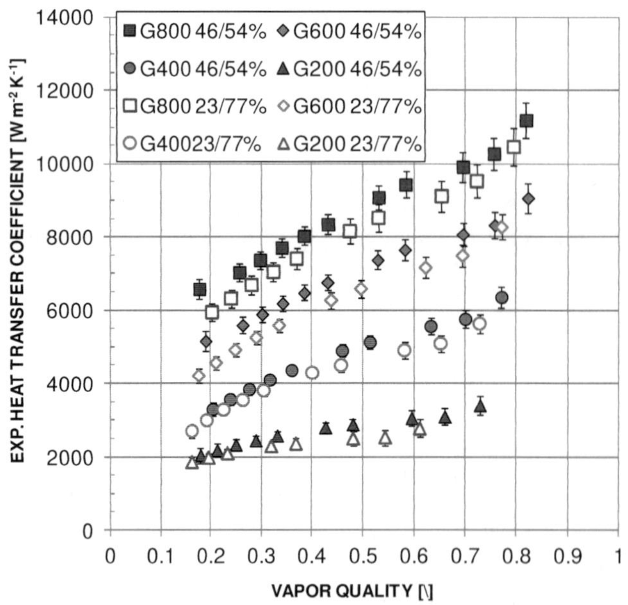

Figure 3. Experimental local heat transfer coefficients versus vapor quality for R32/R1234ze(E) mixtures at 23/77% and 46/54% mass compositions.

For both tested mixtures, the heat transfer coefficient was found to increase with vapor quality and mass velocity although at low mass velocity and vapor quality it does not vary significantly. From Figure 3, it can be noticed the influence of R32 mass fraction on heat transfer coefficient values of the mixture. Indeed, as the mass fraction of R32 is increased from 23% to 46%, heat transfer coefficients are enhanced on average by about 7% at G = 800 kg $m^{-2}s^{-1}$, by 7.4% at G = 600 kg $m^{-2}s^{-1}$,by 8% at G = 400 kg m^{-2} s^{-1} and by 13% at G = 200 kg m^{-2} s^{-1}. Heat transfer coefficients of 46/54% are higher than those of 23/77% mixture for all mass velocities due to the greater liquid thermal conductivity and smaller temperature glide.

In Figs. 4 and 5, heat transfer coefficients of the two R32/R1234ze(E) mixtures inside 0.96 mm inner diameter microchannel are compared to those of pure components R32 and R1234ze(E) at the same working conditions. At x = 0.3 (Fig. 4) the heat transfer coefficients of the 46/54% mixture are roughly the same of those of R1234ze(E) for all the mass velocities. At higher vapor qualities (x = 0.6), as shown in Figure 5, the condensation heat transfer coefficients of pure R1234ze(E) are higher than those of the 46/54% mixture by about 4% at G = 800 kg $m^{-2}s^{-1}$, by 2% at G = 600 kg m^{-2} s^{-1}, by 6.5% at G = 400 kg m^{-2} s^{-1} and by 22% at G = 200 kg m^{-2} s^{-1}.

With regard to Figs 4 and 5, the heat transfer coefficients of R32 are always higher than those of R32/R1234ze(E) mixtures and pure R1234ze(E). The increase of the heat transfer coefficient of R32 compared to that of R32/R1234ze(E) mixture at 46/54% composition ranges from 20 to 28% at x = 0.3 and from 35 to 46% at x = 0.6.

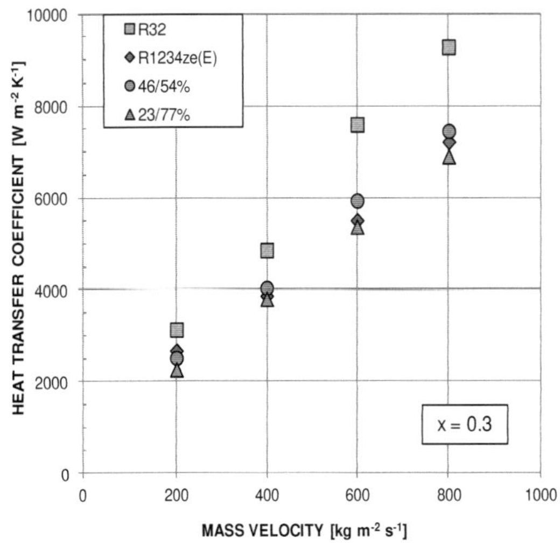

Figure 4. Heat transfer coefficients of R32/R1234ze(E) mixtures and pure components in the microchannel at $x = 0.3$.

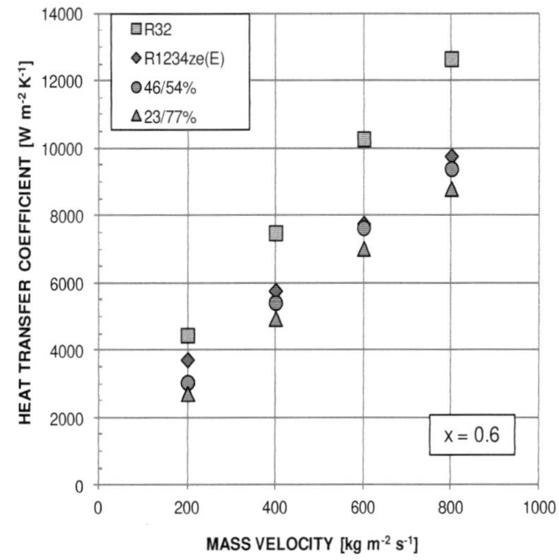

Figure 5. Heat transfer coefficients of R32/R1234ze(E) mixtures and pure components in the microchannel at $x = 0.6$.

6. Prediction of the heat transfer coefficient

The experimental heat transfer coefficient data taken in the 0.96 mm diameter microchannel have been compared to predictions of Cavallini et al. [10] model . This model was originally proposed to predict condensation heat transfer coefficients of pure fluids inside horizontal smooth tubes with internal diameter greater than 3 mm. As underlined in Matkovic et al. [5], this model can be reasonably applied in the case of R32 and R1234ze(E) during condensation inside a 0.96 mm microchannel at mass velocities equal or higher than 200 kg $m^{-2}s^{-1}$. In Figure 6 predicted heat transfer coefficients are plotted versus experimental ones for mass velocities ranging from 200 to 800 kg $m^{-2}s^{-1}$. It is evident that Cavallini et al. [10] model overestimates the experimental data with an absolute deviation e_{AB} equal to 10.9% and a standard deviation of about 6.1%.

As reported in Del Col et al. [11], during condensation of zeotropic mixtures, the effect of the additional mass transfer resistance cannot be neglected. The heat transfer mechanisms occurring during condensation of zeotropic mixtures are the same as for pure fluids, convective condensation and film condensation during stratified flow, with the difference that the condensation process is non-isothermal and there is sensible heat to be removed both in the liquid and vapor phase. The sensible heat removed from the vapour is more significant due to the low heat transfer coefficients associated with this process. Diffusion resistances in both the vapour and the liquid phases are present due to the continuous change in the compositions of both phases, and between them the mass transfer resistance in the vapour phase is the key point for calculations. When considering mixtures, models developed for pure vapor condensation can be applied if complete mixing in both liquid and vapor phase is reached and overall equilibrium is maintained, with a correction term through which the additional resistance of the mixture due to the mass transfer is properly accounted for.

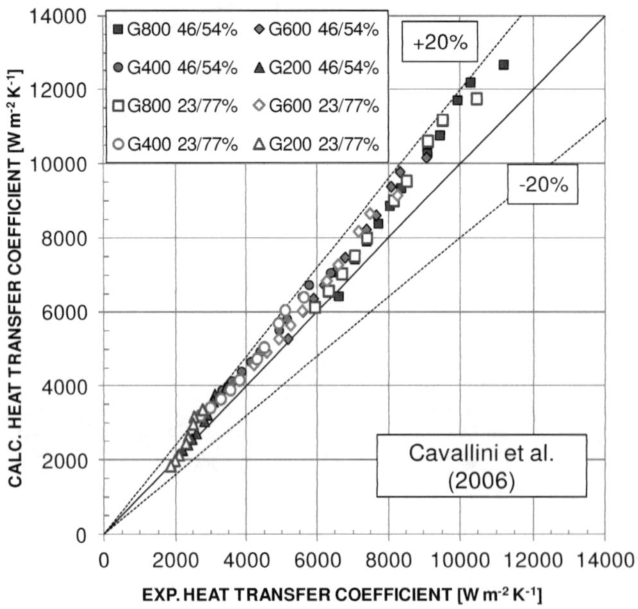

Figure 6. Experimental data of R32/R1234ze(E) mixtures at 23/77% and 46/54% mass compositions compared to values calculated with Cavallini et al. (2006) model.

The Cavallini et al. (2006) model can be applied to zeotropic mixtures using the Bell and Ghaly [12] and the Silver correction [13]. In this procedure the heat transfer coefficient of the mixture is calculated as a function of the condensate heat transfer coefficient α_f (calculated with the model for pure fluids but using thermophysical properties of the mixture), of the heat transfer coefficient of the vapor phase flowing alone in the duct α_g (calculated with the Dittus-Boelter equation), and of the ratio between the sensible flow rate heating the vapor and the total heat flow rate $\partial q_{sg}/\partial q_t$, as reported in Equation (7):

$$\alpha_{mix} = \left(\frac{1}{\alpha_f} + \frac{\partial q_{sg} / \partial q_t}{\alpha_g} \right)^{-1} \tag{7}$$

The ratio of the sensible to total heat duty can be expressed with good approximation as follows:

$$\frac{\partial q_{sg}}{\partial q_t} \approx x c_{pg} \left(\frac{\mathrm{d}T}{\mathrm{d}h} \right) \tag{8}$$

If the ratio dT/dh remains approximately constant during condensation the ratio can be rewritten as a function of the temperature glide ΔT_{GL} and the enthalpy change of isobaric condensation Δh_{mix}:

$$\frac{\mathrm{d}T}{\mathrm{d}h} \approx \frac{\Delta T_{GL}}{\Delta h_{mix}} \tag{9}$$

The comparison between the experimental data and heat transfer coefficients predicted by Cavallini et al. (2006) corrected with Silver-Bell-Ghaly approach is shown in Figure 7. Predicted data are in good agreement with experimental ones, with an absolute deviation e_{AB} of 7.4 % and a standard deviation of 5.4%.

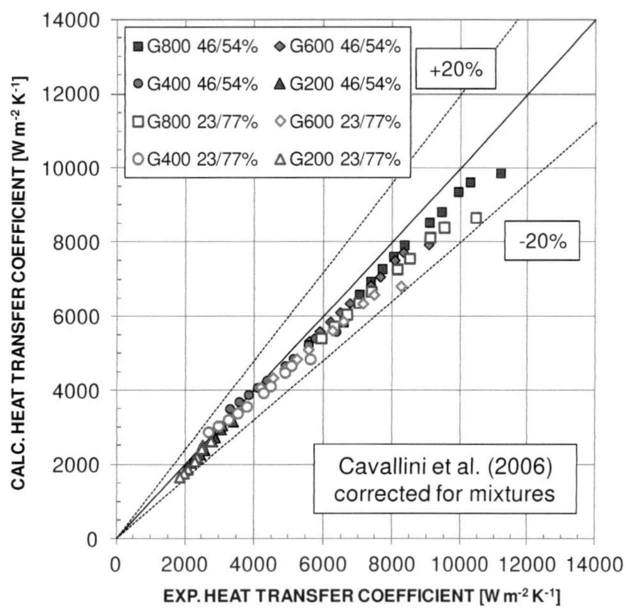

Figure 7. Experimental data of R32/R1234ze(E) mixtures at 23/77% and 46/54% mass compositions compared to values calculated with Cavallini et al. (2006) model coupled with Bell-Ghaly-Silver correction.

7. Conclusions

This work reports experimental heat transfer coefficients during condensation of two R32/R1234ze(E) zeotropic blends at different mass compositions (23/77% and 46/54%), inside a 0.96 mm diameter horizontal smooth microchannel. These zeotropic mixtures display a temperature glide of about 11 K and 8 K, respectively.

Condensation tests have been carried out with mass velocity ranging between 200 kg m^{-2} s^{-1} and 800 kg m^{--2} s^{-1} at a pressure of 13.7 bar and 17.7 bar, respectively, for the 23/77% and the 46/54% mixture. The heat transfer coefficient of both mixtures increases with vapor quality and mass velocity. When the mass fraction of R32 is increased from 23% to 46%, the heat transfer coefficient increases by 7% at G = 800 kg $m^{-2}s^{-1}$, by 8% at G = 400 kg $m^{-2}s^{-1}$ and 13% at G = 200 kg $m^{-2}s^{-1}$ due to better properties (higher thermal conductivity) and smaller temperature glide. With comparison to pure components, at mass velocities ranging from 200 kg m^{-2} s^{-1} and 800 kg m^{--2} s^{-1} the R32/R1234ze(E) mixture at 46/54% composition shows heat transfer coefficients which are lower than those of R1234ze(E) at high vapor qualities but comparable at low vapor qualities.

The experimental heat transfer coefficients have been compared with those predicted by the Cavallini et al. (2006) model coupled with the Silver-Bell-Ghaly correction to account for the additional mass transfer resistance. This model shows a good accuracy in predicting experimental data, displaying an absolute deviation of 7.4% and a standard deviation of 5.4%.

Nomenclature

c_p	[J/(kg K)]	Specific heat	q	[W]	Heat flow rate
d_i	[m]	Hydraulic diameter	q'	[W/m^2]	Heat flux
e_{AB}	[-]	Absolute deviation	t	[°C]	Temperature
G	[kg/(m^{2}s)]	Mass velocity	T	[K]	Temperature
h	[J/kg]	Specific enthalpy	x	[-]	Thermodynamic vapour quality
$\dot{m}$	[kg/s]	Mass flow rate	X	[-]	Mass fraction
p	[Pa]	Pressure	z	[m]	Position

Greek symbols

α	[W/(m^2 K)]	Heat transfer coefficient	λ	[W/(m K)]	Thermal conductivity
Δh	[J/kg]	Specific enthalpy difference	μ	[μPa s]	Dynamic viscosity
ΔT	[K]	Temperature difference	ρ	[kg/m^3]	Density

Subscripts

bub	Bubble	*in*	Inlet	*s*	Sensible
dew	Dew	*l*	Liquid	*SAT*	Saturation
f	Film	*mix*	Mixture	*t*	Total
g	Vapor	*MS*	Measuring section	*water*	Water
GL	Glide	*ref*	Refrigerant	*w*	Wall

References

[1] Hossain M A, Onaka Y, Afroz H M M and Miyara A 2013 Heat transfer during evaporation of R1234ze(E), R32, R410A and a mixture of R1234ze(E) and R32 inside a horizontal smooth tube *Int. J. of Refrigeration* **36** 465-477

[2] Kondou C, Mishima F and Koyama S 2015 Condensation and evaporation of R32/R1234ze(E) and R744/R32/R1234ze(E) flow in horizontal microfin tubes *Science and Technology for the Built Environment* **21** 564-577

[3] Wang L, Dang C and Hihara E 2012 Experimental and theoretical study on condensation heat transfer of nonazoetropic refrigerant mixture R1234yf/R32 inside a horizontal smooth tube *Proceedings from International Refrigeration and Air Conditioning Conference at Purdue*

[4] Del Col D, Bortolin S Cavallini A and Matkovic M 2011 Effect of cross sectional shape during condensation in a single square minichannel *International Journal of Heat and Mass Transfer* **54** 3909-3920

[5] Matkovic M, Cavallini A, Del Col D and Rossetto L 2009 Experimental study on condensation heat transfer inside a single circular minichannel *International Journal of Heat and Mass Transfer* **52** 2311-2323

[6] Lemmon E W, Huber M L and McLinden M O 2013 NIST Standard Reference Database 23: Reference Fluid Thermodynamic and Transport Properties-REFPROP Version 9.1 National Institute of Standards and Technology Standard Reference Data Program NIST Gaithersburg MD

[7] Akasaka R 2013 Thermodynamic property models for the difluoromethane (R-32)+trans-1,3,3,3-tetrafluoropropene (R-1234ze(E)) and difluoromethane+2,3,3,3-tetrafluoropropene (R-1234yf) mixtures *Fluid Phase Equilibria* **358** 98-104

[8] Del Col D, Bortolato M and Bortolin S 2014 Comprehensive experimental investigation of two-phase heat transfer and pressure drop with propane in a minichannel *International Journal of Refrigeration* **47** 66-84

[9] Del Col D, Bortolato M, Azzolin M and Bortolin S 2015 Condensation heat transfer and two-phase frictional pressure drop in a single minichannel with R1234ze(E) and other refrigerants *International Journal of Refrigeration* **50** 87-103

[10] Cavallini A, Censi G, Del Col D, Doretti L, Matkovic M, Rossetto L and Zilio C 2006 Condensation in horizontal smooth tubes: a new heat transfer model for heat exchanger design *Heat Transfer Engineering* **27** 31-38

[11] Del Col D, Cavallini A and Thome J R 2005 Condensation of zeotropic mixtures in horizontal tubes: new simplified heat transfer model based on flow regimes *Journal of Heat Transfer* **127(3)** 221

[12] Bell K J and Ghaly M A 1973 An approximate generalized design method for multicomponent/partial condenser *AIChE Symposium Series* **69** 72-79

[13] Silver L 1947 Gas cooling with aqueous condensation *Transactions of the Institution of Chemical Engineers* **25** 30-42

Experimental investigation on predictive models for motive flow calculation through ejectors for transcritical CO_2 heat pumps

G Boccardi[1], G Lillo[2], R Mastrullo[2], A W Mauro[2], L Saraceno[1], M Pieve[1] and R Trinchieri[1]
[1] Energy Department Production, Conversion and Use of Energy Division, Laboratory of Development of Chemical Processes and Thermofluidynamic,
ENEA, C.R.Casaccia - Via Anguillarese 301, 00123 Roma (Italy)
[2] Department of Industrial Engineering, Federico II University of Naples, p.le Tecchio, 80 - 80125, Napoli (Italy)
e-mail: raniero.trinchieri@enea.it

Abstract. Nowadays, air conditioning systems, especially those used in residential and office buildings, contribute largely to the energy consumptions and to the direct and indirect emissions of greenhouse gases. Carbon dioxide (CO_2) is an interesting option to replace traditional HFCs in vapor compression systems, due to its environmentally friendly characteristics: zero ODP and extremely low GWP. In the case of heat pumps, the use of ejection systems for the expansion phase can contribute to recovery a fraction of the mechanical energy otherwise dissipated as friction, bringing to significant benefits in terms of performance. Currently, at the laboratory DTE-PCU-SPCT of the research center ENEA (Casaccia) in cooperation with the Industrial Engineering Department of Federico II University of Naples, a project is in progress, in order to evaluate experimentally the effect of several ejectors geometries on the global performance of a CO_2 heat pump working with a transcritical cycle. As a part of this project, measurements of the motive flow mass flow rate have been carried out, in transcritical CO_2 conditions. The ejector sizing is a crucial point for the balancing of components and the correct operation of the CO_2 heat pump and therefore the availability of reliable calculation methods for the motive flowrate would be useful.
This paper presents the results obtained by a comparison between the new experimental data and the predictions of some predictive semi-empirical correlations available in the open literature for transcritical CO_2 conditions. Their predictions are analyzed as a function of the main physical parameters of the process to assess their reliability compared to the experimental data. Based on these indications and of the available experimental data, a new semi-empirical correlations and a calculation method based on the hypothesis of isentropic and choked two-phase flow are presented.

1. Introduction

Recently, an experimental campaign on a 30 kW invertible air-to-water heat pump, that uses CO_2 as a working fluid and designed for residential heating and cooling and hot water production, was conducted at ENEA's Casaccia Research Center. The choice of CO_2 as refrigerant is suggested by many positive features (non-flammability, non-toxicity, environmental compatibility), which make it a viable alternative to synthetic refrigerants banned by the latest F-gas regulation [1].

As it is well known, CO_2 heat pumps have significantly higher COPs than similar machines using synthetic refrigerants in domestic hot water production applications (characterized by low gas-cooler inlet water temperatures and high water temperature lift). On the contrary, in residential heating and cooling applications, CO_2 heat pumps may have considerably lower performances than conventional machines [2], due to different gas-cooler operating conditions (high water temperature at gas-cooler inlet and lower temperature lift). In these conditions, the energy dissipated in the lamination phase increases, with consequent reduction of the COP. To reduce such losses, instead of the usual expansion devices, an ejector system can be used to recover part of the expansion work [3].

Many experimental studies on two-phase ejector expansion systems in transcritical carbon dioxide refrigeration cycles have been carried-out. Among others, Elbel et al. [4] made a comparison between a transcritical CO_2 ejector system and a conventional expansion valve system. Results show improvements of COP and cooling capacity of respectively 7% and 8% when using two-phase ejector. In transcritical refrigeration cycles, Lucas et Koehler [5] estimate an improvement up to 17% of the COP with the ejector cycle compared to the maximal COP with expansion valve cycle.

Unlike refrigeration machines, the heat pump works with highly variable operative conditions. For this reason, the use of a single ejector may not bring the desired benefits in all operating conditions. In this regard, the use of several ejectors, working in parallel and conveniently sized, can maximize the efficiency of the system in every operative condition.

The heat pump tested at ENEA facilities had a multi-ejector expansion pack as throttling device.

One of the objectives of the experimental campaign was the development of a thermofluidynamic simulation model calibrated on the experimental data to determine criteria and maps for the ejector sizing. In order to improve the model, correlations to predict efficiency and motive mass flow rates through the ejector are required.

In literature, some methods and semi-empirical correlations for the calculations of the motive mass flow rate were found: one is proposed by Lucas et al. [6]; another is based on the experimental investigation of a R744 multi-ejector by (Banasiak et al [7]; a third one is suggested by Martin et al. [8] for short tube orifices used as expansion devices in CO_2 refrigerant systems. In this paper, the experimental data have been used to calculate the mass flow rate through ejectors by the above mentioned semi-empirical correlations. Besides a new semi-empirical correlation was developed and evaluated by means of a comparison between the literature correlations [6-8] and the experimental data. Finally, an original correlation is proposed, based on the hypotheses of isentropic and choked two-phase flow.

2. Literature survey of existing predictive methods specifically developed for CO_2

In a CO_2 heat pump, the ejectors typically operate with inlet pressures higher than critical value (73.77 bar); moreover, the outlet pressures depend on the evaporator pressure and are usually lower than 40 bar. In this situation, it is reasonable to suppose that the motive flow is choked, Lucas [6], Martin [8]; consequently, the ejector throat pressure is equal to the p_{son} and the mass flow rate only depends on the input conditions.

The accurate prediction of the two-phase mass flow rate in sonic condition is very difficult due to incomplete knowledge of the complex thermal-fluid dynamic phenomena that occur between the two phases. In particular, the sound velocity in two-phase flow can change quickly with the vapour quality [9]. Its calculation is still a matter of discussion M.-S. Chung [10]. Moreover, specific behaviors, issues as metastable conditions and large slip ratio should be taken into account. The equation for calculating the dischargeable mass flow rate "G_r" through a throttling device having a geometric seat area "A" is usually defined as:

$$G_r = k_d \, G_t \, A \qquad (1)$$

where G_t is the theoretical max flux in an ideal (isentropic) nozzle and k_d is the two-phase "discharge coefficient".

Reliable methods for the calculation of mass flow rate when the inlet conditions are supercritical are not available in the literature. This essentially depends on the fact that the available methods use physical quantities related to the vapour state, which are not present in supercritical conditions; e.g., the ISO method [11], derived from the previous HEM [12,13] method, is not recommended when the input pressure is greater than 50% of the critical thermodynamic pressure of the fluid.

In the open literature there are only a few semiempirical correlations based on experimental data to calculate G for CO_2 in supercritical inlet conditions.

Lucas et al [6], propose a method based on the data of an experimental campaign [16] on a multi-ejector heat pump using CO_2 as refrigerant.

Lucas method assumes that the expansion in the converging driving nozzle is isentropic and, consequently, the driving mass flow rate can be calculated by equations:

$$u_{th} = \sqrt{2 \cdot (h_{in} - h_{th})} \tag{2}$$

$$G_{th} = A_{th} \cdot \rho_{th} \cdot u_{th} \tag{3}$$

The density and the enthalpy at the throat can be estimated knowing the throat entropy (equal to the inlet entropy for isentropic flow) and the pressure at the throat.

Correlating experimental data with eq. (2) and (3) with an iterative method, Lucas suggests calculating the ratio η_c between the throat pressure and the inlet pressure with the next equation:

$$\eta_c = \frac{p_{th}}{p_{in}} = 0.0871942 \cdot \frac{p_{in}}{p_{cr}}^{0.9519907} \cdot \frac{\rho_{in}}{\rho_{cr}}^{2.348013} + 0.39387 \tag{4}$$

For the kind of data set used by Lucas (e.g. inlet pressure supercritical, evaporation pressure lower than 40 bar, inlet temperature equal or higher than the CO_2 critical temperature (30.98 °C)), p_{th} corresponds to the choked pressure [8].

Banasiak et al [7] investigated the performance of a multi-ejector for R744 vapour compression units too; they proposed the following correlation for calculation of mass flow rate

$$G = \frac{\pi}{4}\phi^2 \left[A\rho_{in}^{\ 2} + B\rho_{in} + C\frac{p_{in}}{p_{cr}}^2 + D\frac{p_{in}}{p_{cr}} + E \right] \tag{5}$$

where *A,B,C,D,E* are coefficients adjusted for every ejector.
For example, for a ϕ_{th} = 0.7 mm, these coefficients are [7]: *A*=1.71938 10^{-1} m^4 kg^{-1} s^{-1}, *B* = -6.06326 10^1 m s^{-1}, *C*= 4.55787 10^3 kg m^{-2} s^{-1}, *D*= 4.98027 10^4 m^{-2} kg s^{-1}, and *E*= -5.46798 10^4 m^{-2} kg s^{-1}.

Martin et al. [8] conducted an experimental investigation on short tube orifices used as expansion devices in CO_2 refrigerant systems; they suggested a model with dimensionless numbers:

$$G = \phi^2 \sqrt{\rho_{in}\, p_{in}}\, A \left(\frac{L}{\phi}\right)^B \left(\frac{p_{in}}{p_{cr}}\right)^C \left(\frac{T_{in}}{T_{cr}}\right)^D \left(\frac{\phi}{\phi_{cha}}\right)^E \tag{6}$$

where *A,B,C,D,E* are parameters optimized to reach a minimal deviation between measurement and calculation. In the article the values of the coefficients obtained from experimental tests for a short tube are reported, with ϕ=1 mm and L=20 mm: *A*= 0.41, *B*= -0.015, *C*=0.9, *D*=-5, *E*=0.36.

We have tried to find a new semi-empirical correlation to obtain better mass flow rate predictions, using the experimental data available. In particular, in addition to dimensionless parameters already used in the previous correlations, a new geometric parameter was introduced to take into account the ejector geometries.

$$G = \phi^2 \frac{\pi}{4} k \left(A \left(\frac{p_{in}}{p_{cr}} \right)^B \left(\frac{\rho_{in}}{\rho_{cr}} \right)^C \left(\frac{T_{in}}{T_{cr}} \right)^D \left(\frac{L}{\phi_m} \right)^E + F \right) \tag{7}$$

A, B,C, D, E and F are calculated by a mathematical solver that reach the minimum MSE value and $k=10^5$ is a constant used to optimize the solver process.

Another correlation is proposed, considering that the saturation pressure, calculated with the isentropic expansion hypothesis from the input conditions, could be an indicator of the ejector output conditions. In fact:

- during the expansion at pressures lower than 40 bar in the convergent of the motive nozzle, saturation conditions are often reached [8]
- it can be assumed that, in these situations, the sonic velocity is reached at low vapour quality [10]
- the flow through orifices and nozzles can be assumed isentropic

The pressure thus calculated, called p_{ise}, has been correlated to the measured flow rate; the investigations have allowed the identification of a simple correlation expressing the mass flow rate as function of the parameter $\eta_{ise}=p_{ise}/p_{in}$ and the ejector diameter ϕ_{ej}:

$$G = \phi_{ej}^{\ 2} \frac{\pi}{4} \left[a + b\ \eta_{ise} \right] \tag{8}$$

Where *a* and *b* are function of ϕ_{ej} and are determined by experimental data.

$$a = \left(A + B\phi_{ej} + C\phi_{ej}^{\ 2} \right) \tag{9}$$

$$b = \left(D + E\phi_{ej} + F\phi_{ej}^{\ 2} \right) \tag{10}$$

3. Experimental setup

3.1 Test facility

The experimental data used in this paper were obtained testing a 30 kW CO_2 air-water heat pump having a multi-ejector expansion pack as throttling device. The tests were performed using the experimental facility "Calorimetro Enea" at ENEA (Casaccia) research center. This climatic chamber allows testing air-to-water reversible heat pump, with thermal capacity until 50 kW, according to UNI EN 14511/2011 [15].

The experimental set consists of an alternative semi-hermetic compressor (CP) driven by an inverter, a plate heat exchanger (GC), a finned coil (EVAP), a plate internal heat exchanger (IHE), an electronic valve (EEV) and a multi-ejector expansion pack (EJEC). Figure 1 shows the multi-ejector CO_2 system including four different ejector geometries, with throat diameters from 0.7 mm to 2.0 mm.

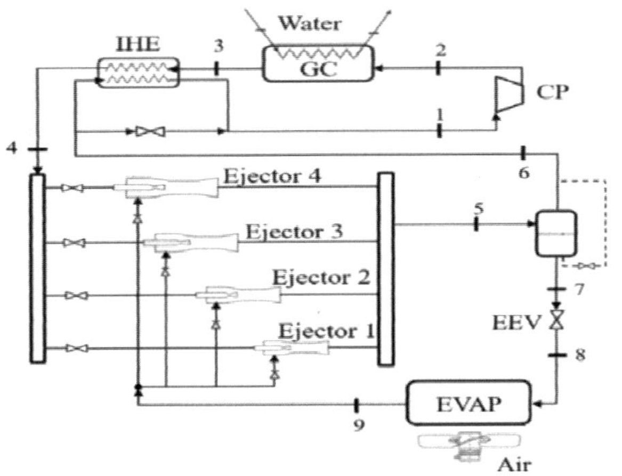

Figure 1. Layout of multi ejector CO_2 system

The heat pump control system can active each ejector independently, depending on boundary conditions, with 15 different configurations; an ejector schematic is shown in figure 2.

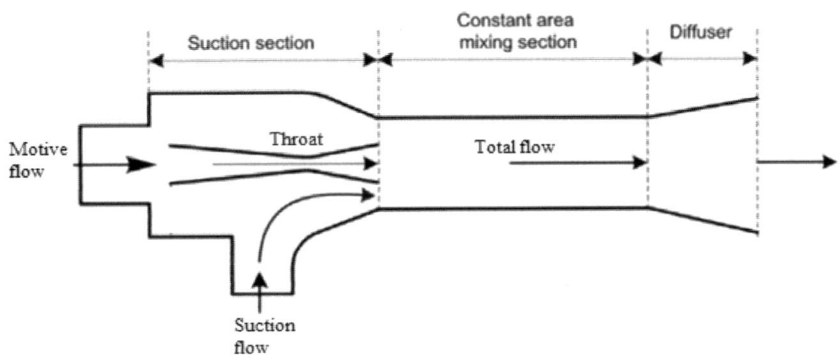

Figure 2. Schematic of the ejector system

Temperature and pressure sensors are installed at the inlet and outlet of each component. K-type and J-type thermocouples were placed to measure the temperatures and piezoelectric transmitters to measure the pressure. An electromagnetic transmitter measures the water volumetric flow rate. The electrical power is measured by a wattmeter. CO_2 mass flow rate is measured with a Coriolis flow meter. All measurement instruments are characterized by high accuracy, according to [15].

Table 1 summarizes the instrument specifications and their uncertainty, as indicated by the manufactures.

Table 1. Measurement instrument and calibrated uncertainties.

Measurement	Range/Unit	Uncertainty
Temperature (K-type)	0/150 °C	± 1.1 °C
Temperature (J-type)	-40/80 °C	± 1.1 °C
Pressure	0-60/0-100/0-160 bar	0.08%
Water volumetric flow rate	0/3.33x10^{-3} m^3/s	0.02% of reading
Electrical power	0/25 kW	Precision class 0.5
CO_2 mass flow meter	0/0.95 kg/s	± 0.10% of reading

3.2 Experimental procedure
All tests were run according to [15]. The boundary conditions in terms of air temperature, relative humidity and water temperature and mass flow rate were fixed and kept constant during the tests. In addition, it was possible to set the outlet evaporator and inlet compressor superheating. Furthermore, it was possible to set manually the chosen ejectors configuration; tests here analyzed were done with ejectors 2, 3, 4 worked one by one. The trends of the thermodynamic measured variables were monitored via a software designed specifically. The main thermodynamic parameters were recorded and processed using Matlab software. The instabilities generated by the control system regulation are limited in according to [15].

Table 2 shows the motive nozzle throat section diameters and the available experimental data subdivided by the inlet temperature range. All tests are with inlet pressure higher than critical. The outlet pressures are between 20÷40 bar for all the tests and they are determined by typical operating conditions of heat pumps in heating applications.

Table 2. Motive nozzle throat sections diameters and experimental test distribution

Ejectors [#]	nozzle throat diameter [mm]	Test		
		$T_{in}<T_{cr}$	$T_{in}>T_{cr}$	Tot
1	0.7	-	-	-
2	1.0	18	0	18
3	1.4	15	4	19
4	2.0	7	12	19

Figure 3 shows the tests input conditions in the T-s diagram. Most of the ejector 4 tests have $T>T_{cr}$, while the ejectors 2 and 3 tests are almost all with $T<T_{cr}$.

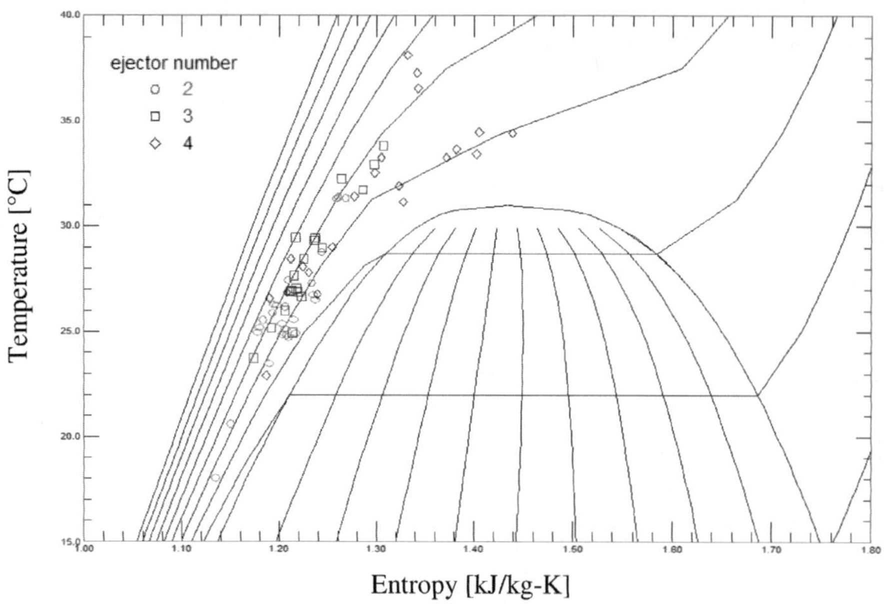

Figure 3. Test input conditions for the three ejectors in the T-s diagram

4 Results and discussion
The mass flow rate calculated applying Lucas's correlation (2), (3), (4) with the coefficients proposed in [5] for p_{th} calculation, figure 4, are affected by high errors (MSE =37.5 10^{-5}). Using the solver to optimize coefficients of (4) respect to experimental data, predictions are improving, but maintaining a high value of MSE =21.18 10^{-5}.

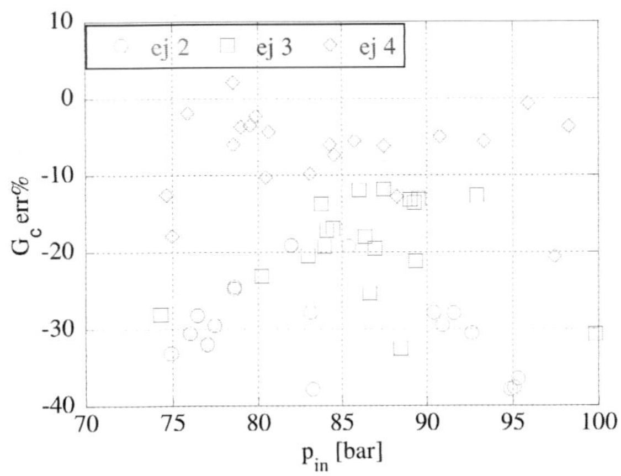

Figure 4. G_c error% vs p_{in} - Lucas (2,3,4)

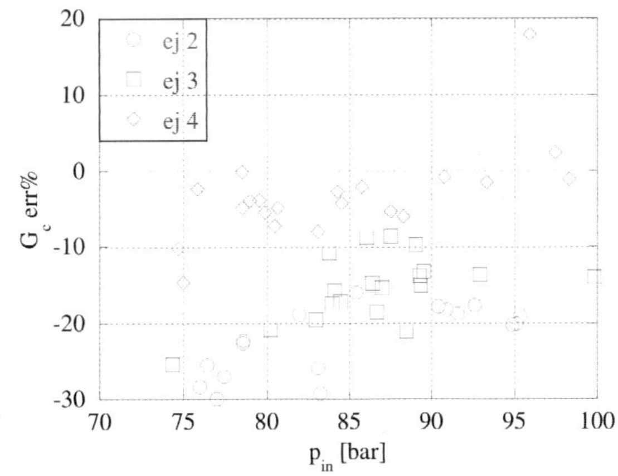

Figure 5. G_c error% vs p_{in} - Lucas (2,3,4) with coefficients recalculated by solver

The recalculated coefficients in (4) are: 0.86535, -0.50933, -0.26973 and the constant term is 0. Figure 5 shows mass flow rate perceptual errors as a function of the ejector p_{in}: predictions are almost always underestimated, but for ejector 4 are closest to the optimal values in comparison with the predictions for the other ejectors. Figure 6 depicts the percentage errors for Banasiak correlation (5) optimized with new experimental data (*A*=0.07561, *B*=30.66, *C*=-3468.23, *D*=47016.37, *E*=-53737.9); MSE improves to 13.11 10^{-5} and error values are between -20 and 20%. The three ejector predictions are almost parallel and errors increase by passing ejector 2 to ejector 4, remarking an uncaptured effect of the size of the system in this type of correlation. To understand these results we would like to point out that, in the spirit of this study, we have tried to use this correlation for the three ejectors simultaneously while the author has suggested that the coefficients for each geometry should be defined. Indeed, using the correlation individually for each ejector, as indicated by Banasiak, the obtained MSEs for ejectors 2 and 3 are good, about 2 and 4 10^{-6} respectively, and the calculated coefficients are close together. Instead, prediction for ejector 4 showed a relatively high MSE, 9.8 10^{-5} and very different coefficients.

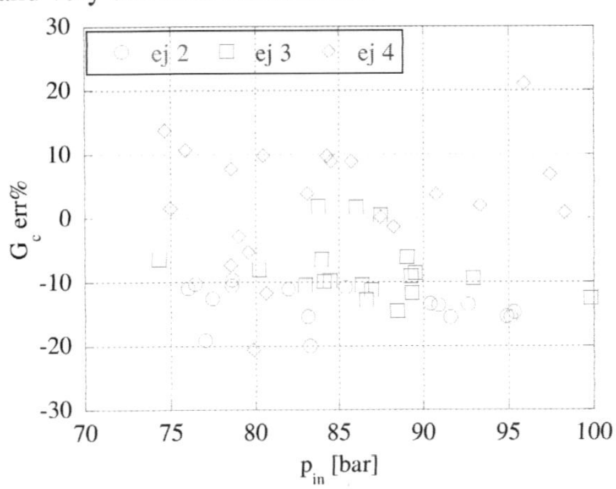

Figure 6. G_c error% vs p_{in} - Banasiak (5) with coefficients recalculated by solver

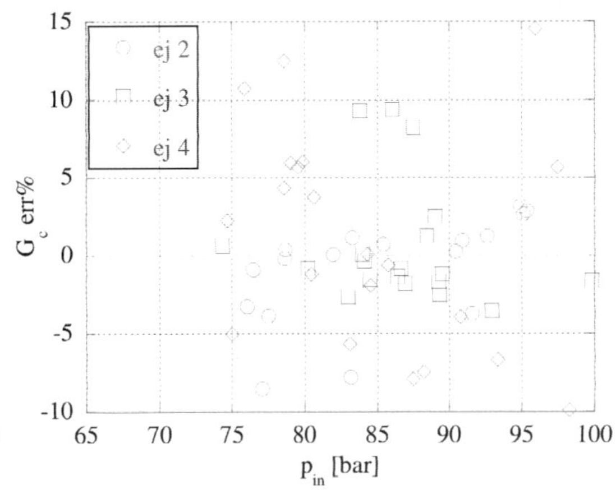

Figure 7. G_c error% vs p_{in} - Martin (6) with coefficients recalculated by solver

In Martin's correlation (6), developed for short tubes, there are two terms related to geometry. To fit the ejector geometry, the term ϕ/ϕ_{ch}, related to E coefficient, was excluded and the other, L/ϕ, was calculated by setting L=nozzle length and ϕ= average nozzle diameter. Martin's correlation predictions are good with MSE = 4.83 10^{-5} with A=1.04901, B=0.72643, C=1.50484, D=-1.11549; E was not calculated. Almost all errors are between -10 and 10% and there are no particular differences for the three ejectors (figure 7).

Figure 8 depicts the predictions of the new correlation (7) where the parameter L/ϕ is the same introduced for Martin's correlation and coefficient are: A= 2.7415, B= 0.4084, C=0.0590, D=0.05048, E=-0.2360 and F=-2.3640. MSE value is 4.08 10^{-5} and most errors are between -7 % and 6% and do not seem to be particularly affected by the ejector geometry.

Fig. 9 shows the relationship between the specific mass flow rate and the η_{ise} for the three ejectors with the relative linear fit function.

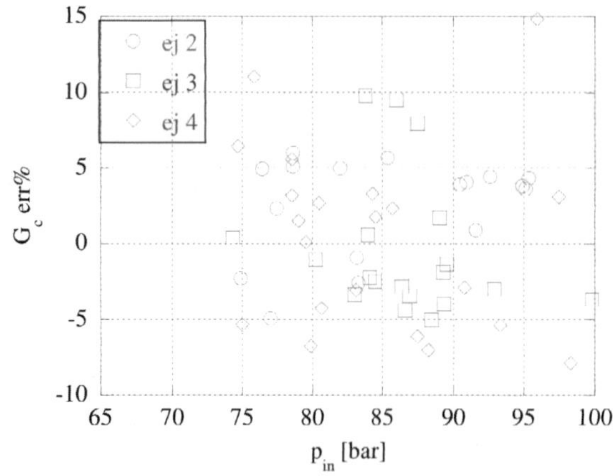

Figure 8. G_c error% vs p_{in} - new correlation (7) with coefficients calculated by solver

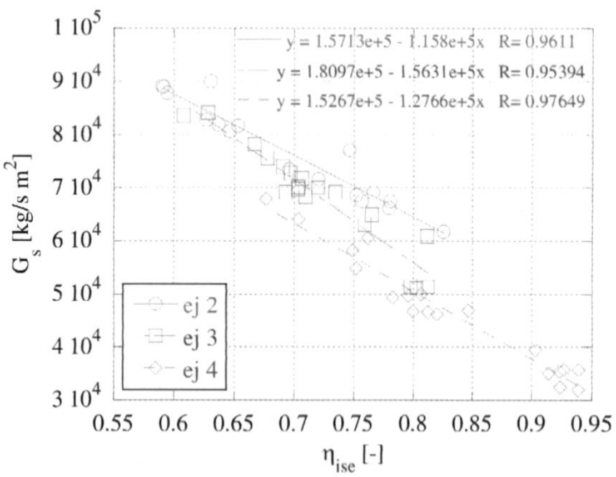

Figure 9. G_s vs η_{ise}

By interpolating the terms of the three linear functions, we obtained the coefficients A=-60017, B=329680, C= -112530, D=203090, E= -473400, F= 155440 to be inserted into (9) and (10) to complete the correlation (8), that expresses the flow rate according to the η_{ise} and the ejector diameters.

Figure 10 depicts the G errors of this method as function of pin for the three ejectors. The MSE=4.02 10^{-5} is almost equal to that of the previous correlation. The most errors are between -5 % and 11% and do not seem to be particularly affected by the ejector geometry.

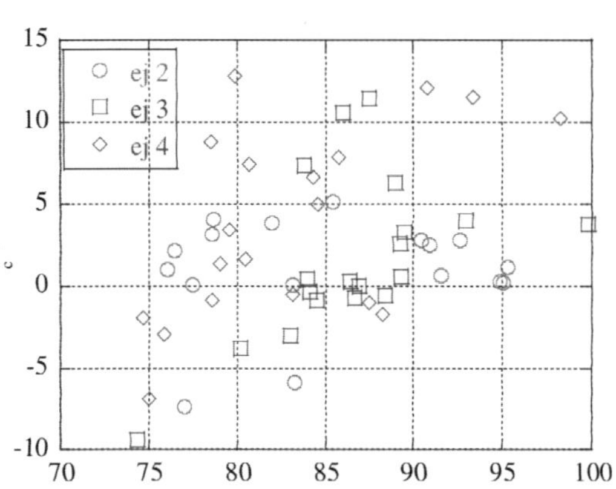

Figure 10. G_c error% vs p_{in} - new method with coefficients calculated by solver

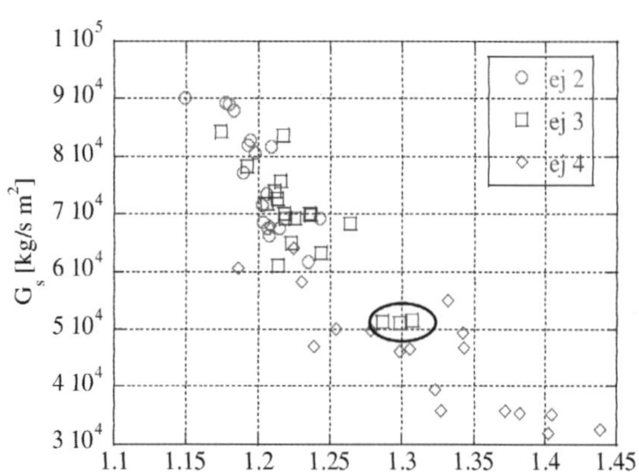

Figure 11. G_s vs Entropy$_{in}$

IOP Conf. Series: Journal of Physics: Conf. Series **923** (2017) 012018 doi:10.1088/1742-6596/923/1/012018

In keeping with eq. (1), calculations have shown how correlations without a parameter related to ejector geometry have worse predictions.

The correlations analyzed, except for (8), use the transcritical conditions (p_{cr}, ρ_{cr}, T_{cr}) to look for representative parameters of the mass flow rate or to compute η_c (4). Also the reliability of calculation method (8) depends on input conditions because they may affect the validity of its initial hypotheses.

In these situations, different input conditions, even considering that expansion is exiting from the transcritical zone, may be a reason for the different reliability of correlations. By going to check the test input conditions (figure 3 and figure 11) we noticed that almost all the tests performed with the ejector 4 were characterized by different values of some physical inlet quantities compared to the other ejectors; the most of experimental points relative to ejectors 2 and 3 are almost in the same range of entropy (1.16÷1.28 kJ /kg) and temperature (23÷32 °C) while tests of ejector 4 have s=1.2÷1.43 kJ /kg and T_{in}=23÷43 °C.

This could explain because:

- the correlation (5), used individually for each ejector, has good predictions for ejectors 2 and 3 and the calculated coefficients are close together while its reliability is much worse for ejector 4.
- the three tests circled in the figure 11, have worse predictions for that geometry (ejector 3) with every correlations.

5 Conclusions

In order to improve the knowledge about ejector performance, experimental tests were carried-out at the DTE-PCU-SPCT laboratory of the ENEA (Casaccia) research center. A complete heat pump system with multi-ejector pack has been tested in a climatic chamber. Transcritical motive nozzle inlet conditions, with pressure range between 74 bar and 98 bar, have been investigated. Therefore, new experimental data in terms of motive nozzle mass flow rate have been obtained. The results have been elaborated using three existing semi-empirical correlations, Lucas et al. [5], Banasiak et al. [7] and Martin et al[8] in order to find the best methods to assessing the motive nozzle mass flow rate. Moreover, a new semi-empirical correlation and a new calculation method have been proposed and evaluated. We observed that:

- The semi-empirical correlations available in literature seem to be very dependent on experimental test from which they have been deduced (Lucas, Banasiak) and require dedicated coefficients for each specific geometry.
- Semi-empirical correlations (Martin, new correlation) that provide a parameter related to the ejector geometry can be used with the same coefficients for different ejectors.
- The method calculating the mass flow rate as a function of η_{ise} and ϕ_{th} of ejectors is interesting and has comparable performance to the new correlation.
- The semi-empirical correlation of Martin, the new correlation and the isentropic method are able to predict the ejector 4 mass flow rate with an acceptable accuracy, while the correlation of Banasiak, which is optimal for the individually treated ejectors 2 and 3, has significantly worse predictions for ejector 4. Instead, Lucas's correlation has the best performance for ejector 4.
- Elaborations performed on the new correlation showed that the introduction of the geometric parameter involves an improvement of more than 50% of the MSE.
- There is a dependence on the reliability of the isentropic method from the test input conditions.

The objectives pursued in the future of the research will concern:

- The identification and study of phenomena occurring during expansion and their correlation with the evolving mass flow rate.
- The improvement of the semi-empirical correlations and the isentropic method.

In particular, it will be evaluated the possibility of identifying methods and/or correlations to be used with reference to CO_2 input conditions.

Nomenclature

A	area	(m^2)
G	mass flow rate	(kg/h)
h	specific enthalpy	(J/kg)
k_d	discharge coefficient	(-)
T	temperature	(°C)
u	velocity	(m/s)
s	specific entropy	(kJ/kg K)
p	pressure	(bar)
L	length	(m)
greek letters		
ϕ	diameter	(m)
ρ	density	(kg/m^3)
η	pressure ratio	(-)

subscripts

c	calculated
cr	thermodynamic critical conditions
cha	chamfer
ise	isentropic
in	inlet conditions
m	average value
r	actual
s	specific
t	theoretical
th	throat conditions
son	sonic (choked) flow condition

References

[1] Regulation (EU) No 517/2014 of the European Parliament and the Council of 16 April 2014 fluorinated greenhouse gases and repealing Regulation (EC) No 842/2006 *Off. J. Union*

[2] Minetto S, Cecchinato L, Brignoli R, Marinetti S and Rossetti A *Int. J. Refrigeration* vol **63** pp 237-50

[3] Trinchieri R, Boccardi G, D'Annibale F, Franchi A, Lattanzi A, Morlacca M, Saraceno L, Scotini A, Simonetti L, Petroni G, Lillo G 2016 Caratterizzazione sperimentale del funzionamento di un sistema ad eiettori multipli in una pompa di calore aria-acqua a CO_2 *Report RdS/2015/170*

[4] Elbel S and Hrnjak P 2008 *Int. J. Refrigeration* vol **31**(3) pp 411–22

[5] Lucas C, Koehler J. 2008 *Int. J. Refrigeration* vol **35**(6) pp 1595-603

[6] Lucas C, Koehler J, Schroeder A, Tischendorf C *Int. J. Refrigeration* vol **36** (3) pp 881-91 http://dx.doi.org/10.1016/j. ijrefrig.2012.10.035

[7] Bansiak K, Hafner A, Kriezi E E, Madsen K B, Birkelund M, Fredslund K and Olsson R 2015 *Int. J. Refrigeration* vol **57** pp 265-76

[8] Martin K, Rieberer R and Hager J 2006 *Int. Refrigeration and Air Conditioning Conf.* Paper 781 http://docs.lib.purdue.edu/iracc/781

[9] Fisher H G, Forrest H S, Grossel S S, Huff J E, Muller A R, Noronha J A, Shaw D A, Tilley B J 1992 Emergency relief system design using DIERS technology *The DIERS Project Manual* AIChE, New York

[10] Chung M -S, Park S –B, Lee H-K, 2004 *J. of sound and vibration* vol **276** pp 13-26

[11] ISO/DIS4126-10 2008 Safety devices for protection against excessive pressure – Part 10 Sizing of safety valves for gas/liquid two-phase flow

[12] Leung, J C, 1986, *AIChE J.* vol **32** pp 1743–46

[13] Leung J C 1995 *Int. Symp. on Runaway Reactions and Pressure Relief Design* Boston pp 367-393

[14] Lucas C, Koehler J, 2012 *Int. J. Refrigeration* vol **35**(6) pp 1595-1603

[15] UNI 2011 Norma UNI EN 14511 Air conditioners, liquid chilling packages and heat pumps with electrically driven compressors for space heating and cooling Part 3: Test methods

IOP Conf. Series: Journal of Physics: Conf. Series **923** (2017) 012019 doi:10.1088/1742-6596/923/1/012019

Effects of Force Fields on Interface Dynamics, in view of Two-Phase Heat Transfer Enhancement and Phase Management for Space Applications

P Di Marco, G Saccone

DESTEC, University of Pisa, Pisa, Italia

E-mail: p.dimarco@ing.unipi.it

Abstract. On earth, gravity barely influences the dynamics of interfaces. For what concerns bubbles, buoyancy governs the dynamics of boiling mechanism and thus affects boiling heat transfer capacity. While, for droplets, the coupled effects of wettability and gravity affects interface exchanges. In space, in the lack of gravity, rules are changed and new phenomena come into play. The present work is aimed to study the effects of electric field on the shape and behaviour of bubbles and droplets in order to understand how to handle microgravity applications; in particular, the replacement of gravity with electric field and their coupled effects are evaluated. The experiments spread over different setups, gravity conditions, working fluids, interface conditions. Droplets and bubbles have been analysed with and without electric field, with and without (adiabatic) heat and mass transfer across the interface. Furthermore, the results of the 4 ESA Parabolic Flight Campaigns (PFC 58, 60, 64 & 66), for adiabatic bubbles, adiabatic droplets and evaporating droplets, will be summarized, discussed, and compared with the ground tests.

1. State of the art

Boiling and evaporation are ubiquitous phenomena. They are relevant for all the industrial applications involving heat transfer mechanisms, most of all: electrical power production. In both phenomena, gravity has a relevant role: it is responsible of bubble detachment during boiling [1], [2] and "flattens" the droplets [3] assisting heat transfer in both processes. The present study wants to extend and use these mechanisms, which are the most efficient heat transfer methods, to space applications. Phase separation and thermal management of space stations and satellites are becoming of growing interest and the reduction of size and weights calls the increase of systems performances and efficiencies [4].

The lack of gravity in such applications can be balanced by the use of a tunable electric field (EF), which creates body forces that substitute gravity. Many published works are available in literature proving the viability of this solution for microgravity applications [5][6][7][8][9][10][11][12].

1.1. Boiling

Boiling heat transfer has been widely studied in the past, in particular, several studies have addressed and speculated on the mechanisms leading to critical heat flux (CHF), the limiting condition of all boiling heat transfer processes, generally associated with the formation of an irreversible dry spot responsible of a drastic reduction of the local heat transfer coefficient, which can lead to catastrophic burnout of the boiling surface. Rosenhow [2] and Zuber [1] are two of the milestones in the explanation of boiling mechanics and limits. In the present study we want to underline the importance of the electric field in increasing the limiting value of CHF, thus increasing safety and capacity of the heat transferred from a surface.

By considering the formation of vapor jets, their velocity and the competition between vapor, moving upwards, and liquid, moving downwards, Zuber [1] proposed a correlation ($K = 0.131$):

$$q''_{CHF} = K h_{lv} \rho_v \left[\frac{g\sigma(\rho_l - \rho_v)}{\rho_v^2} \right]^{0.25} \quad 1$$

Nowadays several techniques can be adopted to enhance CHF beyond the limits predicted by these correlations. Specifically, CHF can be enhanced using passive or active techniques, such as micro-structured surfaces and electric field, respectively.

It is well established that in CHF the dynamics of the contact line plays an important role, and surface conditions, e.g. roughness, wettability and porosity play a role too. As we mentioned, buoyancy determines the efficiency of the boiling process creating bubbles detaching forces; however, it has been shown that the application of an external electric field can improve the boiling heat transfer performance increasing bubble detaching forces and modifying the wettability of the fluid-surface system [13] [14][15][16][17].

1.2. Drop evaporation

When a sessile droplet is exposed to a static electric field, its shape is defined by a balance between surface tension, electrostatic and gravitational forces. The surface tension component tends to make the droplet spherical, the electrostatic force elongates the droplet towards the electrode and the gravitational "flattens" the droplet. The droplet shape and its evaporation rate are strictly linked to triple line movement and wettability. Force fields (namely, a static electric field and the gravitational one) do have a role on triple line dynamics, wettability and heat transfer [18]. Studies showed that different modes of evaporation are possible, depending on the triple line dynamics [19][20][21][22] and thus depending on the hysteresis of the contact angle [23]. A drop can evaporate with a constant contact area overtime, the triple line is pinned and the evaporation is characterized by a decrease of the contact angle during the evaporation. A drop can also evaporate with a constant contact angle overtime; in this case evaporation occurs with a recession of the triple line [24].

2. Interface stability under the action of the electric field

One important first step in deriving an expression for the volumetric electric force (f_e''') that acts on a fluid has been proposed by Landau and Lifsitz [25]:

$$f_e''' = \rho_f E - \frac{\varepsilon_0 E^2}{2} \nabla \varepsilon_R + \frac{\varepsilon_0}{2} \nabla \left(\rho \left(\frac{\partial \varepsilon_R}{\partial \rho} \right)_T E^2 \right) \quad 2$$

Only the first term (electrophoresis) depends on the sign of the electric field, but it is present only when free charge (ρ_f) buildup occurs. It doesn't play a role in the present study, where dielectric fluids are involved. The second and third terms are termed dielectrophoresis and electrostriction. The former one depends on the gradient of the dielectric constant (ε_R, which may occur either from thermal gradients and/or from phase discontinuities) and the latter one is due to the gradient of E^2. According to continuum mechanics, Eq.2 can be expressed as the divergence of a stress tensor. $\overline{\overline{T}}_e$ is the Maxwell stress tensor, that is:

$$f_e''' = \nabla \overline{\overline{T}}_e \quad 3$$

The resulting force acting on a surface S ($\overline{n}$, its unit vector) is:

$$F_e = \iint_S \overline{\overline{T}}_e \cdot \overline{n} \, dS \quad 4$$

Thus, the momentum balance on a bubble in the direction perpendicular to the surface can be written as

$$F_\sigma + F_b + F_p + F_e = 0 \qquad 5$$

where the different terms are defined as:

Forces	Bubbles	Droplets	
Surface Tension	$F_\sigma = -\pi D\sigma \sin\theta$	$F_\sigma = -\pi D\sigma \sin\theta$	6
Buoyancy	$F_b = (\rho_f - \rho_g)gV$	$F_b = (\rho_g - \rho_f)gV$	7
Pressure	$F_p = \frac{\pi D^2}{4}\left(\frac{2\sigma}{R_0} - \rho_f gH\right)$	$F_p = \frac{\pi D^2}{4}\left(\frac{2\sigma}{R_0} + \rho_f gH\right)$	8
Electrical	$F_e = \int_S \overline{\overline{T_{e,g}}} \cdot \overline{n}\, dS$	$F_e = \int_S \overline{\overline{T_{e,g}}} \cdot \overline{n}\, dS$	9

Table 1: Force balance of Eq. 5 extended for both bubbles and droplets

Difficulties in calculating the Maxwell stress tensor arise from the evaluation of the local electric field, which is affected by bubble/droplet size and shape and, in turn, it influences bubble/droplet size and shape. Clearly, the term E^2 and its gradient are the dominating terms in the Maxwell stress tensor which determines the forces associated with the electric field.

The application of an external electric field, results in an increase in pressure and surface tension, obtained respectively by an increase in the height of the shape, a higher curvature radius at its top and a larger contact angle. This behavior has been experienced in all cases: bubble or droplet, micro or normal gravity. The electric field always "calls" the object towards the direction of higher potential (see Figure 1, 2, 3, 4).

Figure 1: bubble with and without EF (3.33 MV/m) normal gravity field

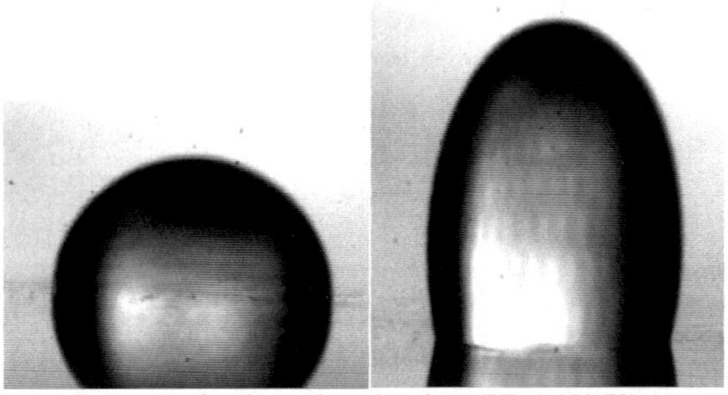

Figure 2: droplet with and without EF (1.25MV/m) normal gravity field

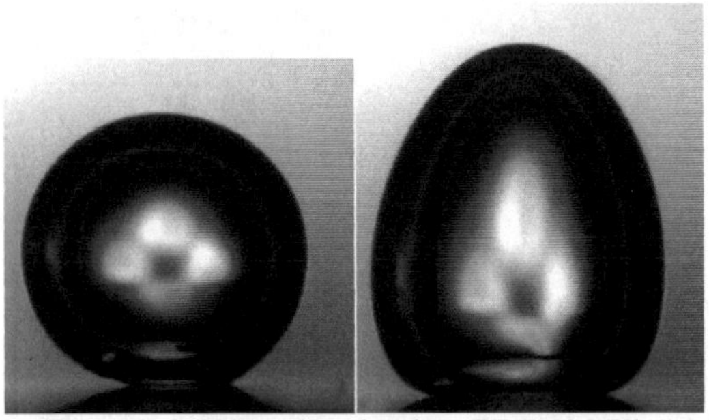

Figure 3: bubble with and without EF (3.33 MV/m) microgravity (58th PFC 2013 – 60th PFC 2014)

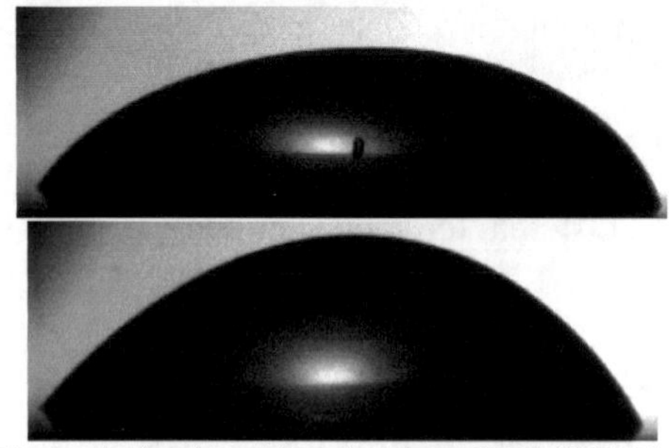

Figure 4: droplet with and without EF (0.833MV/m) microgravity (64th PFC 2016 – 66th PFC 2017)

A last use of mathematics is here necessary to evaluate the force balance in the radial direction. This force balance has been proposed initially for droplets [26], but found its use also to understand bubble behavior. Considering a slice of a resting droplet (see Figure 5) as control volume, a balance between pressure and surface tension is achieved. Forces are shown with arrows.

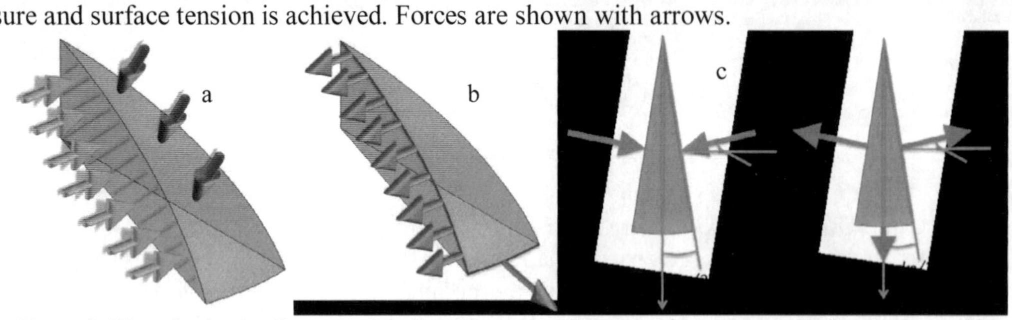

Figure 5: Slice of a droplet. Pressure of the liquid in orange. Pressure of air in green. Surface tension in blue.

The pressure, and therefore, spreading driving force can be written as:

$$F_p = 2\iint \left(p_f(z) - p_g(z)\right)(\vec{n}\cdot\vec{r})\,dr\,dz = \int_0^H\int_0^{r(z)} \left(p_f(z) - p_g(z)\right) d\phi\,dr\,dz \quad (10)$$

The surface tension, and therefore, restraining force can be written as:

$$F_\sigma = 2\,\sigma\,L\,\sin\left({}^{d\phi}/_{2}\right) + \sigma\,\cos\vartheta\,r\,d\phi = \sigma\,L\,d\phi + \sigma\,\cos\vartheta\,r\,d\phi \quad (11)$$

Finally, the equilibrium of the slice $d\phi$ is:

$$\int_0^H\int_0^{r(z)} \left(p_f(z) - p_g(z)\right) dr\,dz = \sigma\,L + \sigma\,r\cos\vartheta \quad (12)$$

or:

$$F_p = F_\sigma \quad (13)$$

Integration of the first term of this balance has to be carried on numerically, point by point. In this study it has been performed from the top to the base. It is worth noting that in the restraining force (Eq. 11) the second term can be either positive or negative (see the direction of blue arrows in Figure 5d), depending on the value of the contact angle. Again, in presence of electric field, we will derive the resulting electric force from the algebraic sum of the others.

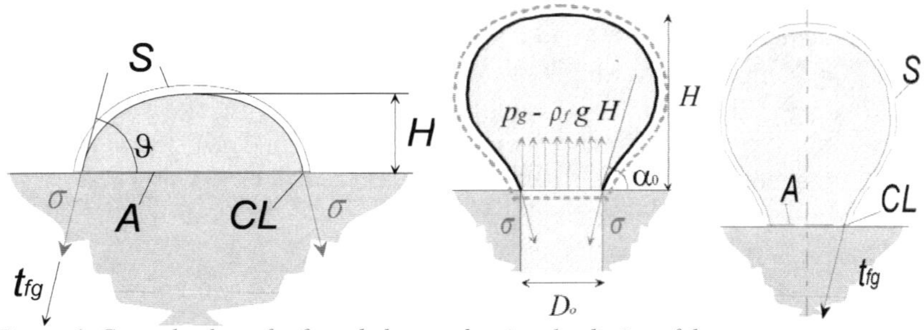

Figure 6: Control volume for force balances showing the diction of the parameters

3. Experimental apparatus

Due to the nature of the experiments, different experimental setups will be here described.

3.1. Bubbles setup

For what concerns bubbles (see Figure 7), it consists in a bubble generator, comprising a flat stainless steel plate with a circular orifice of 0.5 mm internal diameter drilled on it. The orifice was fed with nitrogen by means of a gas flowrate controller. In this way, continuous trains of detaching bubbles, or individual bubbles growing very slowly can be generated. The test cell is an aluminum box filled with the test liquid (FC-72). An electric potential up to 25 kV dc could be applied to a washer-shaped electrode (10x4 mm, 1 mm thickness), located 6 mm above the plate and centered with the orifice axis.

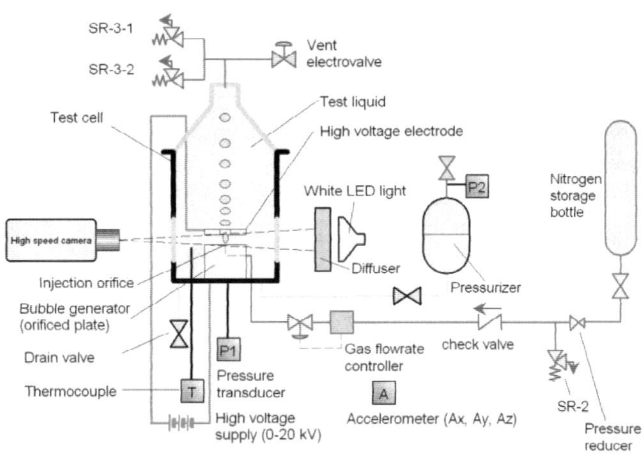

Figure 7: Experimental setup for bubbles (58th PFC 2013 – 60th PFC 2014)

3.2. Droplet setup

For droplets instead (see Figure 8), the experiment's configuration consists of the same test cell provided with a stainless steel plate surmounted by a washer shaped electrode, 6 mm far from the surface, connected with a power supply generating up to 7.5 kV. A syringe connected to a micrometric slide, moved by a stepper motor, assures constant volumetric flow rate through an orifice drilled on the plate. Different working fluids have been tested (Water, FC72, HFE7100).

3.3. Data processing

Images are acquired with a high-resolution camera, equipped with microscopic lens, at 500 fps for bubbles and 50 fps for droplets. The definition achieved was about 5 μm/pixel.

The profile and all relevant parameters are derived via a dedicated software implemented in MATLAB, with the aid of the Image processing toolbox. At first, the background is subtracted to reduce noise, the contour is identified with a threshold (Canny method) and the contact angle is measured by polynomial regression of the points of the profile close to the contact line. The number of pixel of the regression curve used is function of the height of the drop, while the degree of the polynomial curve is 2 and all points are weighted in function of the distance from the contact point.

The radius of curvature at drop top also is calculated fitting the 7 highest points of the profile with a parabola, while the drop volume is calculated with Guldin theorem, specific for rotational bodies. So on, all geometrical parameters of interest are derived from the image processing code.

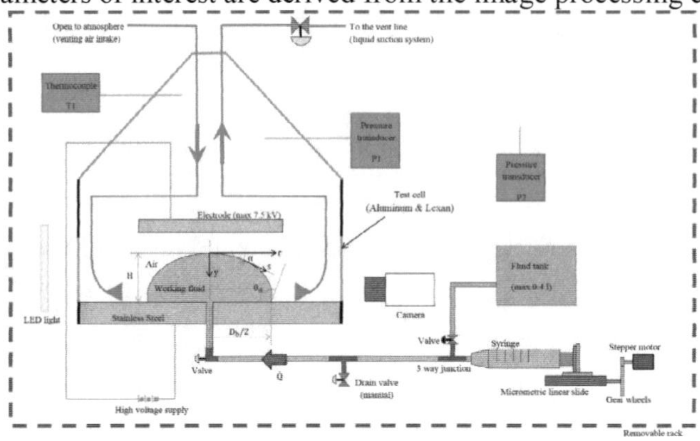

Figure 8: experimental setup for droplets (64th PFC 2016)

4. Results and discussion

A complete review of all the results is behind the purpose of this work. In the present paper we would like to summarize all the findings showing common aspects, achieved goals and particular experienced behaviors. In general, this work can confirm that:

- The electric field exerts a negative radial force (see Figure 10,Figure 12). It means that it shrinks bubbles and droplets. This force is responsible of:
    1. elongation in the vertical direction (see Figure 1, 2, 3, 4),
    2. bubble departure with reduced diameter in microgravity (see Figure 13),
    3. when dielectric fluids are used, formation of a liquid jet; when water is used, dielectric breakup (see Figure 15 and 16).
- The electric field exerts a negative vertical force. It means that pushes bubbles and droplets against the surface (Figure 9, Figure 11).

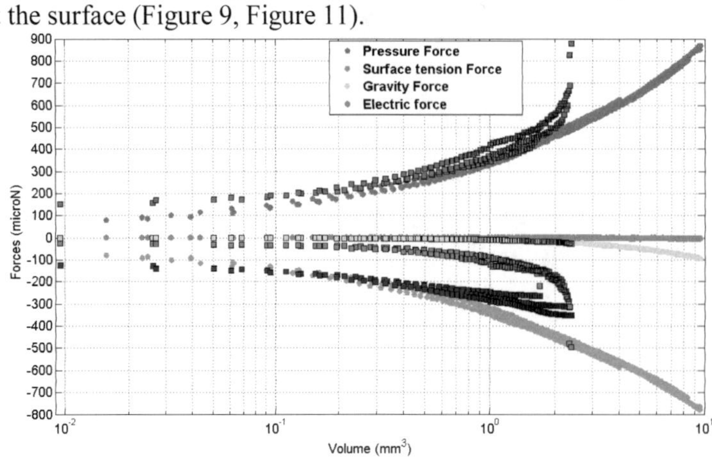

Figure 9: Vertical force balance for the 0 kV (circles) and 7.5 kV (squares with black border) cases. Droplet normal gravity.

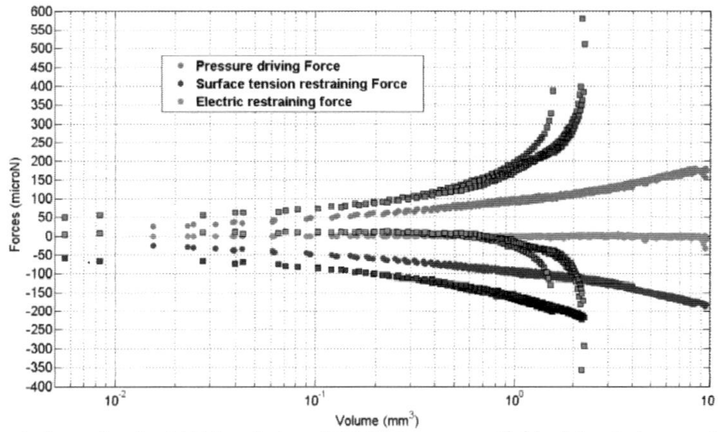

Figure 10: Radial force balance for the 0 kV (circles) and 7.5 kV (squares with black border) cases. Droplet normal gravity.

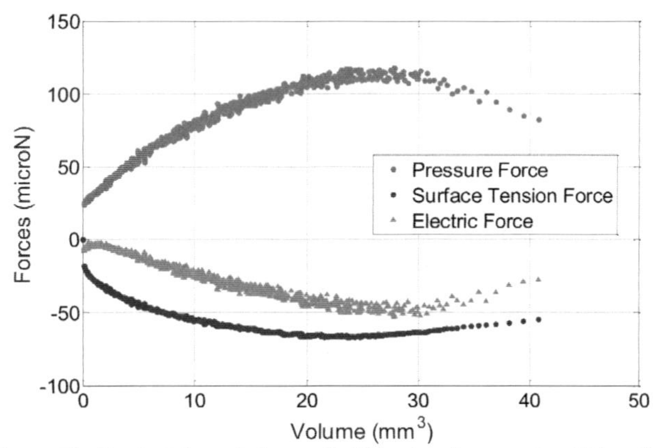

Figure 11: Vertical force balance. Bubbles in microgravity, EF applied

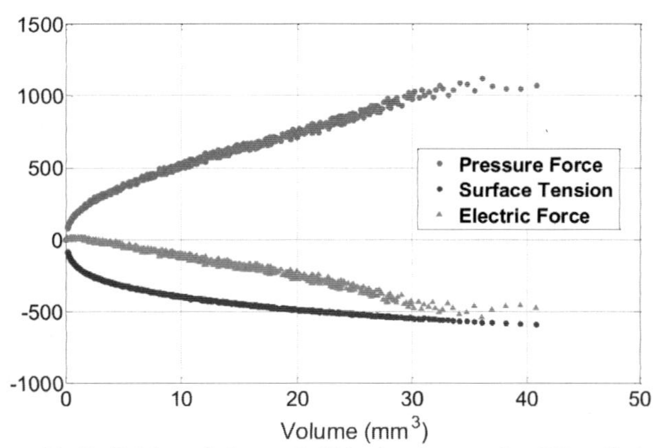

Figure 12: Radial force balance. Bubbles in microgravity, EF applied

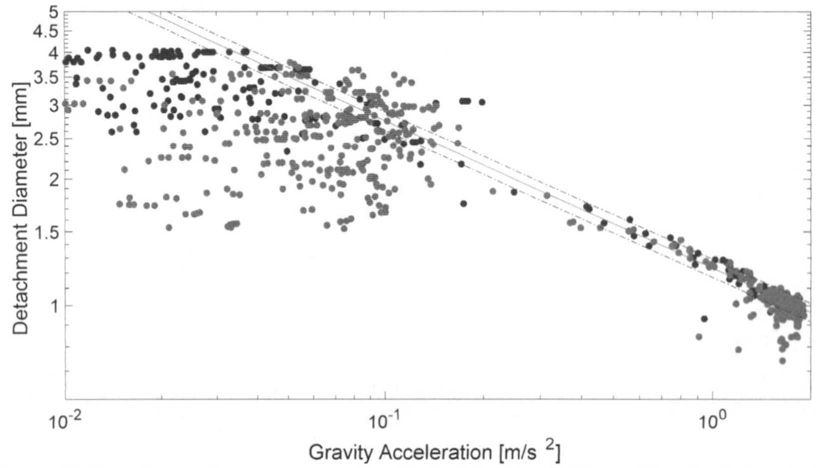

Figure 13: Detachment diameter vs gravity with (red dots) and without (blue dots) electric field.

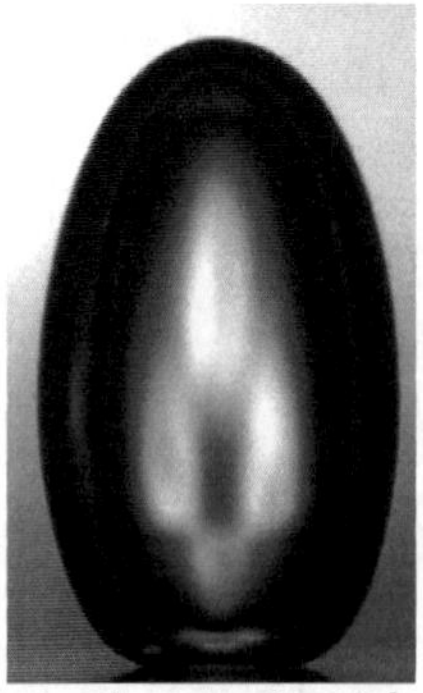
Figure 14: bubble in microgravity, right before departure

Figure 15: Taylor cone in microgravity and liquid jet, HFE7100

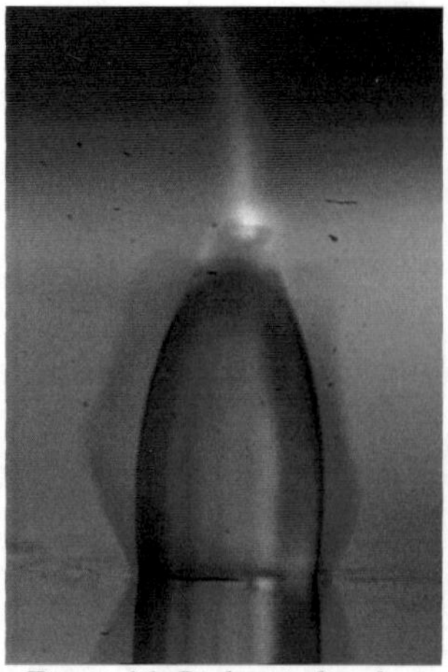
Figure 16: Dielectric breakup, Water

5. Heat and mass transfers with electric field

Introducing Eq. 2, we affirmed that the electric force is a volume force. Thus, the resulting electric force depends on the volume. By analyzing a droplet withdrawal, we found some relevance for what concerns droplet evaporation assisted by electric field. Looking at Figure 17, we can verify that the vertical electric force presses the droplet against the surface, but its effect vanishes with vanishing volume.

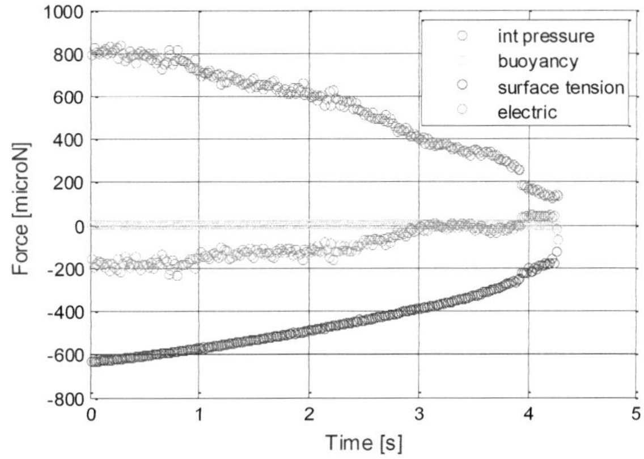

Figure 17: Vertical force balance, droplet withdrawal (reducing its volume with time), microgravity, with electric field

By only considering this fact, we should honestly say that we do not necessarily expect an increase in drop evaporation heat transfer; further studies are currently ongoing to better asses this. The enhancement of the heat transfer might also be induced by the flow of the "corona wind", which is caused by the gas ionisation near the electrode [27][10].

For what concerns boiling, a study has been performed showing an increase of about 27% in CHF and heat transfer coefficient (HTC, see Figure 18Figure 19).

Here, again, when the EF is applied, an "electric wind" attracts the bubbles towards the centre of the electrode (direction of rising voltage). We can speculate that this electro-convection effect is responsible of the increase in HTC and CHF. Furthermore, it is worth noting that bubbles are shrunk with respect to the normal case (Figure 20). This is due to the electric field force, which, when can distribute around the interface properly (undisturbed by other bubbles) creates the radial force that shrinks the bubbles. It is probably the case when the irreversible dry spot, coinciding with CHF, generates: it is smaller and the electric field always acts in order to reduce its size due to the action of radial forces.

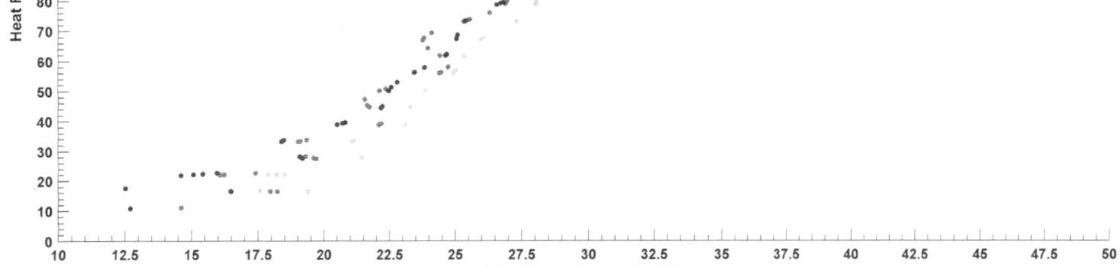

Figure 18: boiling curve of FC72 in pool boiling, with and without EF

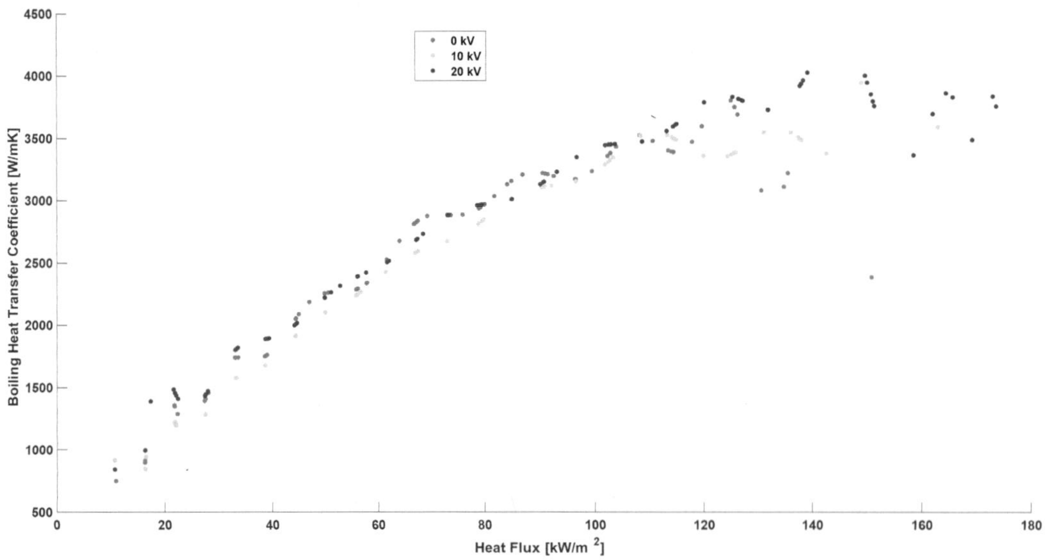

Figure 19: heat transfer coefficient (HTC) of FC72 in pool boiling, with and without EF

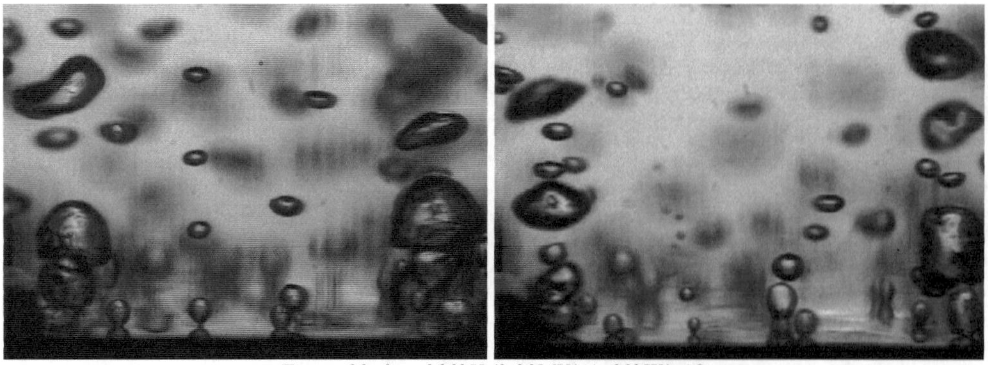

Figure 20: 0 and 20kV (3.33MV/m), 30kW/m^2

In general, the electric field enhances boiling heat transfer coefficient for low heat fluxes (nucleate boiling region) and for high heat fluxes (close to CHF and at CHF). And the reason of CHF suppression when the electric field is applied is due to the force balances discussed previously. Figure 21 shows a recorded experiment performed at a heat flux close to CHF at 0kV, switching the EF on and off. We can affirm without any doubt that the electric field is quenching the dryspot and bringing back the surface to the fully developed nucleate boiling regime.

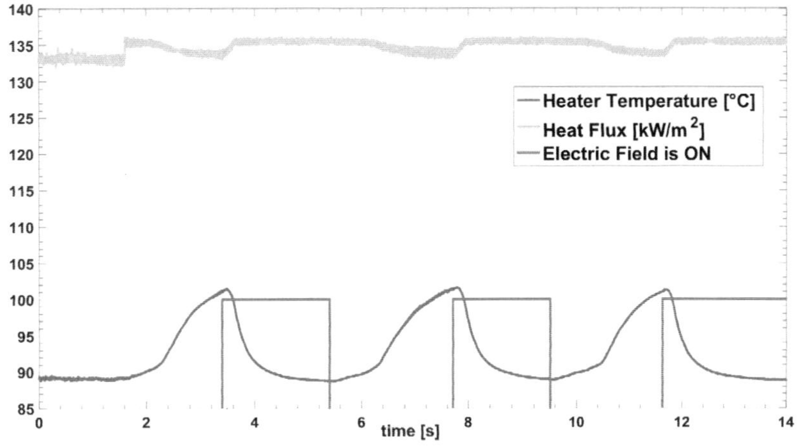

Figure 21: time history of CHF suppression induced by EF application

6. Conclusions

The effects of the electric field on interfaces (bubbles and droplets) have been evaluated. From the activity carried on at DESTEC, University of Pisa, we found that the electric field always acts by shrinking the interface. This can lead either to bubble departure, liquid jet or simply to a rise in the height of the interface.

For what concerns heat and mass transfer in droplets, even if the electric force is negligible, due to small volumes, the "corona wind" may assist in creating a strong flow of air, which can promote heat transfer and replace gravitational natural convection. The effect of electroconvection has been found of particular concern for bubbles during boiling experiments. In general, for bubbles, we have to mention that the EF provides a reliable and repeatable departure diameter also in microgravity.

In all cases the electric field normally concurs with gravity in assisting heat transfer, but it will be the main source of phase separation and management when gravity lacks.

References

[1] Zuber N, 1959, Ph.D. thesis, Research Laboratory, Los Angeles and Ramo-Wooldridge Corporation, University of California, Los Angeles, CA

[2] Rohsenow W. M., 1952, Transactions of ASME, Vol.74, pp. 969-976

[3] Brutin D, Carle F, Academic Press, Oxford, 2015, Pages 383-393, ISBN 9780128007228, https://doi.org/10.1016/B978-0-12-800722-8.00025-4

[4] Zhang H, Mudawar I, 2008 11th Intersociety Conference on Thermal and Thermomechanical Phenomena in Electronic Systems, Orlando, FL, 2008, pp. 949-959. doi: 10.1109/ITHERM.2008.4544370

[5] Di Marco P, Grassi W, 1993, J. Enhanced heat transfer 1:99-114

[6] Di Marco P, Grassi W, 1996, Proc. of Eurotherm Seminar 48. Eds. D. Gorenflo, D. Kenning, C. Marvillet, D. Paderborn, Pisa: ETS, 1:255-264

[7] Di Marco P., Grassi W, 2007, Multiphase Sci. Technol. 19.2, pp. 141-165.

[8] Di Marco P, Kurimoto R, Saccone G, Hayashi K, Tomiyama A, Experimental Thermal and Fluid Science, 2013, Vol.49, P.160, DOI: 10.1016/j.expthermflusci.2013.04.015.

[9] Di Marco P, Kurimoto R, Saccone G, Hayashi K, Tomiyama A, 2014, 15th Int. Heat Transfer Conference, Kyoto, Japan, paper 8960, pp. 1-11.

[10] Vancauwenberghe V, Di Marco P, Brutin D, Colloids and Surfaces A: Physicochemical and Engineering Aspects, Volume 432, 5 September 2013, Pages 50-56, ISSN 0927-7757, https://doi.org/10.1016/j.colsurfa.2013.04.067.

[11] Gibbons MJ, Howe CM, Di Marco P, Robinson AJ, 2016, J. Phys.: Conf. Ser. 745 032066, 7th

European Thermal-Sciences Conference (Eurotherm2016), doi:10.1088/1742-6596/745/3/032066
[12] Di Marco P, Pedretti F, Saccone G, 2013 Transfer, Fluid Mechanics, and Thermodynamics. In: Eighth World Conference on Experimental Heat Transfer, June 16 20, 2013, Lisbon, Portugal.
[13] Sharifi P, 2011, PhD Thesis, Southern Illinois University at Carbondale, UMI Number: 3460461.
[14] Zaghdoudi MC, Lallemand M, 1997, Proc. Int. Symp. on the physics of heat transfer in boiling and condensation, Moscow, May 1997, pp. 335-340.
[15] Zaghdoudi MC, Lallemand M, 2005, Arabian Journal for Science and Eng. 30, 189–212.
[16] Bonjour E, Verdier J, Weil L, 1962, AIChE-ASME 5th Nat. Conference, Houston, USA
[17] Allen PHG, Karayiannis T 1995 Heat Recovery Systems & CHP, 5 pp. 389-423
[18] Saccone G, Kurimoto R, Bucci M, Di Marco P, Tomiyama A, 2015, 7th European-Japanese Two-Phase Flow Group Meeting 12 pages
[19] Picknett RG, Bexon R, 1977, J. Colloid Interface Sci. 61 (2), 336 350.
[20] Bourge`s-Monnier C, Shanahan MER, 1995, Langmuir 11 (7), 2820 2829.
[21] Shanahan MER, 2001a Langmuir 17, 3997 4002.
[22] Shanahan MER, 2001b Langmuir 17, 8229 8235.
[23] Kulinich SA, Farzaneh M, 2009, Appl. Surf. Sci. 255, 4056 4060.
[24] Brutin D, Sobac B, Academic Press, Oxford, 2015, Pages 25-30, ISBN 9780128007228, https://doi.org/10.1016/B978-0-12-800722-8.00003-5.
[25] Landau LD, Lifšitz EM, 2nd edn. Pergamon, New York (1984)
[26] Saccone G, 2014, Master Thesis, University of Pisa – CEA Saclay. https://etd.adm.unipi.it/t/etd-05212014-193232/
[27] Ahmedou SAO, Rouaud O, Food Bioprocess Technol. 2 (2009) 240–247.

IOP Conf. Series: Journal of Physics: Conf. Series **923** (2017) 012020 doi:10.1088/1742-6596/923/1/012020

Numerical simulation of rivulet build up via lubrication equations

N Suzzi[1] **and G Croce**[1]

[1]DPIA - Dipartimento Politecnico di Ingegneria e Architettura - Universitá di Udine - Via delle Scienze - 33100 - Udine (UD) - Italy

E-mail: giulio.croce@uniud.it

Abstract. A number of engineering problems involve the evolution of a thin layer of liquid over a non-wettable substrate. For example, CO2 chemical absorption is carried out in packed columns, where post-combustion CO2 flows up while liquid solvent falls down through a collection of corrugated sheets. Further application include, among others, in-flight icing simulations, moisture condensation on de-humidifier fins, fogging build up and removal. Here, we present a development of an in-house code solving numerically the 2D lubrication equation for a film flowing down an inclined plate. The disjoining pressure approach is followed, in order to model both the contact line discontinuity and the surface wettability. With respect to the original implementation, the full modeling of capillary pressure terms according to Young-Laplace relation allows to investigate contact angles close to $\pi/2$. The code is thus validated with literature numerical results, obtained by a fully 3D approach (VOF), showing satisfying agreement despite a strong reduction in terms of computational cost. Steady and unsteady wetting dynamics of a developing rivulet are investigated (and validated) under different load conditions and for different values of the contact angles.

1. Introduction

The evolution of a thin layer of liquid over a non-wettable surface is a complex phenomenon, involving transition between still/moving droplet, rivulet and continuous film patterns, with physical discontinuities induced by moving contact lines. On the other side, a number of engineering problems requires the knowledge of the liquid distribution over a solid substrate. For example, in-flight icing on aircraft surface is driven by the evolution of a water layer fed by super-cooled drops from the cloud. The liquid distribution, which depends on the cloud water content and the fluid flow field and temperature, influences the location and shape of ice accretion areas, which in turn greatly affect the aircraft aerodynamics. Fogging and defogging of a car windshield or heat transfer on dehumidifier fin surfaces are also affected by condensation and evaporation phenomena with droplets growth, coalescence and motion over solid substrates. Several works concerning these topics are available in literature [1–3].

Even chemical absorption of post-combustion carbon dioxide involves the evolution of a thin layer of liquid. In fact, gas CO_2 flows up while a liquid solvent, which has to capture carbon, flows down through a packed column, composed by a collection of structured corrugated sheets. The absorption process is more efficient when the packed layers are covered by a continuous film, rather than drops or rivulets. Both numerical and experimental literature are available, but it is clearly difficult to look inside the packed columns, making the numerical approach much more

IOP Conf. Series: Journal of Physics: Conf. Series **923** (2017) 012020 doi:10.1088/1742-6596/923/1/012020

suitable. Moreover, most of the above mentioned problems are inherently multiscale: a packed column may have a diameter of 10 m, while the characteristic dimension of structured packing is of the order of 20 cm and the film height may be less than a millimeter [4]; thus, a full CFD approach is typically limited to the investigation of a single facet of the problem: most analysis, in fact, are focused on the hydrodynamic behavior of a liquid along plane surfaces. In particular, Hoffmann et al. [5] validated the Volume Of Fluid (VOF) method for a film flowing down an inclined plate with walls at the lateral boundaries, constructing a two fluid model (gas-liquid) and comparing numerical results with experimental evidences. Singh et al. presented a number of numerical results in [4], where the authors considered the same test case of [5] and adopted a fully 3D approach (VOF) as well, looking for the influence of solvent properties and flow rate. There are other literature works involving absorption and distillation through structured packing [6–8], which focus on numerical simulation of a falling film driven by gravity and shear using VOF method.

Here, we are following a different approach. In order to reduce the computational effort for the simulation of the film dynamics, we consider the 2D lubrication equation, solved numerically by means of Finite Volume Method (FVM). The idea is to look for an efficient, yet accurate model that could be coupled with a full CFD of the core flow, in order to investigate simultaneously both the large scales of the device (the full packed column, or the full wing or nacelle for a in-flight icing problem) and the smaller scale of the film.

Lubrication theory has been widely used in open literature for describing droplet spreading on non-wettable surfaces or flowing film dynamics [9–16], but the small-slope approximation when calculating free surface curvature does not allow to investigate contact angle higher than 20°.

In the present work, the implementation of the capillary pressure according to the membranal theory allows to investigate equilibrium contact angle up to 60°, thus making it suitable for a larger number of practical applications.

An in-house, fully implicit code was built and used for the current simulations. Precursor film together with disjoining pressure model allow for the modeling of the surface wettability (i.e. equilibrium contact angle), as also made in [9, 10, 12, 16]. The FVM solver code is first validated with literature numerical results involving rivulet build up and then applied to the case of a gravity driven film down an inclined plate, looking for useful information about wetting dynamics inside packed columns for CO_2 absorption applications.

2. Mathematical model

Consider a thin film of height h flowing along an inclined plate under the action of both gravity and shear, α being the plate inclination with respect to the horizontal and $\boldsymbol{\tau}$ the shear applied by an external gas flow. Let h_∞ and u_∞ be the undisturbed film height and velocity.

The lubrication theory describing the evolution of a thin layer of liquid over a solid substrate offers:

$$\frac{\partial h}{\partial t} = -\nabla \cdot \boldsymbol{Q} = -\nabla \cdot \left(-\frac{\nabla p}{3\,\mu} h^3 + \frac{\boldsymbol{\tau}}{2\,\mu} h^2 \right) \tag{1}$$

which is valid for low values of Reynolds number, $\mathrm{Re} = \rho\, u_\infty\, h_\infty / \mu \leq 1$.

The pressure field in Eq. (1) is given by hydrostatic, capillary and disjoining contributions:

$$p = \rho\, g\, (h\, \cos\alpha - x\, \sin\alpha) - \sigma\, \kappa - \Pi \tag{2}$$

where x represents the plate downhill direction, κ is the free surface curvature and Π the disjoining pressure.

The small-slope approximation leads to $\kappa = \nabla^2 h$, which is mostly used in literature works involving 2D simulations of thin liquid layers.

The disjoining pressure, related to inter-molecular forces between liquid and solid substrate, was defined by Schwartz and Eley [10] as:

$$\Pi = B\left[\left(\frac{\delta}{h}\right)^n - \left(\frac{\delta}{h}\right)^m\right] \tag{3}$$

$$B = \frac{(n-1)(m-1)}{n-m}\frac{\sigma}{\delta}(1-\cos\theta_e) \tag{4}$$

where $1 < m < n$, δ is the precursor film thickness and θ_e is the equilibrium contact angle. Zhao and Marshall, [12], observed that values of δ larger than the physical one lead to inaccurate results. Thus, the following correction factor was introduced in Eq. (4):

$$f = 1 + 6.069\,\delta + 161.7\,\delta^2 - 1547\,\delta^3 + 5890\,\delta^4 \tag{5}$$

Disjoining pressure allows for the modeling of the contact line dynamics, but the allowed contact angle is still limited due to the small-slope approximation.
The proper equation defining the capillary pressure can be obtained writing the membranal equilibrium in the normal direction of the free surface, see Appendix, which gives:

$$\kappa = \frac{1}{r_x}\left[\frac{1+\left(\frac{\partial h}{\partial x}\right)^2}{1+\left(\frac{\partial h}{\partial x}\right)^2+\left(\frac{\partial h}{\partial y}\right)^2}\right]^{1/2} + \frac{1}{r_y}\left[\frac{1+\left(\frac{\partial h}{\partial y}\right)^2}{1+\left(\frac{\partial h}{\partial y}\right)^2+\left(\frac{\partial h}{\partial x}\right)^2}\right]^{1/2} \tag{6}$$

where r_x and r_y can be calculated according to Young-Laplace equation:

$$r_x = \frac{\frac{\partial^2 h}{\partial x^2}}{\left[1+\left(\frac{\partial h}{\partial x}\right)^2\right]^{3/2}}\ ;\ r_y = \frac{\frac{\partial^2 h}{\partial y^2}}{\left[1+\left(\frac{\partial h}{\partial y}\right)^2\right]^{3/2}} \tag{7}$$

The fourth order, non-linear PDEs, Eqs. (1) and (2), are solved numerically on the 2D domain of the inclined plate. The following relation:

$$\kappa = \frac{1}{r_x} + \frac{1}{r_y} \tag{8}$$

is used in current simulations for calculating the capillary pressure instead of Eq. (6). However, two versions of the solver FVM code are available, with both Eqs. (6) and (8) implemented, but the simplified relation, Eq. (8), is chosen since the two versions lead to similar results but the computational cost is lower if Eq. (8) is implemented.
The case of gravity and shear acting in the same direction is considered, allowing to correlate the average velocity u_∞ and the thickness h_∞ of the undisturbed film according to [16]:

$$u_\infty = \frac{\rho\, g\, \sin\alpha}{3\,\mu}\, {h_\infty}^2 + \frac{\tau}{2\,\mu}\, h_\infty \tag{9}$$

3. Space and time discretization

The 2D spatial domain $L_x \times L_y$ is discretized using a structured, orthogonal grid composed by $n_x \times n_y$ elements of variable size. The convective term $\nabla \cdot \boldsymbol{Q}$ of Eq. (1) is discretized using Finite Volume Method, via a first order upwind scheme. Hence, the flux $\boldsymbol{Q}$ is decomposed into two contributions: $\boldsymbol{Q}^+ = \boldsymbol{u}^+ h$ refers to the the wave propagation with positive velocity, while $\boldsymbol{Q}^- = \boldsymbol{u}^- h$ refers to the wave propagation with negative velocity. This discretization ensures

that the pressure gradient on each element face is computed with a centered difference. More details can be found in [16]. The capillary pressure field, Eqs. (7) and (8), is discretized using a second order centered scheme.
An implicit Euler time integration is chosen due to the restrictive stability constraint on the explicit time step, $\Delta t \propto \Delta x^4$. On the structured grid used in the present work, the computational time can be greatly reduced following an Alternate Direction Implicit scheme, which allows to solve two pentadiagonal algebraic systems at each time step instead of one sparse system. The ADI factorization proposed by Witelski and Bowen [17] for numerical resolution of high-order non-linear diffusion equations, which allows to treat explicitly the mixed derivatives of h arising from the capillary pressure gradient, is adopted. Further details on the discretization can be found in [16, 17].

4. Problem setup

The dynamics of a gravity driven film flowing down an inclined plate are investigated, being the simplest configuration for better understanding the hydrodynamics through structured packing, [4]. The case of gravity driven film ($\boldsymbol{\tau} = 0$) is presented.
Along the upper boundary of the domain, $x = 0$, an inlet section of length $L_{in} \in [0, L_y]$ is defined, where the entering film flux is settled to the undisturbed one, $\boldsymbol{Q_{in}} \cdot \hat{\boldsymbol{n}} = u_\infty h_\infty$, and the pressure $p_\infty = \rho g h_\infty \cos\alpha + \Pi_\infty$ is imposed; outside the inlet section, the entering flux is equal to zero as well as the pressure. Zero flux entering the domain and zero pressure gradient, $\boldsymbol{Q_{in}} \cdot \hat{\boldsymbol{n}} = 0$ and $\nabla p \cdot \hat{\boldsymbol{n}} = 0$, are imposed at the outlet boundary, $x = L_x$.
Two different set of boundary condition on the lateral boundaries $y = 0,\ L_y$ can be used: symmetry boundary condition, $\boldsymbol{Q} \cdot \hat{\boldsymbol{n}} = 0$ and $\nabla p \cdot \hat{\boldsymbol{n}} = 0$, for the simulation of a plate of indefinite extension along y direction, or wall boundary condition, $\boldsymbol{Q} \cdot \hat{\boldsymbol{n}} = 0$ and free surface inclination equal to equilibrium contact angle θ_e, for the simulation of a domain bounded by solid end-walls at $y = 0,\ L_y$. The following equation:

$$\left.\frac{\partial h}{\partial y}\right|_0 = -\frac{\sqrt{1+\left(\frac{\partial h}{\partial x}\right)^2}}{\tan\theta_e}, \quad \left.\frac{\partial h}{\partial y}\right|_{L_y} = +\frac{\sqrt{1+\left(\frac{\partial h}{\partial x}\right)^2}}{\tan\theta_e} \tag{10}$$

is implemented for imposing contact angle at the lateral boundaries, with $\frac{\partial h}{\partial x}$ calculated by means of a second order centered scheme, using the available values of h in the grid elements next to the domain boundary.
The dry initial condition, i.e. film height equal to the precursor film thickness $h = \delta$ over the whole domain, is used for the current simulations, in order to focus on the wetting dynamics.

5. Validation

First of all, the disjoining pressure model, Eqs. (3), (4) and (5), is validated for a 1D film flowing down an inclined plate. Simulations are run imposing: plate inclination equal to $\alpha = 60°$; equilibrium contact angle in the range $\theta_e \in [30°, 60°]$; precursor film thickness $\delta = 0.05\, h_\infty$; spatial discretization step equal to $\Delta x = 0.05\, h_\infty$. The simulated liquid is 1−methylepiperazine $0.31x$ MPZ, which physical properties are listed in table 1.
Figure 1(a) shows the shape of the capillary ridge, which is self-similar in time and typically appears close to the moving contact line: according to literature [12], higher values of the imposed equilibrium contact angle lead to higher ridge dimensions. In figure 1(b) the dynamic contact angle is computed for different film flow rates and fitted using the Tanner-Hoffman-Voinov formula [12, 18]:

$$\theta^3 = {\theta_e}^3 \left(1 + C\,\frac{\mu\, u_\infty}{\sigma}\right) \tag{11}$$

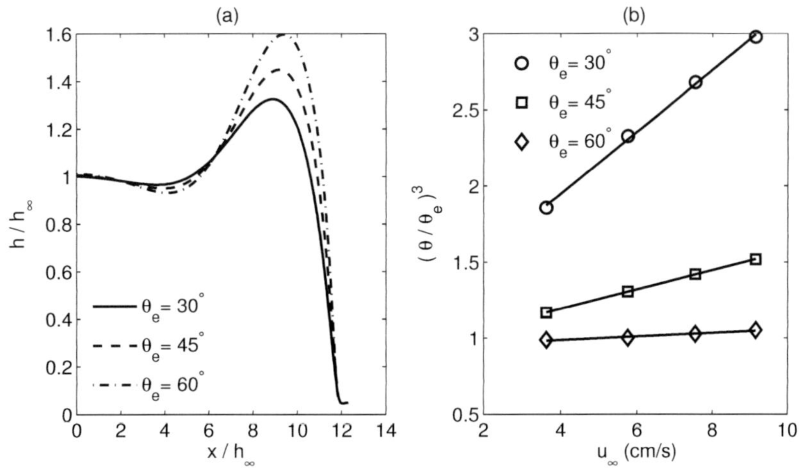

Figure 1. 1D gravity driven film: capillary ridge shape for different equilibrium contact angles θ_e, $Q_{in} = 1\,\mathrm{cm}^2/\mathrm{s}$, $h_\infty = 1.09\,\mathrm{mm}$ (a); dynamic contact angle θ vs undisturbed film velocity u_∞ and imposed equilibrium contact angle θ_e: numerical results (symbols) and Eq.(11) (lines) (b).

where C is the fitting parameter. According to the validation procedure found in [12], it can be stated that surface wettability and contact line dynamics are correctly modeled using disjoining pressure and assuming a thin precursor film, since Eq. (11) well describes the behavior of the dynamic contact angle θ as a function of the undisturbed film velocity u_∞. Moreover, equilibrium contact angle up to $\theta_e = 60°$ had not been investigated (and thus validated) before using disjoining pressure and lubrication theory.

The solver code is then validated with numerical results found in literature, involving rivulets build up. The test case of Singh et al. [4] is chosen for the simulations: $6 \times 5\,\mathrm{cm}^2$ plate, inclined with an angle $\alpha = 60°$ to the horizontal; film flux per unit length $Q_{in} = 1\,\mathrm{cm}^2/\mathrm{s}$ entering through a confined inlet section $L_{in} = 2\,\mathrm{cm}$ at the top of the plate, $\dot{V}_{in} = Q_{in}\,L_{in} = 2\,\mathrm{cm}^3/\mathrm{s}$; 1−methylepiperazine $0.31x$ MPZ liquid simulated (see properties in table 1). It is important to point out that the chosen liquid is used as solvent in packed columns for carbon capture [4].

Thanks to the symmetrical configuration of the problem, only half of the plate is simulated, with symmetry condition imposed at the lateral boundaries $y = 0,\, L_y$, i.e. an array of parallel rivulets with $2\,L_y$ spacing is simulated. Inlet section is implemented as described in section 4. The precursor film thickness is set to $\delta = 0.025\,h_\infty$. A grid dependence analysis suggested to impose spatial discretization step, which is kept uniform over the computational domain for the rivulet build up simulations, of about $0.075 - 0.1\,h_\infty$, depending on the imposed equilibrium contact angle, which is varied in the range $\theta_e \in [20°, 60°]$.

Figure 2, which shows how the stationary liquid layer distribution changes with the surface wettability, clearly proves that the higher the imposed equilibrium contact angles, the narrower the shape of the rivulet, according to the numerical results of Singh et al. [4].

A proper validation is possible computing the wetted area A_{wet}, which is defined as the portion of plate covered by a layer of liquid with height higher than the precursor film one, and the interface area A_{int}, which is the area of the liquid-gas free surface in the wetted region. In figures 3(a) and 3(b) the resulting trends of A_{wet} and A_{int} as a function of θ_e are compared with the numerical results of Singh et al. [4], obtained by means of a fully 3D model: the correspondence is satisfactory for all the investigated values of θ_e.

According to [19], the liquid pressure of a stable rivulet flowing over an inclined plate is independent of the plate downhill direction x, thus the rivulet profile can be calculated solving

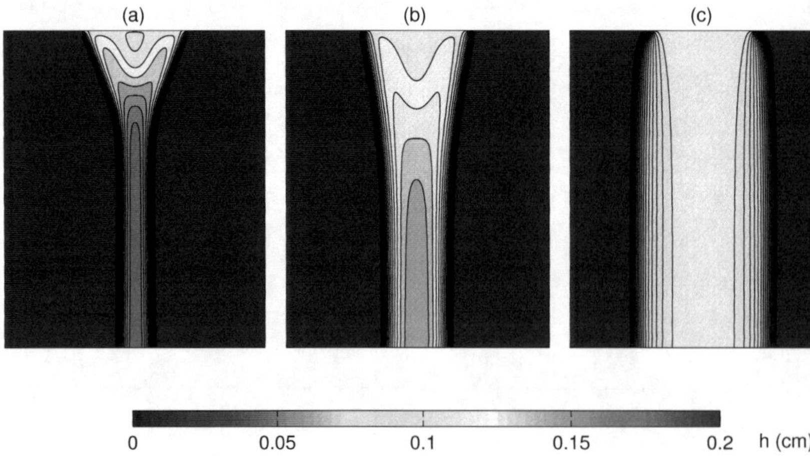

Figure 2. Stationary rivulet distribution over the inclined plate for: $\theta_e = 60°$ (a), $\theta_e = 40°$ (b), $\theta_e = 20°$ (c). $\alpha = 60°$, $\dot{V}_{in} = 2\,\mathrm{cm}^3/\mathrm{s}$.

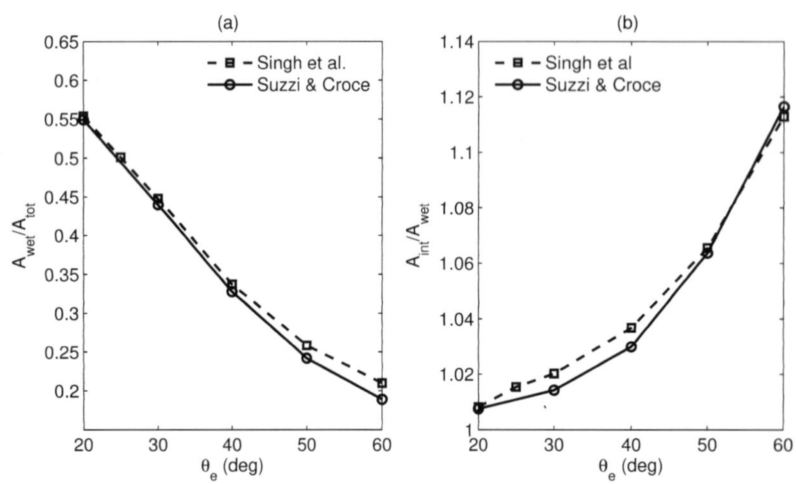

Figure 3. Normalized wetted area of the rivulet as a function of the equilibrium contact angle (a). Normalized interface area of the rivulet as a function of the equilibrium contact angle (b). $\alpha = 60°$, $\dot{V}_{in} = 2\,\mathrm{cm}^3/\mathrm{s}$.

the following initial value problem:

$$p = \rho\, g\, h\, \cos\alpha - \sigma\,\frac{\partial^2 h}{\partial y^2}\left[1 + \left(\frac{\partial h}{\partial y}\right)^2\right]^{-3/2} = \mathrm{const} \tag{12}$$

$$h\big|_{y_0} = 0,\ \left.\frac{\partial h}{\partial y}\right|_{y_0} = \tan\theta_e \tag{13}$$

The solution of Eq. (12) together with the initial conditions is iterated: the value of the rivulet pressure p is updated at each iteration, until the liquid flow rate, which can be calculated integrating the film flux per unit length

$$Q = \frac{\rho\, g\, h^3\, \sin\alpha}{3\,\mu} \tag{14}$$

along the rivulet length, equals the imposed one $\dot{V}_{in} = Q_{in}\, L_{in}$. The resulting profile is then compared with the numerical one, obtained cutting the stationary liquid layer distribution at $x = L_x$. The validation for $\theta_e = 60°$ is shown in figure 4, which shows almost a perfect correspondence between numerical results and Eq. (12).

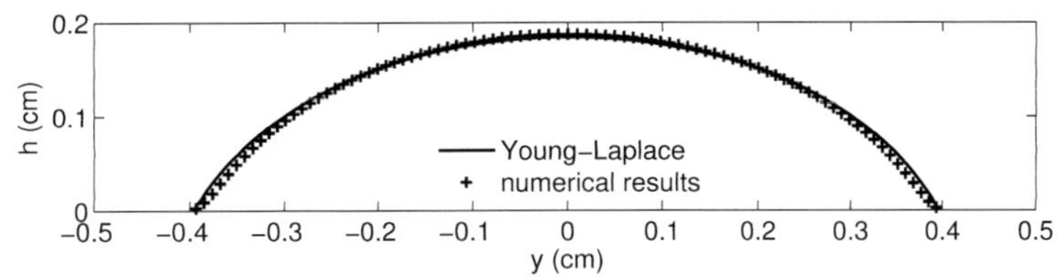

Figure 4. Comparison between rivulet profiles from solution of Eq. (12) and numerical simulation of rivulet build up. $\theta_e = 60°$, $\alpha = 60°$, $\dot{V}_{in} = 2\,\mathrm{cm^3/s}$.

6. Application to CO_2 absorption

Liquid	$\rho\,(\mathrm{kg\,m^{-3}})$	$\mu\,(\mathrm{Pa\,s})$	$\sigma\,(\mathrm{N\,m^{-1}})$	Ka
$0.31x$ MPZ	981.31	0.03642	0.03480	14.8
$0.41x$ MPZ	962.20	0.02348	0.03589	24.6
$0.51x$ MPZ	946.41	0.01336	0.03437	49.7

Table 1. Physical properties of 1−methylepiperazine (MPZ) at $T = 25°\mathrm{C}$ and $p = 1\,\mathrm{atm}$.

The first step for understanding the hydrodynamics of absorption/distillation processes through structured packing is the analysis of the wetting dynamics of an inclined plate covered by a thin liquid layer. As explained in [4], the wetting phenomenon is dictated by several parameters, making crucial to focus on identifying proper non-dimensional numbers. The effect of solvent properties is analyzed through Kapitza number:

$$\mathrm{Ka} = \sigma \left(\frac{\rho}{\mu^4\, g}\right)^{1/3} \tag{15}$$

The flow characteristics are studied in terms of Weber number:

$$\mathrm{We} = \frac{\rho\, {u_\infty}^2\, h_\infty}{\sigma} \tag{16}$$

which represents the ratio between liquid inertia and surface tension force. The effect of surface wettability is obviously given by the equilibrium contact angle θ_e. The wetted area A_{wet}, which is here normalized to the plate area A_{tot}, is used for describing the wetting process, since it is the parameter to be optimized when designing the corrugated sheets composing the packed column.

Similarly to the approach of Singh et al. [4] and Hoffmann [5], an inclined plate is considered: the lateral boundaries are treated as walls, which implementation is explained in section 4; the whole top boundary of the plate represents the inlet section; the case of gravity driven flow is considered. The effect of θ_e, Ka and We on wetting dynamics are investigated, with equilibrium contact angle and Weber number up to 60° and 0.25 respectively. The Reynolds number is limited to $\mathrm{Re} \leq 5.58$, according to lubrication theory validity. The liquids listed in table 1 are simulated. The choice of these liquids is not casual, since they are used as solvent in CO_2 capture through packed column [4]. A $3 \times 3\,\mathrm{cm}^2$ plate, inclined at an angle $\alpha = 60°$ to the horizontal, is considered, but only half of the plate is simulated, thanks to the symmetry of the problem. The imposed precursor film thickness is equal to $\delta = 0.025\,h_\infty$. The spatial discretization step is not uniform over the computational domain, since a finer grid is required close to the wall for a better description of the liquid behavior, especially when low contact angles are imposed, leading to higher slopes.

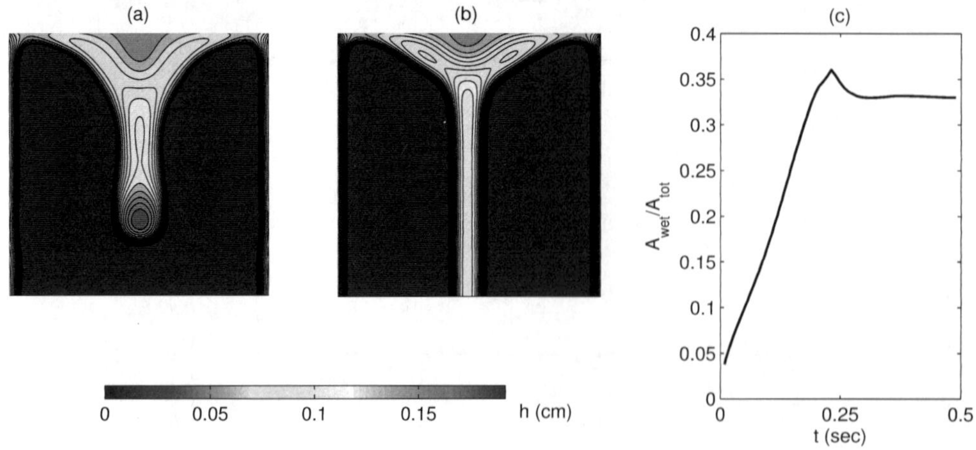

Figure 5. Liquid layer distribution over the inclined plate at $t = 0.19\,\mathrm{s}$ (a) and stationary distribution, $t = 0.49\,\mathrm{s}$ (b). Normalized wetted area as a function of time (c). $\theta = 60°$, $\mathrm{Ka} = 49.7$, $\mathrm{We} = 0.0736$.

Figures 5(a) and 5(b), which refer to the same simulation, show the distribution of liquid 1−methylepiperazine $0.51x$ MPZ over the plate at two different times. It can be noticed that the lateral walls capture some liquid due to capillary effect, perturbing the undisturbed flow. If the inlet flow rate is too low, this phenomenon leads to the growth of dry-patches on the plate, with the liquid forming two rivulets close to the walls and one rivulet in between. Similar results were obtained experimentally by Hoffman [5] and numerically by Singh et al. [4]. In figure 5(c) the plate wetted area is shown as a function of time. The wetting process can be clearly divided into two regimes: first the liquid bulk, driven by gravity, spreads in downhill direction; after the bulk reaches the outlet section, the contact line dynamics are mainly driven by the capillary force and thus influenced by the surface wettability.

The dependence of the plate wetted area on Weber number, Kapitza number and equilibrium contact angle is shown in figures 6(a) and 6(b). Results refer to the stationary solution of Eqs. (1) and (2). In particular, figure 6(a) suggests that the effect of fluid inertia, i.e. of We, on wetting process grows with the Kapitza number, since the whole coverage of the plate, $A_{wet}/A_{tot} = 1$, is reached for higher values of the Weber number. Thus, liquid spreading over the inclined plate is mainly driven by viscous, surface tension and gravity forces only for low Ka. Figure 6(a) also gives some information about the critical flow rate, defined as the minimum flow rate leading to

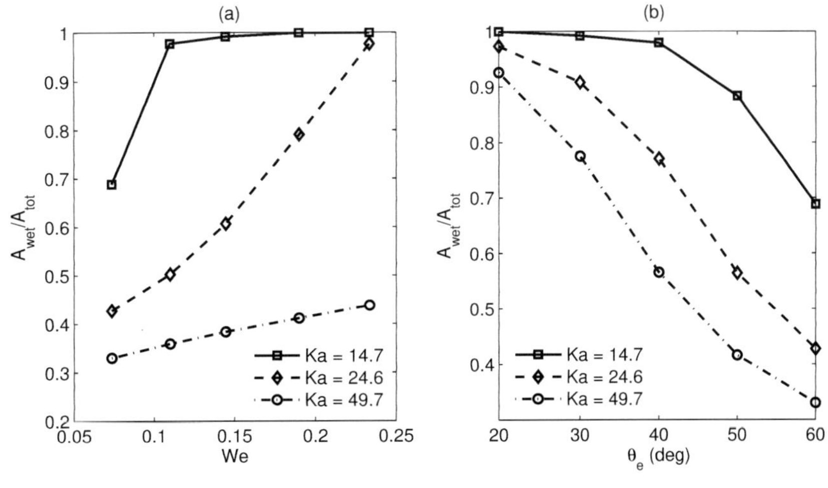

Figure 6. Normalized wetted area at the stationary liquid distribution as a function of Weber number, $\theta_e = 60°$ (a); as a function of equilibrium contact angle, We = 0.0736 (b). Fluids listed in table 1 are simulated, plate inclination is fixed to $\alpha = 60°$.

whole coverage of the plate: the critical value of Weber number (which defines univocally the critical flow rate) can be traced as a function of the Kapitza number of the liquid. The definition of the critical flow rate is crucial, since it defines the most suitable condition (maximum wetted area) for absorption process through structured packing. The surface wettability also has a great impact on wetting process, as shown in figure 6(b): higher values of the equilibrium contact angle lead to lower stationary wetted area, with low changes in contact angle leading to significantly differences in liquid behavior. Similar results were obtained by the rivulet build up simulations, figure 3(a).
Finally, it is important to point out that the computational cost of the presented simulations is limited, thanks to the implemented ADI method, which allows to reach time step Δt up to 10^6 the explicit one. As an example, Singh et al. [4] conducted VOF simulations on a similar setup, running on 128 cores in parallel with 144 h required for a single simulation, while in the present work simulation time is 4 times lower using a single core.

7. Conclusion

An implicit FVM code was implemented in order to solve the 2D lubrication equation describing the evolution of a thin layer of liquid. It was shown that the model is able to compute the film height over the whole domain and capture the contact line dynamics, thanks to the disjoining pressure and the assumption of a thin precursor film. Thus, the 2D approach provides a full analysis of the wetting of a solid substrate, without loss of information comparing to a 3D approach. The fully implementation of capillary pressure allows to investigate values of the equilibrium contact angle up to 60°, which had never been analyzed using lubrication theory before. The code was first successfully validated with literature results for the rivulet build up over an inclined plate. Then, the wetting dynamics of a film flowing down an inclined plate were studied, focusing on the effect of surface wettability, liquid properties (Ka) and flow characteristics (We). The chosen setup and the simulated liquids are directly involved in CO_2 absorption through structured packing. Other numerical works involving distillation/absorption process through packed columns are available in literature, but a fully 3D approach was used by the authors. The advantage of the presented 2D approach is that a whole analysis of the wetting

phenomenon is possible with a gain in terms of computational costs of around two orders of magnitude.
The implemented model also allows to simulate shear driven films, involved in a number of engineering problems. More complex geometries, which are required for fully understanding the liquid behavior over the corrugated sheets involved in structured packing, can also be simulated using the solver code, even in the present structured grid form. It was not possible to simulate high Kapitza number solvents, as done in [4,5], because the low viscosity leads to higher values of the Reynolds number, which are not allowed by the current assumption of neglecting liquid inertia. However, lubrication theory is still valid, keeping liquid inertia terms during the integration of Navier-Stokes equations along plate normal direction.

Appendix

The mathematical procedure for obtaining Eq. (6), which allows to define properly the disjoining pressure term, is derived in this Appendix.

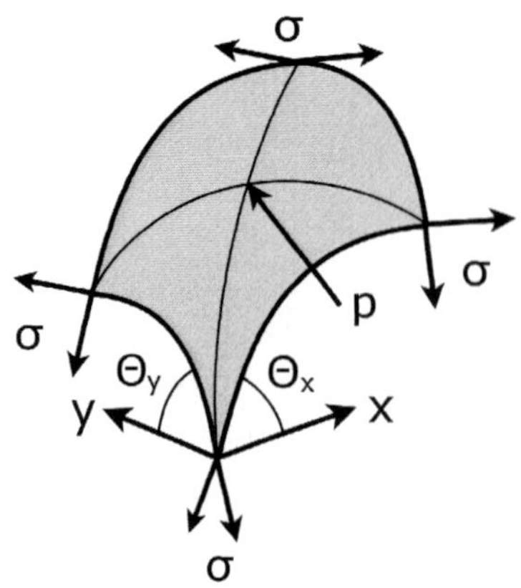

Figure A1. Forces acting on a control volume of free surface.

Consider the portion of free surface shown in figure A1. Assuming low stretching of the control surface (i.e. the corner angles are close to 90°), the membranal equilibrium along the normal direction $\hat{\boldsymbol{n}}$ between the surface tension force and the capillary pressure gives:

$$p\,dA = \sigma\,\sin\left(d\beta_x\right)\,\cos\Theta_y\,r_y\,d\beta_y + \sigma\,\sin\left(d\beta_y\right)\,\cos\Theta_x\,r_x\,d\beta_x \tag{A.1}$$

where $r_{x,y}$, $\Theta_{x,y}$ and $d\beta_{x,y}$ are the radii of curvature, the free surface slopes and the arc angles, calculated along the Cartesian directions.
The free surface area dA, subjected to the capillary pressure p, can be expressed as:

$$dA = \left(r_x\,d\beta_x\right)\,\left(r_y\,d\beta_y\right)\,\sqrt{1-\sin^2\Theta_x\,\sin^2\Theta_y} \tag{A.2}$$

Assuming $\sin d\beta_{x,y} \simeq d\beta_{x,y}$ and dividing both LHS and RHS of Eq. (A.1) for $\sigma\,dA$ lead to:

$$\kappa = \frac{p}{\sigma} = \frac{1}{r_x}\,\frac{\cos\Theta_y}{\sqrt{1-\sin^2\Theta_x\,\sin^2\Theta_y}} + \frac{1}{r_y}\,\frac{\cos\Theta_y}{\sqrt{1-\sin^2\Theta_x\,\sin^2\Theta_y}} \tag{A.3}$$

Using the definitions of the free surface slopes $\Theta_{x,y}$:

$$\tan\Theta_x = \frac{\partial h}{\partial x},\ \tan\Theta_y = \frac{\partial h}{\partial y} \tag{A.4}$$

the terms $\cos\Theta_{x,y}$, $\sin\Theta_{x,y}$ can be expressed as a function of $\frac{\partial h}{\partial x}$, $\frac{\partial h}{\partial y}$ and Eq. (A.3) recast, giving Eq. (6).

References

[1] De Candido E, Croce G and D'Agaro P 2012 *Heat Transfer Eng.* **33** 1130
[2] Croce G, Candido E D, Habashi W G, Munzar J, Aube M, Baruzzi G S and Aliaga C 2010 *J. Aircraft* **47** 1283
[3] Bourgault Y, Habashi W G, Dompierre J and Baruzzi G S 1999 *Int. J. Numer. Methods Fluids* **29** 429
[4] Singh R K, Galvin J E and Sun X 2016 *Chem. Eng. Sci.* **142** 244
[5] Hoffmann A, Ausner I, Repke J U and GWozny 2005 *Comput. Chem. Eng.* **29** 1433
[6] Subramanian K and Wozny G 2012 *Int. J. Chem. Eng.* **2012** 13
[7] Repke J U, Ausner I, Paschke S, Hoffmann A and Wozny G 2007 *Chem. Eng. Res. Des.* **85** 50
[8] Xu Y Y, Zhao M, Paschke S and Wozny G 2014 *Ind. Eng. Chem. Res.* **53** 7797
[9] Sellier M 2015 *Int. J. Multiphase Flow* **71** 66
[10] Schwartz L W and Eley R R 1998 *J. Colloid Interface Sci.* **202** 173
[11] Weidner D E, Schwartz L W and Eres M H 1997 *J. Colloid Interface Sci.* **187** 243
[12] Zhao Y and Marshall J S 2006 *J. Fluid Mech.* **559** 355
[13] Lemaitre C, Hemon P and de Langre E 2007 *J. Wind Eng. Ind. Aerodyn.* **95** 1259
[14] Diez J A and Kondic L 2002 *J. Comput. Phys.* **183** 274
[15] Mata M R and Barozzi A L 2011 *J. Comput. Phys.* **230** 6334
[16] Suzzi N and Croce G 2017 *J. Phys. Conf. Ser.* **796** 012038
[17] Witelski T P and Bowen M 2003 *Appl. Numer. Math.* **45** 331
[18] Blake T D 2006 *J. Colloid Interface Sci.* **299** 1
[19] El-Genk M S and Saber H H 2001 *Int. J. Heat Mass Transfer* **44** 2809

IOP Conf. Series: Journal of Physics: Conf. Series **923** (2017) 012021 doi:10.1088/1742-6596/923/1/012021

Micro Droplets of non-Newtonian Solutions in Silicone Oil Flow through a Hydrophobic Micro Cross-Junction

B Rostami and G L Morini

DIN – Alma Mater Studiorum Università di Bologna, Laboratorio di Microfluidica, Via del Lazzaretto 15/5, Bologna 40131, Italy

E-mail: behnam.rostami2@unibo.it

Abstract. In this paper the generation of non-Newtonian droplets of aqueous Xanthan gum solution (0.3, 0.5 wt%) in a silicone oil flow through a micro cross-junction is experimentally analyzed. A commercial glass cross-junction microchip with hydrophobic walls has been employed to study the droplet generation mechanism. The cross-section of the channel is stadium-shaped, the width of the junction varies between 195 to 390 µm while the height of the channel is fixed at 190 µm. Tween 20 (2 wt%), as a surfactant, has been added to the dispersed phase to avoid the coalescence of the droplets and to enhance the droplet formation. With the aim to follow the time evolution of the droplets inside the channel a specific experimental setup has been implemented. The post-processing of the recorded images has been carried out by means of an "in-house" Matlab code. The typical flow patterns obtained by imposing different flow rates at the inlets of the cross-junction have been observed. The effect of the continuous and dispersed phase flow rates as well as the concentration of Xanthan gum solution on the main droplet characteristics has been studied in detail.

1. Introduction

Nowadays microdevices are commonly used to generate mono-dispersed droplets. A series of simple microfluidic devices have been proposed for droplet generation like T-junctions [1], co-flow devices [2], flow-focusing devices [3] and microchannel terraces [4]. Among flow-focussing devices micro cross-junctions have received a special attention for the generation of mono-dispersed droplets. In these devices, the continuous flow is introduced by means of two lateral junction branches in order to squeeze and shear off the interface of an immiscible liquid used as the dispersed phase and to introduce it inside the junction through its central microchannel.

Mono-dispersed droplets with low size variation are highly desirable in many technical applications such as DNA and blood analysis [5], chemical reactions [6] and drug discovery [7]. It has been demonstrated that the main governing and controlling parameters which enable the control of the droplet size generated in a micro cross-junction are the properties of the liquids introduced in the junctions, their inlet flow rates and the geometry of the micro-junction

Although squeezing regime was observed and considered as the main mechanism in droplet formation by Garstecki *et al*. [8], additional mechanisms, like dripping and jetting can play a role on the droplet generation in these microdevices. In fact, as demonstrated by the experimental observations, all these three mechanisms can be met by varying the continuous phase flow rate Q_c at the entrance of the micro-junction for a fixed flow rate of the dispersed phase (Q_d). By focusing the attention on the location of the droplet generation inside the junction, it is possible to distinguish: (i) droplets formed at the cross-junction (DCJ) and (ii) droplets formed downstream of the channel (DDC), by increasing the inlet dispersed phase flow rate Q_d. Most of the works appeared in literature were devoted to the analysis of the droplet generation in cross-junctions characterized by low inlet flow rates for which the squeezing mechanism becomes predominant. It will be demonstrated in this paper that, especially when non-Newtonian fluids are involved, dripping and jetting flow patterns become more and more important with respect to squeezing and they cannot be ignored.

Despite the importance of non-Newtonian fluids in industrial and pharmaceutical fields, the investigation of their behaviour in micro cross-junction during droplet generation can be nowadays considered disregarded. In order to start to fill this gap in this work an aqueous solution of Xanthan gum with low concentration (0.3-0.5 wt%) has been used as dispersed flow. These aqueous Xanthan gum solutions are typical shear thinning fluids with properties similar to blood plasma. The effects of the elasticity of these solutions as well as the effect of the channel dimension on the droplet generation was studied by Steinhaus *et al.* [9] by considering a Newtonian fluid as carrier fluid. Hong and Cooper-White [10] utilized Carbopol dispersions in order to study the droplet formation and breakup mechanism in presence of non-Newtonian dispersed flows. By the use of Carbopol dispersion which shows shear thinning and elastic properties, they showed that the droplet size was notably influenced by the viscosity ratio between the dispersed and the continuous phase. Husny and Cooper-White [11] have evidenced the presence of elongated filaments during the droplet formation with non-Newtonian dispersed flows due to the elasticity of the droplets as well as the presence of secondary droplets (satellites) close to the main droplet. They highlighted that the droplet polydispersity and the number of secondary droplets are a function of viscosity ratio. Also Lee *et al.* [12] studied the effect of viscoelasticity of the dispersed phase on the breakup dynamics. Arratia *et al.* [13, 14], using polyacrylamide solutions with different molecular weights and different elasticity values, studied in detail the filament thinning and droplet breakup in presence of these non-Newtonian flows. As evidenced by this short literature survey, the researchers have concentrated their attention on the role played by the elasticity and the molecular weight of non-Newtonian fluids on the droplet characteristics but other aspects was not deeply investigated up to now.

In this paper an experimental setup has been designed in order to analyze the generation of non-Newtonian droplets in a silicone oil continuous flow within a commercial glass micro cross-junction. Aqueous Xanthan gum solutions with two different concentrations (0.3 and 0.5 wt%) have been considered as dispersed phase. Tween 20 (2 wt%) has been added to the non-Newtonian solution to avoid the coalescence of the droplets and to enhance their formation. A wide range of volumetric flow rates at the inlet for both continuous and the dispersed phase have been tested in order to be able to detect different flow patterns including squeezing, dripping and jetting regimes. More in detail, the continuous phase (silicone oil) flow rate (Q_c) varies from 0.3 to 10 ml/h; the range of dispersed phase (Xanthan gum solution) flow rate (Q_d) has been varied from 0.01 to 2.1 ml/h. The non-dimensional droplet diameter has been defined and plotted as a function of flow rate ratio and the Capillary number to study the effect of the main physical governing parameters on the droplet features. In addition, the effect of Xanthan gum concentration in aqueous solution has been analyzed.

2. Experimental setup

The experimental apparatus used in this work for the analysis of the characteristics of the droplets generated in a glass micro cross-junction is shown in figure 1.

An inverted microscope (1, *Nikon Eclipse TE2000-U*) with a double system of illumination by means of a Mercury lamp (2a, *Nikon C-SHG1*) from the bottom and a COB LED lamp (2b, *100W, 9000 Lumens*) from the top is the main part of the test rig. An air immersion lens with a numerical aperture $NA = 0.25$ and magnification $M = 10x$ (3, *Nikon CFI DS 10X*) is selected in order to visualize the droplets. A high-speed camera (4, *Olympus I-speed II*) connected to the inverted microscope is used to acquire the droplet images; a LCD monitor (5) and Olympus I-speed II software (6) makes possible a run-time visualization of the flow within the micro-device. The volumetric flow rate of silicone oil (continuous phase) and of the Xanthan gum solution (dispersed phase) are controlled by two independent syringe pumps (7, *Cole-Parmer Version Hills* and 8, *Harvard Apparatus PHD4400 Programmable*) and introduced at the three inlets of the junction (11) using two gastight Hamilton syringes (9 and 10). The relative uncertainty of the imposed volumetric flow rate for continuous and dispersed phase is equal to ± 0.5% and ± 0.35%, respectively, based on the accuracy of the syringe pumps.

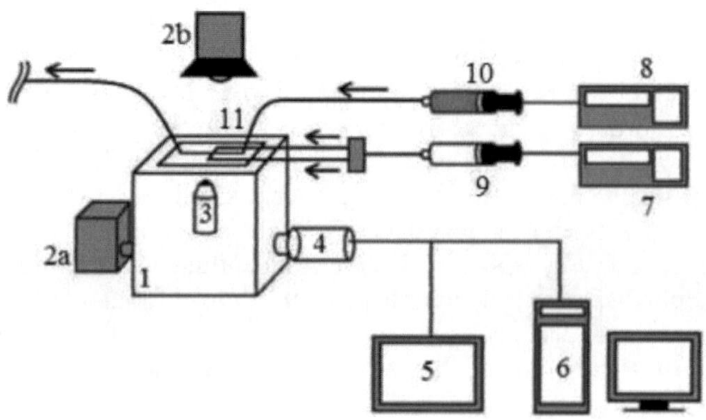

Figure 1. The experimental setup

The generation of non-Newtonian emulsions has been obtained by using a commercial glass micro cross-junction (*Dolomite Microfluidics Co.*). The stadium shape of the cross-section, the channel width variation through the junction and the hydrophobic inner walls are the main features of this junction. The cross-junction (figure 2a) is characterized by a cross-section restriction at the junction in which the width of the channel varies from W_w = 390 μm far from the junction to W_j = 195 μm at the junction (figures 2b and 2c). The restriction ratio defined as the ratio of these two parameters ($R = W_j/W_w$) is equal to 0.5. The height of the channel (H) is uniform and equal to 190 μm. The aspect ratio, i.e., the ratio between the channel height and width H/W, is equal to $A_j = 0.97$ at the junction and $A_w = 0.49$ out of it. In order to avoid the coalescence of the emulsions in an organic phase, a hydrophobic coating is required.

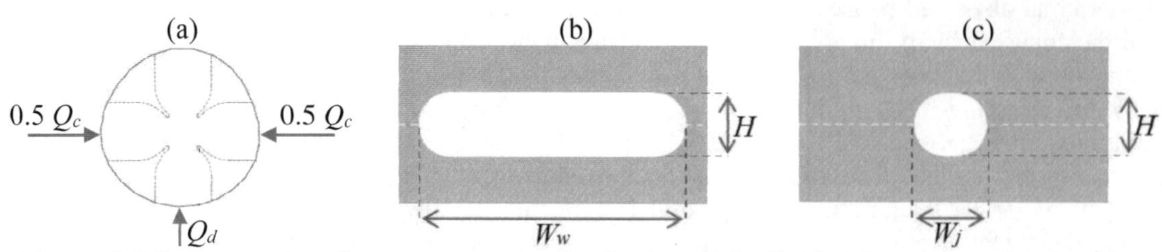

Figure 2. Micro cross-junction geometry; the distribution of the fluids (a); cross-section far from the junction (b) and at the junction (c)

The main physical parameters involved in the droplet generation in the micro cross-junction are the volumetric flow rate Q, density ρ, viscosity μ and the interfacial tension σ of the continuous and dispersed phase. By using the Buckingham-Pi theorem it is possible to demonstrate that a complete set of non-dimensional parameters for the analysis of the droplet generation in a cross-junction is given by the flow rate ratio $\alpha = Q_d/Q_c$, the dimensionless numbers Capillary number ($Ca = \mu u/\sigma$) where u (= Q/A) is the fluid velocity, Reynolds number ($Re = \rho u W/\mu$) and by the channel aspect ratio (A_w or A_j).

The geometry of the channel shown in figure 2 clarifies in which way the continuous and the dispersed phase are introduced in the junction.

The silicone oil (Sigma Aldrich, viscosity 20 cSt, density 0.95 g/ml @ 25 °C) has been employed as the continuous phase in the tests. Tween 20 (Sigma Aldrich, density 1.095 g/ml @ 25 °C) is used as surfactant added to the Xanthan gum solution in small concentrations (2 wt%). Tween 20 (miscible in water) is a non-ionic detergent widely used in biochemical applications. The Hydrophilic-Lipophilic Balance (HLB) value of this surfactant is equal to HLB = 16.7 and it can be proficiently used for O/W emulsions.

The density of continuous and dispersed phases has been measured by using a precise analytical balance (RADWAG AS 220.R2). The measured values of density for the fluids used in this work are reported in Table 1.

The viscosity of silicone oil is measured using a Cannon-Fenske Viscometers and it is shown in table 1. It is evident that the viscosity of silicone oil measured in the Lab is very close to the viscosity value declared by the manufacturer (19 cP @ 25 °C).

A Brookfield DV-II + viscometer with RV spindles has been used for the characterisation of the non-Newtonian solutions. The aqueous solutions of Xanthan gum exhibited a shear thinning property for various shear rates. The power-law model $\mu = r\dot{\gamma}^{n-1}$ can be used for the characterization of the non-Newtonian solution used in these tests where r is the consistency index (cP s^{n-1}), $\dot{\gamma}$ the shear rate (s^{-1}) and n the flow index of the fluid. The values of the consistency index and of the flow index associated to the Xanthan gum solutions is reported in table 1 with the correlation coefficient (R^2).

Surface tension of the working fluids with air has been measured by means of a force tensiometer (KSV Sigma 700). Each value of the surface/interfacial tension is the average value of a ten-time repeated experiment.

Table 1. Physical properties of the working fluids with Tween 20 (2 wt%)

Fluid	Density (g/ml)	Viscosity (cP)			Surface tension (mN/m)
		r	n	R^2	
Silicone oil	0.9480 ± 5.24 E-4	18.07 ± 0.078	1.00	-	19.30 ± 0.005
0.3 wt% Xanthan gum	1.0041 ± 0.00010	2562.5	0.27	0.999	26.44 ± 0.31
0.5 wt% Xanthan gum	0.9993 ± 0.00339	10538	0.16	1.000	27.95 ± 0.12

The interfacial tension between silicone oil and Xanthan gum solutions has been measured by using the same force tensiometer and the values are presented in table 2.

Table 2. Interfacial tension of the silicone oil with Xanthan gum solutions with Tween 20 (2 wt%)

Solution	Interfacial tension (mN/m)
0.3 wt% Xanthan gum	7.4246 ± 0.38
0.5 wt% Xanthan gum	6.4518 ± 0.14

3. Results

As explained before, by means of the two syringe pumps the flow rate of silicone oil and Xanthan gum solutions have been varied at the inlets of the cross-junction in order to investigate the effect of the flow rate ratio on the characteristics of the droplets at the outlet of the junction. In order to study the geometrical characteristics of the droplets, the acquired images have been post-processed by using a Matlab code by means of which the geometry of each droplet (diameter, length, position of the edge with respect to the centre of the junction) is reconstructed [16].

3.1. Flow pattern at low flow rates of the continuous phase (< 2 ml/h)

By varying the volumetric flow rate of the continuous phase (silicone oil) from 0.5 to 2 ml/h a series of visualisations has been recorded for values of the dispersed phase flow rate ranging between 0.01 and 2.1 ml/h.

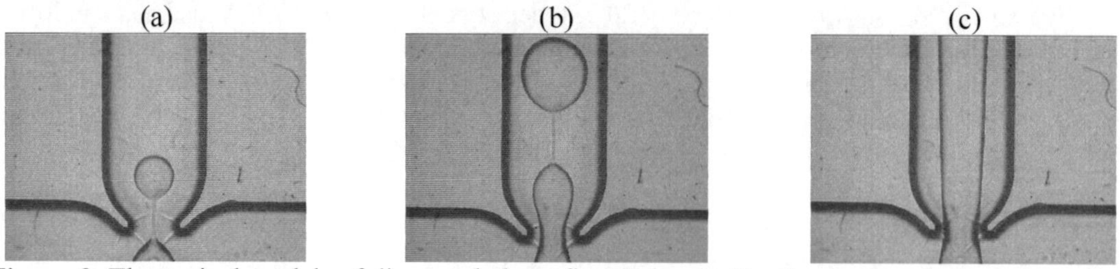

Figure 3. The typical models of dispersed phase flow 0.5 wt% Xanthan gum solution where Q_c = 1.5 ml/h and Q_d = 0.015, 0.2 and 1.65 ml/h, respectively; DCJ (a), DDC (b) and PF (c)

The main mechanism of the droplet formation for low values of Q_c (< 2 ml/h) is squeezing.

In figure 3 the typical flow patterns observed by varying Q_d from 0.015 ml/h to 1.65 ml/h for a fixed value of Q_c equal to 1.5 ml/h are shown. It is evident that droplets are generated close to the centre of the junction when low values of the flow rate ratio (α) are imposed (see figure 3a); in this case the detachment point of the neck connecting the droplet to the source of the dispersed phase takes place close to the centre of the junction. In this case the main mechanism responsible of the droplet generation is squeezing which means that the pressure drop force is the dominant one: this flow pattern will be called DCJ (Droplet generated at the Cross Junction) in this paper. By increasing α the detachment point shifts downstream along the outlet of the junction and an elongated filament links the droplet to the core of the dispersed phase (see figure 3b); this flow pattern is called here DDC (Droplet generated Downstream of the Channel). Finally, for very high values of the flow rate ratio (α) a stratified parallel flow (called PF here) is observed and the droplet generation is stopped (see figure 3c). Both DCJ and DDC regimes are linked to the squeezing mechanism.

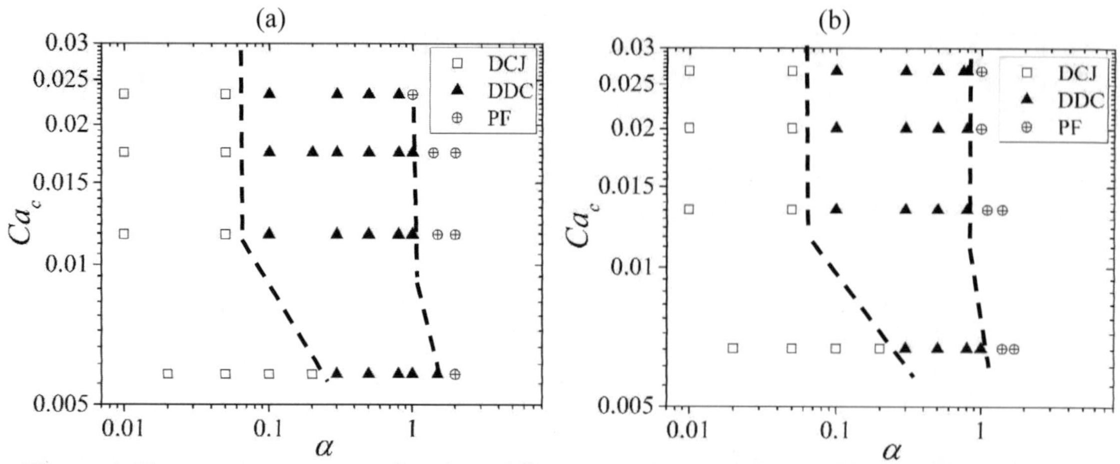

Figure 4. Flow pattern maps as a function of flow rate ratio α and the Capillary number linked to the continuous phase (Ca_c) for a 0.3 wt% Xanthan gum solution (a) and 0.5 wt% Xanthan gum solution (b)

In figure 4 the correlation among these three typical flow regimes (DCJ, DDC and PF), the flow rate ratio (α) and the Capillary number based on the continuous phase (Ca_c) is shown by means of a 2D flow pattern map obtained for two different concentrations of Xanthan gum solution (0.3 and 0.5 wt%). By observing figures 4a and 4b, it is evident how, by fixing the value of Ca_c, an increase of α leads to the transition from DCJ to DDC to PF regime. On the contrary, an increase in Ca_c for a fixed value of α is not able to put in evidence any transition. This confirms that both DCJ and DDC regimes are squeezing regimes in which α plays an important role. However, it is also evident how for low values of Ca_c, the extension of the DCJ region increases which means that the droplet can be regularly generated at the centre of the junction for a larger range of flow rate ratios. This result is in agreement with the observations of Liu and Zhang [17] that in a micro cross-junction at low values of the continuous Capillary number ($0.002 \leq Ca_c \leq 0.008$) observed numerically, by using Lattice Boltzmann method, the same evolution of the droplet regimes from DCJ to DDC to PF when the flow rate ratio was increased.

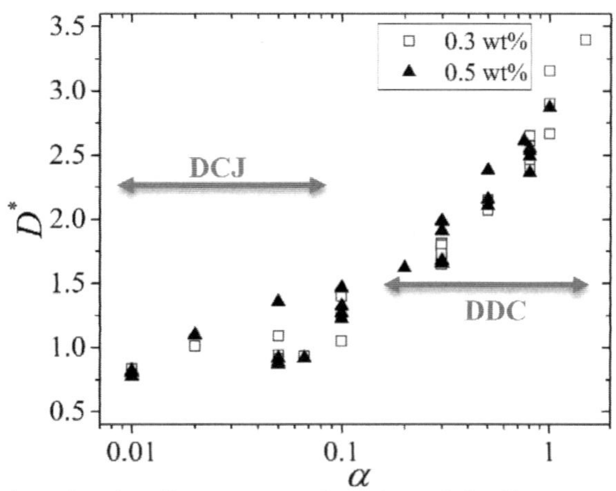

Figure 5. Dimensionless droplet diameter as a function of the flow rate ratio for two different concentrations of Xanthan gum solutions

By comparing figure 4a with figure 4b, it seems that the concentration of Xanthan gum from 0.3 wt% to 0.5 wt% weakly influences the flow map. However, in order to have a clearer understanding about the effect of Xanthan gum concentration on the transition from one regime to another one, more experimental runs are needed with different values of concentrations. The dimensionless diameter of the droplets D^*, defined as the ratio of the square root of the multiplication of horizontal (D_h) and vertical (D_v) diameter of the droplet and the height of the channel $D^* = \sqrt{D_h D_v}/H$, is shown in figure 5 as a function of the flow rate ratio α; Q_c varies from 0.5 to 2 ml/h while the range of variation of Q_d is from 0.01 to 2.1 ml/h.

It is clear that an increase in α determines an increase of the dimensionless droplet diameter. Moving from DCJ to DDC flow patterns, the slope shown in figure 5 tends to change; the data quoted in figure 5 underline that the dimensionless diameter of the droplets is more sensible to a variation of the flow rate ratio in DDC regime with respect to the DCJ regime. Once more, the Xanthan gum concentration seems to have a weak effect on the droplet dimensions by comparing open and closed symbols in figure 5.

3.2. Flow pattern at high flow rates of the continuous phase (>2 ml/h)

When the flow rate of the continuous phase becomes larger than 2 ml/h, the squeezing mechanism tends to become less important in the droplet generation and dripping and jetting regimes become predominant. When the flow rate of the continuous phase is low, due to the low shear forces the control of the droplet formation and breakup is linked to the combined effect of interfacial tensions and pressure drop forces (squeezing regime) and the typical interface is represented in figure 6a. In the case presented in figure 6a the central channel of the cross-junction, where the dispersed phase (Xanthan gum solution) flows, is almost blocked at the cross-junction. The increase in Q_c (Ca_c) leads the shear forces to play a more important role in governing the mechanism of the droplet formation. Dripping regime is obtained when shear forces and the interfacial tensions are of a similar order of magnitude; in this case, in presence of a non-Newtonian flow a filament is originated between the dispersed core at the junction inlet and the droplet.

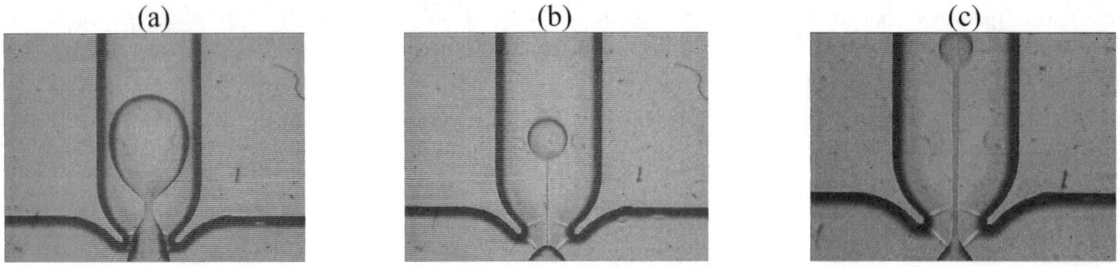

Figure 6. Figure showing Squeezing (a), Dripping (b) and Jetting (c) regimes where Q_c = 0.3, 3 and 8 ml/h and Q_d = 0.1 ml/h, respectively for 0.2 wt% Xanthan gum solution

If the flow rate of the continuous phase is further increased, the filament becomes longer and longer and the droplet dimensions tend to be blocked by the reduced flow of dispersed phase flowing within the filament, as evidenced by the comparison between figure 6b and figure 6c (Jetting regime). It is interesting to highlight that, in presence of non-Newtonian fluids like Xanthan gum solutions, a very long and slender filament can be observed before the droplet breakup which is due to the high viscosity of the dispersed phase. This filament is transformed in a satellite droplet after the breakup. The dimensions of the satellite droplets are correlated to the length of the filament before the breakup.

Figure 7 confirms that the dimensionless droplet diameter decreases when the flow rate of the continuous phase is increased (which means decreasing of flow rate ratio α). The results shown in figure 7 are obtained by varying the flow rate of the dispersed phase from 0.01 to 2.1 ml/h and the flow rate of the continuous phase from 0.3 to 10 ml/h. These results are in agreement with the literature [17,18]. Garstecki *et al*. [18] presented the dimensionless length of the aqueous droplets within a micro T-junction. They observed that at high values of flow rate ratio (α> 1) the size of the droplet increases linearly with α. On the contrary, the droplet length tends to reduce its dependence on the flow rate ratio for $\alpha < 1$. Here the same trend is observed in figure 7 but, due to the properties of the non-Newtonian dispersed phase, the change of slope is located at $\alpha = 0.1$. The results of figure 7 highlight that in order to obtain larger droplets the micro-junction has to work with large flow rate ratio (i.e. at low flow rates of the continuous phase or at large flow rates of the dispersed phase). In addition, the dimensionless droplet diameter becomes more sensitive to the variation of the flow rate ratio for $\alpha > 0.1$ where the squeezing mechanism becomes predominant. In addition, if the flow rate of the continuous phase is increased (for a fixed flow rate value of the dispersed phase), the time needed for the droplet formation decreases and a weak dependence of the droplet dimensions on Q_d is observed.

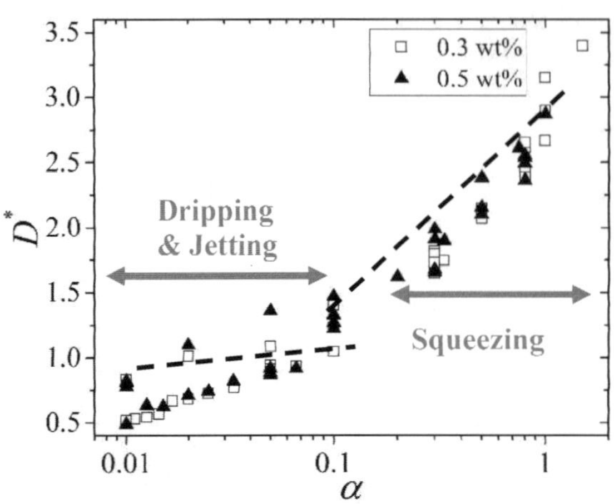

Figure 7. Dimensionless droplet diameter as a function of the flow rate ratio (α) for different Xanthan gum concentrations

In figure 8 the role of the Capillary number linked to the continuous phase on the droplet dimensions is highlighted. By increasing the Capillary number Ca_c the mechanism of the droplet formation changes from squeezing to dripping to jetting with a reduction of the droplet dimensions. At low Capillary numbers $Ca_c \leq 0.01$ the squeezing regime is observed (see figure 6a). When Ca_c is increased, dripping and jetting regimes start, where a trade-off between shear forces and interfacial tensions controls the droplet formation. The large viscosity of the Xanthan gum solution justifies the filament formation evidenced in figures 6b and 6c and, as a consequence, the reduced dimensions of the droplets. In dripping and jetting regimes shear forces are predominant and the droplet dimensions are more dependent on Ca_c than on α as evidenced by the comparison of figures 7 and 8 for $D^* < 1$.

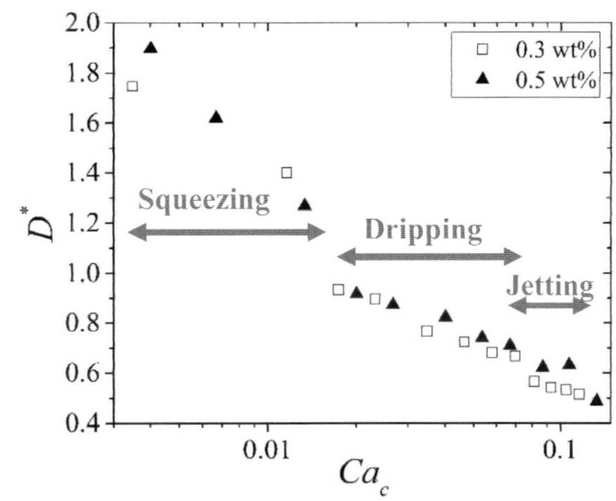

Figure 8. Dimensionless droplet diameter as a function of the Capillary number of the continuous phase

Figure 9 shows the effect of the Xanthan gum concentration on the droplet dimensions. Although the rheological properties of the dispersed phase solutions are different, the droplet diameter variation is not very large for a fixed values of the flow rate ratio. As a rule of thumb, the higher the Xanthan gum concentration is, the larger droplets are generated. Higher values of Xanthan gum concentration mean higher viscosity values.

It's worth mentioning that these results are in disagreement with the case in which Newtonian droplets are formed.

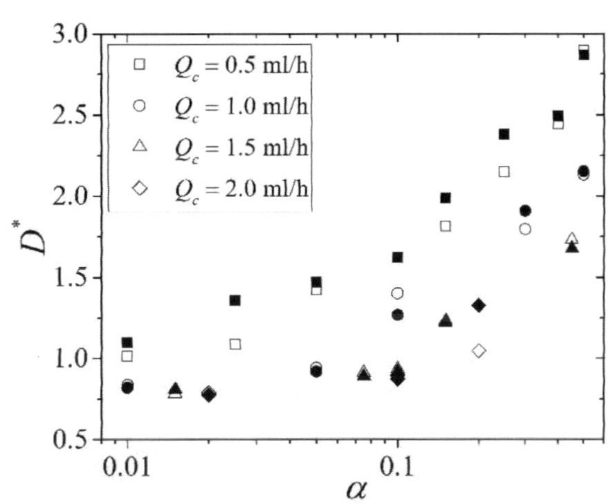

Figure 9. Effect of Xanthan gum concentration on droplet dimensionless diameter for various flow rates while open symbols stand for 0.3 wt% and closed symbol represent 0.5 wt%

In Newtonian fluids larger droplets with less viscous solutions are generated due to the higher viscous pressure in the thread which results in longer filaments [19, 20]. On the other hand, an increase of the dispersed phase viscosity is responsible of a larger shear force which reduces the droplet dimensions [19]. However, more experimental data obtained with different concentrations of Xanthan gum are needed in order to be able to put a spotlight on the rheological effects of the non-Newtonian dispersed phase on the droplet size.

4. Conclusion

In this work the generation of non-Newtonian droplets of aqueous Xanthan gum solutions in silicone oil by using a commercial glass micro cross-junction has been experimentally analyzed. A surfactant (Tween 20) has been added to the non-Newtonian solutions in order to improve the stabilization of the emulsions. The geometrical characteristics of the droplets were analyzed by means of a specific code developed in Matlab. When Q_c is lower than 2 ml/h, the squeezing mechanism is responsible of the droplet generation. Two different flow patterns can be observed linked to the squeezing mechanism: DCJ (droplet formed at the cross-junction) and DDC (droplet formed downstream of the channel). For large values of the flow rate ratio ($\alpha > 2$) the droplet generation is ceased and a stratified parallel flow (PF) is observed. By increasing the flow rate associated to the continuous phase ($Q_c > 2$ ml/h), the squeezing mechanism tends to be replaced by dripping and jetting. In dripping and jetting regimes an elongated microthread connecting the droplet to the core of the dispersed flow at the inlet of the junction is observed which is responsible of the formation of a smaller satellite droplet. The experimental data confirm that the dimensionless droplet diameter in dripping and jetting regimes depends weakly on the flow rate ratio (α) and they are a function of the Capillary number. On the contrary, in squeezing regime Capillary number loses its importance and α influences significantly the dimensions of the droplets. In fact, in squeezing regime since the junction cross-section is almost blocked, the droplet size depends on channel size and Q_d only. If the Xanthan gum concentration is increased, an increase in the viscosity of the dispersed phase is introduced and an increase in the droplet size is evidenced at the same Capillary number (Ca_c) and flow rate ratio α.

References

[1] Thorsen T, Roberts R W, Arnold F H and Quake S R 2001 *Phys. Rev. Lett.* **86** 4163-4166.
[2] Umbanhowar P B, Prasad V and Weitz D A 2000 *Langmuir* **16** 347-351.
[3] Anna S L, Bontoux N and Stone H A 2003 *Appl. Phys. Lett.* **82** 364-366.
[4] Sugiura S, Nakajima M and Seki M 2004 *Ind. Eng. Chem. Res.* **43** 8233-8238.
[5] Xiang X, Chen L, Zhuang Q, Ji X and He Z 2012 *Biosens. Bioelectron.* **32** 43-49.
[6] Song H, Chen D L and Ismagilov R F 2006 *Angew. Chem. Int. Edit.* **45** 7336-7356.
[7] Dittrich P S and Manz A 2006 *Nat. Rev. Drug. Discov.* **5** 210-218.
[8] Garstecki P, Fuerstman M J, Stone H A and Whitesides G M 2006 *Lab. Chip.* **6** 437-446.
[9] Steinhaus B, Shen A Q and Sureshkumar R 2007 *Phys. Fluids.* **19** 073103 (1-13).
[10] Hong J S and Cooper-White J 2009 *Korea-Aust. Rheol. J.* **21** 269-280.
[11] Husny J and Cooper-White J J 2006 *J. Non-Newton. Fluid.* **137** 121-136.
[12] Lee W, Walker L M and Anna S L 2011 *Macromol. Mater. Eng.* **296** 203- 213.
[13] Arratia P E, Cramer L A, Gollub J P and Durian D J, 2009 *New. J. Phys.* **11** 115006 (1-19).
[14] Arratia P E, Gollub J P and Durian D J 2008 *Phys. Rev. E.* **77** 036309 (1-6).
[15] Shui L L, van den Berg A and Eijkel J C T 2009 *Lab. Chip.* **9** 795-801.
[16] Pulvirenti B, Rostami B, Puccetti G and Morini G L 2015 *J. Phys.: Conf. Ser.* **655** 012028.
[17] Liu H and Zhang Y 2011 *Phys. Fluids.* **23** 082101.
[18] Garstecki P, Fuerstman M J, Stone H A and Whitesides G M 2006 *Lab. Chip.* **6** 437-446.
[19] Timgren A, Trägårdh G and Trägårdh C 2009 *Chem. Eng. Sci.* **64** 1111-1118.
[20] Zhang X and Basaran O A 1995 *Phys. Fluids.* **7** 1184-1203.

Fluid-flow pressure measurements and thermo-fluid characterization of a single loop two-phase passive heat transfer device

A Ilinca[1], D Mangini[2], M Mameli[3], D Fioriti[3], S Filippeschi[3], L Araneo[4], N Roth[1], M Marengo[2]

[1] University of Stuttgart, Pfaffenwaldring 31, 70569 Stuttgart, Germany

[2]University of Brighton, Cockcroft Building, Lewes Road, Brighton BN2 4GJ, UK

[3] University of Pisa, Largo Lucio Lazzarino 2, 56122 Pisa, Italy

[4]Politecnico di Milano, Energy Department - Via Lambruschini, 4A - 20156 - Milan, Italy

E-mail: D.Mangini@brighton.ac.uk

Abstract. A Novel Single Loop Pulsating Heat Pipe (SLPHP), with an inner diameter of 2 mm, filled up with two working fluids (Ethanol and FC-72, Filling Ratio of 60%), is tested in Bottom Heated mode varying the heating power and the orientation. The static confinement diameter for Ethanol and FC-72, respectively 3.4 mm and 1.7mm, is above and slightly under the inner diameter of the tube. This is important for a better understanding of the working principle of the device very close to the limit between the Loop Thermosyphon and Pulsating Heat Pipe working modes. With respect to previous SLPHP experiments found in the literature, such device is designed with two transparent inserts mounted between the evaporator and the condenser allowing direct fluid flow visualization. Two highly accurate pressure transducers permit local pressure measurements just at the edges of one of the transparent inserts. Additionally, three heating elements are controlled independently, so as to vary the heating distribution at the evaporator. It is found that peculiar heating distributions promote the slug/plug flow motion in a preferential direction, increasing the device overall performance. Pressure measurements point out that the pressure drop between the evaporator and the condenser are related to the flow pattern. Furthermore, at high heat inputs, the flow regimes recorded for the two fluids are very similar, stressing that, when the dynamic effects start to play a major role in the system, the device classification between Loop Thermosyphon and Pulsating Heat Pipe is not that sharp anymore.

Keywords: Pulsating Heat Pipe, Loop Thermosyphon, Fluid flow Analysis, Local pressure measurements.

1. INTRODUCTION

As modern computer chips and power electronics become more powerful and compact, the need of more efficient cooling systems increases day by day. In the present scenario, Pulsating Heat Pipes (PHPs) are relatively new wickless two-phase passive heat transfer device that aim at meeting all the present and future thermal requirements [1][2]. Although the peculiar advantages of this emergent technology, such as its compactness, the possibility to dissipate high heat fluxes and the ability to work also in microgravity conditions, the PHPs governing phenomena are quite unique and not completely understood [3].

Since the Single Loop PHP (SLPHP) can be considered the basic constituent of a multi-turn Pulsating Heat Pipes, its full thermo-fluidic characterization is fundamental for the complete description of the PHP or Thermosyphon working principles. At present, several studies contribute to the SLPHP knowledge [4][5][6][7] but further work is needed to understand the pressure evolution within PHPs coupled with the fluid flow visualization. For these main reasons, a novel SLPHP could provide vital information when performing simultaneous measurements with different refrigerant fluids on visualization of the fluid flow with a high-resolution camera, detection of the fluid temperature distribution during operation and the fluid pressure evolution both in the heated and in the cooled section. Considering that the inner diameter of the tube of 2 mm is slightly larger than the static confinement diameter for the FC-72 of 1.7 mm, the SLPHP can work theoretically as a thermosiphon. In static conditions, the FC-72 resides at the bottom of the device due to the gravity assistance and the lack of the capillary effects.

The aim of the work is to provide useful information on the basic phenomena involved, improving the PHP understanding. It is intended to demonstrate that a two-phase heat transfer device filled with two different fluids, FC-72 and Ethanol, each of them with static confinement diameter slightly under and slightly above the critical inner diameter in static conditions, have similar working mode during operation. This arise an open point in the literature on the need of more accurate criteria for the definition of the limits between Loop Thermosyphons [8] and Pulsating Heat Pipes. Finally, such experimental analysis provides accurate information for the validation of PHP numerical codes.

2. EXPERIMENTAL APPARATUS AND PROCEDURE

The basic features of the SLPHP are shown in Fig. 1a. The evaporator and the condenser are made of copper tubes (Inner diameter 2 mm, Outer diameter 4 mm) in order to minimize the thermal resistance between the working fluid and respectively the heat source and heat sink. The loop is completed by two transparent vertical tubes (110 mm axial length, 2mm ID,), allowing high-speed visualization of the two-phase flow, The loop components are connected by means of brass joints and sealed by vacuum epoxy (Henkel Loctite® 9492).

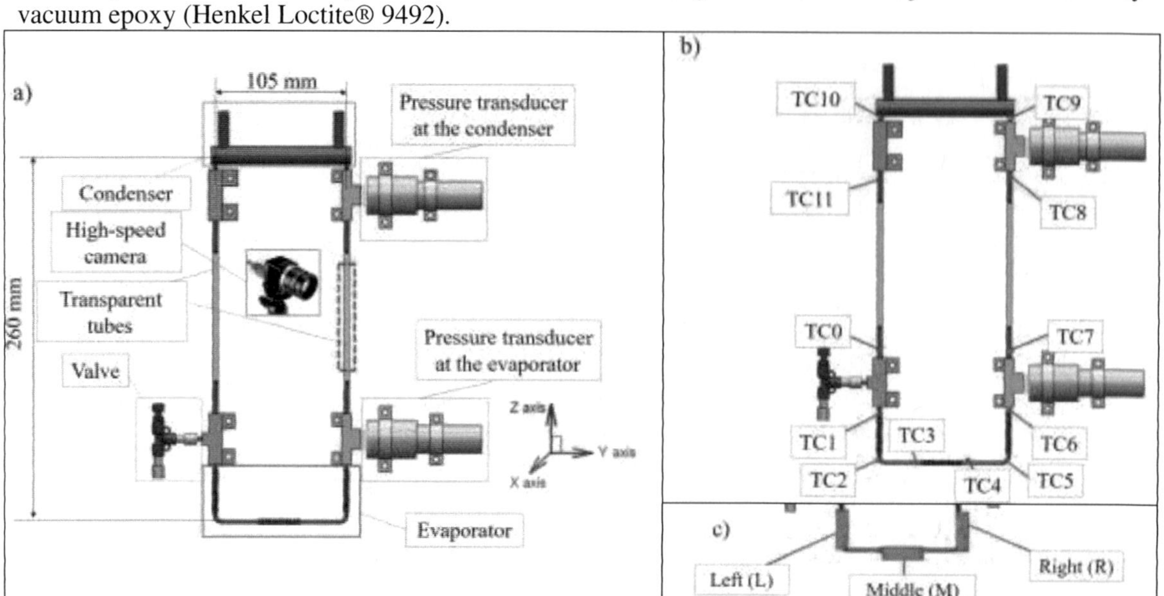

Figure 1. a) Main components of the SLPHP; b) Thermocouple positions; c) Heating elements positions

Two high-accurate pressure transducers (Keller® PX33, 1 bar Absolute, Accuracy 0.05% FSO), mounted just before and after one of the transparent inserts, measure the pressure drop along the adiabatic section.
The condenser section is embedded inside a mini shell and tube heat exchanger, directly connect to a thermal bath (Lauda® A300), recirculating water at 20 °C ±1 °C.

Three heating wires (Thermocoax®, Single core 1Nc Ac), are mounted in the evaporator section, providing a wall to fluid heat flux of 6.5 W/cm^2 at 10 W. The three heaters are controlled independently with a Pulse Width Modulation (PWM) control system, so as to vary the heating distribution along the heated zone. As shown in Fig. 1c, two heating elements (Heater L and Heater R) are positioned just above the 90° curves at the evaporator, in such a way, that by applying different power on them, the device is heated-up non-symmetrically with respect to the gravity field, while the Heater M is mounted at the center of the evaporator. As already demonstrated [9] for a similar multi-turn two-phase passive heat transfer device, when heating up the device non-uniformly at the evaporator with peculiar heating configurations, a circulation in a preferential direction is established, with a subsequent improvement of the overall thermal performance. Twelve T-type thermocouples measure the external wall temperature both in the heated and in the cooled region of the single loop (see Fig. 1c), while other two thermocouples monitor the environmental temperature during tests.

A data acquisition system (NI-cRIO-9074®, NI-9264®, NI-9214®, 2xNI-9205®, NI-9217®, NI-9472®) records the output of the thermocouples (at 10 Hz) and the pressure transducers (at 100 Hz). The high-speed camera (Ximea® USB3 XIQ-093, resolution 1280x1024 pixel) is connected to an ultra-compact PC (NUC® Board D54250WYB) able to store images up to 100 fps and it is synchronized via software with the pressure signals.

The device is vacuumed by means of an ultra-high vacuum system (Varian® DS42 and TV81-T) down to 0.3 mPa and then it is partially filled up with the working fluid with a volumetric ratio of 0.6 ± 0.03 (corresponding to 1.45 ml). Finally, the micro-metering (Upchurch Scientific® UP-P-447) valve that connects the SLPHP to the device is closed, guarantying a leak down to 10^{-8} mbar l/s. The fluid itself is previously degassed within a secondary tank, by continuous boiling and vacuuming cycles [10], in such a way to extract incondensable gases before the filling operation. Ethanol and FC-72 are choose as working fluids for their thermodynamic properties and for the fact that their critical diameter is very close to the inner diameter of the tube. The device filled up with Ethanol will act as a PHP; while with FC-72 will act as a compact loop thermosiphon.

Finally, a rotating structure was used in order to hold the device and incline it at different pivot angles, in order to aid for a faster start-up and promote the stabilization of the circulation. The pivot angle is defined and measured, referring to the laboratory fixed reference XYZ of Fig. 1a, as the rotation of the device around X-axis, counter-clockwise as seen in the figure.

3. EXPERIMENTAL RESULTS AND PROCEDURE

The experimental campaign is carried out in order to point out:

- the effect of the heating distribution on the overall thermal performance;
- the operational regimes in terms of fluid motion;
- the relationship between the pressure drop measurements and the flow patterns observed through the transparent insert;
- the uncertain limit between Loop Thermosyphon and Pulsating Heat Pipe in dynamic conditions for fluids whose static confinement diameter value is lightly lower and higher respectively than the inner diameter value of the tube.
- the effect of the pivot angle on the stabilization of the fluid circulation.

Tests are performed while maintaining the environmental temperature at 20°C ±1°C, inclining the device at four different pivot angles (vertical, 30°, 60° and 90°). The heating power is increased providing a global heating power levels of 1 W, 3 W, 6 W, 12 W and 18 W. For all the above levels, except 1 W, different heating configurations are tested: the heating power is split among the three heaters, to vary the heating distribution during tests. Pseudo-steady state conditions can be reached in approximately 3 minutes, due to the low thermal inertia of the system. Nevertheless, all the heating configuration tested are kept constant for 15 minutes, to ensure that the device reaches the pseudo-steady state conditions for all the cases tested. A video sequence (20 seconds at 100 fps) is recorded during each tested combination of heat input power, pivot angle and working fluid. The video acquisition starts 13 minutes after each heat input power variation in order to wait for the pseudo-steady state conditions. The equivalent thermal resistance (R_{eq}) is evaluated as follows (Eq.1):

$$R_{eq} = \frac{\bar{T}_e - \bar{T}_c}{\dot{Q}_{tot}} \quad (1)$$

Where: $\dot{Q}_{tot}$ is the global heating power, which is the sum of the power of the three heaters: the Heater L ($\dot{Q}_L$), Heater M ($\dot{Q}_M$) and Heater R ($\dot{Q}_R$), as pointed out in the Eq. 2:

$$\dot{Q}_{tot} = \dot{Q}_L + \dot{Q}_M + \dot{Q}_R \quad (2)$$

Since each of the three heating elements, independently controlled, can provide a different heating power and for each of them two temperatures are measured as shown in fig. 1b, the averaged T_e is calculated as pointed out in the Eq. 3:

$$\bar{T}_e = \frac{\dot{Q}_L}{\dot{Q}_{tot}} \max(T_1; T_2) + \frac{\dot{Q}_M}{\dot{Q}_{tot}} \max(T_3; T_4) + \frac{\dot{Q}_R}{\dot{Q}_{tot}} \max(T_5; T_6) \quad (3)$$

$\bar{T}_c$ is the time average of TC9 and TC10 when pseudo-steady state conditions are reached.

A first test campaign is performed with Ethanol, with the device vertically oriented in Bottom Heated Mode, i.e. with the evaporator section below the condenser. For each heating configurations tested the overall temporal evolution of the temperatures and of the pressure recorded in the transparent section are shown in Fig. 2. Only 1 W is not sufficient to activate a self-sustained two-phase flow within the loop: the temperatures at the evaporator increases, while the pressure, close to the saturation pressure of ethanol at 20°C, exhibits a flat trend. When increasing the power at 3 W (Configuration 3a in Fig. 2a), the pressure signal points out some peaks, while the temperatures starts to decrease. In such period, the heating power activates a self-sustained two-phase flow motion, improving the heat transfer between the heated and the cooled sections of the device. However, at 3 W only partial start-ups are detectable: periods in which the two-phase flow oscillates between the heated and the cooled section are alternating with stop-over periods, where the fluid does not move, thus hindering the heat exchange. This is valid for all the configurations tested with a global heating input of 3 W (from 3a to 3e in Fig. 2a), independently from the heating configuration. When the total heating power of 6 W is provided to the heater M only (Configuration 6a) a full start-up is detected: after a sudden peak of pressure at the evaporator, the temperatures in the heated zone decrease abruptly to 35 °C.

a)

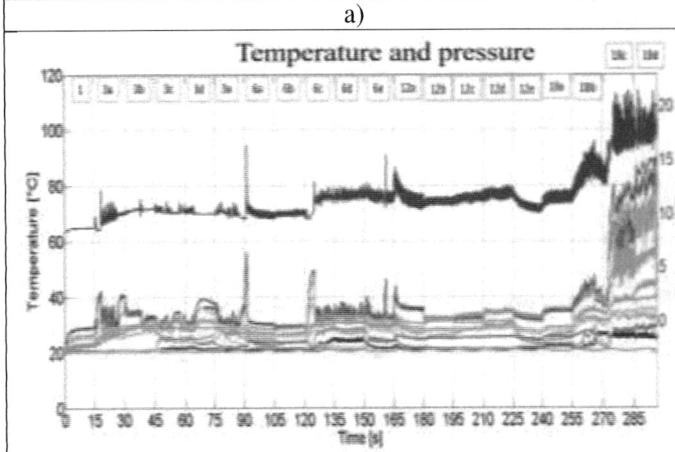

b)

Configuration	HEATER A [W]	HEATER B [W]	HEATER C [W]
1	0	1	0
3a	0	3	0
3b	1	1	1
3c	0	1	2
3d	0	1,5	1,5
3e	1,5	1,5	0
6a	0	6	0
6b	2	2	2
6c	1	2	3
6d	0	3	3
6e	3	3	0
12a	0	12	0
12b	4	4	4
12c	2	4	6
12d	0	6	6
12e	6	6	0
18a	9	9	0
18b	0	9	9
18c	4	6	8
18d	6	6	6

Figure 2. a) Temperatures and pressure evolution for Ethanol in vertical position;
b) Heating configurations sequence

This results in an improvement of the device thermal performance: the equivalent thermal resistance (R_{eq}) decreases down to 1 K/W, as observable in figure 4a. Interestingly, during the configuration 6a, a circulation of the two-phase flow that spontaneously follow a preferential direction was observed through the two transparent sections. The fluid visualization in Fig. 3 shows that an annular flow in the left transparent section flows continuously upwards from the evaporator to the condenser, while a slug/plug flow that preferentially falls from the condenser is detectable in the right transparent insert. Therefore, the TC10 and the TC11 show a higher value than the TC8 and the TC9, since the hot fluid from the evaporator is pushed preferentially through the left side of the loop. This circulation is obtained for the heating configurations 6a and 6b. Such two-phase flow motion in a preferential direction, ,improves the overall performance by continuously refreshing the hot section with a fluid flow that comes from the condenser with a lower temperature.

a) Clockwise motion

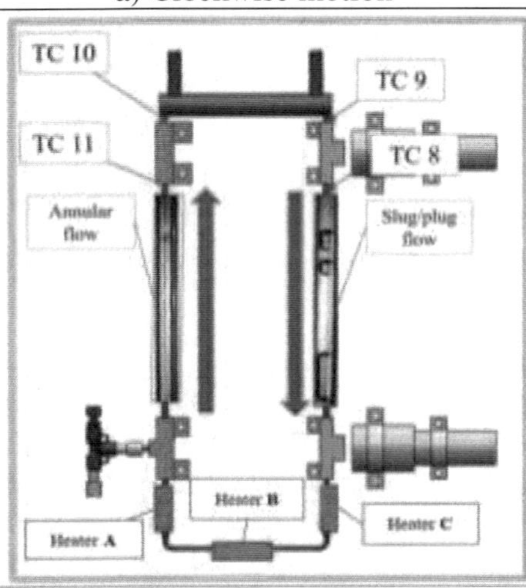

c) Temperature and pressure

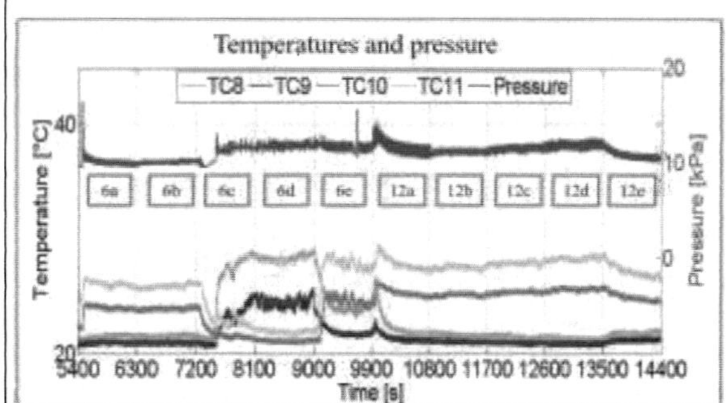

b) Anti-Clockwise motion

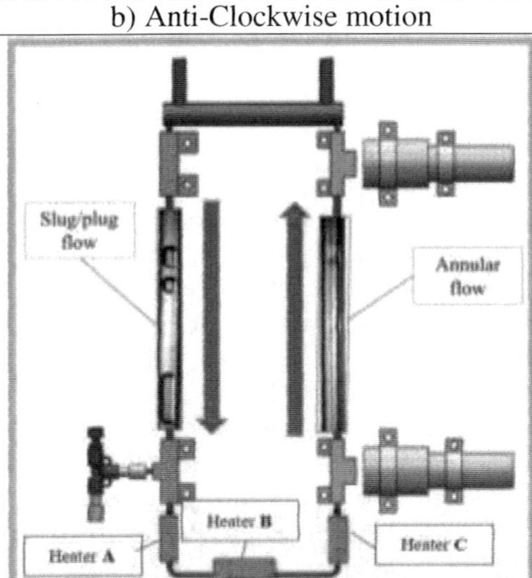

d) Test sequence with direction color coding

Config		Power Input (W) L	M	R	Flow direction
6a		0	6	0	
6b		2	2	2	
6c		1	2	3	
6d		0	•3	3	
6e		3	3	0	
12a		0	12	0	
12b		4	4	4	
12c		2	4	6	
12d		0	6	6	
12e		6	6	0	

Figure 3. Flow direction: a) clockwise circulation, green color in table d); b) Anti-clockwise, blue color in table d); c) temperatures and pressure in the condenser at 6 W and 12 W.

Therefore, the R_{eq} is every time close to 1 K/W. It is worthwhile to note that such circulation is achieved even when the system is heated up with a symmetrical configuration. This could be explained because the SLPHP is very sensitive to small hydraulic non-symmetries and to the higher thermal inertia of the components in the right side, such as the two stainless steel pressure transducers with a mass of 240 g each one. After switching from configuration 6b to 6c, the flow circulation reverses quickly visible from the increase of the temperature from TC8 and TC9. Now the annular flow that pushes upwards the fluid from the evaporator to the condenser is visible through the transparent section on the right side, while a slug/plug that goes preferentially from the evaporator to the condenser in the left one, as "anti-clockwise motion" (highlighted in the blue rectangular shape in Fig.3). This sudden variation of the circulation direction is due to the peculiar distribution of heating power provided by the configuration 6c: the local heating power provided to the Heater R is higher than the power dissipated by the Heater L and M. Therefore, the fluid is pushed preferentially through the right channel, changing the overall circulation direction. Therefore, the TC8 and the TC9 temperatures are for such configuration higher than the TC10 and the TC11. The Anti-Clockwise circulation is also observable during tests performed with the Configuration 6d: since the Heater L is off, the fluid is not pushed through the left side of the loop to the condenser. Changing from the Configuration 6d to the Configuration 6e, the fluid returns to move almost instantaneously following a Clock-wise orientation. During Configuration 6d, the Heater R is off, and therefore the fluid is not able to be pushed through the right side of the loop to the condenser. Interestingly, for such global heating power level of 6 W, the fluid motion is extremely sensitive to the peculiar heating distribution provided: a variation of the circulation direction is easily reachable providing a proper local heating power distribution in the heated region.

Increasing the global heating power from 6 W to 12 W (from the Configuration 12a to the Configuration 12e), the two-phase flow follows every time a "Clock-wise" direction, independently from the peculiar heating configuration at the evaporator. The TC10 and the TC11 are in fact every time higher than the TC9 and TC8 (Fig.4) and the annular flow that pushes continuously the fluid from the heated to the cooled region is always visible in the left transparent section. This could happen because, when increasing the global heat power input to dissipate, the fluid is pushed more vigorously from the heated to the cooled section of the device. The more heating power is provided, the higher the void fraction in the up header and the higher quantity of fluid in the down-comer. It is more difficult for the vapor expansion to contrast liquid column head and momentum. Therefore, once the circulation is established in the loop in a preferential direction for such heating power input, such inertial effect "dampens" the effect of the peculiar heating distribution provided at the evaporator, maintaining its direction for all the configurations tested.

Increasing further the global heating power from 12 W to 18 W, the temperatures at the evaporator increase abruptly, synonym of a thermal crisis (dry-out). The tests will be stopped as soon as the temperatures at the evaporator will reach 90°C in order to preserve the hardware integrity. At 18 W, in vertical position, the system is not able to switch the fluid flow direction between the different configurations tested. As soon as the configuration 18a is provided, the flow starts to oscillate without a preferential direction. The configuration 18a, being in contrast with respect to the "Clock-wise circulation" achieved for all the previous Configurations tested with a global heating power of 12 W, starts initially to push the fluid from the right section. Nevertheless, it is not able to change completely the fluid circulation: the two-phase flow starts to oscillate, and both the two transparent inserts are characterized by an oscillating semi-annular flow. At high heating powers, any attempt to force a flow reversal by changing the heating distribution, when a net flow circulation is already established, fails due to the combination of inertial effect and phase distribution. Therefore, only flow instabilities are detectable, that establish an oscillating flow, thus decreasing abruptly the overall performance. Therefore, the R_{eq} increases when a global heating power of 18 W is provided to the device (Fig.4a).

a) Ethanol, Vertical	b) Additional test up to 18 W
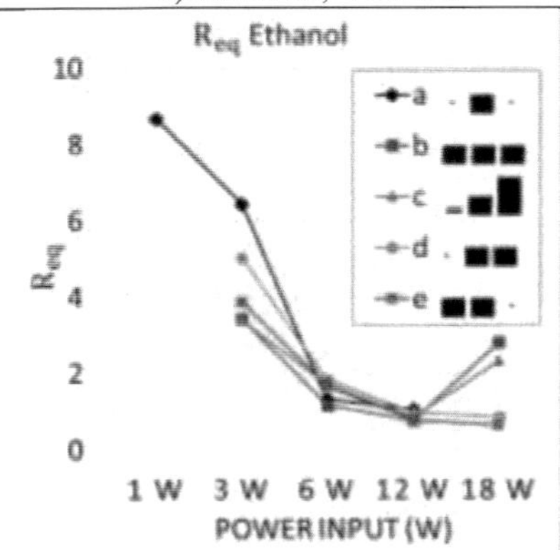	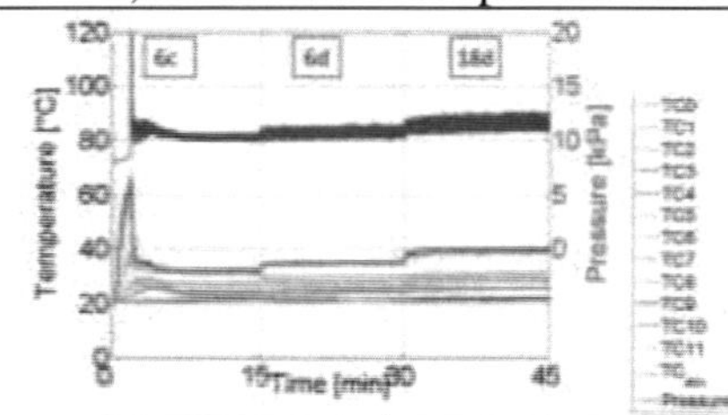

Configuration	Power input L	M	R	Total
12c	6	6	0	12
12d	0	12	0	12
18a	0	18	0	18

Figure 4. a) Equivalent Thermal Resistance for all the heating distribution tested for long test with Ehanol in vertical position and b) Test performed providing to the device up to 18 W solely at the heater M.

a)

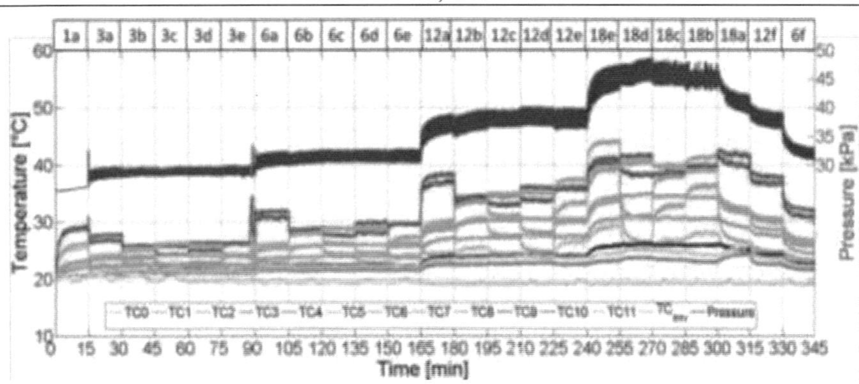

b)

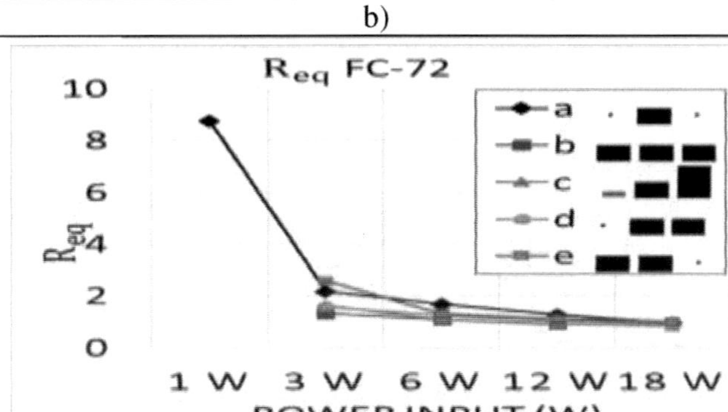

Figure 5. a) Temperatures and pressure evolution and b) Equivalent Thermal Resistance for all the heating distribution, test sequence with FC-72, 30° pivot angle

In order to check whether such thermal crisis is related to the relatively high power level, an additional test is performed, in which 18 W are dissipated solely from the Heater M (Fig. 4b), after the initiation of a self-sustained fluid flow circulation.

The maximum temperatures reached with 18 W are approximately 40°C. The heat exchange is also stable: the temporal evolution of the temperatures during such test reside in a narrow range of temperatures after the start-up, and a fluid circulation in a preferential direction is detectable through the transparent inserts. These results point out that the thermal crisis is not strictly related to the relatively high-power level but on the thermal history.

Similar results were obtained testing the device filled with FC-72 in vertical position, where already at 12 W, the system is not able to change the fluid flow direction between the 12d and 12e tested configurations resulting in a sudden increase of temperature and stop of the experiment.

Furthermore, testing the device filled with FC-72 fluid at a pivot angle of 30° for all the configurations from Fig. 2b, reveals that the pivot angle indeed helps to stabilize the fluid circulation, thus improving the overall performance (Fig. 5). From 6 W up to 18 W, the R_{eq} is every time close to 1 K/W.

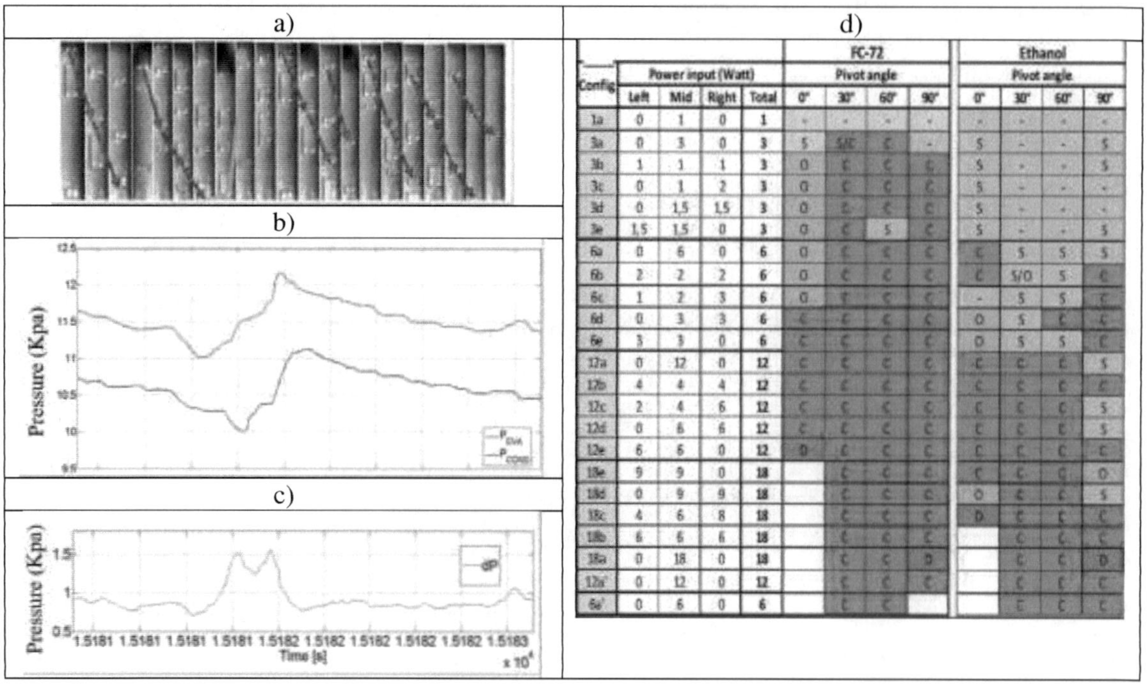

Config	Power input (Watt)				FC-72 Pivot angle				Ethanol Pivot angle			
	Left	Mid	Right	Total	0°	30°	60°	90°	0°	30°	60°	90°
1a	0	1	0	1	-	-	-	-	-	-	-	-
3a	0	3	0	3	S	S/C	C	-	S	-	-	S
3b	1	1	1	3	O	C	C	C	S	-	-	S
3c	0	1	2	3	O	C	C	C	S	-	-	-
3d	0	1,5	1,5	3	O	C	C	C	S	-	-	-
3e	1,5	1,5	0	3	O	C	S	C	S	-	-	S
6a	0	6	0	6	O	C	C	C	C	S	S	S
6b	2	2	2	6	O	C	C	C	C	S/O	S	C
6c	1	2	3	6	O	C	C	C	-	S	S	C
6d	0	3	3	6	C	C	C	C	O	S	C	C
6e	3	3	0	6	C	C	C	C	O	S	S	C
12a	0	12	0	12	C	C	C	C	C	C	C	S
12b	4	4	4	12	C	C	C	C	C	C	C	C
12c	2	4	6	12	C	C	C	C	C	C	C	S
12d	0	6	6	12	C	C	C	C	C	C	C	S
12e	6	6	0	12	O	C	C	C	C	C	C	C
18e	9	9	0	18		C	C	C	C	C	C	O
18d	0	9	9	18		C	C	C	O	C	C	S
18c	4	6	8	18		C	C	C	O	C	C	C
18b	6	6	6	18		C	C	C		C	C	C
18a	0	18	0	18		C	C	O		C	C	O
12a'	0	12	0	12		C	C	C		C	C	C
6a'	0	6	0	6		C	C			C	C	C

Figure 6. Visualization and pressure drop synchronization of: a) the images taken with high speed camera, b) pressure measurements at the evaporator (red line) and at the condenser (blue line) and c) the pressure drop calculated; d) flow pattern map.

Fig. 6d resumes all the flow motions observed for all the pivot angles and fluids tested in a visually intuitive but also complete manner.

FC-72 working fluid has shown better thermal performances over the Ethanol since is it has a prompter start up reaching wider heat transfer capability range. The low boiling point of this liquid ensures the heat discharge to the condenser at low temperatures. An increase in pivot angle improves the flow conditions and broadens the operative range of the device for both fluids. This is due to the significant gravitational assistance and the particular geometry: the evaporator section is no longer horizontal, thus circulation is promoted also by applying power at the middle heater. Once the circulation is activated the performance of the system is very similar for both fluids reaching

IOP Conf. Series: Journal of Physics: Conf. Series **923** (2017) 012022 doi:10.1088/1742-6596/923/1/012022

temperatures in the range from 35 to 45 degrees for 12W and 18 W. The inertia of the fluid in this mode is so strong that the system is not sensitive at sudden changes neither to the distribution of the heat input. The fluid is rotating preferentially counter-clockwise because the configurations had been selected to promote the circulation in a preferential direction.

In Fig 6.d the legend is explained as:
" " no symbol = untested heating power level;
"-" = no fluid motion in the transparent tube;
"S" = start-up, shut down (unstable): flow may be stratified or weakly oscillating, with partial activation or deactivation of at least one heated section visible from the temperature readings;
"O" = oscillating (stable): a strong movement of fluid oscillating back and forth;
"C" = circulating (stable): a strong movement of fluid in one specific preferential direction;
"D" = dry-out: when excessive heat power input causes a thermal crisis in at least one evaporator section.

A synchronization between the images recorded in the transparent section and the pressure signals is performed and highlighted in Fig. 6.b. If a slug/plug flow is oscillating in the transparent section, the relationship between the fluid flow motion and the pressure measurements is clearly observable: when the two-phase flow is coming from the condenser section, the pressure decreases suddenly (sometimes of more than 500 Pa in less than 0.3 seconds) both at the evaporator and at the condenser. Nevertheless, when the flow is pushed from the heated to the cooled region during such flow reversal, a sudden increase of pressure is measured both at the evaporator and at the condenser.
It is observed that the peaks of pressure are more pronounced closer to the evaporator due to the expansions of the vapor bubbles, while at the condenser are smoother (Fig 6.b).

The pressure drop is also calculated, simply subtracting the pressure at the evaporator to the pressure measured at the condenser (Fig. 6c). The pressure difference is characterized by some peaks when fluid flow accelerations are detectable through the transparent section. Nevertheless, the static pressure difference depends on three main terms [11], as shown in Eq. 4:

$$\Delta P = \Delta P_f + \Delta P_a + \Delta P_g \tag{4}$$

where $\Delta P_f, \Delta P_a, \Delta P_g$ are relatively the friction, acceleration and static head terms. The hydrostatic pressure is dependent by the void fraction along the tube in which the pressure measurements are performed, that continuously change in time during PHP operations and it is also dependent by the peculiar flow pattern observed. Further analysis is needed to calculate the magnitude of the three components in Eq. 4.

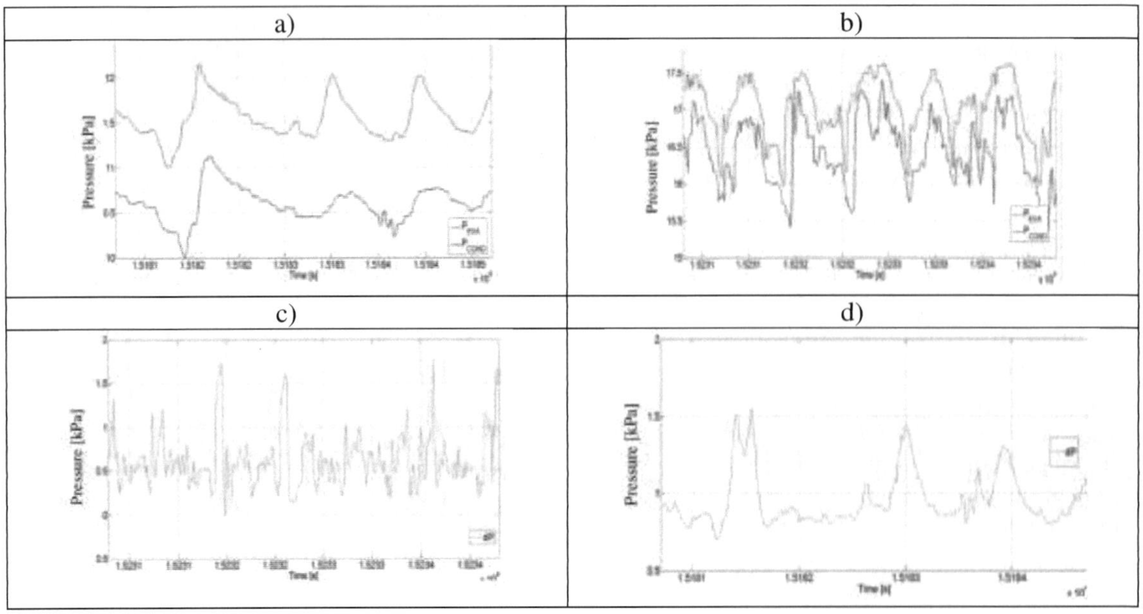

Figure 7. Pressure recorded just after the heated section (red line) and just before the cooled region (blue line) a) slug/plug flow; b) annular flow and Pressure drop measurements (green line) c) slug/plug flow; d) annular flow.

For instance, having the possibility to test the device in microgravity by means of parabolic flights, the hydrostatic pressure is not more present. The acceleration term may be estimated from the liquid velocity calculated by post-processing of the images recorded in the transparent section. In this way, it may be theoretically possible to calculate the pressure related to the skin friction, and to correlate it with respect to the peculiar flow pattern observed.
In the case of an annular flow, it is hard to find a relationship between the pressure measurements and the images recorded by the high-speed camera. In fact, it is difficult to track the annular flow evolution solely with the high-speed camera, since there are no liquid menisci to track as in the slug/plug flow case. The static head is nearly zero because the very small quantity of fluid in the tube, while the vapor acceleration term plays a major role. However, some aspects regarding the fluid-dynamic of the annular flow can be seen looking at the pressure signals during tests. The pressures measured when an annular flow pattern is shown in figure. 7b and 7d (pressure drop) are completely different with respect to the slug/plug flow case (Fig. 7a and Fig. 7c respectively). The oscillations have a higher frequency in time for the annular flow.

4. CONCLUSIONS

A Novel Single Loop Pulsating Heat Pipe, filled up with pure ethanol or FC-72 (Volumetric Filling Ratio = 60%) with an inner diameter value of 2 mm is tested with different heating power distribution and tilting angles. Three heating elements, mounted in strategic points of the evaporator, are controlled independently, in such a way to heat up the device varying the heating distribution. Furthermore, two highly accurate pressure transducers measure the pressure just at the edges of one of the transparent inserts.

Results point out that peculiar heating configurations stabilize the two-phase flow motion in a preferential direction, thus improving the overall performance and it was demonstrated that two fluids, FC-72 and Ethanol, with diameter respectively slightly below and above the critical value, behave respectively as a thermosiphon or as a PHP in static conditions, but show similar working modes during dynamic operation. Moreover, an increase in inclination, improves the flow conditions and broadens the operative range of the device for both fluids.

The pressure measurements recorded just after the heated and before the cooled section highlights that the pressure evolution is strictly dependent on the flow pattern observed. Qualitatively, each flow pattern always exhibits a pressure difference that can be correlated to fluid velocity and acceleration, static head terms, void fraction, fluid density and friction. When a slug/plug flow is observed through the transparent section between the two pressure transducers, the difference of pressure reaches 1.5 kPa and the pressure oscillations exhibit a lower frequency. When annular flow is detected, the pressure variations between the heated and the cooled section are less than the slug/plug flow case, but with higher frequency.

Such high speed visualization coupled with the highly accurate pressure measurements performed in this work, regarding the liquid film evolution, could provide useful information for updating and validating the actual lumped parameters models [12][13] and the emerging Computational Fluid Dynamic simulations on PHPs [14][15].

5. NOMENCLATURE

FC-72 Perfluoro-hexane
Heater-L the heater on the Left side
Heater-M the heater in the Middle, central position
Heater-R the heater on the Right side
P_{EVA} Fluid Pressure measured close to the evaporator
P_{COND} Fluid Pressure measured close to the condenser
PHP Pulsating Heat Pipe
pivot angle the device rotation angle around its X-axis, counter-clockwise
Q'_L power of the Left heater
Q'_M power of the Median heater
Q'_R power of the Right heater
Q'_{TOT} total power of the three heaters
R_{eq} equivalent thermal resistance
SLPHP Single Loop Pulsating Heat Pipe
T_c average temperature at the Condenser
T_e average temperature at the Evaporator
TC_{env} environment thermocouple or temperature
TC_N thermocouple number N ($0 \leq N \leq 11$) or its recorded temperature
ΔP total pressure difference
ΔP_f friction pressure difference
ΔP_a acceleration pressure difference
ΔP_g static gravitational pressure difference

6. REFERENCES

[1] Akachi H. 1990 US Patent 4.921.041.
[2] Akachi H. 1993 US Patent 5.219.020.
[3] Zhang Y. and Faghri A. 2008 *Heat Transfer Eng.* **29** 20-44.
[4] Khandekar S. and Groll M. 2004 *Int. J. of Therm.Sci.* **43** 13-20.
[5] Khandekar S., Gautam A. P. and Sharma P. 2009 *Int. J. of Therm. Sci.* **48** 535–546.
[6] Spinato G., Borani N. and Thome J. R. 2015 *Energy* **90** 889–899.
[7] Spinato G., Borani N. and Thome J. R. 2016, *Int. J. of Therm. Sci.* **102** 78–88.
[8] Mangini D, Mameli M, Fioriti D, Filippeschi S, Araneo L and Marengo M 2017, *Appl.Therm. Eng.* **126** 1029-1043.
[9] Mameli M, Mangini D, Vanoli G, Filippeschi S, Araneo L and Marengo M 2016 . *Energy* **112** 562–573.

[10] Henry C.D, Kim J and Chamberlain B 2004, *3rd Int.Symp. on Two-Phase Flow model. and Exp.* Pisa, Italy
[11] Collier G and Thome J R 1994, Oxford Science Publications, third edition.
[12] Nekrashevych I and Nikolayev V. S 2017 *Appl. Therm. Eng.* **117** 24-29.
[13] Manzoni M, Mameli M, de Falco C, Araneo L and Marengo M 2016 , *Int. J. for Num.Meth. in Fl.* **82(7)** 375–397.
[14] Jiaqiang E, Zhao X, Deng Y and Zhu H, 2015. *Appl. Therm. Eng.* **93** 166–173.
[15] Pouryoussefi S M. and Zhang Y 2015 *Appl. Therm. Eng.* **98** 617-627.

IOP Conf. Series: Journal of Physics: Conf. Series **923** (2017) 012023 doi:10.1088/1742-6596/923/1/012023

Modelling the pressurization induced by solar radiation on above ground installations of LPG pipeline systems

M Leporini[1], A Terenzi[2], B Marchetti[3], G Giacchetta[1], F Polonara[1], F Corvaro[1] and R Cocci Grifoni[4]

[1]Dipartimento di Ingegneria Industriale e Scienze Matematiche (DIISM), Università Politecnica della Marche, via Brecce Bianche, Ancona, Italy
[2]Saipem S.p.A., via Toniolo 1, 61032, Fano (PU), Italy
[3]Facoltà di Ingegneria, Università degli Studi E-Campus, via Isimbardi 10, 22060 Novedrate (CO), Italy
[4] School of Architecture and Design, Università di Camerino, Camerino (MC), Italy

E-mail:m.leporini@univpm.it

Abstract. Pipelining Liquefied Petroleum Gas (LPG) is a mode of LPG transportation more environmentally-friendly than others due to the lower energy consumption and exhaust emissions. Worldwide, there are over 20000 kilometers of LPG pipelines. There are a number of codes that industry follows for the design, fabrication, construction and operation of liquid LPG pipelines. However, no standards exist to modelling particular critical phenomena which can occur on these lines due to external environmental conditions like the solar radiation pressurization. In fact, the solar radiation can expose above ground pipeline sections at pressure values above the maximum Design Pressure with resulting risks and problems. The present work presents an innovative practice suitable for the Oil & Gas industry to modelling the pressurization induced by the solar radiation on above ground LPG pipeline sections with the application to a real case.

1. Introduction

Liquefied Petroleum Gas (LPG) is a mixture of light hydrocarbons (simply propane or butane), gaseous at standard conditions (15°C and 1 bara) and maintained in the liquid state by increased pressure or lowered temperature [1]. Because of its high heating values, high purity, cleanness of combustion and easy of handling, it is used as a multipurpose material in the industrial chemical, commercial, residential, transportation and other sectors of economy [2].

A high-pressure pipeline system is an environmentally friendly way of transporting large volumes of LPG. In fact, it is characterized by less energy consumption and exhaust emissions than other modes of transportation (such as road and ship). In 1940, it was constructed the first LPG pipeline from the Panhandle field to other regions of Texas and other regions of the USA in association with the Phillips Petroleum Company [3]. Worldwide, there are over 350000 kilometres of pipelines transporting refined petroleum products (including LPG) and about 8000 kilometres of single phase pipelines what transport LPG (propane or butane) fluids. The advantage of pipeline transportation of LPG is that it can deliver a constant and steady supply of LPG without the need for intermediate storage along the distribution route. Conveyance of LPG by pipelines continues to grow in the world as the policy recognizes the environmental benefits of this type of transportation. Knowledge of the flow behaviour of the fluid under different operating conditions will assist researchers and engineers in designing these pipelines [4].

IOP Conf. Series: Journal of Physics: Conf. Series **923** (2017) 012023 doi:10.1088/1742-6596/923/1/012023

The literature relating to pipelines transporting fluids similar to LPG from a thermodynamic point of view suggests that the hydraulics of the gas phase transport is much less efficient with respect to liquid/dense phase, since it is disadvantaged by the low density (and consequently large pipe diameter) and high pressure drops. To transport LPG efficiently by pipeline, the fluid must be transported in dense phase (which is defined by pressure values greater than cricondenbar). From a safety point of view, LPG is intentionally aromatized by refineries in order to detect gas leaks [5]. In fact, though LPG is non-toxic, it can be very dangerous if a partial or complete rupture of an LPG pipeline occurs. Pinhole or full-bore leaks can result in an accidental release, which can cause issues related to evaporation, vapour cloud propagation and dispersion. An enormous amount of gasification and energy release occurs with a pressure drop if an LPG pipeline is damaged for any reason, with possible catastrophic consequences for both the line and the surrounding [6]. In fact, when LPG pipeline ruptures occur, two conditions can manifest simultaneously: a particular flow behavior within pipeline and the formation of a vapour cloud extending beyond the leak location. For what concerns the flow after the rupture, in a LPG fluid pipeline (which is a HVP, i.e. high vapor pressure pipeline), differently from what happens for low vapor pressure fluid pipeline (LPV), all the fluid within the line drains off irrespective of elevation profile. Unlike LVP, for which the elevation profile helps to prevent the total fluid drainage, all the LPG contained in a section of a pipeline boils and vaporizes upon pipeline like. Instead, the second condition can lead to a boiling liquid expanding vapour explosion (BLEVE). Boiling liquid expanding vapor explosion (BLEVE) is the most dangerous scenario in the transportation of LPG [7]. BLEVE occurs when LPG (or other similar liquid) in a damaged pipeline gasifies quickly, causing extensive damage. The released gas ignites if this occurs in a quick and intense manner. Furthermore, BLEVE can create widespread fires or flares by a source and the resulting explosions can cause thermal radiation disasters [8]. Concerning the mechanism by which a BLEVE occurs, several authors have analyzed it and proposed theoretical treatments ([9][10][12]) but very few studies investigate the causes of initiation of leaks. In addition, there are different codes that industry follows for the design, construction and operation of LPG facilities. The codes used by the Oil & Gas industry are the main standards related to the pipeline transmission, such as ANSI/ASME B31.4 and B31.8, CSA (Canadian Standard Association) Z662-03 and BS-EN (BSI/European Code) 1442, BS-EN12252 and other practices for engineers provided by the UL International Institution of Gas Engineers and Managers (IGEM). According to the aforementioned international standards, the main consideration to be done during the design and operation of LPG pipelines are the same for the other HVP product pipelines: hydraulics, material fracture control properties, pumping, operation procedures, etc. However, no practices are provided in order to take into account phenomenon that can lead to a formation of leaks and fractures. In fact, various mechanisms can cause the initiation of fractures in a LPG pipeline: operating scenarios that lead the fluid pressure above the Design Pressure (such as water hammer, restart after a shut-down or ramp-up, pigging), the depressurization which can determine temperatures of the pipeline construction material too low (lower than the minimum design temperature) as consequence of liquid pools accumulation within dips (as normally is in hilly terrains) [11], corrosion, the heat radiation from a fire, the pressurization induced by the solar heat radiation, and other phenomenon. In particular, the pressurization induced by the solar radiation can lead the pipeline to operating pressures above the Design Pressure (DP) and this can generates leaks, fractures and most catastrophic scenarios. The design of a LPG pipeline depends on the characteristics of the product it contains and takes into account the fluid vapour pressure at the maximum temperature within the line. This is particularly important in hot countries where heat from solar radiation can cause the liquid within the pipeline to expand, exerting greater pressure [13]. The pipe construction material and the pipeline wall thickness are chosen also in function of the maximum allowable operating pressure; also for this reason, it is very important in the Oil & Gas industry to properly calculate the maximum pressures experienced by the transported fluid and then by the pipeline.

In the present paper, the effect of the solar radiation on the heat transfer on above ground pipeline sections is investigated. The solar radiation on the pipeline sections is modelled in the 1-D thermo-hydraulic simulator OLGA (version 2014.4) through a new method that can be used as standard in Flow Assurance studies. In the past, this method has been validated against field data for a natural gas system

IOP Conf. Series: Journal of Physics: Conf. Series **923** (2017) 012023 doi:10.1088/1742-6596/923/1/012023

and it has been extended to an LPG system in the present study [14]. OLGA is a transient one-dimensional commercial multiphase flow simulator, which is used by the Oil & Gas industry for the multiphase flow design of wells and pipelines. The thermodynamics of the fluid is modelled by the commercial code PVTsim Nova version 3 and Peng-Robinson equation of state is used [15].

2. Governing Equations and Methodology

The pipeline hydraulic calculations are carried out trough OLGA code by solving a set of coupled first-order, nonlinear, one-dimensional partial differential equations. OLGA is a modified two-fluid model, which is able to treat water as a separate third phase in addition to oil and gas phases. In fact, separate continuity equations for the gas (Eq. 1), oil in liquid layer (Eq. 2) and liquid droplets (Eq. 3) are applied; these are coupled through interfacial mass transfer. Only two momentum equations are used, one for the continuous liquid phase (Eq. 4) and one for the combination of gas and possible liquid droplets (Eq. 5). The velocity of any entrained liquid droplets in the gas phase is given by a slip relation. One mixture energy-conservation equation is applied (Eq. 6); both phases are at the same temperature. This yields six conservation equations to be solved (if the water option is not taken into account): three for mass, two for momentum, and one for energy [16].

$$\frac{\partial}{\partial t}\left(V_g \cdot \rho_g\right) = -\frac{1}{A}\frac{\partial}{\partial z}\left(A \cdot u_g \cdot \rho_g \cdot V_g\right) + \Psi_g + G_g \tag{1}$$

$$\frac{\partial}{\partial t}\left(V_l \cdot \rho_l\right) = -\frac{1}{A}\frac{\partial}{\partial z}\left(A \cdot u_l \cdot \rho_l \cdot V_l\right) - \Psi_g \frac{V_l}{V_l + V_d} - \Psi_e + \Psi_d + G_l \tag{2}$$

$$\frac{\partial}{\partial t}\left(V_d \cdot \rho_l\right) = -\frac{1}{A}\frac{\partial}{\partial z}\left(A \cdot u_d \cdot \rho_l \cdot V_d\right) - \Psi_g \frac{V_d}{V_l + V_d} + \Psi_e - \Psi_d + G_d \tag{3}$$

$$\begin{aligned} \frac{\partial}{\partial t}\left(V_l \cdot \rho_l \cdot u_l\right) = & -V_l\left(\frac{\partial p}{\partial z}\right) - \frac{1}{A}\frac{\partial}{\partial z}\left(A \cdot \rho_l \cdot V_l \cdot u_l^2\right) + -\lambda_l \frac{1}{2}\rho_l \cdot \left|u_l\right| \cdot u_l \frac{P_l}{4A} + \lambda_i \frac{1}{2}\rho_g \cdot \left|u_r\right| \cdot u_r \cdot \frac{P_i}{4A} + V_l \cdot \rho_l \cdot g.\cos\vartheta + \\ & -\Psi_g \cdot \frac{V_l}{V_l + V_d} \cdot u_a - \Psi_e \cdot u_i + \Psi_d \cdot u_d - V \cdot_l d \cdot \left(\rho_l - \rho_g\right) \cdot g \cdot \frac{\partial V_l}{\partial z} \sin\vartheta \end{aligned} \tag{4}$$

$$\begin{aligned} \frac{\partial}{\partial t}\left(V_g \cdot \rho_g \cdot u_g + V_d \cdot \rho_l \cdot u_d\right) = & -\left(V_g + V_d\right)\frac{\partial p}{\partial z} - \frac{1}{A}\frac{\partial}{\partial z}\left(A \cdot \rho_g \cdot V_g \cdot u_g^2 + A \cdot \rho_l \cdot V_d \cdot u_d^2\right) + \\ & -\lambda_g \frac{1}{2}\rho_g \cdot \left|u_g\right| \cdot u_g \frac{P_g}{4A} - \lambda_i \frac{1}{2}\rho_g \cdot \left|u_r\right| \cdot u_r \cdot \frac{P_i}{4A} + \left(V_g\rho_g + V_d\rho_l\right) \cdot g.\cos\vartheta + \Psi_g \cdot \frac{V_d}{V_l + V_d} \cdot u_a + \Psi_e \cdot u_i - \Psi_d \cdot u_d \end{aligned} \tag{5}$$

$$\begin{aligned} & \frac{\partial}{\partial t}\left[m_g\left(e_g + \frac{1}{2}u_g^2 + gL\right) + m_l\left(e_l + \frac{1}{2}u_l^2 + gL\right)z + m_d\left(e_d + \frac{1}{2}u_d^2 + gL\right)\right] = \\ & = -\frac{\partial}{\partial z}\left[m_g u_g\left(H_g + \frac{1}{2}u_g^2 gL\right) + m_l u_l\left(H_l + \frac{1}{2}u_l^2 + gL\right) + m_d u_d\left(H_d + \frac{1}{2}u_d^2 + gL\right)\right] + H_s + U \end{aligned} \tag{6}$$

where the subscripts *g*, *l*, *i*, and *d* indicate gas, liquid, interface, and droplets, respectively. The subscript *e* indicates the droplet entrainment. In addition, ρ is the fluid density, u the fluid velocity, z the longitudinal coordinate, p the pressure, A the pipe cross-sectional area, Ψ_g is the mass transfer rate between the phases, Ψ_e and Ψ_d are the entrainment and deposition rates, V is the volume fraction, G_f is the possible mass source of the phase f, ϑ is the pipe inclination with the vertical and P is the wetted perimeter by each phase. e is the internal energy per unit mass, L the elevation, H_s the enthalpy from mass sources, and U the heat transfer from pipe walls. OLGA uses a finite difference scheme on a staggered grid for the spatial discretization.

The basic thermal coupling in OLGA considers radially symmetric heat transfer between the fluid inside a section of a flowpath and the ambient through a set of concentric wall layers (which can represent the pipe wall, insulation material, etc). The inner pipe wall temperature T_{iw} is related to the environment temperature T_{amb} and can be calculated based on the heat flux through the pipe wall and to the environment. OLGA sets up a system of dynamic equations describing the heat flow and heat storage in the pipe wall, which determine the average temperature in each wall layer. The heat transfer process between the pipeline internal fluid and the environment is due to different contributions:

- internal convection between fluid and pipe wall;
- conduction through pipe wall layers;
- external conduction through soil for pipeline buried sections, or external convection for the above ground sections.

The temperature time trend during a transient calculation and/or along the pipeline length is obtained, starting from specified initial conditions, by considering both the fluid cooling by expansion (Joule-Thompson effect) and the heat exchange between fluid inside the pipeline and the external environment through the pipe wall. The heat transfer coefficient between a pipe wall surface and a fluid is calculated by recognized engineering relations and the Nusselt number, Nu, is calculated by the following equations (where Re is the Reynolds number and Pr is the Prandlt number):

- for turbulent flow, $\mathrm{Re}>10^4$, according to Sieder-Tate correlation [17]:

$$Nu = 0.027\,\mathrm{Re}^{0.8}\,(\mathrm{Pr})^{\frac{1}{3}} \tag{7}$$

- for transition zone, $2300<\mathrm{Re}<10^4$, according to Ramm [17]:

$$Nu = 0.027\,\mathrm{Re}^{0.8}\,(\mathrm{Pr})^{\frac{1}{3}}\left(1-\frac{6\cdot 10^5}{\mathrm{Re}^{1.8}}\right) \tag{8}$$

- for laminar flow (natural convection), $\mathrm{Re}<2300$:

$$Nu = \max\left(0.184(Gr\,\mathrm{Pr})^{\frac{1}{3}}, 3.66\right) \tag{9}$$

where Gr is the Grashof number and the equations to calculate Gr number depend on the pipe orientation (vertical or horizontal). If the inequality $(\mathrm{Gr}/\mathrm{Re}^2) \ll 1$ is satisfied, free convection effects may be neglected and Nu =f(Re, Pr). Conversely, if $(\mathrm{Gr}/\mathrm{Re}^2) \gg 1$, forced convection effects may be neglected and Nu= f(Gr,Pr). The combined effects of free and forced convection is considered when $(\mathrm{Gr}/\mathrm{Re}^2) \approx 1$. The transient conduction through the pipe wall layers is simulated integrating the Fourier's equation for a cylindrical geometry, along the radial direction:

$$\rho_w c_{p_w}\frac{\partial T}{\partial t} = \frac{1}{r}\frac{\partial}{\partial t}\left(r k_w \frac{\partial T}{\partial r}\right) \tag{10}$$

where r is the radial coordinate, ρ_w the density of w-th pipe wall layer, c_{pw} the heat capacity of w-th pipe wall layer, k_w the thermal conductivity of w-th pipe wall layer. The equation system for temperatures in the wall layers is solved simultaneously with the energy balance equation (Eq. 6), thus ensuring an implicit coupling between the fluid temperature and the temperature in the wall layers.

The external heat transfer coefficient depends on the considered environment. Modelling for above ground segments is implemented by simulating the external forced convection through an outer wall heat transfer coefficient, h_{amb}, computed by OLGA for convection in water or air or user given and based on the type of ambient medium (in this case fluid properties and velocity of the ambient fluid must be provided). Of course, above ground pipeline sections are subjected also to the thermal radiation effects but OLGA does not implement a model for the simulation of the solar radiation. Maybe the reason is that under normal operating conditions, the exposure of the above ground portions of the pipeline to high intensity solar radiation does not pose any major design problems. In fact, the flowing fluid in the pipeline, by carrying away most of the incident heat flow, will level down the temperature peaks and the subsisting differential thermal expansion is easily accommodated by well-established design techniques. However, other operating scenarios like routine maintenance, emergency repairs or pumping power loss, may require that pipeline valves be shut off to isolate specific sections of the pipeline. As a consequence, the fluid enclosed in these sections, if subject to high heat fluxes, may experience considerable temperature rises [18]. The expanding fluid may cause internal pressures close to the pipeline design pressures with risks of leaks and bursts. This is particular set of conditions for pipeline operation, often not considered by conventional design analyses and neglected by most current specifications. Here below, the methodology developed to simulate this condition is presented.

OLGA implements a model to simulate the electrical heating of the lines (ELECTRICHEAT option). Electric heat tracing or electrical heating, is in fact a system widely used in the Oil & Gas field to maintain or raise the temperature of pipes, to prevent hydrate formation and wax deposition in oil production, to protect pipes from freezing, etc. [19]. The types of electric heat trace cables are four[20]: self-regulating (the cables vary their heat output based on the heat requirements of the system and can be cut to the exact length needed), constant wattage (the cables are used when high heat output or high temperature exposures are required; they can maintain temperatures up to 300°F), mineral insulated trace (it is used when there is even higher heat output, higher exposure temperatures, or resistance to environmental corrosives or abrasives), and series resistance and skin tracing (they are typically used in pipeline applications requiring long runs of pipe). OLGA allows the simulation of fourth kind of electric heat tracing method, with a constant heat absorbed along the pipeline. In order to simulate by OLGA the pressurization induced by the solar heat radiation, the authors suggest a method that uses the aforementioned "ELECTRICHEAT" option in OLGA. The solar radiation impact is simulated by imposing the relevant thermal power input over the pipeline exposed surface, and assuming an absorption coefficient function of the relevant paint colour. As the ELECTRICHEAT option takes heat absorbed by thermal radiation in terms of W/m, heat absorbed by the pipe is calculated as follows:

$$\Omega = \beta \alpha D_e \tag{11}$$

where Ω is the absorbed solar radiation power per unit length, β the pipeline external coating absorption coefficient, α the solar radiation power. Solar radiation impact calculations must be carried out by considering daily variation of absorbed radiation power per unit length.

3. Results and Discussion

The model presented by the authors for simulating the pressurization induced by thermal solar radiation through OLGA code is applied to a practical case in order to show the useful practice that can be adopted in flow assurance studied.

A LPG consortium planned a large-scale development to import Liquefied Petroleum Gas from the American market in an Asian country. A LPG transmission pipeline, called the "LPG Pipeline", is a major component of the development; it is about 500km long. During the design phase of the line, two possible hydraulic configurations have been studied; they are shown in

Figure **1**. The first one (Alternative 1) presents a first 24" pipeline long approximately 402km, running from KP0 to a tie-in where LPG is delivered to a plant for direct usage while the remaining LPG is transported through a 20" line to the delivery point. The second one (Alternative 2) considers a 20" constant diameter pipeline with one intermediate pumping station. The pipeline is considered buried and only some above ground sections are subjected to the solar thermal radiation.

With reference to the hydraulic configurations reported in

Figure **1**, the following above ground pipe sections are studied for pressurization induced by solar radiation, with reference to a long term shutdown: 24" pump suction pipe at head pumping station, 24" pump discharge pipe at head pumping station, 24" pipe at intermediate pigging station, 20" pipe at intermediate pigging station, bypass lines in one 24" MLV (Mainline Valve Station) station, bypass lines in one 20" MLV station, 20" pump suction pipe at head pumping station, 20" pump discharge pipe at head pumping station, 20" pipe at intermediate pump station (pump suction and discharge sides) and

IOP Conf. Series: Journal of Physics: Conf. Series **923** (2017) 012023 doi:10.1088/1742-6596/923/1/012023

bypass lines in one 20" MLV station. Based on the above, 6" and 8" above ground pipe sizes can be analyzed to evaluate pressurization induced by solar radiation.

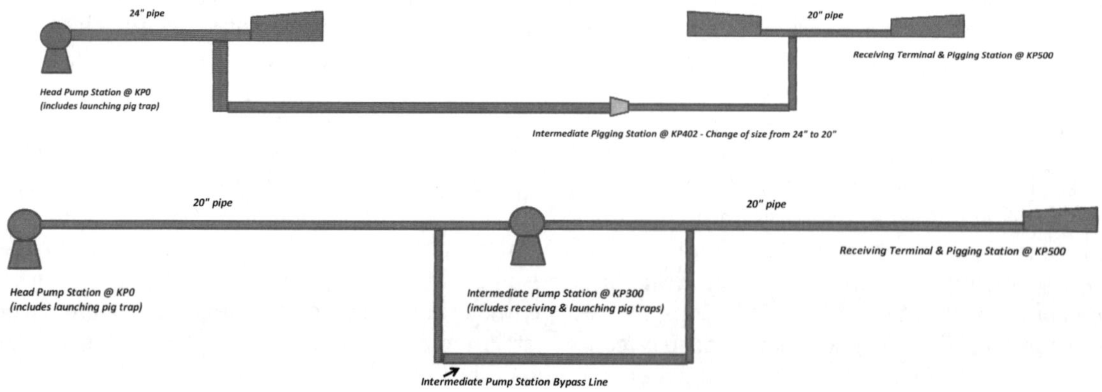

Figure 1. LPG Pipeline. Hydraulic Configurations Investigated.
Alternative 1 (upper) and Alternative 2 (lower).

The above pipes are considered isolated sections. Simulations must be performed to verify the investigated sections in terms of maximum fluid pressure induced by solar radiation, in order to avoid values above the maximum Design Pressure and the possibility of initiation of ruptures. The study is performed considering two reference external air temperatures [21]: 40°C (relevant to the warmest day of the year) and 50 °C (verification for extreme maximum ambient temperature scenario). Starting from these values two different daily temperature trends are developed according to a typical day air temperature evolution, while an average wind velocity equal to 2 m/s has been assumed. The solar radiation trend has been developed considering the following input data [22]: a peak value at noon equal to 1200 W/m^2, a monthly average direct normal irradiance of 7.5 kWh/m^2/day, a sunshine at 6:00 am and the sunset at 8:00 pm. The absorption coefficient, β, is assumed 0.75 for bare pipe and 0.3 for pipes with white painting [23].

Simulations must be performed to verify the investigated sections in terms of maximum fluid pressure induced by solar radiation, in order to avoid values above the maximum design pressure on the basis of the technical data and ambient conditions. Pressurization can be studied in general starting from the maximum operating pressure, i.e. 101 bar (except for the head pumping station suction pipe, for which the maximum operating pressure is 31 bar at 0 °C). Carbon steel is considered as pipe construction material. Initial temperatures are:

- -5°C for the suction pipe of head pumping station (in this case the initial pressure is 31 bar)
- 0°C for the discharge pipe of head pumping station
- 35°C for the discharge pipe of intermediate pumping station
- 25°C for the suction pipe of intermediate pumping station

The LPG fluid studied is composed mainly by propane (60.6 molar percentage) plus small fractions of ethane, butane and pentane. Figure 2 shows the phase envelope of fluid used, obtained by PVTsim Nova code by Calsep.

The summary of the solar radiation simulation scenarios for the isolated sections investigated is reported in Table 1. Polyisocyanurate is considered as insulation material and an insulation thickness (IT) of 30 mm has been calculated in order to avoid sweating.

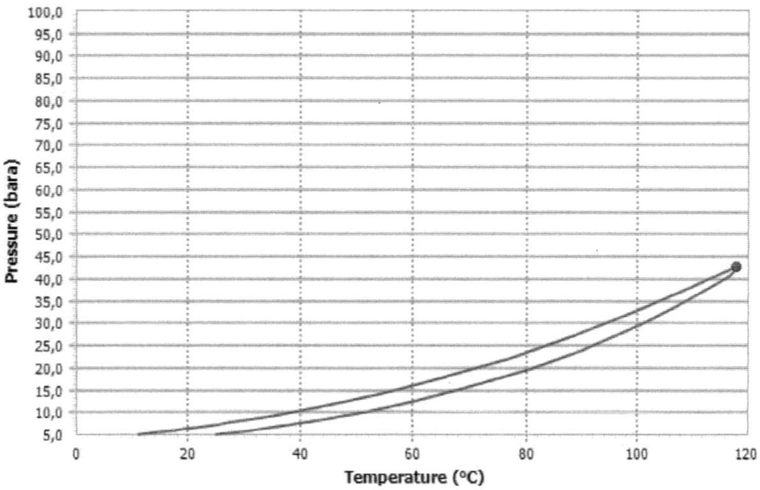

Figure 2. LPG Fluid Used in the Present Study.

Table 1. Summary of Above Ground Sections Calculation Scenarios.

Scenario	Pipe ND (in)	Initial P (bar)	Initial T (°C)	Status of Pipe	IT (mm)	Coating Paint	Maximum Ambient Temperature (°C)
1. Head pump station suction line	24	31	-5	Bare	-	-	40
2. Head pump station suction line	24	31	-5	Insulated	30	White	40
3. Head pump station suction line	24	31	-5	Insulated	30	White	50
4. Head pump station discharge line	24	101	0	Bare	-	-	40
5. Head pump station discharge line	24	101	0	Insulated	30	White	40
6. Head pump station discharge line	24	101	0	Insulated	30	White	50
7. Head pump station suction line	20	31	-5	Bare	-	-	40
8. Head pump station suction line	20	31	-5	Insulated	30	White	40
9. Head pump station suction line	20	31	-5	Insulated	30	White	50
10. Head pump station discharge line	20	101	0	Bare	-	-	40
11. Head pump station discharge line	20	101	0	Insulated	30	White	40
12. Head pump station discharge line	20	101	0	Insulated	30	White	50
13. Intermediate pump station suction line	20	101	25	Insulated	30	White	40
14. Intermediate pump station suction line	20	101	25	Insulated	30	White	50
15. Intermediate pump station discharge line	20	101	35	Insulated	30	White	40
16. Intermediate pump station discharge line	20	101	35	Insulated	30	White	50

Table 2 shows the results of the study in terms of time needed to reach the design pressure in the above ground pipe sections and fluid temperature at Design Pressure (DP) condition, with reference to a long-term shutdown.

Table 2. Solar Radiation Study Results. Time to Reach Design Pressure and Fluid Temperature at Design Pressure Condition.

Scenario	Initial Pressure (bar)	Initial Temperature (°C)	DP (bar)	Time to Reach DP (h)
1. Head pump station suction line	31	-5	56	0.5
2. Head pump station suction line	31	-5	56	5.0
3. Head pump station suction line	31	-5	56	4.7
4. Head pump station discharge line	101	0	148.1	2.6
5. Head pump station discharge line	101	0	148.1	5.2
6. Head pump station discharge line	101	0	148.1	4.1
7. Head pump station suction line	31	-5	56	0.6
8. Head pump station suction line	31	-5	56	4.8
9. Head pump station suction line	31	-5	56	4.5
10. Head pump station discharge line	101	0	148.1	1.6
11. Head pump station discharge line	101	0	148.1	4.0
12. Head pump station discharge line	101	0	148.1	3.5
13. Intermediate pump station suction line	101	25	148.1	9.5
14. Intermediate pump station suction line	101	25	148.1	7.5
15. Intermediate pump station discharge line	101	35	148.1	10.2
16. Intermediate pump station discharge line	101	35	148.1	8.5

From the previous results referred to solar radiation induced pressurization in isolated sections, it can be concluded that during a prolonged shutdown, all the investigated above ground sections could have a pressure raised up to the maximum limit of the rating sizing. The isolated sections subject to an excessive pressurization must be protected by Thermal Relief Valves (TSV) which have a set pressure equal to the rating sizing design pressure.

It has been shown that the modelling of the solar radiation effect proposed by authors can help engineers to properly design safety equipment, by investigating all the technical solutions before and during the engineering phase of a pipeline (and process equipment) project. In fact, if the investigated pipeline sections are not protected by TSVs, the pressure can reach values above the design pressure and the mechanical integrity of the pipe can be compromised.

4. Conclusions

The present work examines a particular set of conditions for pipeline operation, often not considered by conventional design analyses: the pressurization induced by solar radiation which can lead above ground pipeline sections to operating pressures above the design pressure and consequent leaks, fractures and most catastrophic scenarios. The solar radiation on the pipeline sections is modelled in the 1-D thermo-hydraulic simulator OLGA through a new method that can be used as standard in Flow Assurance studies. Through the application of the above model to a practical pipeline design case, it was shown how it works, the steps to follow and the various capabilities. The target of this study is to provide indications for the engineering analysis of such kind (HVP fluids) of pipelines.

5. References

[1] Mohitpour M, Jenkins A and Babuk T 2006 *Proc. Int. Pipeline Conference ASME (Calgary)* pp 531-538
[2] World LP Gas Association (WLPGA) 2015 *Guidelines for Good Safety Practices in the LP Gas Industry*
[3] TSHA (The Texas State Historical Association) 2002 *The Handbook of Texas Online*
[4] Lasode OA, Popoola TO and Prasad B 2013 *Proc. World Academy of Science, Engineering and Technology* n° 78 p 386
[5] Zhang ZX, Wang GX, Massarotto P, Rudolph V 2006 *Energy Convers. Manage* **47** pp 702-715
[6] Bozkurt M, Kulahci Y, Zor F, Kapi E 2008 *J Burn Care Res* **29** p 897-901
[7] Stawczyk J 2003 *J. Hazard. Mater.* B96 pp 189-200
[8] Birk A M 1996 *J Loss Prevent Proc* **9** pp 173-181
[9] Kumar P 2013 *Burns* **39** 1479–87
[10] Casal J, Arnaldos J, Montiel H, Planas-Cuchi E and Vı´lchez J A 2001 In *M. Fingas (Ed.), Handbook of hazardous spills* (New York: McGrawHill)
[11] Terenzi A, Bonato E, Maggiore L 2016 *Proc PSIG Annual Meeting (Vancouver)* pp 1-13
[12] Reid R C 1979 *Science* **203** pp 1263-65
[13] Abbasi T, Abbasi S A 2007 *Journal of Hazardous Materials* **141** pp 489-519
[14] Terenzi A, 2010 *Journal of Pipeline Engineering* **9** p133-143
[15] Peng DY and Robinson DB 1976 *Ind. Eng. Chem. Fundam.* **15** pp 59-64
[16] Bendiksen K H, Malnes D, Moe R, Nuland S 1991 *SPE Prod Eng* **6** pp 171-180
[17] Bird RB, Stewart WE, Lightfoot E N 1960 *Transport Phenomena* (New York: John Wiley & Sons)
[18] Murat Tunç M, Taner Özkaynak F 1985 *Appl Math Model* **9** pp 16-20
[19] Urdahl O, Boernes A H, Kinnari K J, and Holme R *Proc. Offshore Technology Conference* (Houston) pp 161-167
[20] Baen PR 1993 Proc. Petroleum and Chemical Industry Conference (*St Louis*) pp 65-73
[21] DeMenocal PB, Rind D 1993 *J GEOPHYS RES* **98** pp 7265-72
[22] Perry RH, Green DW 1973 *Perry's Chemical Engineers' Handbbok* (McGraw-Hill Book Company)
[23] Siegel R, Howell JR 1992 *Thermal Radiation Heat Transfer* (Hemisphere Publishing Corporation)

IOP Conf. Series: Journal of Physics: Conf. Series **923** (2017) 012024 doi:10.1088/1742-6596/923/1/012024

Magnetohydrodynamic flow and heat transfer around a heated cylinder of arbitrary conductivity

A Tassone, M Nobili and G Caruso

Department of Astronautical, Electrical and Energy Engineering-Nuclear Section, "Sapienza" University of Rome, Corso Vittorio Emanuele II, 244, 00186 Roma, Italy

E-mail: alessandro.tassone@uniroma1.it

Abstract. The interaction of the liquid metal with the plasma confinement magnetic field constitutes a challenge for the design of fusion reactor blankets, due to the arise of MHD effects: increased pressure drops, heat transfer suppression, etc. To overcome these issues, a dielectric fluid can be employed as coolant for the breeding zone. A typical configuration involves pipes transverse to the liquid metal flow direction. This numerical study is conducted to assess the influence of pipe conductivity on the MHD flow and heat transfer. The CFD code ANSYS CFX was employed for this purpose. The fluid is assumed to be bounded by rectangular walls with non-uniform thickness and subject to a skewed magnetic field with the main component aligned with the cylinder axis. The simulations were restricted to $Re = (20, 40)$ and $M = (10, 50)$. Three different scenarios for the obstacle were considered: perfectly insulating, finite conductivity and perfectly conducting. The electrical conductivity was found to affect the channel pressure penalty due to the obstacle insertion only for $M = 10$ and just for the two limiting cases. A general increment of the heat transfer with M was found due to the tendency of the magnetic field to equalize the flow rate between the sub-channels individuated by the pipe. The best results were obtained with the insulating pipe, due to the reduced electromagnetic drag. The generation of counter-rotating vortices close to the lateral duct walls was observed for $M = 50$ and perfectly conducting pipe as a result of the modified currents distribution.

1. Introduction

The flow of an electrically conductive fluid in the presence of an applied magnetic field deviates from the ordinary hydrodynamic behavior due to the arising of induced currents in the fluid bulk, which generate a volumetric Lorentz force that reduces the flow mean velocity and drastically modifies its features. Enhanced corrosion rates, turbulence suppression and additional pressure drops are some of the effects caused by the transition to a magnetohydrodynamic (MHD) flow regime [1][2].

MHD flows are of interest for many industrial applications, including the design of breeder blankets for nuclear fusion reactors [3]. Since the tritium breeder material is lithium, eutectic alloys like LiPb have been considered in the past for their excellent thermal properties as working fluids. To ensure the required cooling of the breeding zone, one of the strategies adopted is to insert pipes transverse to the main flow direction [4].

The bounded flow past a cyrcular cylinder is a classic case studied in hydrodynamics and recently it has been investigated in a MHD perspective. The blockage ratio (β) and the obstacle offset from the duct centerline (G/d) are the most important geometric parameters and, together

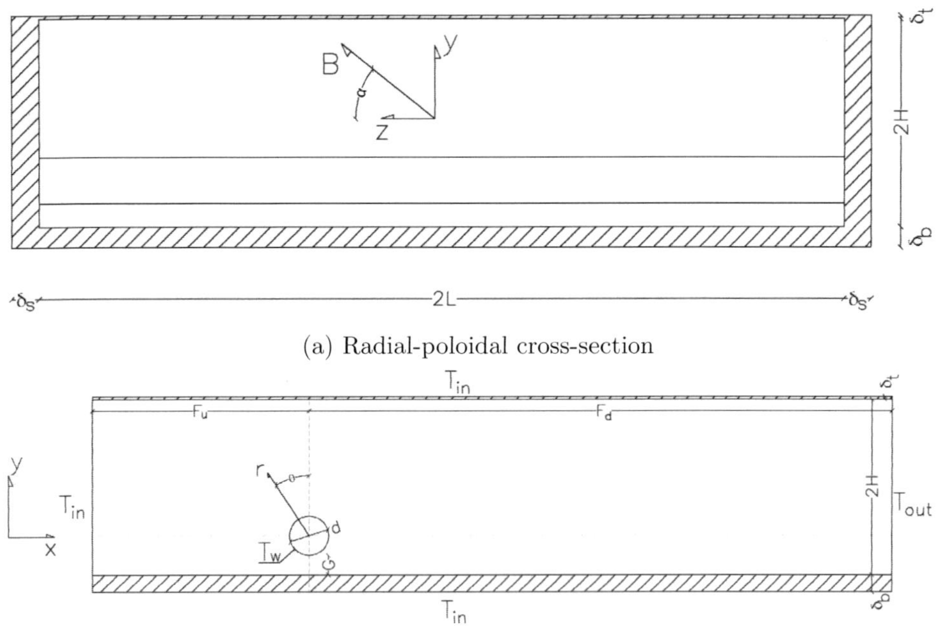

(a) Radial-poloidal cross-section

(b) Toroidal-poloidal cross-section

Figure 1: Problem geometry

with the magnetic field orientation respect to the main stream direction, define the flow. The particular case of magnetic field transverse to the flow but aligned to the obstacle axis has been extensively studied in recent years for its importance in fusion blanket engineering [5][6].

Tassone et al. [7] investigated the flow dynamics and heat transfer for the bounded MHD flow past an heated, electrically insulating, cylinder. Common features encountered in the most recent blanket designs, i.e. skewed magnetic field with the dominant component aligned with the obstacle axis and duct walls of non-uniform thickness, were considered [8]. This work aims to integrate those results by investigating the influence of the cylinder electrical conductivity on the flow pattern and heat transfer mechanisms.

2. Problem formulation

The geometry considered in this work is shown in Figure 1 and a complete overview of the problem geometrical parameters is available in Table 1. A rectangular duct accommodates the obstacle; it is defined by a toroidal half-length L and poloidal half-length H. The duct walls have non-uniform thickness (δ) and satisfy the thin-wall approximation ($\delta \ll L$). The cylinder is defined by blockage ratio $\beta = d/2H$ and offset from the duct centerline G/d, where G is the distance between the cylinder bottom and the closest duct wall. For finite conductivity simulations, inner diameter d_i and wall thickness δ_o were defined for the obstacle. The upstream and downstream lengths (F_u, F_d) identify the cylinder radial position in the duct.

The applied magnetic field $\mathbf{B} = (0, B_y, B_z)$ is uniform and constant in the channel region. The toroidal component is assumed to be dominant with the field inclination on the toroidal axis (α) being equal to 16°. The wall temperature of the external cylinder surface is assumed to be $T_w = 573\,K$. A constant temperature difference with the fluid at the channel inlet is considered such that $\Delta T = T_w - T_{in} = 30\,K$.

To model the behavior of a MHD flow, the Navier-Stokes equations must be combined with the Maxwell ones. The resulting set involves a non-linear and bidirectional coupling between the

IOP Conf. Series: Journal of Physics: Conf. Series **923** (2017) 012024 doi:10.1088/1742-6596/923/1/012024

Table 1: Test case geometry specifications

Duct par. [mm]		Cylinder par. [mm]		Wall thickness [mm]		Wall conductance $\times 10^2$	
L	117.00	d	13.500	δ_t	1.00	c_t	1.25
H	30.25	d_i	8.000	δ_b	6.00	c_b	7.50
F_u	74.25	G/d	0.500	δ_s	10.00	c_s	10.00
F_d	202.50	β	0.223	δ_o	2.75	c_o	$0 - 3.73 - \infty$

fluid velocity and the magnetic field. This phenomenon is mediated through the induced field, which is generated according to the Faraday's law of induction by the currents harbored inside the conductive fluid. A simplifying assumption, called the inductionless approximation, can be employed if the intensity of such field is negligible, thus to consider **B** as being determined only by the boundary conditions. This is accomplished if the magnetic Reynolds number $R_m = \mu\sigma u_0 L \ll 1$; where μ is the magnetic permeability, σ the electrical conductivity, u_0 and L velocity and length characteristic values [9].

Considering an incompressible fluid with constant thermophysical properties and a laminar inductionless MHD flow, the dimensionless governing equations can be derived as follows

$$\nabla \cdot \mathbf{u} = 0 \tag{1}$$

$$\frac{\partial \mathbf{u}}{\partial t} + (\mathbf{u} \cdot \nabla)\mathbf{u} = -\nabla p + \frac{1}{Re}\nabla^2 \mathbf{u} + N(\mathbf{J} \times \mathbf{B}) \tag{2}$$

$$\frac{\partial T}{\partial t} = (\mathbf{u} \cdot \nabla)T = \frac{1}{Pe}\nabla^2 T \tag{3}$$

$$\nabla^2 \phi = \nabla \cdot (\mathbf{u} \times \mathbf{B}) \tag{4}$$

Here, $\mathbf{u}, \mathbf{B}, \mathbf{J}, \phi, T$ and p represents the velocity, magnetic field, current density, electric potential, temperature and pressure scaled by the average inlet velocity u_0, the magnitude of the magnetic induction B_0, $j_0 = \sigma u_0 B_0$, $\phi_0 = d u_0 B_0$, the difference between the local temperature and T_{in} divided by ΔT and, finally, $p_0 = \rho u_0^2$. Lengths are scaled by the external cylinder diameter d, unless otherwise specified, which is also chosen as the length scale of the system for dimensionless parameters. An additional source term in (3) would be formally required to represent the Joule heating. However, in the formulation it has been neglected since it can be demonstrated that its influence for most liquid metal flows is negligible [5]. After solving the set (1-4), the Ohm's equation ($\mathbf{J} = -\nabla\phi + \mathbf{u} \times \mathbf{B}$) is employed to compute the current density from the electric potential distribution.

Reynolds (Re) and Péclet (Pe) numbers, together with the interaction parameter (N), appear in the governing equations and are defined as

$$Re = \frac{u_0 d}{\nu}, \qquad Pe = \frac{u_0 d}{\alpha_t}, \qquad N = \frac{M^2}{Re} \tag{5a,5b,5c}$$

where α_t is the thermal diffusivity. The Hartmann number (M) is a measure of the magnetic field intensity and, therefore, a fundamental MHD parameter

$$M = B_0 d \sqrt{\frac{\sigma}{\rho\nu}} \tag{6}$$

Table 2: Thermophysical properties of LiPb and Eurofer

	Lithium-Lead [10]	Eurofer97 [11]
ρ [kg m^{-3}]	$9.857 \cdot 10^3$	$7.695 \cdot 10^3$
σ [S m^{-1}]	$7.932 \cdot 10^5$	$1.259 \cdot 10^6$
k [W m^{-1} K^{-1}]	12.831	30.060
ν [m^2 s^{-1}]	$2.332 \cdot 10^{-7}$	n.a.
α_t [m^2 s^{-1}]	$6.885 \cdot 10^{-6}$	$7.193 \cdot 10^{-6}$

A fifth parameter, the wall conductance ratio (c), is needed to represent the influence of the electrical boundary conditions on the MHD flow features

$$c = \frac{\sigma_k}{\sigma}\frac{\delta_k}{L} \tag{7}$$

the quantities σ_k and δ_k refer to the wall electrical conductivity and thickness. Different thicknesses were considered for the duct walls, whose conductance ratios are reported in Table 1. The electrical conductivity of the obstacle material was changed each time to simulate insulating, conductive and perfectly conducting pipe.

A local Nusselt number is defined to assess the heat transfer between the obstacle and surrounding fluid employing the temperature gradient normal to the cylinder surface A

$$Nu_w = \frac{d}{T_w - T_{bulk}} \left.\frac{\partial T}{\partial r}\right|_A \tag{8}$$

the bulk temperature of the fluid is computed as the average value on a cylindrical surface S at $r = d$ from the center of the obstacle, weighted on the velocity distribution

$$T_{bulk} = \frac{\iint_S uT\,dS}{\iint_S u\,dS} \tag{9}$$

Accordingly, the average Nusselt number is obtained by

$$Nu = \frac{1}{A}\iint_A Nu_w\,dA \tag{10}$$

The pressure drop penalty due to the obstacle can be defined as the normalized difference between the measured drop and the one for the unperturbed channel (Δp_{2D}), calculated considering the pressure gradient at the outlet

$$p_o = \frac{\Delta p - \Delta p_{2D}}{\Delta p_{2D}} \tag{11}$$

3. Numerical strategy

The commercial CFD code ANSYS CFX 15 was employed to perform this study. The MHD model implemented in CFX relies on the inductionless approximation and, therefore, the governing equations outlined in section 2. The eutectic alloy LiPb was modeled with constant thermophysical properties evaluated at a reference temperature $T_{ref} = 558\,K$ (see Table 2). Correlations developed by Jauch et al. were employed for this purpose [10]. The resulting fluid is characterized by $Pr = 0.034$. The same treatment was applied to the solid domain, which was modeled employing Eurofer97 steel according to the correlations outlined by Mergia & Boukos [11]. The only exception was constituted by the electrical conductivity that was modified to obtain the conductance ratio required to simulate different obstacles.

Typical velocities in the breeding zone for separate cooling fusion blanket ranges from 0.1 mm/s^{-1} to 5 mm/s^{-1} [4]. Accounting for the stabilizing effect of the magnetic field, the resulting low Reynolds number flow would be steady and laminar. For the purpose of this work, a range $Re = [20, 40]$ was considered. A constant mean velocity $u_0$was employed as initial condition for a 2D simulation that produced the fully developed flow velocity profile. This was assumed as the channel inlet BC, whereas at the outlet a zero pressure setting is specified. At the duct walls, no-slip BC for velocity is enforced. The range $Pe = [0.68, 1.36]$ for the Péclet number is derived from u_0 according to 5b.

Three scenarios were considered in the study according to the obstacle material: ideal dielectric, perfect conductor and Eurofer. For these test cases, simulations were performed for the specified Re range and for $M = [10, 50]$. Accompanying simulations for $M = 0$ were performed to highlight the difference with ordinary flows. Electrical boundary conditions must be specified to solve the MHD governing equations. On the duct external surfaces the normal current density flux is zero, as well as the electric potential gradient

$$\frac{\partial \phi}{\partial n} = 0 \qquad (12)$$

Moreover, the BC (12) is employed on the cylinder surface to model the dielectric obstacle. For the duct walls and the finite conductivity obstacle, equation (4) must be solved to obtain the wall electric potential. This is accomplished by coupling the fluid and solid domain, assuming the potential and current density conservation at the interface

$$\phi = \phi_w, \qquad \mathbf{J_n} = \mathbf{J_{n,w}} \qquad (13a,13b)$$

For the perfectly conducting obstacle, BCs (13a-13b) were also employed assuming $\sigma_o = \infty$. The obstacle/fluid interface is assumed at a fixed temperature $T_w = 573\,K$, as well as the inlet, where $T_{in} = 543\,K$. The solid domain external surfaces are adiabatic.

A preliminary study was performed to ensure the result independence from the grid resolution. Simulations were carried out for the insulating obstacle case at $Re = 20$ and $M = 10$ employing five meshes with increasing number of nodes on the cylinder circumference and along the radial direction, ranging from $5 \cdot 10^5$ to $4 \cdot 10^6$ elements. A maximum 2% result divergence for the meshes considered was found [7].

4. Results and discussion

The flow transition from hydrodynamic to MHD behavior is described by the velocity contours presented in Figure 2. For the fully developed flow, the Lorentz force $\mathbf{F_L} = \mathbf{J} \times \mathbf{B}$ opposes the fluid movement and thus balances the driving pressure gradient in the duct center. The viscous forces are confined in thin boundary layers close to the walls that, depending on the relative orientation with the magnetic field, can have thickness $\delta \propto M^{-1}$ (perpendicular, *Hartmann layer*) or $\delta \propto M^{-1/2}$ (parallel, *Shercliff layer*). Since in the present study no wall is perfectly aligned with the magnetic field ($\alpha \gg M^{-1/2}$), the Hartmann layer behavior is observed throughout the duct [12]. Moreover, an internal layer parallel to the magnetic field lines is formed across the core in fully developed condition. This feature carries the bulk of the flow rate and connects two opposite duct corners where high velocity jets are located. This structure formation can be explained by the detachment of the Shercliff layer jets, which would be present for a toroidal magnetic field at $y/H = \pm 1$, from the associated walls and its smearing over the core. The top jet is promoted thanks to the nearby low conductivity wall: those currents whose

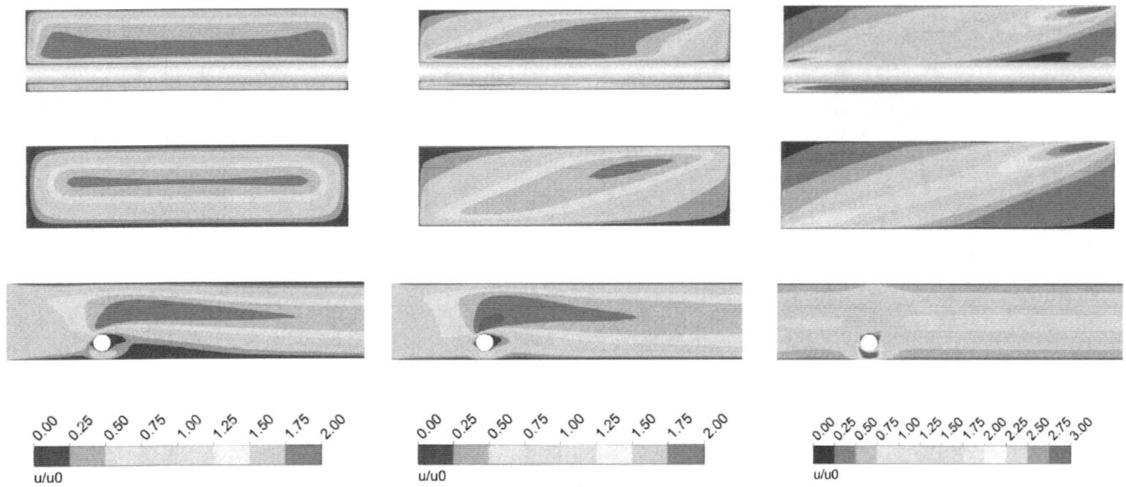

(a) $M = 0$ (b) $M = 10$ (c) $M = 50$

Figure 2: Velocity contour comparison for $Re = 20$. Top and middle row: toroidal-poloidal view at cylinder center ($x/d = 0$) and $x/d = 10$; bottom row: poloidal-radial view at $z/L = 0$. Note that the velocity scale for $M = 50$ covers an higher range. The cylinder is perfectly insulating ($c_o = 0$).

close their path through this wall have a reduced intensity and, therefore, generate a decreased electromagnetic drag. The reversed phenomenon is experienced by the bottom jet [13].

The cylinder offset breaks the duct symmetry in two sub-channels of uneven cross-section, thus the lower, smaller, channel offers a greater hydraulic resistance and carries just $6 - 9\%$ of the total mass flow rate in the hydrodynamic case for Re considered. Increasing M, a stream shift to this channel is observed and, for $M = 50$, its flow rate increases to $\simeq 28\%$ of the total. This phenomenon can be explained by the electromagnetic drag introduction in the duct that, even for low M, it is dominant over the hydraulic losses and it is larger for the top channel due to the lower resistivity experienced by the currents closing through it.Moreover, the sub-channels are linked by the lateral walls and such phenomenon allows currents to pass through both, electrically coupling them, and further compensating the flow rate imbalance [2]. The magnetic field extends the range of stability of the creeping flow regime, which is found for all the cases considered except $M = 10, Re = 40$, where, however, the steady vortices observed in the cylinder wake are partially suppressed [7].

For a fully developed flow, the induced currents are confined to a plane perpendicular to the main flow direction and the flow is essentially 2D [9]. Introducing the obstacle creates velocity gradients in both the radial and poloidal direction thus, from (4), potential differences. The appearance of radial currents is observed, together with a pressure drop increase, and the MHD flow becomes 3D.

The obstacle pressure drop penalty (p_o) is influenced mainly by the flow mean velocity u_0 and the magnetic field intensity B_0, which both appear in the fully developed pressure gradient expression ($dp/dx \propto \sigma u_0 B^2$). The 3D pressure drop term is as well function of these parameters and influenced by the geometrical features of the model, as such β and G/d [5]. However, this component usually features a weaker dependence on B_0 compared with the 2D one, which means that, for constant Re, the increase in M reduces the pressure drop penalty because $\Delta p \simeq \Delta p_{2D}$. Keeping M constant instead, the pressure penalty increases with Re but, for $M \to \infty$, this increment becomes negligible [7].

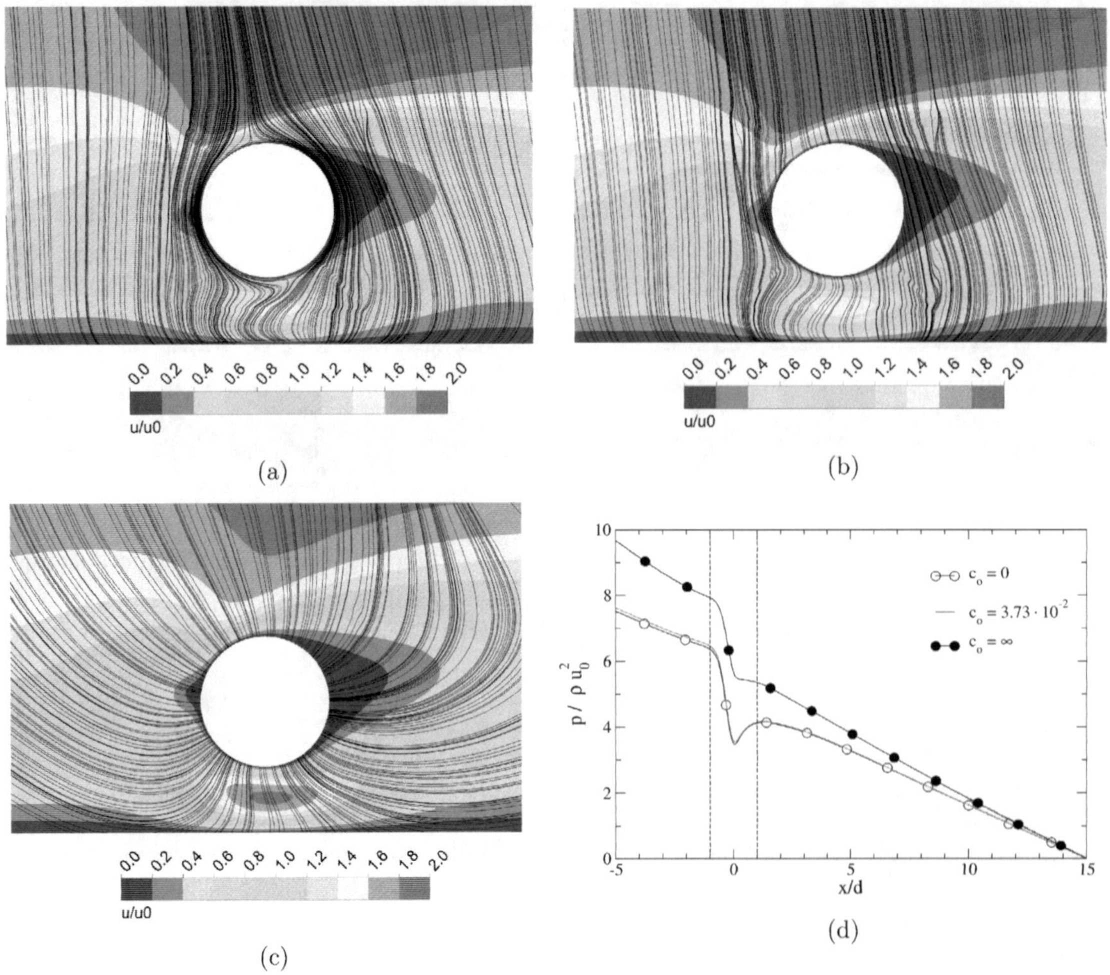

Figure 3: Influence of obstacle conductivity on current path in the fluid domain and channel pressure drop: a) $c_o = 0$, b) $c_o = 3.73$, c) $c_o = \infty$, d) pressure profile along the radial line at $z, y = 0$. Test case $M = 10, Re = 20$. Dotted lines in Figure 3d mark cylinder position.

The obstacle conductivity influences the current paths in the surrounding area in turn affecting the flow pattern and the heat transfer. In Figure 3a, the perfectly insulating surface of the cylinder forbids the currents to penetrate inside the obstacle, forcing them to close through the thin Hartmann layers. It can be observed that the path perturbation introduced by the obstacle extends farther downstream than the wake, which is quickly suppressed by the magnetic field. Conversely, a conductive surface offers a less resistant path than the boundary layer for the currents that, therefore, will tend to close through the cylinder wall. In Figure 3b, this behavior is marked by the streamlines skipping through the boundary layer to enter the solid domain (not represented) and being "attracted" by the obstacle. This phenomenon is evident in Figure 3c where passing through the perfectly conducting cylinder is so favorable that all the currents generated, both upstream and downstream, are collected by the obstacle. This current distribution tends to push away the flow from the duct center toward the poloidal wall at $z/L = -1$ for the top sub-channel and the wall at $z/L = 1$ for the bottom one. For $c_o = \infty$, a slight increase in the channel pressure drop is observed due to these phenomena, whereas for the

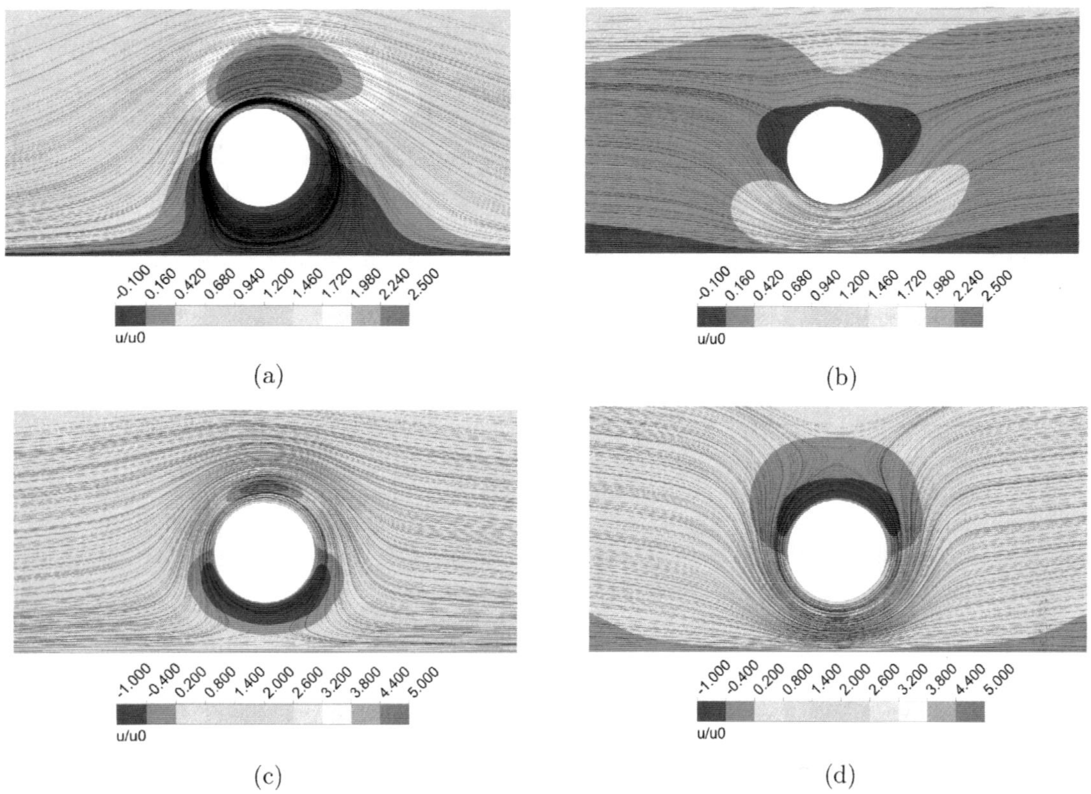

Figure 4: Near-wall vortex formation for perfectly conducting obstacle ($c_o = \infty$) and $Re = 20$. Left: streamline and dimensionless velocity contour at $z/L = -0.95$ and; right: at $z/L = 0.95$. Top row: results at $M = 10$ and; bottom row: at $M = 50$.

intermediate value the obstacle influence on the current distribution is not significant enough to observe a deviation from the insulating case (see Figure 3d). This behavior is mirrored by an increase in the pressure penalty, which for $M = 10$ reaches 40%. Conversely, no significant effect on the pressure penalty is found for $M = 50$ (see Table 3).

Another interesting feature of the $c_o = \infty$ case is the formation of near-wall counter-rotating vortices around the obstacle and close to the poloidal walls. In Figure 4, it can be seen how for $M = 10$ the high velocity jet localized close to the wall at $z/L = -1$ in the top sub-channel is accompanied by a slow, clockwise rotating, vortex that spans across the height of the bottom sub-channel. The vortex is formed by the same localized electromagnetic drag that shifts the bottom jet to the wall at $z/L = 1$. Close to the opposite wall, the flow in the top sub-channel is suppressed but the braking force is not strong enough to provoke a flow inversion. This can be explained by the involved currents in the top sub-channel closing through a longer (thus more resistive) path in the liquid metal. Instead, for the $M = 50$ case the Lorentz forces are strong enough to trigger the counter-clockwise vortex. Due to the angular momentum conservation and the jet peak velocity being proportional to the magnetic field intensity, the vortices shrink closer to the obstacle for increasing M.

The effect of the obstacle conductivity on the heat transfer is less dramatic than on flow dynamics and it is found to lose importance with the increase of the intensity of the applied magnetic field. For $M = 10$ and $c = \infty$, the flow pattern modifications sustain an increase $\Delta Nu = +3.1\%$ compared with the $c = 0$ case for $Re = 20$. This increment can be ascribed

IOP Conf. Series: Journal of Physics: Conf. Series **923** (2017) 012024 doi:10.1088/1742-6596/923/1/012024

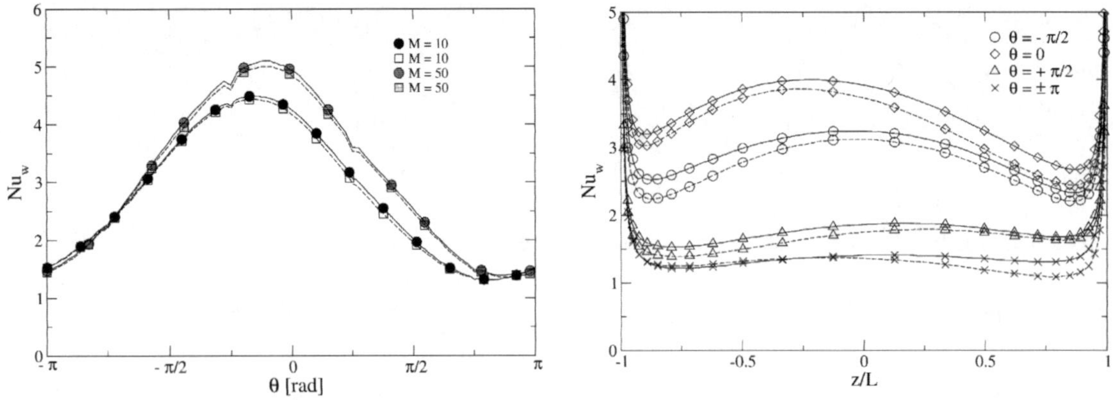

(a) (b)

Figure 5: Influence of obstacle conductivity on local Nusselt number: a) plot along cylinder outer circumference at $z/L = 0$ for increasing M, b) plots along lines parallel to cylinder axis for four relevant azimuthal coordinates (top: $-\pi/2$), front: 0, bottom: $\pi/2$, back: $\pm\pi$) at $Re = 20$. Solid and dotted lines refer respectively to $c = \infty$ and $c = 0$.

Table 3: Pressure penalty and average Nusselt number for different obstacle conductivity. The label E stands for the Eurofer case ($c_o = 3.73 \cdot 10^{-2}$).

	M		10			50	
	c	0	E	∞	0	E	∞
p_o [%]	$Re = 20$	14.074	15.377	34.975	10.918	11.319	12.683
	$Re = 40$	21.128	22.105	40.741	13.356	10.686	14.613
Nu	$Re = 20$	2.5549	2.5427	2.6340	3.1440	3.1325	3.1513
	$Re = 40$	2.6019	2.6046	2.6634	3.1502	3.0893	3.1219

to higher peak velocity for the obstacle jets (compensating for the additional drag in the core) and mass flow rate in the bottom sub-channel (see Figure 5). For $Re = 40$, a similar behavior is observed with a smaller difference between the limiting cases ($\Delta Nu = +2.4\%$). The finite conductivity test case was found to have slightly worse or equal Nu compared with $c_o = 0$. For $M = 50$, the mass flow rate carried by the bottom sub-channel increases from $\sim 10\%$ to $\sim 25\%$ in the Re range considered and the Nusselt number rises accordingly for all the simulations [7]. However, the insulating cylinder is found to perform better compared with the conductive one due to the reduced electromagnetic drag experienced by the fluid. For $c = \infty$, the near-wall vortices sustain the heat transfer bringing the Nu even on par with the insulating case at $Re = 20$ and limiting the net loss at $\Delta Nu = -0.9\%$ for $Re = 40$.

5. Conclusions

The influence of obstacle conductivity on flow dynamics and heat transfer for a bounded MHD flow past a heated cylinder was investigated. Three cases for the obstacle wall conductance ratio (c_o) were considered: perfectly insulating ($c_o = 0$), finite conductivity and perfectly conducting ($c_o = \infty$). The considered range of the Re number was low enough to allow the performing of only steady-state simulations. However, due to the complex magnetic field topology and the

List of symbols

B	magnetic induction [T]	α	magnetic field inclination [°]
d/d_i	outer/inner cyl. diameter [m]	α_t	thermal diffusivity [m^2/s]
H	poloidal half-length [m]	β	blockage ratio
J	current density [A/m^2]	δ	wall thickness [m]
L	toroidal half-length [m]	κ	thermal conductivity [W/mK]
p	pressure [Pa]	μ	magnetic permeability [H/m]
T	temperature [K]	ν	kinematic viscosity [m^2/s]
u	velocity [m/s]	ρ	density [kg/m^3]
r, θ, z	cylindrical coordinate system	σ	electrical conductivity [S/m]
x, y, z	radial, poloidal, toroidal axis	ϕ	electric potential [V]
G/d	offset from centerline	c	wall conductance ratio
M	Hartmann number	N	interaction parameter
Re	Reynolds number	R_m	magnetic Reynolds number
Pe	Péclet number	Pr	Prandtl number
p_o	obstacle pressure penalty [%]	Nu	average Nusselt number

non-uniform wall conductance ratio for the bounding walls, a full 3D computational domain, constituted of solid and fluid bodies, was required to solve the MHD governing equations. Two Hartmann numbers ($M = 10, 50$) were considered.

For low M, the pressure penalty was found to increase with c_o since the obstacle acts like a magnet on the current streamlines, triggering the arise of stronger 3D MHD effects that results in heavier pressure drops. Since the pressure gradient due to the 2D flow is proportional to B_0^2, this effect loses importance for an increasing M and already at $M = 50$ is negligible. No significant difference was found regarding the total pressure drop between the insulating and finite conductivity case.

The influence on the flow dynamics exerted by the modified Lorentz force distribution is significant for the $c_o = \infty$ case. The main feature is the generation of two counter-rotating vortices close to the poloidal walls at $M = 50$. These structures seem to have little repercussion on the overall flow behavior all but buffering the heat transfer losses due to the higher electromagnetic drag. The perfectly conducting case is found to perform slightly better than the finite conductivity scenario in these conditions, even though inferior to the insulating one.

References

[1] Kirillov I R, Reed C B, Barleon L and Miyazaki K 1995 *Fusion Eng. Des.* **27** 553–569
[2] Smolentsev S *et al.* 2010 *Fusion Eng. Des.* **85** 1196–1205
[3] Abdou M *et al.* 2015 *Fusion Eng. Des.* **100** 2–43
[4] Del Nevo A, Martelli E, Agostini P, Arena P, Bongiovì G, Caruso G, Di Gironimo G, Di Maio P, Eboli M, Giammusso R *et al.* 2017 *Fusion Eng. Des.*
[5] Hussam W K and Sheard G J 2013 *Int. J. Heat Mass Transf.* **67** 944–954
[6] Dousset V 2009 *Numerical simulations of MHD flows past obstacles in a duct under externally applied magnetic field* Ph.D. thesis Coventry University
[7] Tassone A, Nobili M and Caruso G 2017 *Proc. of CHT-17 7th Int. Symp. on Adv. in Comp. Heat Transfer* (in pubblication)
[8] Tassone A, Caruso G, Del Nevo A and Di Piazza I 2017 *Fusion Eng. Des.* (in pubblication)
[9] Müller U and Bühler L 2013 *Magnetofluiddynamics in channels and containers* (Springer Science & Business Media)
[10] Jauch U, Karcher V, Schulz B and Haase G 1986 *Thermophysical properties in the system Li-Pb*
[11] Mergia K and Boukos N 2008 *J. Nucl. Mater.* **373** 1–8
[12] Shercliff J 1981 *Zeitschrift für Angewandte Math. und Physik (ZAMP)* **32** 546–554
[13] Tao Z and Ni M 2015 *Sci. China Physics, Mech. & Astron.* **58** 1–18

A multiscale numerical algorithm for heat transfer simulation between multidimensional CFD and monodimensional system codes

A Chierici[*], **L Chirco**[*], **R Da Vià**[*†], **S Manservisi**[*] **and R Scardovelli** [*]
[*]Department of Industrial Engineering
University of Bologna
Via dei Colli 16, 40136 Bologna (BO), Italy
E-mail: [†]roberto.davia2@unibo.it

Abstract. Nowadays the rapidly-increasing computational power allows scientists and engineers to perform numerical simulations of complex systems that can involve many scales and several different physical phenomena. In order to perform such simulations, two main strategies can be adopted: one may develop a new numerical code where all the physical phenomena of interest are modelled or one may couple existing validated codes. With the latter option, the creation of a huge and complex numerical code is avoided but efficient methods for data exchange are required since the performance of the simulation is highly influenced by its coupling techniques. In this work we propose a new algorithm that can be used for volume and/or boundary coupling purposes for both multiscale and multiphysics numerical simulations. The proposed algorithm is used for a multiscale simulation involving several CFD domains and monodimensional loops. We adopt the overlapping domain strategy, so the entire flow domain is simulated with the system code. We correct the system code solution by matching averaged inlet and outlet fields located at the boundaries of the CFD domains that overlap parts of the monodimensional loop. In particular we correct pressure losses and enthalpy values with source-sink terms that are imposed in the system code equations. The 1D-CFD coupling is a defective one since the CFD code requires point-wise values on the coupling interfaces and the system code provides only averaged quantities. In particular we impose, as inlet boundary conditions for the CFD domains, the mass flux and the mean enthalpy that are calculated by the system code. With this method the mass balance is preserved at every time step of the simulation. The coupling between consecutive CFD domains is not a defective one since with the proposed algorithm we can interpolate the field solutions on the boundary interfaces. We use the MED data structure as the base structure where all the field operations are performed, allowing the algorithm to be used with all the numerical codes where a MED duplicate of the field solution can be created. The MEDmem libraries come with the Salome platform and include basic methods for handling meshes and fields. We provide results that show the consistency of the proposed algorithm.

1. Introduction

Numerical code coupling allows scientists and engineers to perform simulations of complex systems. These systems can involve several physical phenomena which can occur on different physical scales. When systems with a great number of components are considered, a complete Computational Fluid Dynamic (CFD) simulation cannot be performed because of the its huge computational cost. In order to reduce it, large systems are then usually simulated by using monodimensional codes. If necessary, a better insight of the phenomena occurring in some components of the system can be achieved by coupling a CFD code with the monodimensional

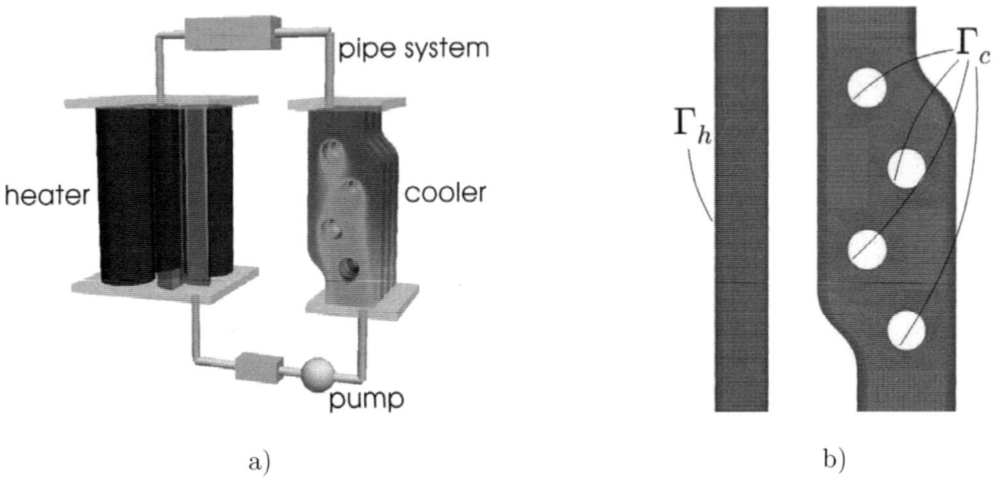

Figure 1. Sketch of the simulated system with the power (heater) and cooling (cooler) assembly a) and boundary surfaces where uniform heat flux is imposed b).

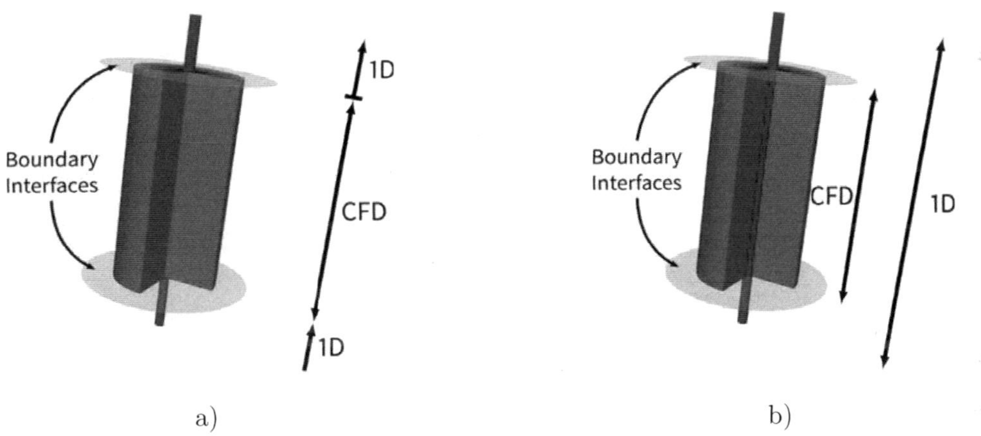

Figure 2. Representation of decomposed (a) and overlapping (b) domain coupling strategies for multiscale simulation.

code. In this way the required computational cost is still acceptable and the solution of the system code can be improved by using the correction of the CFD solution.

In this work we couple a monodimensional code with a CFD code in order to simulate a closed circuit. A sketch of the simulated domain is presented in Figure 1: the circuit contains assemblies of power and cooling units that are connected with an upper and a lower pipe system. The power and cooling units are simulated using the CFD code while the flow inside the pipe system is simulated with the monodimensional code. The numerical coupling of this multiscale simulation can be achieved in two different ways that are represented in Fig. 2. With the decomposed domain defined in Fig. 2 a), the system of interest is divided into several discrete domains that do not overlap and are connected through boundary interfaces. Each domain is simulated with either the system code or the CFD one. The coupling between the different codes is performed via appropriate boundary conditions, as reported in Fig. 3 a). Following the flow direction, the mass flow rate computed on the outlet boundary of the system code is imposed as inlet boundary condition for the CFD code, while the mean pressure calculated on the CFD inlet boundary is imposed as boundary condition on the system code outlet section. A monodimensional domain

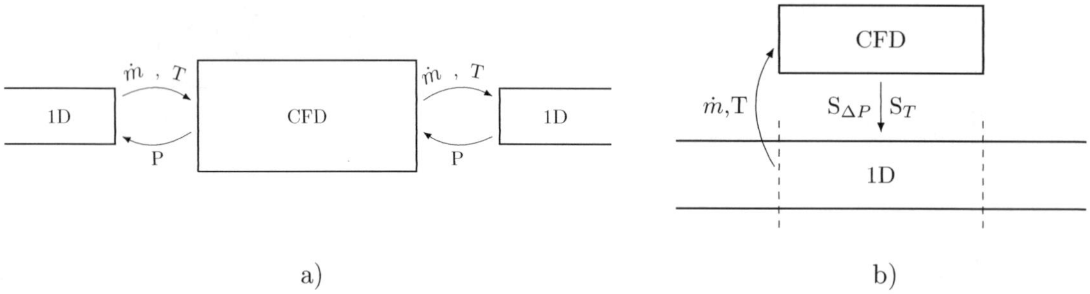

Figure 3. Representation of boundary conditions for numerical code coupling with decomposed (a) and overlapping (b) domain strategies.

that overlaps the CFD domains is instead used with the overlapping domain strategy. In this case the coupling between the two codes is obtained with appropriate boundary conditions when information are passed from the system code to the CFD one. The CFD solution is then used to improve the system code one: this can be performed by calculating some source terms that are then imposed in the system code equations or by modifying some system code parameters [1],[2],[3],[4]. Some performances of domain decomposition and domain overlapping methods are evaluated in [2]. These papers show that the convergence of the decomposed domain strongly depends on the physics of the simulation. In systems driven by an imposed pressure drop on boundary, the mass flow rate is often imposed iteratively from a code to another since the pressure may be discontinuous through the boundary interfaces. Otherwise, with the overlapping domain method the total mass flow rate is continuous across every interface between the monodimensional and CFD domains.

2. Coupling Techniques

In this section we describe the coupling procedure adopted for the multiscale power cooling system simulation. Numerical coupling needs an exchange of information between two or more codes and, in order to perform these operations, interfaces that allow us to access and manipulate input/output solutions are required. For a system code the interface should be able to modify the solution and to access the source terms on every elements at each mesh point of the overlapping monodimensional domain. For a CFD code, the interface may act on boundary points or over points distributed on a sub volume of the domain. This interface should contain a mesh and the solution defined at each mesh point. In the present work, for the realization of the coupling interfaces, we use the MED data format that comes with the SALOME platform. This data structure allows to store numerical fields and meshes within C++ classes. It is then possible to store the required data in the computer memory, without reading and writing operations on external files, obtaining a faster numerical code coupling procedure. The coupling procedure can be divided into 3 main steps: generating the mesh duplicate in the MED format, handling the solution in the data structure and imposing the solution on the target code. From the original solution one need mean integral values (when coupling a CFD and a system code) or field interpolations on a different domain discretization (when coupling two CFD codes). The data transfer is realized by C++ routines designed in an appropriate way for each code which can compute field mean integral values using Gauss integration and point wise field interpolations. In particular, point wise field interpolation routines are based on a search point algorithm, a reverse mapping algorithm (from global coordinate to canonical finite element coordinate system) and an interpolation algorithm based on the classical finite element interpolation. These routines work on MED data format and can be used with all the numerical codes where this format has been adopted for input/output and mesh storage data.

Property	Symbol	Value	Unit
Viscosity	μ	0.00184	Pa s
Density	ρ	10340	Kg/m^3
Thermal conductivity	λ	10.76896	W/(m K)
Heat specific capacity	C_p	146	J/(Kg K)

Table 1. Physical parameters used in the CFD simulations.

In the case of system code solution correction we indicate with λ a variable whose system code solution can be improved with the CFD one. We label with λ_{1D} and λ_{CFD} the values of λ from the system and the CFD code. In the latter case λ_{CFD} is calculated as a mean integral value. At each generic time step n a source term $\mathcal{S}^n$ is imposed on a element or a point of the system code domain. This source term is calculated as

$$\mathcal{S}^n = \mathcal{S}^{n-1} - \omega_\lambda(\lambda_{CFD} - \lambda_{1D}) \,, \tag{1}$$

where ω_λ is an under relaxation parameter smaller than 1. In this way, when $\lambda_{1D} = \lambda_{CFD}$ the source term $\mathcal{S}$ does not impose any correction. In our simulation λ can be the pressure loss across the CFD domain and across the monodimensional overlapping domain or the fluid enthalpy on boundary interfaces.

In the case we need to correct the CFD code or imposing boundary conditions we need to model a solution profile from a single point value. As described in [5] and references therein, various methods are available. Above all, the imposition of a uniform flat profile is the easiest method since it doesn't require the solution of an additional equation or a modification of the numerical code. Sometimes in the case of the Navier Stokes equation this method can lead to an overestimation of the pressure losses inside the CFD component as in the region near the inlet boundary strong velocity gradients take place [2].

3. Numerical Modelling

The operative fluid is Lead Bismuth Eutectic (LBE), whose physical properties are reported in Table 1, while in Table 2 the main geometrical parameters of the simulated domain are reported. We remark that this fluid is characterized by a low value of the Prandtl number, in particular Pr = 0.025. As the fluid flow is turbulent, we solve the Reynolds Averaged Navier Stokes equations

$$\boldsymbol{\nabla} \cdot \mathbf{u} = 0 \,, \tag{2}$$

$$\rho \frac{\partial \mathbf{u}}{\partial t} + \rho(\mathbf{u} \cdot \boldsymbol{\nabla})\mathbf{u} = -\boldsymbol{\nabla} P + \boldsymbol{\nabla} \cdot \left[\mu \left(\boldsymbol{\nabla}\mathbf{u} + \boldsymbol{\nabla}\mathbf{u}^T\right) - \rho\overline{\mathbf{u}'\mathbf{u}'}\right] + \rho\mathbf{g} \,, \tag{3}$$

$$\rho C_p \left(\frac{\partial T}{\partial t} + (\mathbf{u} \cdot \boldsymbol{\nabla})T\right) = \boldsymbol{\nabla} \cdot \left[\lambda \boldsymbol{\nabla} T - \rho C_p \overline{\mathbf{u}'T'}\right] + Q \,, \tag{4}$$

Table 2. Geometrical parameters of the simulated domain.

1D pipe	length	1.5 m
	radius	0.01 m
Heater	length	0.2 m
	inner radius	0.004 m
	outer radius	0.0303 m
Cooler	length	0.2 m
	cooling rods radius	0.01 m
	width	0.05 m

where $\mathbf{u}$, P and T are the mean velocity, mean pressure and mean temperature. The unknown Reynolds stresses $\rho\overline{\mathbf{u}'\mathbf{u}'}$ and the turbulent heat flux $\rho C_p\overline{\mathbf{u}'T'}$ are modeled with the following logarithmic four parameter turbulence model

$$\frac{\partial K}{\partial t} + \mathbf{u}\cdot\nabla K = \nabla\cdot\left[\left(\nu+\frac{\nu_t}{\sigma_k}\right)\nabla K\right] + \left(\nu+\frac{\nu_t}{\sigma_\varepsilon}\right)\nabla K\cdot\nabla K + \frac{P_k}{e^K} - C_\mu e^\Omega , \tag{5}$$

$$\frac{\partial \Omega}{\partial t} + \mathbf{u}\cdot\nabla \Omega = \nabla\cdot\left[\left(\nu+\frac{\nu_t}{\sigma_\varepsilon}\right)\nabla \Omega\right] + 2\left(\nu+\frac{\nu_t}{\sigma_\varepsilon}\right)\nabla K\cdot\nabla \Omega + \\ + \left(\nu+\frac{\nu_t}{\sigma_\varepsilon}\right)\nabla \Omega\cdot\nabla \Omega + \frac{c_{\varepsilon 1}-1}{e^K}P_k - C_\mu\left(c_{\varepsilon 2}f_{exp}-1\right)e^\Omega \tag{6}$$

$$\frac{\partial K_\theta}{\partial t} + \mathbf{u}\cdot\nabla K_\theta = \nabla\cdot\left[\left(\alpha+\frac{\alpha_t}{\sigma_\theta}\right)\nabla K_\theta\right] + \left(\alpha+\frac{\alpha_t}{\sigma_\theta}\right)\nabla K_\theta\cdot\nabla K_\theta + \frac{P_\theta}{e^{K_\theta}} - C_\mu e^{\Omega_\theta} , \tag{7}$$

$$\frac{\partial \Omega_\theta}{\partial t} + \mathbf{u}\cdot\nabla \Omega_\theta = \nabla\cdot\left[\left(\alpha+\frac{\alpha_t}{\sigma_{\varepsilon\theta}}\right)\nabla \Omega_\theta\right] + 2\left(\alpha+\frac{\alpha_t}{\sigma_{\varepsilon\theta}}\right)\nabla K_\theta\cdot\nabla \Omega_\theta + \\ + \left(\alpha+\frac{\alpha_t}{\sigma_{\varepsilon\theta}}\right)\nabla \Omega_\theta\cdot\nabla \Omega_\theta + \frac{c_{p1}-1}{e^{K_\theta}}P_\theta + \frac{c_{p2}}{e^K}P_k - \left(c_{d1}-1\right)e^{\Omega_\theta} - c_{d2}C_\mu e^\Omega . \tag{8}$$

The variables ν_t and α_t represent the eddy viscosity and the eddy thermal diffusivity, while K is the natural logarithm of the turbulent kinetic energy k, Ω is the natural logarithm of k specific dissipation ω, K_θ is the natural logarithm of mean squared temperature fluctuations k_θ and Ω_θ the natural logarithm of k_θ specific dissipation, ω_θ. For details on the benefits of a logarithmic turbulence model and on the formulation of the eddy viscosity and eddy thermal diffusivity on can refer to [6],[7],[8],[9] and [10]. In the present turbulence model we adopt a realizability constrain where the eddy viscosity is limited from above by a value that depends on the local values of the mean strain rate tensor modulus and of the turbulent kinetic energy k [11],[12]. The system of equations (2-8) is solved with an in-house finite element code. The equations are upwinded with a SUPG formulation [13]. In the system (5-8), the cross diffusion terms

$$\left(\nu+\frac{\nu_t}{\sigma_\varepsilon}\right)\nabla K\cdot\nabla K \quad , \quad 2\left(\nu+\frac{\nu_t}{\sigma_\varepsilon}\right)\nabla K\cdot\nabla \Omega + \left(\nu+\frac{\nu_t}{\sigma_\varepsilon}\right)\nabla \Omega\cdot\nabla \Omega$$

$$\left(\alpha+\frac{\alpha_t}{\sigma_\theta}\right)\nabla K_\theta\cdot\nabla K_\theta \quad , \quad 2\left(\alpha+\frac{\alpha_t}{\sigma_{\varepsilon\theta}}\right)\nabla K_\theta\cdot\nabla \Omega_\theta + \left(\alpha+\frac{\alpha_t}{\sigma_{\varepsilon\theta}}\right)\nabla \Omega_\theta\cdot\nabla \Omega_\theta$$

are seen as additional convective terms. We introduce the fictitious velocities $\mathbf{u}_K$, $\mathbf{u}_\Omega$, $\mathbf{u}_{K\theta}$, $\mathbf{u}_{\Omega\theta}$ and rewrite the cross diffusion terms as

$$\mathbf{u}_K\cdot\nabla K \quad , \quad \mathbf{u}_\Omega\cdot\nabla\Omega \quad , \quad \mathbf{u}_{K\theta}\cdot\nabla K_\theta \quad , \quad \mathbf{u}_{\Omega\theta}\cdot\nabla\Omega_\theta .$$

For the system (5-8) the SUPG upwind terms are calculated using the modified velocity field $\tilde{\mathbf{u}} = \mathbf{u} - \mathbf{u}_i$, with i=$K, \Omega, K_\theta, \Omega_\theta$. In this way some diffusion is added also in the direction of sharp gradients of the turbulence variables, which means near the solid boundaries. In the domain regions where the turbulence variable gradients are small, if compared to the velocity field, the SUPG formulation is equivalent to the standard one.

4. Results

In this section we show the results we obtained from the simulation of the multiscale system. The power cooling unit consists of a cylindrical annulus, where a uniform heat flux is applied on the inner surface, a cooling device, an additional heat exchanger, a volume component, a pump and a pipe system. The cylindrical annulus, hereafter labeled as *heater* component, and

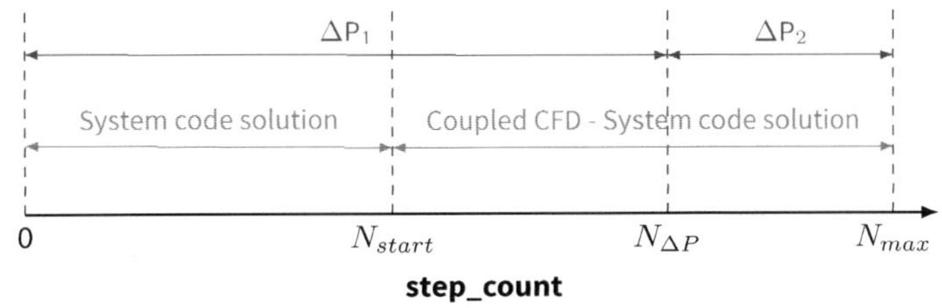

Figure 4. Schematic representation of the transient simulation.

the cooling device, labeled as *cooler* component, are simulated with the CFD code and with an equivalent pipe component in the system code. The additional heat exchanger and the pump are modelled as point wise components within the system code simulation. We simulate a transient situation as reported in Fig. 4, where N indicates the time step. The simulation starts from a stationary solution of the CFD problem. For the first N_{start} time steps the system code reaches a stationary solution based on the stationary CFD one, i.e. correcting the pressure losses and the enthalpy values. From iteration N_{start} we couple the solution of the CFD and system codes and reach a new stationary solution. At time step $N_{\Delta P}$ we increase the pressure gain on the pump component. The simulation is performed for four different values of the under relaxation parameters, namely ω_p for the pressure losses and ω_T for the fluid enthalpy, that are 0.01, 0.05, 0.2 and 0.5. The system is driven by the pressure gain imposed on the pump component. For the CFD components, on the inlet boundaries we impose the mass flow rate and the temperature that are read from the system code solution on the boundaries interfaces. For the turbulence variables we impose uniform values calculated as

$$K_{in} = \ln(1.5(Iv_{mid})^2) \quad , \quad \Omega_{in} = \ln\left(\frac{e^{K_{in}}}{L}\right)$$
$$K_{\theta,in} = K_{in} - \ln(Pr) \quad , \quad \Omega_{\theta,in} = \Omega_{in} - \ln(Pr)\,,$$

where I is the turbulence intensity, $I = 0.16Re_h^{1/8}$, and L is a characteristic length, $L = 0.07D_h$. Re_h and D_h represent the Reynolds number and the hydraulic diameter, while v_{mid} is the averaged velocity component in the normal direction to boundary interface. On wall surfaces, for the turbulent variables we set boundary conditions in accordance with their near wall behavior [7], while on the outlet boundaries we impose homogeneous Neumann boundary conditions. On the boundaries Γ_h and Γ_c, as reported in Fig. 1, we impose a uniform heat flux so that, in stationary condition, the temperature difference between the mean values on the heater inlet and the cooler outlet is zero. In order to test the field interpolation functions, the

Table 3. Mean velocity and bulk temperature values sections, Γ^1_{out} and Γ^2_{in}, at two different time steps: t_1 and t_2.

	$v_{m,\Gamma^1_{out}}$	$v_{m,\Gamma^2_{in}}$	$\Delta v_{m,frac\%}$	$T_{b,\Gamma^1_{out}}$	$T_{b,\Gamma^2_{in}}$	$\Delta T_{b,frac\%}$
t_1	0.10957	0.10944	0.234	578.4	578.34	0.01
t_2	0.14984	0.14961	0.138	577.14	577.1	0.007

simulations are performed considering the heater as a whole component or as split in two regions, labeled by HEAT1 and HEAT2, characterized by a different domain discretization. The latter case leads to a domain decomposition coupling method, when passing information from the first half of the heater component to the consecutive one. In the following, when

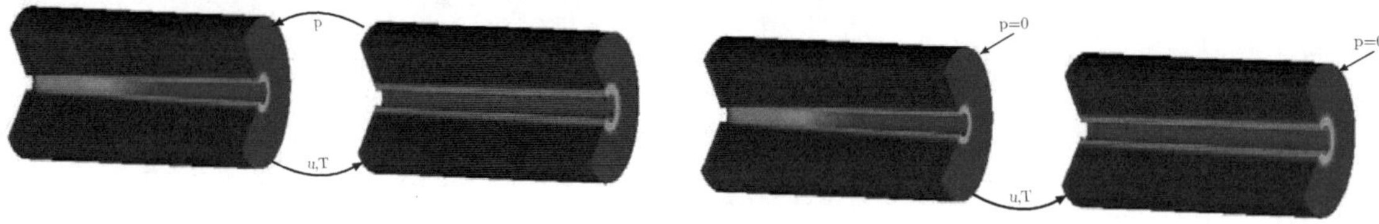

a) b)

Figure 5. Two different split configuration of the heater, with backward projection (a) and without projection (b) of the pressure field.

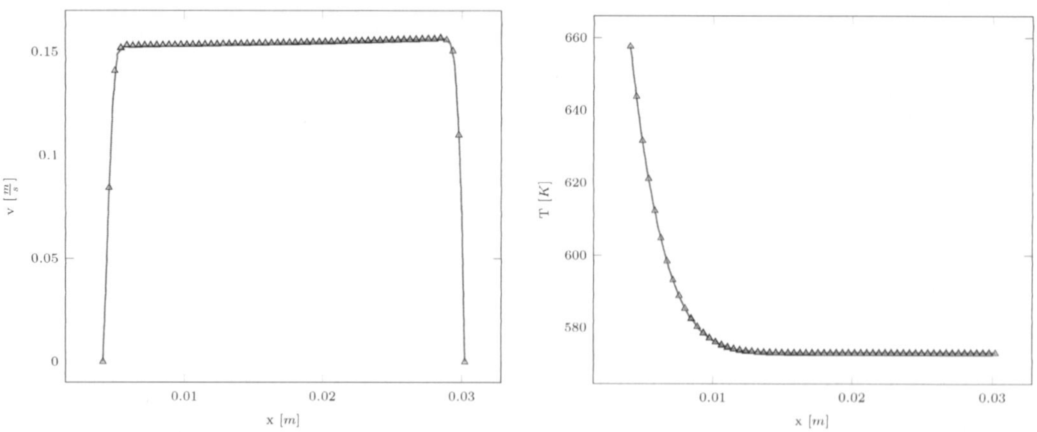

a) b)

Figure 6. Velocity a) and temperature b) profiles on the outlet section of heater HEAT1 Γ^1_{out} (solid lines) and the interpolated values (triangle markers) on the inlet section of heater HEAT2 Γ^2_{in}.

dealing with the heater component split in two parts, we will refer to Γ^1_{in}, Γ^1_{out}, Γ^2_{in}, Γ^2_{out} as the inlet and outlet boundary sections of HEAT1 and HEAT2. As the fluid physical properties do not depend on the pressure field, we can choose whether to perform or not a backward pressure field interpolation from Γ^2_{in} to Γ^1_{out}, as represented in Fig. 5. In fact we need to calculate the overall pressure drop across the heater component. If we perform the backward pressure field interpolation then $\Delta P_{CFD} = P_{\Gamma^1_{in}} - P_{\Gamma^2_{out}}$, otherwise $\Delta P_{CFD} = (P_{\Gamma^1_{in}} - P_{\Gamma^1_{out}}) + (P_{\Gamma^2_{in}} - P_{\Gamma^2_{out}}) = \Delta P_{HEAT1} + \Delta P_{HEAT2}$, where on the outlet boundaries Γ^1_{out} and Γ^2_{out} an arbitrary pressure value can be set as boundary condition. The backward pressure field interpolation can be avoided when a uniform pressure field can be imposed on the outlet boundary, for example in the case of fully developed flow. In Fig. 6 the velocity and temperature fields on the outlet sections of HEAT1 are reported together with the interpolated values that are imposed as inlet boundary conditions on the inlet section of HEAT2. In order to estimate the interpolation field accuracy we report in Table 3 the mean velocity and bulk temperature values of both original and interpolated fields. We also report the fractional difference between the original and the interpolated values, namely $\Delta v_{m,frac\%} = (v_{m,HEAT1} - v_{m,HEAT2})/v_{m,HEAT1}$ and $\Delta T_{b,frac\%} = (T_{b,HEAT1} - T_{b,HEAT2})/T_{b,HEAT1}$. For the velocity field interpolation the difference $\Delta v_{m,frac\%}$ is small ($< 0.25\%$), while for the temperature field interpolation the difference is even smaller ($< 0.01\%$). For the case of split heater component we use a domain decomposition coupling method between HEAT1 and HEAT2 with and without backward pressure field interpolation. In the following, we will label with case 1 the heater component as a whole component, with case 2 the simulations with the heater split into two parts and with backward pressure field projection, while with case 3 the simulations with the heater component split into two parts but without the pressure coupling between HEAT1 and HEAT2. In Fig. 7

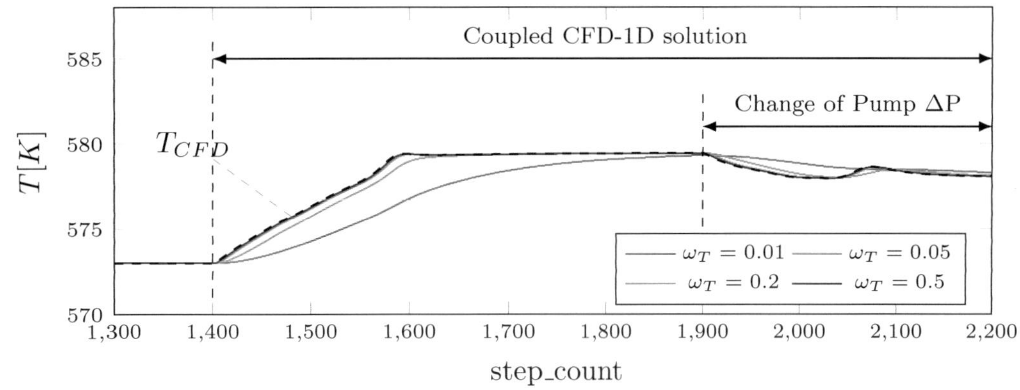

a)

b)

Figure 7. Behavior of mean temperature a), on the outlet boundary of the heater component, and of pressure losses b) across the heater component, as a function of the step count.

Table 4. Values of the mass flow rate with non-split heater (case 1) and with split heater (case 2 and 3)

	Case 1	Case 2	Case 3
Mass flow rate [kg/s]	3.2496	3.2502	3.2502

we report a comparison of the CFD and system code results of case 1, for the heater component. The comparison is performed between the values obtained with the four different values of both ω_p and ω_T, as a function of the step count. In particular, in Fig 7 a) we compare the mean integral temperature calculated on the outlet boundary of the heater component with the point

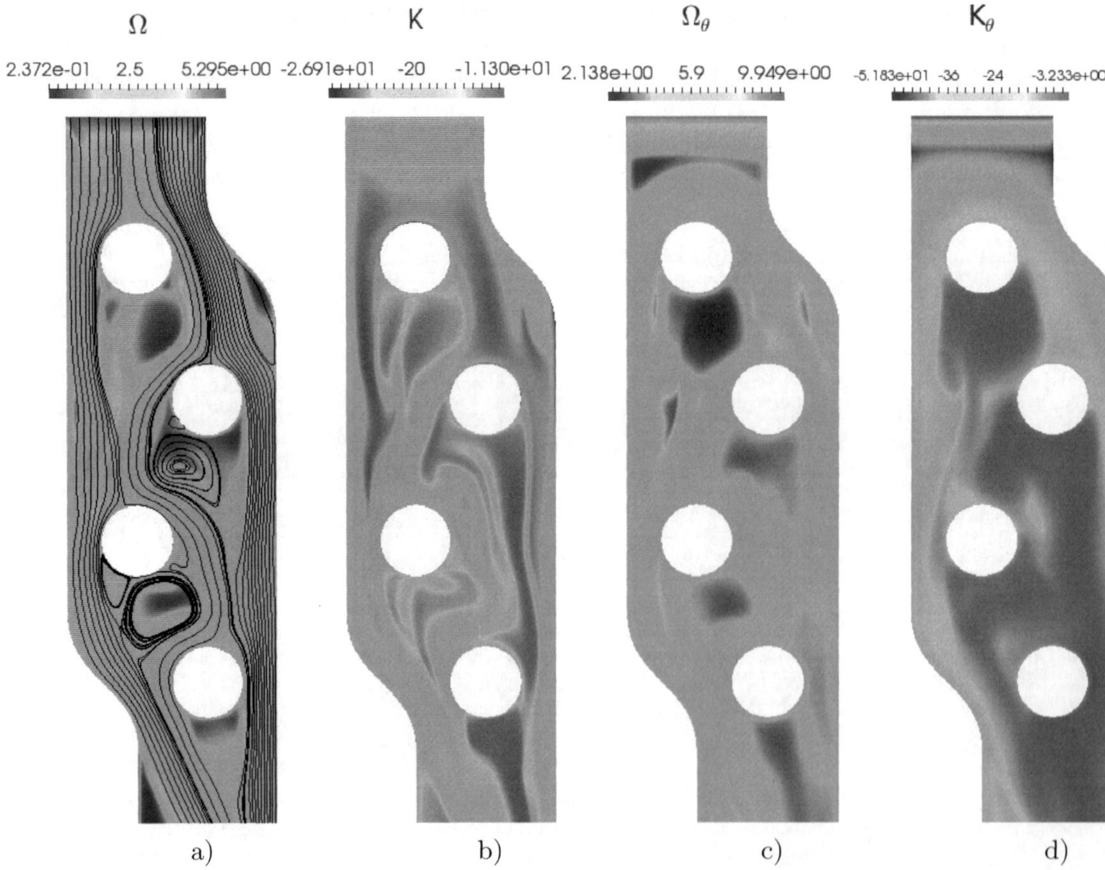

Figure 8. Stationary turbulence variable fields for the cooler component. From left to right, the fields of Ω, K, Ω_θ and K_θ.

wise value coming from the system code solution. We can see that as the under relaxation parameter increases the convergence between CFD and system code solutions is reached with a smaller number of iterations. A stationary value is reached after the coupling of CFD and 1D codes, before changing the value of the pressure gain on the 0D pump element, and also after the change of ΔP. The value $\omega_T = 0.01$ leads to a slow numerical convergence between the CFD and 1D solutions. In order to evaluate the convergence of the pressure losses, in Fig. 7 b) we show the behavior of the parameter ΔP_{frac}, $\Delta P_{frac} = |\Delta P_{CFD} - \Delta P_{1D}|/\Delta P_{CFD}$, for the different values of ω_p. As for the temperature case, the convergence between the values of pressure losses across the CFD domain, ΔP_{CFD}, and the ones across the monodimensional domain, ΔP_{1D}, is faster as the under relaxation parameter increase. As a final comparison between case 1, case 2 and case 3, we report in Table 4 the values of the stationary mass flow rate obtained after the change of the pump pressure gain. We can see that there is a slight difference between the value obtained with the heater as a whole component and the values obtained splitting the heater in the two components HEAT1 and HEAT2. In particular the mass flow rates for case 2 and case 3 are the same, which means that, for this particular simulated system, a uniform pressure value can be set as outlet boundary condition on HEAT1 without altering the final result. In Fig. 8 we report the stationary turbulence variable fields of the cooler component. In particular, in Fig. 8 a), the streamlines are reported with the field of the logarithmic value of turbulent kinetic energy specific dissipation rate, Ω. We can observe the presence of some recirculating regions behind the cooling surfaces. These regions are also characterized by a large value of the logarithmic mean squared temperature fluctuations K_θ. The use of the turbulence limiters proposed in [11] increased the solution stability of the dynamical turbulence system of equations, preventing anomalous growth of the turbulence variables in the recirculation regions.

5. Conclusions

In the present work we presented a coupling algorithm that can be used to perform both multiscale and multiphysics simulations. The proposed algorithm has been tested simulating a complex system, involving several CFD domains and a closed monodimensional pipe system. Both the overlapping and decomposed domain methods have been used in order to realize, respectively, a coupling between a system code and a CFD one and a coupling between two consecutive CFD domains. The data exchange between different codes requires the creation of proper interfaces that are used to access and manipulate code solutions. We used the MED data format, that comes with the MEDmem libraries, for the creation of these interfaces, and the code coupling procedure is realized without using reading/writing procedures on external files. In this way, by using the MEDmem libraries, the coupling operations are performed within the computer memory, allowing a faster numerical code coupling execution. The system of interest has been simulated with various configurations, in order to test both the overlapping and the decomposed domain methods. In particular, for the overlapping domain method, we tested four different values of the under relaxation parameters used to correct the pressure losses and the temperature values of the system code solution. Within the tested range of parameters values, the convergence between the system code solution and the CFD one was found to be faster as the under relaxation parameters increase, without stability problems. For the decomposed domain method we focused our attention on a CFD component and considered it as a whole component and as two different components connected through a boundary interface. The results obtained with the CFD component split in two parts showed only a slight difference with respect to those obtained with the CFD component as a whole component. We can conclude that the proposed algorithm can be used to perform multiscale and multiphysics simulations.

References

[1] Cerroni D 2015 *Multiscale multiphysics coupling on a finite element platform* Ph.D. thesis University of Bologna
[2] Grunloh T P and Manera A 2016 *Annals of Nuclear Energy* **90** 422–32
[3] Bavière R, Tauveron N, Perdu F, Garré E and Li S 2014 *Nucl. Eng. Des.* **277** 124–37
[4] Tenchine D *et al.* 2012 *Nucl. Eng. Des.* **245** 140–52
[5] Prince J A 2015 *Coupled 1D-3D simulation of flow in subway transit networks* Ph.D. thesis Imperial College London
[6] Ilinca F, Hetu J F and Pelletier D 1998 *Computers and Fluids* **27** 291–310
[7] Da Vià R, Manservisi S and Menghini F 2016 *Int. J. Heat Mass Tran.* **101** 1030–41
[8] Manservisi S and Menghini F 2014 *Int. J. Heat Mass Tran.* **69** 312–26
[9] Manservisi S and Menghini F 2014 *Nucl. Eng. Des.* **273** 251–70
[10] Cerroni D, Da Vià R, Manservisi S, Menghini F, Pozzetti G and Scardovelli R 2015 *J. of Phys.: Conf. Series* **655** 012046
[11] Park C H and Park S O 2005 *Int. J. Comput. Fluid D.* **19** 79–86
[12] Durbin P A 1996 *Int J. Heat Fluid Fl.* **17** 89–90
[13] Bochev P B, Gunzburger M D and Shadid J N 2004 *Comput. Methods in Appl. Mech. Eng.* **193** 2301–23

IOP Conf. Series: Journal of Physics: Conf. Series **923** (2017) 012026 doi:10.1088/1742-6596/923/1/012026

An optimal control method for fluid structure interaction systems via adjoint boundary pressure

L Chirco[1], **R Da Vià**[1] **and S Manservisi**[1]

[1] Department of Industrial Engineering DIN
Laboratory of Montecuccolino
University of Bologna
Via dei Colli 16, Bologna 40136 (BO), Italy

E-mail: leonardo.chirco2@unibo.it

Abstract. In recent year, in spite of the computational complexity, Fluid-structure interaction (FSI) problems have been widely studied due to their applicability in science and engineering. Fluid-structure interaction systems consist of one or more solid structures that deform by interacting with a surrounding fluid flow. FSI simulations evaluate the tensional state of the mechanical component and take into account the effects of the solid deformations on the motion of the interior fluids. The inverse FSI problem can be described as the achievement of a certain objective by changing some design parameters such as forces, boundary conditions and geometrical domain shapes.

In this paper we would like to study the inverse FSI problem by using an optimal control approach. In particular we propose a pressure boundary optimal control method based on Lagrangian multipliers and adjoint variables. The objective is the minimization of a solid domain displacement matching functional obtained by finding the optimal pressure on the inlet boundary. The optimality system is derived from the first order necessary conditions by taking the Fréchet derivatives of the Lagrangian with respect to all the variables involved. The optimal solution is then obtained through a standard steepest descent algorithm applied to the optimality system. The approach presented in this work is general and could be used to assess other objective functionals and controls. In order to support the proposed approach we perform a few numerical tests where the fluid pressure on the domain inlet controls the displacement that occurs in a well defined region of the solid domain.

1. Introduction

Optimization has always been used to improve the performance of engineering devices. Nowadays several approaches to optimization are available, such as single and multi-objective, adjoint or sensitivities based method, evolutionary algorithms and many others. In literature there are several works dealing with this subject, for a quick review the interested reader can see [1, 2, 3, 4, 5] and references therein. Fluid-Structure Interaction (FSI) problems can be defined as a system where the fluid flow changes the tensional state of a solid structure that is left free to move and the solid deformation has an important effect on the fluid flow. Examples of this type of systems are quite common in engineering, like wind turbines, man-made drones and in the study of biological systems such as hemodynamics. In literature these topics are investigated deeply and the interested reader can see [6, 7, 8, 9, 10, 11, 12, 13, 14]. Many attempts to apply optimization techniques to FSI problems can be found, for example in [15], where the authors propose a solution method for the problem of optimizing a non-linear aeroelasticity system in a steady-state flow by using a sensitivity method. The work in [16] deals with general shape optimization methods based on design sensitivity analysis.

In this work we refer to adjoint based methods, which have been proven to be a good approach for the optimal control of complex problems in which Computational Fluid Dynamics simulations can be performed on the system of interest, [15, 16]. Moreover these methods have a solid mathematical background and the existence of local optimal solutions can be proven for many interesting cases, [16, 17]. We are interested in a monolithic approach to the solution of the FSI system leading to a stable and well defined solution in a finite element setting [18, 19]. Inside this framework we study a pressure boundary optimal control problem applied to the monolithic FSI system. The objective of the control is the matching of a displacement field in a particular region of the solid domain. This is accomplished by changing the pressure on the fluid inlet boundary which alters the fluid flow profile and thus deforms the shape. This method can be useful in many industrially relevant applications. For example, given a pipe containing a pressurized liquid and the maximum deformation that the pipe can sustain without damaging, it is necessary to know in advance the highest admissible pressure that guarantees both the preservation of the pipe as well as a high liquid flow rate. A possible application in the biomedical field concerns stents, which are metal or plastic tube inserted into arteries. Since their flexibility is limited by the material properties, it can be useful to know what range of inner fluid pressure can be tolerated without damaging the stent itself. In the next section we derive the optimality system which consists of the state, the adjoint system and the control equation. In order to solve the optimality system we propose a simple steepest descent algorithm. In Section 3 we report some numerical results obtained by the implementation of the optimal control algorithm in a finite element parallel code designed for multiphysics simulations.

2. Mathematical formulation

In this section we discuss the mathematical formulation of the FSI problem together with the optimality system. We briefly recall some notations about functional spaces used in this paper, for a detailed description one can consult [20, 21]. We use standard notation $H^s(\Omega)$ for the Sobolev spaces with norm $\|\cdot\|_s$ ($H^0(\Omega) = L^2(\Omega)$ and $\|\cdot\|_0 = \|\cdot\|$). Let $H^s_0(\Omega)$ be the space of all functions in $H^s(\Omega)$ that vanish on the boundary of the bounded open set Ω and $H^{-s}(\Omega)$ be the dual space of $H^s_0(\Omega)$. The trace space for the functions in $H^1(\Omega)$ is denoted by $H^{1/2}(\Gamma)$.

Let us consider a domain $\Omega \subset R^N$ which consists of a structure domain Ω_s and a fluid domain Ω_f so that $\Omega = \Omega_s \cup \Omega_f$. The deformed solid domain $\Omega_s(\mathbf{l})$ is expressed through the solid displacement field $\mathbf{l}$ as

$$\Omega_s(\mathbf{l}) = \{\mathbf{x} \in R^3 \,|\quad \mathbf{x} = \mathbf{x}_0 + \mathbf{l}\}\,, \tag{1}$$

where the vector $\mathbf{x}_0$ defines the initial solid domain $\Omega_s(\mathbf{0})$. For our steady state FSI problem we consider the interaction of a Newtonian fluid, whose behavior is described by the Navier-Stokes equations, with an hyperelastic incompressible St.Venant-Kirchhoff material. The mathematical model in strong form of the FSI problem is then the following

$$\nabla \cdot \mathbf{v} = 0 \qquad on \quad \Omega_f\,, \tag{2}$$

$$\rho_f(\mathbf{v} \cdot \nabla)\mathbf{v} - \nabla \cdot \mathbf{T} = 0 \qquad on \quad \Omega_f\,, \tag{3}$$

$$\nabla \cdot \mathbf{S}(\mathbf{l}) = 0 \qquad on \quad \Omega_s\,, \tag{4}$$

with $\mathbf{v}$ being the fluid velocity field and ρ_f its density. The constitutive relations for the fluid stress tensor $\mathbf{T}$ in the Newtonian incompressible case and for the solid Cauchy stress tensor $\mathbf{S}$

read

$$\mathbf{T}(p_f, \mathbf{v}_f) := -p_f\mathbf{I} + \mu_f(\boldsymbol{\nabla}\mathbf{v} + \boldsymbol{\nabla}\mathbf{v}^T)\,, \tag{5}$$
$$\mathbf{S}(\mathbf{l}) := \lambda_s(\boldsymbol{\nabla}\cdot\mathbf{l})\mathbf{I} + \mu_s\boldsymbol{\nabla}\mathbf{l}\,, \tag{6}$$

where p_f is the fluid pressure, μ_f the dynamic viscosity of the fluid while λ_s and μ_s are the solid Lamé parameters. In the following we study only the incompressible case that occurs when $\lambda_s \to \infty$. In this case we introduce the solid pressure $p_s = -\lambda_s(\boldsymbol{\nabla}\cdot\mathbf{l})$ and write

$$\mathbf{S}(\mathbf{l}) := -p_s\mathbf{I} + \mu_s\boldsymbol{\nabla}\mathbf{l}\,, \tag{7}$$

where the displacement $\mathbf{l}$ is subject to the incompressible constraint $\boldsymbol{\nabla}\cdot\mathbf{l} = 0$. To complete the system (2-4) it is necessary to define the appropriate boundary and interface conditions, which are

$$\begin{aligned}
\mathbf{v} &= \mathbf{v}_0 && on \quad \Omega_{fd}\,,\\
\mathbf{l} &= \mathbf{l}_0 && on \quad \Omega_{sd}\,,\\
\mathbf{T}\cdot\mathbf{n} &= 0 && on \quad \Omega_{fn}\,,\\
\mathbf{S}\cdot\mathbf{n} &= 0 && on \quad \Omega_{sn}\,,\\
\mathbf{T}\cdot\mathbf{n} &= \mathbf{S}\cdot\mathbf{n} && on \quad \Gamma_i\,,\\
\mathbf{v} &= 0 && on \quad \Gamma_i\,,
\end{aligned} \tag{8}$$

where Ω_{fd} and Ω_{sd} are the surfaces where Dirichlet boundary conditions are imposed for the fluid velocity and solid displacement, Ω_{fn} and Ω_{sn} are the surfaces where standard homogeneous outflow boundary conditions are imposed for both the fields. By Γ_i we denote the interface between the solid and fluid domain, $\Gamma_i = \Omega_s \cap \Omega_f$. When using a monolithic approach for the numerical solution of the FSI system the conditions (8) are automatically set, including the equilibrium of normal stresses that has to be imposed on the interface. However it is very common to solve FSI problems by using dedicated solvers for the solid and fluid domain. This solving approach is called *segregated* and has the drawback that the coupling conditions are not implicitly part of the solver but have to be imposed with an iterative loop.

The first step to obtain the optimality system with the Lagrange multipliers method is to choose an objective functional for the control problem. In this work we study a solid domain displacement matching problem where the control is the pressure on the fluid inlet boundary. The objective or cost functional can be defined as follows

$$\mathcal{J}(\mathbf{l}, p) = \frac{1}{2}\int_{\Omega_d} w(\mathbf{l} - \mathbf{l}_d)^2 d\Omega + \int_{\Gamma_c} \beta\, p_c^2\, d\Gamma\,, \tag{9}$$

The non-negative function w is a weight function of the coordinate $\mathbf{x}$ that can be used to impose the objective on a specific part of the domain and $\mathbf{l}_d$ is the desired value of the solid displacement. The functional is completed by a regularization term involving the regularization parameter β, which is needed to obtain a control function p_c in the space of square integrable functions $L^2(\Omega_c)$. If a too high value of β is chosen the control becomes too smooth and the objective cannot be achieved well, while a lack of regularization leads to convergence issues in the numerical solution of the problem. To obtain the optimality system we write the full constrained Lagrangian of the problem which is composed of the cost functional and state equations multiplied by the

IOP Conf. Series: Journal of Physics: Conf. Series **923** (2017) 012026 doi:10.1088/1742-6596/923/1/012026

appropriate Lagrangian multipliers

$$\begin{aligned}\mathcal{L}(p, \mathbf{v}, \mathbf{l}, \widehat{\mathbf{l}}, p_a, \mathbf{v}_a, \widehat{\mathbf{l}}_a, \widehat{\mathbf{s}}_a, \boldsymbol{\beta}_a) &= \mathcal{J}(\mathbf{l}, p) + \int_{\Omega_f} (\boldsymbol{\nabla} \cdot \mathbf{v})\, p_a \, d\Omega + \int_{\Omega_s} (\boldsymbol{\nabla} \cdot \mathbf{v})\, p_a \, d\Omega \\ &+ \int_{\Gamma_c} (\boldsymbol{v}_a \cdot \mathbf{n}) p d\Gamma + \int_{\Omega_f} [\rho_f (\mathbf{v} \cdot \boldsymbol{\nabla})\mathbf{v} + \boldsymbol{\nabla} p - \boldsymbol{\nabla} \cdot (\mu_f \boldsymbol{\nabla} \mathbf{v})] \cdot \boldsymbol{v}_a d\Omega \\ &+ \int_{\Omega_s} \boldsymbol{\nabla} \cdot [\mu_s \boldsymbol{\nabla} \mathbf{l} - \boldsymbol{\nabla} p] \cdot \boldsymbol{v}_a d\Omega + \int_{\Omega_s} \boldsymbol{\nabla}^2 \mathbf{l} \cdot \widehat{\mathbf{l}}_a d\Omega \\ &+ \int_{\Gamma_i} \widehat{\mathbf{s}}_a \cdot \left[(\widehat{\mathbf{l}} - \mathbf{l}) + \frac{\mathbf{v}}{h} \right] d\Gamma + \int_{\Omega_s} \boldsymbol{\beta}_a \cdot \left[\mathbf{v} - h(\mathbf{l} - \widehat{\mathbf{l}}) \right] d\Omega \,. \end{aligned} \tag{10}$$

In (10) we have introduced the auxiliary mesh displacement $\widehat{\mathbf{l}}$, defined only over the solid domain and solution of a Laplace operator. The velocity field $\mathbf{v}$ has been extended to the whole domain Ω as

$$\mathbf{v} = \begin{cases} h(\mathbf{l} - \widehat{\mathbf{l}}) & \text{on } \Omega_s, \\ \mathbf{v} & \text{solution of (2)-(3) on } \Omega_f, \end{cases} \tag{11}$$

with h being a positive constant. It is clear that at steady state conditions $\Omega_s(\mathbf{l}) = \Omega_s(\widehat{\mathbf{l}})$, with $\mathbf{v} = 0$ in the solid. By taking the Fréchet derivatives of the Lagrangian with respect to the adjoint variables the weak form of the state system (2-4) is obtained, together with the correct boundary and interface conditions. When the derivatives are taken with respect to the state variables, after some term rearrangement, the adjoint system in weak form reads

$$\int_{\Gamma_c} (p\beta + \boldsymbol{v}_a \cdot \mathbf{n}) \delta p \, d\Gamma - \int_{\Omega_f} (\boldsymbol{\nabla} \cdot \boldsymbol{v}_a) \delta p \, d\Omega - \int_{\Omega_s} (\boldsymbol{\nabla} \cdot \boldsymbol{v}_a) \delta p \, d\Omega = 0 \qquad \forall \delta p \in L^2(\Omega)\,, \tag{12}$$

$$\int_{\Omega_f} [\rho_f (\delta\mathbf{v} \cdot \boldsymbol{\nabla})\mathbf{v} \cdot \boldsymbol{v}_a + \rho_f (\mathbf{v} \cdot \boldsymbol{\nabla}) \delta\mathbf{v} \cdot \boldsymbol{v}_a + \mu_f \boldsymbol{\nabla} \delta\mathbf{v} : \boldsymbol{\nabla} \boldsymbol{v}_a] d\Omega + \int_{\Omega_f} (\boldsymbol{\nabla} \cdot \delta\mathbf{v}) p_a \, d\Omega + \tag{13}$$

$$\int_{\Omega_s} [\mu_s \boldsymbol{\nabla} \delta\mathbf{v} : \boldsymbol{\nabla} \boldsymbol{v}_a d\Omega + \int_{\Omega_s} (\boldsymbol{\nabla} \cdot \delta\mathbf{v}) p_a \, d\Omega + \int_{\Omega_d} w(\mathbf{l} - \mathbf{l}_d) \delta\mathbf{v} \, d\Omega = 0 \qquad \forall \delta\mathbf{v} \in H^1_{\Gamma_{fd} \cup \Gamma_{sd}}(\Omega)\,.$$

The shape derivatives with respect to the fluid and solid domain have been taken into account and simplified. Since Ω_d and Ω_c are fixed and the system is solved by using a monolithic approach, then the shape derivative contributions appear only in the equation for the adjoint displacement $\mathbf{l}_a$. We do not need the solution of the $\mathbf{l}_a$ adjoint equation to compute the pressure boundary control p and therefore it may be neglected.

Finally, when considering the surface terms in (12) we obtain the following relation for the control pressure p_c on the controlled surface Γ_c

$$p_c = p = -\frac{\boldsymbol{v}_a \cdot \mathbf{n}}{\beta}\,. \tag{14}$$

In order to recover the strong form of the *adjoint system* it is necessary to perform some integrations by parts on the terms where the variations of the state variables are differentiated. After performing the integration by parts, we recover the adjoint state $(\mathbf{v}_a^f, \mathbf{v}_a^s, p_a) \in \mathbf{H}^1_{\partial\Omega^f - \Gamma_i}(\Omega^f) \cap \mathbf{H}^2(\Omega^f) \times \mathbf{H}^1_{\partial\Omega^s - \Gamma_i}(\Omega^s) \cap \mathbf{H}^2(\Omega^s) \times L^2_0(\Omega^f) \cap \mathbf{H}^1(\Omega^f)$, by solving

$$\boldsymbol{\nabla} \cdot \boldsymbol{v}_a^f = 0\,, \tag{15}$$

$$-\rho_l (\boldsymbol{\nabla} \mathbf{v})^T \boldsymbol{v}_a^f + \rho_l [(\mathbf{v} \cdot \boldsymbol{\nabla}) \boldsymbol{v}_a^f] + \boldsymbol{\nabla} p_a - \boldsymbol{\nabla} \cdot (\mu_l \boldsymbol{\nabla} \boldsymbol{v}_a^f) = w(\mathbf{l} - \mathbf{l}_d)\,, \tag{16}$$

$$\boldsymbol{\nabla} \cdot \boldsymbol{v}_a^s = 0\,, \tag{17}$$

$$\boldsymbol{\nabla} \cdot \mathbf{S}(\boldsymbol{v}_a^s) = 0\,. \tag{18}$$

with boundary conditions defined as

$$\begin{aligned}
&\boldsymbol{v}_a^s = \boldsymbol{v}_a^f && && on \quad \Gamma_i\,, \\
&\mathbf{S}(\boldsymbol{v}_a^s)\cdot\mathbf{n} = \mathbf{T}(\boldsymbol{v}_a^f)\cdot\mathbf{n} && && on \quad \Gamma_i\,, \\
&\mu_l(\boldsymbol{\nabla}\boldsymbol{v}_a)\cdot\mathbf{n} = -(\mathbf{v}\cdot\mathbf{n})\,\boldsymbol{v}_a\,, && p_a = 0 && on \quad \Gamma_n^f\,, \\
&\boldsymbol{v}_a = 0 && && on \quad \Gamma_{fd}\cup\Gamma_{sd}\,.
\end{aligned} \tag{19}$$

It is worth noticing the duality between (19) and (8). In fact if a Dirichlet boundary condition for a state variable is set then the corresponding adjoint variable must satisfy the same type of condition in homogeneous form. The adjoint velocity $\boldsymbol{v}_a$ must be continuous and different from zero on the interface since the source term $w(\mathbf{l}-\mathbf{l}_d)$ acts in the solid region and the information has to propagate towards the control boundary Γ_c which is part of the fluid domain. The equilibrium conditions on the interface are automatically satisfied due to the monolithic approach.

In order to solve the optimal control problem iteratively we use the steepest descent method described in Algorithm 1. When the algorithm reaches the minimum the quantity $(\boldsymbol{v}_a\cdot\mathbf{n})/\beta$ is

Algorithm 1 Description of the steepest descent algorithm.

1: **function** FIND OPTIMAL
2: set a state $(\mathbf{v}^0, p^0, \mathbf{l}^0)$ satisfying (2-4) ▷ *Setup of the state - Reference case*
3: compute the functional $\mathcal{J}^0$ in (9)
4: set $r^0 = r_0$
5: **for** $i = 1 \to i_{max}$ **do**
6: Solve the system (12)-(13) to obtain the adjoint state $(\boldsymbol{v}_a^i, p_a^i)$
7: **for** $j = 1 \to j_{max}$ **do**
8: compute the control $p_c^i = p_c^{i-1} - r^{i,j}\,(p_c^{i-1} - \boldsymbol{v}_a^i\cdot\mathbf{n}/\beta)$
9: solve (2-4) for the state $(\mathbf{v}^{i,j}, p^{i,j}, \mathbf{l}^{i,j})$ with the new control p_c^i.
10: compute the new functional $\mathcal{J}^{i,j+1}$ in (9)
11: **if** $\|\mathcal{J}^{i,j+1} - \mathcal{J}^{i,j}\|/\mathcal{J}^{i,j} < toll$ **then**
12: convergence reached ▷ *end of the algorithm*
13: **else if** $\mathcal{J}^{i,j+1} > \mathcal{J}^{i,j}$ **then**
14: set $r^{i,j+1} = 1/2\,r^{i,j}$, $j = j+1$ and go to 8 ▷ *loop on j again*
15: **else if** $\mathcal{J}^{i,j+1} < \mathcal{J}^{i,j}$ **then**
16: set $r^{i,j+1} = r_0$, $i = i+1$ and go to 6 ▷ *loop on i again*
17: **end if**
18: **end for**
19: **end for**
20: **end function**

proportional to the functional gradient. The step length r determines how far, from the current state solution, we are moving, along the gradient direction, by changing the control. In the loop j we solve iteratively the state system by adjusting this parameter and obtain a proper decrease of the functional. When this is accomplished we save the control and iterate again on the main loop i where the adjoint system is solved. This algorithm requires several solutions of the state and adjoint systems in order to find the optimal control. However it does not need a great amount of memory which is limited to a standard CFD simulation.

We implemented this algorithm in an in-house finite element code (FEMuS) which is parallelized by using openMPI libraries and uses a multigrid solver with mesh-moving capability [22, 23, 19]. We have used standard quadratic-linear elements for the velocity and pressure solution, in order to fulfill the *LBB inf-sup* condition and a SUPG stabilization technique. The displacements are approximated with standard quadratic elements as well. In the following we report the results obtained with the proposed algorithm in a few simple test cases. First we give an overview of the not controlled FSI problem, then we report the optimal solutions that have been found for different objective solid domain displacement $\mathbf{l}_d$ and values of β ranging from 10^{-9} to 10^{-11}.

3. Numerical results

In this section we present the results obtained by applying Algorithm 1 to a two-dimensional test case. The reference geometry is shown in Figure 1 where the fluid domain is marked with Ω_f and the solid one with Ω_s. The physical properties are set as $\rho_s = \rho_l = 1000 Kg/m^3$ with $\mu_l/\rho_l = 0.07 Pa\, m^3 s/Kg$ and $\mu_s = 76250 Pa$ so that the flow is not turbulent and the solid can bend easily. The fluid inlet and outlet are marked by the segments AB and EF, with the control

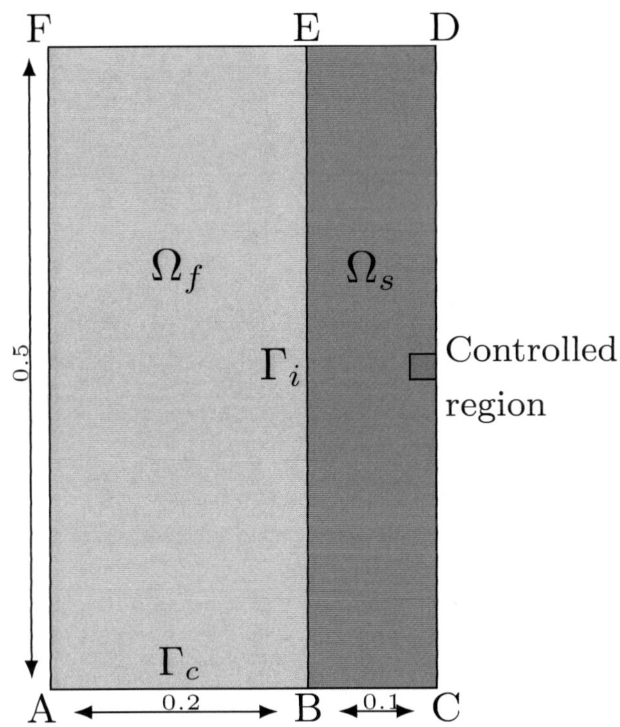

Figure 1. Case study. The region on the left is the fluid and the one on the right is the solid. The segment AB is the fluid inlet, EF the outlet and BE the interface Γ_i. The red square on the right is the controlled region Ω_d.

surface Γ_c along the segment AB. On these boundaries we impose pressure boundary conditions and vanishing tangential velocity, with inlet and outlet pressure p_{EF}, p_{AB} in the uncontrolled case and $p_{AB} = p_c$ for the controlled one. The domain is symmetric with respect to the left boundary AF, while the segment BE lies on the interface Γ_i. The fluid flows from bottom to top on the left side of the domain and the solid bends on the right due to the high pressure values in the fluid region, as imposed by the boundary constraint $\mathbf{T} \cdot \mathbf{n} = \mathbf{S} \cdot \mathbf{n}$. All the boundaries are fixed with the exception of the right segment CD, that may move with fixed endpoints C and D.

We first consider the steady state obtained when the control is not active and the flow is driven by a guess boundary pressure field. From this steady state we then apply an optimal boundary control in order to increase (case A) and decrease (case B) the target desired deformations. In the optimal control cases, in order to simplify the analysis, we set in the controlled region Ω_d a weight function w as follows

$$w = \begin{cases} 1 & \mathbf{x} \in \Omega_d, \\ 0 & \text{otherwise}, \end{cases}$$

where Ω_d denotes the red region around the middle point of the right boundary CD, as shown in Figure 1.

For the initial solution we impose a constant pressure profile. We set $p_{AB} = 16500Pa$ and $p_{EF} = 15000Pa$ for the inlet and outlet pressure, respectively. The displacement, pressure and velocity fields are shown in Figure 2. On the left of Figure 2 the fluid velocity field is

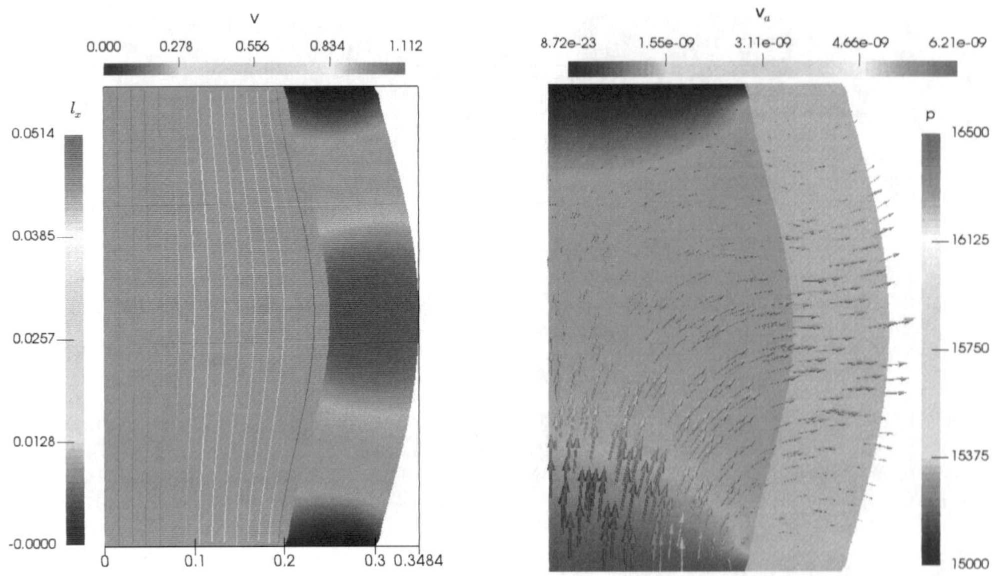

Figure 2. On the left velocity streamlines in the fluid and horizontal component of the displacement field in the solid, with no control. On the right the adjoint velocity and pressure field in the liquid (case (A)).

reported with 15 iso-magnitude lines to better visualize the flow and the horizontal component of the displacement field in the solid region Ω_s is shown in colors. The region that undergoes the maximum displacement is near the horizontal center-line. In Figure 2 one can see clearly the x-coordinate of point-region Ω_d (0.3484) at the steady solution. By subtracting the corresponding value in the reference case, see Figure 1, one recovers the horizontal component of the displacement $l_{\Omega_d,x} = 0.0484$.

In the test cases with optimal control we set the desired displacement $\mathbf{l}_{d1} = (0.07, 0)$ (case A) and $\mathbf{l}_{d2} = (0.03, 0)$ (case B). Therefore, when using $\mathbf{l}_{d1}$ as objective, we aim to reach a higher solid deformation along the x-axis, by increasing the pressure at the fluid inlet. Viceversa, by choosing $\mathbf{l}_{d2}$ as objective we intend to reduce the structure horizontal displacement by reducing the inlet pressure. On the right of the same Figure 2 the liquid pressure is shown in color, while the adjoint velocity field is reported with arrows colored according to its magnitude for the test case A, with $\mathbf{l}_{d1}$ and $\beta = 10^{-11}$. The adjoint flow field enters the domain from the bottom fluid inlet, crosses the interface and exits from the right surface as prescribed by the boundary conditions (19). The control pressure p_c, on the control surface Γ_c, is then related with the adjoint velocity through the scalar equation (14).

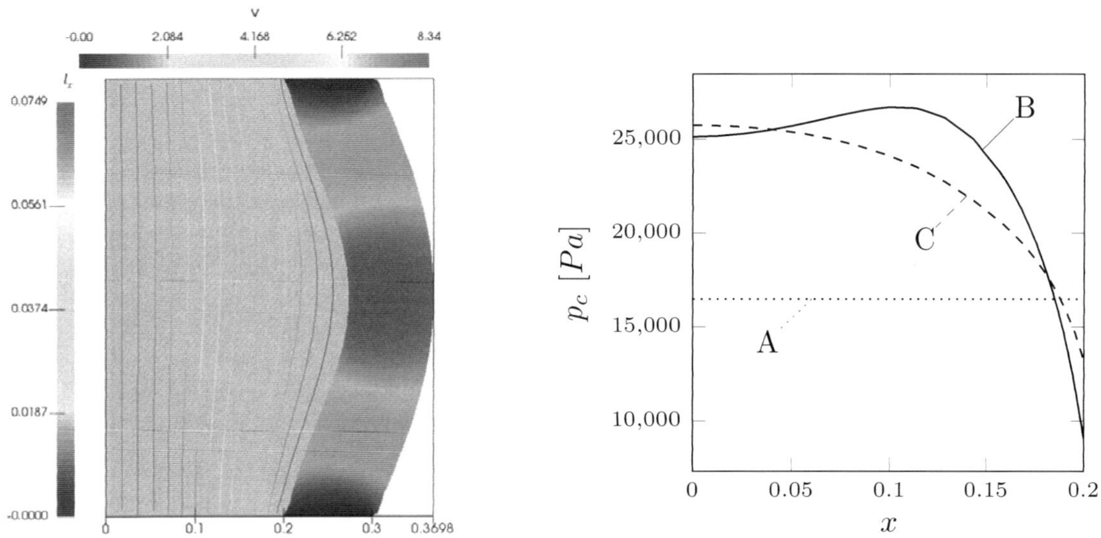

Figure 3. Test case A. On the left velocity streamlines in the fluid and horizontal component of the displacement field in the solid, with $\beta = 10^{-11}$. On the right pressure distribution on the inlet surface, comparison between reference case (A, dotted line), test case with $\beta = 10^{-11}$ (B, solid line) and test case with $\beta = 10^{-10}$ (C, dashed line).

We now report the results of test case A, with objective $\mathbf{l}_{d1}$, in Figure 3. On the left of Figure 3 are reported 15 iso-magnitude lines of the fluid velocity and the solid horizontal displacement in colors, while the x-coordinate of point-region Ω_d can be read on the ticked line. On the right of the same Figure it is shown a plot of the control pressure on the fluid inlet boundary ($y = 0$) for the reference and test case A with two different values of β. In this test case the control increases the inlet pressure, in order to minimize the objective functional (9). As a consequence the fluid velocity also increases, as can be seen on the left of this Figure, in order to satisfy the mass conservation constraint. From the plot on the right we notice that when $\beta = 10^{-10}$ (dashed line) the inlet pressure has a regular, quite parabolic profile, while with $\beta = 10^{-11}$ (solid line) becomes more irregular, moreover in both cases the inlet pressure is significantly higher than in the reference one. In Figure 4 we report some results obtained for test case B with objective $\mathbf{l}_{d2}$. On the left are reported 15 iso-magnitude lines of the fluid velocity, the solid horizontal displacement in colors and the x-coordinate of the point-region Ω_d. On the right of the same Figure the control pressure on the fluid inlet boundary is plotted for test case B, with two different values of β, and for the reference one. In order to reduce the objective functional (9) the control has to reduce the inlet pressure, as it can be seen clearly on the right. In test case B the fluid has reversed the direction of the motion, flowing in the channel from top to bottom and the velocity magnitude is significantly higher than in the reference case. This inversion occurs because the control pressure p_c on the bottom is lower than the one on the top, which is fixed to a positive value ($16500 Pa$) by the boundary conditions. The plot on the right of Figure 4 shows that the inlet pressure reduction obtained with $\beta = 10^{-11}$ (solid line) is more pronounced than the one obtained with $\beta = 10^{-10}$ (dashed line). For a more quantitative description of the results we reported in Table 1 the objective functional (9) as computed in the reference and in the controlled cases with different β. In both test cases the functional has been reduced significantly with $\beta = 10^{-11}$ and 10^{-10}. However, with $\beta = 10^{-9}$ the optimal control algorithm has provided a functional reduction only in the test case B. Also, in Figure 5 one can see a comparison between test case A and B. In this Figure the horizontal (u) and the vertical

IOP Conf. Series: Journal of Physics: Conf. Series **923** (2017) 012026 doi:10.1088/1742-6596/923/1/012026

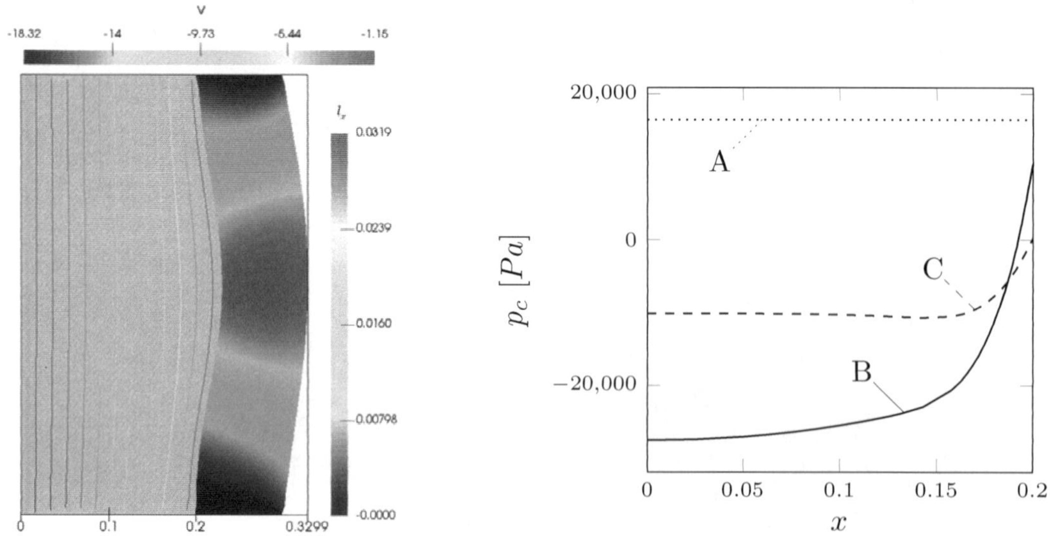

Figure 4. Test case B. On the left velocity streamlines in the fluid and horizontal component of the displacement field in the solid, $\beta = 10^{-11}$. On the right pressure distribution on the inlet surface, comparison between reference case (A, dotted line), test case with $\beta = 10^{-11}$ (B, solid line) and test case with $\beta = 10^{-10}$ (C, dashed line).

Table 1. Objective functionals computed with different β values for test cases with $\mathbf{l}_{d1}$ (Test A) and $\mathbf{l}_{d2}$ (Test B). The reference case with no control is labeled with $\beta = \infty$.

	β	∞	10^{-9}	10^{-10}	10^{-11}
Test A	$\mathcal{J}(\mathbf{l}, p)$	9.30e-04	9.30e-04	1.16e-05	1.29e-07
Test B	$\mathcal{J}(\mathbf{l}, p)$	6.80e-04	7.59e-05	2.48e-05	1.71e-08

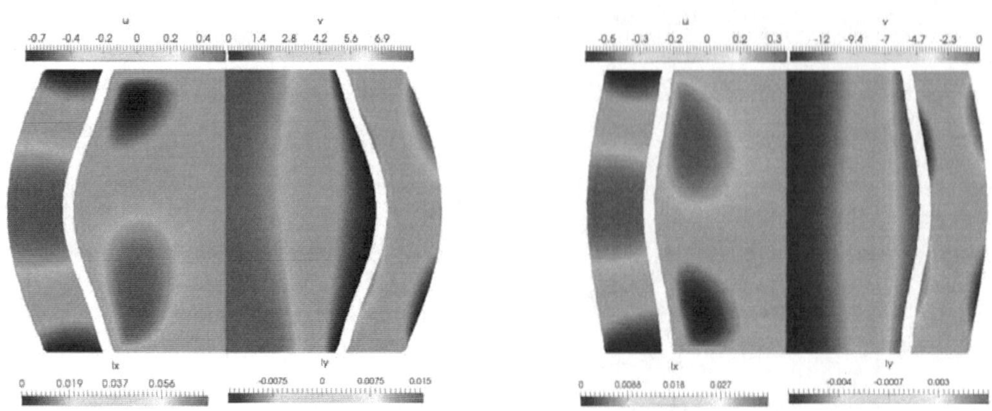

Figure 5. Test case A (on the left) and test case B (on the right). Horizontal (u) and vertical (v) velocity components over the fluid domain Ω_f and horizontal (l_x) and vertical (l_y) displacement components over the solid domain Ω_s.

IOP Conf. Series: Journal of Physics: Conf. Series **923** (2017) 012026 doi:10.1088/1742-6596/923/1/012026

(v) velocity components over the fluid domain Ω_f together with the horizontal (l_x) and vertical (l_y) displacement components over the solid domain Ω_s are reported. It is important to remark the inversion of the direction of the motion, in case B, due to the change of the control pressure.

4. Conclusions

In this work we have studied an optimal control method for a fluid structure interaction problem. The objective is the matching of a displacement field in a particular region of the solid domain by controlling the pressure on the fluid inlet boundary. In order to achieve the coupling between the solid and fluid domain we have used the FSI problem with a monolithic variational formulation that automatically takes into account the interface conditions. Furthermore, we have extended the velocity field to the solid domain to couple adjoint variables and forces on the interface. The optimality system has been obtained by setting to zero the first order Fréchet derivatives of the complete Lagrangian and this system has been solved by using a steepest descent algorithm. The results obtained show the accuracy and robustness of this approach and the possible uses of this method in many industrially relevant applications. In future works we plan to assess this pressure control to more realistic geometries to show the feasibility of this optimization approach in real complex cases. Furthermore this work can provide a good starting point for more complex FSI boundary optimal control problems, for example relying on the inlet velocity profile as control parameter. In that case the optimality condition obtained imposing the first order necessary conditions is a differential equation, which is far more difficult to be solved numerically than the algebraic one obtained considering pressure as control parameter.

References

[1] Gunzburger M D 2003 *Perspectives in flow control and optimization* vol 5 (Siam)
[2] Nocedal J and Wright S 2006 *Numerical optimization* (Springer Science & Business Media)
[3] Rowley C W and Williams D R 2006 *Annu. Rev. Fluid Mech.* **38** 251–276
[4] Kim J and Bewley T R 2007 *Annu. Rev. Fluid Mech.* **39** 383–417
[5] Gunzburger M and Manservisi S 1999 *SIAM Journal on Control and Optimization* **37** 1913–1945
[6] Turek S and Hron J 2006 *Fluid-structure interaction* (Springer) pp 371–385
[7] Formaggia L, Quarteroni A and Veneziani A 2009 *Cardiovascular Mathematics, vol. 1* (Springer, Heidelberg)
[8] Le Tallec P and Mouro J 2001 *Computer Meth. in Appl. Mech. and Eng.* **190** 3039–3067
[9] Bukač M, Čanić S and Muha B 2015 *Journal of Computational Physics* **281** 493–517
[10] Gilmanov A, Le T B and Sotiropoulos F 2015 *Journal of Computational Physics* **300** 814–843
[11] Cerroni D and Manservisi S 2016 *Journal of Computational Physics* **313** 13–30
[12] Calderer A, Kang S and Sotiropoulos F 2014 *Journal of Computational Physics* **277** 201–227
[13] Aulisa E, Cervone A, Manservisi S and Seshaiyer P 2009 *Comm. in Comput.l Phys.* **6** 319–341
[14] Aulisa E, Manservisi S and Seshaiyer P 2005 *Nonlinear Anal., Th., Met. and Appl.* **63** 1445–1454
[15] Maute K, Nikbay M and Farhat C 2003 *Internat. J. for Num. Meth. in Eng.* **56** 911–933
[16] Lund E, Møller H and Jakobsen L A 2003 *Structural and Multidisciplinary Optimization* **25** 383–392
[17] Yan Y and Keyes D E 2015 *Journal of Computational Physics* **281** 759–786
[18] Hron J and Turek S 2006 *Fluid-structure interaction* (Springer) pp 146–170
[19] Cerroni D, Manservisi S and Menghini F 2015 *Internat. J. of Math. Mod. and Meth. in Appl. Sci.* **9** 227–234
[20] Adams R A 1975 *Sobolev spaces* (Academic Press, New York)
[21] Brenner S and Scott R 2007 *The mathematical theory of finite element methods* vol 15 (Springer Science)
[22] Aulisa E, Manservisi S and Seshaiyer P 2006 *Comput. meth. in appl. mech. and eng.* **195** 4604–4616
[23] Bramble J H 1993 *Multigrid methods* vol 294 (CRC Press)

IOP Conf. Series: Journal of Physics: Conf. Series **923** (2017) 012027 doi:10.1088/1742-6596/923/1/012027

CFD analysis of aircraft fuel tanks thermal behaviour

C Zilio[1], G A Longo[1], G Pernigotto[2], F Chiacchio[3], P Borrelli[3] and E D'Errico[3]

[1] University of Padova, Dept. of Management and Engineering , S.lla S. Nicola, 3, I-36100 Vicenza, Italy
[2] Free University of Bozen-Bolzano, Faculty of Science and Technology
Piazza Università, 5 I-39100 Bolzano, Italy
[3] Leonardo, Aircraft Division, Viale dell'Aeronautica, Pomigliano d'Arco, Italy

E-mail: tony@gest.unipd.it

Abstract. This work is carried out within the FP7 European research project TOICA (Thermal Overall Integrated Conception of Aircraft, http://www.toica-fp7.eu/). One of the tasks foreseen for the TOICA project is the analysis of fuel tanks as possible heat sinks for future aircrafts. In particular, in the present paper, commercial regional aircraft is considered as case study and CFD analysis with the commercial code STAR-CCM+ is performed in order to identify the potential capability to use fuel stored in the tanks as a heat sink for waste heat dissipated by other systems. The complex physical phenomena that characterize the heat transfer inside liquid fuel, at the fuel-ullage interface and inside the ullage are outlined. Boundary conditions, including the effect of different ground and flight conditions, are implemented in the numerical simulation approach. The analysis is implemented for a portion of aluminium wing fuel tank, including the leading edge effects. Effect of liquid fuel transfer among different tank compartments and the air flow in the ullage is included. According to Fuel Tank Flammability Assessment Method (FTFAM) proposed by the Federal Aviation Administration, the results are exploited in terms of exponential time constants and fuel temperature difference to the ambient for the different cases investigated.

1. Introduction

In the recent years aeronautical industry has concentrated its design challenges towards a "More Electric Aircraft" concept that implies the reduction of the traditional hydraulic and pneumatic sub-systems and the introduction of electrically driven devices.

As a consequence, aircraft electrical consumption has increased by a factor of five and new engines, whilst more efficient, now produce much more heat. Consequently, a proper thermal management on new aircrafts has become a real challenge. By managing and understanding thermal behaviour early in any aircraft design process, some of the issues related to heat flows towards aircraft components and systems can be abated [1].

Within FP7 research framework, EC launched the Project TOICA (Thermal Overall Conception of Aircraft) with the main target of developing new concepts for improved thermal load management for aircraft components, systems or equipment, which will integrate innovative cooling technologies and products. In particular, TOICA contributed to:

- Improve the overall multidisciplinary conception of aircraft during the architecture phases
- Optimize the overall energy management of the aircraft through a reduction of the aircraft energy consumption
- Reduce thermal constraints on systems and structure, and thermal integrated risks
- Reduce weight and complexity through a fully integrated structure and thermal design of systems.

IOP Conf. Series: Journal of Physics: Conf. Series **923** (2017) 012027 doi:10.1088/1742-6596/923/1/012027

Among the different "Use cases" analyzed by the TOICA consortium, different approaches and methods were investigated to model the fuel tank thermal behaviour. The idea of transferring waste heat to the fuel in the aircraft tanks is relatively old and some pioneering works can be retrieved in the literature. For example, in [2] it was already clearly observed that fuel thermal modeling is a key enabler to assess the capability of the system architecture to use fuel as a heat sink. In that report, a simplified analytical model of the thermal behavior of aviation fuel inside a tank was proposed basing on tests run with a mock-up fuel tank.

Several other analytical models have been proposed in the last decades, thanks to the availability of more sophisticated simulation platforms and the more and more increasing calculation capabilities of computers. However, these methods can be considered suitable only for early-phase trade studies to determine the feasibility of an aircraft [3].

When a more detailed modeling approach is needed, it is possible to consider computational fluid dynamics (CFD) models. Recently, Qian and co-authors [4] coupled the numerical results of FLUENT Volume Of Fluid (VOF) analysis on thermal behavior of jet fuel in the tank with a mono-dimensional dynamic simulation tool (FLOWMASTER). FLOWMASTER was used to build a mono-dimensional simulation of aircraft integrated thermal management, including the convective heat transfer coefficients calculated by using FLUENT. The authors concluded that it is possible to realize thermal simulation of aircraft integrated thermal management without flight tests, in order to provide guidance to the design.

To comply with new civil certification regulation [5], the aircraft fleet fuel tank flammability exposure level must be evaluated. The flammability assessment can be performed using the Federal Aviation Administration (FAA) Fuel Tank Flammability Assessment Method (FTFAM). This method uses Monte Carlo statistical approach to generate flammability data for certain unknown variables over known distributions for a large number of flights. The parameters required to run the flammability assessment can be grouped in the following categories: fuel tank surrounding environment, mission data, fuel properties, and the thermal characteristics of the fuel tank. Once these four components are determined, whether by user input or by Monte Carlo calculations, the model can then determine whether the ullage of the fuel tank for each time increment of flight is flammable or not; and furthermore, the percent of the mission time that the fuel tank is flammable. The flight data, required to perform the Monte Carlo calculations, are: the cruise Mach number, altitude steps and tank ram recovery. The airplane data, required to perform the Monte Carlo calculations, includes: the maximum range of the aircraft, the number of engines, and the Outside Air Temperature (OAT) cutoff limit. Other data that must be fixed by the designer regards fuel tank usage information like the tank full and empty times and the engine start times. Finally, the fuel tank thermal data, required to perform the Monte Carlo calculations, are the temperature differential to both ambient and Total Air Temperature (TAT) as well as a number of exponential time constants for the six cases shown in the Table 1. According to FAA [5], the fuel is charged in the tank at a given uniform temperature. Then the whole tank (including liquid fuel, ullage and solid structure) undergoes the thermal solicitation of the surrounding environment and of any possible internal heat source. Through the analysis of the transient response of the system it is possible to estimate the exponential time constant of the fuel considering different levels of the fuel inside the tank (typically, near full and near empty) with reference to both ground and flight conditions. In addition, for ground conditions, these values must be known with and without the engines running. Hence, there are six exponential time constants required as input, which define how the fuel in the tank heats or cools in response to heat input.

Using experimental values of the exponential time constants is usually unfeasible in the early phase of aircraft design, hence analytical or numerical methods should be employed for estimating the six exponential constants.

The heat and mass transfer processes inside the fuel and the ullage in a wing tank is rather complex. Figure 1, modified from [2], is an attempt to highlight the principal mechanisms.

Table 1. FAA Cases for the fuel tank flammability estimation.

Case		Fuel Quantity
1	Ground-Engine OFF	Near empty[1]
2	Ground-Engine OFF	Near full
3	Ground-Engine ON	Near empty[1]
4	Ground-Engine ON	Near full
5	Cruise-Engine ON	Near empty[1]
6	Cruise-Engine ON	Near full

[1.] Near empty does not refer to the portion of tank, the so called "collector tank", where fuel pump is installed. This part is always full, in normal operation with the engine running.

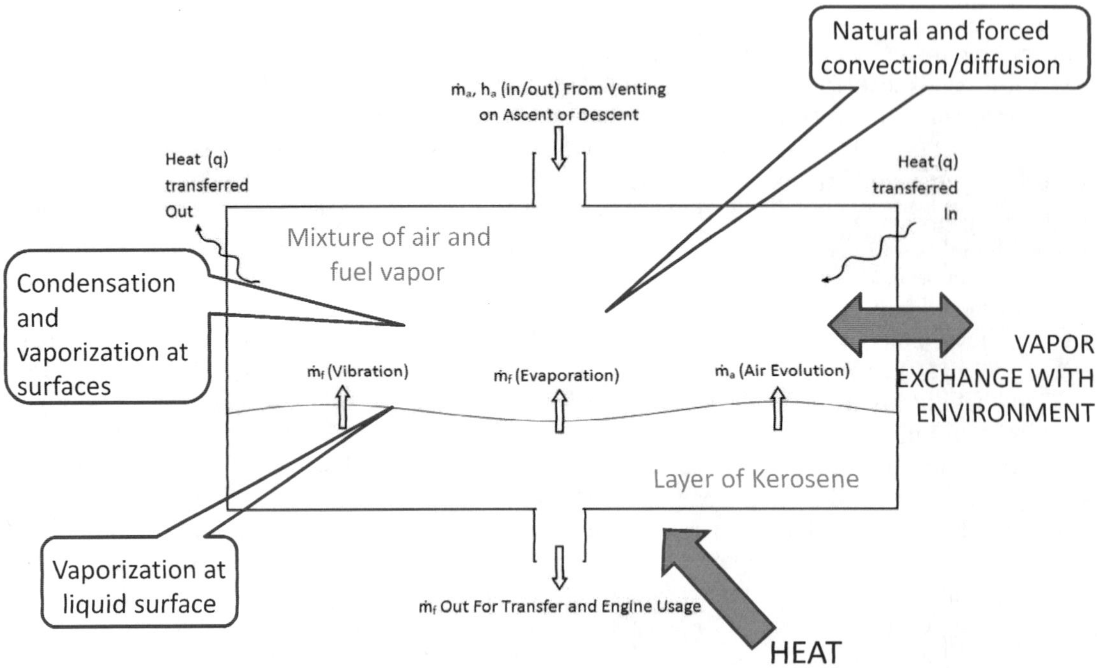

Figure 1. Main heat and mass transfer mechanisms of the fuel in the tank.

Given the rather complex mechanisms inside the tank and the complex interactions of the tank with surrounding components in the wing and with the external ambient, a simple model can hardly properly capture local phenomena that can lead to hazardous conditions in terms of flammability.
CFD is an option to evaluate the time constants requested by FAA regulations and at the same time to properly map local temperatures inside the tank.

In section 2 of this paper, the modeling approach of a portion of the fuel tank of a commercial aircraft is outlined. Furthermore, the relevant boundary conditions for both "ground" and flight" operations are defined according to typical airborne working conditions.
In section 3, the main features of the CFD numerical approach implemented with the software STAR-CCM+ are described in details. An example of the temporal temperature profiles of liquid fuel, ullage and tank structure is reported, in order to exploit the potentiality of the proposed approach to be used in FAA Fuel Tank Flammability Method.

2. Modeling approach and boundary conditions definition

In the present work, reference was made to a regional aircraft. It should be noted that in a regional aircraft, the amount of fuel mass charged in a regional aircraft is lower in comparison to a long haul aircraft. This point can be considered more critical in terms of possible harmful situations as a consequence of an heat input.
In figure 2, an example of a parametric CAD model of the regional turboprop aircraft wing and the fuel tanks is reported.
The studies were performed on a section of fuel tank between three consecutive ribs including the fuel collector tank and the adjacent bay i.e., ribs 4, 5 and 6 in figure 3. The reference tank volume section is reported in figure 4. As it can be seen, the aluminum tank structural parts and the leading edge, including the deicing boot, the hot air bleed pipe and the pipe insulation materials were considered in the simulation.

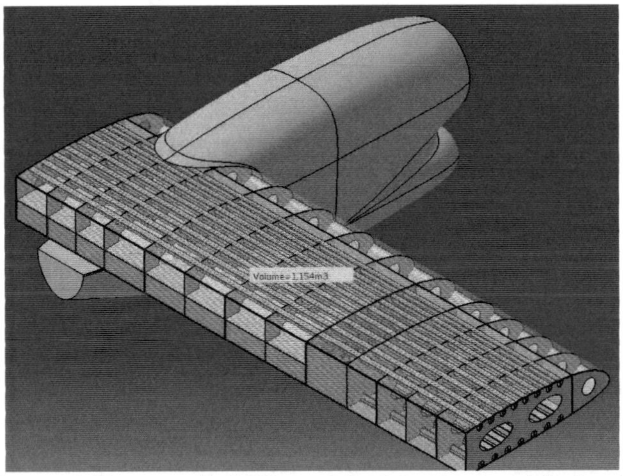

Figure 2. Parametric CAD model of the wing and fuel tank.

Jet A fuel was considered and its thermophysical properties were defined as function of temperature and pressure according to [6].
In flight, the leading edge zone (i.e. front area adjacent to the tank) is vented by air. This comes directly from the wing NACA air intake and it is conveyed through the holes of the ribs along the leading edge (figure 5). The air flow rate has been estimated according to existing regional aircraft specifications while the temperature was set equal to -12.2 °C.

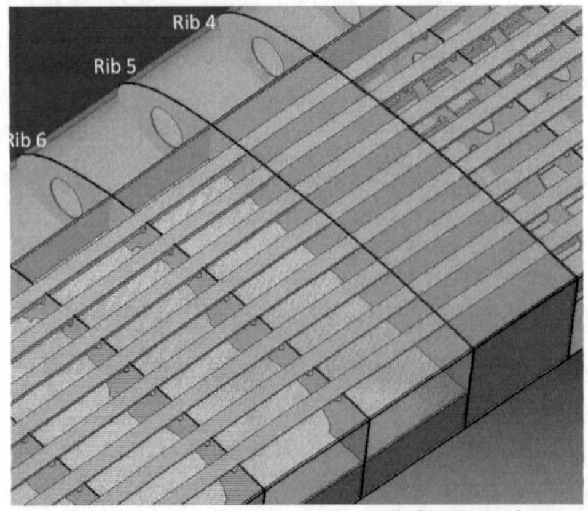

Figure 3. Sketch of the analyzed fuel tank parts.

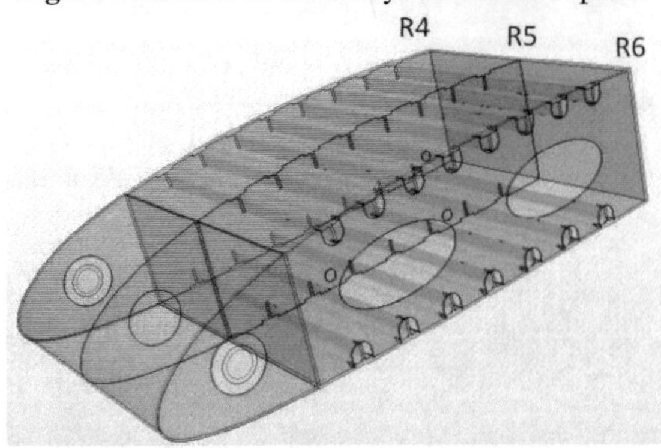

Figure 4. Reference tank volume section.

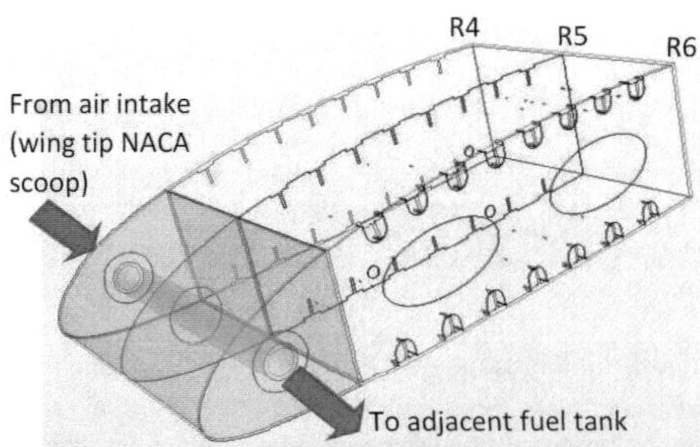

Figure 5. Vent flow-rate for leading edge.

The reduction of the fuel quantity, caused by the engine feeding, is simulated with the exit from rib 4 of a fuel flow rate equal to the engine consumption (see Figure 6). This reduction is compensated by the fuel inlet from the other compartments (adjacent to bay 5-to-6) through the stringers cut-outs of the rib 6. This effect impacts the heat thermal balance because the temperature of the incoming fuel (i.e. coming from the adjacent compartments) is different from the temperature of the fuel leaving the sump tank even if both flow rates are equal.

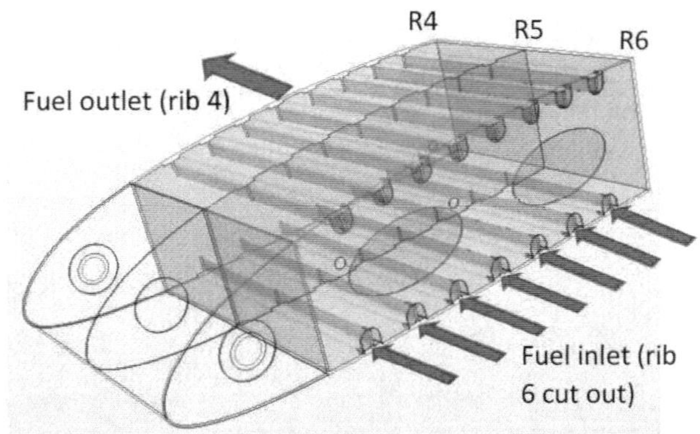

Figure 6. Fuel Flow-rate for Tank Sump - Engine ON Cases.

The environmental solicitation to the fuel tank has been expressed through a single fictitious temperature, named the sol-air temperature, which can be adopted in order to explore a variety of forcing conditions, also in transient studies.
The solar radiation incident on each segment of the skins profiles has been estimated starting from horizontal diffuse solar irradiance and direct normal irradiance recorded in typical weather data file (like for example International Weather Files for Energy Calculations - IWEC).
The chosen calculation model for solar irradiance on tilted surface is the isotropic solar model, which gives good estimations for clear sky conditions (i.e., for days with maximum solar irradiance). The surface of the ground has a reflectivity of 0.2 and the wing profile is oriented towards the equator (e.g., south-oriented for Milan).
For the top skin, an important role in the surface heat balance is played by the infrared radiation towards the sky dome. The top skin profile was split in several segments and for each segment a proper view factor of the sky dome was introduced. The infrared heat transfer coefficient towards the sky dome has been calculated assuming the mean radiant temperature as an average between the OAT and the fictitious sky temperature, assumed as 13 K (tropical regions) or 11 K (temperate regions) lower than OAT.
As regards the surface heat transfer coefficient by radiation h_r, which considers the heat exchanged by radiation with the surrounding environment, it has been calculated considering a mean radiant temperature equal to OAT and is equal to 5 W/(m^2 K). In flight conditions, it has been neglected.
In ground conditions, the convective heat transfer coefficient is 20 W/(m^2 K), which corresponds approximately to the case of wind velocity of 4 m/s. In ground conditions, the surface heat transfer coefficient has been kept constant along the profile.
In flight conditions, the local heat transfer coefficient of the wing skin has been calculated according to [7]. For the remaining parts of the investigated fuel tank i.e., the external remain surfaces (Rib 4, Rib 6, Rear spar) and the bleed pipe in the leading edge, proper surface temperature or local heat transfer coefficients were imposed, basing on typical airborne conditions for both ground and flight operations.

3. Numerical approach

The Jet A thermal behavior in the six FAA conditions reported in Table 1 was studied numerically with the CFD code STAR-CCM+. The volume between Rib 4 and Rib 5 (see figure 3) was considered almost full of Jet A, since this is the normal working condition during standard operation of the aircraft. The liquid fuel level inside the tank part between Rib 5 and Rib 6 can vary, according to the mission profile.
The following Table 2 summarizes the simulation approach used for the structural part (including tank, boot and bleed pipe) and the fluid domain.

Table 2. Main features of the CFD simulation approach.

Domain	Material	Number of cells
Structural parts 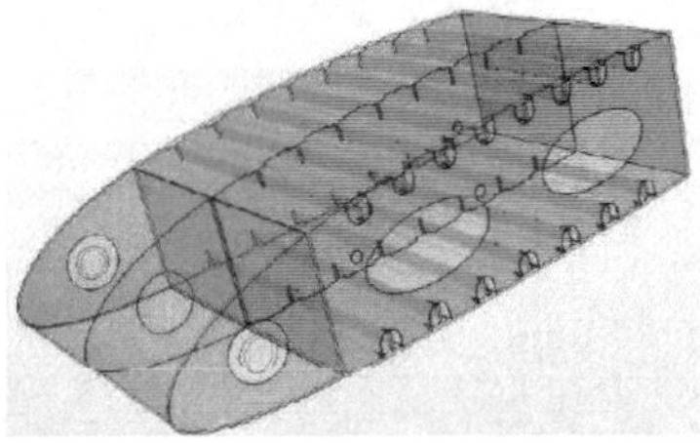	Aluminum	~185000
Boot 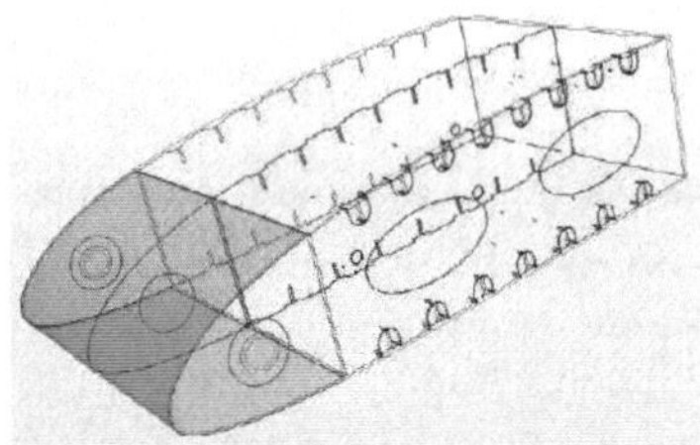	Neoprene	~25000
Insulation Bleed pipe 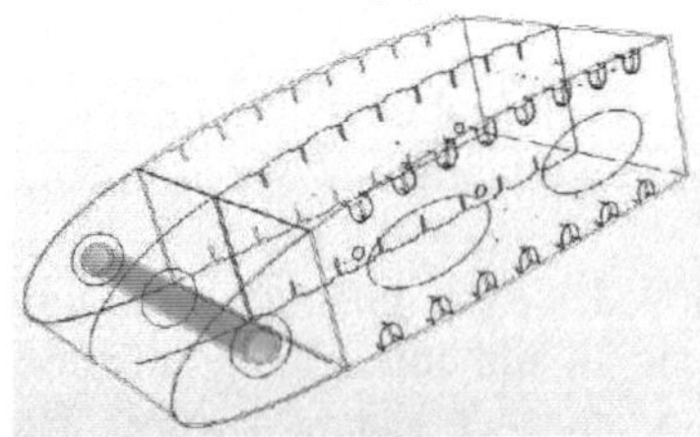	Kevlar	~5000
Fluid parts 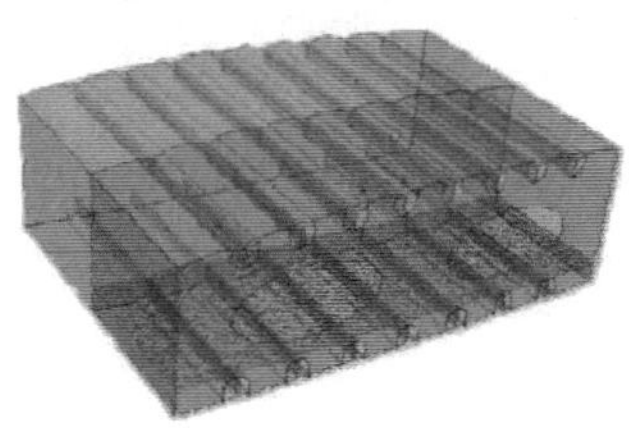	Jet A and air	~675000

Solid parts were studied by adopting the following models: 3D, implicit unsteady (time-step 0.05 s), solid with segregated solid energy, constant density. The surface-to-surface radiation was enabled with gray surfaces hypothesis and view factors calculation.

Regarding fluid domain, the following approach was used: 3D, implicit unsteady (time-step 0.05 s), Volume of Fluid (VOF); multiphase flow with segregated flow and multiphase temperature (two Eulerian phases: liquid Jet A and air). In order to reduce the complexity of the calculations, no multiphase interactions were considered. K-ε turbulence model was imposed and gravity effects were considered.

As mentioned before, relevant Jet A and air properties were considered as a function of local temperature and pressure.

The simulations were run with an High Performance Calculation cluster (HPC) with 180 cores in parallel (Intel Xeon Processor E5620 4C 2.40 GHz 12 MB Cache 1066 MHz 80w).

3.1. Simulation results

Examples of temperature profiles at different times in one longitudinal plane section of the fuel tank are reported in figure 7, while the velocity distribution in the same section after 30 min is reported in figure 8.

In figures 7 and 8, it is possible to appreciate the local temperature values. This feature allows to check the possible subsistence of local hazardous conditions for Jet A fuel flammability.

For example, it is evident that the structural stringers inside the tank have a remarkable effect in the heat transfer inside the fuel. The leading edge is considered to have a relevant impact on the fuel tank boundary conditions.

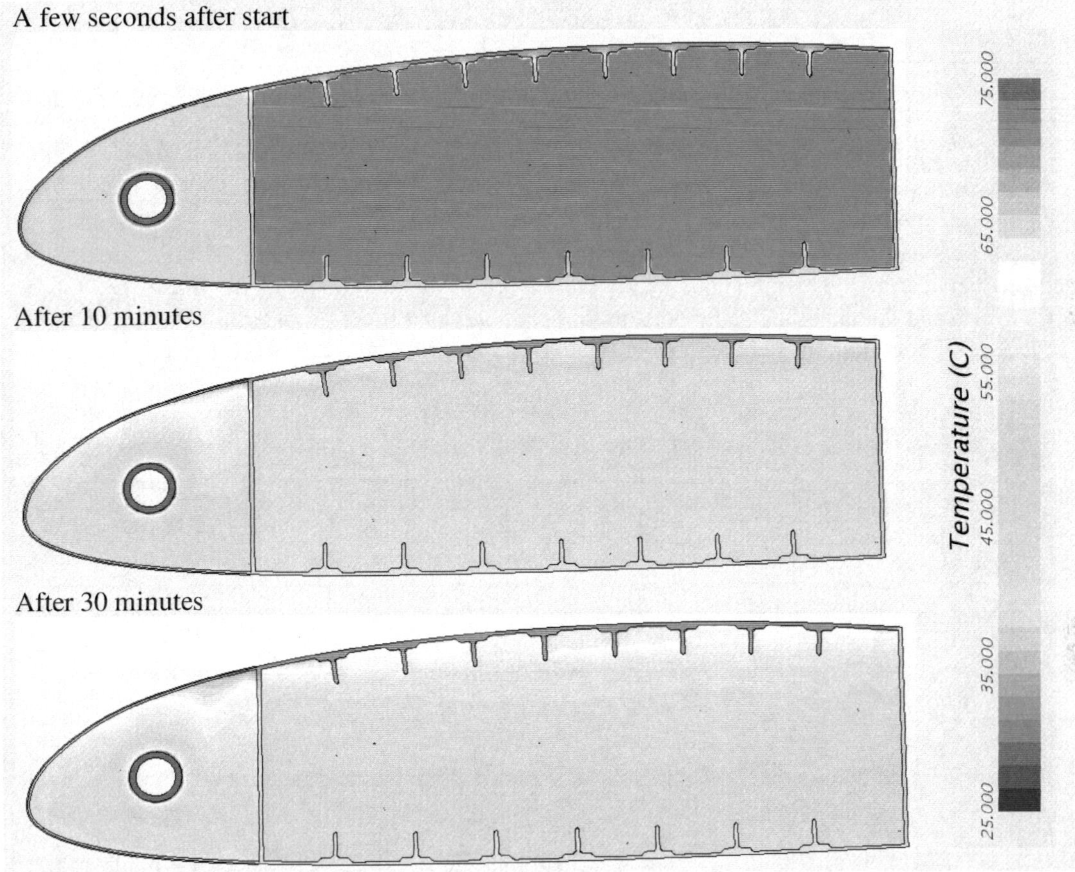

Figure 7. Temperature distribution in one longitudinal section of the fuel tank, including the leading edge (ground operation).

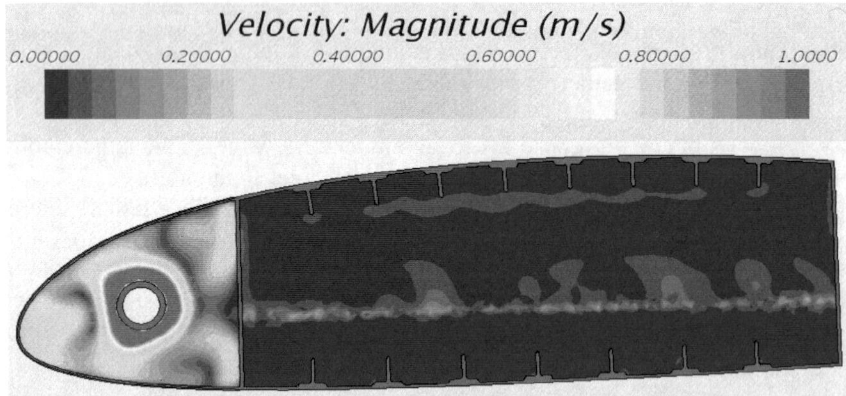

Figure 8. Velocity distribution in a longitudinal section of the fuel tank and leading edge, 30 min after start.

Clearly, the use of the CFD code allows to easily determine the bulk temperature values in order to estimate the exponential time constants as requested by FAA regulation.

In figure 9, the temperature profiles (bulk temperatures of the aluminum structure, of the liquid Jet A fuel in the bay between ribs 4 and 5 and the air temperature in the same bay, of the liquid Jet A fuel between rib 5 to 6 and the air in the same portion) are reported as a function of time with reference to FAA case 5 (i.e. flight conditions, with rib 5 to 6 near empty and rib 4 to 5 almost full of fuel).

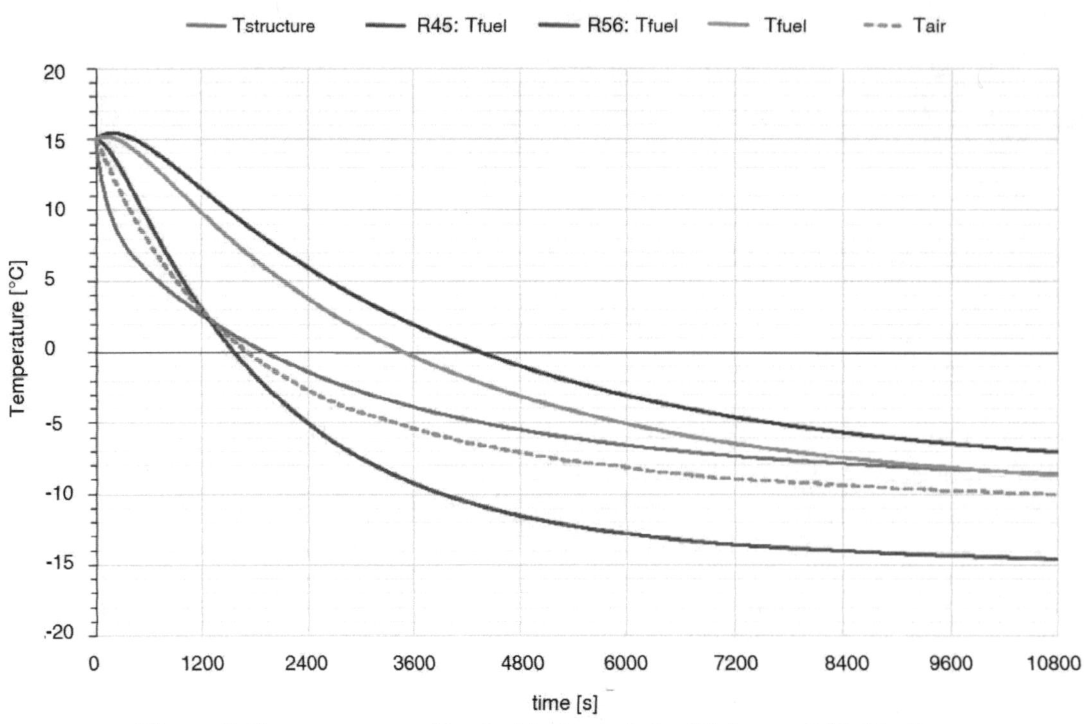

Figure 9. Temperature profiles inside the tank for FAA case 5 (Table 1).

In this case, about 300 kg of Jet A fuel are charged in the tank at 15 °C just before take-off. The cruise pressure is 46450 Pa, the bleed pipe temperature is 200 °C, outside air temperature (OAT) is -26.4 °C and total air temperature (TAT) is -12.2 °C. An internal heat source (2 kW) is located inside the fuel in

the bay between rib 4 and 5. This internal heat source is representative of a generic heat load released within the fuel tank.
According to FAA rules [5], time constants should be evaluated for the fuel contained in Rib 4-to-5 bay (R45) and in Rib 5-to-6 (R56). As expected, given the larger amount of fuel mass present in R45 (almost full), the time constant is larger than for R56 (almost empty). The solid orange line represents the liquid Jet A bulk temperature (i.e. considering both bay R56 and R45), as estimated by STAR CCM+. Dotted orange line represents the ullage bulk temperature. This temperature profile is affected also by the air motion coming from the NACA scopes that permit air flow in order to compensate any possible difference between tank inner pressure and ambient pressure.
The aluminum structure bulk temperature never reaches the ambient air temperature (even after 10800 s, the aluminum is about 19 K warmer that the ambient air). This is a consequence of the internal heat load (pump and fuel) and of the heat rejected by the bleed pipe through the leading edge.
It is interesting to observe that fairly constant bulk fuel temperature (solid orange line in figure 9) is achieved only after about 10800 s (i.e. 3 h) that is a period of time usually longer than the duration of a regional aircraft cruise.
Finally, on the basis of the temperature profiles in figure 9, it is possible to calculate the temperature differential with the ambient temperature (OAT) and the total ambient temperature (TAT) as requested by the FAA Fuel Tank Flammability Method.

4. Conclusions

In this paper a comprehensive numerical approach based on STAR-CCM+ CFD software is proposed for the simulation of fuel thermal behavior inside a portion of the tank of a regional aircraft. In the framework of the European project Thermal Overall Integrated Concept of Aircraft (TOICA), several working conditions were investigated in order to simulate both ground and flight situations with the tank almost full or almost empty of fuel. Particular care was given to the definition of suitable and realistic boundary conditions before running the CFD simulations. The considered cruise scenarios were selected according to the Federal Aviation Administration (FAA) rules.
The CFD tool offers the unique capability of mapping the local temperatures in whole domain, thus exploiting potential hazardous regions. The results indicate that the proposed numerical approach can be used for the estimation of fuel time constants and of the fuel temperature difference to the ambient. This set of data can be used as an input for further assessment of fuel tank flammability by means of statistical analysis (e.g. Monte Carlo method).
Furthermore, the proposed numerical approach is a viable option for the implementation of a thermal simulation of aircraft integrated thermal management without flight tests, in order to provide guidance to the design starting from the aircraft architecture definition.

Acknowledgement

The research leading to these results has received funding from the European Union Seventh Framework Programme (FP7/2007-2013) under grant agreement n° 604981.

References

[1] TOICA Website http://www.toica-fp7.eu
[2] Kosvic T C, Zung L B, Gerstein M 1971 *Tech. Report AFAPL-TR-71-7*
[3] German B J 2012 *J. Propulsion Power* **28** 204 – 10
[4] Su Q, Chang S N, Yang S Y 2016 *IEEE/CSAA Int. Conf. Aircraft Utility System* October 10-12, 2106 Beijing (China) 774 – 79
[5] Summer S M, Hughes W J 2008 *DOT/FAA/AR-05/8*
[6] CRC 2014 *Aviation Fuel Properties Handbook, CRC Report No. 663*
[7] Newton J E, VanFossen G J, Poinsatte P E, DeWitt K J 1988 *NASA Tech. Memorandum 100284 AIAA-88-0287*

IOP Conf. Series: Journal of Physics: Conf. Series **923** (2017) 012028 doi:10.1088/1742-6596/923/1/012028

A priori and a posteriori analysis of the flow around a rectangular cylinder

A Cimarelli[1,2], A Leonforte[2], M Franciolini[1], E De Angelis[3], D Angeli[2] and A Crivellini[1]

[1] DIISM, Università Politecnica delle Marche, Via Brecce Bianche, 60131 Ancona, Italy
[2] DISMI, Università di Modena e Reggio Emilia, Via Amendola, 42122 Reggio Emilia, Italy
[3] School of Engineering, Cardiff University, Cardiff CF24 3AA, UK

E-mail: a.cimarelli@staff.univpm.it

Abstract. The definition of a correct mesh resolution and modelling approach for the Large Eddy Simulation (LES) of the flow around a rectangular cylinder is recognized to be a rather elusive problem as shown by the large scatter of LES results present in the literature. In the present work, we aim at assessing this issue by performing an *a priori* analysis of Direct Numerical Simulation (DNS) data of the flow. This approach allows us to measure the ability of the LES field on reproducing the main flow features as a function of the resolution employed. Based on these results, we define a mesh resolution which maximize the opposite needs of reducing the computational costs and of adequately resolving the flow dynamics. The effectiveness of the resolution method proposed is then verified by means of an *a posteriori* analysis of actual LES data obtained by means of the implicit LES approach given by the numerical properties of the Discontinuous Galerkin spatial discretization technique. The present work represents a first step towards *a best practice* for LES of separating and reattaching flows.

1. Introduction

In Large Eddy Simulation (LES) the large scales of the flow are directly computed while leaving to the subgrid stresses the duty of accounting for the small unresolved scales. Three main elements determine the success of the LES approach: (i) the mesh resolution, (ii) the model for the subgrid stresses and (iii) the numerical scheme. These three aspects are inherently related each other so that a certain turbulence model is well suited for a given mesh resolution and numerical algorithm and not for others [1]. For this reason it is often difficult to find out a best practice for LES. An instructive example of the complexity of the LES approach is given by the flow around a rectangular cylinder. As shown in the BARC project (Benchmark on the Aerodynamics of a Rectangular 5:1 Cylinder), see Bruno et al. [2] and references therein, a large scatter of LES results is present even for very simple statistical quantities. One possible explanation is given by the fact that a clear picture of the combined influence of mesh resolution, turbulence model and numerical scheme on the flow statistics is still missing. In this contribution, we aim at assessing the filtered behaviour of the flow, by analysing for the first time filtered Direct Numerical Simulation (DNS) data of such a system. In particular, in this preliminary work, we aim at assessing the role of the mesh resolution on capturing the main flow features. And, hence, we do not consider the possible coupling of the present results with the modelling approach adopted to represent the stresses at subgrid levels. Finally, we perform an actual LES

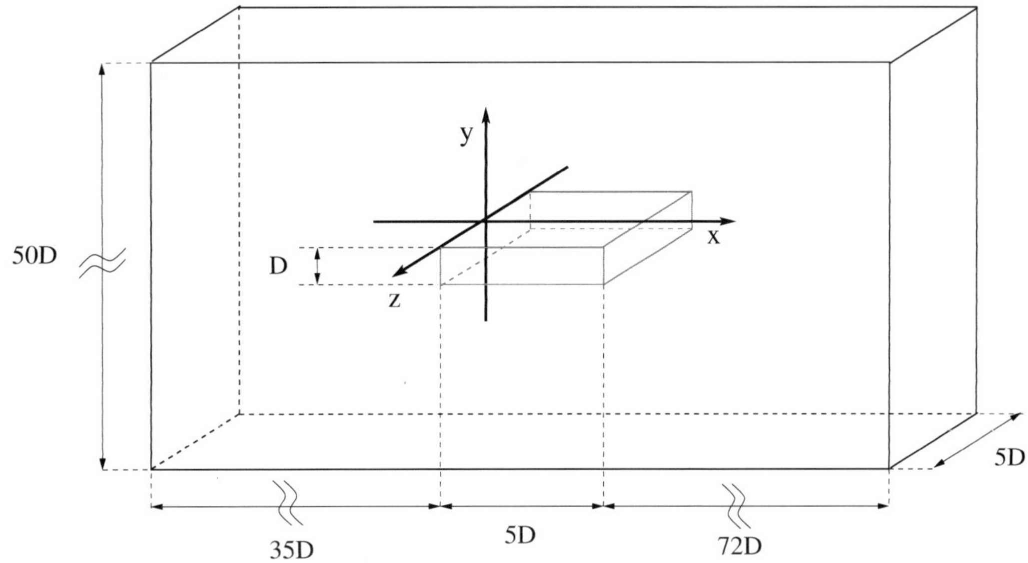

Figure 1. Configuration of the system.

simulation by means of Discontinuous Galerkin (DG) numerical approach in order to prove the results from the DNS.

2. Direct Numerical Simulation

A common practice for the assessment of the behaviour of the resolved field in LES as a function of the filter length employed is to analyse filtered DNS data, the so-called *a priori* analysis. The governing equation numerically integrated when performing a DNS are

$$\begin{aligned} \frac{\partial u_i}{\partial x_i} &= 0 \\ \frac{\partial u_i}{\partial t} + \frac{\partial u_i u_j}{\partial x_j} &= -\frac{\partial p}{\partial x_i} + \frac{1}{Re}\frac{\partial^2 u_i}{\partial x_j \partial x_j} \end{aligned} \tag{1}$$

where $x = x_1$ $(u = u_1)$, $y = x_2$ $(v = u_2)$, $z = x_3$ $(w = u_3)$ are the streamwise, vertical and spanwise directions (velocities), p is the pressure field, and $Re = U_\infty D/\nu$ is the Reynolds number with ν the kinematic viscosity, U_∞ the free-stream velocity and D the rectangle thickness. Unless specifically stated, all the quantities will be reported in a dimensionless form by using D and U_∞ for the lengths and velocities, respectively. A cell-centered finite volume method has been chosen to dicretize the equations by means of the OpenFOAM® open source code [3]. Time integration is performed by means of a second-order backward Euler implicit scheme while convective and diffusive fluxes at the volume faces are evaluated through a second-order central difference scheme. Finally, a pressure-implicit split-operator (PISO) algorithm is used to numerically solve the pressure-velocity coupling. Given the simple geometry of the problem, a structured Cartesian grid is adopted. A laminar free-stream velocity is imposed at the inlet while free-stream boundary conditions (mixed Dirichlet-Neumann conditions) are imposed at the outlet and upper-lower boundaries. Finally, periodic boundary conditions are imposed in the lateral boundaries.

The studied case consists of a flow around a rectangular cylinder whose lengths are $(L_x, L_y, L_z) = (5D, D, 5D)$ at a Reynolds number $Re = 3000$. The extent of the numerical

Case	Δx^+	Δz^+	$N_x \times N_z$	DoF/DoF^{dns}
DNS	6.1	5.4	128×144	1
F1	12.2	10.8	64×72	0.25
F2	12.2	21.6	64×36	0.12
F3	24.4	10.8	32×72	0.12
F4	24.4	21.6	32×36	0.06

Table 1. Parameters of the filtered fields employed for the *a priori* analysis. The values reported of Δx^+ and Δz^+ are averaged in the streamwise directions.

domain is $(\mathcal{D}_x, \mathcal{D}_y, \mathcal{D}_z) = (112D, 50D, 5D)$ and is found to be large enough to not interfere with the flow dynamics, see figure 1. The number of volumes used to discretize the numerical domain is $1.5 \cdot 10^7$ which leads to a near-wall resolution $\overline{\Delta x} = 0.04$, $\Delta y = 0.002$, $\Delta z = 0.034$ where $\overline{(\cdot)}$ denotes the streamwise average along the rectangle length. Although the time integration scheme adopted is implicit, a relatively small constant time step is used, $\Delta t = 0.0023$, which leads to a CFL number less than unity in order to well resolve the small scale unsteadiness of the flow.

3. Filtered fields statistics

In the present section, we assess the behaviour of the LES approach for solving the flow around a rectangular cylinder by means of an *a priori* analysis. The DNS data described in the previous section, are filtered by means of a Gaussian filter,

$$\mathcal{G}_i = \sqrt{\frac{6}{\pi \Delta_i^2}} \exp(-\frac{6r_i^2}{\Delta_i^2}) \tag{2}$$

so that the resolved field can be written as

$$\bar{u}_i(\mathbf{x}, t) = \int \mathcal{G}_i(\mathbf{x}, \mathbf{x}') u_i(\mathbf{x}', t) d\mathbf{x}' \tag{3}$$

and the subgrid stresses as

$$\tau_{ij}(\mathbf{x}, t) = \overline{u_i u_j} - \bar{u}_i \bar{u}_j \ . \tag{4}$$

In order to analyse the role of the filter length on the resolved and subgrid dynamics, we consider 4 different sets of filter widths as reported in table 1. These sets of filter widths has been chosen to understand the effect of reduction of degrees of freedom to reproduce the flow dynamics and to appreciate the effect of the adoption of a different reduction of resolution in the two horizontal directions. Indeed, from F1 to F4 we have a reduction of computational points with respect to DNS from 0.25 to 0.06, respectively. On the other hand, the filter cases F2 and F3 exploit the same total reduction of degrees of freedom with respect DNS, $DoF/DoF^{dns} = 0.12$ but with different resolutions in the streamwise and spanwise directions. It is worth pointing out that filtering is applied only in the horizontal directions since, the vertical one is commonly recognized to not allow for a significant reduction of the degrees of freedom of the problem when a near-wall resolved LES approach is considered.

To assess the effect of filtering, in figure 2 the iso-contours of the resolved turbulent kinetic energy,

$$q = \frac{1}{2} \langle \bar{u}_i' \bar{u}_i' \rangle \tag{5}$$

are shown. Here, $\langle \cdot \rangle$ denotes the ensemble average and, as customary, the prime denotes fluctuating quantities, i.e. $\bar{u}_i' = \bar{u}_i - \langle \bar{u}_i \rangle$. As expected, the overall effect of filtering is to

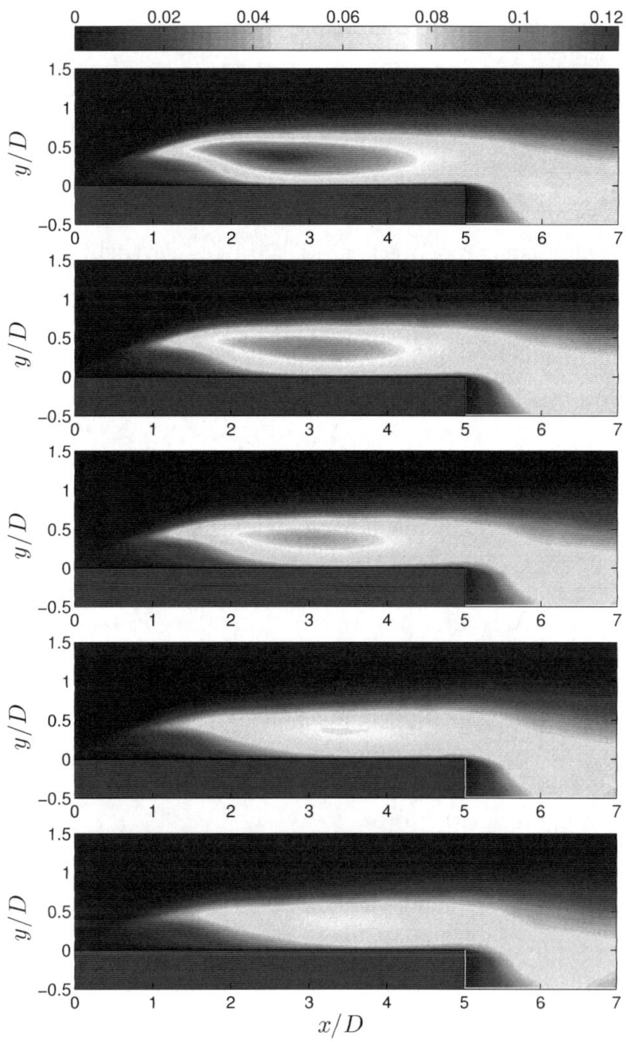

Figure 2. Iso-contours in the $(x - y)$-plane of turbulent kinetic energy. DNS behaviour and the filtering combinations F1, F2, F3 and F4 from top to bottom, respectively.

reduce the amount of resolved energy, see also table 2. In particular, for the F1 case the peak of resolved energy is of the order of 80% the DNS value while for the coarser case F4 is of the order of 59%. Interestingly, the reduction of turbulent kinetic energy is not homogeneous in space thus leading to a modification of the topology of the resolved turbulent intensities. Indeed, moving from F1 to F4 we observe a modification of the shape of the iso-contours and consequently also the position of the maximum intensity is changed. The general trend is a shift towards the wall and in the downstream direction of the region of high turbulence levels, see also table 2. This inhomogeneous effect of filtering is arguably due to the strong inhomogeneity of the flow in terms of spectral distribution of energy, i.e. the energy distribution in the space of scales significantly changes in the different regions of the flow. Hence, the same filter width influences in a different way the energy content in the different regions depending on the amount of energy contained at the small unresolved scales.

From a practical point of view, the *a priori* results shown in figure 2 indicate that the best

Case	q_{max}	$(x/D)_{q_{max}}$	$(y/D)_{q_{max}}$	$q_{max}/q_{max}^{(dns)}$
DNS	0.123	2.57	0.387	1
F1	0.098	2.96	0.383	0.8
F2	0.089	3.07	0.38	0.73
F3	0.078	3.38	0.351	0.63
F4	0.072	3.39	0.35	0.59

Table 2. Scaling of the intensity and position of the peak of turbulent kinetic energy q_{max} as a function of the filter lengths.

compromise between computational costs and reproduction of the main physical processes is the the use of a mesh grid whose resolution is of the order of that of the F2 case. Indeed, we have that the F1 case is certainly able to capture the most energetic flow features but it allows us to reduce the degrees of freedom of the problem to only the 25% those of the DNS. Hence, the mesh properties of the F1 case do not allow for solving this flow at the high Reynolds number regime of the applications. On the contrary, the F4 case allows us to drastically reduce the computational costs being the degrees of freedom of the mesh of the order of 6% those of the DNS. However, the resolved energetics appear to be very poor in terms of both the intensity and the location of the main physical processes. In between these two extreme, the F2 case allows us to reduce the degrees of freedom to the 12% of the DNS and, in the same time, to represent the main physical properties of the flow. It is interesting to point out that the same reduction of the computational costs obtained with a different coarsening in the two wall-parallel directions, as in the F3 case, does not allow to represent the flow features as much with the F2 case.

4. Large Eddy Simulation

Accordingly with the previous results, we perform a Large Eddy Simulation in order to understand if the main outcomes of the *a priori* analysis are confirmed by the *a posteriori* analysis of an actual LES simulation. As anticipated in the introduction, the success of the LES approach is determined not only by the mesh resolution adopted but also by the model for the subgrid stresses used. In the *a priori* analysis shown so far, we considered only the role of the mesh resolution. Actually, the flow around a rectangle and more in general separating and reattaching flows are challenging from a modelling point of view. The reason is given by the simultaneous presence within the flow of laminar, transitional and turbulent regions. These conditions challenge for turbulence models which should be able to switch off when laminar conditions are retained and to smoothly activate when the flow reaches transitional conditions. For these reasons, the choice of the subgrid stress model is of paramount importance when dealing with such kind of flows. Hence, in order to limit our analysis on the effects of resolution, we decide to perform an implicit LES simulation (ILES). This kind of approach consists on resolving the Navier-Stokes equations in a LES grids without the use of an explicit subgrid stress model. The effect of the small unresolved scales is emulated by the numerical dissipation associated with the accuracy of the schemes adopted to solve the equations. This means that the ILES approach automatically predict the laminar regions. Indeed, in the region of the flow where the mesh resolution is fine enough to resolve the flow dynamics the numerical dissipation spontaneously vanishes. Also for these reasons the ILES approach is a good candidate for the present *a posteriori* analysis.

As already demonstrated by several authors, [4, 5], one of the numerical methods particularly prone to perform Implicit Large Eddy Simulations is the Discontinuous Galerkin (DG) method. Indeed, these methods are characterized by a numerical dissipation which mimics the role of

Figure 3. Implicit Large Eddy Simulation of the flow around a rectangular cylinder at $Re = 3000$. Instantaneous iso-surfaces of $\lambda_2 = -1.5$ colored with the distance from the horizontal plane of symmetry of the rectangle.

subgrid-scale models, affecting only the highest wave numbers without spoiling the resolution of the largest resolved scales. It has been also proved [6] that using high order DG discretizations is possible to achieve the same spatial resolution with a reduced number of degrees of freedom (DoF) than standard lower-order methods. In fact, the great geometrical flexibility allows to increase arbitrarily some regions of the fluid domain (for example in the wake of a body) while maintaining very large elements on the farfield.

In this work the DG code MIGALE has been employed to perform the *a posteriori* analysis. The incompressibility constrain is handled through the artificial compressibility flux approach [7], which recovers the hyperbolic nature of the Navier–Stokes equations through a properly modified (only at the mesh element interfaces) Riemann solver, while the viscous numerical fluxes are computed using the BR2 scheme [8]. The simulation has been performed on a computational grid made of 21366 mesh elements using a sixth order polynomial basis. The distribution of the elements and the order of the polynomials are chosen to reproduce the mesh properties of the F2 case shown so far. A Linearly-implicit Rosenbrock-type Runge-Kutta scheme coupled with a low storage matrix-free GMRES solver, preconditioned with a p-multigrid strategy, is used for the time integration. Such high-order implicit scheme allows to integrate efficiently the equations using a large time step size ($\Delta t = 0.05$) without adding sensible numerical dissipation to that of the space discretization.

5. Results of the Implicit Large Eddy Simulation

In this section a preliminary assessment of the Implicit Large Eddy Simulation approach to the flow around a rectangular cylinder is reported. Before addressing the statistics, it is interesting to analyse the instantaneous flow features reproduced by the coarse ILES approach. To this aim we consider the eduction scheme proposed by Jeong et al. [9] which is based on the second largest eigenvalue (λ_2) of the tensor

$$S_{ik}S_{kj} + \Omega_{ik}\Omega_{kj} \qquad (6)$$

where

$$S_{ij} = \frac{1}{2}\left(\frac{\partial u_i}{\partial x_j} + \frac{\partial u_j}{\partial x_i}\right)$$

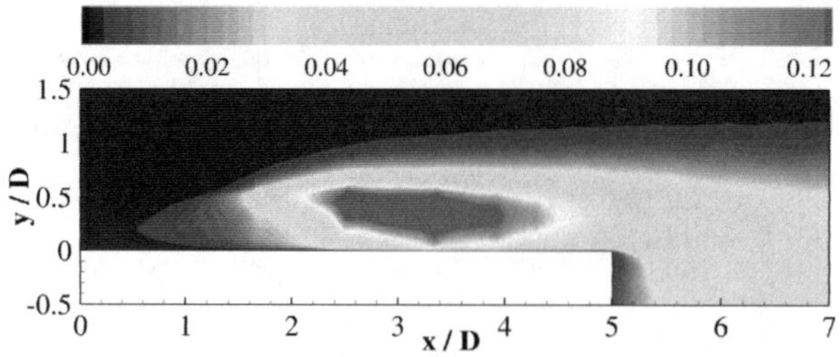

Figure 4. Implicit Large Eddy Simulation of the flow around a rectangular cylinder at $Re = 3000$. Iso-contours in the $(x - y)$-plane of turbulent kinetic energy. The discontinuities observed in the contours are associated with the Discontinuous Glaerkin approach used for the spatial discretization.

$$\Omega_{ij} = \frac{1}{2}(\frac{\partial u_i}{\partial x_j} - \frac{\partial u_j}{\partial x_i}) \tag{7}$$

are the symmetric and antisymmetric part of the velocity gradient tensor. This eduction scheme is known to accurately extract the three-dimensional pattern of vortical structures.

In figure 3, the instantaneous pattern taken by $\lambda_2 = -1.5$ is shown with iso-surfaces colored by the distance from the horizontal plane of symmetry of the rectangle. It is clear that the main turbulent structures of the flow are captured by the ILES approach. In particular, the development of almost two-dimensional spanwise rolls from the initially laminar leading-edge shear layer as a result of the well-known Kelvin-Helmholtz instability is found to be captured. From these spanwise vortices, the formation of hairpin-like structures [10, 11, 12] is also reproduced. These sparsely distributed structures, grow, burst and decay thus giving rise to a fully turbulent flow.

Let us now address how the ILES approach is able to predict the main energetic features of the flow by considering the behaviour of the turbulent kinetic energy. As shown in figure 4, the distribution of turbulent kinetic energy predicted by the ILES recovers the one of the DNS, see the top plot of figure 2 for comparison. Hence, accordingly with the analysis of the turbulent pattern shown so far, the main transitional and production phenomena of turbulence appear to be correctly reproduced by the ILES simulation. The main difference with respect DNS comes from the intensity of the processes. Indeed, we observe a systematic increase of the intensity of the turbulent fluctuations. As an example the maximum intensity of turbulent kinetic energy reached in DNS is $q_{max} \approx 0.123$ while in the ILES is $q_{max} \approx 0.177$.

6. Conclusions

The flow around a rectangular cylinder is recognized to be challenging for numerical simulations adopting turbulence models as shown by several works in literature, see Bruno et al. [2]. The reason is the simultaneous presence of laminar, transitional and turbulent regions within the flow. Here we try to preliminary assess the behaviour of modelling approaches in such a flow system by addressing first the role of the resolution. To this aim we use Direct Numerical Simulation data to quantify the degree of resolution required to resolve the main physical phenomena of the flow in the context of Large Eddy Simulation. In fact, by filtering the DNS data we defined a good compromise between the need of reduction of computational costs and the need of correctly reproduce the flow physics. In order to prove the effectiveness of the *a priori* results, we performed an actual LES simulation. In particular we implemented an implicit LES in order avoid possible modelling effect on the *a posteriori* results that are intended to focus solely to the effect of the mesh resolution. Indeed, the ILES has been performed by means of a Discontinuous Galerkin method whose numerical properties are known to be well-suited for ILES. The *a posteriori* results confirm that to reproduce the main physical phenomena of the flow the wall-parallel resolution should be on average of the order of $\Delta x^+ \approx 12$ and $\Delta z^+ \approx 21$ in the streamwise and spanwise directions, respectively.

References

[1] Cimarelli A and De Angelis E 2014 *Phys. Fluids* **26** 055103
[2] Bruno L, Salvetti M V and Ricciardelli F 2014 *J. Wind Eng. Ind. Aerodyn.* **126** 87–106
[3] Weller H, Tabor G, Jasak H and Fureby C 1998 *Comput. Phys.* **12** 620–631
[4] Moura R C, Sherwin S J and Peiró J 2015 *J. Comput. Phys.* **298** 695–710
[5] Bassi F, Botti L, Colombo A, Crivellini A, Ghidoni A and Massa F 2016 *Eur. J. Mech. B-Fluid* **55** 367–379
[6] Chapelier J B, De La Llave Plata M, Renac F and Lamballais E 2014 *Comput. Fluids* **95** 210–226
[7] Bassi F, Crivellini A, Di Pietro D A and Rebay S 2006 *J. Comput. Phys.* **218** 794–815
[8] Bassi F, Crivellini A, Rebay S and Savini M 2005 *Comput. Fluids* **34** 507–540
[9] Jeong J, Hussain F, Schoppa W and Kim J 1997 *J. Fluid Mech.* **332** 185–214
[10] Hourigan K, Thompson M C and Tan B T 2001 *J. Fluids Struct.* **15** 387–398
[11] Langari M and Yang Z 2013 *Phys. Fluids* **25** 074106
[12] Tenaud C, Podvin B, Fraigneau Y and Daru V 2016 *Int. J. Heat Fluid Fl.*

A Spalart–Allmaras local correlation–based transition model for Thermo–fluid dynamics

V D'Alessandro[1], F Garbuglia[1], S Montelpare[2] and A Zoppi[1]

[1] Dipartimento di Ingegneria Industriale e Scienze Matematiche
Università Politecnica delle Marche
Via Brecce Bianche, 60100 Ancona (AN), Italy
[2] Dipartimento di Ingegneria e Geologia
Università degli Studi "G. D'Annunzio" di Chieti–Pescara
Viale Pindaro 42, 65127, Pescara (PE), Italy

E-mail: v.dalessandro@univpm.it

Abstract. The study of innovative energy systems often involves complex fluid flows problems and the Computational Fluid-Dynamics (CFD) is one of the main tools of analysis. It is important to put in evidence that in several energy systems the flow field experiences the laminar-to-turbulent transition. Direct Numerical Simulations (DNS) or Large Eddy Simulation (LES) are able to predict the flow transition but they are still inapplicable to the study of real problems due to the significant computational resources requirements. Differently standard Reynolds Averaged Navier Stokes (RANS) approaches are not always reliable since they assume a fully turbulent regime. In order to overcome this drawback in the recent years some locally formulated transition RANS models have been developed. In this work, we present a local correlation–based transition approach adding two equations that control the laminar-to-turbulent transition process –γ and $\widetilde{Re_{\theta,t}}$ – to the well–known Spalart–Allmaras (SA) turbulence model. The new model was implemented within OpenFOAM code. The energy equation is also implemented in order to evaluate the model performance in thermal–fluid dynamics applications. In all the considered cases a very good agreement between numerical and experimental data was observed.

1. Introduction

This paper presents the latest developments of a Spalart–Allmaras local correlation based transition model, introduced in [1], for Thermal–Fluid Dynamics applications.

The aim of this work was to contribute to the development of CFD techniques in the prediction of the flow's laminar–to–turbulent transition. Although Direct Numerical Simulation (DNS) or Large Eddy Simulation (LES) are able to predict flow transition, the DNS approach is still inapplicable to the study of engineering interest problems due to the significant computational resources required. On the other hand, the applicability of the LES technique is very limited for the same reason. Moreover standard Reynolds Averaged Navier Stokes (RANS) models are not always reliable because they assume a fully turbulent regime. To overcome this drawback, some locally–formulated RANS transition models have been developed in recent years for general purpose unstructured CFD codes. They can be divided into two main classes: local correlation-based transition models (LCTM), [2, 3], and eddy viscosity phenomenological transition models, [4]. The first class includes the transitional effects by coupling standard RANS

IOP Conf. Series: Journal of Physics: Conf. Series **923** (2017) 012029 doi:10.1088/1742-6596/923/1/012029

models with further transport equations and empirical correlations. The second class of transition models, like the $k - k_L - \omega$ introduced by Walters et al. [4], is based on laminar kinetic energy modelling, see [5]. These models have demonstrated an acceptable accuracy in the simulation of laminar–to-turbulent transition for an ample range of applications. The main drawback of LCTM methods, also named γ–$\widetilde{\mathrm{Re}_{\theta,\mathrm{t}}}$ models, lies in the adoption of empirical correlations that are not typically applicable for a wide range of problems, whereas the $k - k_L - \omega$ technique has not always produced satisfactory results in flow cases characterized by large pressure gradient, see [6].

It should be noted that γ–$\widetilde{\mathrm{Re}_{\theta,\mathrm{t}}}$ models were initially coupled with the SST k–ω turbulence model by its developers, but the γ–$\widetilde{\mathrm{Re}_{\theta,\mathrm{t}}}$ model could be applied to other models too. In this work we present a coupling between the γ–$\widetilde{\mathrm{Re}_{\theta,\mathrm{t}}}$ approach and the Spalart–Allmaras (SA) turbulence model, [7]. In particular we propose an implementation based on the open–source library OpenFOAM. Our source code has been also released with a GNU GPL license to make public all the computational details behind the γ–$\widetilde{\mathrm{Re}_{\theta,\mathrm{t}}}$–SA model; the model and a test case are freely downloadable on `gitHub` at the following address: `https://github.com/vdalessa/gammaReThetatSA`. The main reason behind the choice to investigate the γ–$\widetilde{\mathrm{Re}_{\theta,\mathrm{t}}}$-SA approach is that the SA equation has produced very good results in the computation of external flows. Moreover the model also has lower computational costs than the SST k–ω model.

We have to remark that in this work we face the computation of the flow field past several wind turbine airfoil in transitional regime as well as conjugate heat transfer problems.

The paper is organized as follows. Sec. 2 presents the equations governing the transition model considered here. Sec. 3 briefly describes the discretization technique. Sec. 4 is devoted to the numerical results and Sec. 5 contains the conclusions.

2. Governing equations

The complete set of our flow governing equations can be written as

$$\begin{aligned}
&\boldsymbol{\nabla}\cdot\mathbf{u}=0,\\
&\frac{\partial\mathbf{u}}{\partial t}+\boldsymbol{\nabla}\cdot(\mathbf{u}\otimes\mathbf{u})+\boldsymbol{\nabla}p-\boldsymbol{\nabla}\cdot\left((\nu+\nu_t)\left(\boldsymbol{\nabla}\mathbf{u}+\boldsymbol{\nabla}\mathbf{u}^T\right)\right)=\mathbf{0},\\
&\frac{\partial\mathrm{T}}{\partial t}+\boldsymbol{\nabla}\cdot(\mathbf{u}\mathrm{T})-\boldsymbol{\nabla}\cdot\left(\left(\frac{\nu}{\mathrm{Pr}}+\frac{\nu_t}{\mathrm{Pr_t}}\right)\boldsymbol{\nabla}\mathrm{T}\right)=0,\\
&\frac{\partial\tilde{\nu}}{\partial t}+\boldsymbol{\nabla}\cdot(\mathbf{u}\tilde{\nu})-\mathrm{P}_{\tilde{\nu}}+\mathrm{D}_{\tilde{\nu}}-\frac{c_{b2}}{\sigma}\boldsymbol{\nabla}\tilde{\nu}\cdot\boldsymbol{\nabla}\tilde{\nu}-\frac{1}{\sigma}\boldsymbol{\nabla}\cdot\left((\nu+\tilde{\nu})\,\boldsymbol{\nabla}\tilde{\nu}\right)=0,
\end{aligned}\tag{1}$$

where $\mathbf{u}$ is the velocity vector, $p = P/\rho$ is the pressure divided by the density, T is temperature and d is the distance from the nearest wall; besides ν is the kinematic viscosity and Pr the Prandtl number. The turbulent viscosity, ν_t, needed to take into account the turbulence, is computed according to the $\tilde{\nu}$ variable as

$$\nu_t = f_{v1}\tilde{\nu}, \tag{2}$$

on the other hand the turbulent Prandtl number, $\mathrm{Pr_t}$, is introduced to model the turbulent thermal diffusivity. The production and destruction terms appearing in the $\tilde{\nu}$ transport equation are defined as follows:

$$\begin{aligned}
&\mathrm{P}_{\tilde{\nu}}=\gamma_{\mathrm{eff}}c_{b1}\tilde{S}\tilde{\nu},\\
&\mathrm{D}_{\tilde{\nu}}=\max\left(\min\left(\gamma,0.5\right),1.0\right)\left[c_{w1}f_w\left(\frac{\tilde{\nu}}{d}\right)^2\right].
\end{aligned}\tag{3}$$

The term γ_{eff} in eq. 3 is a term devoted to model the separation-induced transition and it is defined as follows:

$$\gamma_{\text{eff}} = \max\left(\gamma, \gamma_{\text{sep}}\right) \tag{4}$$

with

$$\gamma_{\text{sep}} = \min\left(2.0 \cdot \max\left[0, \left(\frac{\text{Re}_\nu}{3.235\text{Re}_{\theta,c}}\right) - 1\right] \text{F}_{reattach}, 2.0\right) \text{F}_{\theta,t} \tag{5}$$

and

$$\text{F}_{reattach} = \exp\left(-\frac{\text{R}_\text{T}}{20}\right)^4. \tag{6}$$

The following closure functions are now introduced to complete the definition of the SA equation given in eq. 1:

$$\begin{aligned}
f_{v1} &= \frac{\chi^3}{\left(\chi^3 + c_{v1}^3\right)}, & f_{v2} &= 1 - \frac{\chi}{(1 + \chi f_{v1})}, \\
g &= r + c_{w2}\left(r^6 - r\right), & f_w &= g\left[\frac{1 + c_{w3}^6}{g^6 + c_{w3}^6}\right]^{\frac{1}{6}}, \\
\tilde{S} &= \left[\Omega + \min\left(0, S - \Omega\right)\right] + \frac{\tilde{\nu}}{k^2 d^2} f_{v2}, & r &= \begin{cases} r_{max} & \frac{\tilde{\nu}}{\tilde{S}k^2d^2} < 0 \\ \min\left(\frac{\tilde{\nu}}{\tilde{S}k^2d^2}, r_{max}\right) & \frac{\tilde{\nu}}{\tilde{S}k^2d^2} \geq 0 \end{cases},
\end{aligned} \tag{7}$$

where $\chi = \tilde{\nu}/\nu$ is the dimensionless turbulent variable, $\Omega = \sqrt{2\mathbf{W}:\mathbf{W}}$ is the vorticity tensor module, $S = \sqrt{2\mathbf{D}:\mathbf{D}}$ is the strain rate tensor module and $\tilde{S}$ is a function of both the vorticity magnitude, Ω, and $\tilde{\nu}$. The r function is estimated as in [8] to increase the numerical stability, and r_{max} is a constant positive value typically set at 10. Finally, to fully define the SA model, the following standard closure constants are adopted

$$c_{b1} = 0.1355, \quad c_{b2} = 0.622, \quad c_{v1} = 7.1, \tag{8}$$

$$\sigma = 2/3, \quad c_{w1} = \frac{c_{b1}}{k^2} + \frac{(1 + c_{b2})}{\sigma}, \tag{9}$$

$$c_{w2} = 0.3, \quad c_{w3} = 2, \quad k = 0.41. \tag{10}$$

The transport equations needed to model the transition are:

$$\begin{aligned}
&\frac{\partial\gamma}{\partial t} + \boldsymbol{\nabla}\cdot(\mathbf{u}\gamma) = \text{P}_\gamma - \text{D}_\gamma + \boldsymbol{\nabla}\cdot\left[\left(\nu + \frac{\nu_t}{\sigma_f}\right)\boldsymbol{\nabla}\gamma\right], \\
&\frac{\partial\widetilde{\text{Re}_{\theta,\text{t}}}}{\partial t} + \boldsymbol{\nabla}\cdot\left(\mathbf{u}\widetilde{\text{Re}_{\theta,\text{t}}}\right) = \text{P}_{\theta,\text{t}} + \boldsymbol{\nabla}\cdot\left[\sigma_{\theta,\text{t}}\left(\nu + \nu_t\right)\boldsymbol{\nabla}\widetilde{\text{Re}_{\theta,\text{t}}}\right].
\end{aligned} \tag{11}$$

The source terms in the γ equation are defined as:

$$\begin{aligned}
\text{P}_\gamma &= c_{a1} S \left[\gamma \text{F}_{\text{onset}}\right]^{0.5}\left(1 - c_{e1}\gamma\right)\text{F}_{\text{length}}, \\
\text{D}_\gamma &= c_{a2}\Omega\gamma\text{F}_{\text{turb}}\left(c_{e2}\gamma - 1\right),
\end{aligned} \tag{12}$$

in P_γ the term F_{onset} is computed as:

$$\text{F}_{\text{onset}} = \max\left(\text{F}_{\text{onset},2} - \text{F}_{\text{onset},3}, 0\right) \tag{13}$$

with

$$\begin{aligned}
\mathrm{F_{onset,2}} &= \min\left(\max\left(\mathrm{F_{onset,1}}, \mathrm{F_{onset,1}}^4\right), 4\right), \\
\mathrm{F_{onset,3}} &= \max\left(2 - \left(\frac{\mathrm{R_T}}{2.5}\right)^3, 0\right), \\
\mathrm{F_{onset,1}} &= \frac{\mathrm{Re}_\nu}{2.193\mathrm{Re}_{\theta,c}}.
\end{aligned} \tag{14}$$

In eq. 14 the terms Re_ν and $\mathrm{R_T}$ are obtained as follows:

$$\mathrm{Re}_\nu = \frac{Sd^2}{\nu} \qquad \mathrm{R_T} = \frac{\nu_t}{\nu}. \tag{15}$$

Note that the parameter $\mathrm{R_T}$ is redefined because, for the k–ω model, it requires the estimation of ω. The definition based on the viscosity ratio,ν_t/ν, enables the SA equation to be adopted. The aspects concerning the terms: $\mathrm{F_{length}}$ and $\mathrm{Re}_{\theta,c}$ are described in the next subsection. For the destruction term, D_γ, on the other hand, the coefficient $\mathrm{F_{turb}}$ is defined as:

$$\mathrm{F_{turb}} = \exp\left(-\frac{\mathrm{R_T}}{4}\right)^4, \tag{16}$$

As for the source terms in the transport equation for $\widetilde{\mathrm{Re}_{\theta,\mathrm{t}}}$, $\mathrm{P}_{\theta,\mathrm{t}}$, the following equation is adopted:

$$\mathrm{P}_{\theta,\mathrm{t}} = \frac{c_{\theta,t}}{\mathrm{T}}\left(\mathrm{Re}_{\theta,t} - \widetilde{\mathrm{Re}_{\theta,\mathrm{t}}}\right)\left(1 - \mathrm{F}_{\theta,t}\right). \tag{17}$$

In eq. 17 the last term $\mathrm{F}_{\theta,t}$ is defined as:

$$\mathrm{F}_{\theta,t} = \min\left(\max\left(\exp\left(-\frac{|\mathbf{u}|^2}{375\Omega\nu\widetilde{\mathrm{Re}_{\theta,\mathrm{t}}}}\right)^4, 1 - \left(\frac{\gamma - 1/c_{e2}}{1 - 1/c_{e2}}\right)^2\right), 1.0\right). \tag{18}$$

The term T appearing in the source term of the $\widetilde{\mathrm{Re}_{\theta,\mathrm{t}}}$ equation is also defined as follows: $500\nu/|\mathbf{u}|^2$. Finally, the computation of $\mathrm{Re}_{\theta,t}$ in eq. 17 is discussed, together with the $\mathrm{F_{length}}$ coefficient, in the next subsection.
For the turbulence model, the following closure constants are adopted in order to close eq. 11

$$c_{a1} = 2.0, \quad c_{a2} = 0.06, \quad c_{e1} = 1.0, \tag{19}$$
$$c_{e2} = 50, \quad c_{\theta,t} = 0.03, \quad \sigma_f = 1.0, \tag{20}$$
$$\sigma_{\theta,t} = 2.0. \tag{21}$$

2.1. Empirical correlations in the model

Like other γ–$\widetilde{\mathrm{Re}_{\theta,\mathrm{t}}}$ approaches available in the literature, the present model contains three empirical correlations needed to compute $\mathrm{Re}_{\theta,t}$, $\mathrm{Re}_{\theta,c}$ and $\mathrm{F_{length}}$.
$\mathrm{Re}_{\theta,c}$, which appears in eq. 14, is the critical Reynolds number where the intermittency, γ, starts to increase in the boundary layer. This typically occurs upstream from the transition Reynolds number, $\mathrm{Re}_{\theta,t}$. This element relates to the delay between when the turbulence starts and when appreciable turbulence levels are reached within the boundary layer. Note that this last feature is essential to obtain a significant change in the laminar velocity profile. $\mathrm{F_{length}}$, appearing in the production term of the γ transport equation, is an empirical correlation that controls the length

of the transition region. In this paper for $\mathrm{Re}_{\theta,t}$ we adopt a correlation developed by Menter et al. [2]:

$$\mathrm{Re}_{\theta,t} = \begin{cases} \left(1173.51 - 589.428 \cdot Tu + 0.2196/Tu^2\right) F\left(\lambda_\theta\right) & Tu \leq 1.3 \\ 331.5\left(Tu - 0.5668\right)^{-0.671} F\left(\lambda_\theta\right) & Tu > 1.3 \end{cases}, \tag{22}$$

$$F\left(\lambda_\theta\right) = \begin{cases} 1 + \left[12.986\lambda_\theta + 123.66\lambda_\theta^2 + 405.689\lambda_\theta^3\right] \exp\left(-\left(\frac{Tu}{1.5}\right)^{1.5}\right) & \lambda_\theta \leq 0 \\ 1 + 0.275\left[1 - \exp\left(-35\lambda_\theta\right)\right] \exp\left(-\frac{Tu}{0.5}\right) & \lambda_\theta > 0 \end{cases}. \tag{23}$$

It is important to note that the correlations in eq. 22, 23 contain the turbulence intensity Tu. In the framework of the k–ω model, Tu can be computed using the solution to k equation. In this work, we adopt the approach introduced in [9, 10]. Specifically, we establish that $Tu = Tu_\infty$ for all the points of the flow field. Moreover $\mathrm{Re}_{\theta,t}$ is computed by iterating on the value of θ_t, since $\mathrm{Re}_{\theta,t}$ is a function of θ_t itself because of the presence of λ_θ. Differently for $\mathrm{Re}_{\theta,c}$ and $\mathrm{F_{length}}$ we use the correlations introduced by Malan et al. [11]:

$$\mathrm{Re}_{\theta,c} = \min\left(0.615\widetilde{\mathrm{Re}_{\theta,\mathrm{t}}} + 61.5, \widetilde{\mathrm{Re}_{\theta,\mathrm{t}}}\right), \tag{24}$$

$$\mathrm{F_{length}} = \min\left(\exp\left(7.168 - 0.01173\widetilde{\mathrm{Re}_{\theta,\mathrm{t}}}\right) + 0.5, 300\right). \tag{25}$$

2.2. Boundary conditions

Standard boundary conditions are adopted for $\tilde{\nu}$: $\tilde{\nu}_\infty = 3\nu$ at the free stream and $\tilde{\nu} = 0$ at the wall, while the boundary condition for γ at the wall is zero normal flux. At the inlet, the value of γ is 1. The boundary condition for $\widetilde{\mathrm{Re}_{\theta,\mathrm{t}}}$ at the wall is zero flux, while at the inlet $\widetilde{\mathrm{Re}_{\theta,\mathrm{t}}}$ is calculated from the specific empirical correlation based on the inlet turbulence intensity.

It is also very important to note that, in order to capture the laminar and transitional boundary layers correctly, the grid must have a viscous sub-layer scaled first cell height, y^+, of approximately 1 [3]. The value of y^+ is estimated as $y^+ = \frac{u_\tau}{\nu} y_c$, where $u_\tau = \sqrt{\tau_w/\rho}$ is the friction velocity, τ_w is the viscous stress component measured at the wall, and y_c is the height of the cells next to the wall.

3. Numerical solution

The governing equations described in Sec. 2 are solved numerically by adopting a finite volume method (FVM). In particular, we used `simpleFoam`, which is the steady solver for incompressible flows available in the official releases of the `OpenFOAM` (Open-source Field Operation and Manipulation) code [12]. which is unstructured finite volume solver released under the GNU Public License (GPL). The code's object-oriented structure enables users to implement their own models and solvers in the baseline codes with relatively little effort (see for example [13, 14]). `simpleFoam` uses the well-established SIMPLE algorithm [15] for pressure–velocity decoupling and, like the other standard solvers, it is based on a colocated FVM approach. The Rhie–Chow correction is used to remove oscillations in the solutions [16]. For conjugate heat transfer (CHT) problems a fully coupled approach for energy equation is adopted.

For all the computations presented in this paper the diffusive terms and pressure gradients were approximated with second-order accurate central schemes. The convective terms for momentum, energy and turbulence equations were handled with a second order accurate linear–upwind scheme. For the linear solvers a preconditioned bi-conjugate gradient method (PBiCG) with the DILU preconditioner was used to solve the discretized momentum, $\tilde{\nu}$, γ and $\widetilde{\mathrm{Re}_{\theta,\mathrm{t}}}$ equations. A preconditioned conjugate gradient method (PCG) with a diagonal incomplete-Cholesky preconditioner was adopted instead for the pressure, while energy equation a Biconjugate Gradient

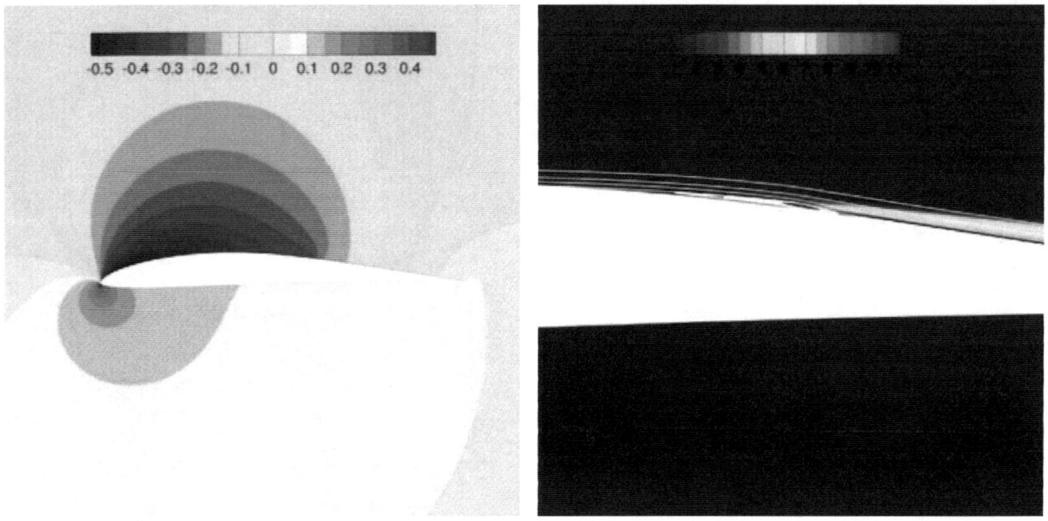

(a) Pressure contour plot (b) $\tilde{\nu}/\nu$ contour plot and streamlines

Figure 1. E387, Re $= 3 \cdot 10^5$. $\alpha = 4°$.

Stabilized (BiCGStab) method with Cholesky preconditioner was used. In particular, a local accuracy of 10^{-7} was established for the pressure, whereas other linear systems were considered as converged when the residuals reached the machine precision.

4. Results

4.1. Eppler 387 airfoil

In this sub-section we consider the flow over an Eppler 387 (E387) airfoil; see Fig. 1 for a representation of the solution. This case was chosen as a test case involving a separation–induced transition. The E387 airfoil is particularly appealing for this purpose because McGhee et al. [17] have published a large experimental dataset also involving the distribution of c_p along the airfoil surface at several Re numbers. Indeed it was used in several paper for validation purposes, [18, 19].

The solutions reported here were computed using a structured computational grid having about $n_c = 2.2 \cdot 10^5$ and a C-shaped domain. The inflow and outflow boundaries were placed at about 10 times the airfoil chord, c, from the airfoil. The cells were clustered near the airfoil surface and the first cell height was arranged in order to obtain $O(y^+) \approx 1$. The effect of the angle of attack, α, on the pressure distribution, for the case at Re $= 3 \cdot 10^5$, can be seen in Figure 2.

For all the reported angles of attack, the results obtained with the well–established XFOIL code [20] (which is a recognized method for use in airfoil design) have also been included. The data presented here were obtained using the default settings for the critical amplification factor, $N_{crit} = 9$. It is very easy to see that XFOIL tends to overestimate the c_p value with respect to the experimental results, while the RANS transition model proposed here is globally very consistent with the experimental findings. The reported results reveal a good agreement between the experimental and numerical data in the context of laminar separation bubble (LSB) simulation.

The force coefficients plotted at Re $= 3 \cdot 10^5$ from Fig. 3, confirm the very good consistency between the numerical and experimental data. In all these figures, for further comparison, the results obtained with the XFOIL code have been included too. Prior to the stall region, the force

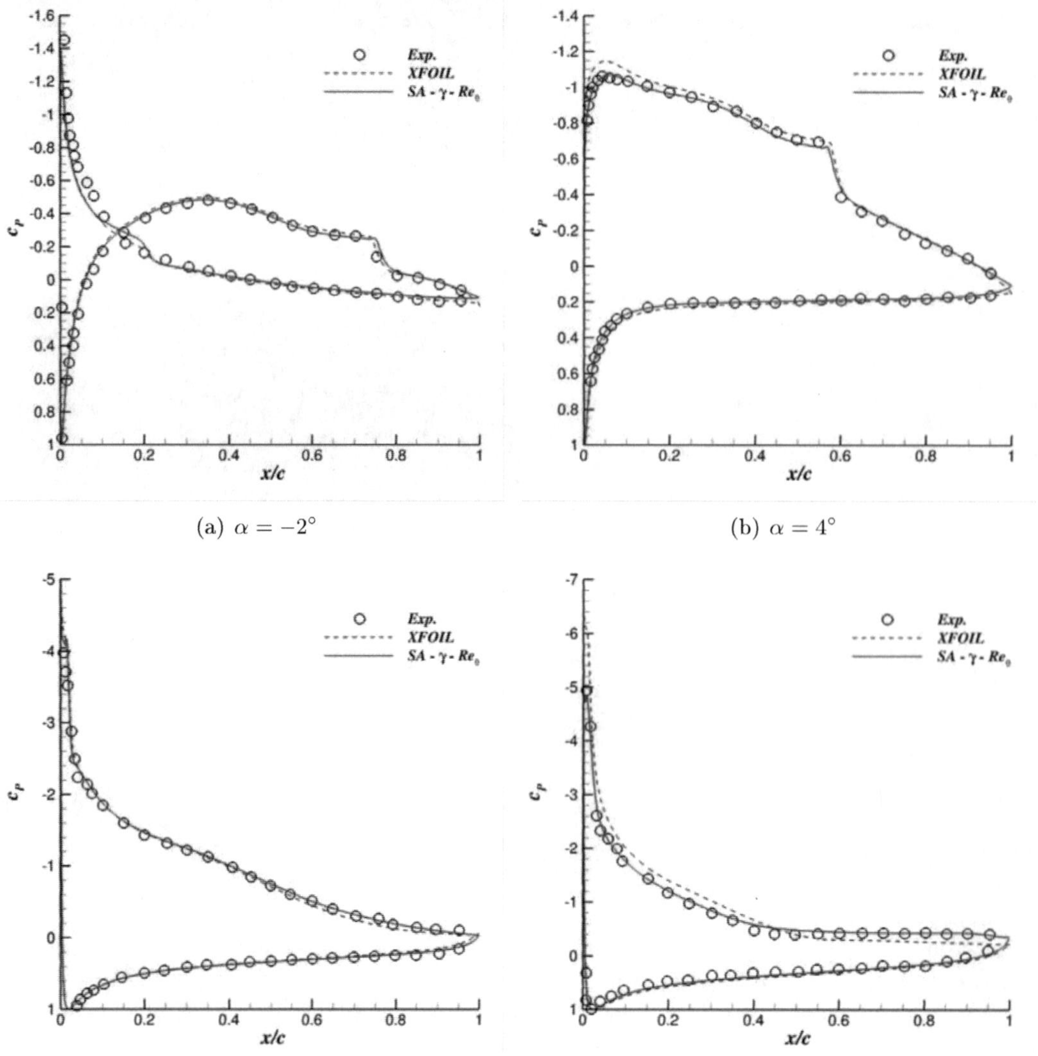

(a) $\alpha = -2°$ (b) $\alpha = 4°$

(c) $\alpha = 10°$ (d) $\alpha = 14°$

Figure 2. E387, $\mathrm{Re} = 3 \cdot 10^5$. Pressure coefficients at several angles of attack.

coefficients computed by the two completely different approaches are clearly always in very good agreement. Our γ–$\widetilde{\mathrm{Re}}_{\theta,\mathrm{t}}$–SA model is able to capture airfoil performance in the stall region very well too.

4.2. SD7003 airfoil

The Selig–Donovan (SD) 7003 airfoil was also investigated at $\mathrm{Re} = 6 \cdot 10^4$ and $\alpha = 4°$, our numerical solutions are depicted in Fig. 4. The results obtained in this work were compared with LES and ILES computations performed by Catalano and Tognaccini, [21] and Galbraith and Visbal, [22] respectively. Very recently Islam et. al. [23] performed RANS computations based on SST k–ω and a local correlation based transition model. We have also compared our results with their ones.

The solutions reported in Fig. 5 were computed using a structured computational grid having

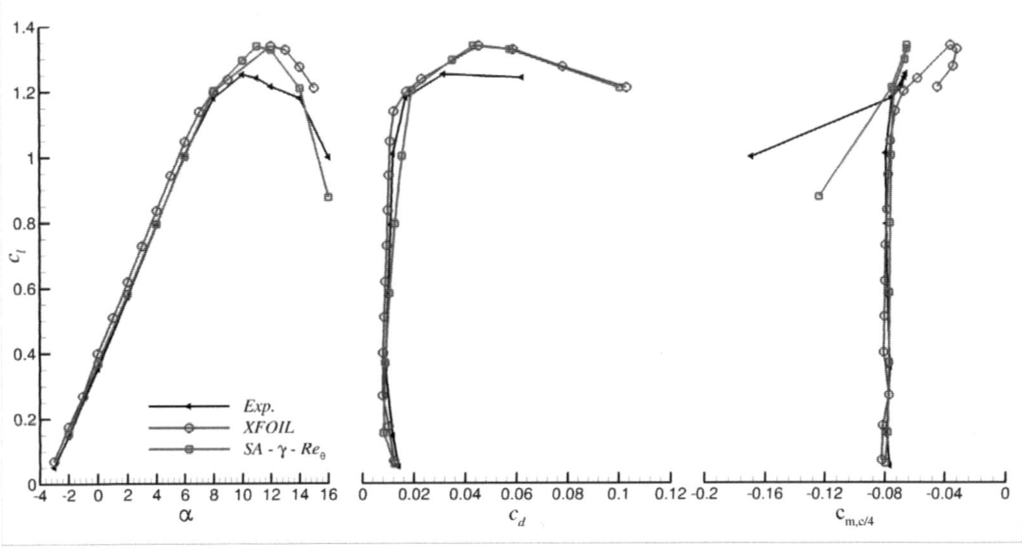

Figure 3. E387, $\mathrm{Re} = 3 \cdot 10^5$. Force coefficients.

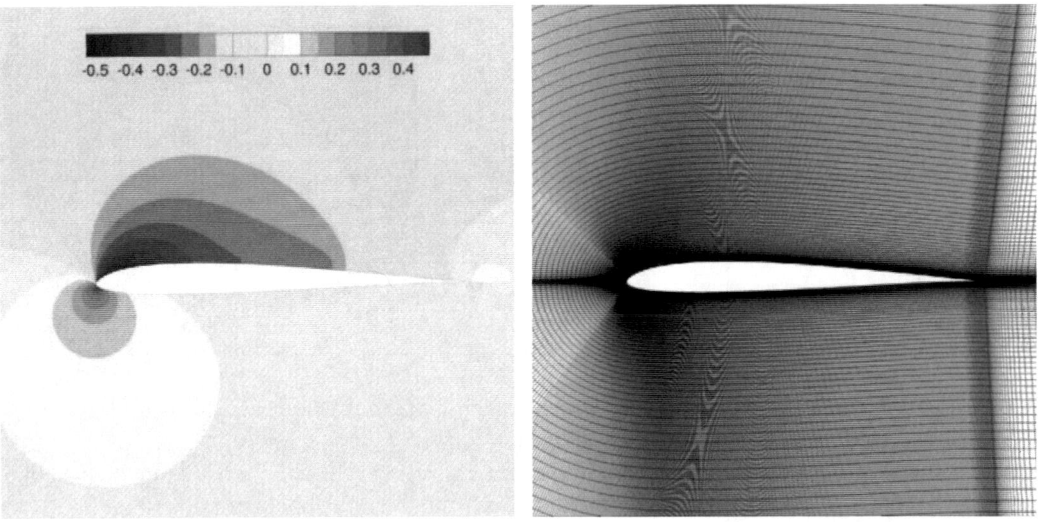

(a) Pressure contour plot (b) Computational grid

Figure 4. SD7003, $\mathrm{Re} = 6 \cdot 10^4$. $\alpha = 4°$.

similar features to that employed for the E387 airfoil. Our pressure coefficient distribution is very close to Islam et. al. [23], however there are noticeable differences in the LSB zone with literature LES and ILES results. This result demonstrates objective evidence of LCTM based transition approaches in the prediction of separation–induced transition phenomena at very low Reynolds numbers. In Fig. 5(b) the skin friction coefficient, C_f, distribution around the airfoil is showed. The LSB is situated between the separation point, x_s, and the and reattachment point, x_r; in Fig. 5(b) is clearly showed that between x_s and x_r we have $C_f < 0$, thus the flow experiences a recirculation region. We have to put in evidence that RANS solutions over-predict the reattachment point significantly when compared with the ILES/LES results. SA and k–ω

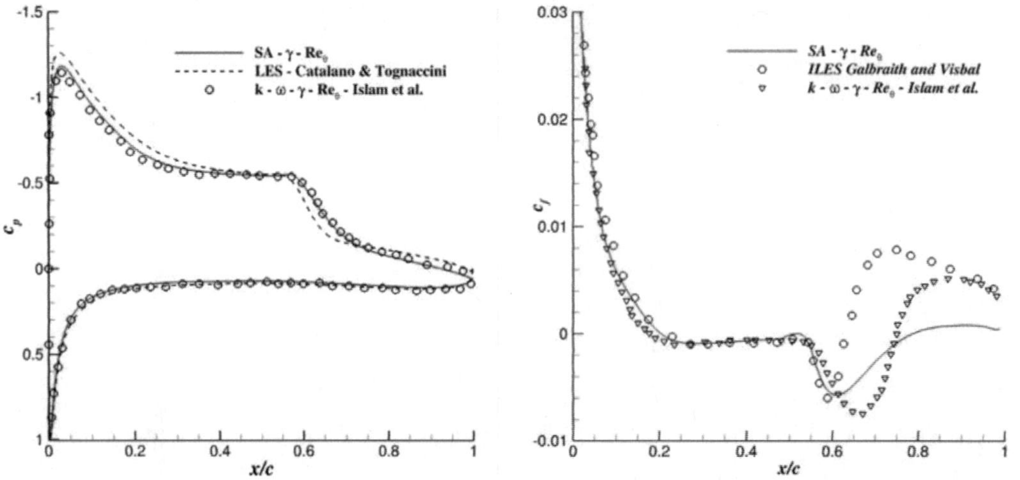

(a) Pressure coefficient (b) Skin friction coefficient

Figure 5. SD7003, $\mathrm{Re} = 6 \cdot 10^4$. $\alpha = 4°$.

Table 1. SD 7003, $\mathrm{Re} = 6 \cdot 10^4$, $\alpha = 4°$

	C_D	C_L	C_M
γ–$\widetilde{\mathrm{Re}_{\theta,t}}$ –SA – Present work	0.0139	0.5320	0.0244
DNS – Carton de Wiart and Hillewaert, [24]	0.0196	0.6020	–
ILES – Galbraith and Visbal, [22]	0.021	0.590	–
γ–$\widetilde{\mathrm{Re}_{\theta,t}}$ –SST – Islam et. al. [23]	0.0255	0.4782	0.0328
k–k_L–ω – Islam et. al. [23]	0.0188	0.5150	0.0387
Exp. – Selig et al. $(\mathrm{Re} = 6.14 \cdot 10^4)$, [25]	0.0166	0.6038	

models show similar x_r values, however it is also evident that their C_f results show sensible differences in the both LSB and turbulent attached region.

In Table 1, aerodynamic coefficients obtained from this research work with are compared with literature data. It can be seen that our approach exhibits a good agreement with k–k_L–ω data published by Islam et al. [23]. Differently DNS/LES based approach finds a better agreement with experimental data especially for what concerns the lift coefficient.

4.3. Conjugate Heat Transfer over a flat plate

In this subsection we present the CHT modeling of a case derived from the commonly used Schubauer and Klebanoff benchmark [26]. The flow problem is a zero pressure gradient flat plate with a very low free-stream turbulence intensity, *i.e.* $Tu = 0.03\%$, so it is used to test transition models in natural transition conditions.

The present calculations were performed on a computational mesh having two domains, one for a fluid and another for a solid. The fluid domain and the computational grid adopted for its geometric discretizazion were the same used in [1]. The solid one was represented by a flat plate with a thickness of 0.015 m; at the fluid interface the solid's grid shares the same cells' distribution used for the fluid domain, along its thickness the mesh elements were clustered

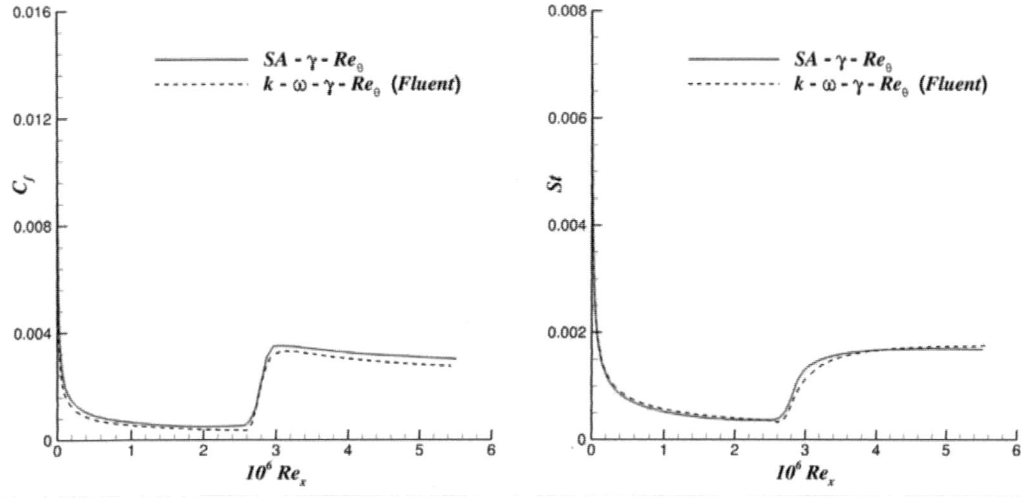

(a) Skin friction coefficient

(b) Local Stanton number

Figure 6. CHT over a flat plate.

near the fluid–solid interface. The bottom side of the plate was cooled, it was established by constant temperature $T = 300\ K$; the incoming wind temperature was fixed to 293.15 K. For the fluid domain we have adopted a Prandtl number equal to 0.71, on the other hand the solid was represented by a glass, *i.e.* $\rho = 2500\ Kg/m^3$, $c = 789\ J/kgK$ and $\lambda = 0.55\ W/mK$.
Our results were obtained using an energy equation coupled CHT solver available within `foam-extend` code which is a fork of the OpenFOAM library (developed with the aim to integrate community contributions). Hence in this paper, in order to perform CHT computations, we have also ported the model described in Sec. 2 to `foam-extend v4.0`.
For the sake of validation we have compared our results with k–ω–γ–$\widetilde{\mathrm{Re}_{\theta,\mathrm{t}}}$ approach available within `Ansys-Fluent v16.0` code. Since k–ω produces the decay of turbulence from the inlet to the flat plate leading edge the free–stream turbulence was increased up to $Tu = 0.2\%$. This element, in some sense, can be also considered a further advantage of the SA–γ–$\widetilde{\mathrm{Re}_{\theta,\mathrm{t}}}$ model. Indeed, our modeling technique fully transports Tu avoiding the calibration process required by the turbulence decay.

In Fig. 6 we present the obtained the results in terms of skin–friction coefficient, $C_f = 2\tau_w/\left(\rho u_\infty^2\right)$, and Stanton number, $St = h/(\rho c_p u_\infty)$. The dimensionless parameters reported in Fig. 6(a) and Fig. 6(b) show very good agreement, thus we can retain the presented CHT approach sufficiently reliable.

5. Conclusions

In this work, we have presented a γ–$\widetilde{\mathrm{Re}_{\theta,\mathrm{t}}}$ RANS model for transitional flows based on the SA equation. The main reasons behind this approach are as follows: (i) the SA model offers very reliable results for external flow applications; and (ii) the SA model has lower computational costs than the SST k–ω equations.

We obtained a good feedback from our model's implementation in predicting airfoil performance, as well as in the computation of a CHT problem on a zero pressure gradient flat plate. In all cases, our model implementation showed that the model is effective in predicting flow feature even in the airfoils stall region. Future research will focus on extending the γ–$\widetilde{\mathrm{Re}_{\theta,\mathrm{t}}}$–SA model to the Detached Eddy Simulation technique.

References

[1] D'Alessandro V, Montelpare S, Ricci R and Zoppi A 2017 *Energy* **130** 402 – 419

[2] Menter F, Langtry R, Likki S, Suzen Y, Huang P and Volker S 2006 *J Turbomach* **128** 413–422

[3] Menter F, Langtry R and Volker S 2006 *Flow Turbul and Combust* **77** 277–303

[4] Walters D and Cokljat D 2008 *J Fluid Eng–T ASME* **130** 1214011–12140114

[5] Mayle R and Schulz A 1991 *ASME J of Turbomach* **119** 405–411

[6] Cutrone L, Palma P D, Pascazio G and Napolitano M 2007 *Int J Heat Fluid Fl* **28** 161 – 177 ISSN 0142-727X

[7] Spalart P and Allmaras S 1994 *La Recherche Aérospatiale* **1** 5–21

[8] Crivellini A, D'Alessandro V and Bassi F 2013 *J Comput Phys* **241** 388 – 415

[9] Medida S and Baeder J 2011 *American Helicopter Society 67th Annual Forum*

[10] Medida S 2014 *Correlation-based Transition Modeling for External Aerodynamic Flows* Ph.D. thesis University of Maryland

[11] Malan P, Suluksna K and Juntasaro E 2009 *47th AIAA Aerospace Sciences Metting* (Orlando, FL)

[12] Weller H, Tabor G, Jasak H and Fureby C 1998 *Comput Phys* **12** 620–631

[13] D'Alessandro V, Montelpare S and Ricci R 2016 *Comput Fluids* **136** 152–156

[14] Silva L and Lage P 2011 *Comput Chem Eng* **35** 2653–2666

[15] Patankar S V 1980 *Numerical heat transfer and fluid flow* Series in computational methods in mechanics and thermal sciences (Washington: Hemisphere Pub. Corp. New York) ISBN 0-07-048740-5 URL http://opac.inria.fr/record=b1085925

[16] Ferziger J and Peric M 1999 *Computational Methods for Fluid Dynamics* (Springer)

[17] McGhee R, Walker B and Millard B 1988 Experimental Results for Eppler 387 Airfoil at low Re numbers in Langley Low Turbulence pressure tunnel Tech. Rep. TM 4062 NASA

[18] Crivellini A and D'Alessandro V 2014 *Int J Heat Fluid Fl* **47** 70–83

[19] Crivellini A, D'Alessandro V, Di Benedetto D, Montelpare S and Ricci R 2014 *J Phys: Conf Ser* **501**

[20] Drela M 2002 XFOIL Users Guide, Versione 6.94 MIT Aeronautics and Astronautics Department

[21] Catalano P and Tognaccini R 2011 *Aerospace Science and Technology* **15** 615 – 626 ISSN 1270-9638

[22] Galbraith V and Visbal M 2008 *Forty-sixth AIAA Aerospace Sciences Meeting and Exhibit* (Reno, NV)

[23] Islam M, Langfeldt F, Furst J and Wood D 2017 *IOP Conf Ser: Mat Sci Eng* **184**

[24] Carton de Wiart C and Hillewaert 2012 *7th Int Conf on Comput Fl Dyn* (Hawaii)

[25] Selig M, Guglielmo J J, Broeren A P and P G 1995 *Summary of low speed airfoil data vol 1* (SoarTech Publications)

[26] Schubauer G and Klebanoff P 1955 Contributions on the mechanics of boundary–layer transition Tech. Rep. NACA–N–3489 NASA

Study of the flow field past dimpled aerodynamic surfaces: numerical simulation and experimental verification

L Binci[1], G Clementi[1], V D'Alessandro[1], S Montelpare[2] and R Ricci[1]

[1] Dipartimento di Ingegneria Industriale e Scienze Matematiche
Università Politecnica delle Marche
Via Brecce Bianche, 60100 Ancona (AN), Italy
[2] Dipartimento di Ingegneria e Geologia
Università degli Studi "G. D'Annuzio" di Chieti–Pescara
Viale Pindaro 42, 65127, Pescara (PE), Italy

E-mail: l.binci@univpm.it

Abstract. This work presents the study of the flow field past of dimpled laminar airfoil. Fluid dynamic behaviour of these elements has been not still deeply studied in the scientific community. Therefore Computational Fluid-Dynamics (CFD) is here used to analyze the flow field induced by dimples on the NACA 64-014A laminar airfoil at $\mathrm{Re} = 1.75 \cdot 10^5$ at $\alpha = 0°$. Reynolds Averaged Navier–Stokes (RANS) equations and Large-Eddy Simulations (LES) were compared with wind tunnel measurements in order to evaluate their effectiveness in the modeling this kind of flow field. LES equations were solved using a specifically developed OpenFOAM solver adopting an L–stable Singly Diagonally Implicit Runge–Kutta (SDIRK) technique with an iterated PISO-like procedure for handling pressure-velocity coupling within each RK stage. Dynamic Smagorinsky subgrid model was employed. LES results provided good agreement with experimental data, while RANS equations closed with k–ω–γ–$\widetilde{\mathrm{Re}}_{\theta,t}$ approach overstimates laminar separation bubble (LSB) extension of dimpled and undimpled configurations. Moreover, through skin friction coefficient analysis, we found a different representation of the turbulent zone between the numerical models; indeed, with RANS model LSB seems to be divided in two different parts, meanwhile LES model shows a LSB global reduction.

1. Introduction

In the recent years the increasing interest of governments to diversification of electrical production source is realized with renewable sources plant installation increase. The wind power sector is one of the most developed, thanks also to the innovative construction techniques that have led to wind turbines size growth and a consequent electricity production optimization [1]. It is clear that the development of high efficiency blades still plays a central role in the design of new generation wind turbine. Laminar airfoils can satisfy thees requirements: indeed thees airfoils operate for a certain angles of attack without increasing their aerodynamic drag. However, such airfoils are often affected by a laminar separation bubble (LSB) which produces a reduction of the aerodynamic performance. It is worth nothing that the adoption of dimples to reduce the drag is credible since the accidental discovery of drag reduction by surface roughening on golf balls. Nevertheless many papers have been devoted to study dimples behavior for bluff body flows and heat transfer enhancement in internal flows; we can not find in the present literature any study that addresses the dimples effect on aerodynamic surfaces for transitional flows. However few researches provide some information about flow structure due to dimples on a channel surface and the influence of dimple depths, [2, 3]. Moreover Lake, [4], in his PhD thesis, conducted experimental tests on a gas turbine blade cascade in order to estimate the best shape and position of the dimples elements. So the main aim of this research is to deeply investigate the dimples effect on aerodynamic surfaces. In particular our efforts have been oriented on low Reynolds number operating airfoils which are of interest for wind energy applications. In particular the impact of dimples on the reduction of LSB was faced. Therefore the flow field past a dimpled NACA 64–014A laminar airfoil was carefully studied. This airfoil was selected because its abrupt pressure recovery and the systematic presence of a LSB. Hence this element allows to evaluate dimples performance about the LSB reduction. The airfoil was studied at $\alpha = 0°$ and $\mathrm{Re} = 1.75 \cdot 10^5$: the dimpled configuration consists in one row of these elements placed at 55 % of chord length with one dimple diameter spacing in the span-wise direction. Firstable Reynolds Averaged Navier–Stokes (RANS) solution were obtained using the well–known commercial CFD code `Star-CCM+`. On the other hand, Large-Eddy Simulations (LES) were performed in order to investigate the flow mechanism in the dimple zone and to reconstruct the flow features in the downstream region. LSB modeling is well recognized challenging task in the CFD community and in the present literature we can find several paper addressing this topic, [5, 6, 7, 8]; however, in the authors knowledge we cannot find any detailed study about LSB reduction modeling using dimples, hence this element can be considered a point of novelty of this paper. Our numerical investigations were validated using experimental data obtained at Environmental Wind Tunnel of the "Università Politecnica delle Marche". The results, presented in this paper, show that dimpled NACA 64-014A airfoil, compared with its un–dimpled version, experiences a LSB reduction.

The paper is organized as follows. Sec. 2 presents the dimples geometric configuration. Sec. 3 briefly describes the flow models and the discretization techniques adopted for their solution. Sec. 4 treats the description of the experimental set–up. Sec. 5 is devoted to the results and Sec. 6 contains the conclusions.

2. Dimple geometric configuration

The dimples geometric features are reported in Fig.1 (values expressed in chord length percent, $\% c$): a spherical cut of $4.7\% c$ as executive diameter and $0.705\% c$ in depth make an elliptic dimple, depending of airfoil local curvature. The dimple axis is placed at $55\% c$ and it is perpendicular to airfoil surface, thus we have a $3.35\% c$ equivalent diameter (ratio between perimeter and π), as reported in Table 1. The dimpled airfoil has only one row of these elements with about one dimple diameter spacing in the spanwise direction.

Table 1. Dimpled NACA 64-014A geometric details; values expressed in chord length percent

x_D	D_{exec}	h	D^*	$\frac{h}{D^*}$
55	4.7	0.705	3.35	0.2107

3. Numerical modeling

For the numerical simulations, RANS and LES approaches were used. For un–dimpled airfoil a RANS 2D model was used, while 3D domain with $14\% c$ spanwise depth is used for dimpled configuration (consisting of two consecutive dimples representation) and LES un–dimpled case.

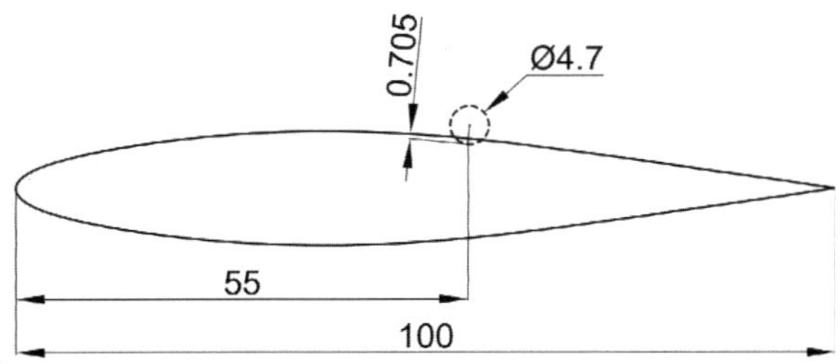

Figure 1. Dimples geometric configuration

A C–shaped domain with hexaedral elements was built. The employed grid has a viscous sublayer scaled first cell height, y^+, of approximately 1. The value of y^+ is estimated as $y^+ = \frac{u_\tau}{\nu} y_c$, where $u_\tau = \sqrt{\tau_w/\rho}$ is the friction velocity, τ_w is the viscous stress component measured at the wall, and y_c is the height of the cells next to the wall which was fixed equal to 10^{-4} c.
The starting 2D domain consists of $1.75 \cdot 10^5$ quadrilateral elements. We have to note that 290 cells were used to discretized the airfoil along the chord length, 180 cells were adopted in the wake region, while 195 in the direction normal to the airfoil. The 3D domain, used for LES computations, was obtained extruding the 2D one in the spanwise direction using 40, uniformly spaced, cells. Thus the total cells' number was $5.6 \cdot 10^6$. The dimples curvature strictly requires a local mesh refinement, thus nodes' number was increased from 290 to 345 for the chord direction and from 40 to 112 for the spanwise direction. In particular the dimples were discretized using 44 elements along its diameter. The obtained computational grid owns $22.1 \cdot 10^6$ cells. Moreover a coarse mesh, having $3.1 \cdot 10^6$ of cells, was also carried out by halving the cells' number in all the x–y–z direction. Fine and coarse computational grids for the dimpled configuration are showed in Fig.2.

For RANS simulations, laminar–to–turbulent transition model available within the commercial CFD software `Star-CCM+` was used. In particular k-ω SST model with γ-$\widetilde{\mathrm{Re}_{\theta,t}}$ transitional approach was chosen. Unfortunately, only first order upwind scheme can be used for convective terms discretization in this solver. In order to overcome the dissipation induced by the flow model and numerics the free–stream turbulent intensity was set about 3 times greater than turbulent intensity of the experimental tests.
LES seeks to calculate the large energetic structures in a turbulent flow and it allows to reduce the computational resources required by DNS. The LES approach filters the Navier–Stokes equations and it inserts a closure approximation for unresolved turbulent structures. Such models are based on the decomposition of the fluid–dynamic quantities in resolved and unresolved components, $[\overline{\mathbf{u}}, \overline{p}]$ and $[\mathbf{u}', p']$, respectively, so that $\mathbf{u} = \overline{\mathbf{u}} + \mathbf{u}'$ and $p = \overline{p} + p'$ [9]. In order to derive a set of equations for $\overline{\mathbf{u}}$ and $\overline{p}$, a filtering procedure is performed over the Navier–Stokes. Filtered equations reads

$$\begin{aligned} &\boldsymbol{\nabla} \cdot \overline{\mathbf{u}} = 0 \\ &\frac{\partial \overline{\mathbf{u}}}{\partial t} + \boldsymbol{\nabla} \cdot (\overline{\mathbf{u}} \otimes \overline{\mathbf{u}}) = -\frac{1}{\rho} \boldsymbol{\nabla} \overline{p} + \boldsymbol{\nabla} \cdot (\nu \boldsymbol{\nabla} \overline{\mathbf{u}}) - \boldsymbol{\nabla} \cdot \mathbf{B} \end{aligned} \tag{1}$$

where $\mathbf{B} = \overline{\mathbf{u} \otimes \mathbf{u}} - \overline{\mathbf{u}} \otimes \overline{\mathbf{u}}$ is the so called subgrid–scale stress tensor; $\mathbf{B}$ model is the effect of the unresolved scales on the resolved one, [10].
Usually the $\mathbf{B}$ tensor is modeled in analogy with Boussinesque approximation by introducing a subgrid–scale viscosity, ν_{sgs}, as follows:

$$-\mathbf{B} + \frac{2}{3} k_{sgs} \mathbf{I} = \nu_{sgs} \left(\boldsymbol{\nabla} \overline{\mathbf{u}} + \boldsymbol{\nabla} \overline{\mathbf{u}}^T \right) \tag{2}$$

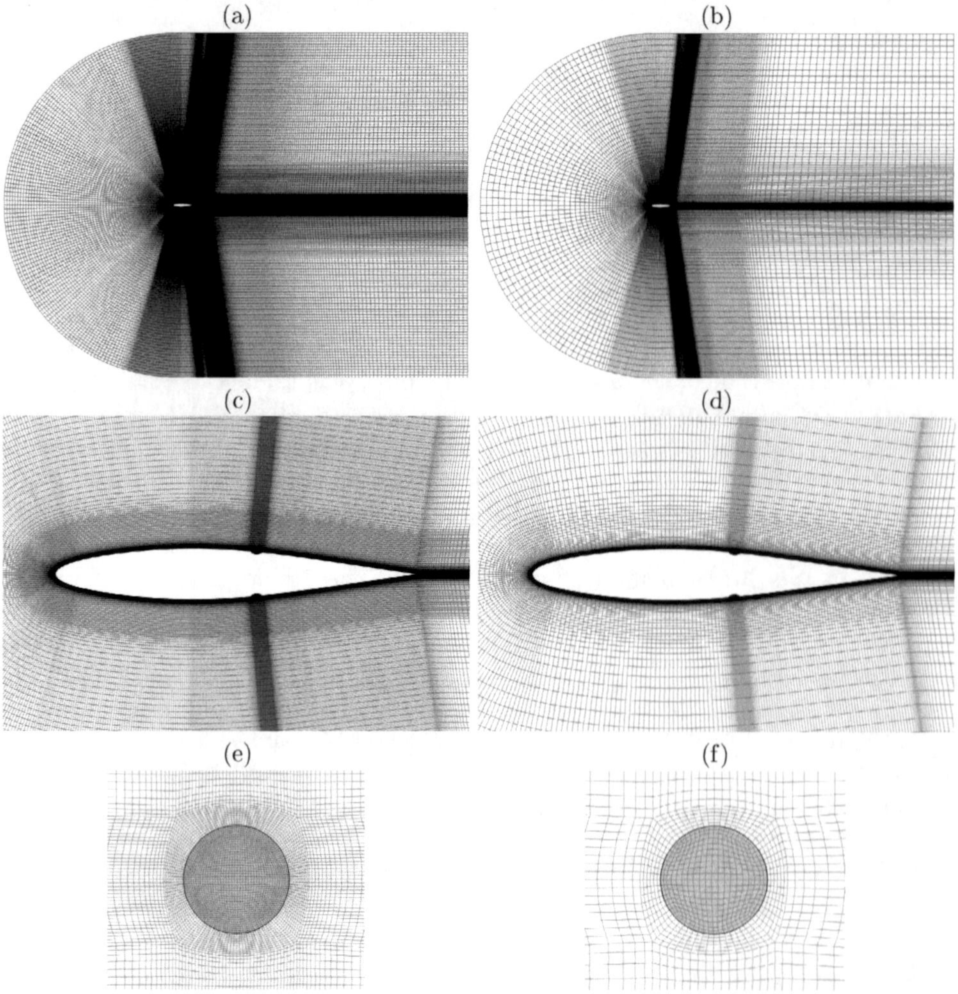

Figure 2. Comparison between fine grid (a,c,e) and coarse grid (b,d,f)

where k_{sgs} is sub–grid kinetic energy and $\mathbf{I}$ the identity tensor. Replacing $\mathbf{B}$ from eq. 2 in eq. 1, we obtain the LES governing equations:

$$\begin{aligned} &\boldsymbol{\nabla} \cdot \overline{\mathbf{u}} = 0 \\ &\frac{\partial \overline{\mathbf{u}}}{\partial t} + \boldsymbol{\nabla} \cdot (\overline{\mathbf{u}} \otimes \overline{\mathbf{u}}) = -\boldsymbol{\nabla} \bar{P} + \boldsymbol{\nabla} \cdot \left[(\nu + \nu_{sgs}) \left(\boldsymbol{\nabla} \overline{\mathbf{u}} + \boldsymbol{\nabla} \overline{\mathbf{u}}^T \right) \right] \end{aligned} \tag{3}$$

with $\bar{P} = \frac{2}{3} k_{sgs} + \bar{p}/\rho$. The sub–grid scales models based on eq. 2 produce the mean (in the statistical sense) dissipation provided by the unresolved eddies. In particular, in this work we have considered dynamic version, [11, 12] of the Smagorinksy subgrid scale model, [10]. Therefore a specifically developed `OpenFOAM` solver named `dirkFoam` has been adopted to solve the flow governing equations. `OpenFOAM` (Open-source Field Operation and Manipulation) code [13] is unstructured finite volume solver released under the GNU Public License (GPL). The code's object-oriented structure enables users to implement their own models and solvers in the baseline codes with relatively little effort (see for example [14, 15, 16]).
The adopted solver has been developed by our group for DNS/LES applications in order to guarantee very low dissipation in time integration, [17, 18]. It adopts a third–order accurate

Singly Diagonally Runge–Kutta (SDIRK) scheme for time integration and for each implicit Runge–Kutta stage, a PISO–like procedure based on Rhie–Chow correction, [19], in the spirit of OpenFOAM PISO implementation is employed, [17, 18]. Moreover all diffusive terms and pressure gradients were approximated with a second–order accuracy, while the convective terms were discretized using high–resolution NVD Gamma scheme, [20]. For what concerns linear solvers, we used a preconditioned bi–conjugate gradient method (PBiCG) with a diagonal incomplete–LU (asymmetric) preconditioner for the momentum equation. For the pressure, on the other hand, we used a preconditioned conjugate gradient method (PCG) with a diagonal incomplete Cholesky preconditioner. The linear system for pressure was solved using a local accuracy of 10^{-6}, whereas the other systems were considered converged when the residuals reached 10^{-9}.Lastly we want also to remark that no wall functions were used, standard inflow/outflow boundary conditions were adopted and the flow periodicity was assumed to be in the span–wise direction.
The computations involving the largest number of grid cells were completed on the GALILEO supercomputing facility at CINECA using 384 CPU–cores. The solutions on the coarser grids were obtained on a small Linux Cluster with 6 AMD Opteron-based nodes for a total of 96 CPU cores operating at 2.6 GHz.

4. Experimental apparatus

Figure 3. (Left) Wing model in EWT of Università Politecnica delle Marche. (Right) Angle of attack rotation system

The experiments were performed at the Environmental Wind Tunnel (EWT) of the "Università Politecnica delle Marche". The EWT test chamber consists of three main sections: the first is used for aerodynamic tests requiring a uniform velocity distribution and a low turbulence level; the second is used to test reciprocal interference effects between slender bodies; the latter is used to test wind effects over buildings, structures, orography models which are subjected to fully developed environmental boundary layers. The wind tunnel is supplied by a fan having a constant rotational speed of 975 RPM, consisting of 16 blades with an adjustable pitch that ensure a regulated wind velocity in the test section between 5 m/s and 40 m/s. Constant Temperature Hot Wire Anemometer (CTA HWA) measurements showed a lack of flow uniformity less than 2.5% and a turbulence intensity level less than 0.3% on an area larger than 90% of the test cross section. Velocity and turbulence measurements are obtained with Dantec CTA system equipped with a pneumatic calibration unit. The wind tunnel is also equipped with a compact heat exchanger able to control temperature variations inside a range of one Celsius degree. NACA 64-014A dimpled and un–dimpled airfoils are tested. The wing models have

(a) (b)

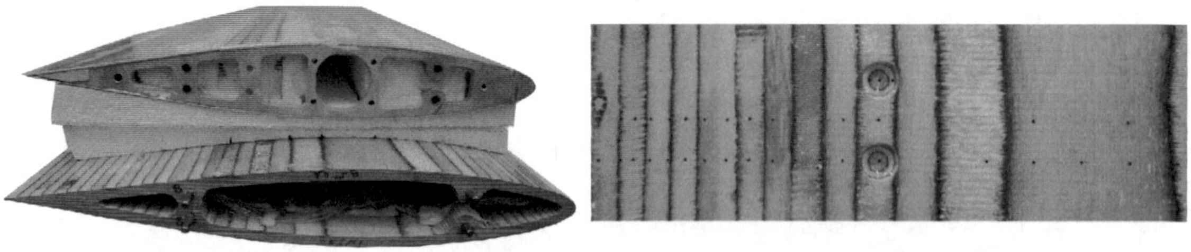

Figure 4. (a) Wing model modules. (b) Pressure holes

a chord and span of 0.45 m and 1.4 m respectively. They are formed by three modules: two modules of wooden template and balsa wood cover and one module formed by tho half birch plywood wooden shells. Pressure measurements were obtained by means of a Games 5266 very low range differential pressure transducer (± 50 Pa operating range) with a typical error below 1% FS BSL. To obtain the airfoil pressure coefficient distribution, this device in sequentially connected to all the pressure taps drilled in the section of the wings. The pressure holes are placed along the wooden shells airfoil section as showed in Fig.4. The original airfoil has one row of pressure taps, while the dimpled one has a row passing through the middle section of a dimple and a row between two consecutive dimples. The models are constrained to the wind tunnel by means of a specific framework in order to avoid mechanical vibrations transmission. The angle of attack is set by means of a zero–backlash cycloidal gear train with a graduated indicator and a Handel, see Fig.3.

5. Results

This section presents a comparison of the pressure distribution as well as drag coefficient obtained from LES and RANS computations of the dimpled and un–dimpled airfoil. Finally computed pressure distributions are compared with experimental data.
In Fig. 5 we show a representation of the vortex structures computed using LES technique and identified using the Q–criterion; it is very easy to note as coarse grid results exhibits a loss in the vortical structures resolution.

5.1. Differences between LES and RANS model

As can be seen in Fig. 6, C_p distributions show that our numerical solutions predict a more abrupt pressure recovery for the un–dimpled airfoil configuration. However RANS and LES approaches provide different results downstream the dimples region in terms of LSB modification. Indeed, as shown in Fig. 7, RANS model produces a sort of splitting of the LSB around the half dimple section, meanwhile LES solution indicates a LSB length global reduction (over 13.6% c), with a restricted attached flow area downstream the dimples.

Furthermore, we have to note that the grid resolution is more significant for the LES approach than for the RANS one, in particular coarse grid overestimates the dimples effect and underestimates C_f value on turbulent reattachment area, as can be seen in Fig. 8(a).

Lastly, Table 2 reports the impact of the different turbulence models here used on the LSB extension prediction. Differently Table 3 shows the computed drag coefficients. In particular, we can note that RANS cases and coarse grid LES solution provide a similar global C_D reduction if compared to un–dimpled configuration. On the other hand, LES model with finer grid provides

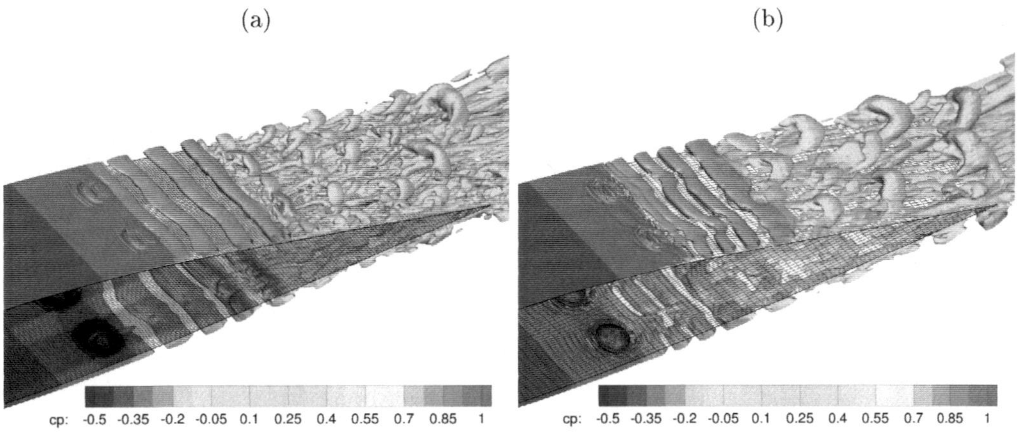

Figure 5. Vortical structures, identified by the Q-criterion ($Q = 0.5u_\infty^2/c^2$), isosurfaces colored by pressure coefficient; (a) fine mesh, (b) coarse mesh.

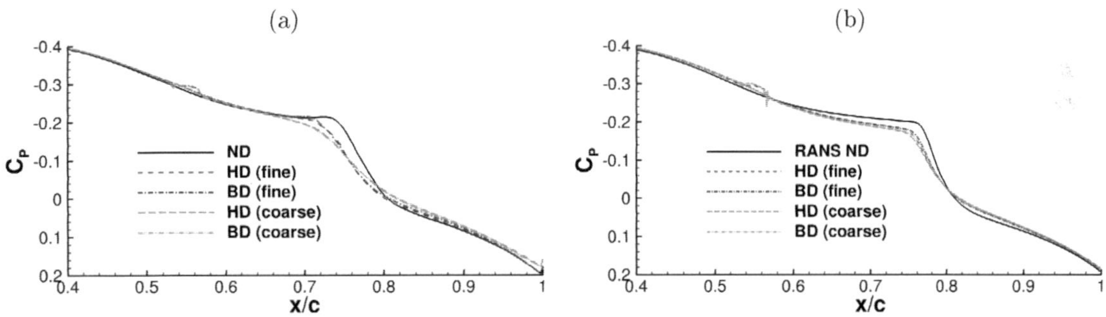

Figure 6. C_p distribution; (a) LES model, (b) RANS model

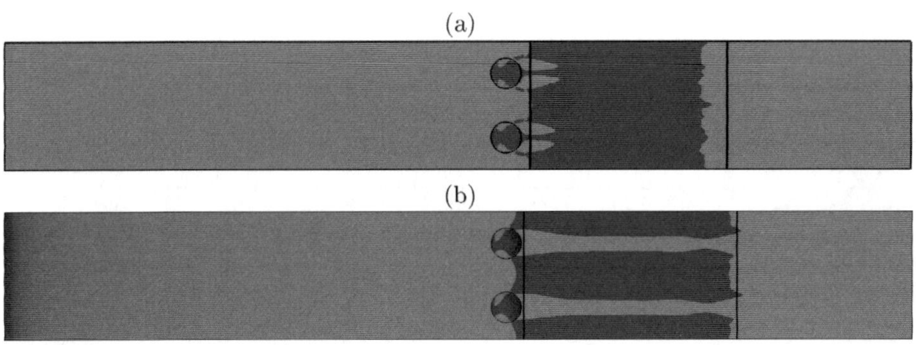

Figure 7. Visualization of positive and negative C_f zones, in red and blue respectively; (a) LES model, (b) RANS model. Blue zones represent the LSB extension of the dimpled airfoil cases, while the vertical black lines indicate the starting-ending flow separation points of the un–dimpled ones.

a minor C_D reduction due to a shear drag increase. Anyway, drag coefficients obtained by RANS model are higher than the LES ones.

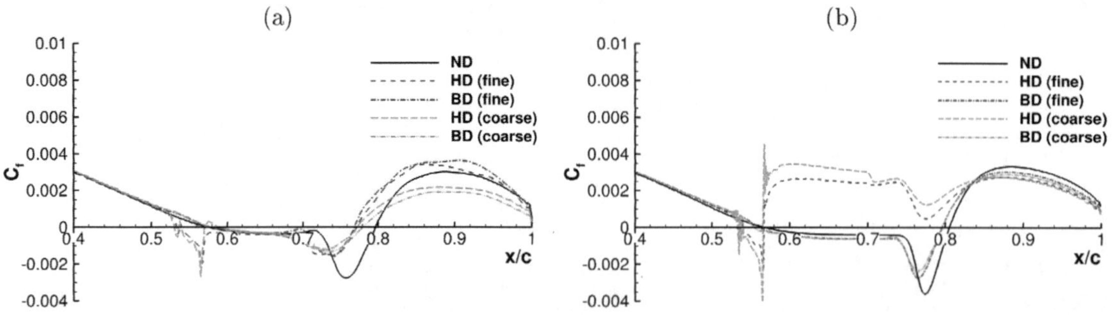

Figure 8. C_f distribution; (a) LES model, (b) RANS model

Table 2. LSB extension; separation and reattachment points expressed in chord length percent. ND = no dimpled airfoil section, HD = half dimple airfoil section, BD = between dimples section.

Section	LES		RANS	
	x_s/c	x_r/c	x_s/c	x_r/c
ND	57.6	79.6	56.9	80.5
HD	53.3	77.4	–	–
BD	58.0	77.0	56.0	79.5

Table 3. Comparison of drag coefficients. C_D = total drag coefficient, $C_{D,p}$ = pressure drag coefficient, $C_{D,f}$ = friction drag coefficient

	$C_D \cdot 10^2$	$C_{D,p} \cdot 10^2$	$C_{D,f} \cdot 10^2$	ΔC_D [%]
LES no dimple	1.188	0.524	0.664	
LES dimple fine grid	1.181	0.465	0.716	-0.59
LES dimple coarse grid	1.155	0.495	0.660	-2.81
RANS no dimple	1.228	0.561	0.667	
RANS dimple fine grid	1.193	0.512	0.681	-2.82
RANS dimple coarse grid	1.192	0.505	0.687	-2.93

5.2. Comparison with experimental data

The pressure coefficient, C_p, distributions of the original and dimpled airfoil are depicted in Fig. 9. Both LES and RANS simulations are in good agreement with experimental data up to 55% c, *i.e.* in the fully laminar zone. In the laminar separation region, the predicted pressure coefficients are not in agreement; however their differences fall inside the experimental uncertainty. Differently in laminar–to–turbulent transition region, as well as in turbulent reattachment one, LES results put in evidence a better fitting with the experimental data if compared with RANS solutions. In particular RANS results are not in good agreement with experiments. As showed in Fig. 10, our dimpled and un–dimpled configurations put in evidence a similar behaviour of the adopted numerical models.

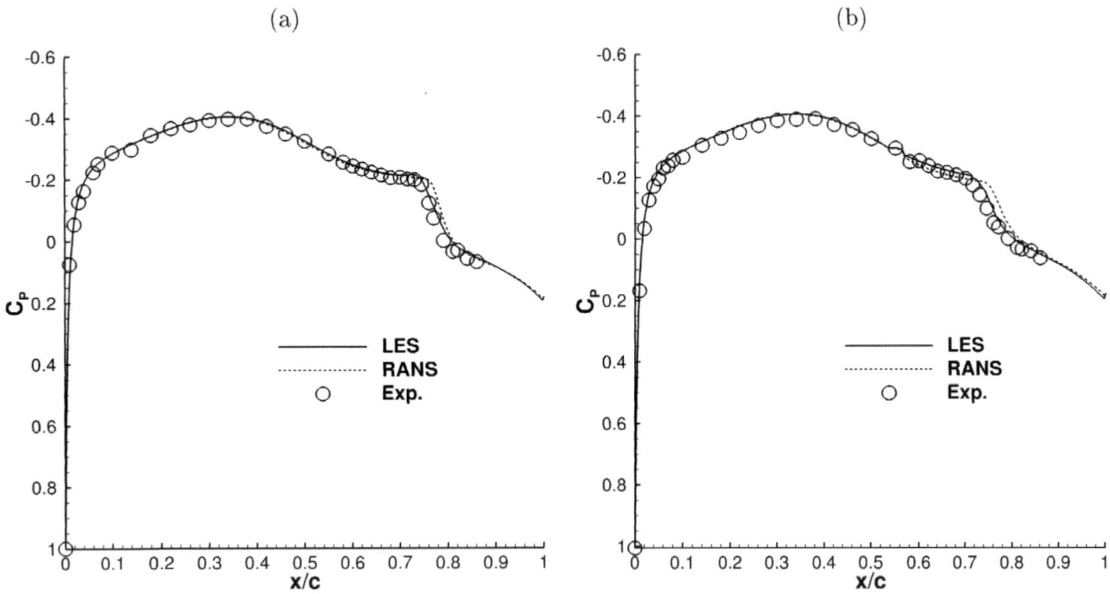

Figure 9. Comparison of C_p with experimental data; (a) un–dimpled airfoil, (b) dimpled airfoil

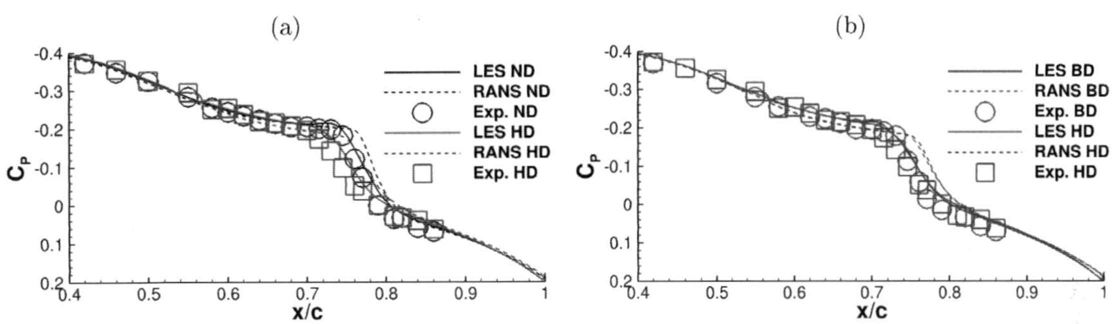

Figure 10. C_p distributions; (a) Comparison of original and dimpled airfoil; (b) differences between the two dimpled airfoil sections. For symbols see caption of table 2

6. Conclusion

In this paper, dimples effect on NACA 64-014A airfoil at $Re = 1.75 \cdot 10^5$ and $\alpha = 0°$ was presented. In particular, the dimples impact on the reduction of the LSB is faced. Therefore, RANS simulations, within a commercial solver, and LES simulations, with a low-dissipative algorithm developed by our group, were used. For LES computations, dynamic version of the Smagorinksy subgrid scale model were used. An experimental measurement were also carried out for validation purposes.

It has been shown that dimples application produces a reduction of the LSB extension. RANS model overestimates the laminar–to–turbulent transition and turbulent reattachment locations of the dimpled and un–dimpled airfoils,while LES model provides better agreement with experimental data for both the cases. Therefore, LES is able to capture the main features of the flow past the dimpled and un–dimpled airfoils. The main drawback of the LES approach is related to its high computational costs and, for this reason, the presented analysis is limited only to a single angle of attack.

In conclusion, dimples application on NACA 64-014A airfoil at $Re = 1.75 \cdot 10^5$ and $\alpha = 0°$

produces a reduction of the LSB extension and a consequent pressure drag decrease. Both numerical models show a total drag reduction, while experimental force measurements are still in processing, with a preliminary good agreement. This is very intresting because dimple application could be the way to reduce the drag by preserving the general blade features of a wind turbine. Indeed, a drag reduction without any lift change will have the double benefit of an increased tangential force and a reduced axial force, i.e. an electric power production increase and a minor mechanical stress for a wind turbine. Future work will be devoted to the evaluation of both the angle of attack and the Re number effects.

7. Acknowledgements

We acknowledge the CINECA Award N. HP10CDBABZ YEAR 2016 under the ISCRA initiative, for providing high-performance computing resources and support.

References

[1] Yu-Fong Huang Xing-Jia Gan P T C 2017 *Renew Energ* **102** 98–106
[2] Ligrani P M, Harrison J L, Mahmmod G I and Hill M L 2001 *Phys Fluids* **13(11)**
[3] Won S Y and Ligrani P M 2007 *Mechl Sci Technol* **21**
[4] Lake J 1999 *Flow Separation Prevention on a Turbine Blade in Cascade at Low Reynolds Number* Ph.D. thesis Air Force Institute of Technology, Ohio
[5] Lee D, Nonomura T, Oyama A and Fujii K 2017 *Int J Comp Fl Dyn* **31** 57–67
[6] Catalano P, Mele B and Tognaccini R 2015 *Comput Fluids* **109** 67–71
[7] Crivellini A and D'Alessandro V 2014 *Int J Heat Fluid Fl* **47** 70–83
[8] Crivellini A, D'Alessandro V, Di Benedetto D, Montelpare S and Ricci R 2014 *J Phys: Conf Ser* **501**
[9] Rogallo R and Moin P 1984 *Annu Rev Fluid Mech* **16** 99–137
[10] Smagorinsky J 1963 *Mon Weather Rev* **91** 99–164
[11] Germano M, Piomelli U, Moin P and Cabot W 1991 *Physics of Fluids* **3** 1760–1765
[12] Lilly D 1991 *Phys Fluids* **4** 633–635
[13] Weller H, Tabor G, Jasak H and Fureby C 1998 *Comput Phys* **12** 620–631
[14] D'Alessandro V, Montelpare S and Ricci R 2016 *Comput Fluids* **136** 152–156
[15] D'Alessandro V, Montelpare S, Ricci R and Zoppi A 2017 *Energy* **130** 402 – 419
[16] Silva L and Lage P 2011 *Comput Chem Eng* **35** 2653–2666
[17] D'Alessandro V, Binci L, Montelpare S and Ricci R 2017 *Comput Phys Commun*
[18] D'Alessandro V, Zoppi A, Binci L and Ricci R 2016 *Int J Comp Meth Exp Meas* **4(4)** 594–603
[19] Ferziger J and Peric M 1999 *Computational Methods for Fluid Dynamics* (Springer)
[20] Jasak H, Weller H and Gosman A 1999 *Intl J Num Meth Fl* **31** 431–449

CFD three dimensional wake analysis in complex terrain

F Castellani[1], D Astolfi[1] and L Terzi[2]

[1] Department of Engineering, University of Perugia, Perugia, Italy
[2] Renvico srl, Via San Gregorio 34, Milano 20124, Italy

E-mail: francesco.castellani@unipg.it

Abstract. Even if wind energy technology is nowadays fully developed, the use of wind energy in very complex terrain is still challenging. In particular, it is challenging to characterize the combination effects of wind flow over complex terrain and wake interactions between nearby turbines and this has a practical relevance too, for the perspective of mitigating anomalous vibrations and loads as well improving the farm efficiency. In this work, a very complex terrain site has been analyzed through a Reynolds-averaged CFD (Computational Fluid Dynamics) numerical wind field model; in the simulation the influence of wakes has been included through the Actuator Disk (AD) approach. In particular, the upstream turbine of a cluster of 4 wind turbines having 2.3 MW of rated power is studied. The objective of this study is investigating the full three-dimensional wind field and the impact of three-dimensionality on the evolution of the waked area between nearby turbines. A post-processing method of the output of the CFD simulation is developed and this allows to estimate the wake lateral deviation and the wake width. The reliability of the numerical approach is inspired by and crosschecked through the analysis of the operational SCADA (Supervisory Control and Data Acquisition) data of the cluster of interest.

1. Introduction

The development of onshore wind energy industry [1] has stimulated a certain number of scientific and technological challenges: the comprehension of flow over terrain at the microscale level [2, 3, 4, 5, 6], the assessment of what happens in the downstream area especially when terrain-induced flow acceleration combines with wake interactions [7, 8, 9, 10] between nearby turbines.

Therefore, the scientific literature about wakes in complex terrain [11, 12, 13, 14] has been recently vastly developing. A general boost has been provided by the widespread diffusion of SCADA control systems and by an increasingly open-minded attitude towards the use of data in wind energy industry [15, 16, 17, 18]. An impressive boost, inspiring also the present work, has been provided by the evolution of precision wind measurements techniques, allowing the reconstruction of the full three dimensional wind field. For example, in [13], it is shown, basing on an experimental drone campaign, that in complex terrain the near-wake of a full scale wind turbine extends up to two rotor diameters and is therefore 35% shorter than in flat terrain. In [19], high-frequency time series measurements obtained with remote scanning systems are employed for the analysis of wakes in complex terrain. Similarly to the case of

[13], it arises that the waked area is smaller than in complex terrain because wake effects are often overruled by the mixing induced by topology effects.

In [20], the three dimensionality of the waked characteristics of a yawed wind turbine immersed in a turbulent boundary layer is studied: in that work, wind tunnel testing, through high-resolution stereoscopic particle image velocimetry system, is used to measure the three velocity components in the turbine wake under different yaw angles and tip-speed ratios. Some remarkable features of the wakes of yawed turbines, such as the asymmetric distribution of the wake skew angle with respect to the wake center, are highlighted. This is due to the formation of a counter-rotating vortex pair in the wake cross-section as well as the vertical displacement of the wake center.

The test case of this study is the upstream turbine of a cluster of 4 turbines, having 2.3 MW of rated power each, sited in Italy in a very complex terrain. In [21], the cluster is studied through Reynolds-averaged CFD modeling and SCADA data analysis. The objective of this combination of methods is inquiring how much one can understand the three-dimensional wind field through a two-dimensional lens: the wind field at hub height, as predicted by the numerical model and as observed at the nacelle of the wind turbines. The main result of [21] is that observing and interpreting judiciously the wind field at hub height, it is possible to identify the main drivers of the performances of the cluster: most of all, the impressive directional distortion of the wakes induced by the terrain.

These results, as well with those of [20], inspire the idea of improving the comprehension of the three-dimensional nature of wakes in complex terrain through numerical modeling. In this work, a novel post-processing method is developed for elaborating the output of a steady state RANS CFD model. This allows to quantify the lateral deviation and the width of the wake and therefore the impact of three-dimensionality on the evolution of the waked area between nearby turbine.

Summarizing, the structure of the work is as follows: in Section 2, the test case wind farm is described. The methods are described in Section 3. Section 4 is devoted to the collection of the results and, finally, conclusions as well as further directions are drawn in Section 5.

2. The Wind Farm

The cluster of wind turbines is depicted in Figure 1; the black turbine is the upstream one for which the wake development has been simulated. The contour lines allow to appreciate the considerable complexity of the terrain. The inter-turbine distances are indicated in units of rotor diameter. The wind turbines have 2.3 MW of rated power each.

The 270° is a very populated sector of the wind rose for this site. In this regime, turbine T11 is under a full wake from turbine T10 and turbine T12 is under the multiple full wake of turbines T10 and T11. In [22], the distribution of the performances of the cluster when the wind blows from 270° is estimated using SCADA operational data: it arises that, at T12 site, a considerable wake recovery is observed, so that turbine T13 is often the most productive turbine of the cluster. In [21], this picture is interpreted as due to the northward distortion of the wake of T11. This is precisely the effect of the complexity of the terrain. Anyway the cited analysis was performed only observing the 2D wind field distribution at hub height; for a fully three

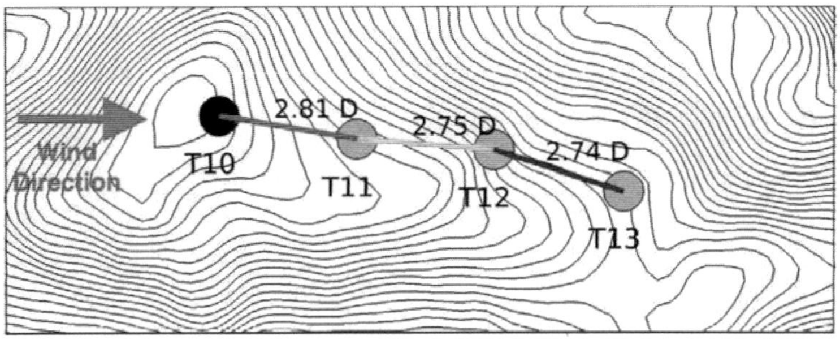

Figure 1. The layout of the test case cluster, where D is the rotor diameter.

dimensional insight it is therefore fundamental to explore the behavior of the wake at different heights from ground.

3. The method

The WindSim numerical tool has been employed [2, 5] for the simulations. The inlet boundary is a logarithmic profile blowing from 270°, and the top of the boundary layer is 250 meters. RANS equations are solved through the General Collocated Velocity (GCV) method, for its capability of robust convergence also in very complex terrain. The complexity of the terrain suggests the optimal choice of the turbulence model: RNG k-ϵ is selected, as for example in [23, 24]. The wake interaction was modeled through the Actuator Disk (AD) approach. The used wake model indeed disregards wake rotation and reproduces the speed deficit applying the thrust on the rotor surface: despite its simplicity such a model [25] can be reliable to reproduce the waked flow in the mid to far wake region.

In [21], a method for defining a wake area and a mean wake line was proposed. It is based on the comparison between the simulated free flow wind field and the simulated wind field where the upstream turbine is included and accounted for through the AD model. All the points where, within a certain tolerance, the wind field with the turbine is lower than the free wind field are included in the wake area. The midpoints of each wake section can be computed and they form a wake line. As an example, see Figures 2 and 3.

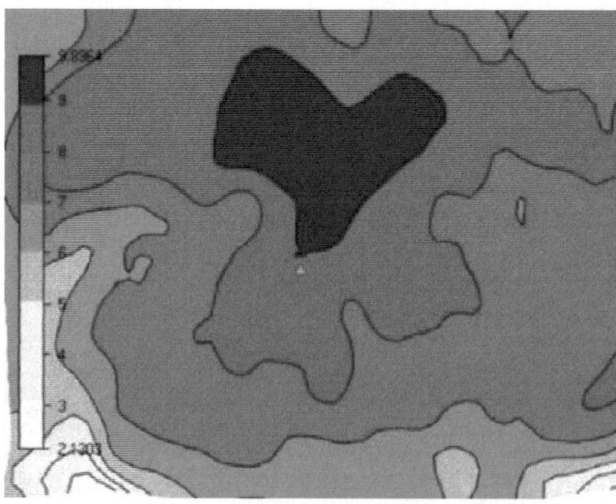

Figure 2. Free flow simulation

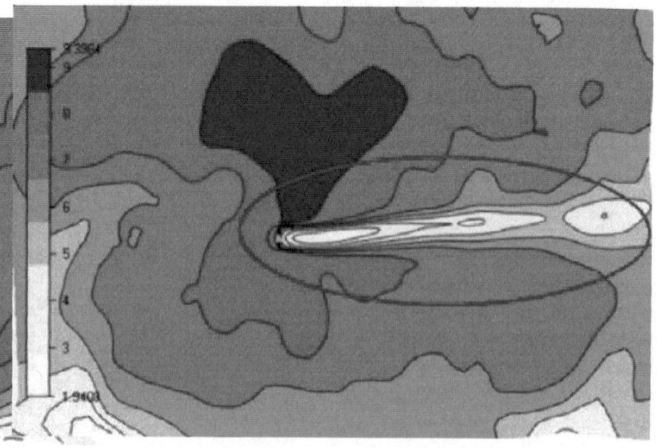

Figure 3. Single wake flow simulation: the area of the cluster in circled in blue.

This method is purely geometric, basing on the idea that the wake deficit should be symmetric. In this work, the definition of wake area is improved through the improvement of the post processing acting on the simulated wind field. The center of the wake in the lateral z-direction is defined by looking at the maximum speed deficit for each streamwise x-coordinate and this can be performed at any height from ground y. The comparison between this method and the one from [21] is shown in Figures 4, 5 and 6. In particular, from Figure 6, the role of the terrain in distorting the wake area arises more clearly.

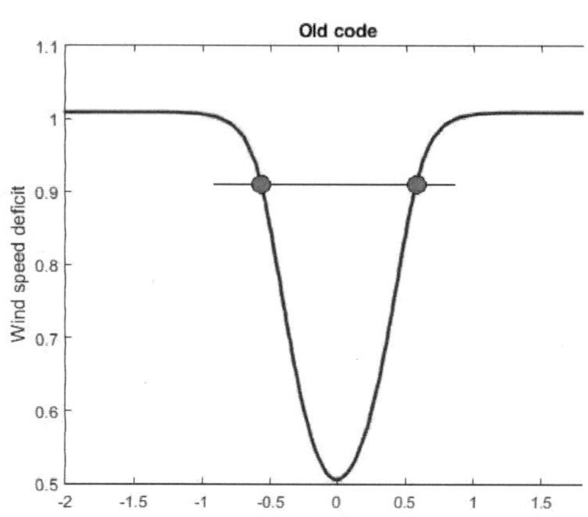

Figure 4. Purely geometrical wake definition

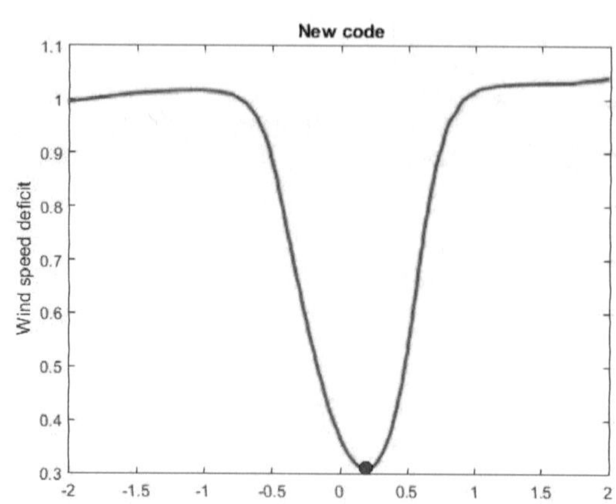

Figure 5. Definition of the wake basing on the speed deficit.

CFD three dimensional wake analysis in complex terrain

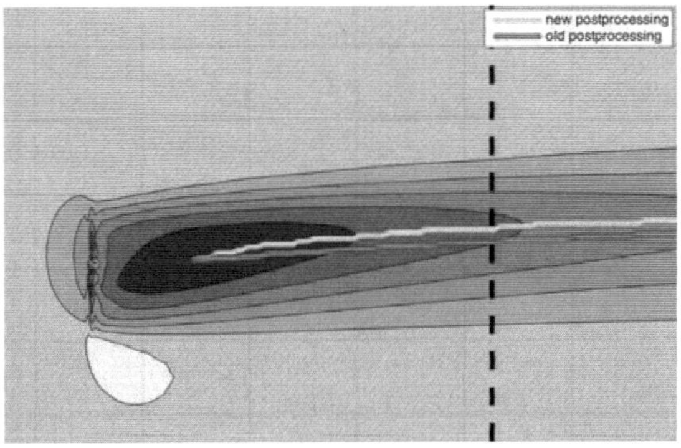

Figure 6. The wake line according to geometrical and optimized post-processing, at hub height.

This approach is employed to reconstruct the full three-dimensionality of the wake, by cutting the wind flow, and therefore the wake, in slices: wind intensity planes parallel to the ground at different heights are considered. For each height, the wake section is reconstructed by individuating, as above, the minimum velocity. As a reference for the flow upstream, the distance of two rotor diameters upstream is taken. The renormalization can be relative (referred to the same height), or absolute (referred to the hub height plane). This method provides as output a surface evolving downstream and up and down and this gives insight into the three-dimensional nature of the wake. Further, in this work, the above method is applied in flat terrain and on the real terrain of the wind farm, in order to appreciate the distortion of the wind flow.

4. Results

The results collected here on refer to the three-dimensional evolution of the wake of turbine T10: therefore, the focus is on the single wake case. The results are shown at two distances downstream. These distances have been selected because they have been considered instructive:

- 2.8 rotor diameters downstream, because it is the position (along the streamwise coordinate) of the turbine T11;
- 8.2 rotor diameters downstream, because it is the position (along the streamwise coordinate) of the turbine T13.

In particular, the position of turbine T11 has been selected because how the wake evolves at that position heavily influences (in a far from trivial way [21]) what happens more downstream.

Three heights are reported: hub height and a lower and a upper (symmetrical) height with respect to the hub (at D/3 distance). The upstream and the downstream profiles (normalized, as explained in Section 3, to v_∞ at hub height) are reported as a function of the lateral displacement.

Figures 7 and 8 refer to the flat case and it arises that, as expected, the profiles are symmetric with respect to the z coordinate. The interesting point is that the maximum deficit arises at hub height, with respect to the two other heights, but this

IOP Conf. Series: Journal of Physics: Conf. Series **923** (2017) 012031 doi:10.1088/1742-6596/923/1/012031

effect decreases proceeding downstream and, at 8.2 diameters, the deficit of the lower profile is a bit higher to that of hub height, while the opposite happen for the higher profile.

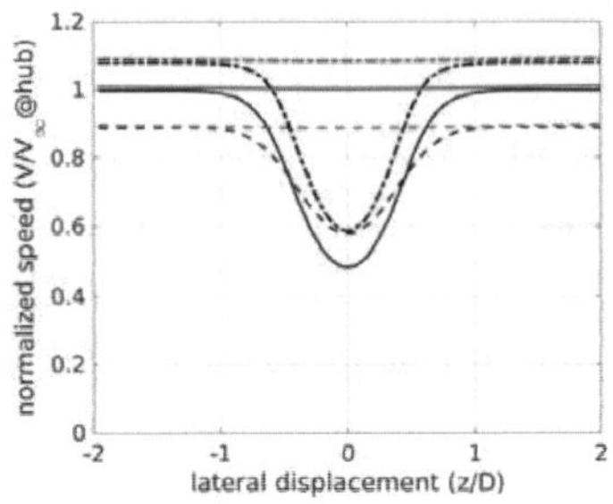

Figure 7. Wake profile 2.8D downstream (Flat terrain case)

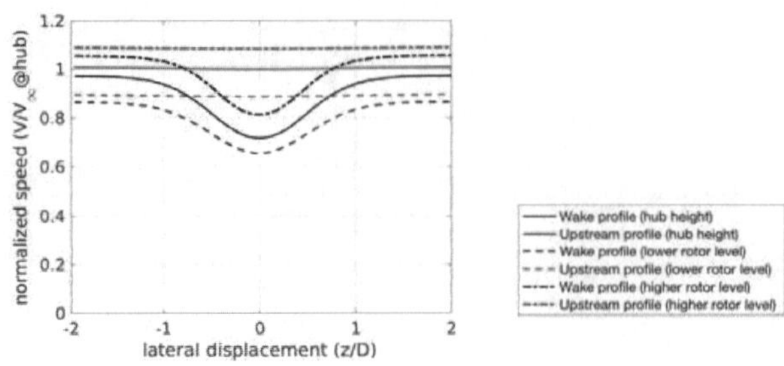

Figure 8. Wake profile 8.2D downstream (Flat terrain case)

In Figures 9 and 10, the results are collected for the simulations on the real terrain. The main feature is that the wake profile isn't symmetric with respect to the lateral displacement and with respect to the height. The former effect is due to the lateral flow distortion, the latter is due to the vertical wind shear. The deformation of the wake becomes more relevant proceeding downstream, and the effect of the topography is so relevant that at T13 site the profile of the wake has a less defined minima (Figure 10). This can be interpreted by looking at the topography of the site (Figure 1): a singularity in the orography occurs, more or less at the streamwise position of turbine T12, and this diffuse the flow horizontally enlarging the width of the wake and deviating its direction. The role of vertical mixing is also important: in the terrain case the width of the wake is higher in all the three investigated levels while in the flat case the width is considerably lower in the upper section of the wake. This picture is confirmed by the final Figures 11 and 12, where the three-dimensional evolution of the wake is represented: the asymmetry of the wake and the role of the terrain and of the wind shear can be fully appreciated.
Another interesting result is that the wake deficit, especially at the beginning of the wake, is stronger for the terrain case; this is also reported in [26] where a similar numerical approach has been used.

IOP Conf. Series: Journal of Physics: Conf. Series **923** (2017) 012031 doi:10.1088/1742-6596/923/1/012031

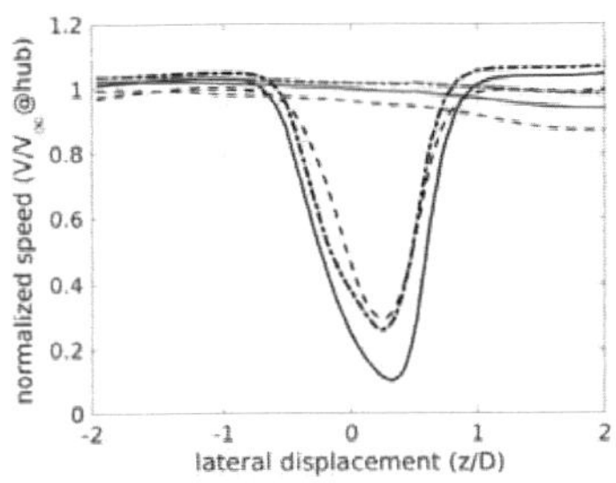

Figure 9. Wake profile 2.8D downstream (Complex terrain case)

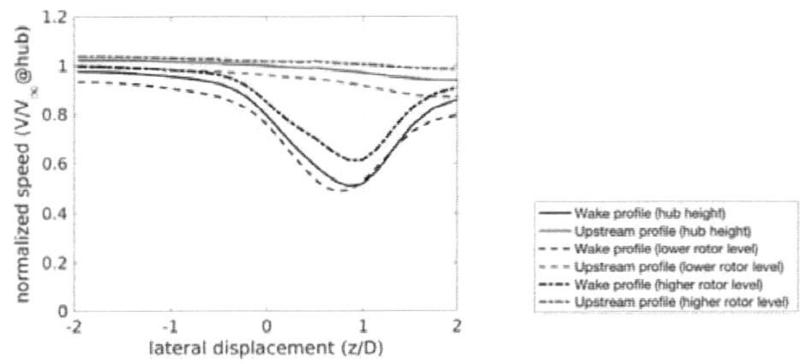

Figure 10. Wake profile 8.2D downstream (Complex terrain case)

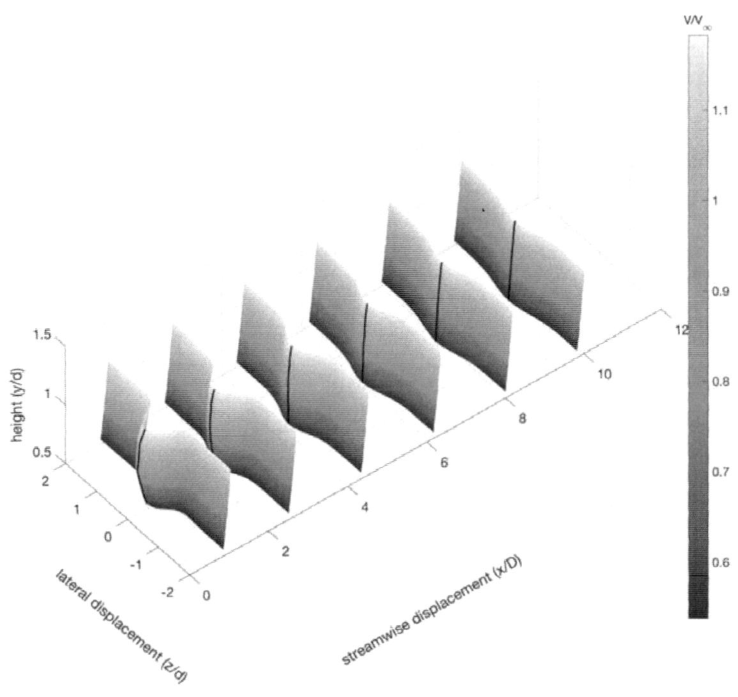

Figure 11. Three dimensional wake analysis with the black line indicating the estimated center (Flat terrain case).

From figure 13 it is also possible to notice how the lateral displacement of the center of the wake has a different behavior according to the height. The lower region, due to the ground influence, tends to be less affected by the general northward deviation induced by the terrain; this also indicate a rotation of the wake volume.
So the general direction of the wake proceeding downstream is defined by the influence of the surface and by the micro-scale inflow defined by the local orography; the effect

of the latter is much more powerful in the upper region of the wake.

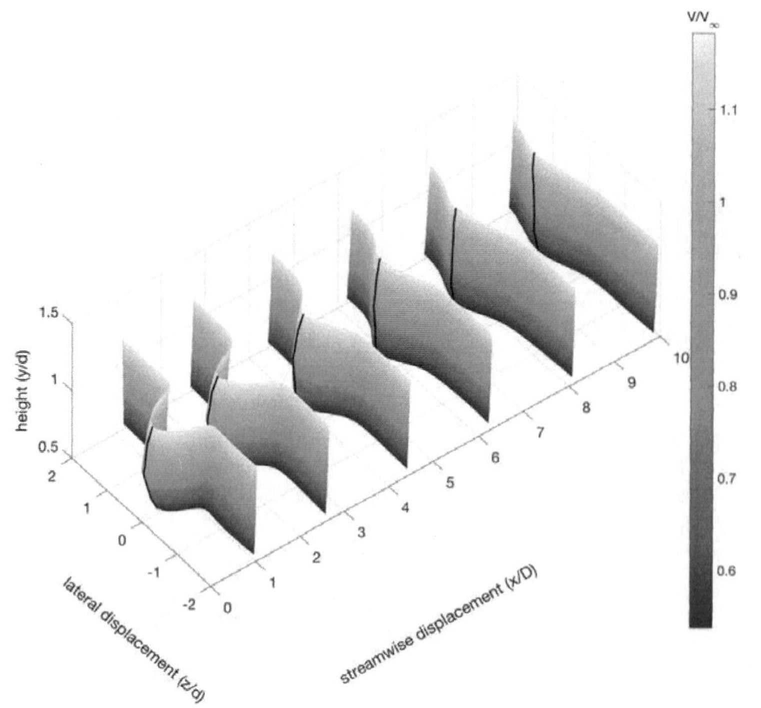

Figure 12. Three dimensional wake analysis with the black line indicating the estimated center (Complex terrain case).

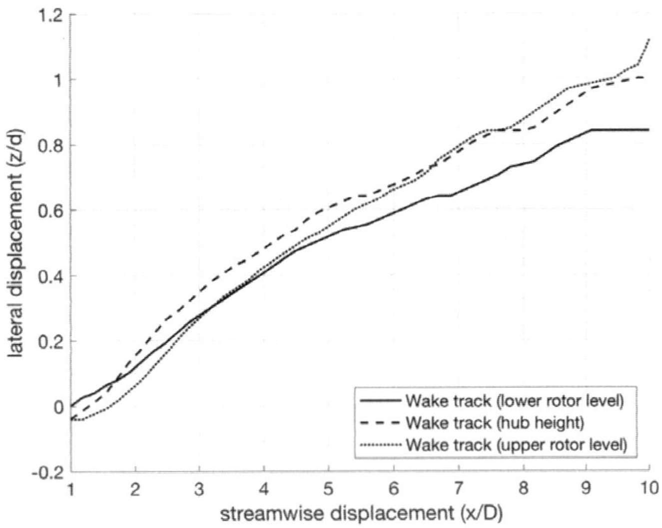

Figure 13. Lateral deviation of the wake in the terrain case at different level from ground (lower= hub-D/3 upper=hub+D/3).

5. Discussion and Further Directions

In this work, wind turbine wakes over a very complex terrain have been analyzed. The inspiration to this work comes in general, from the very recent attention on the comprehension of wake recovery in complex terrain [13, 19], and in particular from the study of a cluster of four 2.3 MW wind turbines, sited in southern Italy in a very complex terrain: in [21], the very relevant role of the terrain in distorting the single wake of the upstream wind turbine has been highlighted. Further, in [21], some insights on the evolution of the waked section proceeding downstream have been provided. The straightforward step ahead in the comprehension of this phenomenon is understanding the three-dimensional nature of the wake and its evolution. For these general and particular reasons, the same test case of [21] has been selected and the evolution of the wake of the upstream turbine has been studied when the wind blows from the 270° direction. The main novelty of this work has been the formulation of a new post-processing method of the results from a RANS CFD simulation: by looking at the speed deficit with respect to the v_∞ at hub height, a wake profile can be defined and the main observation is that the terrain distorts the wind flow so that the profile is severely asymmetric with respect to the lateral displacement. This approach has been applied on the wind field at different heights and this allowed to appreciate the effect of the vertical wind shear in the lateral distortion of the wake. In particular, this method has provided further insight into the role of the orography singularities that the wake crosses moving downstream; such terrain details tend to destroy the main structure of the wake and this is a further explanation of the performances of the cluster. Actually, for example in [22] it was noticed that the far downstream T13 wind turbine is the best performing of the cluster when the wind blows from 270°.

CFD three dimensional wake analysis in complex terrain

This work might be fruitful for several further directions: the yawing behavior of the wind turbines, as observed through SCADA data, can in perspective be put in relation with the three-dimensional evolution of the wake. This study also motivates a considerable step forward: the numerical modeling and the experimental study of the mechanical loads and the fatigue behavior of wind turbines undergoing so complex wind flow conditions‡. Further, this framework can be fruitful to interpret experimental data coming from remote scanning systems (as in [19]) spanning three dimensions.
As a future development of the present work also the possible influence of rotation in the evolution of the three dimensional shape of the wake should be investigated; this could be important especially in the near and mid wake region.

References

[1] Cheng P W 2013 *Transition to Renewable Energy Systems* 241–264
[2] Moreno P, Gravdahl A and Romero M 2003 *European Wind Energy Conference and Exhibition, Madrid* pp 16–19
[3] Lee M, Lee S, Hur N and Choi C 2010 *WIND STRUCT* **13** 375
[4] Rodrigo J, Gancarski P, Arroyo R, Moriarty P, Chuchfield M, Naughton J, Hansen K, Machefaux E, Koblitz T, Maguire E *et al.* 2014 *J PHYS CONF SER* vol 524 (IOP Publishing) p 012105
[5] Dhunny A, Lollchund M and Rughooputh S 2017 *RENEW ENERG* **101** 1–9
[6] Segalini A 2017 *PHILOS T R SOC A* **375** 20160099
[7] Barthelmie R J, Hansen K, Frandsen S T, Rathmann O, Schepers J, Schlez W, Phillips J, Rados K, Zervos A, Politis E *et al.* 2009 *WIND ENERGY* **12** 431–444
[8] Barthelmie R J and Jensen L 2010 *WIND ENERGY* **13** 573–586
[9] Hansen K S, Barthelmie R J, Jensen L E and Sommer A 2012 *WIND ENERGY* **15** 183–196
[10] Bastankhah M and Porté-Agel F 2014 *RENEW ENERG* **70** 116–123
[11] Politis E S, Prospathopoulos J, Cabezon D, Hansen K S, Chaviaropoulos P and Barthelmie R J 2012 *WIND ENERGY* **15** 161–182
[12] Hyvärinen A and Segalini A 2017 *J ENERG RESOUR-ASME*
[13] Subramanian B, Chokani N and Abhari R 2016 *RENEW ENERG* **85** 454–463
[14] Murali A and Rajagopalan R 2017 *J WIND ENG IND AEROD* **162** 57–72
[15] Blanco M I 2009 *Renewable and Sustainable Energy Reviews* **13** 1372–1382
[16] Yang W, Court R and Jiang J 2013 *RENEW ENERG* **53** 365–376
[17] Scappaticci L, Bartolini N, Garinei A, Becchetti M and Terzi L 2017 *International Journal of Renewable Energy Research (IJRER)* **7** 323–329
[18] Bartolini N, Scappaticci L, Garinei A, Becchetti M and Terzi L 2016 *Diagnostyka* **17**
[19] Hansen K S, Larsen G C, Menke R, Vasiljevic N, Angelou N, Feng J, Zhu W J, Vignaroli A, Xu C, Shen W Z *et al.* 2016 *J PHYS CONF SER* vol 753 (IOP Publishing) p 032013
[20] Bastankhah M and Porté-Agel F 2016 *J FLUID MECH* **806** 506–541
[21] Castellani F, Astolfi D, Mana M, Piccioni E, Becchetti M and Terzi L 2017 *WIND ENERGY* **20** 1277–1289
[22] Castellani F, Astolfi D, Terzi L, Hansen K and Rodrigo J 2014 *J PHYS CONF SER* vol 524 (IOP Publishing) p 012142
[23] Peralta C, Nugusse H, Kokilavani S, Schmidt J and Stoevesandt B 2014 *ITM Web of Conferences* vol 2 (EDP Sciences) p 01002
[24] Abdi D S and Bitsuamlak G T 2014 *ADV ENG SOFTW* **75** 30–41
[25] Castellani F and Vignaroli A 2013 *APPL ENERG* **101** 432–440
[26] Nedjari H D, Guerri O and Saighi M 2017 *ENERG CONVERS MANAGE* **138** 224–236

‡ www.softwind.it

IOP Conf. Series: Journal of Physics: Conf. Series **923** (2017) 012032 doi:10.1088/1742-6596/923/1/012032

Investigation of the effect of cylindrical insert devices on laminar convective heat transfer in channel flow by applying the Field Synergy Principle

P Vocale[a], A Mocerino[a], F Bozzoli[a,b], S Rainieri[a,b,*]
[a]Department of Engineering and Architecture, University of Parma, Parco Area delle Scienze 181/A, 43124 Parma – Italy
[b]SITEIA.PARMA Interdepartmental Centre, University of Parma, Parco Area delle Scienze 181/A, 43124 Parma – Italy

E-mail: pamela.vocale@unipr.it

Abstract. The Field Synergy Principle is widely applied to the evaluation of the convective heat transfer mechanism. In fact, as highlighted in literature, the evaluation of the synergy between the velocity and the temperature gradient vectors could provide a better insight on the local convective heat transfer mechanism. In this paper, the field synergy approach is adopted to numerically investigate the fluid dynamic and thermal behaviour of a fully developed flow between parallel plates with asymmetric heating, when cylindrical inserts are present. To better evaluate the influence of the inserts on the convective heat transfer mechanism, different values of the insert diameter are considered, for a given pitch value. The numerical results in terms of Nusselt number point out that the convective heat transfer coefficient decreases as the insert diameter increases. The Field Synergy Principle allows to explain the cause of the convective heat transfer reduction identifying the regions in which the heat transfer mechanism is ineffective: the extent of these areas increases as the insert diameter increases.

1. Introduction

The heat transfer enhancement is one of the main scientific and technological challenge of the last decades; therefore, various kinds of technologies have been developed for improving the efficiency of heat transfer equipment.

Since the duct shape significantly influences the flow behaviour, the most widespread techniques for enhancing the internal convective heat transfer that do not require external power (i.e. passive techniques) are based on the proper conformation of the duct wall [1].

To assess the performance of passive heat transfer enhancement techniques, many experimental and numerical methodologies have been proposed. While the experimental approach generally enables predicting the average performance of the enhanced heat transfer equipment [2-7], numerical investigations allow to evaluate the velocity and temperature fields in the duct, thus providing a useful tool to better understand the causes of the convective heat transfer augmentation [8-15].

Although the performance of enhanced heat transfer equipment can be properly assessed in terms of Nusselt number, Guo et al. [16] proposed a new method to evaluate the heat transfer enhancement, namely the Field Synergy Principle (FSP), which is based on the interaction between the velocity vector and the temperature gradient.

Their analysis reveals that the investigation of the synergy between the two vectors provides a better insight on the local convective heat transfer mechanism. The results presented in [16] point out that the convective heat transfer enhancement can be achieved by reducing the intersection angle between the velocity and the temperature gradient and by increasing simultaneously the local values of both fields.

Since the feasibility of the FSP has been confirmed in many works [17-26], several analyses of passive heat transfer enhancement techniques have been carried out by adopting this approach.

In particular, by numerically investigating the heat transfer mechanism in several geometries (i.e. wavy channels, corrugated ducts, obliquely positioned plate array, circular tube with a coaxially inserted cylindrical bar and parallel plates with inserted square blockages), Tao et al. [17] observed that an enhanced heat transfer surface can reach the best performance for minimum average intersection angle between the velocity vector and the temperature gradient.

The same conclusions were highlighted by He et al. [20] who evaluated the influence of the Reynolds number and the geometric parameters on the dynamic and thermal behaviour of laminar flow in plain plate fin-and- tube heat exchangers.

Guo et al. [24] investigated the heat transfer augmentation in the curved channels by considering different boundary conditions for the thermal problem. Their numerical results confirm that the better synergy between the temperature gradient and the velocity vector corresponds to the higher convective heat transfer rate.

This trend was also confirmed by the analysis of the effect of the irregular boundary conditions on the convective heat transfer mechanism carried out by Zhu et al. [26] who analysed the influence of the irregular boundary conditions on the convective heat transfer coefficient, both in laminar and in turbulent flow regimes.

By applying the FSP to the investigation of heat transfer augmentation in a corrugated tube, Vocale et al. [27] pointed out that this approach allows to identify the zones that do not positively contribute to the convective heat transfer.

More recently, Vocale et al. [28] have presented an analytical investigation on the heat transfer mechanism in non-Newtonian fluids by applying the FSP approach. They highlighted that this approach provides a detailed analysis of the local interaction between temperature and velocity fields, by then offering a promising tool to investigate convective heat transfer in systems where velocity and boundary conditions are totally asymmetric.

The present work is focused on the numerical investigation of the influence of flow insertion devices on the fluid-flow and heat transfer phenomena in parallel plates with asymmetric heating. This geometry is often used in compact heat exchangers or in solar collector. For the purpose of the here presented analysis, cylindrical inserts are considered and their effects are evaluated from the view-point of the FSP, to better evaluate the local heat transfer enhancement.

2. Mathematical model

The geometry here investigated is characterized by a parallel plates channel with cylinders inserted along the flow direction, as shown in figure 1(a), being L, H and W the channel length, height and width, respectively, D_c and P_c the cylinders diameter and pitch, respectively.

Since $W >> L$, the analysis has been carried out considering a two-dimensional parallel plates channel, as shown in Figure 1(b).

A Cartesian coordinate system x, y is introduced, being x the axial coordinate and y the perpendicular one. The centre of the cylinders is positioned on the channel axis (i.e. at $y=H/2$), as depicted in figure 1(b).

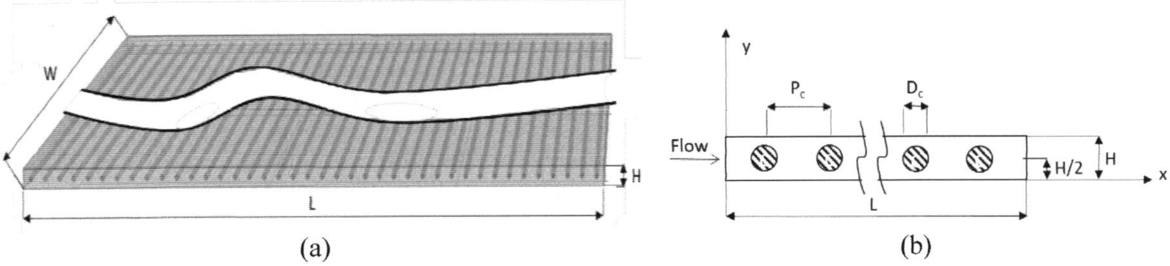

Figure 1. Sketch of the geometry under investigation.

By assuming that the fluid is Newtonian with constant physical properties, the flow is laminar and steady-state the governing equations in Cartesian coordinates are as follows:

$$\frac{\partial u}{\partial x}+\frac{\partial v}{\partial y}=0 \tag{1}$$

$$\begin{gathered}\rho\left(u\frac{\partial u}{\partial x}+v\frac{\partial u}{\partial x}\right)=-\frac{\partial p}{\partial x}+\mu\left(\frac{\partial^2 u}{\partial x^2}+\frac{\partial^2 u}{\partial y^2}\right)\\ \rho\left(u\frac{\partial v}{\partial x}+v\frac{\partial v}{\partial x}\right)=-\frac{\partial p}{\partial y}+\mu\left(\frac{\partial^2 v}{\partial x^2}+\frac{\partial^2 v}{\partial y^2}\right)\end{gathered} \tag{2}$$

$$\rho c_p\left(u\frac{\partial T}{\partial x}+v\frac{\partial T}{\partial y}\right)=\lambda\left(\frac{\partial^2 T}{\partial x^2}+\frac{\partial^2 T}{\partial y^2}\right)+\mu\left[2\left(\frac{\partial u}{\partial x}\right)^2+2\left(\frac{\partial v}{\partial y}\right)^2+\left(\frac{\partial u}{\partial y}+\frac{\partial v}{\partial x}\right)^2\right] \tag{3}$$

where u and v are the components of the velocity field, ρ, μ, c_p and λ are the fluid density, dynamic viscosity, specific heat at constant pressure and thermal conductivity, respectively, p is the pressure and T the fluid temperature.

The governing equations have been solved by considering the following inflow boundary conditions at x=0:

$$u=W, \qquad v=0, \qquad T=T_{in} \tag{4}$$

The boundary conditions for the fluid flow problem have been completed by the no-slip condition on any solid boundary (i.e. at the walls and on the cylinders surface) and by pressure condition at the channel outlet (i.e. $x=L$). Regarding the thermal problem, a prescribed and uniform heat flux q has been applied to the wall at $y=H$ while the wall at y=0 has been considered to be adiabatic as well as the cylinders surface. Outflow boundary conditions has been applied at the channel outlet.

The influence of the inserted cylinders on the heat transfer mechanism can be evaluated in terms of the Nusselt number, which on the non-adiabatic wall is defined as follows:

$$Nu=\frac{hD_h}{\lambda}=\frac{qD_h}{\lambda\left(T_w-T_b\right)} \tag{5}$$

being h the convective heat transfer coefficient, D_h the channel hydraulic diameter, T_w and T_b the wall and bulk temperature averaged over a single representative module of the channel, respectively.

Although the performance of enhanced heat transfer equipment can be properly assessed by means of the Nusselt number, the Field Synergy Principle approach is a promising tool to investigate convective heat transfer, since it provides a detailed analysis of the local interaction between temperature and velocity fields [2].

In literature, it has been demonstrated that the convective heat transfer rate is proportional to the dot product of the velocity vector and the temperature gradient integrated over the entire domain [2,28].

In the here analyzed model, the dot product between the velocity vector and the temperature gradient reads as follows:

$$\vec{v}\bullet\overrightarrow{\nabla T} = u\frac{\partial T}{\partial x} + v\frac{\partial T}{\partial y} \tag{6}$$

The dot product in Eq.(6) can be also defined as a function of the intersection angle between the two vectors:

$$\vec{v}\bullet\overrightarrow{\nabla T} = \left\|\vec{v}\right\|\left\|\overrightarrow{\nabla T}\right\|\cos\beta \tag{7}$$

where β is the intersection angle between the velocity vector and the temperature gradient.

From Eq.(7) it can be deduced that to properly evaluate the dot product (i.e. the convective heat transfer phenomenon), it has to be considered not only the magnitude of the two vectors but also the value of the intersection angle between them, which can be evaluated as follows:

$$\cos\beta = \left(\frac{\vec{v}\bullet\overrightarrow{\nabla T}}{\left\|\vec{v}\right\|\left\|\overrightarrow{\nabla T}\right\|}\right) = \frac{u\frac{\partial T}{\partial x} + v\frac{\partial T}{\partial y}}{\sqrt{u^2+v^2}\sqrt{\left(\frac{\partial T}{\partial x}\right)^2 + \left(\frac{\partial T}{\partial y}\right)^2}} \tag{8}$$

3. Results

Equations (1) to (3), together with their boundary conditions, have been numerically solved by adopting the finite element method, within the COMSOL Multiphysics environment. Both the governing equations have been solved using a direct linear solver and imposing a relative tolerance equal to 1e-06.

To evaluate the effects of inserted devices on the convective heat transfer three values of the cylinders diameter have been considered, namely D_c=0.25 H, D_c=0.375 H and D_c=0.50 H.

The accuracy of the numerical results have been checked by testing different grid sizes for each value of the cylinders diameter considered in the present study. Due to the interest of the present investigation in the hydrodynamically and thermally fully developed flow, for each considered grid size, the value of Nusselt number in the fully developed region has been evaluated.

Representative results of the mesh convergence analysis are shown in figure 2 where the Nusselt number is depicted as a function of the grid elements number N for D_c=0.50 H, together with the adopted mesh.

It has been observed that for a number of elements higher than 50000 the value of the Nusselt number becomes independent on the grid size, as shown in figure 2. Since this trend has been observed for all values of the cylinders diameter here considered, a mesh characterized by at least 50000 elements has been adopted in all numerical runs.

After having checked the accuracy of the numerical solution, several runs have been performed to evaluate the effects of the cylinders diameter on the dynamic and thermal fluid behaviour.

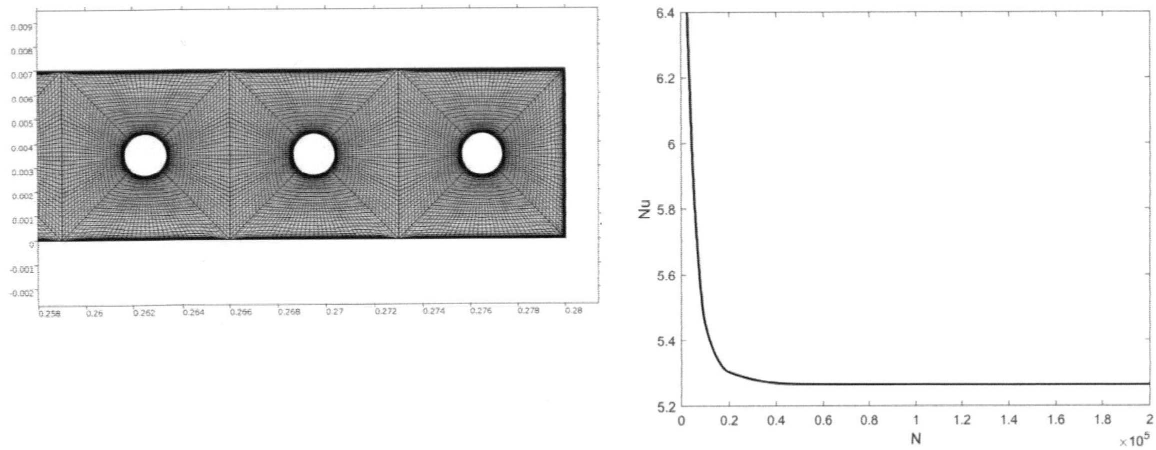

(a) (b)

Figure 2. Sketch of the mesh (a) and mesh sensitivity analysis (b).

As above discussed, the here presented analysis is focused on hydrodynamically and thermally fully developed flow, therefore all the results are presented for the fully developed region.
The influence of the inserted cylinders on the flow pattern and temperature field is presented in Figure 3 where the streamlines and the dimensionless temperature contours for two representative values of the cylinders diameter (i.e. the lowest and the highest values considered in the present analysis) are depicted.
Concerning the flow pattern, due to the symmetric conditions and the low value of the Reynolds number (i.e. Re=100), the flow is characterized by a symmetric behaviour with respect to the channel symmetry axis. It has been also observed that for all values of the here considered cylinders diameter there is interference between two adjacent cylinders [8].
As expected, since only the upper wall is subjected to the heat flux, the fluid presents an asymmetric thermal behaviour, as shown in Figure 3(b) where the temperature contours (normalized with respect to the maximum and minimum values) are depicted.
Due to this behaviour, the interaction between the velocity vector and the temperature gradient is not easily predictable and the Field Synergy Principle approach becomes fundamental to understand heat transfer enhancement mechanisms.
By evaluating the influence of the inserted cylinders in terms of Nusselt number it has been observed that Nu decreases as the cylinders diameter increases. This effect becomes more evident for high values of the cylinders diameter; in fact, for D_c=0.5H the heat transfer mechanism is less efficient than in the parallel plates channel without insertion (i.e. the reference case [29]), as shown in Table 1.
The FSP helps to understand this behaviour. In fact, while the Nusselt number definition is limited to the system's boundary, the Field Synergy Principle coveys information representative of an effect distributed over the whole domain. Therefore, the analysis of the local interaction between the velocity vector and the temperature gradient could suggest suitable geometry modifications, which allow to enhance the convective heat transfer mechanism
On the left side of Figure 4, the velocity vectors (red arrows) and the temperature gradients (blue arrows) in the non-dimensional form are depicted for all the values of the cylinders diameter here considered. It can be observed that, due to the non-symmetric boundary conditions, the two vectors are almost perpendicular over the whole domain. Smaller intersection angles are observable only in the region close to the adiabatic wall.
To better evaluate the influence of the cylinders on the dynamic and thermal fluid behaviour, in Figure 5 the velocity vectors and the temperature gradients for the reference case (i.e. the parallel plates channel without insertion) are reported.

IOP Conf. Series: Journal of Physics: Conf. Series **923** (2017) 012032 doi:10.1088/1742-6596/923/1/012032

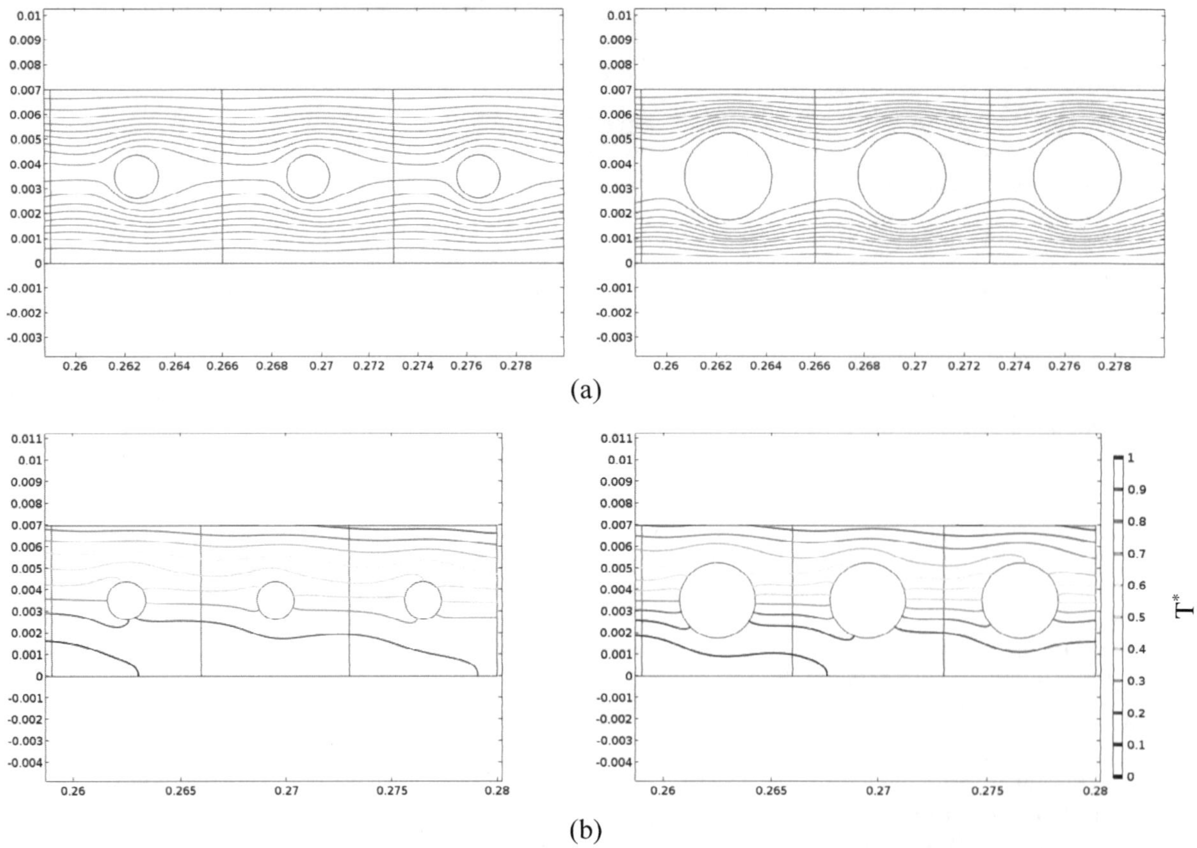

Figure 3. Streamlines and dimensionless temperature contours: (a) D_c=0.25 H; (b) D_c=0.50 H.

Table 1. Nusselt number for the here analysed cases.

	Nu
D_c=0.25 H	5.984
D_c=0.375 H	5.873
D_c=0.50 H	5.407
Reference case	5.385

It has been observed that in the reference case the region in which the two vectors are perpendicular was higher than in cases where the cylinders are inserted in the flow direction As demonstrated in literature [16,17] the higher convective heat transfer coefficient corresponds to a lower value of the intersection angle between the vectors (i.e. the two vectors have to be aligned), therefore, the insertion of the cylinders in the streamwise direction enhances the convective heat transfer coefficient.
On the right side of Figures 4 and 5 the maps of the dimensionless dot product between the velocity vector and the temperature gradient (normalized with respect to the total thermal power and the hydraulic diameter of the channel) are also reported.

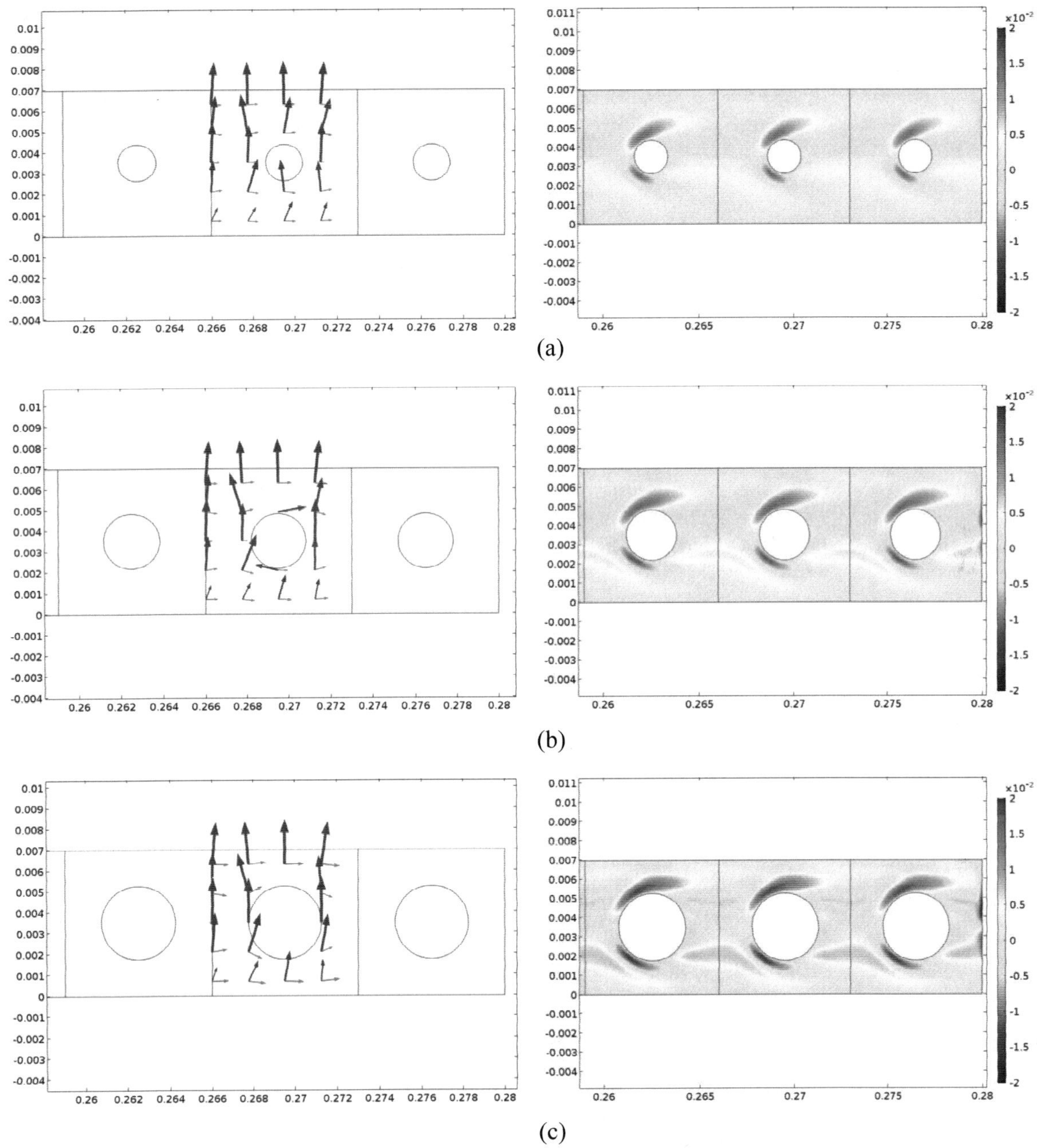

Figure 4. Dimensionless velocity vectors (red arrows), temperature gradients (blue arrows) and dot product: (a) D_c=0.25 *H*; (b) D_c=0.375 *H*; (c) D_c=0.50 *H*.

The maps in Figure 4 show that the region of the channel in which the dot product assumes negative values increases as the cylinders diameter increases. This result can explain the cause of the reduction in the convective heat transfer coefficient that has been observed by evaluating the performance of the heat transfer device in terms of Nusselt number.

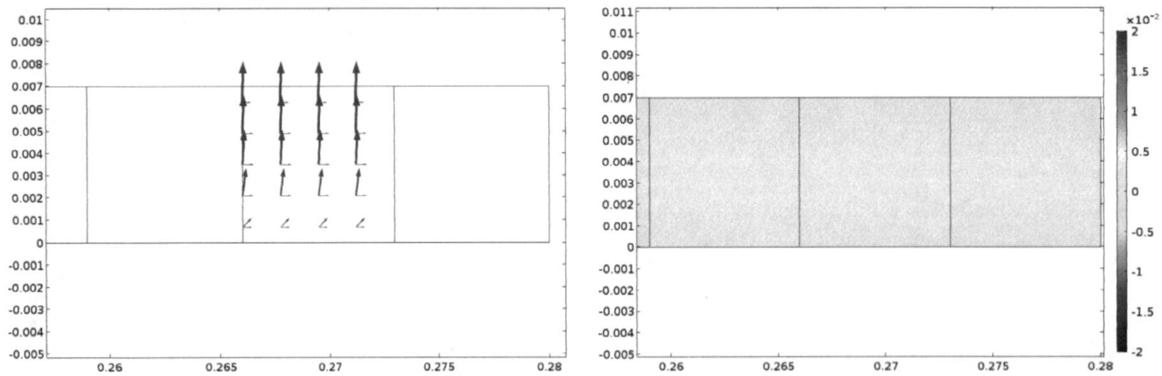

Figure 5. Dimensionless velocity vectors (red arrows), temperature gradients (blue arrows) for the reference case (i.e. parallel plates).

Defining the region in which the dot product assumes positive values as effective heat transfer area e the other one (i.e. the region in which the dot product assumes negative values) as ineffective heat transfer area it is more clear what happens as the cylinders diameter increases. In Figure 6 the effective (red region) and the ineffective (blue region) heat transfer areas are depicted.

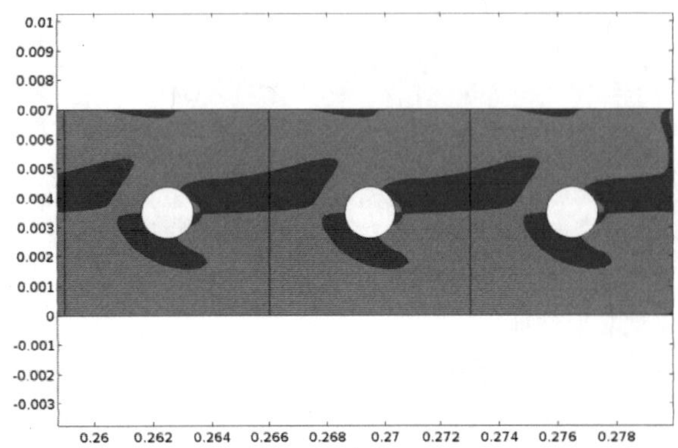

(a)

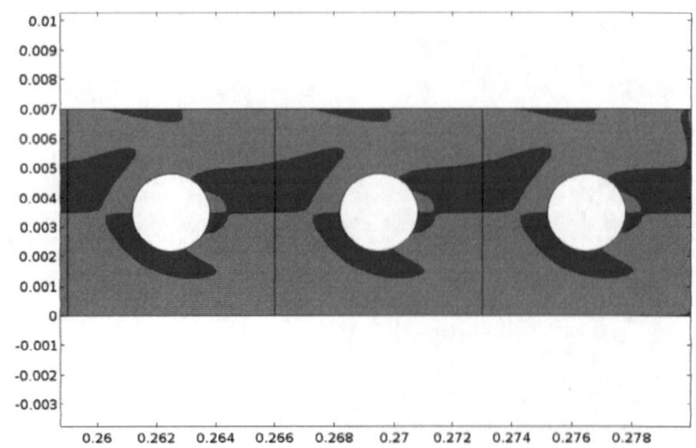

(b)

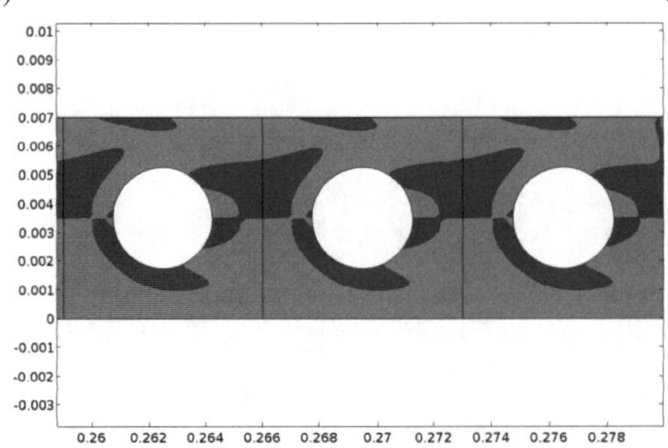

(c)

Figure 6. Effective (red regions) and ineffective (blue regions) heat transfer area: (a) D_c=0.25 H; (b) D_c=0.375 H; (c) D_c=0.50 H.

IOP Conf. Series: Journal of Physics: Conf. Series **923** (2017) 012032 doi:10.1088/1742-6596/923/1/012032

It can be observed that with increasing cylinders diameter the effective heat transfer area decreases and the ineffective one increases, as shown in Table 2.
Therefore, as the cylinders diameter increases the overall thermal performance of the whole system decreases.

Table 2 Effective and ineffective heat transfer area.

	Effective [%]	Ineffective [%]
D_c=0.25 H	76.63	23.37
D_c=0.375 H	67.20	32.80
D_c=0.50 H	57.25	42.75

4. Conclusions

The dynamic and thermal behaviour of a fully developed flow in a channel bounded by two parallel plate walls heated asymmetrically and with cylindrical inserts has been numerically investigated. The influence of the cylinders on the flow pattern and the heat transfer mechanism has been evaluated by considering different values of the cylinders diameter.

The outcomes in terms of Nusselt number point out that the convective heat transfer coefficient decreases with increasing inserts diameter for a given pitch value. The causes of the convective heat transfer reduction have been explained by adopting the Field Synergy Principle approach, which provides a detailed analysis of the local interaction between the velocity and temperature gradient fields.

The numerical results, in terms of dot product between the two vectors, point out that the extent of the region in which the heat transfer mechanism is effective decreases as the cylinders diameter increases, thus reducing the convective heat transfer coefficient.
Therefore, it can be concluded that the Field Synergy Principle approach could be a suitable tool to assess the performance of passive heat transfer enhancement techniques in internal flows, especially when the flows are subjected to asymmetric boundary conditions. Moreover, the analysis of the local interaction between the velocity vector and the temperature gradient could suggest suitable geometry modifications, which allow to enhance the convective heat transfer mechanism.

References

[1] Webb R L 1994 *Principles of Enhanced Heat Transfer* (New York: Taylor & Francis)
[2] Rainieri S and Pagliarini G 2002 *Int. J. Heat Mass Tran.* **45** 4525–36
[3] Rainieri S, Bozzoli F and Pagliarini G 2012 *Int. J. Heat Mass Tran.* **55** 498–504
[4] García A, Solano J P, Vicente P G and Viedma A 2012 *App. Therm. Eng.***35** 196-201
[5] Rainieri S, Bozzoli F, Cattani L and Pagliarini G 2013 *Int. J. Heat Mass Tran.* **59** 353–62
[6] Bozzoli F, Cattani L and Rainieri S 2016 *Int. J. Heat Mass Tran.* **101** 76–90
[7] Harleß A, Franz E and Breuer M 2017 *Int. J. Heat Mass Tran.* **107** 1076–84
[8] Kundu D, Haji-Sheikh A and Lou D Y S 1991 *Numer. Heat Tr. A-Appl.* **19** 345-60
[9] Rainieri S, Bozzoli F, Schiavi L and Pagliarini G 2011 *Int. J. Numer. Method H.* **21** 559–71
[10] Rainieri S, Bozzoli F, Mordacci M and Pagliarini G 2012 *Heat Tran. Eng.* **33** 1120–29
[11] Zachár A 2010 *Int. J. Heat Mass Tran.* **53** 3928–39
[12] Zheng N, Liu W, Liu Z, Liu P and Shan F 2015 *App. Therm. Eng.***90** 232-41
[13] Vocale P, Mocerino A, Bozzoli F and Rainieri S 2016 *J. Phys. Conf. Ser.* **745** 032072
[14] Pagliarini G, Vocale P, Mocerino A and Rainieri S 2017 *J. Phys. Conf. Ser.* **796** 012014
[15] Lin Z M, Wang L B, Lin M, Dang W and Zhang Y H 2017 *App. Therm. Eng.***115** 644-58
[16] Guo Z Y, Li D Y and Wang B X 1998 Int. J. Heat Mass Tran. 41 2221–25

[17] Tao W Q, Guo Z Y and Wang B X 2002 *Int. J. Heat Mass Tran.* **45** 3849–56
[18] Guo Z Y, Tao W Q and Shah R K 2005 *Int. J. Heat Mass Tran.* **48** 1797–1807
[19] Tao W Q, He Y L, Wang Q W, Qu Z G and Song F Q 2002 *Int. J. Heat Mass Tran.* **45** 4871–79
[20] He Y L, Tao W Q, Song F Q, Zang W 2005 *Int. J. Heat and Fluid Flow* **26** 459–473
[21] Ma L D, Li Z Y and Tao W Q 2007 *Int. Comm. Heat Mass Tran.* **34** 269–276
[22] Tao Y B, He Y L, Wu Z G and Tao W Q 2007 *Int. J. Heat and Fluid Flow* **28** 1531–44
[23] Wei L, Zhichun L, Tingzhen M and Zengyuan G 2009 *Int. J. Heat Mass Tran.* **52** 4669–72
[24] Guo J, Xu M and Cheng L 2011 *Int. J. Heat Mass Tran.* **54** 4148–51
[25] Hamid M O A, Zhang B and Yang L 2014 *Energy* **76** 241–253
[26] Zhu X W and Zhao J Q 2016 *Int. J. Heat Mass Tran.* **100** 347–54
[27] Vocale P, Mocerino A, Bozzoli F and Rainieri S 2017 *Proc. of 7th International Symposium on Advances in Computational Heat Transfer, CHT-17* (Napoli, 2017)
[28] Vocale P, Mocerino A, Bozzoli F and Rainieri S *Advances in Mathematical Physics* (*in press*)
[29] Bejan A 1993 *Heat Transfer* (New York: John Wiley & Sons, Inc.)

Numerical analysis of flow resistance and heat transfer in the transitional regime of pipe flow with twisted-tape turbulators

R Rossi[1], L Cattani[3], A Mocerino[4], F Bozzoli[1,3], S Rainieri[1,3], R Caminati[2] and G Pagliarini[1]

[1] CIDEA Interdepartmental Centre, University of Parma, Parco Area delle Scienze 181/A, I-43124 Parma, Italy

[2] Be.Tube S.r.l., Via Cicagna 52, 47522, Cesena (FC), Italy

[3] SITEIA.PARMA Interdepartmental Centre, University of Parma, Parco Area delle Scienze 181/A, I-43124 Parma, Italy

[4] Department of Industrial Engineering, University of Parma, Parco Area delle Scienze 181/A, I-43124 Parma, Italy

E-mail: fabio.bozzoli@unipr.it

Abstract. In this paper, we present the numerical analysis of the fully developed flow and heat transfer in pipes equipped with twisted-tape inserts in the laminar to transitional flow regime. The flow Reynolds number ranges from 210 to 3100 based on the pipe diameter, whereas the Prandtl number of the working fluid, a 40% mixture of water and ethylene glycol, is about 45 at the average film temperature. The numerical study is carried out via Scale Adaptive Simulations (SAS) where the k-ω SST model is employed for turbulence modeling. Using SAS and low-dissipation discretization schemes, the present study shows that it is possible to capture the transition from the laminar regime to the pulsating or pseudo-laminar flow regime induced by the twisted-tape at low Reynolds numbers, as well as the transition to moderate turbulent regime at the higher, yet non-turbulent for smooth pipes, range of Reynolds numbers. Numerical results, validated against experiments performed in a dedicated test rig, show very good agreement with measured data and an increase of the friction factor and Nusselt number in the range of 4 to 7 times and 6 to 15 times, respectively, of the values for an empty pipe.

1. Introduction

The use of heat transfer augmentation techniques is a common practice when heat exchangers operate with high-viscosity fluids and thus at low Reynolds numbers, where the heat transfer is minimal due to the persistence of the laminar flow regime. The urgency on heat transfer enhancement is also driven by the tendency in many industrial, commercial and domestic applications (e.g. heat pumps, air conditioning) towards increasing compactness and hence, progressively reduce the Reynolds number [1].

Performance evaluation of various type of heat transfer passive augmentation mechanism have been the objective of several studies in the present research group, with a focus on the effect of pipe wall corrugations and wall curvature [2, 3]. Turbulators represent another mean of increasing the heat transfer rate in high-Prandtl number fluids and among the numerous types of turbulators available to date, twisted-tapes represent simple and effective devices employed in many industrial applications.

Although several experimental studies and correlations exist for the pressure drop and heat transfer characteristics of twisted-tapes, there are only few numerical studies concerned with the mechanism of heat transfer augmentation at low Reynolds number in the transitional regime. Hong and Bergles [4] reported one of the first available set of pressure drop and heat transfer measurements in pipes with twisted-tape inserts using water and ethylene glycol as working fluids. The flow Reynolds number based on the pipe diameter ranged from 13 to 2460 and the Prandtl number from 3 to 192. An increase of the Nusselt number up to 9 times the empty tube value was reported and a correlation based on the Reynolds and Prandtl numbers and on the tape twist-ratio was provided by the authors, but the onset of transition in the flow regime was not addressed in their study. Moreover, from the plots of friction and heat transfer data a sudden variation in measured values indicating a change in the flow regime was not clearly discernible.

A similar but more recent experimental study was reported by Agarwal and Rja Rao [5], where the pressure drop and heat transfer characteristics for the twisted-tape were investigated for both cooling and heating conditions using Servotherm oil as working fluid (Pr=195-375) up to a Reynolds number of 4000 based on the pipe diameter. The authors reported friction factors and Nusselts numbers up to 9 and 5 times higher, respectively, than in a smooth tube. A significant departure from available correlations for the laminar regime at approximately Re=500 could also be clearly noted from the plots of friction and heat transfer data, indicating the anticipated onset of the unstable regime in the pipe equipped with the twisted tape inserts.

Kaliakatsos et al [6] presented a numerical study of the flow and heat transfer in a pipe equipped with a twisted tape for Reynolds numbers in the range from 500 to 1500 based on the pipe diameter. The numerical model employed a steady and laminar formulation of the flow and energy equations and even though the predicted Nusselt number compared favorably to the correlation proposed by Hong and Bergles [4], the numerical results underestimated the heat transfer rate up to 35% for the highest Reynolds number when compared to the correlation presented in the study of Saha and Dutta [7].

A numerical study in the transitional flow regime from a Reynolds number of 800 to 2000 based on the pipe diameter using a turbulent, yet steady, formulation of the flow and energy equations was presented by Savekar et al. [8], where the k-ω model was used to to predict the flow pressure drop and heat transfer in a pipe with the twisted-tape insert. The comparison of numerical results with the measurements reported by Manglik and Bergles [9] shows increasing divergence in the predicted Nusselt number with increasing Reynolds number, presumably because of the onset of time-dependent perturbations which could be not accounted for by the steady formulation adopted in the analysis.

In this work, we present a numerical study of the friction and heat transfer characteristics for pipes equipped with twisted-tape inserts in the laminar to transitional flow regime. The computations are carried out with the open-source CFD package OpenFOAM®, where the flow and energy equations are solved using unstructured and second-order accurate finite-volume schemes. The numerical study is based on Scale Adaptive Simulations (SAS) where the k-ω SST model is employed for turbulence modeling. Using such model and low dissipation discretization schemes, we show that is possible to capture the laminar to turbulent transition in the flow regime and provide an accurate prediction of the increased pressure drop and heat transfer associated with the twisted-tape inserts. The paper is organized as follows: in Section 2 the numerical model adopted in the analysis is presented. An overview of turbulence modeling is given in Section 3, with particular emphasis on the SAS model. The validation of the numerical model is presented in Section 4, where the estimated friction factor and Nusselt number are compared to measurements performed in a dedicated test rig. Additional insights on the laminar

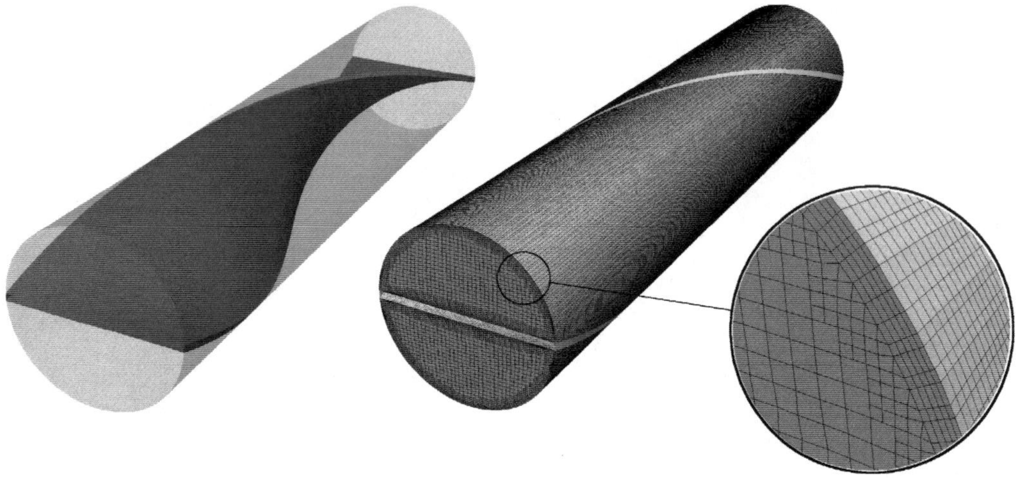

Figure 1. Numerical model of the pipe with twisted-tape insert: left, model geometry; right, computational grid; detail, close-up view of the near-wall mesh structure and resolution.

to turbulent flow transition and on the heat transfer augmentation mechanism are provided in Section 5. Conclusions and final remarks are reported at the end of the paper.

2. Numerical model

The numerical model adopted in the present study, shown in Fig. 1, consists of a single periodic module of a smooth pipe equipped with the twisted-tape, corresponding to a 180 deg rotation of the turbulator, where fully-developed flow conditions are assumed for the velocity and temperature fields. The tape is characterized by a twist-ratio $y = H/D$, where H is the pitch for 180-deg rotation of the tape and D the pipe diameter, of 4.44. In the present model, the tape is in perfect contact with the pipe wall, thus forming two separate half-sections in the pipe. The flow Reynolds number, based on the pipe diameter, ranges from 210 to 3100, whereas the Prandtl number of the working fluid, evaluated at film temperature, is 45. Because of the short length of the periodic module with respect to the pipe diameter, the variation of the flow bulk temperature is minimal and thus the temperature dependence in the properties of the working fluid is neglected in the simulations.

The following RANS formulation of flow and energy equations are solved in the simulations using the open-source finite-volume code OpenFOAM®:

$$\frac{\partial U_i}{\partial x_i} = 0 \tag{1}$$

$$\frac{\partial U_i}{\partial t} + \frac{\partial U_i U_j}{\partial x_j} = -\frac{1}{\rho}\frac{\partial P}{\partial x_i} + \nu \frac{\partial^2 U_i}{\partial x_j \partial x_j} - \frac{\partial}{\partial x_j}(\overline{u_i u_j}) \tag{2}$$

$$\frac{\partial T}{\partial t} + \frac{\partial U_j T}{\partial x_j} = \alpha \frac{\partial^2 T}{\partial x_j \partial x_j} - \frac{\partial}{\partial x_j}(\overline{u_j \theta}), \tag{3}$$

where uppercase and lowercase denote mean and fluctuating values, respectively. The $\overline{u_i u_j}$ and $\overline{u_j \theta}$ terms in Eq. (2) and (3) represent the Reynolds stress tensor and the turbulent heat flux, respectively, and require modeling in the framework of RANS simulations. In the present work, the turbulent heat flux is approximated using the eddy-diffusivity model:

$$\overline{u_j\theta} = -\alpha_t \frac{\partial T}{\partial x_j} = -\frac{\nu_t}{Pr_t}\frac{\partial T}{\partial x_j}, \tag{4}$$

where ν_t and $Pr_t = \nu_t/\alpha_t$ denote the eddy-viscosity and the turbulent Prandtl number, respectively, and α_t the eddy-diffusivity. The standard value for wall-bounded turbulent flows of Pr_t of 0.9 is employed in the simulations, whereas the the eddy-viscosity is obtained via the $k-\omega$ SST model described in the next section.

A hex-dominant grid, also shown in Fig. 1, is employed in the computations, where boundary-layers cells aligned with the pipe surface are used to resolve the flow and thermal boundary layer at the wall. A constant and isotropic grid resolution of about $2.3\times10^{-2}D$, where D is the pipe diameter, is employed in the core region of the flow whereas the grid is isotropically refined near the tape and pipe walls to twice the resolution of the core region. The total cell count for the present grid is about 1.4M. The time-integration of the flow and energy equations is performed using a second-order implicit time-stepping scheme, whereas a limited linear scheme and a corrected linear scheme are used, respectively, for the convective and diffusive terms of the governing equations.

The boundary conditions for the velocity and temperature field at the pipe wall are of no-slip and uniform heat flux, respectively, whereas the no-slip condition and zero-flux are employed on the twisted-tape surface, meaning the heat conduction through the thickness of the tape is neglected. In the streamwise direction, the fully-developed condition for the flow field is achieved by imposing a cyclic condition between the period sections of the pipe, where the computed velocity profile at the outflow section is simply copied at the inlet section of the computational domain and the flow driven by an externally imposed constant pressure gradient.

In the case of the energy equation, however, the temperature field must be rescaled between the inlet and outlet sections of the computational domain due to the rise in the bulk temperature of the flow following from the imposed heat flux at the pipe wall. This is achieved in the present numerical model by modifying the energy equation (3) following the method proposed by Patankar et al. [10]. Figure 2 shows the validation of the implemented rescaling technique for the case of an empty pipe against the theoretical values of the friction factor and Nusselt number factors in the laminar regime $f = 64/Re$ and $Nu = 4.364$ as well as against the correlations proposed by Petukhov [15] for the turbulent regime:

$$f = (0.79\log(Re) - 1.64)^{-2} \tag{5}$$

$$Nu = \frac{(f/8(Re-1000)Pr)}{(1+12.7(f/8)^{1/2}(Pr^(2/3)-1))}, \tag{6}$$

The agreement between numerical results and theoretical values in the laminar regime is excellent and that between numerical results and Eqs. (5) and (6) also very good in the turbulent regime.

3. Turbulence modeling

The target flow regime of the present analysis is particularly challenging from the simulation standpoint because of the transition mechanism from the stable laminar regime to the fully developed turbulent flow at the highest Reynolds number. Such mechanism is in fact controlled by the balance between the stabilizing effect of the swirling motion induced by the twisted tape and the growth of perturbations associated with the onset of the turbulent flow regime [9]. Therefore, the onset of time-dependent perturbations, even at large-scales, cannot be accounted

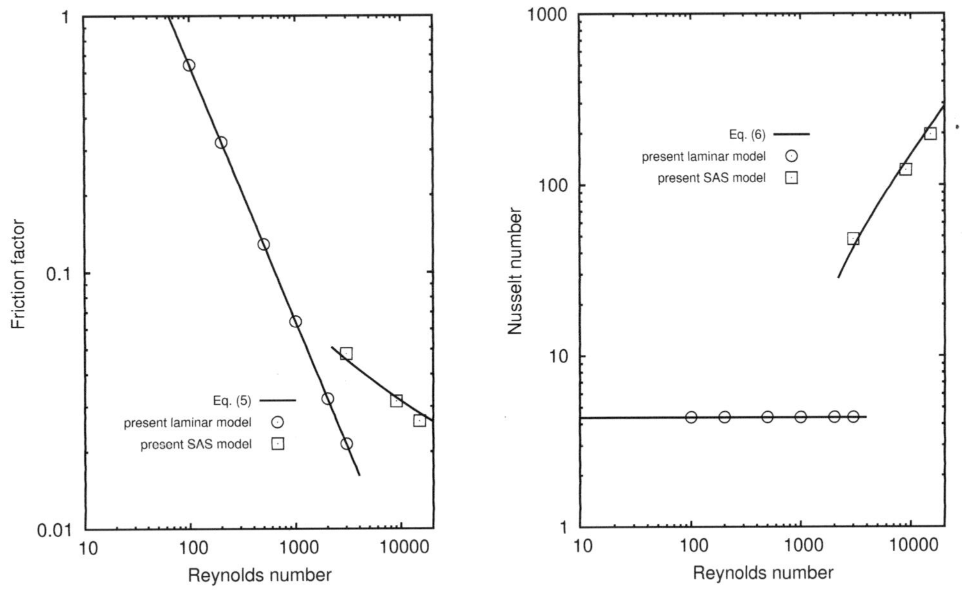

Figure 2. Validation of numerical model for laminar and turbulent flow in an empty pipe: left, friction factor; right, Nusselt number.

for in the framework of standard Unsteady-RANS (URANS) simulations. URANS operates in fact on a "separation of scales" basis and can resolve only unsteady effects characterized by very low frequencies, which are typically associated with abrupt changes in the geometry or time-dependent boundary conditions.

In the SAS formulation employed in this work, the turbulence model is formulated in a way to operate in Scale-Resolving Simulation (SRS) mode, where the model automatically balances the contribution of modelled and resolved parts of the turbulent stresses [11]. One of the more interesting features of the SAS approach is that for unstable flows as that induced by the twisted tape in the transitional regime, the model changes smoothly from a time dependent scale-resolving model to a steady scale-separating model, based on the specified time step.

The eddy-viscosity ν_t required to close the RANS formulation of governing equations (2) and (3) is obtained in the SAS model via the following relation:

$$\nu_t = c_\mu^{1/4}\sqrt{k}L, \tag{7}$$

where k is the turbulent kinetic energy and L is the integral length-scale introduced in the SRS formulation. The integral length-scale is computed via a modified form of the $k-\omega$ SST model of Menter [12], where an additional source term is introduced in the original transport equation for the turbulence frequency ω (not reported here for sake of brevity):

$$Q_{SAS} = \max\left[\rho\zeta^2 S^2\left(\frac{L}{L_{\nu\kappa}}\right)^2 - C_{SAS}\frac{2\rho k}{\sigma_\Phi}\max\left(\frac{1}{k^2}\frac{\partial k}{\partial x_j}\frac{\partial k}{\partial x_j}, \frac{\partial 1}{\partial \omega^2}\frac{\partial \omega}{\partial x_j}\frac{\partial \omega}{\partial x_j}, 0\right)\right], \tag{8}$$

where ζ, σ_Φ and C_{SAS} are constants, S is the strain rate magnitude and $L_{\nu\kappa}$ the Von-Karman length-scale. As in the standard RANS or scale-separating formulation of the $k-\omega$ SST, the model is able to account for the viscosity-affected near-wall region of the flow without the use

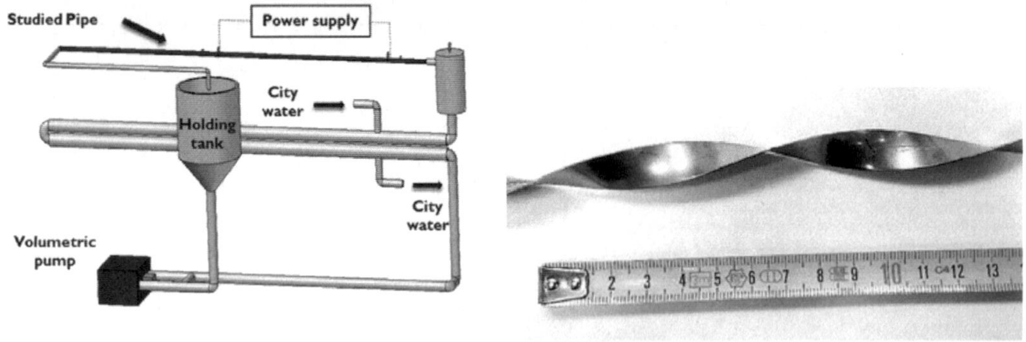

Figure 3. Experimental setup for measurements of friction factor and Nusselt number in the pipe flow with twisted-tape insert: left, schematic view of the test rig; right, twisted-tape employed in the test-rig.

of extra damping functions provided that the grid resolution is sufficient. In the present work, the grid resolution in wall-units at the first cell-centroid next to the wall ranges from $y^+=0.24$ to $y^+=0.85$ for the lowest and highest Reynolds number cases, respectively, and the viscous sublayer is therefore well-resolved. However, the thickness of the conductive thermal sublayer scales with the $Pr^{-1/2}$, leading to a near-wall resolution for the thermal field with the same grid of about $y^+=1.6$ to $y^+=5.7$ for the lowest and highest Reynolds number cases, respectively. Nonetheless, the validation study presented in Fig. 2 shows that the reduced grid resolution in the case of the energy equation is not such to limit the accuracy of present results.

The $k-\omega$ SST SAS model has been tested in numerous complex flow configurations [13] and proved to be effective in presence of instabilities associated with large separated zones or vortical instabilities, as in the case of pipe flows with twisted-tape inserts. Note that results for the friction factor and Nusselt number obtained with a laminar formulation of the numerical model are also reported for the Reynolds number range from 210 to 910 for sake of comparison with the SAS model.

4. Experimental validation of the numerical model

The validation of the numerical model for the pipe with the twisted-tape insert is carried out by comparing the predicted friction factor and Nusselt number factors with measurements performed in a dedicated test rig, shown in Fig. 3. The experimental tests are performed by adopting a stainless steel type AISI 304 tube fitted with the twisted tape. The pipe is 3 m long and presents a wall thickness of 1 mm and an internal diameter of 13.5 mm. The test section is inserted horizontally in a loop, which is completed by a secondary heat exchanger fed with city water enabling to keep the working fluid temperature constant at the tube's inlet.

The working fluid is conveyed by a volumetric pump to a holding tank from which it enters the test section equipped with fin electrodes connected to a power supply (type HP 6671A) working in the ranges 0−8 V and 0−220 A. This setup allows investigating the heat transfer performance of the tube under the prescribed condition of heat generated by Joule effect in the wall. By neglecting the changes in the local wall resistivity due to the cold working process, changes in the geometry and the effect of heat conduction in the wall, this heating condition well approximates the constant heat flux boundary condition at the fluidwall interface. The whole length of the heat transfer section was thermally insulated to minimize the heat exchange to the environment. The wall temperature was measured through 40 type T thermocouples, previously calibrated and connected to a multichannel ice point reference (type KAYE K170- 50C). Regarding the wall temperature, the sensors are attached at different circumferential locations to the external tubes surface and at different axial locations along the heated section. The inlet temperature is measured by a thermocouple probe placed on the tubes wall upstream the start of the heating section. The bulk temperature at any location in the heat transfer section was calculated from the power supplied to the tube, which is assumed distributed uniformly per unit length over the heat transfer surface area, and by taking into account for the heat losses through the insulation. The outlet temperature is also monitored by placing a thermocouple at the end of the heated section.

The heat losses towards the environment are estimated from a preliminary calibration of the apparatus. Volumetric flow rates are obtained by measuring the time needed to fill a volumetric flask placed at the outlet of the test section. The data acquisition system consisted essentially of a high precision multimeter (type HP 3458A) connected to a switch control unit (type HP 3488A) driven by a Personal Computer. Pressure drops throughout the section fitted with twisted tapes are measured in isothermal conditions by a Rosemount-3051S differential pressure transducer. A 40% mixture of water and ethylene glycol is used as working fluid in the Reynolds number range 210-3100. In the temperature range characterizing the experimental conditions the Prandtl number of the working fluid varied in the range 44-51.

Figure. 4 shows the comparison of numerical results with measured data from the test rig and with correlations available from the experimental study of Agarwal and Raja Rao [5]:

$$f = 19.48(Re)^{-0.6519}(y)^{-0.6281} \tag{9}$$

$$Nu = 0.725(Re)^{0.568}(y)^{-0.788}(Pr)^{1/3}, \tag{10}$$

The agreement with measurements is very good for the Nusselt number, whereas the numerical model under-predicts to some extent the friction factor. The trend in the predicted pressure drop and heat transfer characteristics is overall in good agreement between the numerical model and the reference data, where the measured friction factor and Nusselt number factors are found in very good agreement with the existing correlation in the literature. Moreover, the comparison of numerical results obtained with the laminar and the SAS model suggests that the onset of the transitional regime affects the heat transfer rate to a greater extent than the pressure drop.

5. Flow and temperature fields analysis

The overall good agreement between the computed friction factor and Nusselt number against the experimental measured and reference data demonstrates the capability of the present numerical model to predict the characteristics of the pipe flow with the twisted-tape insert from the laminar to the transitional regime. In this section, such characteristics are illustrated and discussed in more detail with the help of contour plots and time-histories of the computed velocity and temperature fields.

In Fig. 5, the contour plots of velocity components and temperature on a cross-section located halfway of the periodic module are shown for increasing values of the Reynolds number. At Re=496, the instantaneous axial velocity u_z/U_b features two loci of stable and fast moving fluid on the outer side of the pipe half section with respect to the bending direction of the twisted-tape. The high-speed flow is pushed by centrifugal forces acting outwards from the inner side of the pipe, where slow moving fluid resides. With the increase in the Reynolds number, the contour plots show clearly how the flow instability originates from the low speed flow, where the negative radial pressure gradient has a destabilizing effect promoting the growth of perturbations in the flow. The flow instability propagates from the inner side to the outer side of the pipe half section with further increase in the Reynolds number, eventually determining the breakdown of the high-speed flow regions at the highest values.

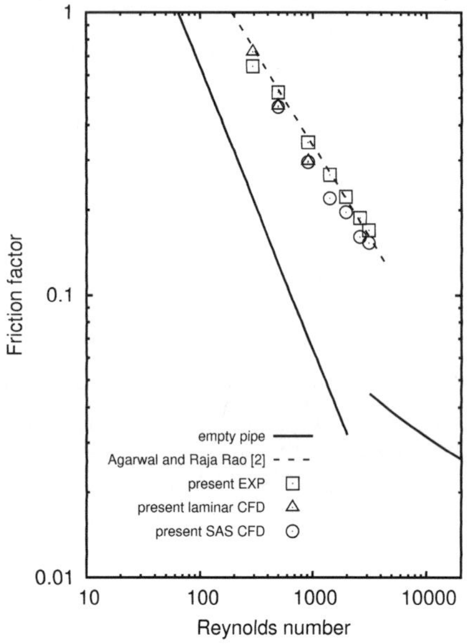

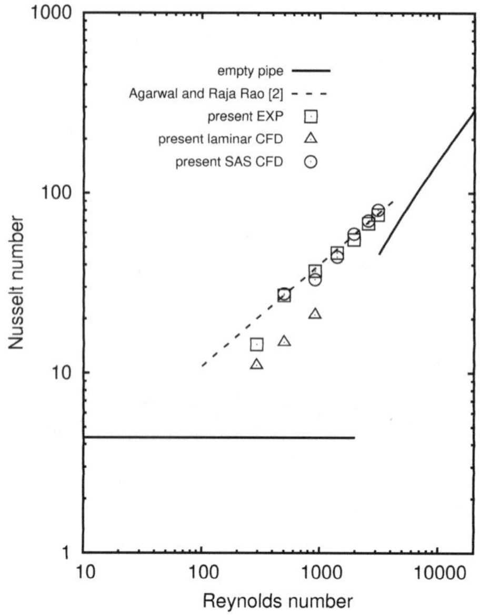

Figure 4. Validation of numerical results against measured data and available correlations for pipe flow with twisted-tape insert: left, friction factor; right, Nusselt number.

The contours of instantaneous swirl velocity $(u_x^2 + u_y^2)^{\frac{1}{2}}/U_b$ are also reported in Fig. 5. In the fully stable flow regime, the magnitude of the swirl flow is about 20% of the flow bulk velocity and the swirl flow is almost evenly distributed across each of the pipe half sections, with the peak in swirl velocity located about halfway between the pipe wall and the twisted tape surface along the radius. Because of the increasing dominance of centrifugal forces with increasing Reynolds numbers, the magnitude of the swirl flow reaches a maximum of about 35% of the bulk velocity at Re=3100. Moreover, the peak in swirl velocity moves towards the pipe wall while the uniform distribution of the swirl flow is replaced by several local maxima due to the flow instability taking place within the pipe.

The change in the flow structure due to the onset of the flow instabilities is reflected in the contour plots of normalized temperature $(T - T_w)/(T_w - T_b)$ in Fig. 5, where T_w is the average temperature of the pipe wall and T_b the flow bulk temperature. Within the stable flow regime at Re=496, the temperature distribution in each of the pipe half section features a cold fluid region corresponding to the high-speed flow in the outer part of the pipe as well as a secondary cold spot located in the inner region of the pipe occupied by the flow moving at low speed. As the Reynolds number is increased and the flow instability takes place, the two cold spots start to blend together thanks to the enhanced mixing associated with the onset and growth of perturbations in the flow. Once the turbulent mixing takes place the cold fluid is also able to penetrate the contact edge between the twisted-tape and the pipe, a region where the heat transfer rate is significantly decreased in the laminar and stable regime.

In order to better highlight the onset of the transition regime in the flow, Fig. 6 presents the time-history of the spanwise velocity component for increasing Reynolds numbers from a probe located at the center of one of the pipe half sections. At Re=496, the flow appears to be perfectly stable after the the initial oscillations due to the simulation start up from a zero velocity initial conditions, indicating that perturbations induced by the twisted-tape are damped

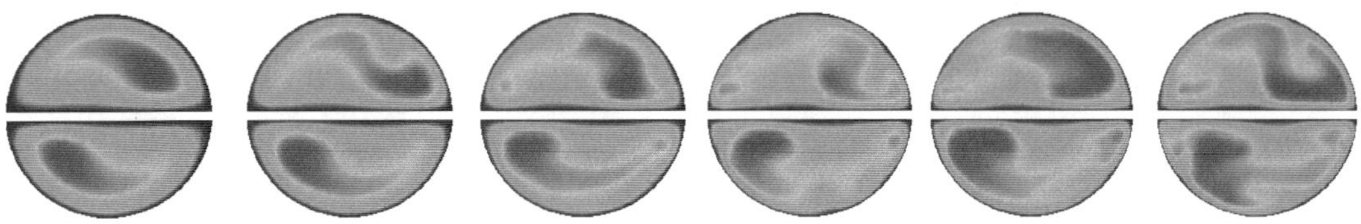

(a) Normalized axial velocity

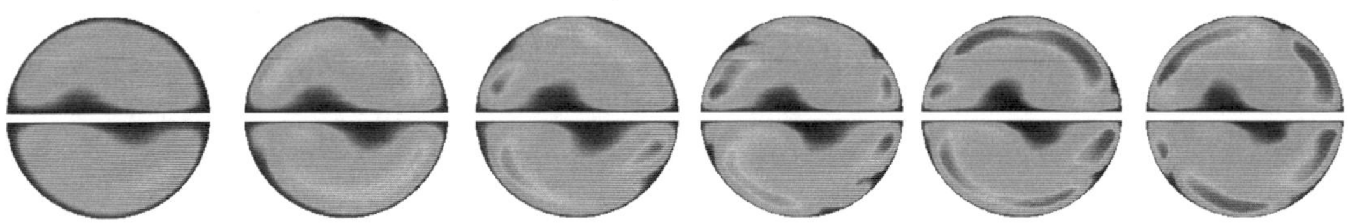

(b) Normalized swirl velocity

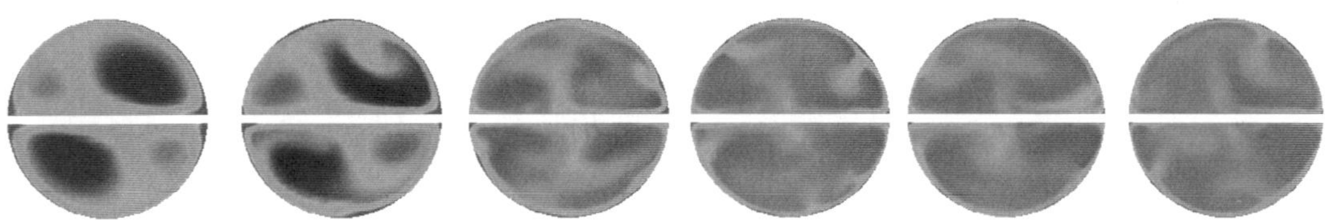

(c) Normalized temperature

Figure 5. Contours of flow and temperature field variables for increasing values of the Reynolds number: from left to right; Re=496, Re=910, Re=1400, Re=1940, Re=2580, Re=3100; from top to bottom; contours of normalized axial velocity $u_z/U_b = 0 \div 1$, contours of normalized swirl velocity $(u_x^2 + u_y^2)^{\frac{1}{2}}/U_b = 0 \div 0.35$, contours of normalized temperature $(T - T_w)/(T_w - T_b) = -1.25 \div 0$.

within the flow. By increasing the Reynolds number to 910, the recorded time history presents weak instabilities that are still dissipated within the simulated time window.

At Re=1400, however, an initial train of periodic oscillations with period $T = 4H/U_b$, where U_b is the flow bulk velocity, develops at the simulation start up anticipating the transition to a turbulent regime at $15H/U_b$. When the Reynolds number is further increased, the onset of transition occurs shortly after the start up of the simulation and the recorded time history exhibits fluctuations of increasingly larger amplitude.

6. Conclusions

In this paper, we presented a numerical study of the fully-developed flow in a pipe with twisted-tape inserts in the laminar to transitional flow regime. The simulations were carried using the

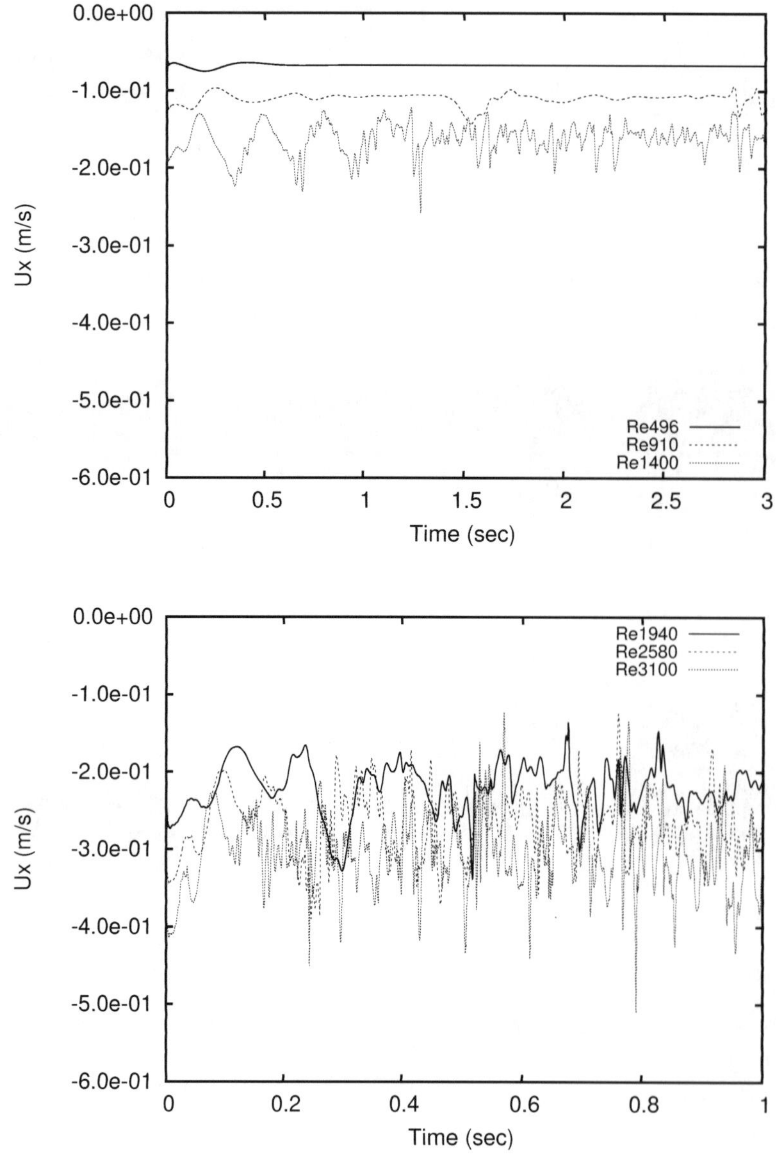

Figure 6. Time history of the spanwise velocity from a probe monitor located at the center of one of the two pipe half-sections: top, Re=469, 910, 1400; bottom, Re=1940, 2580, 3100.

open source code OpenFOAM®, based on the unstructured finite-volume scheme method, and turbulence modeling performed in the framework of Scale-Adaptive Simulations (SAS), where a modified formulation of the $k - \omega$ SST model is used to close the flow and energy RANS equations. Using SAS and low-dissipation discretization schemes, the present study shows that it is possible to capture the transition from the laminar regime to the pulsating or pseudo-laminar flow regime induced by the twisted-tape at low Reynolds numbers, as well as the transition to moderate turbulent regime at the higher, yet non-turbulent for smooth pipes, range of Reynolds numbers.

The numerical model has been validated against experimental measurements carried out in a dedicated test rig as well as against available correlations for the present flow setup, showing general good agreement with the reference data. The friction factor and Nusselt number, characterizing the pressure drop and heat transfer rate of the flow, are found increased up to 4 to 7 times and 6 to 15 times, respectively, of the values for an empty pipe, indicating that twisted-tapes are able to enhance the heat transfer performance without compromising the overall efficiency due to a significant increase in the requested pumping power.

A detailed analysis of flow and temperature fields across the transitional regime shows that the onset and growth of turbulent fluctuations occur between a Reynolds number between 910 and 1400 based on the pipe diameter. Fully-turbulent flow conditions are established beyond Re=1940, where the heat transfer rate approaches that provided by available correlations for turbulent flows in empty pipes.

Since the SAS formulation operates at a computational costs significantly lower than Direct Numerical Simulations (DNS) or Large-Eddy Simulations (LES), the present work shows that the present model as a good candidate to perform numerical simulations in similar flow conditions for other type of devices for passive heat transfer enhancement.

Acknowledgments
This work was partially supported by the Emilia-Romagna Region (POR-FESR 2014-2020: "Progetto HEGOS - nuove pompe di calore per l'Harvesting EnerGeticO in Smart buildings" and "Progetto Nuovi paradigmi per la progettazione, costruzione ed il funzionamento di macchine e impianti per l'industria alimentare").

References

[1] Reay D A, Hesselgreaves J E and Law R 2001 *Compact Heat Exchangers: Selection, Design and Operation* (Oxford: Pergamon press)
[2] Rainieri S, Bozzoli F and Pagliarini G 2012 *Int. J. Heat Mass Transf.* **55** 498
[3] Rainieri S, Bozzoli F, Cattani L and Pagliarini G 2013 *Int. J. Heat Mass Transf.* **59** 353
[4] Hong S W and Bergles A E 1976 *J. Heat. Trans.-T ASME* **98** 251
[5] Agarwal S K and Raja Rao M 1996 *Int. J. Heat Mass Transf.* **39** 3547
[6] Kaliakatsos D, Cucumo M, Ferraro V, Mele M, Galloro A and Accorinti F 2016 Int. J. Heat Technol. **34** 172
[7] Saha S K and Dutta A 2001 *J. Heat. Trans.-T ASME* **123** 417
[8] Savekar A, Jangid D, Gurjar M, Patil V and Sewatkar C M 2015 Proc. $2^{n}d$ Int. Conf. on Fluid Flow, Heat Mass Transf. *(Ottawa)* p 143
[9] Manglik R M and Bergles A E 1993 *J. Heat. Trans.-T ASME* **115** 881
[10] Patankar S V, Liu C H and Sparrow E M 1977 *J. Heat. Trans.-T ASME* **99** 180
[11] Menter F R and Egorov Y 2010 *Flow Turbul. Combust.* **85** 113
[12] Menter F R 1994 *AIAA J.* **32** 1598
[13] Menter F R and Egorov Y 2010 *Flow Turbul. Combust.* **85** 139
[14] Manglik R M and Bergles A E 1993 *J. Heat. Trans.-T ASME* **115** 890
[15] Petukhov B S 1970 *Adv. Heat Transf.* **6** 503

IOP Conf. Series: Journal of Physics: Conf. Series **923** (2017) 012034 doi:10.1088/1742-6596/923/1/012034

Numerical analysis of heat conduction problems on 3D general-shaped domains by means of a RBF Collocation Meshless Method

R Zamolo[1] **and E Nobile**[1]

[1] Dipartimento di Ingegneria e Architettura, Università degli Studi di Trieste, via A. Valerio 10, 34127 Trieste, Italy

E-mail: riccardo.zamolo@phd.units.it

Abstract. A Collocation Meshless Method based on Radial Basis Function (RBF) interpolation is employed to solve steady state heat conduction problems on 3D domains of arbitrary shape. The set of points required by the numerical method is generated through a novel and simple technique which automatically produces a distribution with variable point density and which adapts to each specific geometry. Numerical results are systematically compared to the corresponding analytical solutions considering several combinations of parameters; convergence tests have also been carried out. The favorable properties that will be outlined suggest that this approach can be an effective and flexible tool in the numerical simulation of heat conduction problems with complex 3D geometries.

Keywords:
Meshless methods, Local RBF, 3D node/point generation, Octree, Poisson equation

1. Introduction

The simulation of engineering relevant problems with standard numerical approaches (Finite Element and Finite Volume methods among all) requires a spatial discretization of the domain; such phase, known as meshing, can require a valuable fraction of the entire simulation time, especially when dealing with complex 3D geometries.

In recent years, several *meshless* approaches have been proposed [1–5] to avoid the need of a mesh: only a set of points is required to approximate the unknown physical quantities and the field equations, with slight constraints on the distribution of points; these approaches have been successfully applied to a wide range of problems, from heat transfer to solid mechanics.

Meshless methods employing Radial Basis Function (RBF) interpolation [6–11] will be considered; in particular, Hardy's Multiquadric (MQ) [12] can be used as RBF; such MQ-RBF meshless approaches are becoming more and more popular in the field of numerical simulations because of their efficiency, accuracy and flexibility [13–16].

The original MQ-RBF method [13, 14] employed a global support for the RBF interpolation, which leads to spectral convergence and insensitivity to point locations, at the price of dealing with fully populated matrices [16]; this approach becomes clearly unfeasible as the number of unknowns N grows, because of excessive time and memory consumptions. Futhermore the choice of interpolation parameters had strong influence on the numerical results.

Instead of a globally supported interpolation, most of recent MQ-RBF meshless approaches [17–21] use a local support for the RBF interpolation, which allows the use of efficient solvers at the price of a loss of spectral convergence and greater sensitivity to point locations; for these reasons a locally supported strategy will be employed in this work.

MQ-RBF meshless methods require a distribution of points which covers the whole domain and its boundary; higher point density is required where the solution exhibit large gradients because of the local support, as in a traditional mesh-based method, but without any other constraint on the distribution. In this work the point distributions will be generated by a novel and simple technique which is based on an Octree space partitioning algorithm [22] to obtain a variable point density which mets a defined spacing function; then a refinement process is employed to adapt the distribution to the domain boundaries. This refinement process also smooths the local spacing discontinuities due to the Octree algorithm.

Several computations will be carried out considering two 3D geometries and various combinations of parameters for a *Poisson* equation (steady state heat conduction) in order to outline the properties of this numerical approach; computed solutions will be compared to the corresponding analytical solutions and FEM solutions. These tests showed that this approach can be an accurate and flexible tool in the numerical simulation of heat conduction problems on complex-shaped 3D domains.

2. Governing equation and boundary conditions

Let us consider the following 3D Poisson equation in the unknown temperature field T:

$$-\nabla^2 T = q \tag{1}$$

defined on the domain Ω; equation (1) is representative of steady state heat conduction problems with internal heat generation q, in the case of a constant $k = 1$ thermal conductivity.

Two different kinds of boundary conditions (BC) will be considered; *Dirichlet* BC (prescribed temperature) on the Dirichlet boundary Γ_D:

$$T = \bar{T} \tag{2}$$

and *Neumann* BC (prescribed heat flux) on the Neumann boundary Γ_N:

$$\frac{\partial T}{\partial \mathbf{n}} = \mathbf{n} \cdot \nabla T = \bar{f} \tag{3}$$

where $\mathbf{n}$ is the exterior normal to the boundary $\Gamma = \partial\Omega$; obviously we have $\Gamma_D \cup \Gamma_N = \Gamma$.

Since the computed solution will be compared to the corresponding analytical solution T_{an}, we will set $\bar{T} = T_{an}$ and $\bar{f} = \mathbf{n} \cdot \nabla T_{an}$; the domain Ω and the terms q and T_{an} will be defined case by case.

3. Point distributions

A point distribution is defined as a set of N points $\mathbf{x}_i \in \Omega$ whose local spacing is weakly defined as the mean distance from $\mathbf{x}_i$ to its nearest neighbors.

The point distribution process is composed by two consecutive phases:

- Octree phase: an Octree algorithm is used to generate a distribution of points whose local spacing is close to a prescribed spacing function $\bar{s}(\mathbf{x})$; starting from the smallest box bounding Ω, the algorithm recursively partitions the space in 8 octants till the local partitioned box, centered in $\mathbf{x}_i$, has a size $l \leq \bar{s}(\mathbf{x}_i)$: this point is then inserted if $\mathbf{x}_i \in \Omega$;

- Refinement phase: since the Octree algorithm produces a distribution inside Ω but not on its boundary Γ, a simple point refinement process is employed to move the points near Γ onto it; this is accomplished moving the points according to the reciprocal repulsion of points: each point is subjected to the radial repulsion force of the closest 14 points.
  The repulsion force magnitude is chosen to be the following:

$$F(r) = \omega\bar{s}\left[4\left(\frac{r}{\bar{s}}\right)^2 + 1\right]^{-2} \tag{4}$$

  where r is the distance between the points and $\omega = 0.5$ is a relaxation parameter.
  This refinement phase is iterated 30 times over the whole point set, and whenever the distance from a point $\mathbf{x}_i$ to the boundary is less than the local spacing $\bar{s}(\mathbf{x}_i)$, the point is projected onto Γ; 30 iterations are typically enough to obtain a boundary distribution whose local spacing is comparable to $\bar{s}$ (there are no "holes" or clustered patches on the boundary). The refinement phase also smooths the local spacing discontinuities inside Ω due to the Octree algorithm.

4. Numerical method

4.1. Local RBF interpolation

RBF interpolation approximates the field T around $\mathbf{x}$ through the following expansion [9]:

$$T(\mathbf{x}) = \sum_{j\in J_n(\mathbf{x})} a_j\varphi(\|\mathbf{x}-\mathbf{x}_j\|) + \mathbf{b}\cdot\mathbf{x} + c \tag{5}$$

where $J_n(\mathbf{x})$ represents the indices j of the n points $\mathbf{x}_j$ closest to $\mathbf{x}$ in the usual Euclidean norm. $T(\mathbf{x})$ is therefore a linear combination of n radial functions φ centered at the n field points $\mathbf{x}_j$ plus a linear polynomial in $\mathbf{x}$.

There are many possible choices for φ [10]; Hardy's Multiquadric (MQ) has been chosen since it seems to offer the better results if an appropriate *shape factor* ε is employed [15, 16]:

$$\varphi(r) = \sqrt{1+(\varepsilon r)^2} \tag{6}$$

The coefficients a_j, $\mathbf{b}$ and c can be formally computed writing equation (5) for the $m \le n$ closest points $\mathbf{x}_i$, $i \in J_n(\mathbf{x})$ which do not lie on the Neumann boundary Γ_N:

$$T(\mathbf{x}_i) = T_i \tag{7}$$

where T_i is the unknown temperature in $\mathbf{x}_i$.

The following additional relations are needed [3] because of the linear polynomial in (5):

$$\sum_{j=1}^{n} a_j = 0 \quad , \quad \sum_{j=1}^{n} x_j a_j = 0 \quad , \quad \sum_{j=1}^{n} y_j a_j = 0 \quad , \quad \sum_{j=1}^{n} z_j a_j = 0 \tag{8}$$

If any of the neighbor points $\mathbf{x}_i$, $i \in J_n(\mathbf{x})$ lies on the Neumann boundary, for each of these $n-m$ Neumann points the corresponding BC (3) must be satisfied:

$$\mathbf{n}\cdot\nabla T(\mathbf{x}_i) = \bar{f}_i \tag{9}$$

where $\nabla T(\cdot)$ is given by the gradient of equation (5) with $\nabla\varphi(r_j) = \varepsilon^2\mathbf{r}_j/\varphi(r_j)$, $\mathbf{r}_j = \mathbf{x}-\mathbf{x}_j$.

Collecting the n coefficients a_i, the m unknown values T_i and the $n-m$ Neumann boundary contributions $\bar{f}_i$ in column vectors $\mathbf{a}$, $\mathbf{T}$ and $\bar{\mathbf{f}}$, respectively, the interpolation system, in compact form, is the following:

$$\mathbf{G}\begin{Bmatrix}\mathbf{a}\\ \mathbf{b}\\ c\end{Bmatrix}=\begin{Bmatrix}\mathbf{T}\\ \mathbf{0}\\ \bar{\mathbf{f}}\end{Bmatrix} \tag{10}$$

4.2. Rescaled Multiquadrics

For each local MQ-RBF interpolation, the shape parameter has been rescaled as $\varepsilon = \bar{s}_{max}\bar{\varepsilon}/d_k$ to avoid ill-conditioned interpolation matrices where the prescribed spacing function $\bar{s}$ is small; $\bar{\varepsilon}$ is the modified shape parameter (called shape parameter from now on), $\bar{s}_{max}$ is the maximum spacing within the domain and d_k is the local subdomain size, defined as the maximum distance between the points $\mathbf{x}_i$, $i \in J_n(\mathbf{x_k})$ near $\mathbf{x_k}$.

4.3. Collocation

Equation (1) with RBF approximation (5) becomes:

$$\nabla^2 T(\mathbf{x}) = \sum_{j\in J_n(\mathbf{x})} a_j \nabla^2\varphi(\|\mathbf{x}-\mathbf{x}_j\|) = -q(\mathbf{x}) \tag{11}$$

where the 3D Laplacian of the RBF φ is $\nabla^2\varphi(r_j) = [2\varphi(r_j)+1/\varphi(r_j)][\varepsilon/\varphi(r_j)]^2$, $r_j = \|\mathbf{x}-\mathbf{x}_j\|$.

Writing equation (11) for a generic point $\mathbf{x}_k$ near $\mathbf{x}$ gives:

$$\mathbf{a}^T\mathbf{L}(\mathbf{x}_k) = -q(\mathbf{x}_k) \tag{12}$$

where $\mathbf{L}(\mathbf{x}_k)$ is the column vector of the Laplacian of the RBF φ evaluated in $\mathbf{x}_k$ for each of the n neighbors $\mathbf{x}_j$. Finally, recalling $\mathbf{a}^T$ from equation (10) we have:

$$\begin{Bmatrix}\mathbf{T}^T & \mathbf{0} & \bar{\mathbf{f}}^T\end{Bmatrix}[\mathbf{G}^T]_a^{-1}\mathbf{L}(\mathbf{x}_k) = -q(\mathbf{x}_k) \tag{13}$$

where $[\mathbf{G}^T]_a^{-1}$ is the left $(n+4)\times n$ submatrix of $[\mathbf{G}^T]^{-1}$. Writing equation (13) for each point $\mathbf{x}_k$ which does not lie on the boundary gives the following linear system:

$$\mathbf{A}_{II}\mathbf{T}_I = -(\mathbf{q}+\mathbf{A}_{ID}\mathbf{T}_D+\bar{\mathbf{f}}_N) \tag{14}$$

where the subscripts I, D, and N refer to *Internal*, *Dirichlet* and *Neumann* contributions, respectively; since $\mathbf{A}_{II}$ is nonsymmetric, sparse and very large, an iterative solver has to be used. The biconjugate gradient stabilized method (BiCGSTAB) [23] with diagonal preconditioning has been employed with a relative tolerance of 1E-8, which converged in less than 150 iterations for each case that has been considered.

4.4. Error norm and FEM results

The comparison between the computed solution T_{comp} and the corresponding analytical solution T_{an} will be done computing the normalized root-mean-square norm of the error (NRMSE):

$$\text{NRMSE} = \sqrt{\frac{1}{V(\Omega)}\int_\Omega \left(\frac{T_{comp}-T_{an}}{T_{max}-T_{min}}\right)^2 d\Omega} \tag{15}$$

where $V(\Omega)$ is the volume of Ω; the 3D integral has been approximated by numerical quadrature using the prescribed spacing function $\bar{s}^3$ as quadrature weight.

Finite Element results have been obtained using COMSOL Multiphysics® FEA software using tetrahedral meshes with linear elements (4 nodes); the number of mesh nodes is kept closest possible to the number of unknowns N of the corresponding meshless solution, as well as the element sizing which is chosen to be closest possible to the prescribed spacing function $\bar{s}(\mathbf{x})$. Dirichlet BC have been prescribed, while different BC had negligible influences on the error.

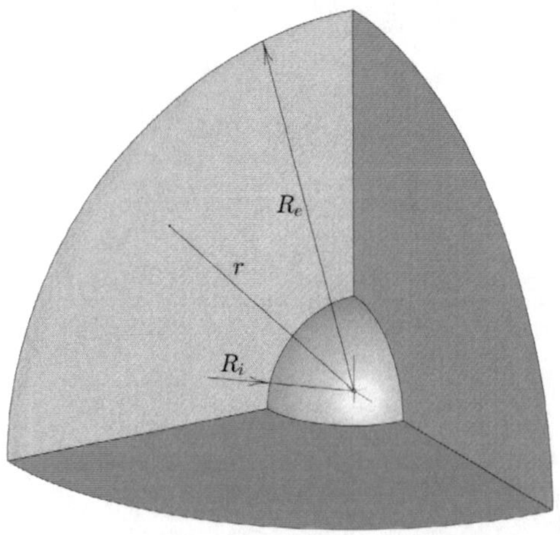

Figure 1. Test case A: geometry.

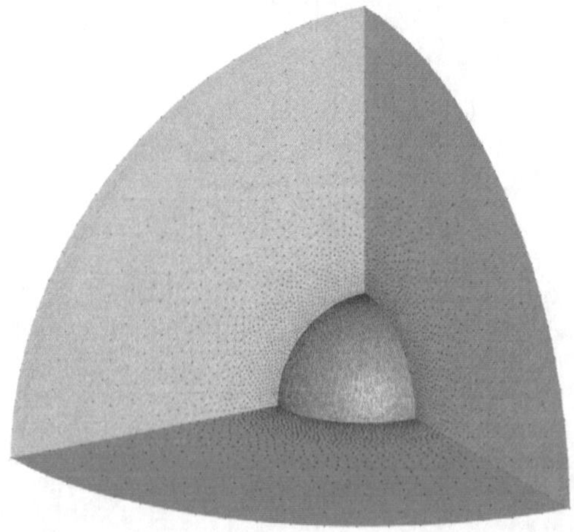

Figure 2. Test case A: boundary point distribution with $N \approx 40$k points.

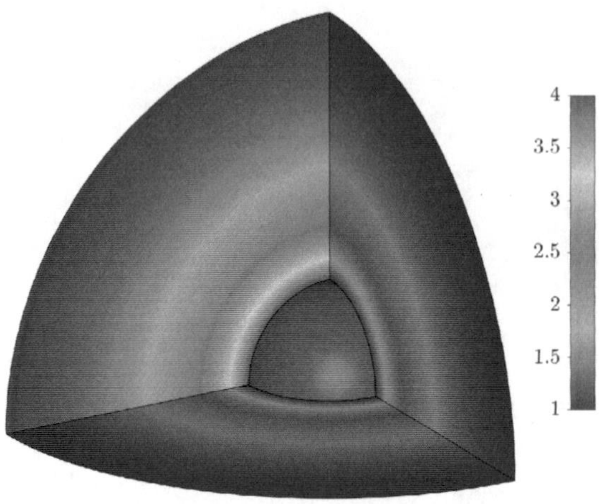

Figure 3. Test case A: surface plot of the solution.

5. Numerical results and discussion

5.1. Test case A: 1/8 of a spherical shell

The geometry of the problem is reported in Figure 1 where the center of the spherical shell is also the origin for the spherical r coordinate; the ratio R_i/R_e is chosen to be 0.25.

The chosen analytical solution is the following:

$$T_{an} = R_e/r \tag{16}$$

which is harmonic: $q = -\nabla^2 T_{an} = 0$; the prescribed spacing function is chosen to be $\bar{s} \propto r^2$. A surface plot of this solution is reported in Figure 3, while a boundary point distribution for $N \approx 40$k points can be seen in Figure 2.

Boundary conditions

- A1: Dirichlet BC on the whole boundary;

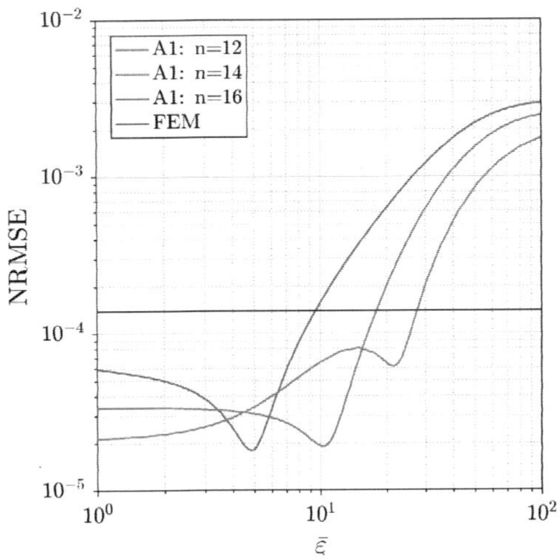

Figure 4. Test case A1: error vs shape parameter with Dirichlet BC, $N \approx 100$k.

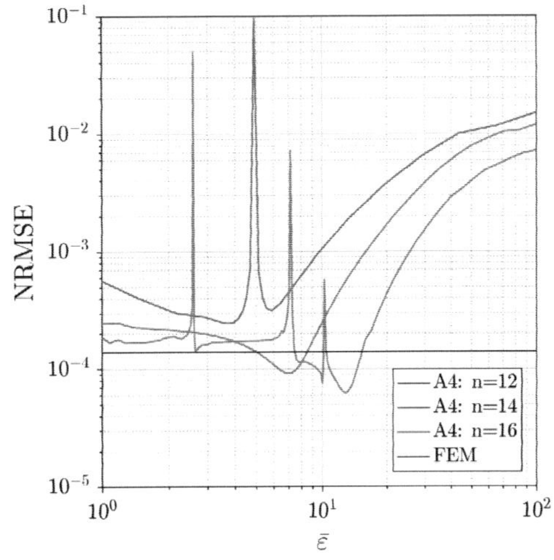

Figure 5. Test case A4: error vs shape parameter with Mixed BC, $N \approx 100$k.

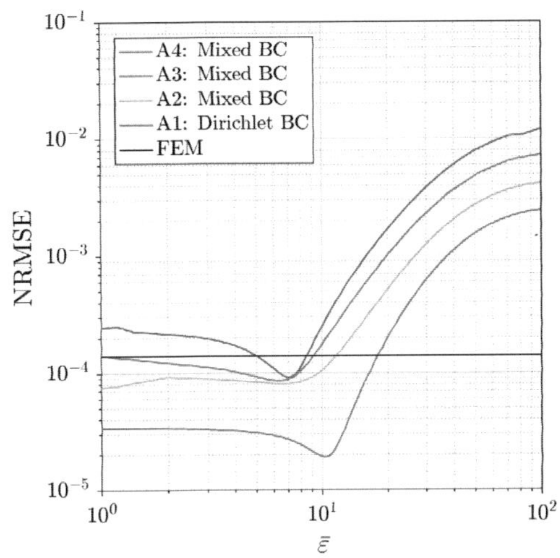

Figure 6. Test case A: error vs shape parameter with $n = 14$, $N \approx 100$k.

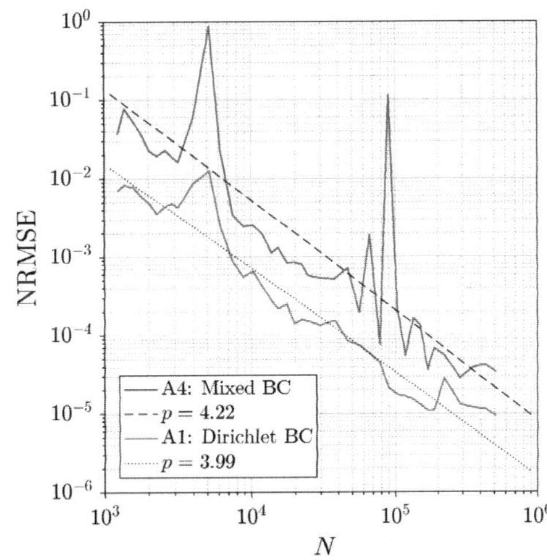

Figure 7. Test case A: convergence curves with $n = 12$ and $\bar{\varepsilon} = 5$.

- A2: Neumann BC on one flat face, Dirichlet BC elsewhere;
- A3: Neumann BC on two flat faces, Dirichlet BC elsewhere;
- A4: Neumann BC on three flat faces, Dirichlet BC elsewhere.

First of all we investigated the influence of the shape parameter $\bar{\varepsilon}$ in the range $[1, 100]$ on the error, considering three cases with n=12, 14 and 16 local interpolation points and two different BC, A1 and A4, Figures 4 and 5, with $N \approx 100$k points. These figures show that with Dirichlet BC, for each n there exists a value of the shape parameter which minimizes the error; for $n = 16$ the error has a local minimum which is not global in the considered range. For all the values of n, there exists a value of $\bar{\varepsilon}$ below which the error is lower than the FEM. With Mixed BC, Figure

IOP Conf. Series: Journal of Physics: Conf. Series **923** (2017) 012034 doi:10.1088/1742-6596/923/1/012034

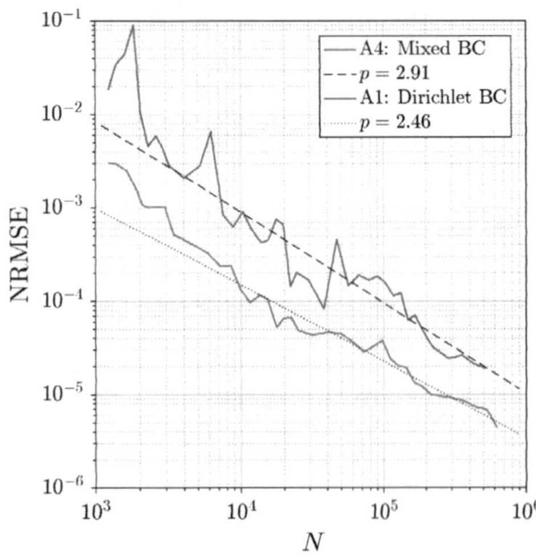

Figure 8. Test case A: convergence curves with $n = 14$ and $\bar{\varepsilon} = 5$.

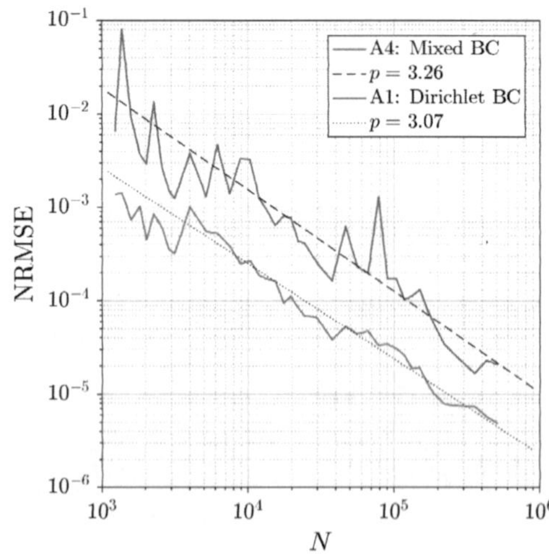

Figure 9. Test case A: convergence curves with $n = 16$ and $\bar{\varepsilon} = 5$.

5, an irregular and peaked behaviour is observed, except for the $n = 14$ case; this undesired feature is therefore due to the imposition of Neumann BC on a portion of the boundary: further analyses have to be carried out in order to understand the influence of local point distribution and shape factor when dealing with Neumann BC within the local MQ-RBF interpolation. In addition to this irregular behaviour, the errors are comparable with FEM only for small ranges of $\bar{\varepsilon}$ around 10.

Figure 6 shows the influence of $\bar{\varepsilon}$ on the error for all the considered BC with $n = 14$ and $N \approx 100$k points: a behaviour with minimum error around $\bar{\varepsilon} = 10$ is visible for all the curves, with growing error passing from Dirichlet BC (A1) to Mixed BC (A4); the errors for this case with $n = 14$ are comparable with the FEM for a wide range of $\bar{\varepsilon}$ and for each BC.

Figures 7-9 show convergence curves for three cases with n=12, 14 and 16 local interpolation points with $\bar{\varepsilon} = 5$. We first note that for every n there is a difference of one order of magnitude in the error passing from Dirichlet BC (A1) to Mixed BC (A4), while the irregular behaviour of the curves is probably due to the nature of the point distribution process (Octree phase) which has discontinuous dependence from the prescribed spacing function $\bar{s}$. In particular the case $n = 12$ presents strong discontinuities, while the cases $n = 14$ and $n = 16$ have a similar and more regular behaviour with estimated order of accuracy[1] p varying from 2.5 to 3.3.

5.2. Test case B: hemisphere with cylindrical through hole

The geometry of the problem is reported in Figure 10 where r is the distance from the vertical axis and z starts from the center of the hemisphere; the ratio R_i/R_e is chosen to be 0.5.

The chosen analytical solution is the following:

$$T_{an} = \exp\left(k_z\left(\frac{z}{H} - 1\right)\right) \exp\left(k_r \frac{R_i - r}{R_e - R_i}\right) \tag{17}$$

with $k_z = 1.5$ and $k_r = 1.0$; T_{an} has therefore exponential growth in the direction of the upper circular edge located at $z = H = \sqrt{R_e^2 - R_i^2}$, as can be seen in the surface plot of Figure 12.

[1] For 3D cases the order of accuracy p is defined by: NRMSE $\sim N^{-p/3}$.

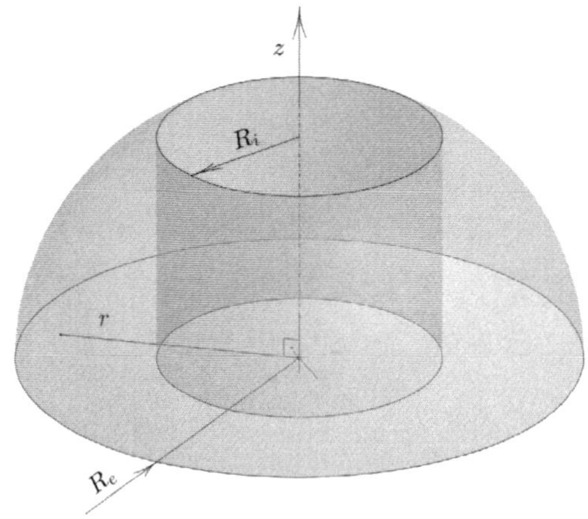

Figure 10. Test case B: geometry.

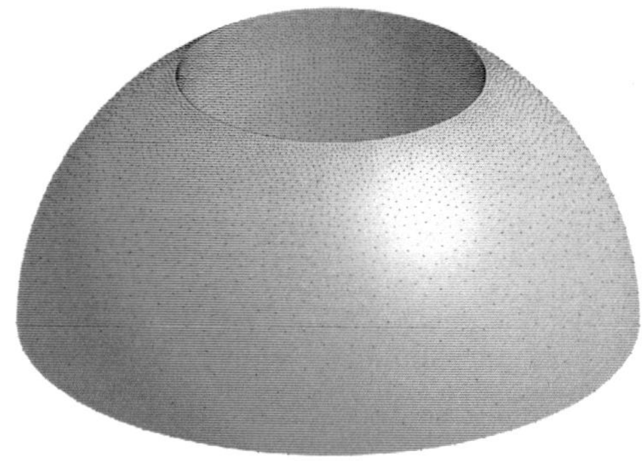

Figure 11. Test case B: boundary point distribution with $N \approx 50$k points.

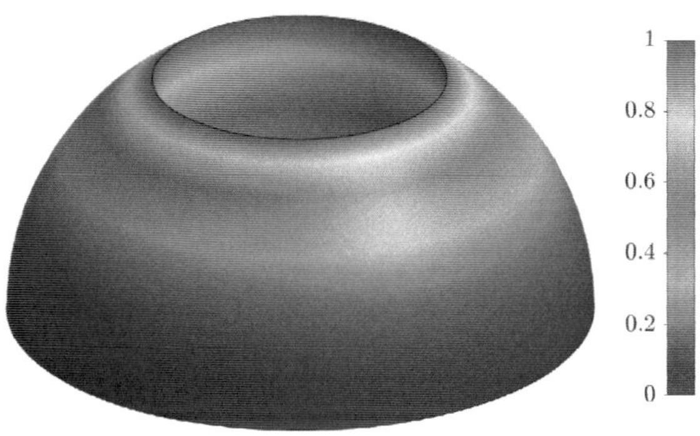

Figure 12. Test case B: surface plot of the solution.

The prescribed spacing function is chosen to be $\bar{s} \propto T_{an}^{-1}$: a boundary point distribution for $N \approx 50$k points can be observed in Figure 11.

The internal heat generation q is therefore:

$$q = -\nabla^2 T_{an} = -\left[(C_z^2 + C_r^2) + C_r/r\right] T_{an} \tag{18}$$

where $C_z = k_z/H$ and $C_r = -k_r/(R_e - R_i)$.

Boundary conditions

- B1: Dirichlet BC on the whole boundary;
- B2: Neumann BC on the spherical surface, Dirichlet BC elsewhere.

Again, we first investigated the influence of the shape parameter $\bar{\varepsilon}$ in the range $[1, 100]$ on the error, considering three cases with n=12, 14 and 16 local interpolation points and two different BC, B1 and B2, Figures 13 and 14, with $N \approx 100$k points. Dirichlet BC error curves have very similar behaviour for different n, with no local minimal points in the considered range for $\bar{\varepsilon}$; below $\bar{\varepsilon} = 10$ the error is lower than the FEM, even one order of magnitude less than FEM error for $\bar{\varepsilon}$ around 1. With Mixed BC, Figure 14, the error curves have shifted towards large errors,

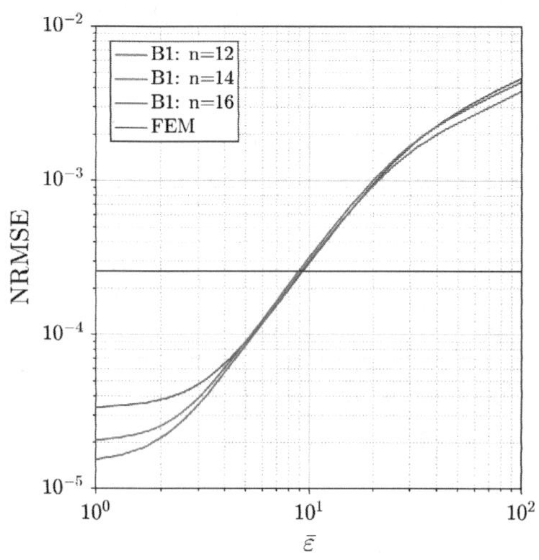

Figure 13. Test case B1: error vs shape parameter with Dirichlet BC, $N \approx 100$k.

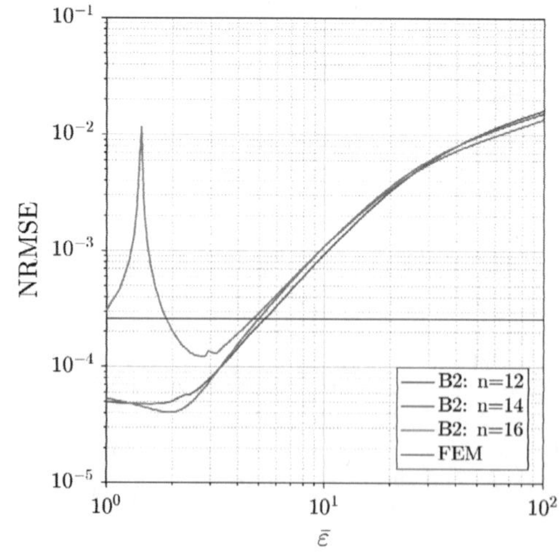

Figure 14. Test case B2: error vs shape parameter with Mixed BC, $N \approx 100$k.

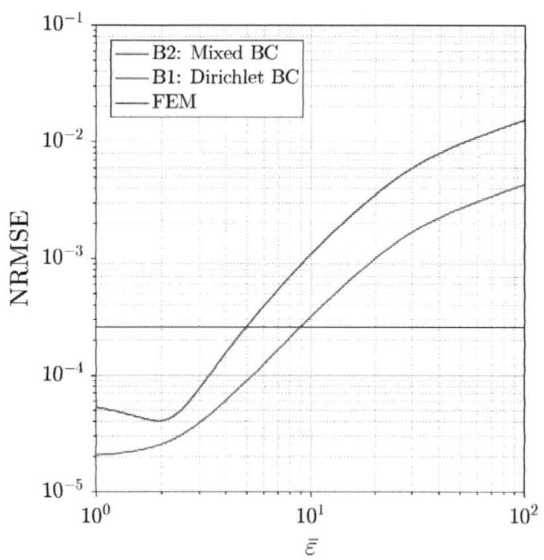

Figure 15. Test case B: error vs shape parameter with $n = 14$, $N \approx 100$k.

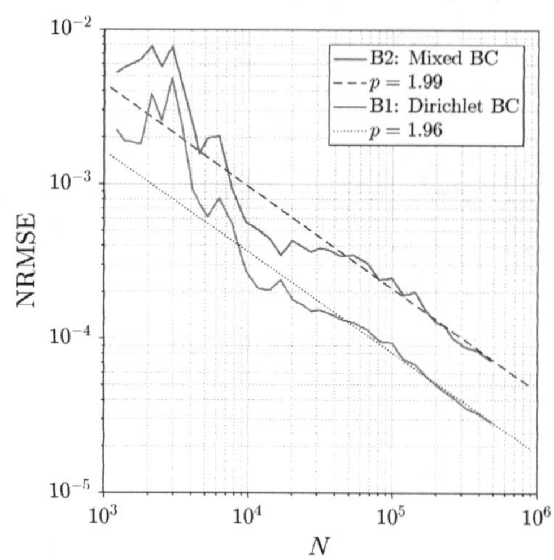

Figure 16. Test case B: convergence curves with $n = 12$ and $\bar{\varepsilon} = 5$.

while a peak is observed for the $n = 16$ case around $\bar{\varepsilon} = 1.3$; for $n = 12$ and 14 the errors are lower than FEM only for $\bar{\varepsilon} < 5$.

Figure 15 shows the influence of $\bar{\varepsilon}$ on the error for both Dirichlet and Mixed BC with $n = 14$ and $N \approx 100$k points: a decreasingly error for small $\bar{\varepsilon}$ is visible for both curves, with growing error passing from Dirichlet BC (B1) to Mixed BC (B2); the errors for this case $n = 14$ are comparable to the FEM for a wide range of $\bar{\varepsilon}$ and for both BC.

Figures 16-18 show convergence curves for three cases with n=12, 14 and 16 local interpolation points with $\bar{\varepsilon} = 5$. Again we note that for every n there is a difference of a factor 5 in the error passing from Dirichlet BC (B1) to Mixed BC (B2), while the irregular behaviour of the curves is less evident than the test case A. Furthermore for large N the convergence curves become more

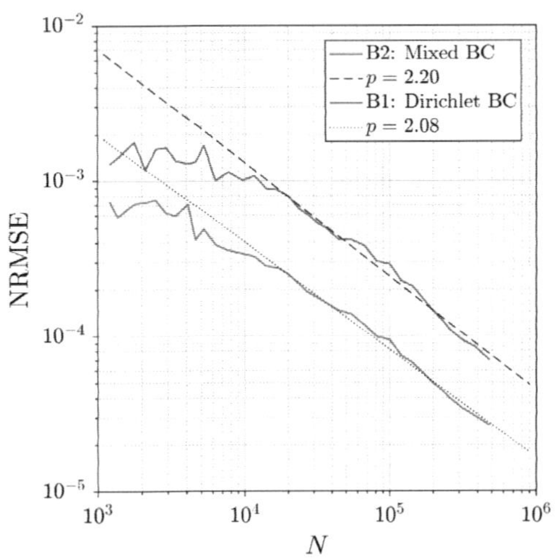

Figure 17. Test case B: convergence curves with $n = 14$ and $\bar{\varepsilon} = 5$.

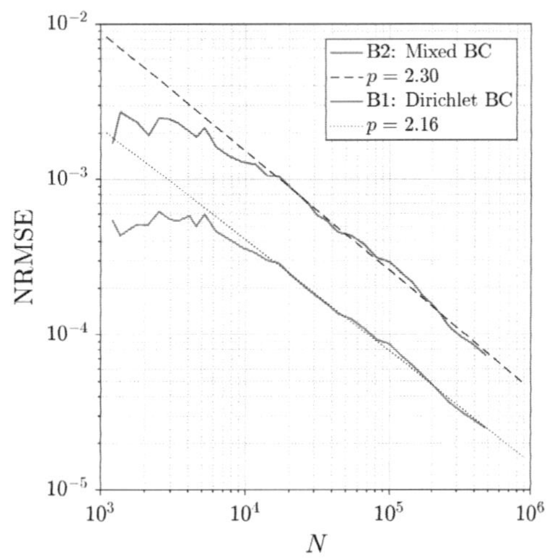

Figure 18. Test case B: convergence curves with $n = 16$ and $\bar{\varepsilon} = 5$.

and more regular, allowing an estimation of the order of accuracy p between 2.0 and 2.3.

5.3. CPU times
Among all phases of the proposed approach, the point distribution process is the most time consuming; its cost is approximately 1 second/50k points[2] and its complexity is linear in N. The local interpolation phase has a cost which is the half of the previous phase, approximately 1 second/100k points in the case of $n = 14$ and is again linear in N; the solution phase through BiCGSTAB algorithm took the same amount of time of the local interpolation phase for N around 500k and $n = 14$. We remark that the point distribution process and the local interpolation phase are not parallelized in any way, while BiCGSTAB is automatically parallelized on multicore architectures within MATLAB®.

6. Conclusions and future work
In this work a Collocation Meshless Method, based on Multiquadric Radial Basis Function interpolation, is used for the numerical simulation of steady state heat conduction problems (Poisson equation) on complex-shaped 3D domains. A simple and flexible process is employed to generate a distribution of points with a prescribed spacing; points are automatically generated inside the domain and on its boundary.

Two test cases of practical relevance have been considered, and for each case several computations are carried out for various combinations of parameters (total number of points N, shape factor $\bar{\varepsilon}$, number of local interpolation points n, boundary conditions). Most of these tests revealed good accuracy (comparable with linear FEM) and good convergence properties (more than second order in space); excellent results are obtained with $n = 14$ (the asymptotic number of edges per node in a tetrahedral mesh with recursive baricentric partitioning is exactly 13 [24], corresponding to the choice $n = 14$).

These positive results confirm that the coupling of the local RBF meshless discretization with the point generation process here presented can therefore be extremely effective in the numerical simulation of heat conduction problems on general-shaped 3D domains of engineering relevance.

[2] On a modern laptop (quad-core Intel® i7 2.6GHz).

Further investigations will be conducted considering a variable number of local interpolation points n and a variable shape factor $\bar{\varepsilon}$, different strategies for shape factor rescalation and improved point distribution techniques.

References

[1] Liu G 2003 *Meshfree Methods: Moving Beyond the Finite Element Method* (CRC Press) ISBN 978-1-4200-8209-8
[2] Li H and Mulay S 2013 *Meshless Methods and Their Numerical Properties* (CRC Press) ISBN 978-1-4665-1747-9
[3] Fasshauer G 2007 *Meshfree Approximation Methods with MATLAB* (World Scientific) ISBN 978-981-270-633-1
[4] Belytschko T, Krongauz Y, Organ D, Fleming M and Krysl P 1996 *COMPUT METHOD APPL M* **139**(1–4) 3–47
[5] Nguyen V, Rabczuk T, Bordas S and Duflot M 2008 *MATH COMPUT SIMULAT* **79**(3) 763–813
[6] Franke C and Schaback R 1998 *APPL MATH COMPUT* **93**(1) 73–82
[7] Fasshauer G 1997 *In: Surface Fitting and Multiresolution Methods* ed Mehaute A L, Rabut C and Schumaker L (Vanderbilt University Press) pp 131–138
[8] Zerroukat M, Power H and Chen C 1998 *INT J NUMER METH ENG* **42**(7) 1263–1278
[9] Golberg M, Chen C and Bowman H 1999 *ENG ANAL BOUND ELEM* **23**(4) 285–296
[10] Larsson E and Fornberg B 2003 *COMPUT MATH APPL* **46**(5–6) 891–902
[11] Fornberg B and Flyer N 2015 *ACTA NUMER* **24** 215–258
[12] Hardy R 1971 *J GEOPHYS RES* **76**(8) 1905–1915
[13] Kansa E 1990 *COMPUT MATH APPL* **19**(8-9) 127–145
[14] Kansa E 1990 *COMPUT MATH APPL* **19**(8-9) 147–161
[15] Franke R 1982 *MATH COMPUT* **38**(157) 181–200
[16] Sarra S and Kansa E 2009 *Multiquadric Radial Basis Function Approximation Methods for the Numerical Solution of Partial Differential Equations* (*Advances in Computational Mechanics* vol 2) ed Atluri S N (Tech Science Press)
[17] Šarler B 2007 *From Global to Local Radial Basis Function Collocation Method for Transport Phenomena* vol 5 (Springer) chap 14, pp 257–282
[18] Šarler B and Vertnik R 2006 *COMPUT MATH APPL* **51**(8) 1269–1282
[19] Mavric B and Šarler B 2015 *INT J NUMER METHOD H* **25**(6) 1488–1510
[20] Wang Z, Huang Z, Zhang W and Xi G 2015 *NUMER HEAT TR B-FUND* **67**(4) 320–337
[21] Waters J and Pepper D 2015 *NUMER HEAT TR B-FUND* **68**(3) 185–203
[22] Aluru S 2004 *Quadtrees and Octrees* (Chapman and Hall/CRC) chap 19, pp 19/1–19/26
[23] der Vorst H A V 1992 *SIAM J SCI STAT COMP* **13**(2) 631–644
[24] Plaza A and Rivara M 2002 *J COMPUT APPL MATH* **140**(1-2) 673–693

IOP Conf. Series: Journal of Physics: Conf. Series **923** (2017) 012035 doi:10.1088/1742-6596/923/1/012035

Anisotropy effects on convective heat transfer and pressure drop in Kelvin's open-cell foams

M Iasiello [1*], N Bianco[*], W K S Chiu[#] and V Naso[*]
* Dipartimento di Ingegneria Industriale, Università degli studi di Napoli Federico II, P.le Tecchio 80, 80125, Napoli, Italy

Department of Mechanical Engineering - University of Connecticut, 191 Auditorium Road, 06269, Storrs (CT), USA

[1] E-mail: marcello.iasiello@unina.it

Abstract. Open-cell foams have the potential to offer heat transfer enhancement in many applications, such as heat exchangers, solar air receivers, porous burners, thanks to their high heat transfer surface to volume ratio and to tortuosity, that promotes the internal flow mixing. Their microstructural transport features are affected by foam anisotropy, and could play an important role not only because manufacturing processes can stretch the cells along a preferential direction, but also because modern techniques allow for the customization of the foam cell shape. However, structural anisotropy in open cell foams has been scarcely investigated. In this paper, anisotropy effects on convective heat transfer and pressure drop in Kelvin's open-cell foams are analyzed. Since Kelvin's model is geometrically regular, anisotropy has been investigated by stretching the foam along three orthogonal directions, at equal cell volume. Governing mass, momentum and energy equations have been solved using a finite element method. Results are presented for different fluid inlet velocities and cell sizes along three orthogonal stretching directions. They show that anisotropy affects the velocity and temperature fields, and, consequently, the permeability, inertial factor and volumetric heat transfer coefficients.

Nomenclature

a, b	Ellipsoid axes half lengths (m)
C_p	Heat capacity at constant pressure (J/kg K)
d	Diameter (m)
h	Heat transfer coefficient (W/m^2 K)
Hg	Hagen number
k	Thermal conductivity (W/m K)
L	Foam length (m)
Nu	Nusselt number
p	Pressure (Pa)
$\mathbf{q}$	Heat flux (W/m^2)
Re	Reynolds number
S	Heat exchange surface (m^2)
T	Temperature (K)
$\mathbf{u}$	Velocity vector (m/s)
V	Volume (m^3)
x, y, z	Rectangular coordinates (m)

Greek symbols

ε	Porosity
μ	Dynamic viscosity (kg/m^2·s)
ρ	Density (kg/m^3)

Subscript

c	Cell
f	Fluid
h	Hydraulic
in	Inlet
m	Maximum
s	Solid
v	Volumetric

Other symbols

<>	Volume average

1. Introduction

For applications in which enhancing heat transfer is a primary aim, open-cell foams are a promising class of materials because of their weight, high effective thermal conductivity, mixing-flow characteristic and high heat exchange surface to volume ratio. They are used in various applications, such as volumetric solar receivers [1, 2], heat sinks [3-5], phase change materials [6, 7] and porous burners [8].

Flow and temperature fields in open-cell foams can be obtained experimentally, numerically or analytically. Since the scales of the problem are very small, the last two techniques allow one to carry out analysis at these scales. Due to the complex geometry, thermal behavior of open-cell foams is analyzed by employing the Volume Averaging Technique (VAT) [9], in which the whole foam is treated as an equivalent medium with variables averaged over a Representative Elementary Volume (REV). The local temperature difference between the two phases can be taken into account with a Local Thermal Non-Equilibrium (LTNE) model, with energy equations written for each phase and coupled with the so-called volumetric heat transfer coefficient.

The volumetric heat transfer coefficient, together with permeability and inertial factor for the momentum equation, can be obtained from experiments [10] or predictions. The latter can be performed by simulating the single-phase flow that passes through the foam, whose geometry needs to be modeled. The most accurate approach uses tomographic scans of real foam samples. However, a less costly approach models the geometry by making reference to geometrical models that represent the foam structure as accurately as possible.

The first models were proposed by Du Plessis and Masliyah [11] and Lu et al. [12]. In the former, the foam geometry was treated as three mutually perpendicular struts and the pressure drop was predicted. In the latter, the geometry was modeled as various orthogonal cylinders. During the years, the aim has been to describe the foam geometry as accurately as possible. To the authors' knowledge, the most accurate foam cell model was proposed by Weaire and Phelan [13], that reduces the surface energy by a 0.3% when compared to the Kelvin's foam model (that was the most accurate model until then). The Weaire and Phelan's model was used by Boomsma et al. [14], Kopanidis et al. [15] and by Cunsolo et al. [16]. The Kelvin foam model was used by [17-19]. Iasiello et al. [18] compared the Kelvin's foam model with a tomography-based model, while Lucci et al. [19] showed that a better accuracy is obtained if the Kelvin's model is built up at equal surface to volume ratio. The comparison between Weaire and Phelan's and Kelvin's models carried out by Cunsolo et al. [16] exhibited similar Nusselt numbers and friction factors, especially for high porosities. The Kelvin's foam model is computationally lighter since the unit cell consists of one polyhedron, whereas the Weaire and Phelan's model contains eight polyhedrons.

Some of the studies carried out using Kelvin's foam model took into account the effects of microstructural parameters, since they can strongly affect transport parameters. The effects of structural parameters, such as angulation and tapering, were analyzed by Pusterla et al. [20], in order to optimize heat transfer features. Ambrosio et al. [21] showed that the accuracy of the model is improved when the strut shape is accounted for. It is also widely known that, because of viscosity and gravity effects, a slight anisotropy stretches a foam in a certain direction during the foaming process. The foam cell anisotropy can be also induced with modern manufacturing techniques, like 3D printing [22], that allow for the customization of foam cells. Ranut et al. [23] investigated anisotropy in real foams and showed that the anisotropy effect is inversely proportional to PPIs, since increasing PPI makes the foam cells more regular.

Anisotropy effects in a Kelvin-based foam are analyzed in the present study. A 0.94 porosity foam is reconstructed by using Surface Evolver [24] and simulations of a stream of air under uniform solid/fluid boundary heat flux are carried out by means of a finite element scheme. Various fluid velocities, cell sizes and stretching factors are analyzed in order to evaluate anisotropy effects caused by the foam cell stretching along three orthogonal axes.

2. Geometrical model

The foam generation process is described in the following. An open-cell foam with 0.94 porosity is generated with the free-to-use software Surface Evolver [24]. The procedure is the same as that used in [18] and [25]. After importing the Kelvin's foam in the COMSOL Multiphysics software, the open-cell foam is modified in order to analyze anisotropy effects. Since the starting structure is perfectly isotropic, a sphere of volume $V_c = 4/3\pi(d_c/2)^2$ and diameter d_c can be inscribed through the cell. At equal volume, the cell is stretched along one of the principal (*x*, *y* or *z*) directions; thus, the cell is better represented as an ellipsoid rather than as a sphere. Under the hypothesis that the two minor axes of the stretched ellipsoid are equal, the cell volume is expressed as follows:

$$V_c = \frac{4}{3}\pi\left(\frac{d_c}{2}\right)^3 = \frac{4}{3}\pi ab^2 \tag{1}$$

where a is the half length of the axis along the stretching direction and b is the half length of the other two axes. It is worth noting that the elliptical nature of foam cells has been already discussed in the literature by defining two cell sizes that are the axes of the ellipse that approximates the foam cell structure [26]. The parameter a in equation (1) is varied in order to appreciate the foam anisotropy; on the other hand, the length of minor axes is obtained from the following equation:

$$b = \left[\left(\frac{d_c}{2}\right)^3 \Big/ a\right]^{1/2} \tag{2}$$

A resume of the stretching process is presented in figure 1. After stretching the foam cell, a boundary box is built up in order to distinguish the fluid domain from the solid domain. The latter is replaced by a boundary condition. Ficticious inlet and outlet regions are built up in order to make lead effects negligible, with the length prescribed depending on how much the foam is stretched.

3. Mathematical model

Mass, momentum and energy equations for the fluid phases (air) are solved under the hypothesis of laminar, steady and incompressible flow, negligible buoyancy effects and uniform thermophysical properties:

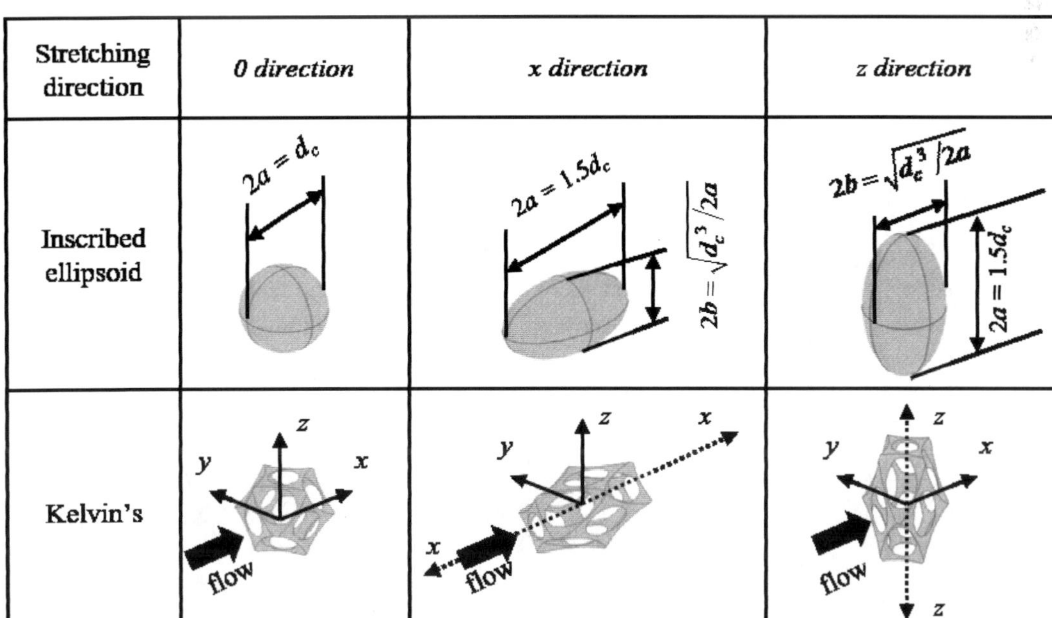

Figure 1. Stretching process in the *x* direction and *z* direction (*0* direction refers to an isotropic cell).

$$\nabla \cdot \mathbf{u} = 0 \tag{3}$$

$$\rho\, \mathbf{u} \cdot \nabla \mathbf{u} = -\nabla p + \mu\, \nabla^2 \mathbf{u} \tag{4}$$

$$\rho\, C_p\, \mathbf{u} \cdot \nabla T = k\, \nabla^2 T \tag{5}$$

where **u** is the air velocity vector, ρ the density, p the pressure, μ the dynamic viscosity, C_p the heat capacity at constant pressure, T the temperature, and k the thermal conductivity.

Boundary conditions of the problem are the same as those in [18] and [25]. At the inlet section, an air uniform temperature (T = 293.15 K) plug flow is assumed. At the walls of the computational domain, symmetry conditions are employed, *i. e.* slip condition with no shear stresses and a zero temperature gradient condition. At the solid/fluid boundary, a no-slip condition with uniform entering heat flux is assumed. Finally, atmospheric pressure is assumed at the outlet section, with an outflow condition for heat transfer. Since both geometry and boundary conditions have symmetrical features, only a quarter of the cell is analyzed.

Governing equations with boundary conditions are solved with the Galerkin-based finite element method by means of the commercial code COMSOL Multiphysics. A tethraedral mesh is employed in the computations, with a number of elements proportional to the cell size. Grid convergence has been verified in terms of heat transfer coefficients and pressure drop, by employing the numerical grid with deviations less than 10^{-3} when compared to two finer grids. A convergence criterion of 10^{-6} has been adopted.

It is worth noticing that few experimental data on convective heat transfer in open-cell foams are available in the literature, and, to the authors' knowledge, none of them refers to anisotropic foams. Comparisons among various foam morphological parameters, presented by Iasiello et al. [25], exhibited deviations lower than about 30% in the worst case, with velocities far larger than those in the present paper. One can, therefore, argue a better accuracy of the model at the velocities investigated in the present study. The model had been already validated for an isotropic cell in previous studies by the same authors [18, 25].

4. Results and Discussion

Simulations are carried out for cell diameters equal to 1.0, 2.0, 3.0 mm; air inlet velocities equal to 0.5, 1.0, 1.5, 2.0 m/s, and stretching factor in the range $1a$ - $1.5a$. The imposed heat flux is equal to 1000 W/m^2, so that temperature variations are small enough to ensure the reliability of the uniform properties assumption. Scaled velocity fields for stretching along x and y directions, with stretching factors equal to $1a$, $1.2a$ and $1.4a$, are presented in figure 2. Five slices are considered, at x/L = 0, 0.25,

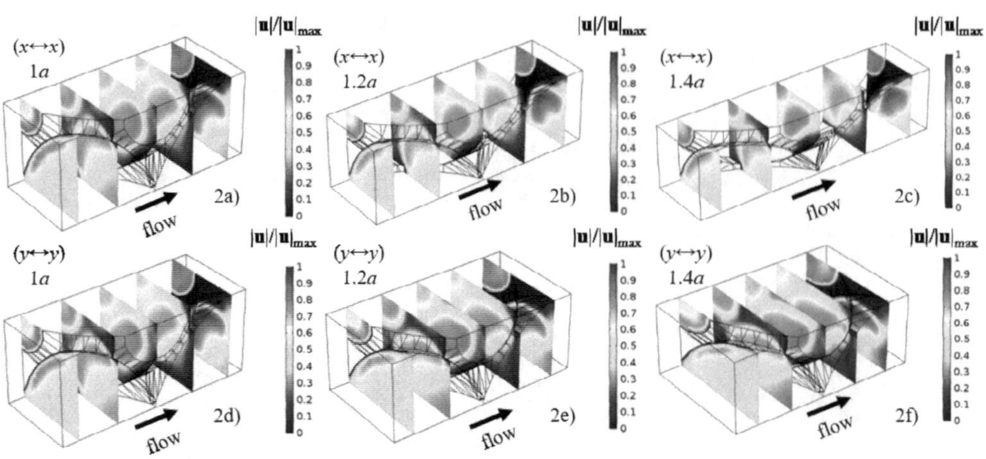

Figure 2. Scaled velocity slices for stretching along the x direction: a) $1a$; b) $1.2a$; c) $1.4a$ and along the y direction: d) $1a$; e) $1.2a$; f) $1.4a$.

0.50, 0.75 and 1.0 in order to make the comparisons reliable. As to the x direction stretching ($x\leftrightarrow x$), figures 2a, 2b and 2c exhibit slight differences in the shape of the velocity profile along the flow direction. One can also notice that, at the same x/L, the larger the stretching factor the more uniform the velocity profile in the cross section, since the decrease in the area available for fluid flow increases the fluid velocity in the core region. For the stretching along the y direction ($y\leftrightarrow y$) figures 2d, 2e, 2f show that velocity profiles are almost unaffected by the increase in the scaled velocity. It is likely due to the opposite effects of the stretching: the fluid velocity is decreased by the increase in the cross section available to the fluid whereas it is increased by the decrease in the domain length along x axis.

Iasiello et al. [27] compared various characteristic lengths for forced convection, and showed that the best scaling parameters for open-cell foams are the ratio of the cell size to the porosity and the hydraulic diameter. The predicted dimensionless pressure drop, convective heat transfer coefficient and volumetric convective heat transfer coefficient are presented in the following, using the hydraulic diameter as the scaling parameter. With reference to the hydraulic diameter

$$d_h = 4\varepsilon/S/V \tag{6}$$

with ε the porosity, S/V the ratio of the heat transfer surface to the total foam volume [28], the Reynolds number, Re, is defined as

$$\mathrm{Re} = \frac{\rho\ |\mathbf{u}|_{\mathrm{in}}\ d_h}{\mu} \tag{7}$$

where $|\mathbf{u}|_{\mathrm{in}}$ is the air inlet velocity, that is equal to the ratio of the volumetric flow rate to the inlet section of the domain. In order to scale pressure drop and heat transfer coefficients, the Hagen number, Hg, and Nusselt number, Nu, are employed:

$$\mathrm{Hg} = \frac{\Delta p}{d_c}\frac{\rho\ d_h^{\ 3}}{\mu^2} \tag{8}$$

$$\mathrm{Nu} = \frac{h\ d_h}{k} = \frac{|\mathbf{q}|}{\langle T\rangle_s - \langle T\rangle_f}\frac{d_h}{k} \tag{9}$$

where Δp is the pressure drop between the entrance and exit sections, $|\mathbf{q}|$ is the imposed heat flux, $\langle T\rangle_s$ is the solid/fluid boundary average temperature and $\langle T\rangle_f$ is the volumetric average fluid temperature. Finally, reference is also made to a volumetric heat transfer coefficient, h_v, and to its dimensionless form, the volumetric Nusselt number, $\mathrm{Nu_v}$:

$$h_v = h\frac{S}{V} = \mathrm{Nu_v}\frac{d_h^{\ 2}}{k} \tag{10}$$

It is worth noticing that the hydraulic diameter depends on both porosity and specific surface area. This means that scaling variables into their dimensionless form is meaningful, since scaled variables aren't proportional to their dimensional form.

We can also notice that S/V slightly increases with the stretching factor (S/V at $1.5a$ is 1.05 times S/V at $1a$); thus the hydraulic diameter, as well as the dimensionless numbers based on it, vary. On the contrary, they don't vary when the stretching direction changes, the same stretching factor remaining the same. The pressure drop per foam length and the heat transfer coefficient, for a stretching factor equal to $1a$, as a function of the cell diameter, at different inlet velocities, as well as a function of the inlet velocity, at different cell diameters, are presented in figures 3a and 3b, respectively. Both figures exhibit an increase in pressure drop and heat transfer coefficient at decreasing cell size and increasing inlet velocity. One can also remark that pressure drop and heat transfer coefficient are more affected by the cell size than by the inlet velocity.

The Hagen number as a function of the Reynolds number, for all the investigated values of inlet velocity and cell size and for stretching along the x direction and along the y direction, is presented in

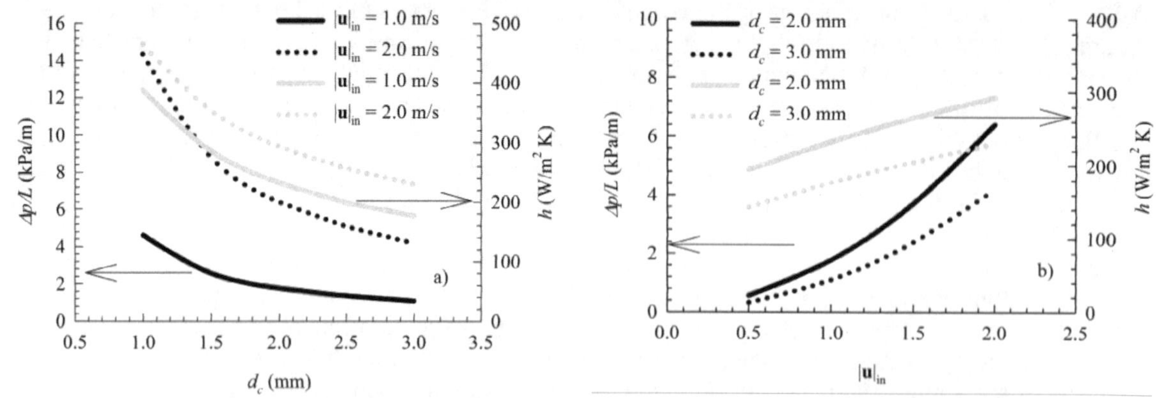

Figure 3. Pressure drop per foam length and heat transfer coefficient *vs.*:
a) cell size, at different inlet velocities; b) inlet velocity, at different cell sizes.

Figs. 4a and 4b, respectively. Data for the stretching along x are more scattered than those along y. They are correlated by the power laws reported in the figures, whose regression coefficients are rather different. Similar differences were also found between results for the stretching along the y direction and for the stretching along the z direction, not reported in this paper.

The pressure drop per foam length as a function of the stretching factor, for stretching along the three directions as well as for different cell diameters and inlet velocities, is presented in figures 5a and 5b, respectively. Both figures point out that the pressure drop increases at smaller cell diameter and larger inlet velocity. One can also remark a higher dependence of the pressure drop on the stretching factor when the foam is stretched along the x direction as well as a stronger effect when it is stretched along y and z directions. Since it is well known that the flow can hydrodynamically and thermodynamically develop in a foam [25, 29] and taking into account that stretching a foam along the x direction makes the domain longer, as depicted in figure 2, the flow is more prone to hydrodynamically develop, with a reduction in the pressure drop. On the other hand, stretching along the y and z directions makes the foam domain shorter along the flow direction, thus making the hydrodynamic development less likely and enhancing the pressure drop.

The Hagen number as a function of the stretching factor, for stretching along the three directions as well as for different cell diameters and inlet velocities, is presented in figures 6a and 6b, respectively. A slight increase with the stretching factor is remarked for y and z directions, while a strong decrease is found for higher stretching factors when the foam is stretched along the x direction. This likely occurs because the non-dimensionalization involves the hydraulic diameter, that depends on the

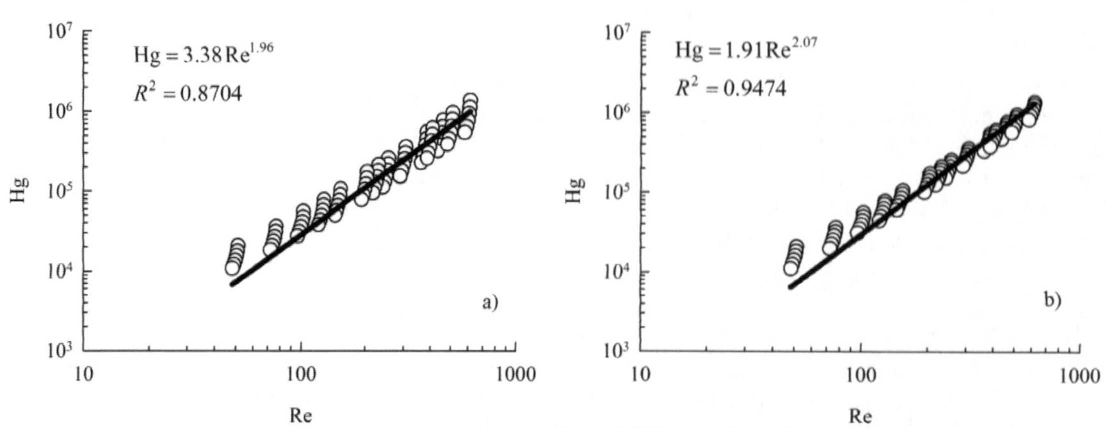

Figure 4. Hagen number *vs.* Reynolds number, for stretching along:
a) the x direction; b) the y direction.

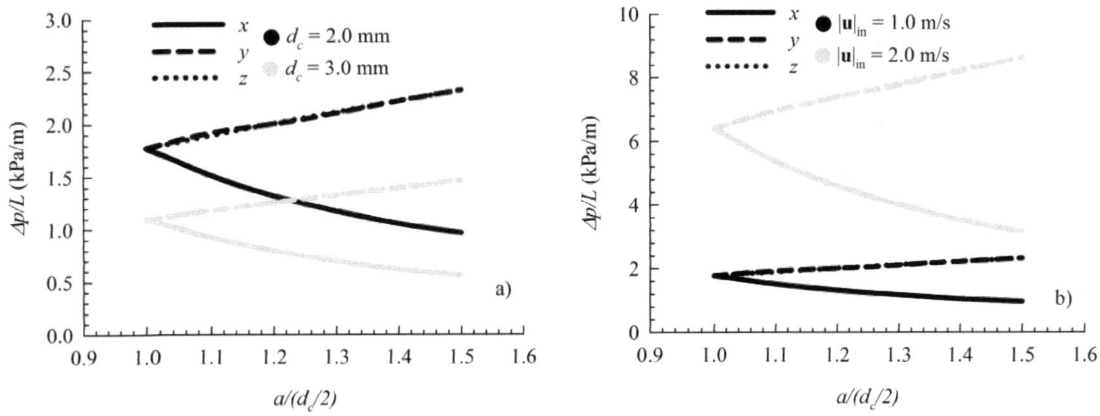

Figure 5. Pressure drop per foam length *vs.* stretching factor, for stretching along the three directions: a) $|\mathbf{u}|_{in}$ = 1.0 m/s and d_c = 2.0 mm, 3.0 mm; b) d_c = 2.0 mm and $|\mathbf{u}|_{in}$ = 1.0 m/s, 2.0 m/s.

specific surface area, which, in turn, slightly increases with the stretching factor.

The heat transfer coefficient as a function of the stretching factor, for stretching along the three directions as well as for different cell diameters and inlet velocities, is presented in figures 7a and 7b, respectively. Both figures show that the heat transfer coefficient increases when the cell diameter is smaller and when the velocity is larger. As far as the stretching effect is concerned, we can notice that the heat transfer coefficient is practically unaffected by the stretching factor for stretching along the y and z directions, whereas it decreases at increasing stretching factors along the x direction. The above different behaviors can be caused by variations in the velocity slices due to developing effects for the stretching along the x direction (see figure 2) that, on the contrary, do not occur when the foam is stretched along the y and z directions, thus making the heat transfer coefficient independent of the stretching factor.

The Nusselt number as a function of the stretching factor, for stretching along the three directions as well as for different cell diameters and inlet velocities, is presented in figures 8a and 8b, respectively. Since the heat transfer coefficient is made dimensionless through the hydraulic diameter, depending on S/V, the Nusselt number decreases at increasing stretching factor, which is larger when the foam is stretched along the x direction.

The volumetric heat transfer coefficient as a function of the stretching factor, for stretching along the three directions as well as for different cell diameters and inlet velocities, is presented in figures 9a and 9b, respectively. Both figures show that the volumetric heat transfer coefficient is almost

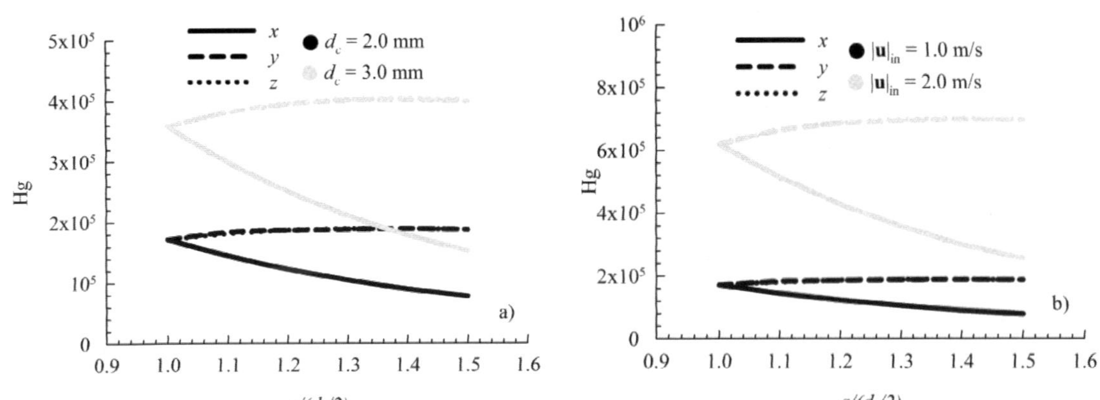

Figure 6. Hagen number *vs.* stretching factor, for stretching along the three directions: a) $|\mathbf{u}|_{in}$ = 1.0 m/s and d_c = 2.0 mm, 3.0 mm; b) d_c = 2.0 mm and $|\mathbf{u}|_{in}$ = 1.0 m/s, 2.0 m/s.

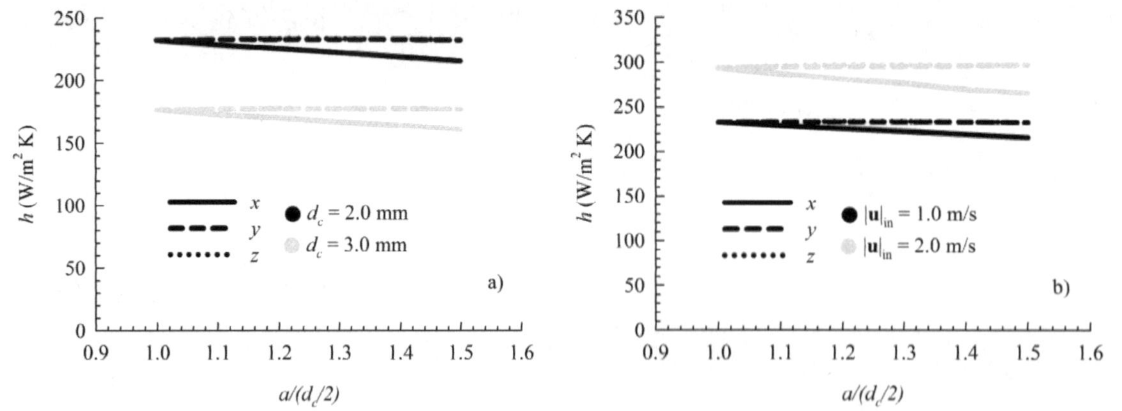

Figure 7. Heat transfer coefficient *vs*. stretching factor, for stretching along the three directions: a) $|\mathbf{u}|_{in} = 1.0$ m/s and $d_c = 2.0$ mm, 3.0 mm; b) $d_c = 2.0$ mm and $|\mathbf{u}|_{in} = 1.0$ m/s, 2.0 m/s.

independent of the stretching factor along the *x* direction, whereas it increases when the foam is stretched along *y* or *z* directions. This can be physically explained by the following considerations. In the volumetric heat transfer coefficient, that can be viewed as the heat rate per foam volume, the imposed heat flux doesn't vary, whereas the specific surface area increases with the stretching factor. When the foam is stretched along the *y* or *z* direction, the temperature difference $<T_s>_s$ - $<T_f>_f$ is nearly independent of the stretching factor because of the negligible developing effects (see figure 2), and the heat rate per foam volume increases. Consequently the volumetric heat transfer coefficient increases as well. On the contrary, when the foam is stretched along *x* direction the increase in the heat rate per foam volume is balanced by the increase in $<T_s>_s$ - $<T_f>_f$ caused by developing effects (see figure 2).

The volumetric Nusselt number as a function of the stretching factor, for stretching along the three directions as well as for different cell diameters and inlet velocities, is presented in figures 10a and 10b, respectively. Similarly to the Nusselt number, also the volumetric Nusselt number decreases with the stretching factor, since it is inversely proportional to *S*/*V*.

5. Conclusions

Anisotropic effects on convective heat transfer and pressure drop in Kelvin's foams have been investigated numerically by stretching a foam along three orthogonal directions. Pressure drop per unit length, heat transfer coefficients, volumetric heat transfer coefficients, and their dimensionless forms

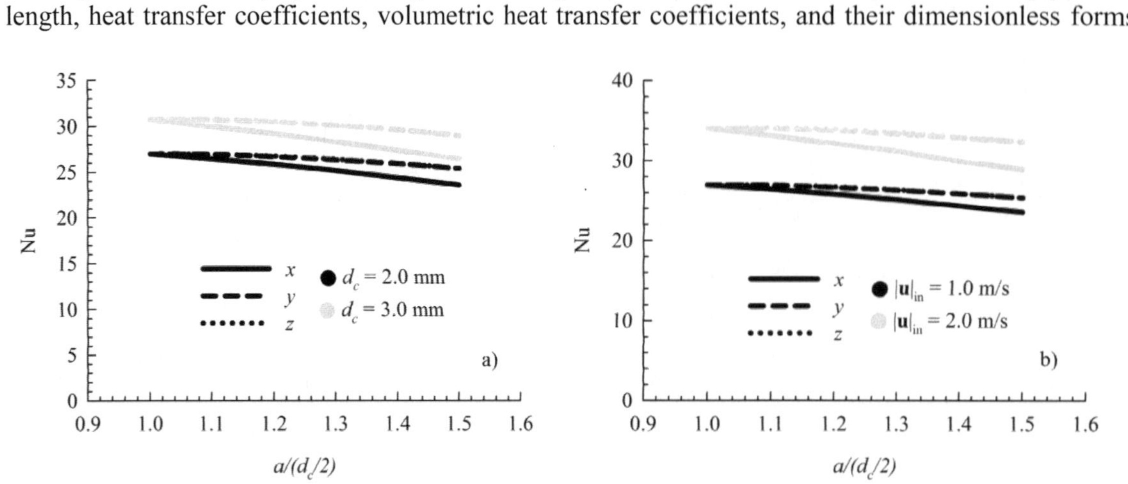

Figure 8. Nusselt number *vs*. stretching factor, for stretching along the three directions: a) $|\mathbf{u}|_{in} = 1.0$ m/s and $d_c = 2.0$ mm, 3.0 mm; b) $d_c = 2.0$ mm and $|\mathbf{u}|_{in} = 1.0$ m/s, 2.0 m/s.

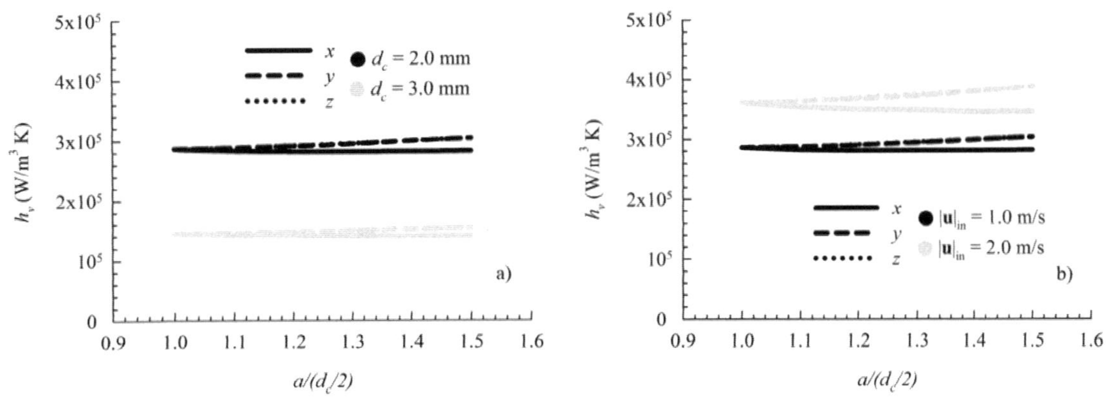

Figure 9. Volumetric heat transfer coefficient *vs*. stretching factor, for stretching along the three directions: a) $|\mathbf{u}|_{in}$ = 1.0 m/s and d_c = 2.0 mm, 3.0 mm; b) d_c = 2.0 mm and $|\mathbf{u}|_{in}$ = 1.0 m/s, 2.0 m/s.

are presented, for different cell diameters, inlet velocities and stretching factors.

Results show that anisotropy of the foam affects the coefficients depending on the stretching direction, in particular hydrodinamic and thermal development. Pressure drop and heat transfer coefficient decrease with the stretching factor when the foam is stretched along the flow direction. In the orthogonal directions, pressure drop increases with the stretching factor whereas heat transfer coefficients are unaffected by the stretching factor. The increase of the surface to volume ratio at increasing stretching factors makes the dimensionless numbers dependent on the stretching factor. They generally decrease when the stretching factor increases, more markedly when the foam is stretched along the flow direction than in the orthogonal directions.

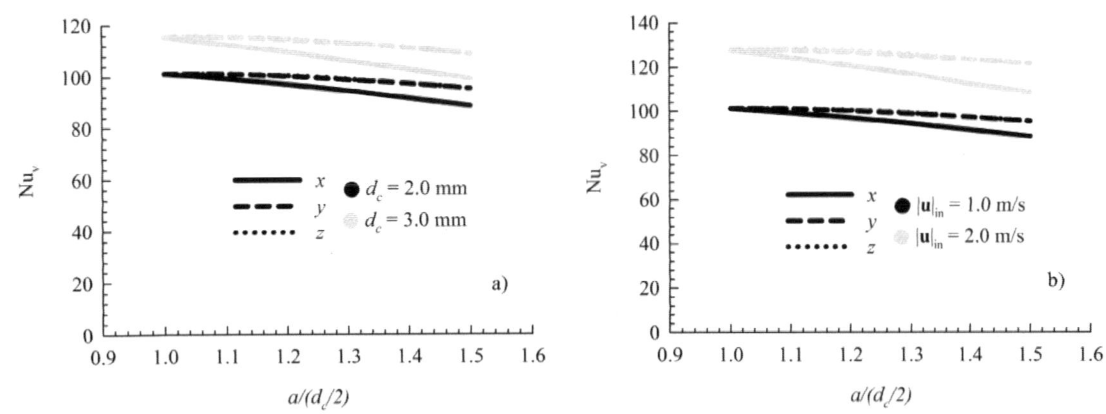

Figure 10. Volumetric Nusselt number *vs*. stretching factor, for stretching along the three directions: a) $|\mathbf{u}|_{in}$ = 1.0 m/s and d_c = 2.0 mm, 3.0 mm; b) d_c = 2.0 mm and $|\mathbf{u}|_{in}$ = 1.0 m/s, 2.0 m/s.

6. References

[1] Avila-Marin A L 2011 *Solar energy* **85** 891-910.
[2] Wu Z, Caliot C, Flamant G and Wang Z 2011 *Solar Energy* **85** 2374-85.
[3] Bhattacharya A and Mahajan R L 2002 *ASME J Electronic Packaging* **124** 155-63.
[4] Hsieh W H, Wu J Y, Shih W H and Chiu W C 2004 *Int. J. Heat Mass Transfer* **47** 5149-57.
[5] Andreozzi A, Bianco N, Iasiello M and Naso V 2017 *J. Phys.: Conf. Series* **796** 012002
[6] Mancin S, Diani A, Doretti L, Hooman K and Rossetto L *Int. J. Thermal Sciences* **90** 79-89.
[7] Di Giorgio P, Iasiello M, Viglione A, Mameli M, Filippeschi S, Di Marco P, Andreozzi A and Bianco N 2017 *J. Phys.: Conf. Series* **796** 012032.
[8] Bouma P H and De Goey L P H 1999 *Combustion and Flame* **119** 133-43.
[9] Vafai K 2015 *Handbook of porous media* (Boca Raton: CRC Press).

[10] Kamiuto K 2005 *Int. Comm. Heat Mass Transfer* **32** 947-53.
[11] Du Plessis J P and Masliyah J H 1988 *Transport in Porous Media* **3** 145-61.
[12] Lu T J, Stone H A and Ashby M F 1998 *Acta Materialia* **46** 3619-35.
[13] Weaire D and Phelan R 1994 *Phil. Mag. Letters* **69** 107-10.
[14] Boomsma K, Poulikakos D and Ventikos Y 2003 *Int. J. Heat Fluid Flow* **24** 825-34.
[15] Kopanidis A, Theodorakakos A, Gavaises E and Bouris D 2010 *Int. J. Heat Mass Transfer* **53** 2539-50.
[16] Cunsolo S, Iasiello M, Oliviero M, Bianco N, Chiu W K S and Naso V 2016 *ASME J. Heat Transfer* **138** 022601.
[17] Wu Z, Caliot C, Flamant G and Wang, Z 2011 *Int. J. Heat Mass Transfer* **54** 1527-37.
[18] Iasiello M, Cunsolo S, Oliviero M, Harris W M, Bianco N, Chiu W K S and Naso V 2014 *ASME J. Heat Transfer* **136** 112601.
[19] Lucci F, Della Torre A, Montenegro G, Kaufmann R and Eggenschwiler P D 2017 *Int. J. Heat Mass Transfer* **108** 341-50.
[20] Pusterla S, Barbato M, Ortona A and D'Angelo C 2012 *Int. J. Heat Mass Transfer* **55** 7902-10.
[21] Ambrosio G, Bianco N, Chiu W K S, Iasiello M, Naso V and Oliviero M 2016 *Appl. Thermal Engineering* **103** 333-43.
[22] Ortona A, D'Angelo C, Gianella S, and Gaia D 2012 *Materials Letters* **80** 95-8.
[23] Ranut P, Nobile E and Mancini L 2014 *App. Thermal Engineering* **69** 230-40.
[24] Brakke K A 1992 The surface evolver *Exp. Math.* **1** 141-65.
[25] Iasiello M, Cunsolo S, Bianco N, Chiu W K S, Naso V 2017 *Int. J. Thermal Sciences* **111** 129-37.
[26] De Jaeger P, T'Joen C, Huisseune H, Ameel B and De Paepe M 2011 *J. Appl. Phys.* **109** 103519.
[27] Iasiello M, Cunsolo S, Bianco N, Naso V, Oliviero M, Harris W M and Chiu W K S 2014 *Proc. 3d Int. Conf. Computational Methods for Thermal Problems (Bled, SL)* Napoli: Massimo Giannini Editore) 117-20.
[28] Dietrich B, Schabel W, Kind M and Martin H 2009 *Chem. Eng. Science* **64** 3633-40.
[29] Dukhan N and Suleiman A S 2014 *Transport in Porous Media* **101** 229-46.

IOP Conf. Series: Journal of Physics: Conf. Series **923** (2017) 012036 doi:10.1088/1742-6596/923/1/012036

CFD simulations of power coefficients for an innovative Darrieus style vertical axis wind turbine with auxiliary straight blades

F Arpino[1], G Cortellessa[1], M Dell'Isola[1], M Scungio[1], V Focanti[2], M Profili[2] and M Rotondi[2]

[1] Department of Civil and Mechanical Engineering, University of Cassino and Southern Lazio, via G. di Biasio 43, 03043 Cassino (FR), Italy
[2] Arken S.p.A. - Gymnotus S.a.s., via Stazione 182, 03013 Ferentino (FR), Italy

E-mail: f.arpino@unicas.it

Abstract. The increasing price of fossil derivatives, global warming and energy market instabilities, have led to an increasing interest in renewable energy sources such as wind energy. Amongst the different typologies of wind generators, small scale Vertical Axis Wind Turbines (VAWT) present the greatest potential for off grid power generation at low wind speeds. In the present work, Computational Fluid Dynamic (CFD) simulations were performed in order to investigate the performance of an innovative configuration of straight-blades Darrieus-style vertical axis micro wind turbine, specifically developed for small scale energy conversion at low wind speeds. The micro turbine under investigation is composed of three pairs of airfoils, consisting of a main and auxiliary blades with different chord lengths. The simulations were made using the open source finite volume based CFD toolbox OpenFOAM, considering different turbulence models and adopting a moving mesh approach for the turbine rotor. The simulated data were reported in terms of dimensionless power coefficients for dynamic performance analysis. The results from the simulations were compared to the data obtained from experiments on a scaled model of the same VAWT configuration, conducted in a closed circuit open chamber wind tunnel facility available at the Laboratory of Industrial Measurements (LaMI) of the University of Cassino and Lazio Meridionale (UNICLAM). From the proposed analysis, it was observed that the most suitable model for the simulation of the performances of the micro turbine under investigation is the one-equation Spalart-Allmaras, even if under the conditions analysed in the present work and for TSR values higher than 1.1, some discrepancies between numerical and experimental data can be observed.

1. Introduction

The interest in developing wind energy technology is growing significantly since its potential in power generation is unaffected by political and economic instabilities [1-4]. The increasing prices of fossil derivatives, in fact, is pushing industrialized countries to the diversification of energy sources, also raising awareness to environmental issues related to the exploitation of traditional non-renewable energy reserves. Amongst the different typologies, vertical axis wind turbines (VAWT) are very promising for small scale power generation at low wind speeds, but more attention by scientific community and dedicated industry is needed in order to develop low cost reliable wind energy converters [5, 6]. One of the aerodynamic issues of the VAWT is its continuously variable angle of attack of the blades with respect to the incoming wind [7] that causes high turbulence on the blades when they pass in the leeward side of the turbine, lowering the overall efficiency. The characterisation of the turbine performances by means of on-field analysis is practically unfeasible because of the high variability of the meteorological conditions and local wind speeds. In this context one of the most powerful tool for the wind turbine characterisation and optimisation is the Computational Fluid Dynamics (CFD) approach. The CFD technique [8-10], anyway, cannot be used without appropriate experimental campaigns aimed to validate the numerical data [11-14]. In order to obtain robust and reliable data from the experiments at constant testing conditions, it is essential the use of a wind tunnel equipment, which allows a velocity uniformity of the flow in correspondence of the measurement section in order to get the more accurate data possible with the corresponding measurement uncertainty [15, 16]. Wind tunnel experiments on Darrieus wind turbine were carried out by Khadir and Mrad [17] and Bianchini, Balduzzi, Rainbird, Peiró, Graham, Ferrara and Ferrari [18] who conducted experiments on different NACA airfoils in order to study the aerodynamic behaviour of the blades, while CFD investigations of static and dynamic performance on a Darrieus type wind turbine in relation to the blade profile, was recently performed by [19, 20]. In the present work, CFD analysis of the performances of an innovative straight-blades Darrieus-style configuration of vertical axis micro wind turbine was performed using the open source CFD toolbox OpenFOAM adopting a moving mesh approach for the turbine rotor. The simulated data were reported in terms of dimensionless power coefficient. The results from the simulations were validated against the data obtained from experiments on scaled models of the same VAWT configuration, conducted in a closed circuit open chamber wind tunnel facility available at the Laboratory of Industrial Measurements (LaMI) of the University of Cassino and Lazio Meridionale (UNICLAM).

2. Materials and methods

2.1. Characteristics of the turbine blades

The proposed micro wind turbine configuration is composed of three pairs of airfoils placed at 120° each other. Each pair is composed of two airfoils, whose features are reported in Figure 1. The blades are a modified version of the DU 06-W-200 airfoil [21]. The angle of attach, α is equal to 13° for both the main and auxiliary airfoils; the ratio between main and auxiliary airfoils chord lengths, C_1/C_2, is equal to 0.57; the distance between the two airfoils, y, is equal to $0.29\times C_1$ while the longitudinal distance between the airfoils, x, is equal to $0.38\times C_1$. In order to fit the micro turbine model into the wind tunnel measurement section, it was scaled and printed in PLA (polylactic acid) through a Fused Deposition Modelling (FDM) technique using a 3D printer. Referring to Figure 1, the main dimensions of the proposed rotor are the following: main and auxiliary airfoil chord lengths, C_1 and C_2, equal to 47.2 mm and 27 mm, respectively; distance between the two airfoils, y, equal to 13.8 mm; longitudinal distance between the airfoils, x, equal to 18.2 mm; angle of attach, α, equal to 13° for both the main and auxiliary airfoils. The height and diameter of the micro turbine model are equal to 150 mm and 200 mm, respectively.

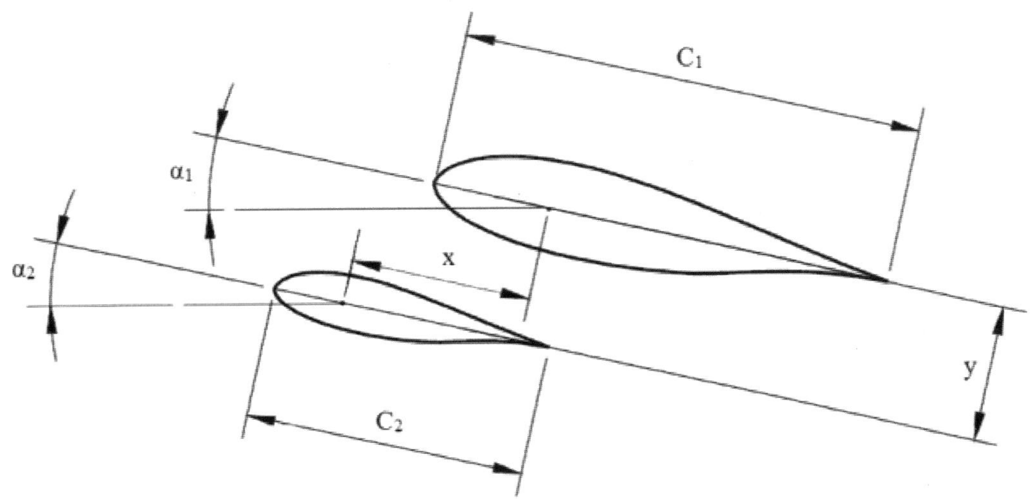

Figure 1. Scheme of the two-bladed design of the micro wind turbine under investigation.

2.2. CFD modelling setup

The numerical CFD software used for the simulation of the wind turbine performance is the open source OpenFOAM toolbox. In our simulations the default linear interpolation scheme has been adopted while as a linear solver the GAMG - Geometric agglomerated algebraic multigrid solver) with a tolerance value equal to 10^{-6}. A moving mesh technique was adopted for the reproduction of the fluid-airfoil interaction in correspondence of the turbine rotor. The simulations were conducted in a bi-dimensional domain configuration using a time step value necessary to obtain a rotation of the moving mesh of 1°; as observed during the simulations, two full rotations (720°) are necessary, in order to reach a stable periodic state. A detail of the mesh adopted is reported in Figure 2. A fully unstructured grid also in non-rotating region has been used in order to obtain acceptable values of the skewness factor and aspect ratio. In particular, the grid is composed of about 88000 triangular elements. The maximum and minimum grid dimensions were set to 10^{-2} and 10^{-4}, respectively. The grid was refined in correspondence of the airfoil surfaces using a refinement factor of 1.1. At the rotating domain interface, the static and the rotating grids present the same dimension of elements. With these characteristics of the grid, the y^+ value was maintained under about 5 and then no wall functions were adopted. Different turbulence models were evaluated for the numerical simulations, and after a preliminary analysis of the performances of each one, the two-equations SST k-ω and the one-equation Spalart-Allmaras models were chosen for the complete simulations since they showed the best overall capability of reproducing the experimental measurements. As boundary condition at the inlet section for the turbulence model a value equal to 1% of the turbulence intensity has been imposed. This value was measured in our laboratories by a PIV system.

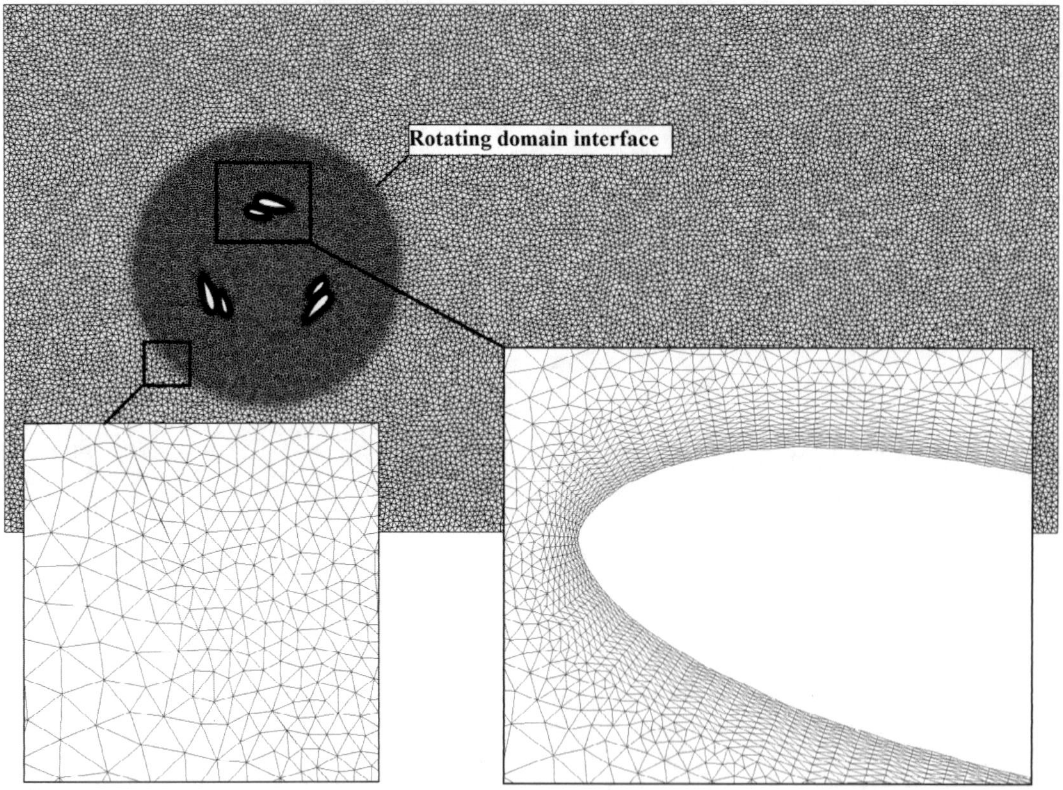

Figure 2. Computational domain adopted for the numerical simulations and detail of the grid in correspondence of the blade surfaces and at the rotating domain interface.

2.3. Wind tunnel testing setup

The experimental investigations were carried out in the wind tunnel of the Laboratory of Industrial Measurements (LaMI) of the Department of Civil and Mechanical Engineering (DICeM) at the University of Cassino and Lazio Meridionale. In Figure 2 a scheme of the wind tunnel (on the left) and the 3D printed model of the turbine (on the right) are reported.

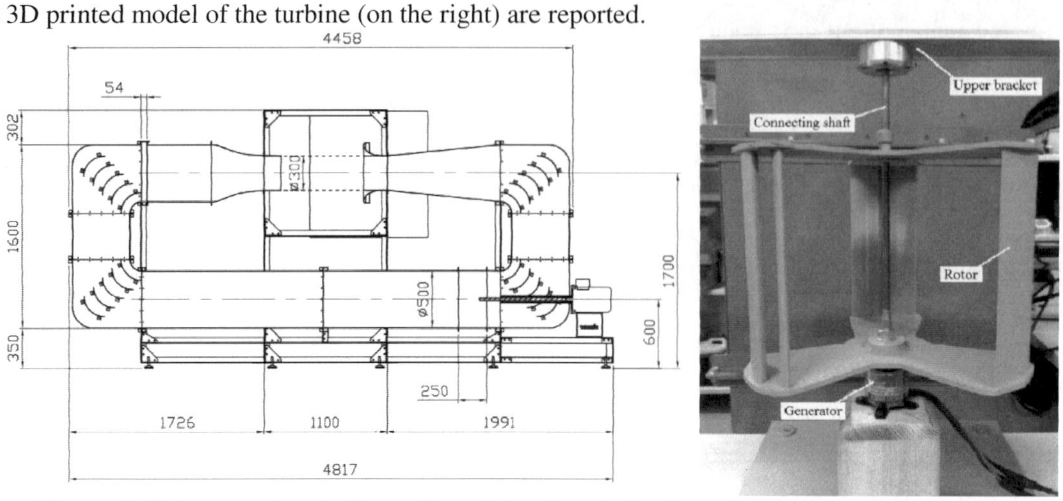

Figure 3. Scheme and main dimension of the closed circuit-open chamber wind tunnel (left panel) and 3D printed model of the turbine under investigation (right panel).

The wind tunnel works in the range 0.2 – 30 m/s at a maximum temperature of 80°C, providing good characteristics of response time, repeatability and flow uniformity. The performance analysis of the

IOP Conf. Series: Journal of Physics: Conf. Series **923** (2017) 012036 doi:10.1088/1742-6596/923/1/012036

micro turbines was made by using a brushless electric motor, acting as generator, directly connected to the rotor axis, while the variable load was simulated by means of electric resistors connected on the three phases of the brushless motor. The wind turbine characteristic curve was obtained point by point varying, for a fixed wind speed value imposed in the wind tunnel, the resistance on each phase, and measuring the electric power and the rotational speed, once stable operating conditions were achieved for each point. Additional details about the wind tunnel characteristics, the instrumentation used for electric power measurements, and the testing procedure can be found in Scungio, Arpino, Focanti, Profili and Rotondi [22].

2.4. Dimensionless performance parameters

For the performance evaluation of the VAWTs under investigation, the following non dimensional parameters were used. Tip Speed Ratio (TSR), defined as the ratio between the tip speed of the blade and the wind speed:

$$TSR = \frac{\omega \cdot R}{V} \tag{1}$$

where ω is the rotational speed in (rad/s), R is the radius of the rotor in (m) and V is the wind speed in (m/s).

Power Coefficient:

$$C_P = \frac{P}{\frac{1}{2} \cdot \rho \cdot V^3 \cdot A} \tag{2}$$

where P is the power in (W), ρ is the air density in (kg/m^3), and A is the swept area in (m^2).

2.5 Measurements uncertainty analysis

Uncertainty affecting measurements has been estimated taking into account type A and type B contributions and applying the well-known uncertainty propagation law . In particular, type A uncertainty has been evaluated *a posteriori*, through the statistical analysis of data obtained by repeated measurements, while type B is evaluated *a priori*, by applying the uncertainty propagation law to equations (1) – (4). The calculation of the combined uncertainty, u, has been made on the two extreme points and on the middle point of the measured power-rpm curve, taking as a final value the average of the three evaluated. The expanded uncertainty, U, was obtained multiplying the combined uncertainty by the coverage factor, k [15, 23] as follows:

$$U_y = k \cdot u_y \tag{3}$$

By fixing a value for the coverage factor equal to 2, i.e. fixing a probability of 95% that the measured value falls within the defined range, and normalizing the extended uncertainty to the corresponding measured quantities, the relative expanded uncertainty values reported in Table 1 were obtained:

Table 1. - Relative expanded uncertainty values for TSR and CP evaluated for three wind speeds as an average on three points for each power-rpm curve.

	$U_{r,TSR}$	$U_{r,CP}$
U_r (%)	2.01	11.29

IOP Conf. Series: Journal of Physics: Conf. Series **923** (2017) 012036 doi:10.1088/1742-6596/923/1/012036

A detailed discussion about the equations used for the estimation of the above mentioned propagation law of the uncertainty can be found in Scungio, Arpino, Focanti, Profili and Rotondi [22].

3. Results and discussion

In this section the results in terms of the dimensionless performance parameters presented in the previous sections are reported. In Figure 4a the power coefficient, CP, obtained from the numerical simulations and from the experimental measurements in the wind tunnel is showed as a function of the tip speed ratio (TSR), while in Figure 4b the torque on the turbine shaft as a function of the rotational angle is reported for the two turbulence models considered. The power coefficient – TSR curves were measured/calculated for a wind speed equal to 12 m/s. For the numerical simulations, the SST k-ω and the Spalart-Allmaras turbulence models were considered, as stated previously, while the experimental curves have been constructed point by point, and each point was obtained as average value of a sufficiently large number of acquisitions in the wind tunnel. Looking at Figure 4a it can be seen that both the turbulence models considered tend to overestimate the power coefficient for the higher TSR values (i.e. for the lowest loads applied to the turbine generator in the wind tunnel measurements), while for higher loads (i.e. lower TSR), the experimental data are well reproduced from both the turbulence models. As a general remark, anyway, the Spalart-Allmaras model tends to match the experiments better than the SST k-ω. The discrepancy between the experimental and numerical results for TSR>1 could be attributed to two different reasons. The first regards 3D effects not considered in the present numerical simulation and the second one could be due to the characteristic of the brushless electric motor which present a strong efficiency reduction in correspondence of TSR>1 (RPM>1000) as showed in Figure 4 (c).

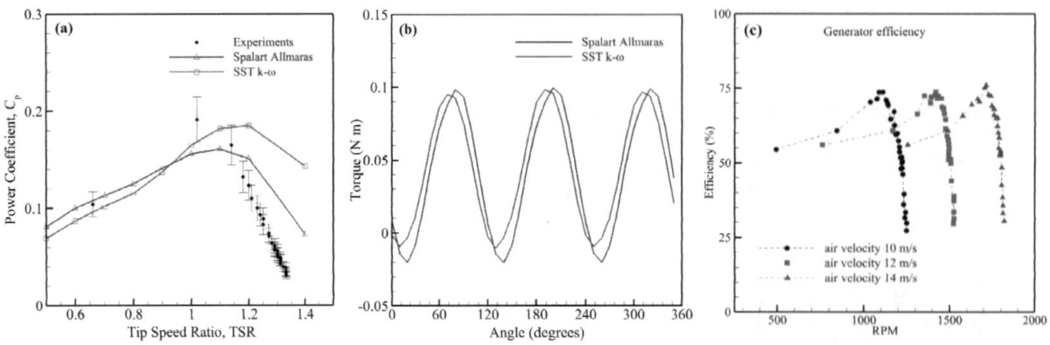

Figure 4. Power coefficient as a function of the Tip Speed Ratio (a) torque value as a function of the rotational angle of the turbine rotor with respect to the incoming wind flow (b), and generator efficiency as a function of RPM (c). Wind speed of 12 m/s.

In Figure 4b, the torque as a function of the rotational angle of the turbine with respect to the incoming wind flow is reported for both the turbulence models under examination. As can be seen, the results from the two models are almost overlapped, with only a slightly higher amplitude of the torque values for the Spalart-Allmaras model with respect to the SST k-ω. In Figure 5 the x-velocity and pressure fields are showed for the same rotational angle of the turbine rotor, for a fixed TSR value of 1, for the Spalart-Allmaras model (a), (b) and for the SST k-ω model (c), (d). As can be seen, the fields result very similar close to the blade surfaces, while in the wake flow the Spalart-Allmaras model tends to reproduce zones with lower and higher velocity values with respect to the SST k-ω model.

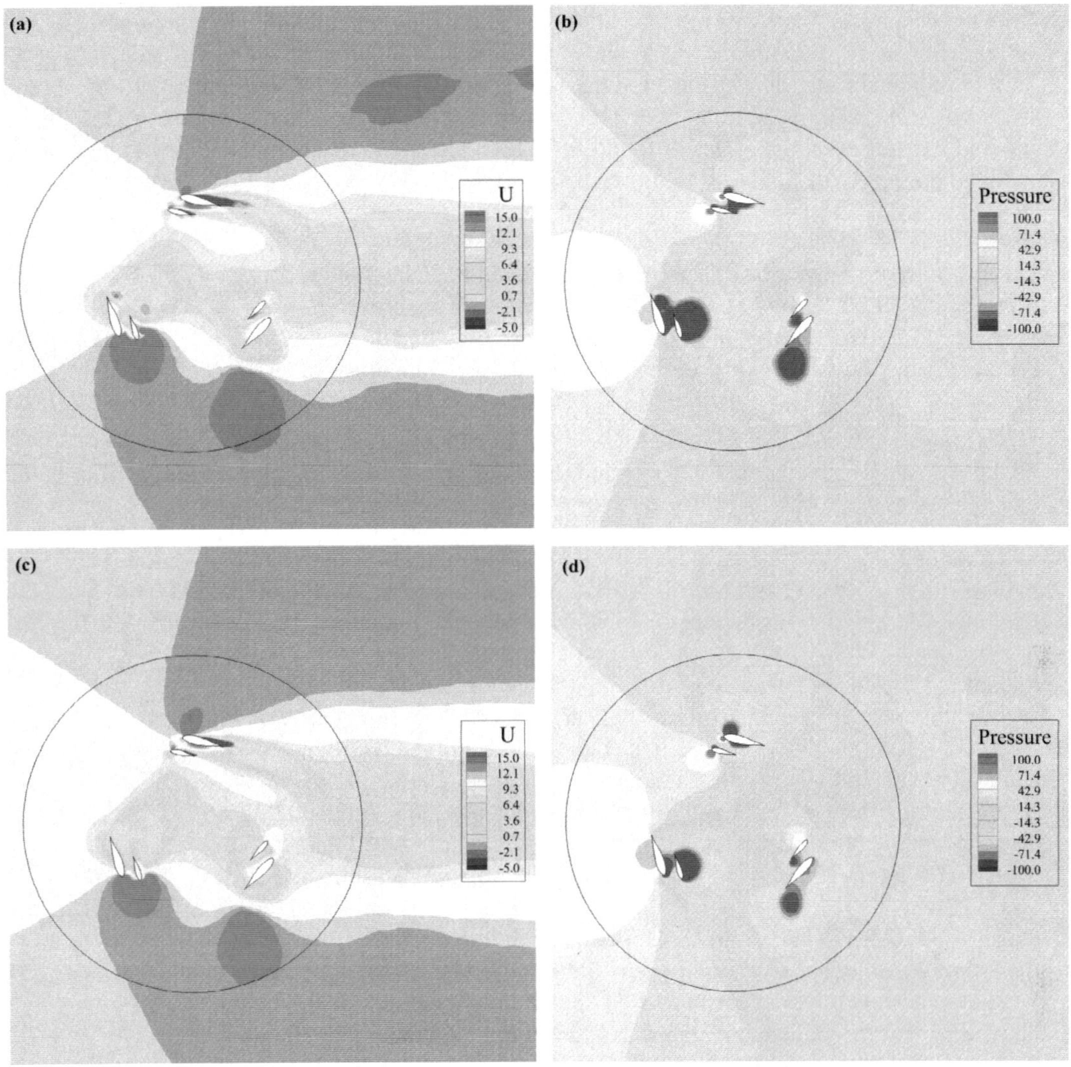

Figure 5. x-velocity and pressure fields for a fixed value of TSR (1.0) in the same rotor angle for both the turbulence models analysed at wind speed of 12 m/s: SST $k-\omega$ (a), (b); Spalart-Allmaras (c), (d).

4. Conclusions

In the present paper, an experimental campaign was carried out at the Laboratory of Industrial Measurements (LaMI) of the Department of Civil and Mechanical Engineering at the University of Casino and Lazio Meridionale, in order to analyse the performance of an innovative micro Vertical Axis Wind Turbine (VAWT). The performance analysis was conducted by means of a numerical CFD approach using the open source OpenFOAM toolbox by using different turbulence models and adopting a moving mesh approach. The results from the numerical CFD toolbox were compared to experimental data measured in a wind tunnel, reproducing a scaled model of the micro turbine through a Fused Deposition Modelling (FDM) 3D printer. The dynamic performance of the micro turbine was analysed in terms of power coefficient, CP evaluated as a function of the tip speed ratio, TSR. From the measured and numerical data, it was found that the most suitable models for the simulation of the performances of the micro turbine under investigation are the one-equation Spalart-Allmaras and the two-equations SST k-ω. In particular, between these two models the Spalart-Allmaras resulted able to better reproduce the experimental data, even if for the highest TSR values both the turbulence models tend to overestimate the measured power coefficients. Looking at the torque values as function of the rotational angle of the turbine, instead, the two models reproduce almost the same results. In conclusion, from the presented experimental campaign, it was found that for the simulation of the power coefficient-TSR curves of a vertical axis micro wind turbine, the moving mesh simulation approach with the Spalart-Allmaras turbulence model is able to give satisfactory results, even if some discrepancies between numerical and experimental data can be found for TSR values higher than 1.1, under the conditions analysed in the present work.

References

[1] Tjiu W, Marnoto T, Mat S, Ruslan M H and Sopian K 2015 *Renew. Energ.* **75** 50-67
[2] Akwa J V, Vielmo H A and Petry A P 2012 *Renew. Sust. Energ. Rev.* **16** 3054-64
[3] Monteiro J P, Silvestre M R, Piggott H and André J C 2013 *J. Wind Eng. Ind. Aerod.* **123** 99-106
[4] Bottasso C L, Campagnolo F and Petrović V 2014 *J. Wind Eng. Ind. Aerod.* **127** 11-28
[5] Balduzzi F, Bianchini A, Carnevale E A, Ferrari L and Magnani S 2012 *Appl. Energ.* **97** 921-9
[6] Bianchini A, Ferrara G and Ferrari L 2015 *Energ. Convers. Manage.* **89** 690-707
[7] Bazilevs Y, Korobenko A, Deng X, Yan J, Kinzel M and Dabiri J O 2014 *J. Appl. Mech-T Asme, Trans.* ASME **81** (8) 081006
[8] Arpino F, Cortellessa G, Dell'Isola M, Massarotti N and Mauro A 2014 *Numer. Heat Tr. A-Appl.* **66** 839-62
[9] Arpino F, Cortellessa G and Mauro A 2015 *Numer. Heat Tr. A-Appl.* **67** 605-31
[10] Arpino F, Carotenuto A, Ciccolella M, Cortellessa G, Massarotti N and Mauro A 2016 *Int. J. Heat Tech.* **34** S512-S8
[11] Arpino F, Cortellessa G and Frattolillo A 2015 *Sol. Energy* **118** 7-19
[12] Massarotti N, Ciccolella M, Cortellessa G and Mauro A 2016 *Int. J. Numer. Method. H.* **26** 1187-225
[13] Scungio M, Arpino F, Cortellessa G and Buonanno G 2015 *Atmos. Pollut. Res.* **6** 351-64
[14] Neft I, Scungio M, Culver N and Singh S 2016 *Aerosol Sci. Tech.* **50** 937-46
[15] Natrella M G 1966 *Experimental statistics* (New York: Wiley-Interscience)
[16] Sparrow E M 1976 *Measurements in Heat Transfer* Washington: E.R.G. Eckert and R.J. Goldstein
[17] Khadir L and Mrad H 2015 *Int. J. Multiphys.* **9** 383-96
[18] Bianchini A, Balduzzi F, Rainbird J M, Peiró J, Graham J M R, Ferrara G and Ferrari L 2015 *Energ. Convers. Manage.***106** 373-84
[19] Zamani M, Maghrebi M J and Varedi S R 2016 *Renew. Energ.* **95** 109-26
[20] Bedon G, De Betta S and Benini E 2016 *Renew. Energ.* **94** 328-40
[21] Claessens M C 2009 *The Design and Testing of Airfoils for Application in Small Vertical Axis Wind turbines* Delft University
[22] Scungio M, Arpino F, Focanti V, Profili M and Rotondi M 2016 *Energ. Convers. Manage.***130** 60-70
[23] Baker R C 2000 *Flow Measurement Handbook - Industrial designs, operating principles, performance and applications* New York: Cambridge University Press

Use of phase change materials during compressed air expansion for isothermal CAES plants

B Castellani, A Presciutti, E Morini, M Filipponi, A Nicolini and F Rossi

CIRIAF, University of Perugia, Via G. Duranti 67, 06125, Perugia ITALY

E-mail: beatrice.castellani@unipg.it

Abstract. Compressed air energy storage (CAES) plants are designed to store compressed air into a vessel or in an underground cavern and to expand it in an expansion turbine when energy demand is high. An innovative CAES configuration recently proposed is the isothermal process. Several methods to implement isothermal CAES configuration are under investigation. In this framework, the present paper deals with the experimental testing of phase change materials (PCM) during compressed air expansion phase. The experimental investigation was carried out by means of an apparatus constituted by a compression section, a steel pressure vessel, to which an expansion valve is connected. The initial internal absolute pressure was equal to 5 bar to avoid moisture condensation and the experimental tests were carried out with two paraffin-based PCM amounts (0.05 kg and 0.1 kg). Results show that the temperature change during air expansion decreases with increasing the PCM amount inside the vessel. With the use of PCM during expansions an increase of the expansion work occurs. The increase is included in the range from 9.3% to 18.2%. In every test there is an approach to the isothermal values, which represent the maximum theoretical value of the obtainable expansion work.

1. Introduction

The increasing share of renewable energy in the existing electricity network leads to several challenges, mainly related to the intermittency and unpredictability of the renewable sources. Due to such a characteristic, there is mismatch between the electricity production and the corresponding consumption, which is currently compensated by spinning reserve and fast start generators. In addition, when the renewable sources are highly available, the produced power exceeds the transmission capacity limits, with a consequent increase of service costs [1]. The introduction of energy storage systems in the above-described scenario provides several beneficial services, allowing the electrical system to run more efficiently [2-3].

There is a wide range of different technologies to store electrical energy: mechanical, electrochemical, chemical, electrical and thermal energy storage systems [4-6]. The most common mechanical storage systems are pumped hydroelectric power plants, compressed air energy storage (CAES), flywheel energy storage and mechanical springs. Electrochemical storage systems consist of various types of batteries (lead acid, NiCd/NiMH, Li-ion, metal air, sodium sulphur, sodium nickel chloride and flow battery). Chemical energy storage focuses on hydrogen and synthetic natural gas (SNG) as secondary energy carriers and, finally, electrical storage systems include double-layer capacitors and superconducting magnetic energy storage [7-14]. Technologies used for high power ranges and energy capacities are pumped hydro storage (PHS) and CAES [15].

CAES plants work by pumping and storing air into a proper reservoir when excess or low-cost electricity is available. Then, during the peak demand periods, the pressurized air is expanded in an expansion turbine to produce electricity. Today only two CAES power plants are in operation worldwide, both based on the diabatic configuration [8]. In the diabatic CAES, heat released during compression is dissipated by cooling and not stored, therefore air must be reheated prior to expansion in the turbine through fuel combustion. To improve efficiency of diabatic CAES, other configurations have been proposed and worldwide other CAES plants are under development and construction. Among the improved or advanced CAES technological proposals, there is the adiabatic CAES, in which heat from the high-temperature compressed air is absorbed, stored and used to reheat the air before expansion. The roundtrip efficiency is over 70 % [16].

The new emerging technology which attempts to overcome the limits of diabatic and adiabatic CAES is isothermal CAES. Isothermal CAES is based on isothermal compression and expansion in situ. In this way the need for fuel and high temperature thermal energy storage is eliminated. Isothermal CAES requires to remove heat continuously during compression and add heat continuously during expansion by means of effective heat transfer [17].

Currently there are no commercial isothermal CAES plants, but some solutions have been proposed. One consists in water spraying into the air-filled chamber of pneumatic cylinders, so that heat is transferred from water to air during expansion or from air to water during compression [18].

Another way to do isothermal air expansion or compression is the use of phase change materials (PCM) as thermal storage media to obtain no temperature change during air transformations. The present paper deals with the experimental investigation on the effects of PCM during compressed air expansion. Tests were carried out in an experimental apparatus, whose main section is constituted by a 100 L pressure vessel in which air is compressed and stored. The experimental campaign has as objectives the temperature decrease monitoring and the evaluation of the obtainable expansion work. Work values were then compared to the theoretical isothermal value.

2. Motivation

Even though CAES technology was historically applied for large-size installations for grid management, greater attention is currently directed to small-size innovative applications [19-23]. Large-size CAES plants require the selection of suitable geological reservoirs, which limit the installed capacity because of geographical and geological constraints [24].

Small-size CAES instead can use custom-fitted above-ground storage vessels, allowing the construction of the plant where the service is requested. This ensures portability and adaptability of the system also for distributed energy productions [25]. Currently, increasing effort is devoted to complete energy evaluation and process optimization studies with the objective of overall performance improvement. Energy and exergy analyses in literature highlighted that micro-CAES systems are very effective systems for distributed energy production applications [23, 26] and they reach high efficiency if based on quasi-isothermal compression and expansion processes [11].

Given the above-mentioned theoretical findings, this paper wants to give an experimental contribution on the advanced highly efficient CAES technology development, especially for small-size applications.

It deals with the evaluation of the effect of PCM during air expansion in obtaining quasi-isothermal conditions. PCM are storage media with a high latent heat value. Latent heat is the energy exchanged in a phase change, during which there is no change of temperature. The use of PCM is proposed in order to control P-v curve during air expansion so that it can resemble or practically approach an isotherm. The phase change of paraffin-based PCM has the effect of both levelling the internal temperature profile during air expansion, getting close to isothermal conditions and increasing the obtainable work.

3. Materials and methods

The experimental procedure consists in storing compressed air in a steel vessel and expanding it in presence of PCM. During expansion, temperature and pressure values are recorded and used for calculation of P-v curve and the extractable useful work. The scheme of the experimental apparatus is shown in Figure 1. It is formed by a compression section, a steel pressure vessel, to which an expansion valve is connected.

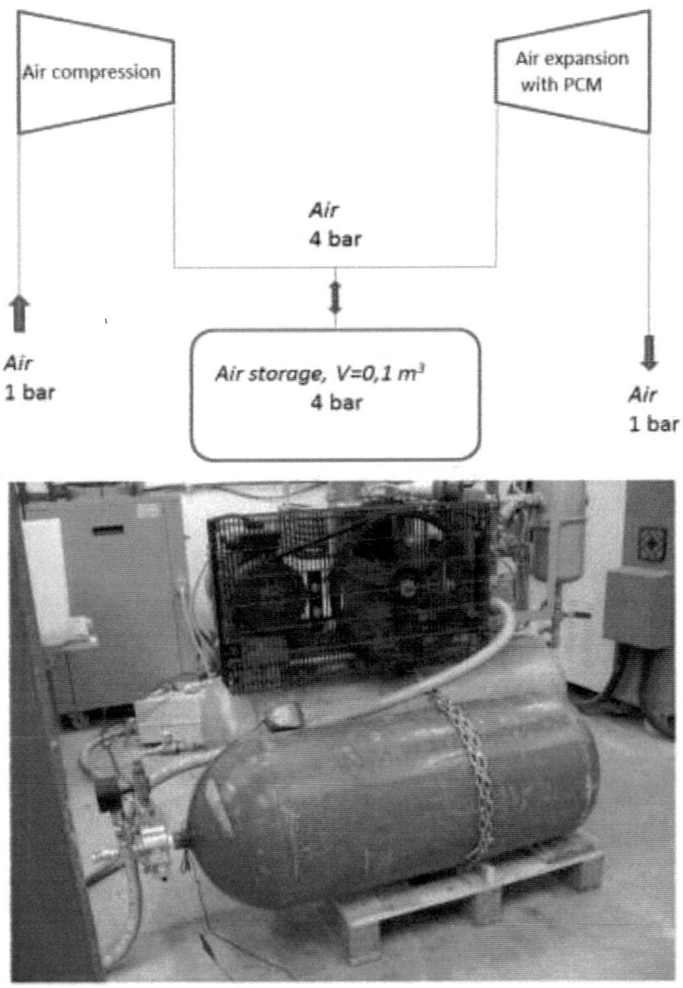

Figure 1. Schematic diagram of the experimental apparatus with the air compressor and high-pressure vessel.

Compression, supplied by Adicomp Italy, is done by an air-cooled two-stage reciprocating compressor. The maximum outlet pressure is 30 bar with a flow rate of 4 Nm^3/h. Compressed air is uploaded in a steel vessel with a total internal volume of 100 l. It has been designed for pressure values up to 200 bar. The reservoir is equipped with two temperature sensors, which measure the temperature inside the vessel in the lower part and in the upper part, and a pressure gauge.

The temperature sensors are mineral insulated type K thermocouples; the accuracy class is 1 with ± 1.5 °C tolerance range. The pressure sensor is a 0.5 accuracy class digital piezo-resistive manometer supplied by Kobold; the pressure measurement uncertainty is ± 0.2 bar. Voltage signals from pressure transducers and temperature sensors are collected every 2 seconds by a software for data acquisition on a personal computer.

Tests were conducted in accordance with the following procedure: i) in case of expansion in presence of PCM, the proper amount of PCM is loaded in the tank (50 g or 100 g), ii) air is compressed and loaded inside of the tank until an internal pressure of 5 bar is reached, iii) internal temperature and pressure are adjusted to achieve the correct initial conditions, iv) the expansion valve is opened to complete the expansion of the air.

In this investigation, a turbine to recover the expansion work was not used. In absence of the turbine, air expansion takes place by opening the expansion valve. The effect of PCM on the expansion is estimated by monitoring the values of temperature and pressure, which are used at a later stage for the calculation of the expansion work.

The PCM used in the experimental investigation is a paraffin-based material and was supplied by Rubitherm Technologies GmbH. Physical properties are described in [17]. Physical properties of the used PCM are shown in Table 1. The reported value of the heat storage capacity is a combination of latent and sensible heat in a temperature range of 11 °C to 26 °C, as stated by the supplier.

The calculated amount of PCM is settled on the vessel's internal wall.

Table 1. PCM: list of physical properties.

Properties	
Material	Paraffin
Melting Area [°C]	17-19
Congealing area [°C]	19-17
Heat storage capacity [kJ/kg]	250
Specific heat capacity [kJ/kg K]	2
Density solid [kg/l]	0,88
Density liquid [kg/l]	0,77
Heat conductivity [W/m K]	0,2
Volume expansion [%]	12

4. Results and discussion

Tests were devoted to determine the effect of PCM on air expansion phase. The initial internal absolute pressure was equal to 5 bar to avoid moisture condensation. The measured air's relative humidity was equal to 65%. Table 2 summarizes experimental conditions for the carried out tests. Temperature and pressure data were used to calculate specific volume through CoolPack Software. Tests were carried out varying both the PCM amount and the initial air temperature.

The theoretical PCM quantity to obtain isothermal conditions was calculated through Equation 1:

$$m_{PCM} = \frac{m_{air} \cdot c_{Pair} \cdot \Delta T_{air}}{C_{PCM}} \quad \text{(Eq. 1)}$$

where

c_{Pair} is air specific heat capacity [kJ/kg K]

C_{PCM} is the PCM heat storage capacity [kJ/kg]

ΔT_{air} is the temperature difference between initial temperature and final temperature [K]

m_{air} is the air inside the vessel at experimental conditions of pressure and temperature [kg]

m_{PCM} is the quantity of PCM necessary for isothermal expansion [kg].

For a 4-bar expansion, the calculated PCM amount is equal to 47 g (0.047 kg, 0.061 l). It was decided to carry out experimental tests with two PCM amounts: 50 g (0.05 kg) equal to the theoretical value and 100 g (0.1 kg), double the theoretical amount. To increase the interfacial area between air and PCM and to reduce the negative effects of the low thermal diffusivity ($1.298 \cdot 10^{-7}$ m^2/s) on the heat transfer, PCM was laid down on the internal wall of the vessel. The internal volume occupied by PCM (0.065 l and 0.13 l respectively) is practically negligible with respect to the total volume of the vessel. Thus this does not affect heat transfer conditions, expansion times and air flows during the expansion.

Internal temperature values are calculated as the average of the two temperature values measured by two thermocouples in two different positions. Different temperature values from 28.6 °C and 19.3°C are investigated, comparing expansions without PCM with expansions in presence of different PCM amounts. Temperature values are all above the phase change temperature of the used PCM.

Table 2. Experimental tests.

N.	PCM (g)	Initial Temperature (°C)	Final Temperature (°C)	ΔT (°C)	Initial Absolute Pressure (bar)
1	-	28.6	3.0	25.6	5.1
2	-	27.2	1.3	25.9	5.0
3	-	26.7	2.1	24.6	5.1
4	-	24.3	1.0	24.3	5.0
5	-	19.3	0.6	18.7	5.1
6	50	28.5	14.9	13.6	5.0
7	50	27.2	15.1	12.1	5.0
8	50	26.6	12.1	14.5	5.0
9	50	24.2	13.2	11.0	5.0
10	50	19.3	10.0	9.3	4.9
11	100	28.4	16.8	11.6	5.0
12	100	27.4	16.6	10.8	5
13	100	26.6	15.3	11.3	5
14	100	24.2	15.4	8.8	4.9
15	100	19.3	14.8	4.5	4.9

The first five tests were carried out without PCM, with an initial temperature of 28.6 °C (test 1), 27.2°C (test 2), 26.7°C (test 3), 24.3°C (test 4) and 19.3°C (test 5). Tests with 50 g and 100 g of PCM were carried out at the same values of initial temperature, so that we can compare test 1 with test 6 (with 50 g of PCM) and test 11 (with 100 g of PCM) and so on.

Table 2 reports, for each test, also the final temperature and the difference between initial and final values. The temperature change during air expansion decreases with increasing the PCM amount inside the vessel. In particular, for tests 1-4 the variation is in the range 24-26°C, while for tests 6-9 with 50 g of PCM the variation is in the range 11-15°C and for tests 11-14 it is included in the range 8-12°C. No substantial temperature differences ΔT are recorded comparing 50 g PCM tests with 100 g PCM tests. For tests at 19.3°C, instead, the temperature variation undergoes a greater decrease, from 18.7°C of test 5 to 9.3°C and 4.5°C of tests 10 and 15 respectively.

Profiles of internal pressure and temperature during air expansion for tests 1, 6 and 11 and tests 5, 10, 15 are shown in Figure 2. Temperature profiles show visibly that in presence of PCM, temperature decrease during air expansion is leveled off and temperature difference at the end of the expansion is lower than that without PCMs. Pressure profiles instead do not show significant dissimilarities.

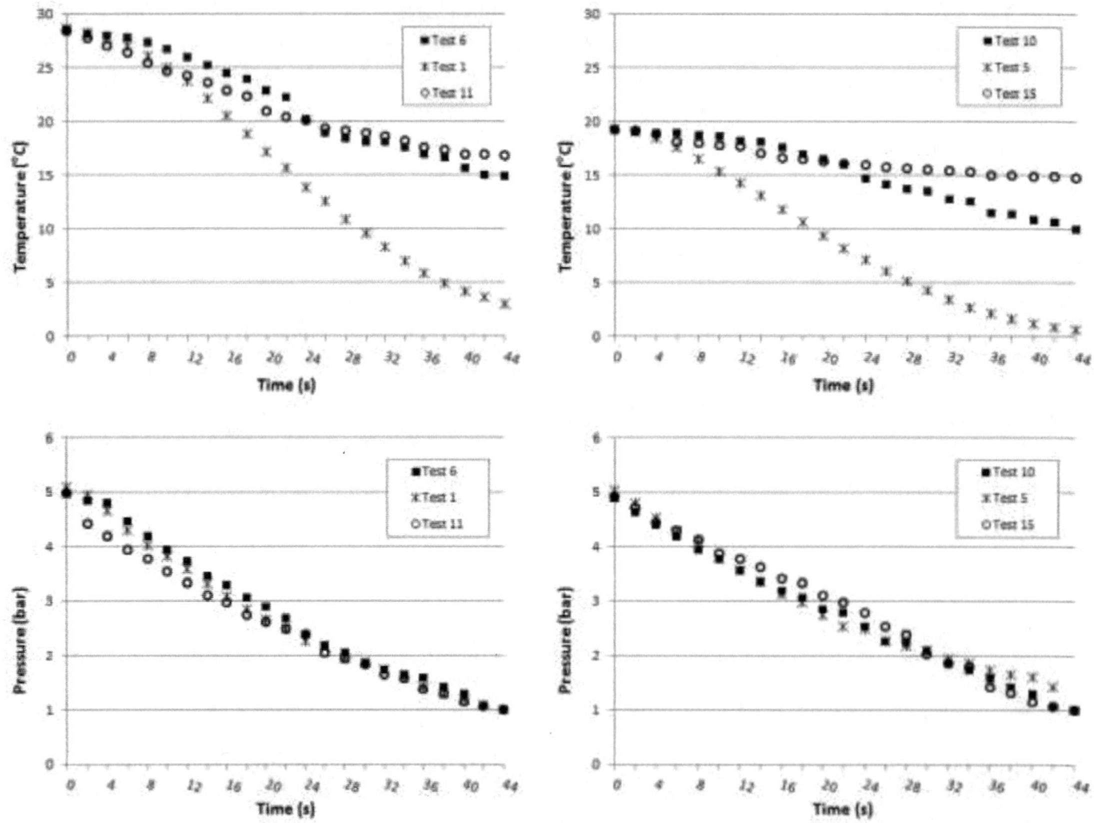

Figure 2. Temperature and pressure profiles.

For every test, the acquired values of pressure and specific volume were used to build the P-v curve and to calculate the obtainable expansion work. Results from the tests were compared to the theoretical adiabatic and isothermal expansions. Adiabatic and isothermal curves on the P-v plane were depicted in accordance to Equation 2 and Equation 3 respectively:

$$P = \frac{k}{v^{\gamma}} \qquad \text{(Eq. 2)}$$

$$P = \frac{R \cdot T}{v} \qquad \text{(Eq. 3)}$$

where:
P is pressure;
v is the specific volume;
k is adiabatic constant;
γ is the adiabatic index calculated as the ratio of the heat capacity at constant pressure to heat capacity at constant volume;
R is gas constant;
T is the internal temperature.

P-v diagrams for 4-bar expansions in tests 1, 7, 13 are depicted in Figure 3. Test 1 without PCM is placed between the adiabatic and the isotherm, while tests 7 and 13 with PCM approach the isothermal curve. The benefit of PCM use in terms of obtainable expansion work is better clarified in Table 3.

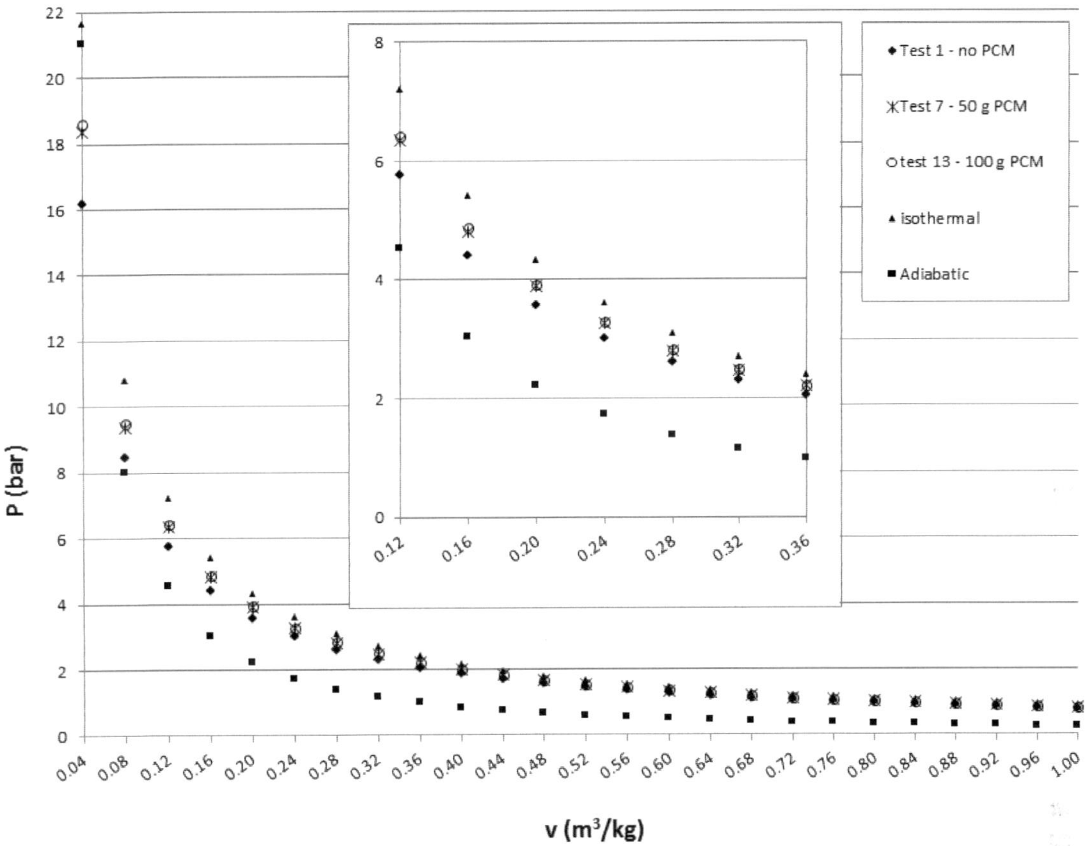

Figure 3. P-v curves – Tests 1-7-13 (28.5 °C) .

Table 3 shows the expansion work for all the tests. In the same row, work values in J/kg from expansions with the same initial temperature are compared. As a reference, also the isothermal work, calculated considering the same initial conditions, is reported.

Together with the expansion work, also the related increase obtained with the use of PCM is reported and expressed as a percentage value with respect to the baseline tests without PCM. Benefits from the use of PCM during expansions are demonstrated by the increase of the expansion work which occurs in every test with PCM. The increase is included in the range from 9.3% in test 6 to 18.2% in test 15. The general trend of the results shows that there is an approach to the isothermal values, which represent the maximum theoretical value of the obtainable expansion work.

Some other observations can be deduced from data in Table 3. The influence of PCM becomes greater with increasing the amount of PCM. Comparing tests with 50 g of PCM with tests with 100 g of PCM, the increase in the expansion work is bigger. This is quite evident at 19.3°C, when the increase with 50 g of PCM is equal to 14.0%, while with 100 g of PCM it is equal to 18.2%.

This result is in contrast with other tests in which PCM were encapsulated [15]. In that experimental investigation, PCM were encapsulated in several disk packets, resulting in an incomplete phase transition with consequent lower thermal storage.

Table 3. Expansion work.

	Test 1	Test 6	Test 11	Isothermal
Expansion work (J/kg)	26.87	29.37	29.68	33.04
Increase (%)		9.3	10.5	
	Test 2	**Test 7**	**Test 12**	**Isothermal**
Expansion work (J/kg)	26.46	29.41	30.46	32.89
Increase (%)		11.1	15.1	
	Test 3	**Test 8**	**Test 13**	**Isothermal**
Expansion work (J/kg)	25.91	29.95	30.05	32.83
Increase (%)		15.6	16.0	
	Test 4	**Test 9**	**Test 14**	**Isothermal**
Expansion work (J/kg)	26.36	29.7	30.56	32.57
Increase (%)		12.7	15.9	
	Test 5	**Test 10**	**Test 15**	**Isothermal**
Expansion work (J/kg)	26.21	29.88	30.97	32.02
Increase (%)		14.0	18.2	

In the cited paper, higher improvements in the expansion work were obtained with a lower amount of PCM. Heat capacity phenomena play a significant role in the complete exploitation of the transition phase and this aspect should be investigated in the future stages of the research.

5. Conclusion

Great efforts are devoted on CAES systems in order to identify novel highly-efficient configurations. Current research is focused also on the study of materials for thermal exchange and critical components. The present work is well integrated in this context, since the objective of the experimental investigation is the testing of PCM in compressed air expansions to approach isothermal conditions.

The testing took place through the use of an apparatus, constituted by a compression section and a pressure storing vessel, from which air expands via an expansion valve. 15 tests were conducted at an initial pressure of 5 bar and for each expansion achieved was calculated by the obtainable work.

Results show that through the use of the PCM a levelling of the internal temperatures occurs and an increase of the work of expansion up to a maximum of 18.2%. Through the use of PCM, P-v curves approach the isotherm, providing near-isothermal conditions for the expansion phase. Results give experimental evidence of the potential benefit of PCM in CAES plants, providing a viable solution to increase the work of expansion, thereby increasing considerably the efficiency of the whole process.

Nevertheless, some aspects should clarified in the future, such as the role of encapsulation. Different methods of encapsulation and deposition of PCM inside the vessel lead to contrasting results. Finally, a future step of the research is the integration of a 4-bar expansion turbine in the experimental apparatus, so that complete energy balance analysis will be carried out.

References

[1] Eyer M, Iannucci JJ, Corey JP 2004 *Energy Storage Benefits and Market Analysis Handbook* SAND2004-6177
[2] Lund H, Salgi G *Energ. Convers. Manage.* **50** 1172
[3] Ippolito MG, Di Silvestr ML, Riva Sanseverino E, Zizzo G, Graditi G 2014 *Energy* **68** 648
[4] Díaz-González F, Sumper A, Gomis-Bellmunt O, Villafáfila-Robles R 2012 *Renew. Sust. Energ. Rev.* **6** 2154

[5] Salata F, Golasi I, Domestico U, Banditelli M, Lo Basso G, Nastasi B, de Lieto Vollaro A 2017 *Energ. Convers. Manage.* 138
[6] Garcia DA, Barbanera F, Cumo F, Di Matteo U, Nastasi B 2016 *Energies* **9** art. no. 96
[7] Kyriakopoulos GL, Arabatzis G 2016 *Renew. Sust. Energ. Rev.* **56** 1044
[8] Kim YM, Favrat D 2010 *Energy* **35** 213
[9] Rossi F, Castellani B, Morini E, Di Giovanna L, Corsi N, Giuliobello M, Nicolini A 2015 *Pr. SASEC2015* (South Africa)
[10] Nataf K, Bradley TH 2016 *Appl. Energ.* **164** 133
[11] Kim YM, Shin DG, Favrat D 2011 *Energy* **36** 6220
[12] Cotana F, Filipponi M, Castellani B 2013 *Appl. Mech. Mater.* **392** 512
[13] Castellani B, Morini E, Filipponi M, Nicolini A, Palombo M, Cotana F, Rossi F 2014 *Sustainability* **6** 6815
[14] Rossi F, Castellani B, Nicolini A 2015 *Energ. Proc.* **82** 805
[15] International Energy Agency – IEA 2014 *Technology Roadmap: Energy storage* (Paris)
[16] Grazzini G, Milazzo A 2008 *Renew. Energ.* **33** 1998
[17] Castellani B, Presciutti A, Morini E, Filipponi M, Nicolini A, Rossi F 2015 *Sustainability* **7** 9773
[18] Tallini A, Vallati A, Cedola L 2015 *Energ. Proc.* **82** 797
[19] Baqari FF, Vahidi B 2013 *Renew. Sustain. Energy Rev.* **21** 365
[20] Zhao P, Dai Y, Wang J 2014 *Energy* **70** 674
[21] Bagdanavicius A, Jenkins N 2014 *Energ. Convers. Manage.* **77** 32
[22] Li Y, Wang X, Li D, Ding Y 2012 *Appl. Energ.* **99** 316
[23] Alami AH 2015 *J Energy Storage* **1** 38
[24] Konrad J, Carriveau R, Davison M, Simpson F, Ting SKD 2012 *Int. J. Environ. Stud.* **69** 350
[25] De Lieto Vollaro R, Faga F, Tallini A, Cedola L, Vallati A 2015 *Energ. Proc.* **82** 645
[26] Minutillo MA, Lavadera L, Jannelli E 2015 *J Energy Storage* **4** 135

Multi-objective Optimization of a Solar Humidification Dehumidification Desalination Unit

M Rafigh [1], M Mirzaeian[1], B Najafi [2*], F Rinaldi[2], and R Marchesi[2]

[1] Dipartimento di Energia, Politecnico di Torino, Corso Duca degli Abruzzi, 24 - 10129 Torino (TO) , Italy
[2] Dipartimento di Energia, Politecnico di Milano, Via Lambruschini, 4, 20156, Milano, Italy

Email: behzad.najafi@polimi.it

Abstract. In the present paper, a humidification–dehumidification desalination unit integrated with solar system is considered. In the first step mathematical model of the whole plant is represented. Next, taking into account the logical constraints, the performance of the system is optimized. On one hand it is desired to have higher energetic efficiency, while on the other hand, higher efficiency results in an increment in the required area for each subsystem which consequently leads to an increase in the total cost of the plant. In the present work, the optimum solution is achieved when the specific energy of the solar heater and also the areas of humidifier and dehumidifier are minimized. Due to the fact that considered objective functions are in conflict, conventional optimization methods are not applicable. Hence, multi objective optimization using genetic algorithm which is an efficient tool for dealing with problems with conflicting objectives has been utilized and a set of optimal solutions called Pareto front each of which is a tradeoff between the mentioned objectives is generated.

1. Introduction

Water demand already exceeds supply in many parts of the world and as the world population continues to rise, so does the water demand. It is expected that by 2025, more than 60% of world's population will have water shortages [1]. Regarding the fact that seawater is one of the most convenient sources of water, desalination can be a practical solution to this concern.

Desalination refers to a process which removes salt and other minerals from water resulting in having fresh water. The required energy for this purpose can be provided by different sources including refrigeration cycle [2], combined cogeneration cycle, heat pump [3] or solar system. Limitation of natural gases and oil in addition to their undesired effect on environment arises more interest in application of renewable sources of energy. Due to the fact that solar energy has a good consistency with climatic conditions in regions which need potable water, it can be considered as an appropriate source of energy. It is noteworthy that among the various desalination technologies in use, only those based on a thermal principle of operation must be considered in combination with solar ponds [4].

Additionally, many of the remote arid areas need low capacity desalination systems, while conventional desalination methods such as MSF, MEE, VC and RO are suitable for large and medium capacity of fresh water production; therefore utilizing mentioned methods for small capacity systems is not economical. Humidification-dehumidification desalination is a suitable choice for production of fresh water when the demand is decentralized [5]. A significant advantage of the HD technology besides its simplicity of operation is that it provides a means for low pressure, low temperature desalination that operates off of waste heat and is potentially very cost competitive [6].

In addition, HD operates at low temperature; therefore choosing solar system as the source of energy is an ideal decision. Solar desalination based on the HD principle results in an increase in the overall efficiency of the desalination plant and is therefore considered as a promising technique for small capacity, solar-driven desalination plants [7]. Several studies have been performed on solar driven HD desalination systems considering different objective functions and design parameters.

Ben Bacha et al. [8] conducted a work including modeling, simulation and experimental validation of a solar HD system and showed that that elevated flow rate at the entrance of evaporation tower, low temperature of water at the entrance of condenser, and hot water recycling can improve operation and production of the system. Soufari et al. [9] optimized the performance of HD desalination using mathematical programming and found the productivity seems to be the best objective function. Al-Hallaj et al. [10] studied the solar-based processes from an economic point of view. The result of the research was that the solar HD process is a suitable replacement to all other forms of solar desalination techniques in the smaller capacity range. Hermosillo et al. [1] analysed a desalination system working under the air humidification and dehumidification principle by comparing mathematical model and experimental results and a sufficient agreement was shown, although no optimization was carried out in this study. Gude et al. [11] studied the advantages and opportunities for enhanced utilization of solar energy using theoretical and experimental approaches. It was indicated that solar energy is best utilized when the source is integrated with energy-efficient desalination process. Zamen et al. [12] instead optimized the cost of a complete solar HD process.

The main purpose of the present project is to optimize the operation of a solar HD desalination process by defining suitable objective functions such as the area of the humidifier and dehumidifier and thermodynamic parameters. Since the objective functions are conflicting, global optimum cannot be achieved by the means of conventional optimization methods. Multi-objective method is a suitable choice in such cases in which both the thermodynamics and economic objectives can be considered simultaneously and a set of optimal solutions, called Pareto solutions can be achieved. As a result, based on the project limits and available investments, the optimal solution for each case can be chosen by the designer. In other words, a reasonable solution to a multi-objective problem is to investigate a set of solutions, each of which satisfies the objectives at an acceptable level without being dominated by any other solution [13]. Multi-Objective optimization has been successfully utilized for optimizing many different thermodynamic systems. Baghernejad et al. [14] successfully used multi-objective evolutionary algorithms in order to find solutions that simultaneously satisfy exergetic as well as economic objectives for an integrated solar combined cycle system. Najafi et al. [15] used multi-objective optimization in order to maximize the performance and minimize the cost of an SOFC-Gas Turbine hybrid cycle coupled to an MSF Desalination unit.

In the present work, multi-objective optimization of the plant is carried out while the required areas of the subsystems of the HD unit, as an indicator of the cost, and also the specific energy of the solar energy source, indicating the energy efficiency, are considered as objective functions. Although the considered system in the present work is a well-stablished technology, the main contribution of the present work, compared to the previous studies, is providing a set of optimal system designs, each of which can be utilized by the user depending on the project requirements and the corresponding available investment. Simulation and optimization procedures have been carried out using MATLAB™.

2. Model Description

The humidification dehumidification process is based on the fact that the capability of air in carrying vapor increases progressively with temperature. The HD unit is the distillation under atmospheric conditions by an air loop saturated with water vapor, and has three main sections: the humidifier (evaporator), dehumidifier (condenser) and solar heat source. A schematic representation of the plant has been demonstrated in Fig. 1.

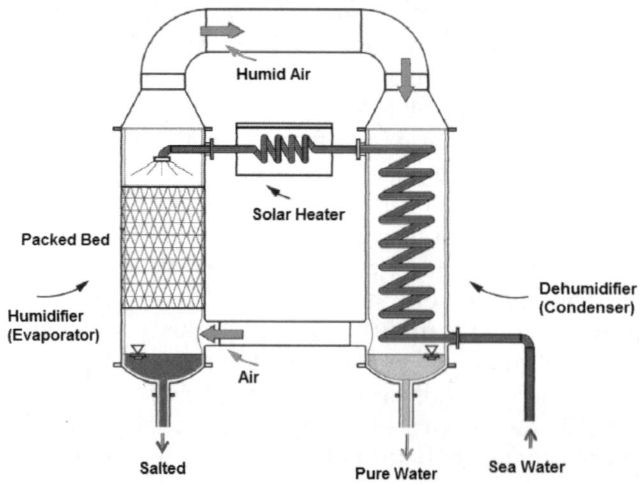

Figure 1. Sketch of solar HD unit with closed air cycle and water [16]

2.1 Solar source

In the solar heater, the sea water leaving the condenser is heated and will enter the evaporator with higher temperature. In comparison with other desalination systems, considering the fact that HD unit requires lower water temperature; solar heater is an appropriate choice for the heat source of this type of units.

2.2. Humidifier

In the humidifier, air and water, where the water has been heated by an external heat source, are in contact and a certain amount of vapor is absorbed by air. The temperature of air will increase and water temperature will decrease during this process.

In the presented model, packing bed evaporator is used. Usually, packings are used when a liquid is in contact with a gas or vapor. It is desired to have the greater liquid surface contact with the vapor or gas. As the liquid flows over the packings, essentially wetting the latter, the area in contact between the two fluids is related to the area of the packing. Packings can also be used in liquid–liquid systems to assist in removing the wake from the dispersed liquid and increasing the heat transfer [16].

3. Mathematical Model

In order to establish the thermal transfers and mass balances of the evaporation tower for the liquid phase, the gas phase and the liquid-gas interface the following hypotheses are made [8]:

- Steady state conditions
- The tower is adiabatic.
- Head losses in the tower column are low and it can be assumed that the density of the air mass flux is constant, since the evaporation rate is small compared to the water flow rate, then the water mass flux remains constant throughout the column.
- The water distribution over the packed bed is uniform; hence there is only a vertical gradient of temperature and humidity.

3.1. Solar source

The rate of heat transferred from the solar collector to water is expressed as:

$$\dot{Q} = \dot{m} C_p \Delta T_w \quad (1)$$

Hence, the energy per unit of time radiated on the surface of the collector is equal to:

$$\dot{Q}_{sol} = \frac{\dot{Q}}{\eta_{\eta_{col}}} \tag{2}$$

η_{col} can be obtained from the empirical relation [17].

$$\eta_{col} = 0.8 - 6.8(\frac{T_{col,in} - T_{amb}}{I}) \tag{3}$$

Finally, the required surface area for collector is given by:

$$A_{col} = \frac{\dot{Q}}{I} \tag{4}$$

3.2. Humidifier

As demonstrated in Fig. 2 for modeling the humidifier, a volume element with height of dz containing three control volumes is considered: 1) C.V.1: total system, 2) C.V.2: water side, 3) C.V.3: air side

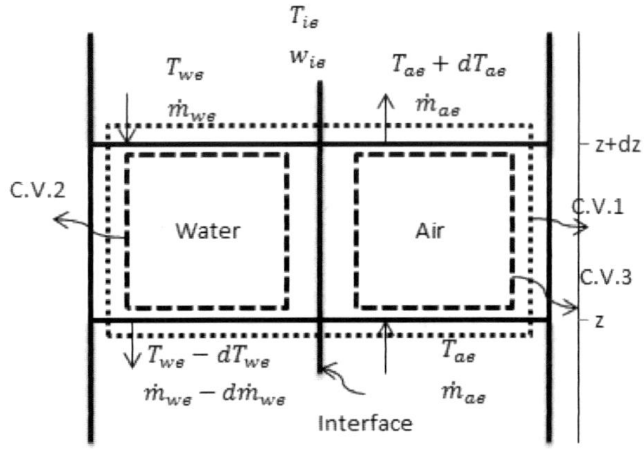

Figure 2. Element of humidifier volume

The mass rate balance for C.V.1 can be written as:

$$d\dot{m}_{we} = d\dot{m}_{ve} = \dot{m}_{ae} d\omega_e \tag{5}$$

The heat rate balance for C.V.2 results in Equation 2.

$$\frac{dT_{we}}{dz} = \frac{h_{we} a_{He}(T_{we} - T_{ie})}{\dot{m}_{we} C_{we}} \tag{6}$$

h_{we}, is obtained from Equation (7) [8].

$$h_{we} = \frac{5900\, \dot{m}_{ae}^{0.5894}\, \dot{m}_{we}^{0.169}}{a_{He}} \tag{7}$$

For C.V.3 the heat and mass rate balances can be expressed respectively as:

$$\frac{d\omega_e}{dz} = \frac{k_{ae} a_{Me}(\omega_{ie} - \omega_e)}{\dot{m}_{ae}} \tag{8}$$

$$\frac{dT_{ae}}{dz} = \frac{h_{ae} a_{He}(T_{ie} - T_{ae})}{\dot{m}_{ae}(C_{ae} + \omega_e C_{ve})} \tag{9}$$

Where h_{ae} and k_{ae} are obtained using Lewis relation [8]:

$$k_{ae} = \frac{2.09\, \dot{m}_{ae}^{0.11515}\, \dot{m}_{we}^{0.45}}{a_{Me}} \tag{10}$$

$$h_{ae} = C_{ae}k_{ae} \tag{11}$$

Finally heat rate balance for interface per unit of surface is as follows:

$$h_{we}a_{He}(T_{we} - T_{ie})dz = h_{ae}a_{He}(T_{ic} - T_{ac})dz + L_{ve}k_{ac}a_{Me}(\omega_{ie} - \omega_e)dz \tag{12}$$

The interface is assumed to be a film of saturated air. Therefore T_{ie} and ω_{ie} are dependent variables. T_{ie} with respect to ω_{ie} can be calculated using an experimental relation [18].

$$\omega_{ie} = f_{exp}(T_{ie}) = 2.19 \times 10^{-6} \times T_{ie}^3 - 1.85 \times 10^{-4} \times T_{ie}^2 + 7.06 \times 10^{-3} \times T_{ie} - 0.077 \tag{13}$$

3.3. Dehumidifier

As demonstrated in Fig. 3 using the same procedure as the one applied on the humidifier, the dehumidifier volume element contains C.V.1 (total system), C.V.2 (water side) and C.V.3 (air side).

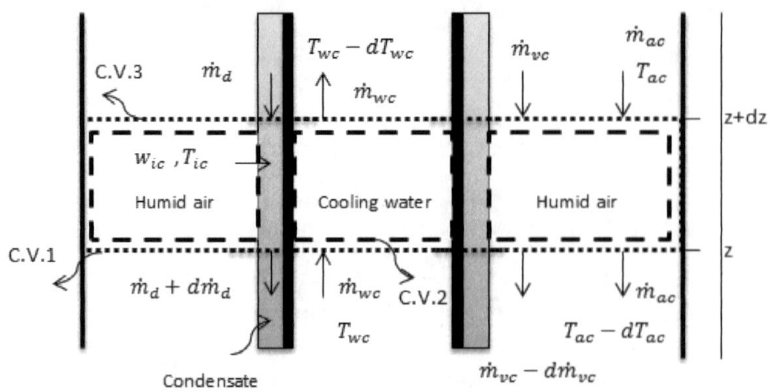

Figure 3. Element of dehumidifier volume

Mass rate balance for C.V.1 can be written as:

$$d\dot{m}_d = d\dot{m}_{vc} = \dot{m}_{ac}d\omega_c \tag{14}$$

The heat rate balance for C.V.2 can be calculated as:

$$\frac{dT_{wc}}{dz} = \frac{h_{wc}a_{Hc}(T_{ic} - T_{wc})}{\dot{m}_{wc}C_{wc}} \tag{15}$$

For C.V.3 heat and mass rate balances can be expressed respectively as:

$$\frac{d\omega_c}{dz} = \frac{k_{ac}a_{Mc}(\omega_c - \omega_{ic})}{\dot{m}_{ac}} \tag{16}$$

$$\frac{dT_{ac}}{dz} = \frac{h_{ac}a_{Hc}(T_{ac} - T_{ic})}{\dot{m}_{ac}(C_{ac} + \omega_c C_{vc})} \tag{17}$$

Finally, for the interface:

$$h_{wc}a_{He}(T_{we} - T_{ic}) = h_{ac}a_{He}(T_{ic} - T_{ac}) + L_{vc}k_{ac}a_{Mc}(\omega_{ic} - \omega_c) \tag{18}$$

And the experimental relation for T_{ic} and ω_{ic}:

$$\omega_{ic} = f_{exp}(T_{ic}) = 2.19 \times 10^{-6} \times T_{ic}^3 - 1.85 \times 10^{-4} \times T_{ic}^2 + 7.06 \times 10^{-3} \times T_{ic} - 0.077 \tag{19}$$

The mean values of h_{ac} and h_{wc} calculated and k_{ac} is obtained from Lewis relation.

$$k_{ae} = \frac{h_{ae}}{C_{ae}} \tag{20}$$

It is important to note that the mass transfer coefficient of the air in the humidifier should satisfy a constraint which is related to the ratio of mass velocity of water to air and also geometrical parameters. For this purpose, the characteristic curve of selected packing is formed as mathematical relations [12].

$$\frac{k_{ae}aV}{L} = (1.222H + 0.3667) \times \left(\frac{\dot{m}_{we}}{\dot{m}_{ae}}\right)^{-0.62} \tag{21}$$

4. Mathematical programming model

Finite element method has been implemented in order to model the air and water behaviour in the humidifier and dehumidifier units. Based on the discretization which is carried out in FEM, the governing equations of each element which are in the format of differential equations are converted into a set of algebraic relations; hence, the water and air properties of each element is determined by solving above mentioned system of equations. In order to perform the FEM discretization both the humidifier and dehumidifier are divided into n and m vertical elements respectively.

4.1. Solar collector equations

The heat transfer rate and surface area of the solar collector can be determined using the following equations:

$$\dot{Q} = \dot{m}_{wc}A_{cross}C_{wc}[T_{we}(1) - T_{wc}(1)] \tag{22}$$

$$A_{col} = \frac{\dot{Q}}{I\left[0.8 - 6.8\left(\frac{T_{col,in} - T_{amb}}{I}\right)\right]} \tag{23}$$

4.2 Humidifier finite difference equations

$$T_{we}(i+1) = T_{we}(i)\left(1 - \frac{\Delta z_e h_{we} a_e}{\dot{m}_{we}C_{we}}\right) + T_{ie}(i)\frac{\Delta z_e h_{we} a_e}{\dot{m}_{we}C_{we}} \tag{24}$$

$$T_{ae}(i+1) = T_{ae}(i)\left(1 + \frac{\Delta z_e h_{ae} a_e}{\dot{m}_{ae}C_e}\right) - T_i(i)\frac{\Delta z_e h_{ae} a_e}{\dot{m}_{ae}C_e} \tag{25}$$

$$\dot{m}_{we}(i+1) = \dot{m}_{we}(i) + \dot{m}_{ae}[\omega_{ae}(i+1) - \omega_{ae}(i)] \qquad 1 \le i < n \tag{26}$$

$$L_{ve}k_{ae}[\omega_{ie}(i) - \omega_{ae}(i)] = h_{we}[T_{we}(i) - T_{ie}(i)] + h_{ae}[T_{ae}(i) - T_{ie}(i)] \qquad 1 \le i < n \tag{27}$$

$$\omega_e = f_{exp}[T_{ae}(i)] \qquad 1 \le i < n \tag{28}$$

$$\omega_{ie} = f_{exp}[T_{ie}(i)] \qquad 1 \le i < n \tag{29}$$

4.3 Humidifier finite difference equations

$$T_{wc}(j+1) = T_{wc}(j)\left(1 + \frac{\Delta z_c h_{wc} a_c}{\dot{m}_{wc}(i)C_c}\right) - T_{ic}(j)\frac{\Delta z_c h_{wc} a_c}{\dot{m}_{wc}(i)C_c} \qquad 1 \le j < m \tag{30}$$

$$T_{ac}(j+1) = T_{ac}(j)\left(1 - \frac{\Delta z_c h_{ac} a_c}{\dot{m}_{ac}C_c}\right) + T_{ic}(j)\frac{\Delta z_c h_{ac} a_c}{\dot{m}_{ac}C_c} \qquad 1 \le j < m \tag{31}$$

$$L_{vc}k_{ac}[\omega_{ic}(j) - \omega_c(j)] = h_{wc}[T_{wc}(j) - T_{ic}(j)] + h_{ac}[T_{ac}(j) - T_{ic}(j)] \qquad 1 \le j < m \tag{32}$$

$$\omega_c(j) = f_{exp}[T_{ac}(j)] \quad 1 \le j < m \tag{33}$$

$$\omega_{ic}(j) = f_{exp}[T_{ic}(j)] \quad 1 \le j < m \tag{34}$$

The constraints regarding the closed air cycle and water flow rate are as follows:

$$\dot{m}_{we}(1) = \dot{m}_{wc} \tag{35}$$

$$\dot{m}_{ae} = \dot{m}_{ac} \tag{36}$$

$$T_{ae}(1) = T_{ac}(1) \tag{37}$$

$$T_{ae}(n) = T_{ac}(m) \tag{38}$$

Another constraint of the system is the pinch temperature difference. Humidifier and dehumidifier towers can be considered as two heat exchangers. An important factor in design of heat exchangers is pinch temperature difference. In practice, the outlet temperature of the hot stream cannot reach the inlet temperature of the cold stream and vice versa.
Pinch temperature difference for the dehumidifier occurs at the inlet or outlet of the tower, but for the humidifier it may occur at each point of the tower. Thus, the following constraints should also be considered [9].

$$T_{ac}(1) - T_{wc}(1) < \Delta T_p \tag{39}$$

$$T_{ac}(m) - T_{wc}(m) < \Delta T_p \tag{40}$$

$$T_{ae}(i) - T_{we}(i) < \Delta T_p \tag{41}$$

The heat recovery can be obtained by

$$Heat\ Recovery = \frac{T_{wc}(1) - T_{wc}(m)}{T_{we}(1) - T_{wc}(m)} \tag{42}$$

Finally, the objective functions are calculated as follows:

$$Specific\ Energy = \frac{\dot{Q}}{\dot{m}_d} \tag{43}$$

Where the distilled water flow rate:

$$\dot{m}_d = \dot{m}_{we}(1) - \dot{m}_{we}(n) \tag{44}$$

And the areas which can represent the cost if the HD:

$$\frac{A}{\dot{m}_d} = \frac{A_e + A_c}{\dot{m}_d} \tag{45}$$

5. Optimization procedure

Optimization of the objective functions based on mentioned parameters is conducted employing multi-objective using genetic algorithm. Most of the real-world problems deal with multiple objectives which are often in conflict with each other. Therefore, any effort to optimize the problem with respect to a single objective, will lead to improper results regarding the rest of objectives. The fact which necessitates using methods which can satisfy all objective functions concurrently. Vividly, a perfect solution which can satisfy all of the objectives at the same time, does not exist. Therefore, a logical alternative for dealing with a multi-objective problem is to determine a set of solutions, each of which satisfies the objectives at a suitable level while not being dominated by any other solution [19]. In case all of the objective functions should be minimized, a feasible solution x is said to dominate another feasible solution y $(x > y)$, if and only if $Z_i(x) \leq Z_i(y)$, for i = 1,…,K and $Z_j(x) < Z_j(y)$ for least one objective function j. A solution is called Pareto optimal if it is not dominated by any other solution in the considered space [19]. Evolutionary algorithms, specifically genetic algorithm (GA), with their unique capabilities for this kind of problems, are promising methodologies for dealing with multi-objective optimization problems [19-20].

6. Results and discussion

Considered input conditions of the regarded case are given in Table 1 [9]. The inlet mass velocity and temperature of water entering the condenser and also the air mass velocity have been imposed as input conditions. Optimization parameters include the collector outlet temperature (with the range of 68-80 °C), Pinch temperature difference (1.0-6.0 °C) and condenser length (with the range of 0.7-1.5 m).

Table 1. Input conditions

Parameter	Value
Inlet water temperature to the condenser (°C)	15
Water mass velocity in to the condenser ($\frac{kg}{m^2.s}$)	0.132
Air mass velocity ($\frac{kg}{m^2.s}$)	0.043

Multi-objective optimization using genetic algorithm is applied in order to find a set of optimal parameters. In this method, the population size is adjusted to 40 and mutation and Pareto fractions are 0.8 and 0.35, respectively. After 80 generation, the optimization procedure was terminated when the average change in the spread of Pareto solutions was less than the considered termination tolerance (10^{-6}). Final Pareto optimal set generated by multi-objective genetic algorithm is depicted in Fig. 4. Parameters of a few of the determined optimal points are given in Table 3.

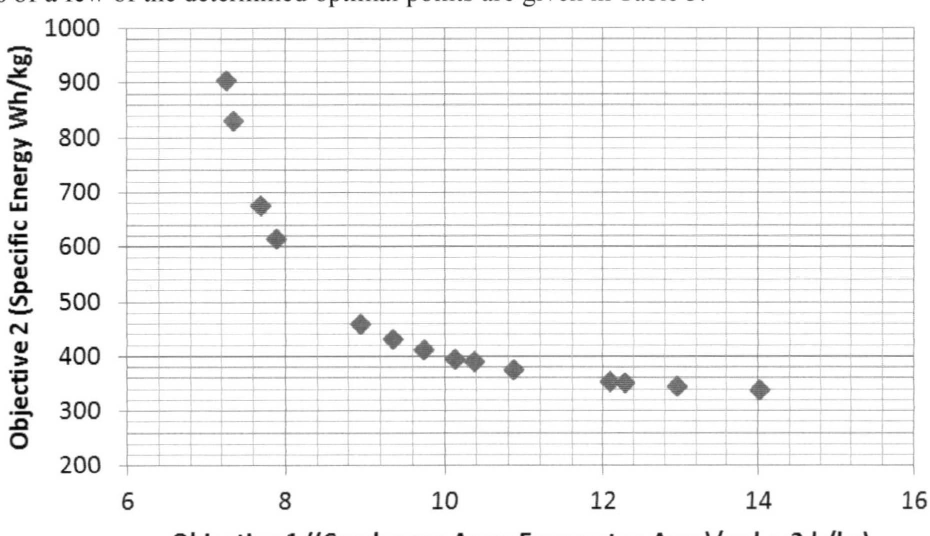

Figure 4. Pareto front

As can be observed in Table. 3 each solution includes the corresponding optimum parameters which include the collector outlet temperature, the pinch point temperature difference and the length of the condenser. For each optimal solution, the value of the considered objective functions which are the required specific energy (required thermal power for production of unit mass flow rate of distilled water) and the required specific area (for a unit mass flow rate of distilled water) are given. Accordingly, the efficiency of the collector, the required area of collector (for a unit mass flow rate of

IOP Conf. Series: Journal of Physics: Conf. Series **923** (2017) 012038 doi:10.1088/1742-6596/923/1/012038

distilled water) and heat recovery percentage for each optimum solution is determined and is also demonstrated in Table 2. As was previously explained, each of the achieved solutions can be chosen based on the requirements of a specific project. It is worth noting that the required specific energy employing HD desalination is significantly larger than that of other desalination methods including reverse osmosis (RO). Nevertheless, HD desalination unit has a notably simpler configuration compared to other desalination systems and consequently required a lower investment which makes it a suitable candidate for small scale applications. Accordingly, the main contribution of the present work, is providing a set of optimal system designs each of which can be selected based on the available capital. Hence, it further facilitates the implementation of HD based small scale solar driven applications.

Table 2. Selection of optimal values of optimization procedure

Optimization Parameters			Objective Functions			
$T_{col-out}$ (°C)	ΔT_p(°C)	L_c(m)	Specific Energy ($\frac{W.h}{kg}$)	$\frac{A}{m_d}(\frac{m^2.h}{kg})$	$\frac{A_{col}}{m_d}(\frac{m^2.h}{kg})$	HR %
68.0	6.0	1.13	340.07	14.02	0.63	44.69
71.0	5.9	1.12	377.72	10.87	0.70	41.68
79.6	1.2	0.96	833.52	7.35	1.53	21.91
72.7	5.5	1.01	413.79	9.74	0.76	39.14
68.8	5.8	1.21	355.45	12.09	0.65	43.46
78.3	2.4	0.97	676.14	7.70	1.24	26.64
74.4	4.9	0.99	461.24	8.91	0.85	36.24

7. Conclusion

In the present work, a humidifier-dehumidifier (HD) desalination plant integrated with a solar heat source was simulated. Specific required energy provided by the solar source, as an indicator of the energy efficiency of the plant, and the areas of the humidifier and dehumidifier, indicating the capital cost of the plant, were taken into account as objective functions. Since the considered objective are in conflict, multi-objective optimization using genetic algorithm was employed to carry out the optimization procedure and a set of optimal solutions each of which is a trade-off between the objectives was achieved. It is worth mentioning that the humidifier-dehumidifier desalination system is a well stablished technology and the corresponding mathematical model is comprehensively studied and validated in the literature. Hence, an experimental validation procedure was not necessary and thus was not conducted. Furthermore, although HD desalination is a less efficient method compared to other desalination technologies, it has a simpler configuration and requires lower investment which makes it a more promising alternative for decentralized small scale applications. In this context, the main contribution of the present work, is providing a set of optimal system designs each of which can be selected by the designer based on the available capital along with the specific requirements and the limitations of the project.

8. Nomenclature

A — Surface area, m^2
a— Specific area, m^2/m^3
C — Specific heat, J/kg.K
FEM—Finite element method
h— Heat transfer coefficient, W/m.K
H— Height of evaporator, m
HD—Humidification dehumidification
I — Total solar energy incident on the collector
MEE—Multi effect evaporation
MSF— Multi Stage Flash
RO—Reverse Osmosis
$\dot{Q}$— Heat transfer rate, W
T — Temperature, K, °C
V— Volume, m^3
VC—Vapor compression
z— Height, m

aperture W/m^2
k— Mass transfer coefficient, kg/m^2.s
L_v— Evaporation latent heat, J/kg
$\dot{m}$— Mass velocity, kg/m^2.s

Greek
η— Efficiency
ω— Specific humidity, kg vapor/kg dry air

Subscripts
a —Air
amb— Ambient
c — Condenser (dehumidifier)
col — Collector
cross— Cross section of HD unit
d — Distillated water
e — Evaporator (humidifier)
i — Interface
H — Heat
in— Inlet
M — Mass
p — Pinch
sol— Solar
v — Vapor
w — Water

References

[1] Hermosillo JJ, Arancibia-Bulnes CA, and Estrada CA 2011 *Solar Energy* **86** 1070-76
[2] Wanga Y and Lior N 2011 *Energ. Convers. Manage* **52** 228–35
[3] Wang Y and Lior N 2011 *Energy* **36** 3878-87
[4] Caruso G and Naviglio A 1999 *Desalination* **122** 225-34.
[5] Bourouni K, Chaibi M and Tadrist L 2001 *Desalination* **137** 167–76.
[6] Parekh S, Farid M, Selman J, and Al-Hallaj S 2004 *Desalination* **160** 167-86.
[7] Mathioulakis E, Belessiotis V and Delyannis E 2007 *Desalination* **203** 346-65.
[8] BenBacha H, Bouzguenda M, Abid MS and Mallej AY 1999 *Renew. Energ.* **18** 349–65.
[9] Soufari SM, Zamen M, and Amidpour M 2009 *Desalination* **237** 305–17.
[10] Al-Hallaj S, Parekh S, Farid MM and Selman JR 2006 *Desalination* **195** 169–86.
[11] Gude VG, Nirmalakhandan N and Deng SH 2011 *Energy* ***36*** 78 - 85.
[12] Zamen M, Amidpour M and Soufari SM 2009 *Desalination* **239** 92–9.
[13] Aminyavari M, Mamaghani AH, Shirazi A, Najafi B and Rinaldi F 2016 *Appl. Therm. Eng.* **108** 833-46
[14] Baghernejad A and Yaghoubi M 2011 *Int. J. of energy Res.* **35** 601–15
[15] Najafi B, Shirazi A, Aminyavari M, Rinaldi F and Taylor RA 2014 *Desalination*, **334** 46-59
[16] Liburd SO 2010 *Solar-driven humidification dehumidification desalination for potable use in Haiti* (Diss. Massachusetts Institute of Technology)
[17] Kalogirou SA 2004 Progr. Energ. Comb. Sci. **30** 231–95
[18] Stocker WF and Jones JW 1982 *Refrigeration and Air Conditioning* (McGraw Hill, Singapore)
[19] Selleri T, Najafi B, Rinaldi F and Colombo G 2013 *ASME J. Thermal Sci, Eng. Appl* **5** 031013
[20] Najafi B, Najafi H and Idalik MD 2010 *J Mech. Eng. Sci.* C **225** 1389-98

Development of a low-cost temperature data monitoring. An upgrade for hot box apparatus

T de Rubeis, I Nardi and M Muttillo

G. Parolini Lab – Dept. of Industrial and Information Engineering and Economics (DIIIE), University of L'Aquila, P.le Pontieri 1, Monteluco di Roio, I-67100, L'Aquila, Italy

E-mail: tullio.derubeis@graduate.univaq.it

Abstract. The monitoring phase has gained a fundamental role in the energy efficiency evaluation of a system. Number and typology of the probes depend on the physical quantity to be monitored, and on the size and complexity of the system. Moreover, a measurement equipment should be designed to allow the employment of probes different for number and measured physical quantities. For this reason, a scalable equipment represents a good way for easily carrying out a system monitoring. Proprietary software and high costs characterize instruments of current use, thus limiting the possibilities to realize customized monitoring.
In this paper, a temperature measuring instrument, conceived, designed, and realized for real time applications, is presented. The proposed system is based on digital thermometers and on open-source code. A remarkable feature of the instrument is the possibility of acquiring data from a high and variable number of probes (order of hundred), assuring flexibility of the software, since it can be programmed, and low-cost of the hardware components. The contemporary use of multiple temperature probes suggested to apply this instrument for a hot box apparatus, although the software can be set for recording different physical quantities. A hot box compliant with standard EN ISO 8990 should be equipped with several temperature probes to investigate heat exchanges of a specimen wall and thermal field of the chambers. In this work, preliminary tests have been carried out focusing only on the evaluation of the prototypal system's performance. The tests were realized by comparing different sensors, such as thermocouples and resistance thermometers, traditionally employed in hot box experiments. A preliminary test was realized imposing a dynamic condition with a thermoelectric Peltier cell. Data obtained by digital thermometers DS18B20, compared with the ones of Pt100 probes, show a good correlation. Based on these encouraging results, a further test was carried out in hot box, comparing the data measured by digital thermometers, Pt100 and T-type thermocouples. In this case also, the analyses show a good correlation between either digital thermometers and analog sensors.
From these results, it is reasonable to foresee that this measuring instrument could help those willing to realize or refurbish a hot box apparatus, and those who want to undertake temperature monitoring.

Keywords: temperature measuring instrument; hot box apparatus; flexible and low-cost instrument.

1. Introduction

The monitoring phases of energy systems allow to assess their real behavior, highlighting possible challenges. Frequently, proprietary software and high costs characterize instruments of current use, thus limiting the possibilities to realize customized monitoring and adversely affecting costs, flexibility, and data integration. For these reasons, customized measurement equipments are rapidly increasing and find application in many fields. Arnold et al. [1] presented an apparatus that allows to monitor and record high-temperature thermocouple measurements of dynamic systems in real time, with wireless network technology, to bridge the gap between a dynamic (moving) sample frame and the static laboratory frame. Gad and Gad [2] discussed a new sensor-based temperature data acquisition system for solar energy applications that allows flexibility and ease of changing the type of sensors and way of recording data. This system is especially suitable for large and remote installations where cost is a key factor in the choice of the measuring system. In the work of Ali et al. [3] the Open Source Building Science Sensors (OSBSS) project, created to design and develop a suite of inexpensive, open source devices based on the Arduino platform for measuring and recording long-term indoor environmental and building operational data, is presented. Revel et al. [4] disclosed a low-cost infrared measurement system, developed to monitor in real time thermal comfort conditions in indoor environments.

For the building sector, which is responsible for more than one third of the total energy use and associated greenhouse gas emissions, both in developed and developing countries [5], temperature monitoring phases represent a fundamental aspect, in order to evaluate energy performance and possible energy efficiency interventions. Indeed, the knowledge of the thermal behavior of buildings is an essential component of the strategies of the policy makers and energy efficiency is a key issue for the energy policies of the most industrialized countries [6].

In this paper, a flexible, low-cost and multi-probe measuring instrument is presented. The system is based on a microcontroller board ATmega2560 [7], that allows to measure and record simultaneously a large number of digital temperature probes, type DS18B20 [8]. The flexible management of the employable sensors is allowed through an optimized algorithm, that manages the microcontroller and automatically recognizes the number of probes really used.

Based on high flexibility and high number of temperature probes, the proposed measuring instrument fits well for laboratory tests in hot box apparatus. Since a hot box allows to assess energy performance of structural members with real dimensions, a large number of temperature probes is needed, in order to evaluate all the heat exchanges. Usually, to simplify the management of these temperature probes, commercial measuring instruments are used, but with high costs and poor flexibility. Martin et al. [9] evaluated the thermal response of thermal bridges through the guarded hot box of the LCCE (Construction Quality Control Laboratory) of the Basque Government in Vitoria-Gasteiz, equipped with a group of 86 T-type thermocouples inside each of the chambers (hot and cold), which measure the temperature of the air and the sample surface. Asdrubali and Baldinelli [10] discussed calibration and experimental procedures of the calibrated hot box of the Department of Industrial Engineering of the University of Perugia, taking into account three standards for calibrating hot boxes: the European EN ISO 8990 [11], the American ASTM C1363-05 [12], and the Russian GOST 26602.1-99 [13]. This calibrated hot box is equipped with 142 T-type thermocouples inside the hot and cold chambers, and 69 differential thermocouples to assess the temperature differences between the metering chamber and the external ambient. Chen and Wittkopf [14] used a calorimetric hot box to measure thermal transmittance of fenestration systems in summer condition. Integrated in the walls are sensors to monitor the heat loss through the walls using a 112- junction thermopile system, made of T-type copper-constantan thermocouple wire. Kus et al. [15] conducted a study to evaluate hygrothermal performance of pumice aggregate concrete (PAC) block walls using a calibrated hot box method, equipped with 52 NiCr–Ni (K) type thermocouples for measuring indoor and outdoor temperatures, surface temperatures of wall blocks, and the hollow walls inside the blocks.

The proposed system has been tested in the guarded hot box of the G. Parolini Lab. of University of L'Aquila, described and employed in previous works [16, 17]; the results obtained have shown good response of the system and good correlation between digital thermometers DS18B20 and analog probes (Pt100 and T-type thermocouple).

2. Methodology

The methodology applied for this work is synthetically shown in figure 1. The first step was to define the concept and all the properties required for the temperature measuring instrument. The following phases focused on the software and hardware design, which allowed the realization of the prototype and its laboratory tests. Finally, experimental analyses have been performed, in order to verify the actual performance of the measuring instrument.

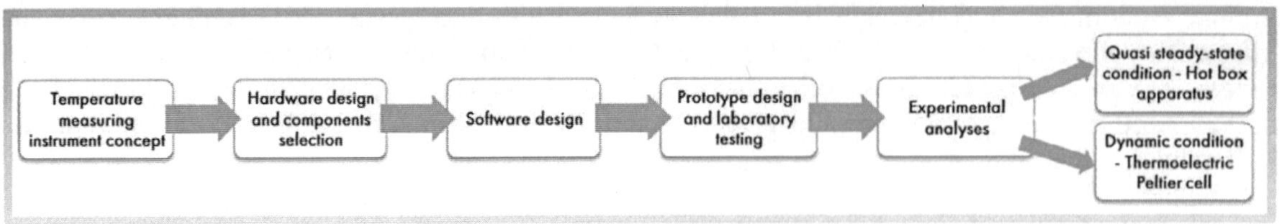

Figure 1. Methodology flow-chart.

2.1 *System description*

The systems that need many analog temperature probes, such as an hot box apparatus, are usually based on data loggers with a limited number of input temperature probes. For this reason, the total cost of the measuring instruments has high incidence. Furthermore, the employment of analog probes determines disturbances introduced by connection cables [18], that represent one of the main problems, due to losses and noise that lie along transmission lines [19]. Therefore, adaptation and conditioning circuits are needed for each probe [20], determining complexity of the measuring system, both for software and hardware level, and a remarkable design work for the choice of the most suitable components. The high number of electronic components determines a complex management of the system.

Nowadays, new technologies based on digital electronic systems allow different solutions for temperature measuring instruments. The employment of digital thermometers solves most of the problems above described for analog solutions, determining cheaper and easier creation and management of the system. In particular, the digital probes selected for the proposed measuring system are the DS18B20, provided by Maxim Integrated. The main characteristics of these probes are: voltage range equal to 3-5.5 V, measurement interval from 218 K to 398 K (-55 °C to 125 °C) with 12 bits resolution, corresponding to increments of 0.0625 K, and an accuracy equal to ±0.5 K. Moreover, digital thermometers can be directly powered through the data line with the Unique 1-Wire protocol and each sensor is univocally identified through a 64 bit code, that is memorized in an internal ROM. The 1-Wire protocol is proprietary and created by Maxim Integrated, and it provides two wires for power and one wire for data bus. Thanks to this protocol, it is possible to connect in parallel many probes, along the same line, in order to reduce the number of cables with respect to analog solutions. These characteristics make digital temperature probes particularly suitable to be used in multi sensors applications, as for instance the hot box apparatus. Moreover, digital sensors are also available with waterproof and moisture-proof properties, allowing also external uses. The general scheme of the designed measuring system is shown in figure 2.

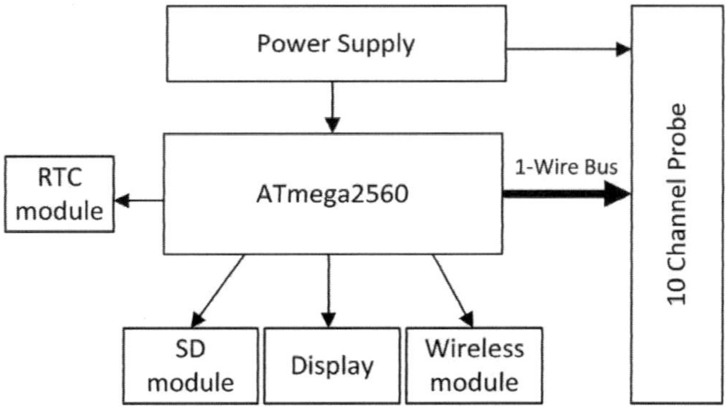

Figure 2. The measuring system architecture.

The system is based on digital sensor interface that allows both data management and signals acquisition. In this prototypal phase, a microcontroller board ATmega2560 has been chosen. The system is equipped with a 20x4 Character LCD Display Module with an HD44780 controller driver to view average temperatures and all error messages, in order to facilitate system debugging without connecting the system to a PC. A card reader permits to save all measured data on a SD memory and it supports up to 32 GB memory device. A real time clock, the low power ds1307 module connected through an I2C protocol, has been also included to keep track of measurements time. Finally, a wireless module was put in order to communicate the data with a remote PC. The software, loaded on the microcontroller, has the capability to manage up to 100 probes through the 1-Wire bus, but with the possibility of extending it with many more probes. In fact, the designed algorithm includes the subdivision of the 10-channel connection as a good compromise for what concerning sensing capability for minimal time sampling and typical application. The probes have an analog-digital converter that converts the read analog temperature values into digital values to be send to the microcontroller by the bus. The maximum conversion time provided by the DS18B20 is 750 ms with a 12 bits resolution. The time requested to start analog-digital conversion and read the measured data is equal to 75 ms and the total time needed to obtain the output is equal to 75 ms. Therefore, this means that each channel has an overall acquisition time of 900 ms, independently of the probes number, because the sensors on the same 1-wire bus receive the data request simultaneously. The designed algorithm has an efficient data management system able to minimize the acquisition time. The proposed algorithm is shown in figure 3.

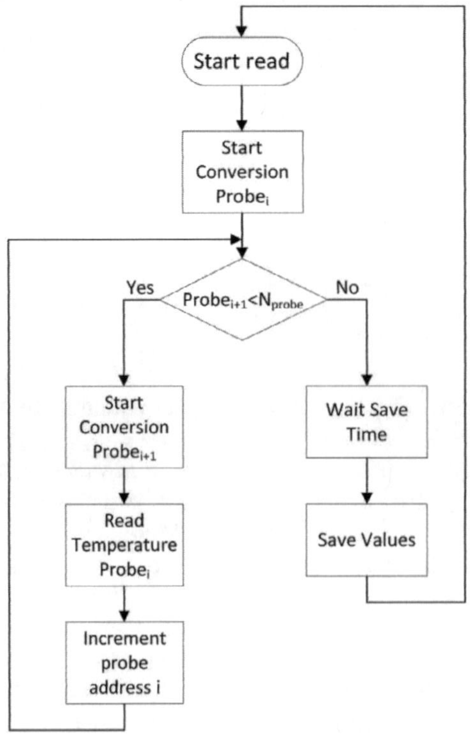

Figure 3. Proposed algorithm to acquire temperature measurements.

The algorithm is able to recognize and identify automatically the probes really plugged to the system, being not necessary to use the full probes capability in some applications.

The measurement system can be used setting continuous or stepped acquisitions. Measurements data are stored in a 32GB SD card that, considering continuous acquisitions, allows to store data for up to few years.

The total power consumption mainly depends on the acquisition time step and on the number of probes. Without considering the LCD module that is an utility not mandatory for the overall functionality, in the worst case of continuous acquisition the fixed current consumption is 150 mA. Beyond it, the variable component depends on the number of sensors. Each one has a current consumption of 1.5 mA when active and a 0.75 mA when inactive. A more detailed description of the measuring instrument is given in [21].

2.2 *Experimental set up*

The proposed measuring system has been verified in laboratory, through two different preliminary tests, in dynamic and quasi steady-state conditions. For the dynamic condition a thermoelectric Peltier cell has been employed, while for the quasi steady-state condition, the test was realized in guarded hot box, neglecting thermal analysis of the specimen wall. Both the tests have been carried out with an overall number of the sensors less than the total number usable. The tests have been realized by using n.12 DS18B20, n.4 Pt100 and n.3 T-type thermocouples.

The first dynamic test has been conducted through the use of a thermoelectric Peltier cell (figure 4) to heat up the water contained in a basin, in order to analyze the dynamic response of the digital thermometers under analysis. Initially, the sensors have been maintained at 292 K (+19.0 °C) (room temperature) for about 15 minutes. After this period, the sensors have been immersed for 15 minutes in the water, previously heated up and maintained at about 328 K (+55.0 °C), thanks to the thermoelectric Peltier cell. At the end of this phase, temperature probes have been taken again to room temperature (292 K).

For this test, the DS18B20 sensors were compared with Pt100 probes.

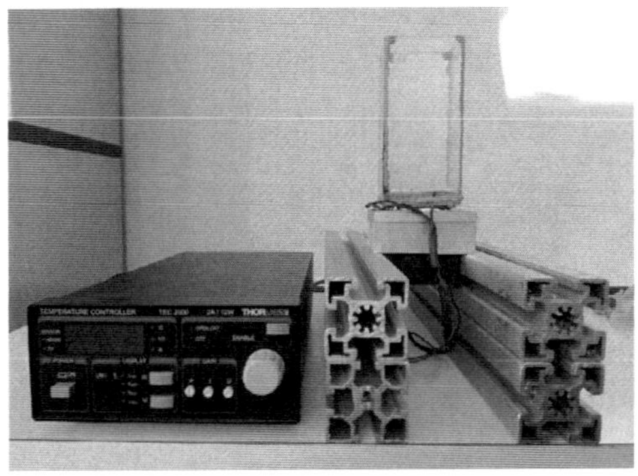

Figure 4. Thermoelectric Peltier cell.

The second test has been carried out imposing quasi steady-state condition in hot chamber (about 292 K (+19.0 °C)), maintained for about 300 minutes. The digital thermometers response (DS18B20) has been compared with the ones of resistance thermometers (Pt100) and thermocouples (T-type). The setup of the sensors installation has been realized following the standard [11] and the thermal model of the specimen wall interposed between hot and cold chamber (figure 5).

Figure 5. Image of the specimen wall. (a) Visible. (b) IR image with thermal model.

The temperature probes have been positioned according to a matrix scheme on the surface of the sample wall, as shown in figure 6. All the probes have been placed where the hot chamber ensures the best thermal uniformity, i.e. the metering chamber. The digital thermometers have been installed with a horizontal spacing equal to 0.75 m and a vertical distance of 0.50 m. Furthermore, thermal compound has been employed to ensure the mounting between temperature probes and sample wall.

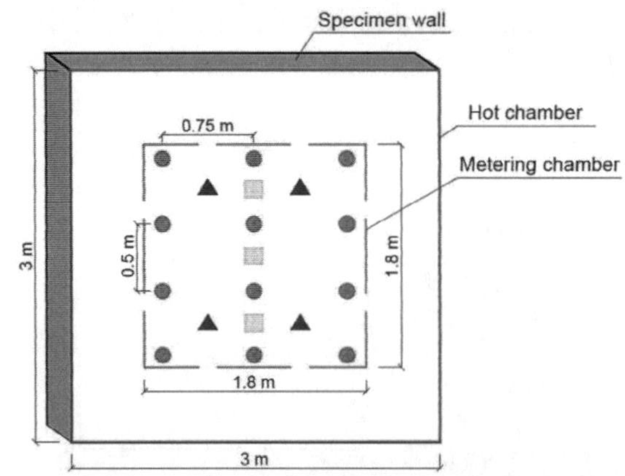

Figure 6. Installation scheme of the temperature probes. Red circles for DS18B20. Blue triangles for Pt100. Green squares for T-type thermocouples.

Both the tests have been carried out with an acquisition time-step equal to one minute. The characteristics of the comparison probes are listed in table 1. The values measured by analog sensors have been acquired through a Delta-T Logger acquisition system.

Table 1. Characteristics of comparison probes.

Type	Measurement interval	Sensitivity
Resistance thermometer (Pt100)	from 233 K to 353 K (-40.0 °C to +80.0 °C)	0.4 Ω /K
Thermocouple (T-type)	from 73 K to 673 K (-200.0 °C to +400.0 °C)	48.2 µV/K

3. Results and discussion

The first dynamic test shown that in quasi steady-state condition all the temperature probes have similar behavior. When temperature changes, digital thermometers follow temperature variations faster than analog probes, i.e. Pt100, as plotted in figure 7.

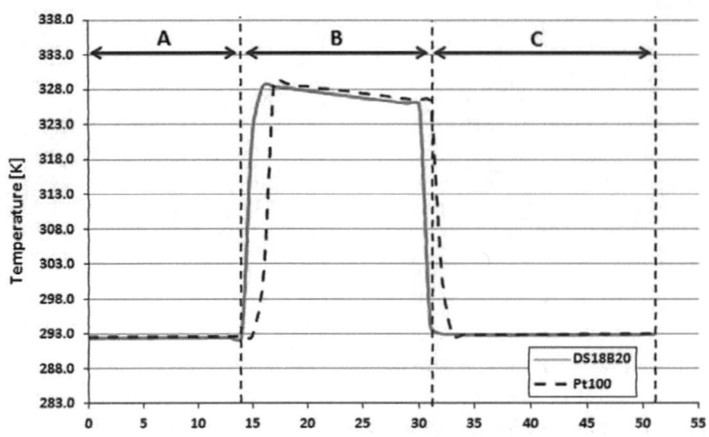

Figure 7. Average temperatures for the dynamic test. (A): quasi steady-state condition. (B): dynamic condition (warming and cooling). (C): quasi steady-state condition.

In this case, by comparing values measured by digital and analog sensors, it can be observed a good correlation (figure 8), with a correlation coefficient (R) equal to 0.895 and a coefficient of determination (R^2) greater than 0.80. Values hooped in figure 7 highlight that, when temperature changes, the digital probes (DS18B20) have a faster reaction with respect to the analog sensors (Pt100), both in warming (blue circle "1") and cooling (orange circle "2").

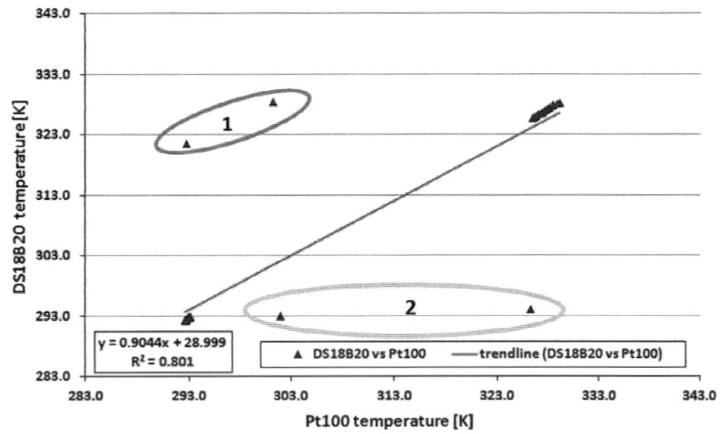

Figure 8. Scatter graph for correlation between digital and analog probes (DS18B20 vs Pt100).

For the second test in quasi steady-state condition, the results obtained show that temperature trend measured by digital thermometers are close to those measured by analog probes, as plotted in figure 9.

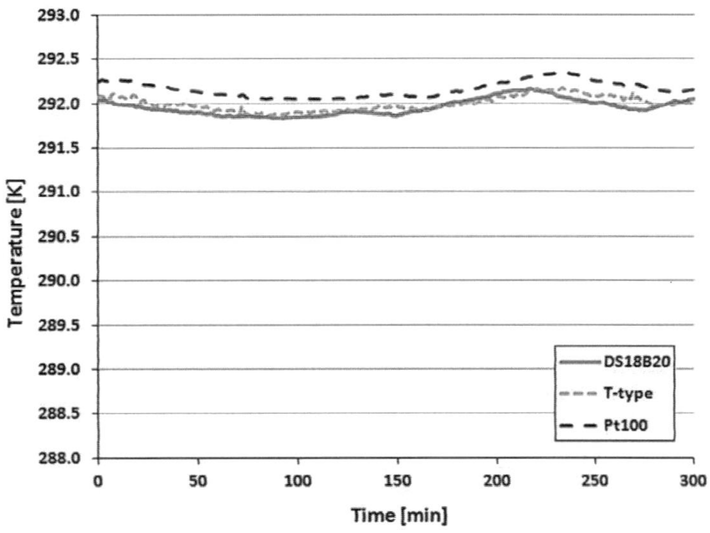

Figure 9. Average temperatures for the test in quasi steady-state condition.

Also in this case, the comparison between digital and analog sensors highlights a good correlation: the correlation coefficients (R) are equal to 0.805 for the comparison with Pt100 probes and equal to 0.835 for the comparison with T-type sensors (figure 10).

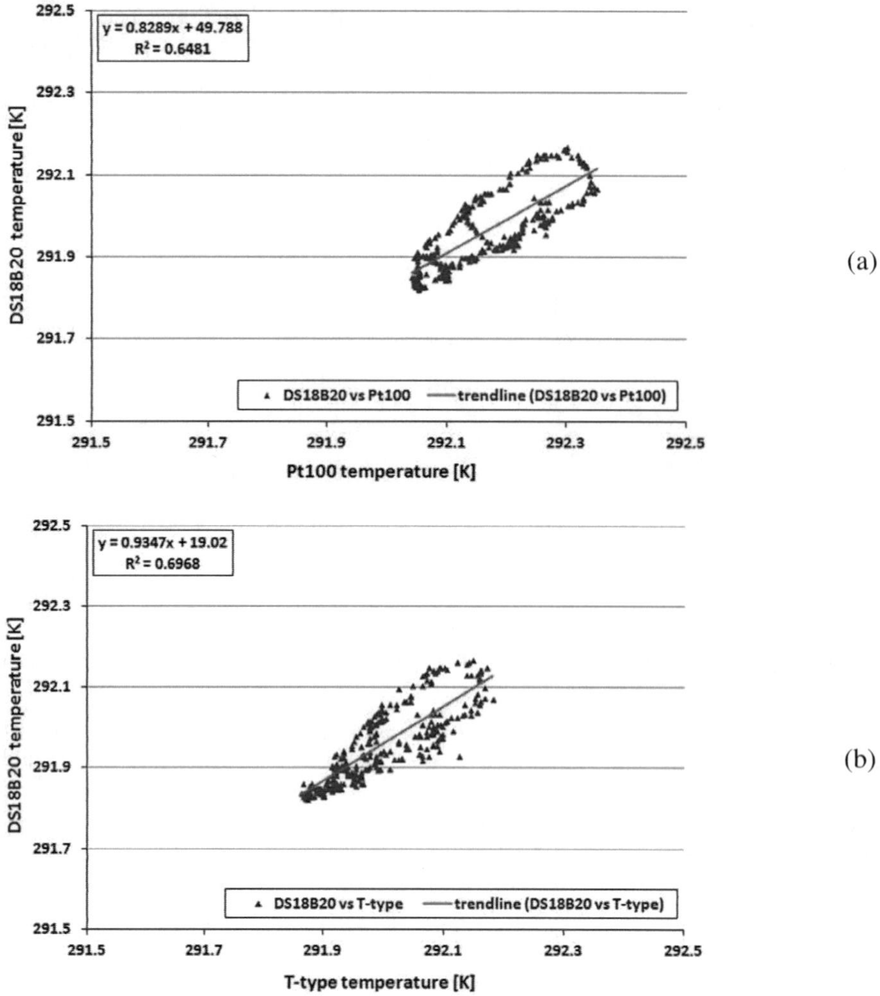

Figure 10. Scatter graph for correlation between digital and analog probes. (a) DS18B20 vs Pt100. (b) DS18B20 vs T-type.

4. Conclusion

In this paper, a new low-cost measuring system based on digital thermometers DS18B20 and on a microcontroller board ATmega2560, is proposed for new applications or refurbishment of hot box apparatus. This system allows to acquire data from high and flexible number of temperature probes, with only one acquisition system. The employment of digital thermometers represents an innovative choice with respect to traditional sensors used in hot box experiments, such as thermocouples and resistance thermometers. This choice allows to achieve advantages for the management of the large number of probes and for the significant reduction in total costs of the instrumentation. Technological reasons, that motivated the creation of this instrument, and hardware and software components have been described.

Preliminary tests have been realized in order to compare digital thermometers with analog sensors traditionally employed in hot box apparatus, such as T-type thermocouples and resistance thermometers Pt100. Two different tests have been carried out. A dynamic test was realized with a thermoelectric Peltier cell to heat up the water contained in a basin, in order to analyze the response of the digital thermometers under analysis. A further test was carried out in hot box. Temperature probes have been installed on the surface of the specimen wall and quasi steady-state condition has been imposed. Both the tests have shown good correlation between digital thermometers and analog sensors.

Therefore, the proposed measuring system could represent an opportunity for those who want to realize or revamp a hot box apparatus. Future development of the proposed system will concern the possibility of implementing a control system of the temperature actuators in hot box, i.e. electric resistances and refrigerating unit.

References

[1] Arnold F, DeMallie I, Florence L and Kashinski D O 2015 *Rev. Sci. Instrum.* **86** 035112.
[2] Gad H E and Gad H E 2015 *Renew. Energ.* **74** 337.
[3] Ali A S, Zanzinger Z, Debose D and Stephens B 2016 *Build. Environ.* **100** 114.
[4] Revel G M, Arnesano M and Pietroni F 2014 *Meas. Sci. Technol.* **25** 085101.
[5] IEA – International Energy Agency, World Energy Outlook 2016, ISSN 2072-5302.
[6] de Rubeis T, Nardi I, Paoletti D, Di Leonardo A, Ambrosini D, Poli R and Sfarra S 2016 *Energ. Convers. Manage.* **128** 261.
[7] ATmega2560, Microcontroller, http://www.atmel.com/Images/Atmel-2549-8-bit-AVR-Microcontroller-ATmega640-1280-1281-2560-2561_datasheet.pdf
[8] DS18B20, Digital Thermometer datasheet, https://datasheets. maximintegrated.com.
[9] Martin K, Campos-Celador A, Escudero C, Gómez I and Sala J M 2012 *Energ. Buildings* **50** 139.
[10] Asdrubali F and Baldinelli G 2011 *Energ. Buildings* **43** 1618.
[11] European StandardENISO 8990, Thermal insulation – Determination of steadystate thermal transmission properties – Calibrated and guarded hot box, 1996.
[12] American Society for Testing and Materials, ASTM C1363-05: Standard Test Method for Thermal Performance of Building Materials and Envelope Assemblies by Means of a Hot Box Apparatus, West Conshohocken, PA, 2005.
[13] Interstate Standard of Russian Federation GOST 26602.1-99, Windows and doors. Methods of determination of resistance of thermal transmission, 1999.
[14] Chen F and Wittkopf S K 2012 *Energ. Buildings* **53** 47.
[15] Kus H, Özkan E, Göcer Ö and Edis E. 2013 *Constr. Build. Mater.* **38** 837.
[16] Nardi I, Paoletti D, Ambrosini D, de Rubeis T and Sfarra S 2015 *J. Phys. Conf. Ser.* **655** 012006.
[17] Nardi I, Paoletti D, Ambrosini D, de Rubeis T and Sfarra S 2016 *Energ. Buildings* **122** 211.
[18] Smith I B 1923 *Transactions of the American Institute of Electrical Engineers* **42** 349.
[19] Kester W, Bryant W and Jung W 1999, *Practical Design Techniques for Sensor Signal Conditioning* ed Walt Kester (Analog Devices) Section 7: Temperature Sensors p 7.1.
[20] De Marcellis A, Depari A, Ferri G, Flammini A, Marioli D, Stornelli V and Taroni A 2008 *IEEE Trans.on Instrum. and Measurements* **57** 1596.
[21] Pantoli L, Muttillo M, de Rubeis T, Nardi I, Stornelli V and Ferri G 2017 *Proceedings*, **1(4)**, 596.

Energetic performance analysis of a commercial water-based photovoltaic thermal system (PV/T) under summer conditions

I Nardi, D Ambrosini, T de Rubeis, D Paoletti, M Muttillo and S Sfarra
DIIIE, University of L'Aquila, P.le E. Pontieri 1, Monteluco di Roio, 67100 L'Aquila, Italy

E-mail: iole.nardi@univaq.it

Abstract. In the last years, the importance of integrating the production of electricity with the production of sanitary hot water led to the development of new solutions, i.e. PV/T systems. It is well known that hybrid photovoltaic-thermal systems, able to produce electricity and thermal energy at the same time with better energetic performance in comparison with two separate systems, present many advantages for application in a residential building.
A PV/T is constituted generally by a common PV panel with a metallic pipe, in which fluid flows. Pipe accomplishes two roles: it absorbs the heat from the PV panel, thus increasing, or at least maintaining its efficiency; furthermore, it stores the heat for sanitary uses.
In this work, the thermal and electrical efficiencies of a commercial PV/T panel have been evaluated during the summer season in different days, to assess the effect of environmental conditions on the system total efficiency.
Moreover, infrared thermographic diagnosis in real time has been effected during the operating mode in two conditions: with cooling and without cooling; cooling was obtained by natural flowing water. This analysis gave information about the impact of a non-uniform temperature distribution on the thermal and electrical performance. Furthermore, measurements have been performed in two different operating modes: 1) production of solely electrical energy and 2) simultaneous production of thermal and electrical energy.
Finally, total efficiency is largely increased by using a simple solar concentrator nearby the panel.

1. Introduction

The ever-growing buildings' energy consumption is a relevant issue for the civil sector.
It is evident that the spread of electric appliances for domestic uses leads to an increase of electric energy consumption. It is also clear that, in the last decades, the living standards are getting higher, and people are reluctant to reduce them. For instance, we're getting used to have televisions in several rooms, to have cooling devices for the hottest seasons, and to have multiple electric devices and lights simultaneously switched on. Simple hacks (like installing occupancy sensors) help to significantly reduce energy needs [1].
Besides these concerns, domestic hot water must be provided, too.
On the whole, in a building, whatever is its intended use, there is the need to face requirement for electric energy and domestic hot water. In some special cases, situations may be even more complex [2]. Fortunately, the civil sector itself can provide the possibility of supply such requirements. In fact, the installation of photovoltaic systems and solar heating collectors on rooftops can help to provide the self-production of the supplies.
The technological solution that can match both requirements is the PV-Thermal (PV/T or PVT) collector. It consists of a PV module that simultaneously serves as electricity producer and thermal absorber, thus it uses solar energy for producing hot water and electricity at the same time. There are many benefits in the use of PV/T. In fact, the rooftop area used is lower than the equivalent needed for

the separate production of hot water and electricity, and such solution is better integrated in the architectural feature of a building, also due to the several available shapes and layout of PV/T collectors.

Different PV/T typologies are commercially available, and several research and review studies have been carried out on these systems.

The PV/T collector concept was firstly introduced in the '70s, as one of the responses to the oil crisis of such period. Although the PV/T technology is quite mature, the effort of the scientific community is to fully characterize its behavior, in terms of electrical and thermal efficiency, and to improve its capabilities.

In the study of Huang [3], an integrated photovoltaic and thermal solar system is proposed, and its performances are compared to a conventional solar water heater. In the paper, the primary-energy saving efficiency is also proposed. Results show that PV/T efficiency is increased if the heat collecting plate constituted the base plate of PV cells, and if it is directly in thermal contact with them.

An experimental prototype was employed in [4], and different numerical models (dynamic 3D or steady-state with 1D, 2D and 3D) for the thermal yield definition were also employed, proving reliable results, but with different calculation times.

The development of the thermal model of the integrated photovoltaic and thermal solar system mentioned before is proposed in [5], and results agree with those proposed in [3].

A review on PV/T collectors is proposed in [6], where different outcomes provided by research studies are compared.

In 2008, the study of [7] gives a review focused on flat -plate PV/T collectors, and the history of PV/T systems is also outlined, together with the main findings from a literature review that covers 30 years of publications.

In 2009, a study [8] compared the results of the total efficiency obtained from simulations carried out on seven design configurations for the absorber collectors of PV/T. Several parameters, like flow rate, solar radiation and ambient temperature, were also considered. As result, the design that performs the highest thermal efficiency is the one having spiral flow, which is one of the simplest investigated.

A work by Kostic et al [9] deals with the influence of a flat aluminum concentrator on energy efficiency of PV/T. The addition of a reflective foil, laid at specific angles in respect of horizontal and vertical plane, ensures the increase of the thermal and electrical energy given by the PV/T. The study proves that both the thermal and electrical efficiency, and the total daily thermal and electrical efficiency decrease with the solar radiation intensity concentration factor.

In 2011, Ibrahim at al [10] proposed a paper on recent advances in PV/T collectors, showing different design configurations and the future development of such systems, with attention to the building integrated photovoltaic systems. A similar approach was followed by Zhang et al [11].

The theme of building integrated photovoltaic and thermal system was deepened in a recent work [12], where an extensive review on the subject is presented.

Another recent review on PV/T systems is given by Al-Waeli et al [13], where thermal and electrical efficiency of PV/T systems assessed in literature are also compared in respect of the different cooling media.

A work authored by Vittorini et al [14] aims at developing a Matlab model that reproduces the thermal balance of modules, calibrated on experimental data.

It is evident that the interest of the scientific community on PV/T is high, because of the great potentiality of such system, and the possibility of integrating this technology in the building skin. This, of course, is an advantage and, possibly, a push to the wider installation of such systems in new buildings.

In this paper, a commercial PV/T collector consisting of a plane PV module of polycrystalline silicon, cooled by water, was investigated. Tests were performed during summer season, under real solar radiation conditions. The employed set up is quite simple, but it allowed to characterize the thermal and electrical efficiency of the collector under different weather conditions. Results are compared with a commercial PV panel, having the same characteristics (displayed in Table 1), showing the better performance of the PV/T system.

2. Materials and methodology

2.1. PV/T and PV setup

Tests and measurements were carried out on a PV/T collector and on a frameless PV mounted on aluminum supports, needed for comparison. Panels were mounted on a flat roof of the Faculty of Engineering in L'Aquila (42°20' N, 13°22'E); they were tilted of 41.5° in respect to the horizontal plane, and were south oriented.
The employed PV/T collector, shown in Figure 1, is certified by IEC EN 61215, IEC EN 61730-2, IEC EN 61730-3 and EN 12975-2. Its technical features are summarized in Table 1.

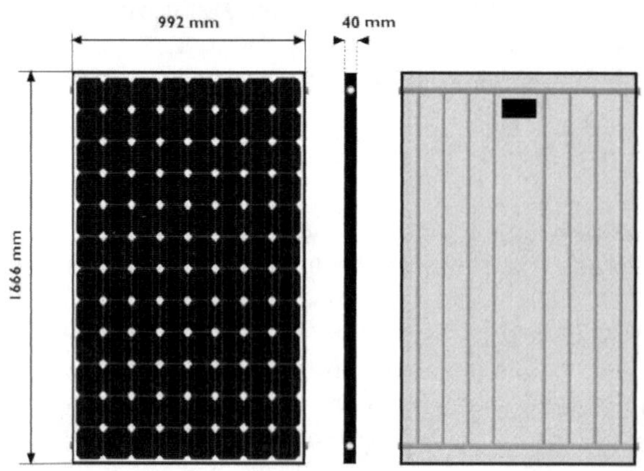

Figure 1. Drawing of the employed PV/T module, with its sizes. Right side shows the absorber layout.

Table 1. PV/T and PV module technical specifications.

	PV/T panel	PV panel
Number of solar cells	60 polycrystalline cells	72 polycrystalline cells
Cell sizes	156 x 156 mm	156 x 156 mm
Module sizes (H x L x W)	1666 x 992 x 40 mm	1985 x 1005 x 4 mm
Pp *	250 W	300 W
Pt**	785 W	-
Gross surface	1.63 m^2	1.99 m^2
Net surface	1.48 m^2	1.99 m^2
Weight	22 kg	27 kg
Liquid content	1.15 l	-
Absorber	Copper harp laser welded on aluminum plate	-

* peak power
** maximum producible thermal power

A simple way to install a PV/T collector is shown in Figure 2.

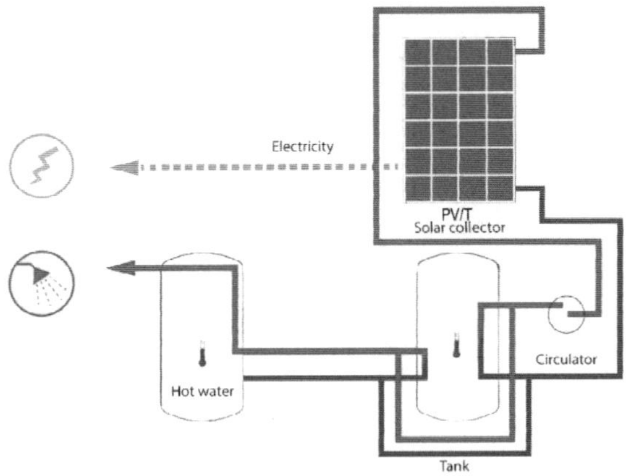

Figure 2. Schematic diagram of a possible installation of a PV/T collector.

In our case, it wasn't possible to realize such layout, because of the lack of a boiler tank and of the possibility of net connection. Therefore, it has been decided to employ the configuration displayed in Figure 3.

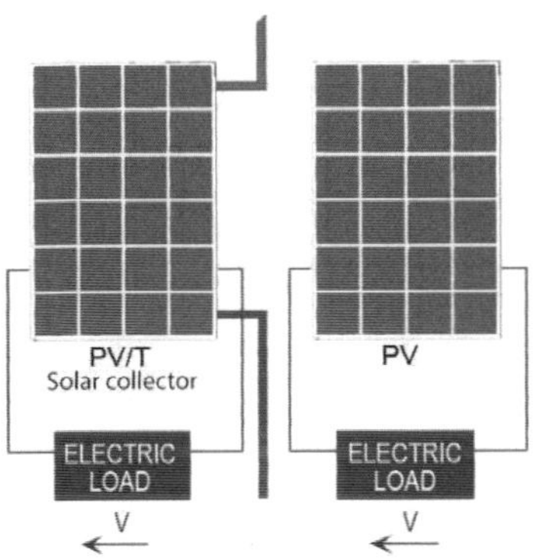

(a)

(b)

Figure 3. Adopted set-up: (a) schematic diagram; (b) picture.

In fact, given the electrical characteristics of the two modules (detailed in Table 2), a proper external electrical load was needed in the set-up, to measure the current and voltage on the load itself, thus having the possibility to measure the electrical power instantaneously produced by the module. The same load was applied to the two modules (PV/T and PV), and it was sized according to the maximum power current that could circulate in them (which is quite the same). In fact, hypothesizing electrical power (P) of about 200 W for each module, the required load, in terms of resistance, can be calculated as the ratio between the power and the square of the maximum current, that is:

$$R = \frac{P}{I_{mp}^2} = \frac{200}{(8.33)^2} \cong 3\Omega \quad (1)$$

Table 2. Electrical characteristics of employed modules (from products datasheet).

	PV/T module	PV module
Module Efficiency	15%	15.04%
Maximum Voltage (Vmp)	30.03 V	36.10 V
Maximum Current (Imp)	8.33 A	8.31 A
Open Circuit Voltage (Voc)	37.68 V	45.10 V
Short Circuit Current (Isc)	8.81 A	8.98 A

Due to the unavailability of a hot water storage tank, it has been decided to have a continuous water flow in the PV/T absorber, directly taken from the aqueduct. This, of course, causes a great penalty on the evaluation of the thermal efficiency of the module, since the module is always cooled a new water rate. This choice may affect the electrical efficiency, too. In fact, as known, the electrical efficiency increases with the decrease of module temperature. The continuous heat exchange from module to water ensures the cooling of the module, whose temperature is lowered and whose electrical performances might be, in a closed circuit, lower.

2.2. Performance evaluation

The performance evaluation of a PV/T collector is expressed by its capability to convert solar energy into thermal and electrical energy.
The thermal efficiency η_{th} of the system can be defined [9,13] as the capability of the device to convert the solar radiation into thermal energy, that is the ratio between solar irradiance on the surface and the thermal energy gained by water, that is:

$$\eta_{th} = \frac{\dot{m} * c * (T_{out} - T_{in})}{G * A} \tag{2}$$

where $\dot{m}$ is the water mass flow rate, c is water specific heat, *Tout* and *Tin* are respectively the outlet and inlet water temperatures, G is the solar radiation on the module whose area is A.
The electrical efficiency η_{el} can be defined as the amount of electric power generated by the PV module in respect of the solar radiation on the module surface. This parameter can be expressed as:

$$\eta_{el} = \frac{I * V}{G * A} \tag{3}$$

where I and V are the current and the voltage, whilst G and A have the meaning expressed before.
The total performance η_{tot} of a PV/T panel can be expressed as the sum of the thermal and electrical efficiencies [13]:

$$\eta_{tot} = \eta_{th} + \eta_{el} \tag{4}$$

To evaluate thermal and electrical efficiency, data regarding ambient air temperature, relative humidity and wind speed were acquired through a weather station placed nearby the modules. Solar radiation G was measured by a pyranometer included in the weather station, whose working principle is a thermopile. Weather station specifications are listed in Table 3. Alternatively, another equipment, already introduced in literature [15], which is based on the voltage difference on PV cell, can be used.

The following evaluations, however, are carried out with respect to the solar radiation.

Table 3. Weather station technical specifications.

	Operating range	Accuracy
Pyranometer	0~2000 W/m^2	ISO 9060 - 2nd class
Temperature	-50~+150 °C	±0.5 °C
Relative humidity	0~100 %	±2%
Cup anemometer	0~50 m/s	±0.1 m/s

Thermocouples were employed for the measurement of inlet and outlet water temperatures.
All quantities were acquired every 15 minutes during the morning.
Different tests were carried out during the summer season; tests differ for the weather conditions that occurred, water mass flow rate set, and for the absence or presence of a flat reflector in front of the modules. This feature allows to increase the solar radiation on the module surface, that, in turn, implies an increase in the generated power. The flat reflector was realized with a cardboard covered with aluminum foil. Tests overview is shown in Table 4.

Table 4. Tests matrix.

	Weather condition	Water mass flow rate in the PV/T [kg/s]	Flat reflector	Thermographic inspection
Test 1	Overcast	0.212	-	
Test 2	Clear sky	0.212	-	
Test 3	Overcast	0.106	-	
Test 4	Clear sky	0.106	-	
Test 5	Clear sky	-	-	✓
Test 6	Clear sky	0.106	✓	✓
Test 7	Clear sky	-	✓	✓

3. Results and discussion

PV and PV/T panels efficiencies are shown in Figure 4 from (a) to (e), together with the recorded solar radiation, against the hour, for the tests carried out.
For the PV/T panel, thermal, electrical and total efficiencies are shown. As foreseen, the thermal efficiency has always low values, due to the employed set-up. In literature, high values (order of 60 % to 90 %) of thermal efficiency are assessed for water cooled PV/T systems [13]. This, of course, affects the total efficiency.
Measured electrical efficiency of both the PV and the PV/T panel agree with those expected from manufacturers and available from previous studies. As a general remark, PV/T electrical efficiency is always slightly higher than the PV panel one. Such efficiencies have swinging values when the measured solar radiation varies rapidly. Such variations are, sometimes, due to clouds passing through. It's the case of Test 2: the weather condition was favorable, in fact as a whole the sky was clear, and solar radiation reaches a peak of about 1067 W/m^2 (the highest value of all campaigns); nevertheless, moving clouds significantly reduced the solar radiation in the middle of the test, thus affecting the efficiencies, whose trends recall the one of G.
Test 3 and Test 4 were carried out by halving the water mass flow rate. The heat exchange was therefore enhanced, causing the increase of Tout, and, therefore, the (low) increase of η_{th}.
The indirect effect due to this is that electrical efficiency of Test 3 and Test 4 follows the solar radiation behavior. Instead, in Test 1 and Test 2, the electrical efficiency is less influenced by solar

radiation, unless the spikes already discussed for Test 2. Then, it has been decided to assess the electrical efficiency of the PV/T module when it is employed as a simple electricity producer, that is, without the cooling effects due to water circulation, which means it is employed as a simple PV panel (Test 5).

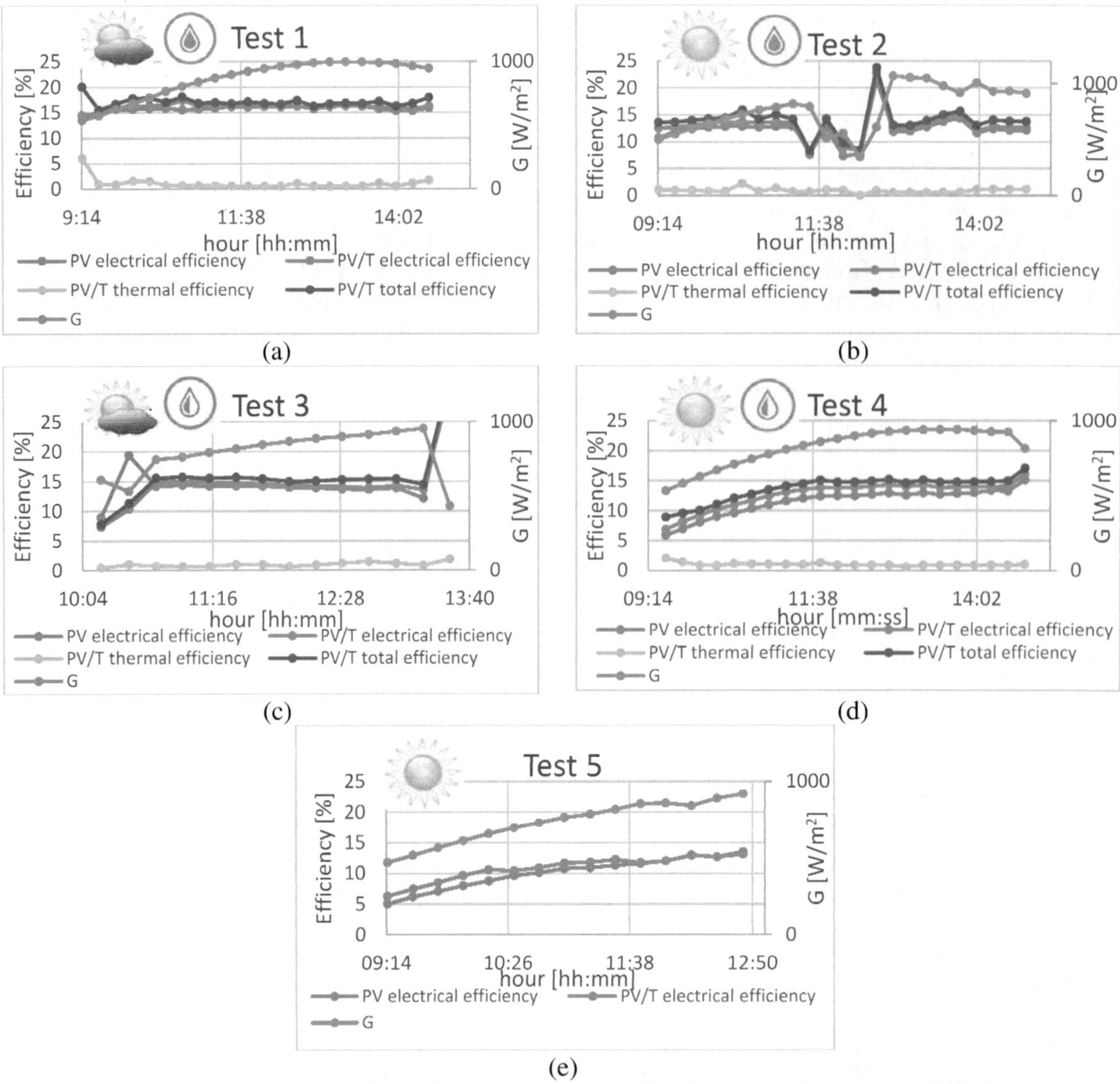

Figure 4. Thermal and/or electrical efficiency of the PV/T and PV panel, and solar radiation against hour. Results of: (a) Test 1; (b) Test 2; (c) Test 3; (d) Test 4; (e) Test 5.

Then, in order to enhance the solar radiation on panels, it has been decided to install a flat reflector in front of them. The conceived system wasn't equipped with a solar tracking device, therefore its position remained fixed for all the measurement duration. Two tests (Test 6 and Test 7) were performed by using this feature, both of them under clear sky. Due to the fixed position of the flat concentrator, the concentrating factor (that is, the ratio between total solar radiation that reaches the panel - sum of reflected and solar radiation- and the solar radiation) was variable along the hours, therefore this parameter has not been measured.

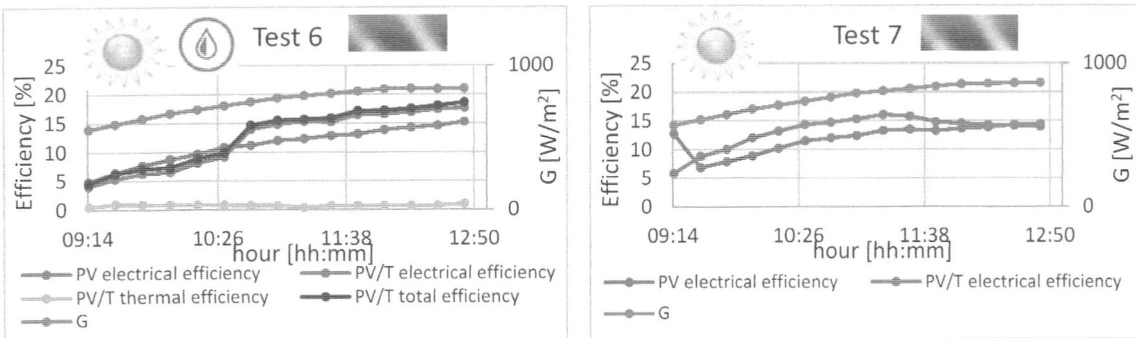

(a) (b)

Figure 5. Thermal and/or electrical efficiency of the PV/T and PV panel, and solar radiation against hour. Results of: (a) Test 6; (b) Test 7.

Electrical efficiency obtained in Test 5 are lower than the ones of Test 6 and Test 7, carried out with the employment of a flat reflector. This implies that, despite the ease of the adopted reflector, its benefits on the efficiency are remarkable. In fact, η_{el} increases of some percentage points (depending on the hour) in Test 6 and Test 7 in respect of results of Test 5.
By observing results of Test 5 and Test 7, it can be seen that, unless a few values, PV/T electrical efficiencies stand upon the PV ones. The difference between these values is higher when a reflector is employed (Test 7). However, when the solar radiation is high (at noon), electrical efficiency of both PV and PV/T module converge in asymptotical way to the module efficiency declared by manufacturers.
By comparing results from Test 6 and Test 7, it can be noted that, in case of water cooling, electrical efficiency gets higher. In this case, for instance, the PV/T electrical efficiency measured at 12:45 a.m. increases from 14% (Test 7) to 18% (Test 6), that is a difference of 28%.
Analogous considerations can be drawn by comparing results from Test 4 and Test 6 (that differ for the reflector absence or presence, respectively), showing the reflector influence on both thermal and electrical efficiency.
These encouraging results might be due to the thermal uniformity of panels, due to cooling.
During Test 5, Test 6 and Test 7, thermographic inspection was carried out to visualize panels' thermal field. Figure 6 shows thermal images taken at 12:30 a.m., i.e. in condition of maximum solar radiation (as deducible from plots of Figure 4 and Figure 5).
Temperature scale is the same for all images. Edge effects are clearly visible on both moduls, while squared contours reveal the positioning of superficial temperature probes (whose results are not subject of the present work).
Temperatures shown in Figure 6(c) are the highest, due to the severe working conditions (absence of water cooling and adoption of flat reflector) whilst in case of absence of both water circulation and reflector, temeperatures are lower. The lowest temperatures, averaged on the surface, are obtained in case of moderate mass flow rate and in presence of reflector. Particularly, image of Figure 6(b) clearly displays cells layout.

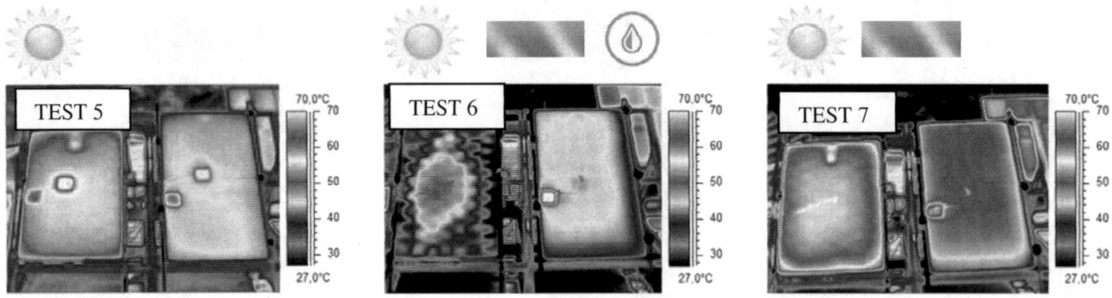

(a) (b) (c)

Figure 6. Thermal images taken at 12:30 a.m. during: (a) Test 5; (b) Test 6; (c) Test 7.

4. Conclusions

In this study, a PV and a PV/T module were investigated during the morning of some summer days, in 7 different conditions.

Electrical and total efficiency of the systems were assessed, by using a set-up specifically conceived for the purpose. Due to the simplicity of the set-up, the PV/T thermal efficiency was significantly underestimated, and its total efficiency, too.

Nevertheless, some relevant outcomes can be drawn from tests:

- Electrical efficiency of both PV and PV/T modules obtained on field agrees with manufacturer expectations;
- PV/T electrical efficiency is always higher than the PV one, whatever the operative conditions are;
- When water flow rate is high, energy efficiency is quite constant against hours, although it seems deeply influenced by rapid solar radiation variation (Test 2);
- By lowering cooling fluid mass flow rate, electrical efficiency trend recalls solar radiation increase (due to its mathematical definition);
- If the PV/T module is employed without cooling (i.e. as a simple PV module), its efficiency is higher than the one of the comparison PV panel, especially in the early hours of the morning. Such difference decreases in the more irradiated hours (Test 5);
- Thermal images, besides the identification of possible damaged or inoperative cells, allow to visually sketch the benefits of different operative conditions;
- Highest temperatures are reached in the set-up with reflector without cooling, followed by the configuration without reflector. The cooling effect due to a moderate water mass flow rate decreases panel temperature, whose distribution, however, is uneven: the central area of the panel shows higher temperatures;
- The employment of simple reflector can significantly (+28%) increase electrical efficiency;
- Panel cooling is therefore advisable to enhance the electrical energy capability of panels.

Future development of this work will deal with the analysis of temperature trends on the centre and on edge of panels, and on the influence of thermal field uniformity on PV and PV/T panels efficiency.

Acknowledgments
Authors are thankful to Eng. Giulia Marinucci for her kind cooperation.

References

[1] de Rubeis T, Muttillo M, Pantoli L, Nardi I, Leone I, Stornelli V and Ambrosini D 2017 *Energy Build.* **152** 24–39

[2] de Rubeis T, Nardi I, Paoletti D, Di Leonardo A, Ambrosini D, Poli R and Sfarra S 2016 *Energ. Convers. Manage.* **128** 261–72

[3] Huang B J, Lin T H, Hung W C and Sun F S 2001 *Sol. Energy* **70** 443–8

[4] Zondag H A, De Vries D W, Van Helden W G J, Van Zolingen R J C and Van Steenhoven A A 2002 *Sol. Energy* **72** 113–28

[5] Tiwari A and Sodha M S 2006 *Sol. Energy* **80** 751–9

[6] Charalambous P G, Maidment G G, Kalogirou S A and Yiakoumetti K 2007 *Appl. Therm. Eng.* **27** 275–86

[7] Zondag H A 2008 *Renew. Sust. Energ. Rev.* **12** 891–959

[8] Ibrahim A, Othman M Y, Ruslan M H, Alghoul M A, Yahya M, Zaharim A and Sopian K 2009 *WSEAS Transactions on Environment and Development* **5** 321–30

[9] Kostic L T, Pavlovic T M and Pavlovic Z T *Applied Energy* **87** 410–6

[10] Ibrahim A, Othman M Y, Ruslan M H, Mat S and Sopian K *Renew. Sust. Energ. Rev.* **15** 352–65

[11] Zhang X, Zhao X, Smith S, Xu J and Yu X 2012 *Renew. Sust. Energ. Rev.* **16** 599–617

[12] Yang T and Athienitis A K 2016 *Renewable and Sustainable Energy Reviews* **66** 886–912

[13] Al-Waeli A H A, Sopian K, Kazem H A and Chaichan M T 2017 *Renew. Sust. Energ. Rev.* **77** 109–30

[14] Vittorini D, Castellucci N and Cipollone R 2017 *App. Energ.* **205** 44–56

[15] Orsetti C, Muttillo M, Parente F R, Pantoli L, Stornelli V and Ferri G 2016 *Procedia Engineering* **0** 0–3

Innovative coupling of cogeneration units with fire tube boilers: thermo-fluid dynamics of the fire tubes

L Cioccolanti[1], A Arteconi[1], C M Bartolini[2] and F Polonara[3]

[1] Università eCampus, via Isimbardi 10, 22060 Novedrate (CO), Italy

[2] Società per il TRAsferimento TEcnologico e Guida all'Innovation Engineering, S.TRA.TE.G.I.E. srl, via Sandro Totti 1, 60131 Ancona, Italy

[3] Dipartimento di Ingegneria Industriale e Scienze Matematiche, Università Politecnica delle Marche, via Brecce Bianche 1, 60131 Ancona, Italy

E-mail: luca.cioccolanti@uniecampus.it

Abstract. Nowadays the thermal energy demand in the industrial sector is usually satisfied by means of fire tube boilers while electricity is supplied from the grid. Alternatively cogeneration units could be adopted for thermal and electrical energy self-production, whilst installing boilers only as back-up units. However, even when cogeneration is profitable, it is not widespread because industries are usually unwilling to accept cogeneration plants for reliability and high investment costs issues. In this work a system aimed at overcoming the above mentioned market difficulties is proposed. It consists of an innovative coupling of a combined heat and power unit with a modified fire tube boiler. In particular, a CFD analysis was carried out by the authors in order to address the most critical aspects related with the coupling of the two systems. More precisely, the following aspects were evaluated in detail: (i) pressure losses of the exhausts going from the prime mover to the boiler due to the sudden cross-section area variations; (ii) thermal power recoverable from the exhausts in the tubes of the boiler; (iii) dependence of the system on the final users' specification.

1. Introduction

Despite the prolonged economic crisis that has affected the world's most advanced countries in the last years, industry remains one of the most energy-intensive sectors worldwide. According to the Energy Information Administration (EIA) [1] indeed in the United States the industry sector accounts for about one third of the energy used in the country while in the EU-28 countries the industry, excluding the energy sector, accounted for more than 25% of the overall final energy consumption in 2015 [2]. Moreover, it will continue to play a key role also in the future since an increase of about 1.2% per year in the industrial sector energy consumption is foreseen in the next decades according to reference case scenarios [3] mainly due to the contributions of the countries outside the Organization for Economic Cooperation and Development (OECD).

As regards the highly industrialized countries energy is used for a wide range of purposes, such as process and assembly, steam and cogeneration, heating and cooling. With respect to the energy supply, when a simultaneous demand of thermal and electrical energy is needed, thermal energy is usually satisfied by means of fire tube boilers while electricity supplied from the grid. Sometimes, cogeneration units are adopted for energy self-production whilst fire tube boilers are also installed for surplus production or as back-up units. However, due to the high investment cost and higher complexity compared to traditional systems, the use of combined heat and power (CHP) units is still limited even when largely profitable. Therefore, also in the industrial sector energy efficiency measures, such as the combined heat and power production, have a very interesting potential in curbing energy consumptions and the related CO_2 emissions.

Taking into account the above mentioned market difficulties, some of the authors patented [4] and designed in cooperation with an Italian engineering company [5] an innovative solution in order to combine the well-known technology of the fire tube boiler with a cogeneration unit. In particular, by means of an adequate conveyor, exhausts from the cogeneration unit are forced into the boiler and intercept a few of its tubes thus recovering the high temperature heat. In this way, it is possible to avoid the auxiliary heat exchanger for the recovery of the exhausts and to reduce of about 10-15% the investment cost of the CHP unit. In addition, even the low temperature heat from the auxiliaries or from the engine cooling system of the prime mover can be recovered to preheat the make-up water of the boiler or the air in the combustion chamber of the boiler in order to reduce the primary energy consumption of the burner. Hence, the proposed solution is a system with a very high flexibility in terms of thermal and electrical power production and lower investment and operating costs compared to the separated configurations of the two systems involved.

Despite the fire tube boilers are by far the most used technology in the industry sector for the steam and heating production, very limited research is published in the literature on such systems. Ortiz [6] developed a dynamic model of a three-pass fire tube boiler in Matlab in order to evaluate the performance of such a system and assist plant engineers in their tasks. On the contrary, Bisetto et al. [7] experimentally investigated and then modelled in Matlab/Simulink a three-pass boiler operating with a single phase energy heating fluid. However, none of them analyzed in detail the pressure drop of the exhausts in the fire tubes which on the contrary is extremely important in order to evaluate the potential coupling with a cogeneration unit. Higher the back pressure of a prime mover indeed lower is its electrical efficiency. For this reason, in the present work authors have carried out a CFD analysis to evaluate in detail the thermo-fluid dynamics performance of several tubes of a boiler specifically designed for such a coupling. After the Introduction, Section 2 reports the detailed description of the innovative system here proposed; then the CFD model to predict the thermo-fluid dynamics performance of the system under investigation is presented in Section 3. Section 4 reports the main results of the analysis while Section 5 draws the conclusions.

2. System description

The proposed solution is schematically shown in Figure 1 where a fire tube boiler is combined with an upward CHP unit, thus to theoretically recover the whole thermal power output of the exhausts from the cogeneration plant.

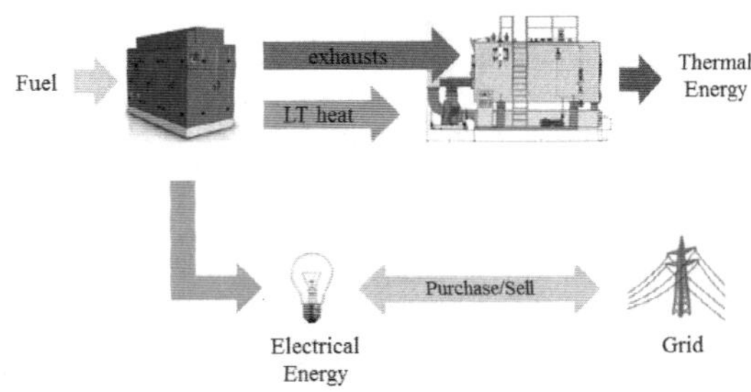

Figure 1. Schematics of the proposed innovative combination of a fire tube boiler with a CHP unit.

Both micro gas turbine and internal combustion engine (ICE) can be used as prime movers. However, a gas turbine has higher air to fuel ratio and lower electrical efficiency compared to an ICE, thus requiring a higher number of fire tubes for the exhausts for the same power range. On the contrary a gas turbine has a lower environmental impact [8,9]. Nevertheless, whenever allowed, ICEs are preferred. Fire tube boilers are widely adopted in industry for hot water and steam production at a rate of few tons per hours. The design of fire-tube boilers consists of a bundle of fire tubes contained in a shell full with water that evaporates to produce steam. Fire-tube boilers are usually distinguished by the number of passes, namely the number of times that the flue gas flows along the length of the vessel transferring its heat. At the end of each pass, the flue gas turns and goes back through the opposite direction of the vessel. In a typical three-pass fire tube boiler, that consists of three sets of horizontal tubes, the internal furnace is placed at the first pass, then the flue gas flows into the second pass and leaves the stack after the third pass.

Since in the proposed solution the exhausts from the CHP unit are directly conveyed into the boiler, the design of the traditional three-pass fire tube boiler is modified accordingly. In a three-pass fire tube boiler the flue gas at the entrance of the third pass has temperatures > 400°C. Even if the exhausts from the CHP unit have temperatures similar to those of the flue gas from the furnace, a separate section to convey the former is needed. Back pressure of a prime mover indeed has to be limited as far as possible and the mixing of the exhausts extremely discouraged.

Figure 2 and 3 report an axonometric projection of the modified fire tube boiler and a cross-section of the front view.

Figure 2. Axonometric projection of the modified fire tube boiler.

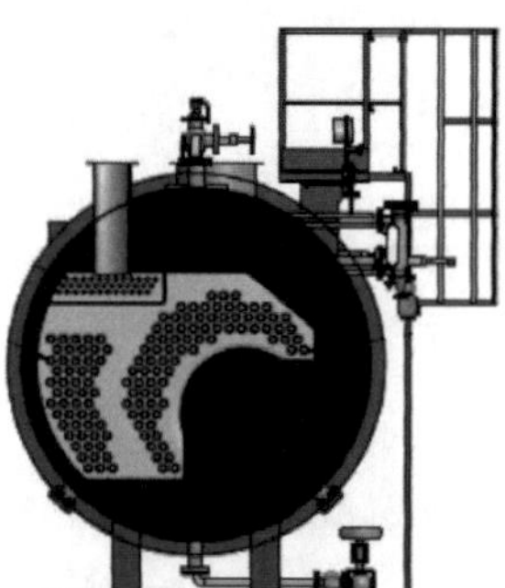

Figure 3. Cross-section of the front view of the modified fire tube boiler.

As clearly shown in Figure 3, the exhausts from the CHP unit have their own entrance and exit. The number and the dimensions of the fire tubes depend on the size of the coupled CHP unit, while the overall dimensions of the boiler do not differ sensibly compared to the standard configuration. Taking into account the different industrial sectors and the market of the two systems, a $5 \cdot 10^3$ $kg \cdot h^{-1}$ third-pass fire tube boiler and a 210 kWe ICE have been considered for the purpose of this analysis. In this way, the operation of the cogeneration plant can be maintained also when the primary burner of the boiler is off. The thermal power of the CHP unit is indeed considerably lower than that of the primary burner, thus allowing the recovering of the heat from the CHP unit also when the steam production is not instantaneously requested. In case of excess, a by-pass valve allows to temporarily discharge the exhausts in the atmosphere. At the same time, the cogeneration plant can be switched off without compromising the life time of the fire tube boiler. The tubes that convey the exhausts from the CHP unit are indeed submerged in the water and their surface temperature is almost constant in steady state conditions if water is in saturated conditions. Thus the operation of the cogeneration plant does not sensibly affect the temperature of the related tubes.

As regards the ICE considered in the present analysis, it is manufactured by MAN Engine [10] with the following technical specifications:

Table 1. Technical specifications of the considered ICE.

Model	MAN E2876 LE302
Fuel	Natural gas
Electrical power output (ISO conditions)	210 kWe
Power input	538 kWt
Thermal power output from the exhausts	143 kWt
Thermal power output from engine cooling	99 kWt
Thermal power output from intercooling (LT + HT)	41 kWt
Electrical efficiency	39.0%
Thermal efficiency	48.9%
Exhausts temperature	510°C
Exhausts flow rate	1101 $kg{\cdot}h^{-1}$

In particular, according to the selected CHP unit, 36 tubes were considered to recover the thermal power output of the exhausts from the CHP unit.

3. CFD model

A three-dimensional model of the fire tubes of the boiler crossed by the exhausts from the CHP unit was developed using a mesh generator. The computational domain was discretized using an unstructured grid of tetrahedral elements in the inlet and outlet plenum chambers and structured hexahedral elements in the tubes. In the x direction which corresponds to the longitudinal direction of the tubes, the grid density is reduced moving towards the centre in order to limit the number of the total cells as a compromise of accuracy and computing efficiency. On the contrary, in proximity of each cross-section area variations a layer refinement has been realized to better evaluate the flow field and the pressure loss of the exhausts. Figure 4 reports the mesh of the considered geometry which consists of about 1’310’000 nodes, almost 2’180’000 cells with an overall volume of 0.365 m^3.

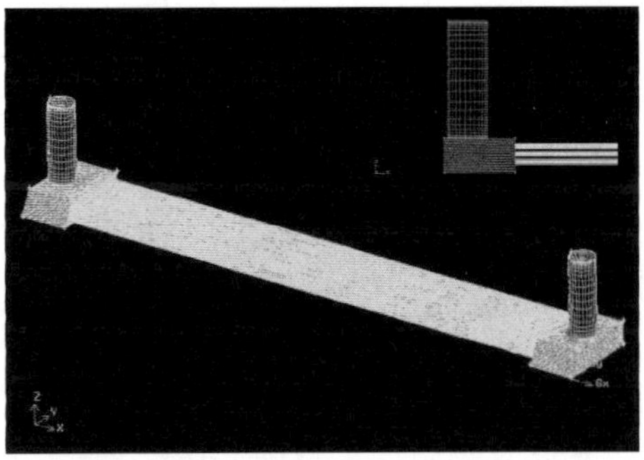

Figure 4. An axonometric projection of the meshed geometry.

IOP Conf. Series: Journal of Physics: Conf. Series **923** (2017) 012041 doi:10.1088/1742-6596/923/1/012041

Calculations were assessed by means of a CFD commercial code. With respect to the CFD model the following assumptions have been considered: (i) there is no heat transfer in the vertical tubes that convey the exhausts from the CHP unit and to the chimney; (ii) flow is at steady state conditions; (iii) fire tubes are completely submerged by water at saturated conditions (at 200°C).

The steady state Reynolds Averaged Navier – Stokes (RANS) equations were solved for the computation of the turbulent flow, using the realizable k-ε turbulence model. The PRESTO! pressure interpolation scheme was adopted for pressure–velocity coupling equations. As regards the boundary conditions, the inlet surface of the vertical tube was set as mass flow inlet while the outlet surface as pressure outlet at atmospheric pressure. The vertical tubes are adiabatic surfaces while all the other tubes were set as wall. The wall was no-slip boundary shear condition.

Steel was used for the tubes while air was considered for the exhausts.

4. Results and discussion

The objectives of the present numerical simulation analysis were to evaluate: (i) the pressure losses of the exhausts from the CHP unit due to the sudden cross-section area variations of the fire tubes inside the boiler; (ii) the thermal power recovered from the exhausts in the tubes; (iii) the dependence of the system on the final users' requests. The CFD analysis indeed was oriented to obtain a first guess solution of the back pressure of the CHP unit and the thermal power that can be recovered from the exhausts in order to estimate the potential of such a coupling prior to a subsequent experimental test campaign. Figures 5a-c report the contours of the static pressure, static temperature and the velocity magnitude of the flow along the tubes.

Contours of Static Pressure (pascal)

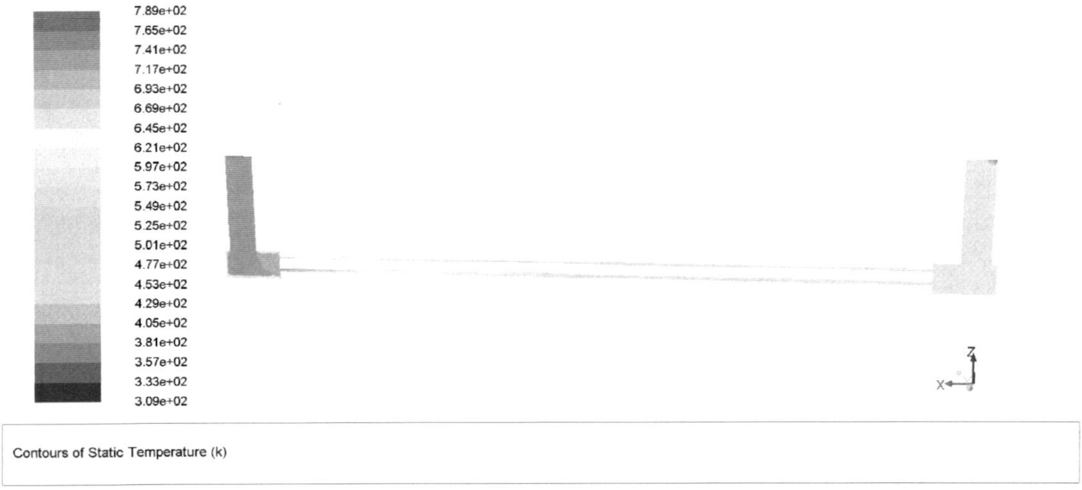

Figures 5. CFD results: a) static pressure; b) static temperature; c) velocity magnitude.

Figure 5a clearly shows that for the considered size of the systems the back pressure is very low and limited to a maximum of about 4.6 mbar which is considerably lower than the maximum back pressure tolerable by the prime mover (50 mbar). In terms of temperatures, the heat transfer in the plenum chambers are also limited and the most part of the temperature reduction occurs along the tubes. For an inlet temperature of the exhausts of about 510°C the outlet temperature of the exhausts is in the range of 220-230°C. This means that the recovered thermal power is about 90 kWt which corresponds to more than 60% of the thermal power input of the exhausts. Of course, the recovered thermal power would be even higher if lower temperatures of the water/steam inside the boiler were considered. In any case, this represents an interesting amount of energy that can be recovered from the CHP unit, thus contributing to operate at valuable overall efficiencies and reaching important Primary Energy Savings (PES) [11]. As regards the velocity of the exhausts, the magnitude was evaluated also in proximity of the plate that connects the entrance plenum chamber with the fire tubes. In particular, the maximum velocity magnitude was about 13.3m·s^{-1} which does not represent an issue in terms of noise and vibrations.

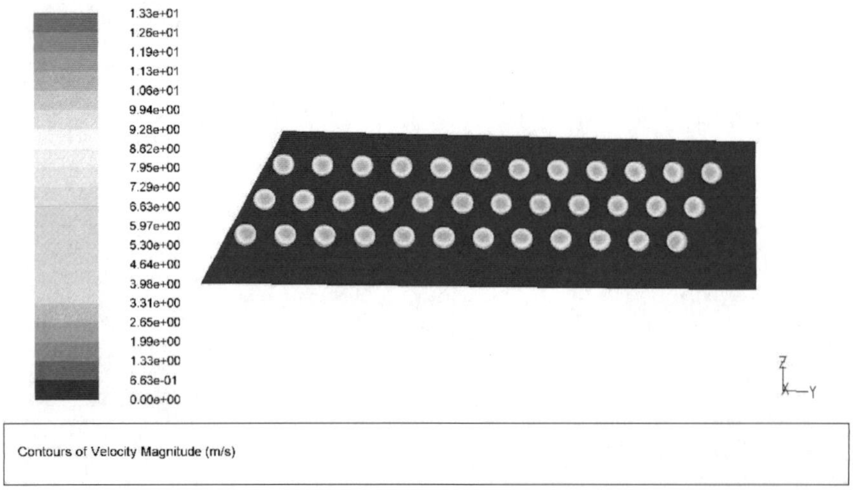

Figures 6. Contours of the velocity magnitude at the entrance of the fire tubes.

Finally, although the boiler was designed to be coupled with a 210 kWe ICE, sensitivity with users' needs was evaluated in terms of different flow rates. For the same size of the boiler indeed a different size of the ICE could be required or even micro-turbine technology preferred. In particular, such analysis showed that the back pressure reaches the maximum tolerable by an ICE with an exhausts flow rate of about 0.712 $kg \cdot s^{-1}$ which corresponds to two times the design flow rate of the proposed configuration. Figure 7 reports the contours of the static pressure for an exhausts flow rate of 0.712 $kg \cdot s^{-1}$.

Figures 7. Contours of the static pressure for a flow rate of 0.712 kgs^{-1}.

Therefore, on the basis of the obtained simulation results some conclusions can be drawn prior to the subsequent experimental phase. The flow field showed a good homogeneous distribution of the exhausts into the different fire tubes. Moreover, since the back pressure and the maximum flow velocity are low even at the nominal operating conditions of the CHP unit, the numbers of the tubes could be also reduced in order to keep the extra-costs of the modified boiler down.

5. Conclusions

Coupling a CHP unit with a fire tube boiler allows to overcome some skepticisms of the industries that are sometimes unwilling to accept cogeneration plants for reliability and high investment costs issues. The proposed solution presents higher complexity compared to the separate configuration of the two systems but allows to reduce the investment cost and assure reliable supply. In this work, the main criticisms arisen from such a coupling were addressed by means of a CFD analysis. In particular, the simulation analysis showed that at nominal operating conditions the back pressure of the ICE is < 5 mbar which is far lower than the maximum tolerable pressure loss by an ICE usually in the range of 50 mbar. Also in terms of temperatures, the fire tubes are able to extract the medium-high temperature heat of the exhausts from the prime mover with outlet temperatures that are close to those of the flue gas from the furnace. From the energetic point of view, for the considered temperatures of the exhausts and the water inside the boiler, the heat recovered in the fire tubes is >60% of the thermal power output of the exhausts from the CHP unit. Finally, in order to appreciate the sensitivity of the proposed configuration of the fired tubes boiler with the users' requests, the pressure losses were evaluated also with varying the mass flow rates of the exhausts from the prime mover. In particular, it was found that pressure losses still remain lower than the admissible level even at a flow rate of about 0.712 $kg \cdot s^{-1}$ which corresponds to two times the flow rate at the design conditions.

Therefore, the present analysis allows to assess that the proposed solution can be adopted to reduce the investment cost of the CHP unit (because the heat exchanger for heat recovery can be avoided) with interesting overall energy performance. Nevertheless, prior to a potential subsequent industrialization and commercial phase experimental analysis are needed to confirm the numerical performance of the designed configuration.

Acknowledgements

The authors wish to thank S.TRA.TE.G.I.E. srl for funding the patent application and also INVITALIA for partially funded the patent development under the program Brevetti+.

References

[1] U.S. Energy Information Administration, https://www.eia.gov/energyexplained (Accessed: 08 May 2017).
[2] European Statistical System – Eurostat, http://ec.europa.eu/eurostat/ (Accessed: 08 May 2017).
[3] U.S. Energy Information Administration https://www.eia.gov/outlooks/ieo/ (Accessed: 08 May 2017).
[4] European Patent Office, https://goo.gl/oE9Jod (Accessed: 08 May 2017).
[5] Società di TRAsferimento TEcnologico e Guida all'Innovation Engineering – S.TRA.TE.G.I.E. srl, http://www.strategiesrl.com/eng/Default.aspx (Accessed: 08 May 2017).
[6] Gutiérrez Ortiz F J 2011 *App. Therm. Eng.* **31** 3463.
[7] Bisetto A, Del Col D and Schievano M 2015 *App. Therm. Eng.* **78** 236.
[8] Comodi G, Cioccolanti L and Renzi M 2014 *Energy* **68** 92.
[9] Caresana F, Comodi G, Pelagalli L, Renzi M and Vagni S 2011 *App. Therm. Eng.* **31** 3552.
[10] MAN Engines, http://www.engines.man.eu/global/en/index.html (Accessed: 08 May 2017).
[11] Comodi G, Cioccolanti L, Renzi M, Vagni S., Pelagalli L and Caresana F 2013 *App. Therm. Eng.* **54** 336.

Thermal analysis of a Phase Change Material for a Solar Organic Rankine Cycle

M Iasiello*, K Braimakis, A Andreozzi* and S Karellas****
* Dipartimento di Ingegneria Industriale - Università degli Studi di Napoli Federico II - Piazzale Tecchio, 80, 80125 Napoli, Italy
** Laboratory of Steam Boilers and Thermal Plants, School of Mechanical Engineering, National Technical University of Athens, Zografou, 15780 Athens, Greece

Email: marcello.iasiello@unina.it

Abstract. Organic Rankine Cycle (ORC) is a promising technology for low temperature power generation, for example for the utilization of medium temperature solar energy. Since heat generated from solar source is variable throughout the day, the implementation of Thermal Energy Storage (TES) systems to guarantee the continuous operation of solar ORCs is a critical task, and Phase Change Materials (PCM) rely on latent heat to store large amounts of energy.
In the present study, a thermal analysis of a PCM for a solar ORC is carried out. Three different types of PCMs are analyzed. The energy equation for the PCM is modeled by using the heat capacity method, and it is solved by employing a 1D explicit finite difference scheme. The solar source is modeled with a time-variable temperature boundary condition, with experimental data taken from the literature for two different solar collectors. Results are presented in terms of temperature profiles and stored energy. It has been shown that the stored energy depends on the heat source temperature, on the employed PCM and on the boundary conditions. It has been demonstrated that the use of a metal foam can drastically enhance the stored energy due to the higher overall thermal conductivity.

1. Introduction

Organic Rankine Cycle (ORC) represents a promising technology with which it is possible to use low-temperature heat to generate electrical power. This low-temperature heat can be obtained from various sources, like a gas turbine or an internal combustion engine [1-7]. ORCs can also be coupled with environmentally friendly renewable energy sources (RES), such as solar. Especially regarding solar ORCs, they can play an important role in decentralized power generation, while they can be especially important in islanded microgrids. Additional applications include cogeneration for community or industrial use.

The intermittent nature of solar radiation can lead to discrepancies between demand and heat source availability. The addition of thermal energy storage (TES) systems makes it possible to redirect excess energy produced during daytime to nighttime or periods of unfavorable meteorological conditions, thus making the power system more efficient, flexible and reliable. A key advantage of solar ORCs over photovoltaic systems is the opportunity to integrate TES equipment to store excess heat energy in thermal rather than electrical form. TES, which is currently cheaper than commercial batteries and has a longer usable lifetime, comes in many varieties.

In the present study, thermal energy is stored in the latent heat of a phase-change material (PCM) [8]. Since latent energy of phase-change transitions is much higher than specific heat capacity, PCM storage has the potential for much higher energy storage density than sensible TES. It has to be observed that PCMs can be used also for mechanical energy systems or for buildings [9, 10]. Another key advantage of latent storage is that the phase transition is a quasi-isothermal process, a feature that facilitates TES integration and control within the ORC plant. The tradeoff, however, is the low thermal conductivity of the storage media which results in an excessive charging and discharging time for thermal storage. Faster response times can be achieved by adding high-conductivity materials (e.g., metals or graphite) within the PCM. Media for latent storage include organic materials (e.g., paraffins and fatty acids), inorganic compounds (e.g., salt hydrates), and various eutectic mixtures. Despite its technical advantages compared to sensible technologies, latent heat storage is more expensive and still at the prototype phase.

The melting point of PCMs can be chosen in order to have the desired operation temperature for the ORC. Zalba et al. [11] provide a review of studies dealing with TES using phase change materials. The authors categorized different materials depending on the following criteria: 1) Phase change materials, 2) Heat transfer analysis, 3) Applications. Wang and Baldea [12] analyzed latent-heat storage using phase-change materials (PCM) as a mean for improving temperature control and energy management in cooling systems. They proposed a novel, systems-centric approach to PCM-based thermal management and established a connection between the quantity and geometric properties of the PCM, the dynamics of the integrated system, and potential energy savings. They focused on composite heat sinks consisting of PCM elements encapsulated in a conductive matrix material as a practical implementation of PCM-enhanced thermal management. Most PCMs present the common problem of low thermal conductivity that strongly limits the energy charging/discharging rates. Therefore, the introduction of highly conductive materials embedded with PCMs has been already proposed in literature in order to improve the heat exchange process [13].

In this paper, a thermal analysis of a PCM for a solar ORC is carried out. Three different types of PCM are herein analyzed. The energy equation for the PCM is modeled by using the heat capacity method, and it is solved by employing a 1D explicit finite difference scheme. The solar source is modeled with a time-variable temperature boundary condition, with experimental data taken from the literature and referred to two different solar collectors. At the PCM/ORC interface, a sensitivity analysis on the boundary conditions is carried out in order to analyze the different cases. Results are presented in terms of temperature profiles and stored energy.

2. Mathematical model

2.1. Description of the storage system

A diagram of the PCM-solar-assisted ORC is presented in Fig. 1. Since the ORC requires a relatively low-temperature heat source [14], a medium-temperature solar collector can be employed for this purpose. Having a constant temperature solar heat source is pretty challenging, thus an integrated system is required. The PCM has the role of storing solar energy and giving it to the ORC when the solar source cannot produce sufficiently high operation temperatures, while a high temperature heat pump can be employed to further integrate the process.

In this paper, it is assumed that the ORC fluid works at a maximum operation temperature of about 90 °C. Thus, the solar collector has to provide an outlet fluid temperature that is in the range between 100 and160 °C. For this purpose, two different solar collectors, a parabolic trough solar collector (39 m^2 made up of three modules in series, with a 19.6 geometric concentration ratio) and an evacuated tubes solar collector (two panels with 6 collector tubes, with 2.76 and 1.82 m^2 gross and net absorber area, respectively) have been employed. The first is referred to an experimental study performed in Tunisia [15], while the second in Japan [16]. The collector fluid outlet temperature profiles are presented in Fig. 2, showing that these vary in the range from 120 to160 °C, with maximum values achieved around midday, when the solar radiation is maximized. Sixth-order polynomial fits were performed on experimental data, in order to obtain analytical function that can be used as a boundary condition for the PCM. The collector gives heat to the PCM, for which the melting point must be in the range of about 110-130 °C, in order to exploit the energy gain due to the phase change. Among various PCMs, two of these were chosen due to their melting points: erythritol and magnesium chloridehexahydrate $MgCl_2$ $6H_2O$. Their thermophysical properties are resumed in Table 1 [17], together with melting characteristics that will be described in the following. For the sake of simplicity, the thermal conductivity and density variation with temperature are neglected in the present study (temperature variations through the space and time are less than about 50 °C), thus solid phase thermophysical properties are employed.

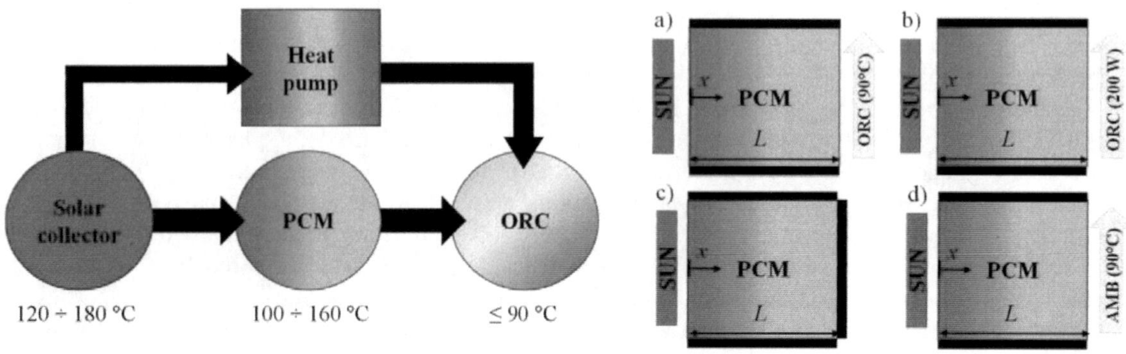

Figure 1.A sketch of the thermal system with boundary conditions employed in the present paper.

Table1.PCMs thermophysical properties.

PCM	k (W/m K)	ρ (kg/m³)	$c_{p,s}$ (kJ/kg K)	$c_{p,l}$ (kJ/kg K)	λ (kJ/kg)	T_m (°C)	ΔT (°C)
Erythritol (A)	0.733	1480	1350	2740	339	122.0	20
$MgCl_2$ $6H_2O$ (B)	0.704	1569	2250	2610	134	121.2	10

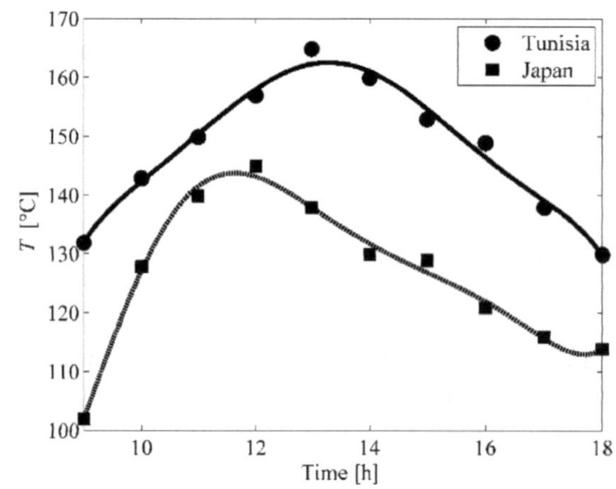

Figure 2. Temperature profiles during the day for Tunisia [15] and Japan [16] heat sources.

2.2. Mathematical model

In order to analyze temperature fields and energy stored, the energy equation of the PCM is solved. Due to temperature gradients, natural convection occurs; but for lower temperature gradients through the domain, it is reasonable to neglect them. The PCM starts from an initial temperature that is below the melting point, and the melting front moves from the heated boundary to the opposite one with a certain velocity. Since this front is not known a priori, a fixed grid technique is used. With this technique, the energy equation is characterized depending on the punctual temperature of the computational domain. This characterization happens by employing the apparent heat capacity method, with which the energy equation is described by assuming that the enthalpy is proportional to the temperature by means of the heat capacity, and the heat capacity is characterized depending on the PCM phase. For a 1D coordinate system, the transient energy equation is as follows:

$$\rho c_p \frac{\partial T}{\partial t} = k \frac{\partial^2 T}{\partial x^2} \tag{1}$$

The heat capacity curve is obtained starting from Direct Scanning Calorimetry (DSC) of the substance, taken from literature [18, 19]. By means of a graphical construction, it is possible to obtain from experiments the latent heat of fusion, the melting temperature, and a temperature difference ΔT in which the melting occurs. Such curves are reported in Fig. 3, together with the graphical construction and the heat capacity vs. temperature curve obtained. The latent heat is taken from the aforementioned literature source, while melting temperature and range are herein derived (Fig. 3). From the heat capacity curve, it is shown that the erythritol has more storage capacities when compared to the magnesium chloride hexahydrate. The heat capacity is described with the following equation:

$$c_p = c_{p,s}\left[1-\alpha(T)\right] + c_{p,l}\alpha(T) + \lambda \frac{\mathrm{d}\alpha}{\mathrm{d}T} \tag{2}$$

where α is the alpha function. The function alpha is also known as the melting fraction, and it has the physical meaning of porosity, since it described how the PCM melts by determining the ratio between the two phases volume. It is equal to 0 for a pure solid and 1 for a pure liquid. In order to make the solid-liquid transition smoother, a sinusoidal function is employed:

$$\alpha(T) = \frac{\sin\left[\frac{\pi}{2\Delta T}\left(T - T_m\right)\right] + 1}{2} \tag{3}$$

valid for $0<\alpha<1$. This operation also avoids derivative discontinuities in the points $T_m = T \pm \Delta T$.

Boundary conditions are presented in the following and resumed in Fig. 1. For $t = 0$, all the solid PCM is at a temperature of 100 °C. At the right boundary of the domain ($x = L = 0.1$ m), different boundary conditions are analyzed, making references to four cases. The first is a Dirichlet boundary condition (Fig. 1a), that can occur for example with very high heat transfer coefficients at the right boundary of the domain. The second and the third (Figs. 1b and 1c) are Neumann conditions, respectively fixed heat flux and adiabatic condition. The first simulates a heat storage system that guarantees the operation of the PCM, while the second simulates a case in which the PCM has the role of storing thermal energy. Finally, the last boundary condition is a Robin one (Fig. 1d), that simulates heat given to a fluid, that might be represent the environment, at a temperature just below the PCM operation point. All these boundary conditions are resumed in Table 2.

In order to solve Eq. 1, a finite-difference explicit scheme is herein employed. Grid convergence has been checked in terms of temperature profiles and stored energy and presented in Fig. 4a, for erythritol with $T_m = 122$ °C and $\Delta T = 20$°C, with adiabatic boundary condition on the right side of the PCM and by using the heat source referred as Tunisia (Fig. 2). In calculating the energy stored, and as it will be done in the rest of the paper, it is assumed that the height of the PCM is equal to the length 0.1 m, obtaining a square PCM. The time step is chosen with references to the numerical stability of the problem. Finally, a comparison with a solution developed with a Galkerin-based finite element scheme by means of the commercial code COMSOL Multiphysics has been performed. It is observed

IOP Conf. Series: Journal of Physics: Conf. Series **923** (2017) 012042 doi:10.1088/1742-6596/923/1/012042

that the finite-element code models the phase change by means of a smoothed Heaviside function. Results in terms of temperature profiles are presented in Fig. 4b, showing an excellent agreement.

Table2. Investigated boundary conditions.

Boundary condition		Mathematical description	
BC_1	Dirichlet	$T = T_\infty$	$T_\infty = 90$ °C
BC_2	Neumann	$-kdT/dx = Q/A$	$Q = 200$ W
BC_3	Neumann	$-kdT/dx = Q/A$	$Q = 0$ W
BC_4	Robin	$-kdT/dx = h(T-T_\infty)$	$T_\infty = 90$ °C

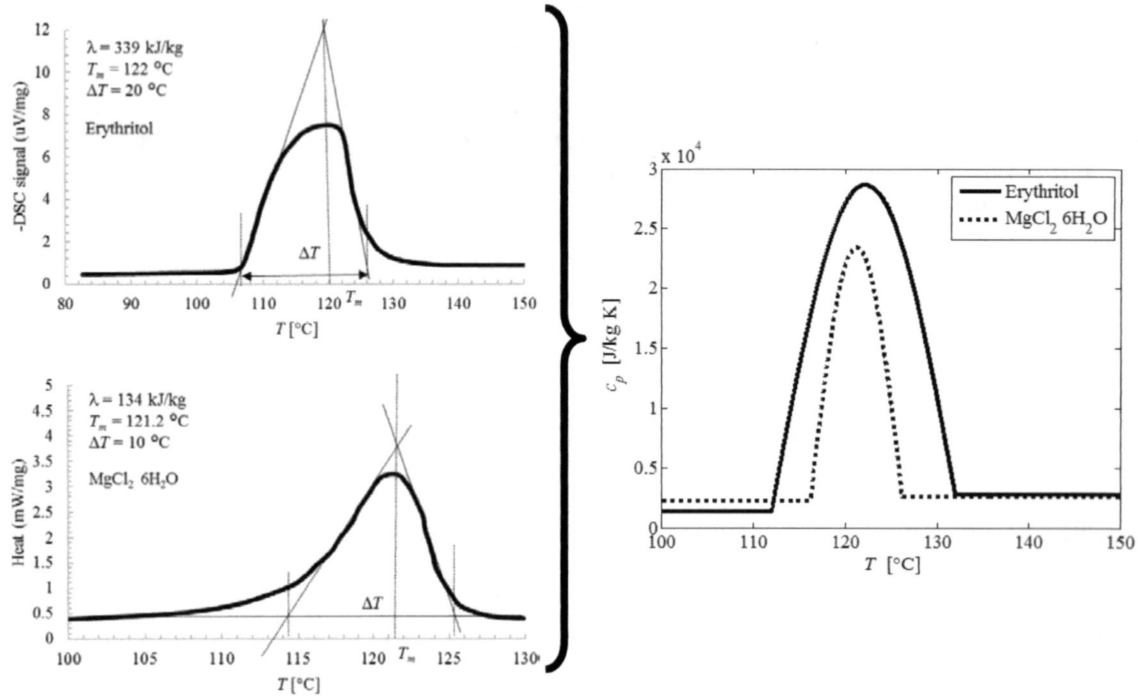

Figure 3. Derivation of the effective heat capacity curves.

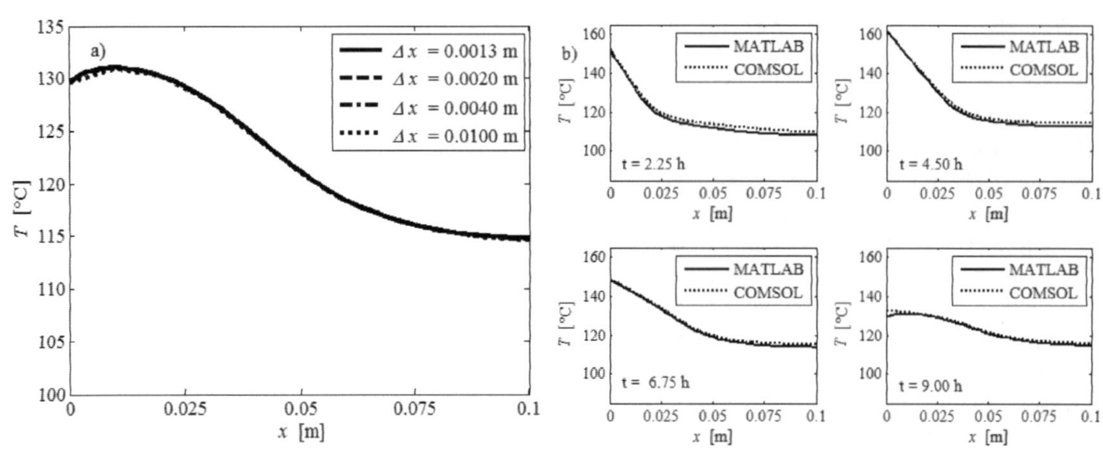

Figure 4.a) Grid independence check and b) comparisons with results obtained with a finite-element scheme for temperature profiles for erythritol, adiabatic boundary condition and Tunisia.

3. Results

Results will be presented in terms of temperature profiles and cumulated energy. The total cumulated energy per unit length *U*, in (J/m), can be calculated from the following relationship:

$$\int_{t_{in}}^{t_{fin}} \mathrm{d}U = \rho A \left(\int_{T_{in}}^{T_m} c_{p,s} \mathrm{d}T + \lambda + \int_{T_m}^{T_{fin}} c_{p,l} \mathrm{d}T \right) \tag{4}$$

where the subscript *in* and *fin* respectively refer to initial and ending conditions and *A* is the PCM area. Since the apparent heat capacity method is herein applied, the energy ΔU cumulated between two time steps referred to *t* and *t+1* is herein derived from the following:

$$\Delta U = U(t+1) - U(t) = \rho \Delta x^2 c_p(t) \left[T(t+1) - T(t) \right] \tag{5}$$

where Δx is the grid size. It is important to observe that the heat capacity is calculated at a time step referred to *t*, since an explicit method is here used. The energy cumulated is obtained by making the sum of consecutive energy accumulations, starting from an initial state with zero-energy.

In Fig. 5, a comparison between two different PCMs (PCM_A denotes Erythritol, while PCM_B denotes $MgCl_2\ 6H_2O$) is presented. Results are presented in terms of temperature profiles for different times, and energy cumulated. The Robin boundary condition is used (BC_4, see Table 2), for climate date referred to Tunisia (Fig. 2). With reference to the temperature profiles, it is shown that the temperature on the left increases in the first half of the simulations that follows what reported in Fig. 2, while a drop occurs for the other parts of the simulations, due to the fact that the sun provides lower energy. For the phase change, it is shown that the heat moves from the left to the right, causing a general increase of the temperature profile, that means phase change. With reference to the cumulated energy, it is shown that there is a high increase at the beginning of the simulations, while a maximum is reached just before the end. The reason of this is that the solar source doesn't provide constant temperature, thus at the end of the transient the reduction of temperature could cause an inversion of heat flux, making energy in the PCM to become lower. By comparing the two PCMs, it is shown that the erythritol melts easier than the magnesium chloride hexahydrate, since the slightly lower heat capacity in the solid phase (Fig. 3) means less resistance to the heat that comes from the solar collector. On the other hand, the erythritol makes possible to accumulate more energy; this happens because the heat capacity is in general higher.

In Fig. 6, different boundary conditions (see Table 2) are compared in terms of temperature profiles and stored energy. The analysis is referred to erythritol (PCM_A) and for climate date referred to Tunisia (Fig. 2). It is shown that the adiabatic condition (BC_3) provides higher temperature values, since the heat comes from the solar collector without any exits, causing temperature increases higher than the other boundary conditions investigated. As obvious, this boundary condition also provides higher stored energy values. On the other hand, the Dirichlet boundary condition (BC_1) provides lower temperature profiles and stored energy. This happens because the locked temperature is a constraint in terms of temperature enhancing through the PCM. With references to the other two BCs, the Robin condition (BC_4) provides a slight increase with respect to the time for the temperature at the right end of the PCM (x = 0.1 m), since the heat that comes from the solar system is higher than the heat that goes to the environment. Similar conclusions can be made for the Neumann BC (BC_2).

In Fig. 7, different solar sources (see Fig. 2) are presented. Results are referred to erythritol (PCM_A) and for the Robin boundary condition (BC_4). The computations in which the solar source named Tunisia are the ones that have higher temperatures and energy stored, since the solar source temperatures are in general higher than that for Japan.

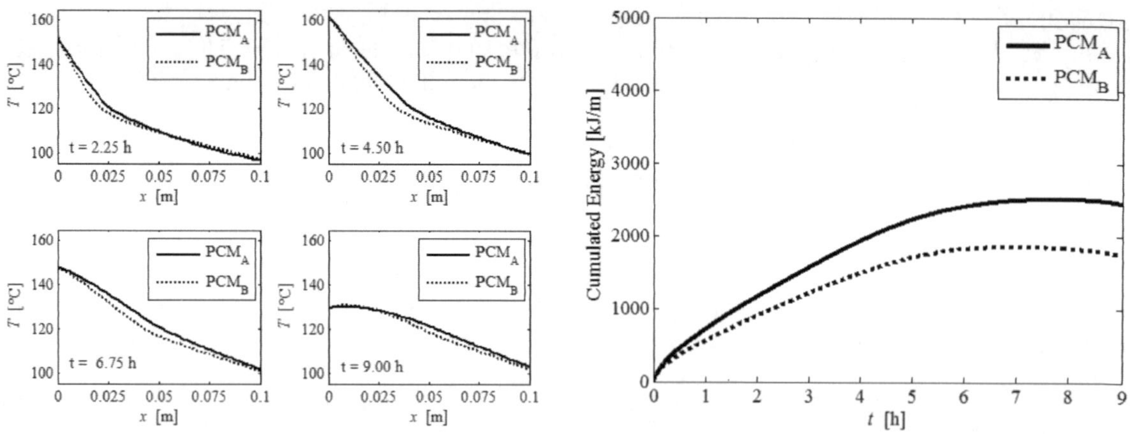

Figure 5. Temperature profiles for different times and cumulated energy for different PCMs, for Robin BC (BC_4) and Tunisia.

It is important to observe that at about midday the energy stored for Japan starts to decrease, since the solar source starts to become lower, as depicted in Fig. 2.

It is widely known that the weakest point, from the thermodynamic point of view, is the relatively low thermal conductivity of the PCM. Indeed, this causes the PCM to have some problems in melting at all [20]. For this purpose, various techniques have been proposed during the years to increase the conductivity [21]. One of these consists into using a metal foam that has the ability of increasing the overall thermal conductivity of the foam, since a high conductivity material with high porosity is put in the PCM. In this paper, simulations for an aluminum foam with 90% of porosity are carried out. In order to take into account the foam in the computational domain, the thermophysical properties are averaged over the computational domain, as follows:

$$\rho = \rho_{PCM} + \rho_{foam} = \varepsilon \rho_{PCM} + (1-\varepsilon)\rho_{foam} \tag{6}$$

$$k = k_{PCM} + k_{foam} = \varepsilon k_{PCM} + \frac{(1-\varepsilon)}{3} k_{Al} \tag{7}$$

$$c_p = c_{p,PCM} + c_{p,foam} = \varepsilon \left\{ c_{p,s}\left[1-\alpha(T)\right] + c_{p,l}\alpha(T) + \lambda \frac{\mathrm{d}\alpha}{\mathrm{d}T} \right\} + (1-\varepsilon) c_{p,\mathrm{Al}} \tag{8}$$

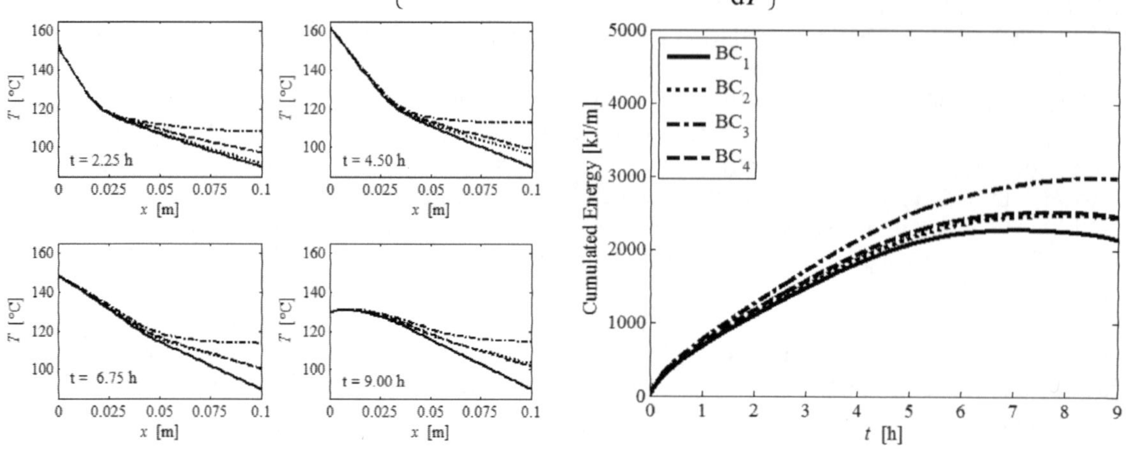

Figure 6. Temperature profiles for different times and cumulated energy for different BCs, for erythritol (PCM_A) and Tunisia. Legend for temperature profiles is the same reported in the figure on the right.

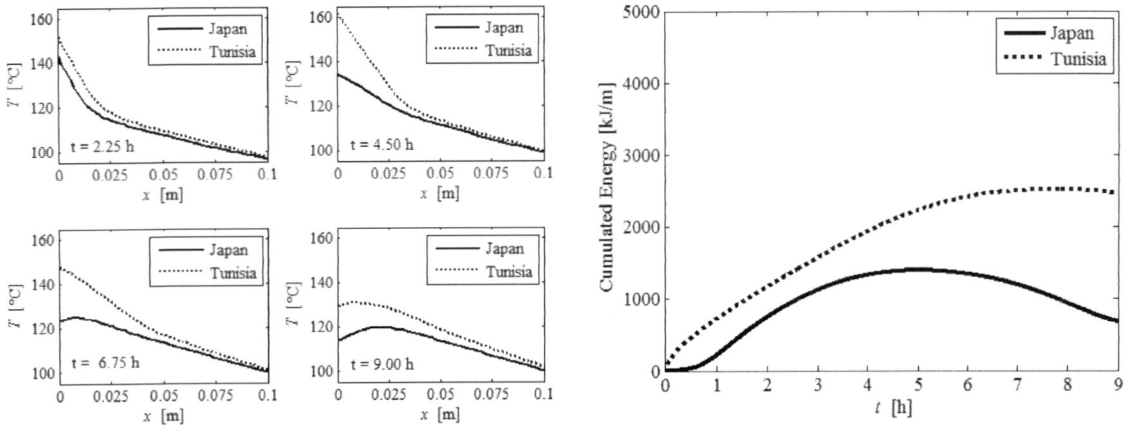

Figure 7. Temperature profiles for different times and cumulated energy for different heat sources, for erythritol (PCM_A) and Robin BC (BC_4).

The aluminum thermophysical properties, denoted with the subscript Al, are evaluated at 400 K. It should be noted that the thermal conductivity relationship for the foam is taken from [22, 23], since it has been demonstrated that a weighted average or a series model is not accurate enough [24, 25]. Density and heat capacity, respectively Eqs. 6 and 8, are taken as the weighted average with respect to the porosity, as usually done for porous media [26]. The porosity is here the ratio between the PCM volume and the total volume. The heat capacity (Eq. 8) is obtained by observing that the PCM heat capacity has been already presented in Eq. 2. Results in terms of temperature profiles and stored energy and referred to PCM_A, BC_3 (adiabatic) and Tunisia, are presented in Fig. 8. It is shown that the metal foam drastically enhances both temperature values and stored energy. This because the higher thermal conductivity causes the melting front to proceed faster from the left side to the right one, obtaining more melting fraction through the foam. However, the overall heat capacity tends to reduce because the foam has lower heat capacity than the PCM, but this effect is overwhelmed by the higher melting fraction. It has to be observed that the foam cumulated energy reaches a maximum at the middle point of the curve. This occurs since the melting process becomes very fast with a foam, then from that peak the temperature heat source reduction (Fig. 2) is as prominent as to make the temperature reducing in the PCM, thus cumulated energy starts to reduce. This effect is also evident with references to the temperature profiles, since from 4.50 h to 9.00 h the profiles are almost uniform, and they tend to reduce following the temperature of the heat source on the left.

4. Conclusions

In this paper, various aspects of a solar-assisted PCM for ORC energy storage have been numerically investigated with the apparent heat capacity method. Since the PCM needs to operate close to 100-120 °C, two different PCMs are analyzed and compared: erythritol and magnesium chloride hexahydrate. Making reference to such temperature range, two solar heat sources from the literature have been chosen and compared. Various boundary conditions were analyzed, in order to investigate different configurations. Finally, a comparison with a PCM-metal foam system has been reported.

Results, presented in terms of temperature profiles and stored energy, show that the latent heat has a primary role in storing energy, since the PCM with an higher latent heat (Erythritol) can store higher amounts of energy when compared to the other investigated PCM (about less than 30%). As expected, higher solar source temperature increases the temperature and the stored energy of the system. Finally, it has been shown that using a metal foam can drastically enhance stored energy, since melting occurs easily due to the relatively high thermal conductivity of the foam. Finally, it is possible to conclude that ORC temperatures can be controlled when the solutions proposed in the paper are employed.

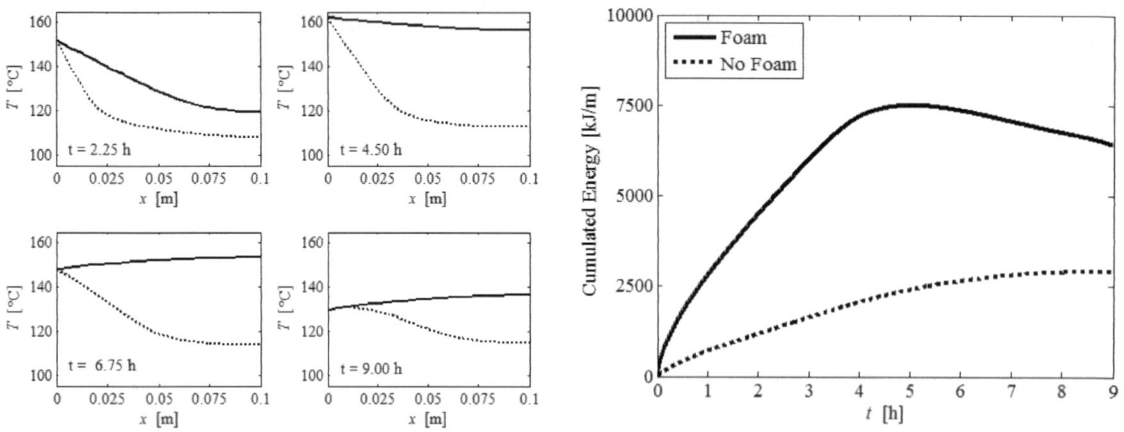

Figure 8. Temperature profiles for different times and cumulated energy with and without aluminum foam, for erythritol (PCM_A), adiabatic BC (BC_3) and Tunisia. Legend for temperature profiles is the same reported in the figure on the right.

5. Nomenclature

A	Area (m^2)	ε	Porosity
c_p	Heat capacity (J/kg K)	λ	Latent heat of fusion (J/kg)
h	Heat transfer coefficient (W/m^2K)	ρ	Density (kg/m^3)
k	Thermal conductivity (W/m K)	*Subscripts*	
L	Length (m)	∞	Environment
Q	Heat rate (W)	*fin*	Final
t	Time (s)	*foam*	Aluminum foam
T	Temperature (°C)	*in*	Initial
U	Energy per unit length (J/m)	*l*	Liquid PCM
x	Axial coordinate(m)	*m*	Melting
Greek symbols		*PCM*	Phase Change Material
α	Alpha function	*s*	Solid PCM

References

[1] Wang M, Wang J, Zhao Y, Zhao P and Dai Y 2013 Thermodynamic analysis and optimization of a solar-driven regenerative organic Rankine cycle (ORC) based on flat-plate solar collectors. *Appl Therm Eng* **50** 816-825.

[2] Franco A 2011 Power production from a moderate temperature geothermal resource with regenerative organic Rankine cycles. *Energy Sustain Dev* **15** 411-419.

[3] Algieri A and Morrone P 2012 Comparative energetic analysis of high-temperature subcritical and transcritical organic Rankine cycle (ORC). A biomass application in the Sibari district. *Appl Therm Eng* **36** 236-244.

[4] Aussant C D, Fung A S, Ugursal V I and Taherian H 2009 Residential application of internal combustion engine based cogeneration in cold climated Canada. *Energy Build* **41** 1288-1298.

[5] Dolz V, Novella R, García A and Sanchez J 2012 HD diesel engine equipped with a bottoming Rankine cycle as a waste heat recovery system. Part 1: study and analysis of the waste heat energy. *Appl Therm Eng* **36** 269-278.

[6] Wang D, Ling X and Peng H 2012 Performance analysis of double organic Rankine cycle for discontinuous low temperature waste heat recovery. *Appl Therm Eng* **48** 63-71.

[7] Wang H, Peterson R and Herron T 2011 Design study of configurations on system COP for a combined ORC (organic Rankine cycle) and VCC (vapor compression cycle). *Energy* **36** 4809-4820.

[8] Zhou D, Zhao C Y and Tian Y 2012 Review on thermal energy storage with phase change materials (PCMs) in building applications. *Appl Energ* **92** 593-605.

[9] Castellani B, Presciutti A, Filipponi M, Nicolini A and Rossi F 2015 Experimental investigation on the effect of phase change materials on compressed air expansion in CAES plants. *Sustainability-basel* **7 (8)** 9773-9786.

[10] Castellani B, Morini E, Filipponi M, Nicolini A, Palombo M, Cotana F and Rossi F 2014 Clathrate hydrates for thermal energy storage in buildings: overview of proper hydrate-forming compounds. *Sustainability-basel* **6 (10)** 6815-6829.

[11] Zalba B, Marìn J M, Cabeza L F and Mehling H 2003 Review on thermal energy storage with phase change: materials, heat transfer analysis and applications. *Appl Therm Eng* **23** 251–283.

[12] Wang S and Baldea M 2013 Temperature control and optimal energy management using latent energy storage. *Ind Eng Chem Res* **52 (9)** 3247-3257.

[13] Fan L and Khodadadi J M 2011 Thermal conductivity enhancement of phase change materials for thermal energy storage: A review. *Renew Sust Energ Rev* **15 (1)** 24-46.

[14] Cataldo F, Mastrullo R, Mauro A W and Vanoli G P 2014 Fluid selection of Organic Rankine Cycle for low-temperature waste heat recovery based on thermal optimization. *Energy* **72** 159-167.

[15] Balghouthi M, Ali A B H, Trabelsi S E and Guizani A 2014 Optical and thermal evaluations of a medium temperature parabolic trough solar collector used in a cooling installation. *Energ Convers Manage* **86** 1134-1146.

[16] Sharma S D, Iwata T, Kitano H and Sagara K. 2005 Thermal performance of a solar cooker based on an evacuated tube solar collector with a PCM storage unit. *Sol Energy* **78** 416-426.

[17] Mehling H and Cabeza L F 2008 *Heat and cold storage with PCM: an up to date introduction into basics and applications* (Berlin: Springer-Verlag Berlin Heidelberg).

[18] Ji H, Sellan D P, Pettes M T, Kong X, Ji J, Shi L and Ruoff R S 2014 Enhanced thermal conductivity of phase change materials with ultrathin-graphite foams for thermal energy storage. *Energ Environ Sci* **7** 1185-1192.

[19] Ushak S, Gutierrez A, Galleguillos H, Fernandez A G, Cabeza L F and Grágeda M 2015 Thermophysical characterization of a by-product from the non-metallic industry as inorganic PCM. *Sol Energ Mat Sol C* **132** 385-391.

[20] Mongibello L, Atrigna M, Bianco N, Di Somma M, Graditi G and Risi N 2017 Experimental test of a hot water storage system including a macro-encapsulated phase change material (PCM). *J Phys Conf Ser* **796** 012030.

[21] Jegadheeswaran S and Pohekar S D 2009 Performance enhancement in latent heat thermal storage system: a review. *Renew Sust Energ Rev* **13** 2225-2244.

[22] Schuetz M A and Glicksman L R 1984 A basic study of heat transfer through foam insulation. *J Cell Plas* **20** 114-121.

[23] Kamiuto K 2008 Modeling of composite heat transfer in open-cellular porous materials at high temperatures *Cellular and Porous Materials: Thermal Properties, Simulation and Prediction*, ed A Öchsner, G E Murch and M J S de Lemos (Weinheim: Wiley-VCH) chapter 6 pp. 165-198

[24] Boomsma K and Poulikakos D 2001 On the effective thermal conductivity of a three-dimensionally structured fluid-saturated metal foam. *Int J Heat Mass Trans* **44** 827-836.

[25] Wang F and Li X 2017 The stagnant thermal conductivity of porous media predicted by the random walk theory. *Int J Heat Mass Trans* **107**5 20-533.

[26] Nield D A and Bejan A 2006 *Convection in porous media* (New York: Springer).

Energy saving opportunities in the refrigerated transport sector through Phase Change Materials (PCMs) application

P Principi[1], R Fioretti[1] and B Copertaro[1]

[1]Università Politecnica delle Marche, Department of Industrial Engineering and Mathematical Sciences, Ancona, Italy

E-mail: b.copertaro@univpm.it

Abstract. Transportation of food products at controlled temperature is a critical task in the transport sector. In fact, whilst there is a need of ensuring both food quality and safety to the global population, its impact in terms of energy consumption and related CO_2 emissions into the atmosphere is becoming increasingly evident. In this regard, Thermal Energy Storage (TES) using Phase Change Materials (PCMs) can be considered as a potential way of reducing the cooling load, energy consumption and related greenhouse gas emissions in refrigerated transport sector. In this paper two different PCM applications are investigated. Specifically, in the first study a PCM (35 °C melting temperature) layer was added to the external side of a refrigerated enclosure wall with the aim of managing the cooling peak (shifting and reducing) and reducing the daily energy rate. Outdoor experimental results showed that the added PCM layer helps to reduce (between 5.55% and 8.57%) and delay (between 4.30 h and 3.30 h) the peak load of incoming heat compared to the reference one. In the second study, the energy performance of a refrigerated chamber with an air heat exchanger containing PCM (5°C melting temperature) was investigated. The study purpose was to reduce the cooling energy consumption during steady state operating conditions and the rate of temperature increase throughout the course of a power failure event. Test results showed that using a PCM air heat exchanger addition, up to 16% of energy can be saved.

Keywords: phase change materials; PCM; refrigerated transport, energy conservation; air heat exchanger

1. Introduction

Globalization of the food system has led to an increase of volume and variety of refrigerated food traded globally. At the same time, during the carriage of perishable foodstuffs it is important to guarantee a constant temperature throughout the entire journey, which determines a high energy consumption and related CO_2 emissions into the atmosphere [1]. It is well known that maintenance of the refrigerated compartment below the ambient air temperature represents a driving force for heat flow towards the refrigerated compartment from the external environment [2]. In this regard the total heat gain, which generally affects a refrigerated system, is determined by the surrounding environment (heat conducted through the insulated envelope), door openings (infiltration load), and product loading (heat that has to be removed from stored products) process. The latter in conjunction with a low efficiency-refrigerating unit, results in an increased energy consumption. Therefore, a solution to this problem could be the use of Thermal Energy Storage (TES) with the addition of Phase Change

Materials (PCMs) [3]. PCM is a substance characterized by a high heat of fusion, capable of storing and releasing large amounts of energy, thereby producing the cooling effect necessary to maintain a constant temperature throughout the journey. As shown in the next section (2. Study background), several authors have investigated various promising applications based on the use of PCM in the refrigeration sector. The main purpose of the paper is to describe two different PCM applications over refrigerated transport systems. The first experimental investigation focused on the addition of PCM to the external compartment walls of a refrigerated container stacked in a container terminal. The aim of this technology is the decrease and phase displacement of the incoming heat load with a consequent reduction of the cooling energy required to maintain the internal temperature at a constant level. The second experimental investigation focused on PCM air heat exchanger addition near the evaporator of a refrigerated enclosure.

2. Study background

Many technical options based on PCM have been developed in order to improve the energy performance of refrigerated transport systems. For example Tassou et al. [4] reported a road transport refrigeration technology consisting of hollow tubes, filled with a eutectic solution, positioned inside the refrigerated compartment. Specifically the operating principle of this equipment is based on the use of latent heat storage capacity of the phase change material. In fact all the cold energy stored during the night by power supply, can be progressively released during the truck journey, thus providing the cooling effect necessary for a specific duration of time. Moreover this technology can be useful in contrasting the infiltration heat load due to door openings of refrigerated vans. Another technology consisting of a PCM application in a typical refrigerated truck compartment was numerically analyzed by Simard and Lacroix [5]. Specifically the latent heat cold storage unit involves parallel plates, filled with an aqueous-glycol mixture, with the purpose of absorbing the incoming heat load and keeping the internal temperature below -8.15 °C for almost 8 consecutive hours. The authors explored all the operational and design parameters that could have a negative effect on the performance of the whole system. Specifically, it was found that a too wide or too narrow plate spacing can result in a reduced heat transfer coefficient and in a complete obstruction of the air flow passage respectively. Still regarding the refrigerated transport sector, a novel refrigeration system for trucks was proposed by Liu et al. [6]. This technology is based on a phase change thermal storage unit (PCTSU), which can be charged with cold energy when the truck is parked and discharged during working hours. Moreover, the authors developed a new and economical PCM with a phase change temperature compatible with the study object. The obtained results demonstrated that the energy cost was 86.4 % lower when the PCTSU was applied to the refrigerated transport system. Another aspect to be taken into account is the capability of PCM in improving the thermal inertia of refrigerated compartment walls. In this way a reduction of cooling energy required to break down the heat load deriving from the external environment can be achieved. This is what the study by Ahmed et al. [7] is based on. Substantially, the authors modified the conventional method of insulation of refrigerated wall enclosure by using PCMs. By comparing two test truck trailer simulators, it was found that the inclusion of copper pipes containing paraffin, allowed an average daily heat flow reduction inside the refrigerated compartment equal to 16.3%. Moreover, for individual walls the peak heat transfer rate reduction was found to range between 11.3% - 43.8%. Glouannec et al. [8] numerically and experimentally analyzed three different multilayer insulation walls of a refrigerated van. They aimed to study the effect of the external environment such as solar irradiation and temperature over energy consumption and thermal behavior. It was found that the wall made with reflective multi-foil insulation and aero gel decreased the external surface temperature by about 9 °C, highlighting its high insulating capacity compared to the reference. On the other hand, considering the wall made with microencapsulated paraffin, thanks to PCM thermal inertia the energy consumption decreased by about 25 % when compared to the standard insulation wall. Moreover a significant decrease in the external surface temperature due to the melting of paraffin was observed. Finally, Tinti et al. [9] using experimental approach, evaluated the incorporation of microencapsulated phase change material into standard polyurethane foam walls designed to improve the thermal performance of refrigerated vehicles. This technology resulted more useful in contrasting all the thermal loads which could occur during the journey of the vehicle such as refrigeration system blackout, compartment doors opening and varying solar irradiation. As proposed in the present study, a PCM air heat exchanger addition near the evaporator of a refrigerated enclosure can lead to an improvement of energy efficiency. In fact in accordance with Chatzidakisn and Chatzidakis [10] the low efficiency of a cooling unit can be determined by higher condenser temperature (more than 32 °C) which generally characterizes the Mediterranean climatic context. This is even truer for refrigerated enclosures which are on the ground or in a ship. Therefore, the aim of this study was the reduction of cooling energy required to run a low efficiency refrigeration unit operating during the summer season.

3. Proposed technologies in refrigerated transport sector

The reduction of energy consumption and environmental impact of refrigerated transport systems can be achieved by the increase of insulation enclosure performance and energy efficiency of the refrigeration units. In the following paragraphs the development of two technologies easily implementable in refrigerated transport systems was investigated.

3.1. PCM thermal shield

3.1.1. Thermal behaviour. The most commonly used insulation material for refrigerated transport systems is polyurethane foam. In fact, its use allows a suitable thermal resistance against the external heat load due to a reduced global heat transfer coefficient. However, an improvement of lightweight wall energy performance can be achieved by increasing its thermal inertia. This goal is currently being realized by adding PCMs. In fact, thanks to their high heat storage capacity, PCMs are able to reduce and delay the heat load determined by solar irradiation and temperature difference between the external and internal environment. Specifically, the term "heat load delay" is intended as the time shift at which the highest external surface temperature and the internal surface temperature values occur. A heat load delay of about eight hours allows the refrigeration unit to work at a higher performance (due to lower environmental temperature) and reduced energy costs. At the same time, a reduction of the peak heat load can be translated into a lower cooling energy amount required to maintain a constant temperature inside the refrigerated compartment.

3.1.2. Technological development. When selecting a PCM there are three main requirements to be considered: a suitable phase change temperature, a large latent heat of fusion and a proper amount, Orò et al. [11]. Specifically the selection of a PCM that properly matches all the above requirements, represents a delicate phase in the refrigerated transport sector. In fact when considering a PCM applied to the external refrigerated compartment walls, it should be compatible with the varying environmental conditions during the journey. In fact a too high or too low phase change temperature could determine a partial exploitation of the PCM's latent heat storage capacity. Strictly linked to the phase change temperature, the latent heat of fusion is another important factor to be considered when selecting a PCM. This latter expresses the amount of energy that a PCM can store or release per unit mass during its phase change process. Therefore the selection of a PCM with a large latent heat of fusion can be translated in a reduced material quantity. This aspect is crucial in the refrigerated transport system. In fact, the optimization of PCM thickness that has to be applied to insulated walls, results in a limited weight to be added at the transportation system. Therefore considering all these advantageous characteristics and in accordance with a previous study [12] the three centimetres thick paraffin wax RT35HC (www.rubitherm.de) was selected Table 1. Moreover, due to the PCM's solid-liquid transition it was encapsulated in polyethylene containers (with dimensions of 0.6 x 0.6 x 0.03) internally divided into 81 small square tanks. The latter, filled by paraffin in liquid phase, were then hermetically closed (three bars of pressure, 170°C) using a thin multilayer film. This packaging system configuration allows a good thermal contact between all the stratigraphic components and a flexibility necessary to contain the PCM volume expansion during its melting phase. Finally the PCM layer was then added to all the external surfaces of a cold room using a wood frame and closed by a metal sheet layer (Figure 1).

3.2. PCM air heat exchanger

3.2.1. Thermal behaviour. The reduction of temperature fluctuations and the enhancement of energy performance of the refrigerated transport system, led to the development of technologies involving the use of PCM inside a refrigerated chamber [13]. In this study a PCM air heat exchanger was applied close to the evaporator of a cold room. Due to the high latent heat, the integration of PCM near the evaporator could determine a prolonged ON-OFF compressor time and could also act as a cooling unit during power failure. In fact, as shown in Figure 2, during the compressor OFF time period, due to the PCM melting phase, the heat load transmitted through the insulated envelope is absorbed by the PCM air heat exchanger, determining a reduction of the rate of compartment air temperature increase. Conversely, during the compressor ON time period, due to the PCM freezing phase, heat is released to the evaporator outgoing air determining a reduction of the rate of compartment air temperature. Therefore as a result, an extension of the ON-OFF compressor time and a reduction of the ON-OFF cycles can be achieved over time. This means that the refrigerating unit energy efficiency can be increased together with a reduction of electric power consumption.

Table 1. Thermophysical properties of PCM used for "*thermal shield*" proposal.

	Value
PCM commercial name	Rubitherm RT35HC
Phase change temperature	35 [°C]
Heat storage capacity	220000 [J/kg]
Specific heat capacity	2000 [J/kg K]
Heat conductivity	0.2 [W/m K]
Volume expansion	12 %

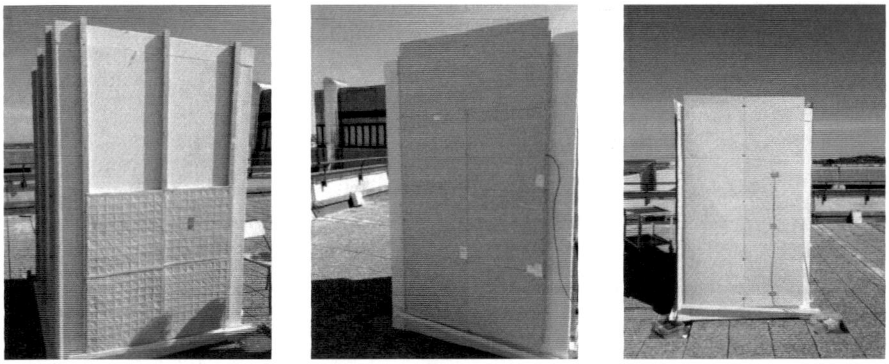

Figure 1. PCM layer addition procedure.

3.2.2. Technological development. The technology developed in this work is based on an air heat exchanger made up of six PCM container units assembled in pairs and located in a polystyrene channel near the evaporator of a refrigerated enclosure Figure 2. Each PCM container unit is characterized by an anodized aluminium and finned surface with internal dimensions of 1.00 m x 0.03 m x 0.14 m (length x height x width), equal to a 4.2•10-3 m^3 in volume. The exchange surface area is equal to 1.44 m^2 per meter of length. Moreover in order to obtain a better heat exchange between the evaporator outgoing air and the PCM air heat exchanger a fan system (electric power of 20 W) was added on the outlet section of the channel. In this study the paraffin wax RT5HC was selected (Table 2). In fact, in accordance with Joybari et al. [14] the selection of PCM with a phase change temperature close to the operating temperature (5 °C), was found to be the most effective choice in terms of improving the energy performance of a refrigerated system and maintaining the compartment air temperature at the desired thermal conditions. After selecting a proper phase change temperature, the next phase consists in choosing the correct amount of PCM. It is well known that thinner of thicker PCM can result in reduced thermal effectiveness. Therefore in this study 19.15 kg of RT5HC were used. This amount of PCM could be useful in contrasting all the heat load transmitted through the insulated envelope during 12 hours of a power failure event. Even if not all the PCM thickness underwent a phase change during steady state operating conditions, an extension of the compressor ON-OFF time and a reduction of start/stop compressor cycle over time were achieved.

Table 2. Thermophysical properties of PCM used for "*air heat exchanger*" proposal.

	Value
PCM commercial name	Rubitherm RT5HC
Phase change temperature	5 [°C]
Heat storage capacity	240000 [J/kg]
Specific heat capacity	2000 [J/kg K]
Heat conductivity	0.2 [W/m K]
Volume expansion	13 %

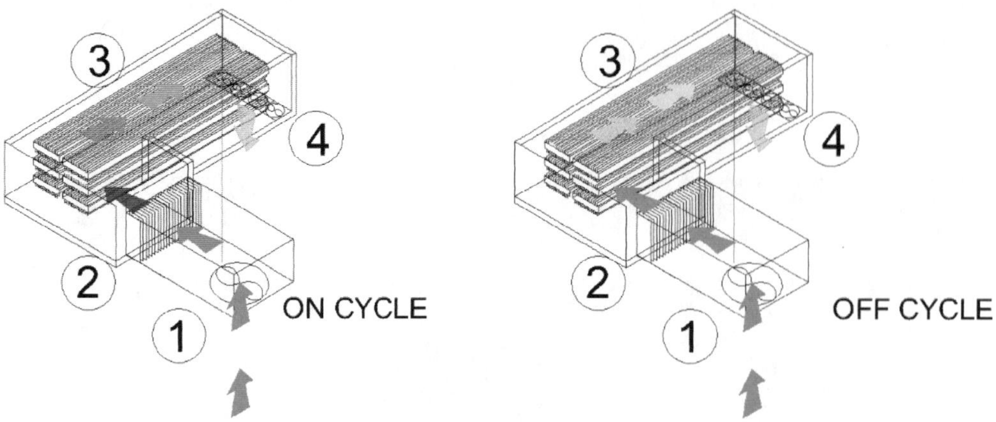

Figure 2. PCM air heat exchanger operating principle: 1) Fan 2) Evaporator 3) PCM air heat exchanger 4) Additional fan system.

4. Measurement and test conditions

The experimental set-up consists of a walk-in temperature controlled mini cold room equipped with a refrigeration unit (1140 W of cooling capacity). The maximum and the minimum operating temperature values of the cold room are equal to 8 °C and 0 °C respectively. The cold room walls are made of galvanized metal sheet, insulated by polyurethane foam which determines a total panel thickness of 10 cm.

4.1. Cold room with PCM thermal shield

Both the novel and the reference cold rooms were tested during a typical Italian summer (August-September 2014 Ancona, Italy) in order to verify the thermal improvement deriving from PCM addition to the refrigerated enclosure envelope. A monitoring system was developed and the real time data were acquired by using Datataker DT500, set to one reading every 5 min. In Figure 3 the location of the Thermal Resistance Detectors (RTDs) "PT100" with a four-wire connection through the cross section of the new and reference composite walls are shown (T_n). Moreover, a fluxmeter (Hukseflux HP01) (Q_1) was used to measure the heat flux passing through the south oriented wall which represented the most affected wall by solar irradiation. Finally, in order to measure the external climatic conditions a meteorological station was employed.

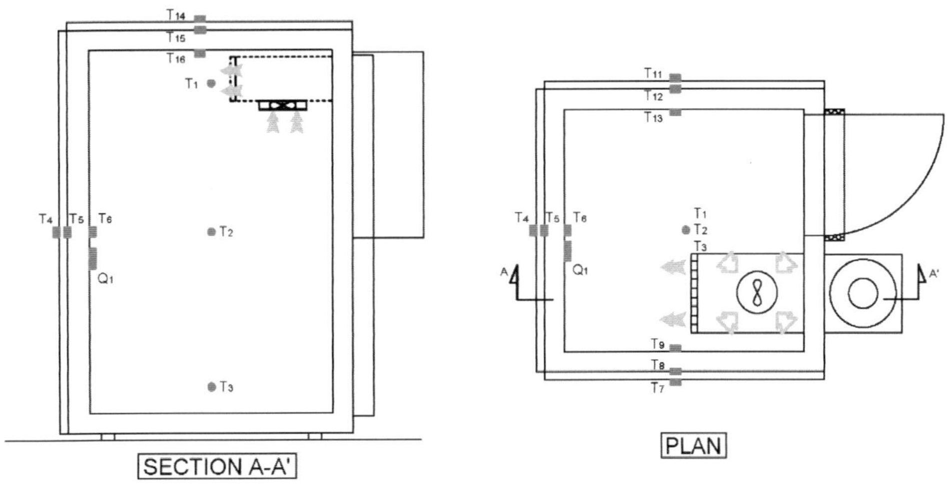

Figure 3. Monitoring system developed for PCM thermal shield experimental tests. (Tn = Sensors of temperature; Q1 = Heat flux meter sensor).

4.2. Cold room with PCM air heat exchanger

In this experiment both the novel and the reference cold room were tested in a laboratory room kept at 32 °C. The aim was to study the thermal improvement deriving from this technology considering the refrigeration unit operating at steady state conditions under high environmental temperature. Therefore, a monitoring system was developed in order to measure the thermal parameters characterizing the PCM air heat exchanger behaviour and its influence over the internal conditions of the refrigerated compartment. Power consumption was tested (wattmeter Christ Elektronik CAC 140). Also in this case data were acquired using Datataker DT500 linked to a computer for monitoring experimental real time data. In Figure 4 the location of the Thermal Resistance Detectors (RTDs) "PT100 with a four-wire connection over the novel and the reference cold room is shown (T_n).

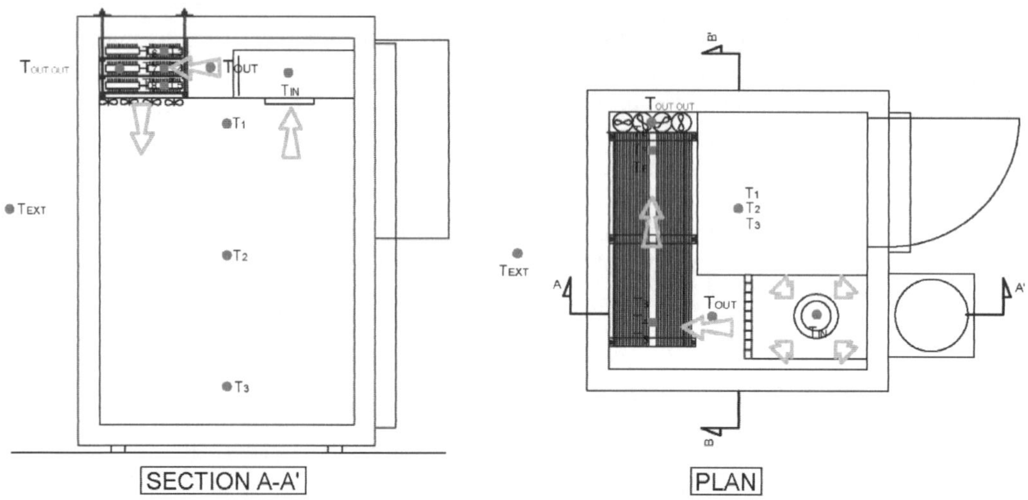

Figure 4. Monitoring system developed for PCM air heat exchanger experimental tests (T_n = Sensors of temperature)

5. Results and discussion

5.1. PCM thermal shield

In this section the main monitoring data of the reference and the novel cold room, referred to one test day, are reported and discussed. Meteorological parameters such as solar irradiation and external environmental temperature affecting the refrigerated enclosure, are also analysed. In Figure 5 it is possible to appreciate that the highest values of solar irradiation (almost 800 W/m^2) and outdoor temperature (almost 28 °C) occurring during the daily midday hours, determine a high thermal heat load and a consequent reduction of refrigeration unit energy efficiency. This means that the maintenance of a constant temperature inside the refrigerated enclosure leads to a high energy consumption as well. Figure 7 and Figure 6 show the comparison of measured heat flux and temperature through the cross section of the novel and reference south oriented wall respectively. Specifically in Figure 6 the time shift of almost five hours is evident, which occurs between the highest value of the external surface ($R_{4\text{-}PCM}$) and the PCM surface ($R_{5\text{-}PCM}$) temperature. This thermal behaviour is due to the increase of the compartment wall thermal inertia, determined by PCM addition. In fact thanks to the PCM high heat storage capacity it is able to store all the incoming heat load at a relatively constant temperature (during its melting phase), determining a reduction and phase displacement of the incoming heat load. Moreover, by comparing the external surface temperature of the reference cold room ($R_{4\text{-}REF}$) with the PCM surface temperature ($R_{5\text{-}PCM}$), it is evident that the PCM heat load storage process leads to a reduction in the maximum temperature value (33 °C instead of 44 °C). Finally Figure 7 shows a reduction and a phase displacement of the heat flux coming into the refrigerated compartment. In fact, as can be observed, the heat transfer rate of the reference cold room peaked at 4:30 PM while the heat flux in the novel cold room peaked at 9:00 PM. This means that a peak delay of almost 4.30 h was achieved. Moreover, the thermal inertia compartment wall improvement determined a thermal heat flux peak reduction equal to 5.55%. In fact, PCM acting as an energy storage medium can absorb and displace the incoming heat flux during its melting phase. This means that the reduced heat load stored during the day reaches the internal refrigerated environment during the night, which is characterised by high refrigeration system performance (due to lower environmental temperature) and reduced energy costs. In this way a reduction in energy or fuel consumption by the refrigeration unit, especially in the daytime, characterised by high external temperatures, can be achieved.

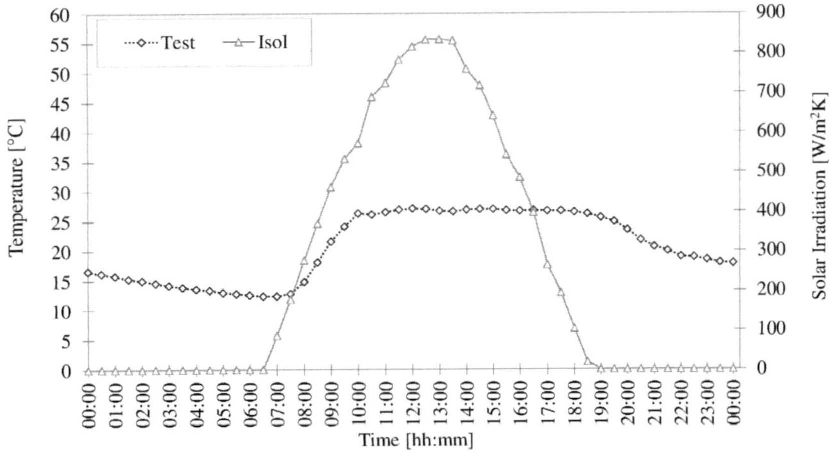

Figure 5. Meteorological parameters such as solar irradiation and external environmental temperature detected during the experimental campaign.

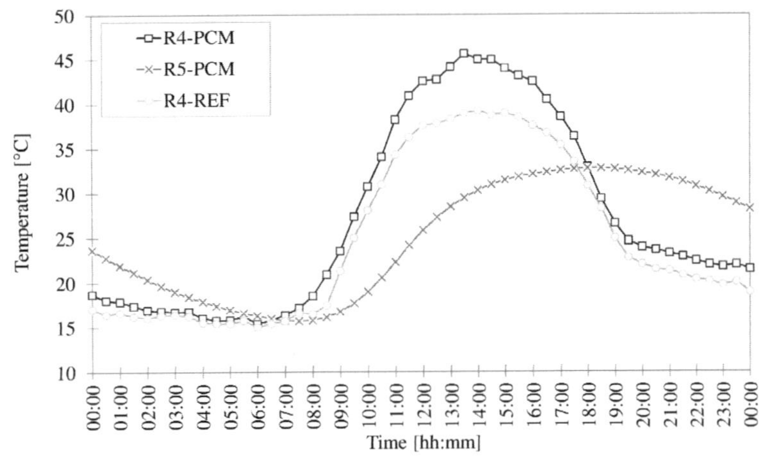

Figure 6. Measured temperature trend through the cross section of the novel and reference cold room south oriented wall.

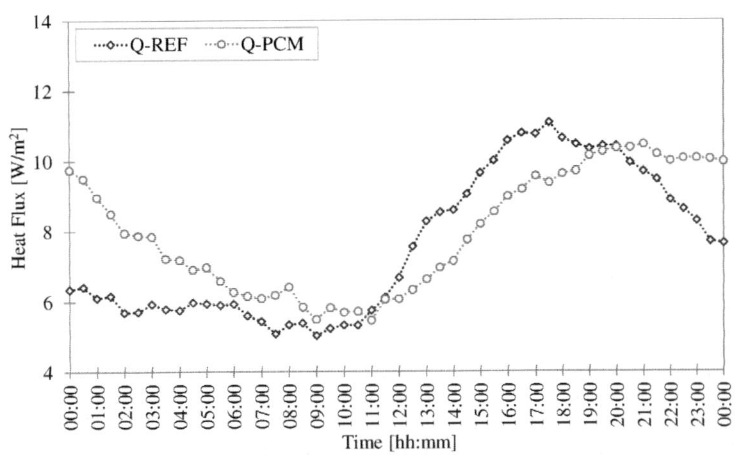

Figure 7. Comparison of thermal flux through the south-oriented wall of novel and reference cold room.

5.2. PCM air heat exchanger

The thermal behaviour of both the novel and reference cold room under steady state operating conditions was studied during a 24 h test. To make the results clear and readable 2h test results are reported and discussed. In Figure 8 the reference and novel cold room compartment air temperature are shown. Moreover both the reference evaporator ($T_{OUT-REF}$) and PCM air heat exchanger ($T_{OUT-OUT-PCM}$) outgoing air temperature are reported. For the ordinary cold room the average compartment air temperature was equal to 3.68 °C. For the novel cold room it was equal to 5 °C. As can be seen from the graph, the number of fluctuations over time was significantly reduced in the PCM added system. These results suggest that the PCM air heat exchanger application, close to the evaporator of refrigeration unit, allowed the increase and stabilization of air temperature leaving the evaporator. In fact, during the compressor OFF time period the heat gain, transmitted through the insulated walls, was absorbed by PCM (thanks to its high storage capacity) avoiding a rapid rise in the compartment air temperature. Moreover, during the compressor ON time period the evaporator outgoing air passing through the PCM air heat exchanger, was heated as a result of PCM freezing process. Consequently, the air temperature leaving the heat exchanger ($T_{OUT-OUT-PCM}$) increased. This means that the lowest value of the Tout out PCM was 2.9 °C higher than the lowest value of the reference evaporator outgoing air temperature ($T_{OUT-REF}$). In this way, the destructive effect of evaporator freezing temperature over fresh products could be avoided [13]. Figure 9 shows the comparison of the compressor power between the reference and the novel cold room. Moreover PCM air heat exchanger surface temperature was reported. From the graph it is immediately evident that thanks to the PCM solid/liquid transition, an increase of the compressor ON/OFF cycle time and a reduction of compressor START/STOP over time was achieved. In fact, during the OFF period, the warmed air passing through the air heat exchanger, released heat at PCM, which changed its state (melted). In this way, thanks to the PCM melting process the compressor STOP period increased by about 7.75 min compared to the reference. In the same way during the ON period, the evaporator outlet air passing through the PCM air heat exchanger absorbed heat from the PCM, which changed its state (solidified). Thanks to the PCM freezing process the compressor START period increased by about 3.17 min compared to the reference. This means that the number of the ON/OFF compressor cycles over time was also reduced (six cycles instead of thirteen). Overall the total electric power consumption of the novel cold room, during 2 h of stable operating conditions, was reduced by about 16% compared to the reference.

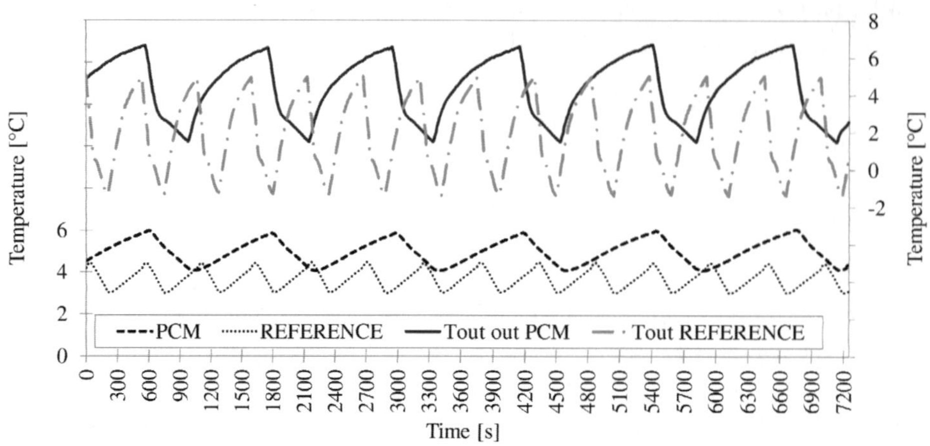

Figure 8. Compartment air temperature with/without PCM, evaporator outgoing air temperature without PCM ($T_{OUT-OUT-PCM}$) and temperature of air leaving the air heat exchanger ($T_{OUT-REF}$).

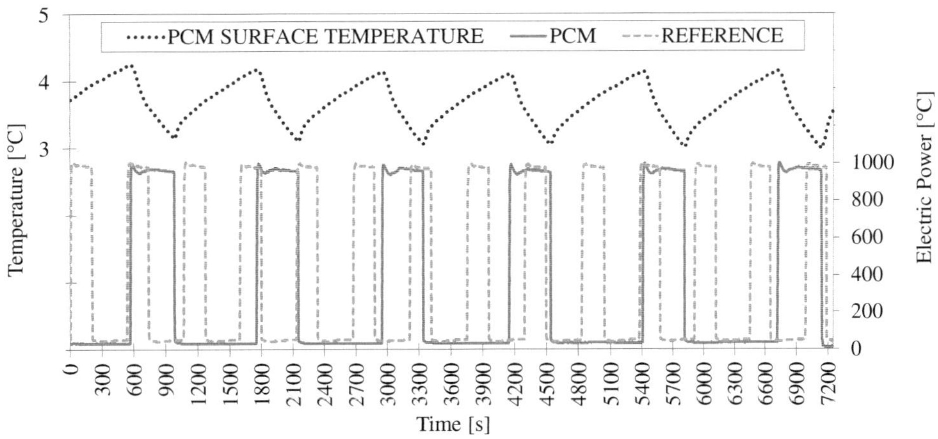

Figure 9. PCM air heat exchanger surface temperature and electrical power consumption with/without PCM.

6. Conclusion

Nowadays, refrigerated transport plays an essential role in the food industry, preserving the quality of foodstuffs and satisfying an increasingly wide range of consumer needs. However, it is well known that the maintenance of a proper temperature along the entire journey determines a high-energy consumption and related greenhouse gas (GHGs) emissions into the atmosphere. Even small improvements in energy performance of refrigerated transport systems can lead to an interesting reduction in overall energy impact on an already high-tech sector. Considering the main heat loads affecting a refrigerated compartment, the application of PCM on two different thermal energy storage systems were assessed. The first analysis was focused on the reduction of cooling energy required to break down the heat load deriving from the external environment (e.g. heat conducted through the insulated envelope). In this context, a PCM layer was added to the external side of a refrigerated container envelope. Afterwards, an experimental campaign was aimed at reducing cooling energy required to run a low efficiency-refrigerating unit. In this way, an air heat exchanger containing PCM was positioned close to the evaporator of a cold room. Results underline that improvement of the envelope thermal inertia, after adding the PCM layer, led to a consistent reduction (5.55%) and phase displacement (between 4.30) of the daily heat load. This means that the refrigeration unit operated with high efficiency due to a reduced temperature gradient between the external and the internal environment therefore requiring a lower amount of energy. Results obtained by positioning a PCM air heat exchanger close to the evaporator highlighted a lower number of compressor ON-OFF cycles (6 cycles instead of 13 cycles in 2 hours) and in an increase of their length. In this way, the refrigerating unit energy efficiency increased and the electric power consumption decreased (about 16%). Finally, it is also worth noting that the technologies presented here can be easily applied to other compartments involved in refrigerated transport systems and should not be considered alternative to each other. In fact, considering a possible research development, it would be interesting to study a refrigerated compartment (travelling by rail, ship or road) which is fully integrated with all the described technologies. The development of energetically "independent" refrigerated compartments able to meet the required cooling energy in different environmental conditions should be strongly considered.

References

[1] Oro´ E, Miro´ L, Farid MM and Cabeza LF 2012 *Int. J. Refrig.* **35** 1709–1714.
[2] Çengel YA and Ghajar A 2010 *Heat and Mass Transfer: fadamental & Applications, fourth ed., Refrigeration and Freezing of Foods* (2 Penn Plaza, New York NY 10121: McGraw-Hill Education) Chapter 17 pp 1–59.

[3] Oro´ E, Miro´ L, Farid MM, Martin V and Cabeza LF 2014 *Int. J. Refrig.* **42** 26–35.
[4] Tassou SA, De-Lille G and Ge YT 2009 *Appl. Therm. Eng.* **29** 1467-1477.
[5] Simard AP and Lacroix M 2003 *Energ. Convers. Manage.* **44** 1605-1624.
[6] Liu M, Saman W and Bruno F 2012 *Appl. Energy.* **92** 336-342.
[7] Ahmed M, Meade O and Medina MA 2010 *Energ. Convers. Manage.* **51** 383–392.
[8] Glouannec P, Michel B, Delamarre, G and Grohens Y 2014 *Appl. Therm. Eng.* **73** 196–204.
[9] Tinti A, Tarzia A, Passaro A and Angiuli R 2014 *Appl. Therm. Eng.* **70** 201–210.
[10] Chatzidakisn SK and Chatzidakis KS 2004 *Int. J. Energy Res.* **28** 887–897.
[11] Oró E, Barreneche C, Farid MM and Cabeza LF 2013 *Renew. Energ.* **57** 130–136.
[12] Copertaro B, Fioretti R, Sangelantoni L and Principi P 2015 *Proc. Int. Conf. on environment and electrical engineering (Rome)* pp 378–383
[13] Marques AC, Davies GF, Maidment GG, Evans JA and Wood ID 2014 *App. Therm. Eng.* **63** 511–519.
[14] Joybari MM, Haghighata F, Moffat J and Sra P, 2015 *Energy and Build.* **106** 111–124.

Recent trends in the development of heat exchangers for geothermal systems

A Franco and M Vaccaro

Department of Energy, Systems, Territory and Constructions Engineering (DESTEC), University of Pisa, Largo Lucio Lazzarino, 56126 PISA (ITALY).

E-mail: alessandro.franco@unipi.it

Abstract. The potential use of geothermal resources has been a remarkable driver for market players and companies operating in the field of geothermal energy conversion. For this reason, medium to low temperature geothermal resources have been the object of recent rise in consideration, with strong reference to the perspectives of development of Organic Rankine Cycle (ORC) technology.
The main components of geothermal plants based on ORC cycle are surely the heat exchangers. A lot of different heat exchangers are required for the operation of ORC plants. Among those it is surely of major importance the Recovery Heat Exchanger (RHE, typically an evaporator), in which the operating fluid is evaporated. Also the Recuperator, in regenerative Organic Rankine Cycle, is of major interest in technology. Another important application of the heat exchangers is connected to the condensation, according to the possibility of liquid or air cooling media availability. The paper analyzes the importance of heat exchangers sizing and the connection with the operation of ORC power plants putting in evidence the real element of innovation: the consideration of the heat exchangers as central element for the optimum design of ORC systems.

1. Introduction

ORC plants have been identified since the beginning of the 1990s as the basic reference technology for increasing the use of medium to low temperature geothermal resources for electricity production purposes. In fact, as it has been largely addressed in the scientific literature in the last twenty years, the medium to low temperature reservoirs, with temperature in the range between 70 °C and 150 °C, are the most diffused all around the world [1, 2].

The potential use of geothermal resources has been a remarkable driver for market players and industry operating in the field of geothermal energy in the last ten years. For this reason, medium to low temperature geothermal resources have been the object of recent rise in consideration, driven by the growth of the geothermal market around the world due to subsidies to ORC applications by the national and local Institutions, mainly in the European Union, together with the development of the geothermal industrial branch in emerging countries of Central America, Africa and Asia.

In recent years, a significant attention to the ORC development and utilization has been addressed by the scientific community and academic researchers and by the companies involved in geothermal industry. Both theoretical and practical solutions based on the optimum matching among geothermal resource characteristics, ORC working fluid, and thermodynamic cycle arrangement have been largely proposed in the last 15 years [3-5].

Notwithstanding the diffused debate, the industrial applications of ORC plants in geothermal energy field seems to be limited only to some specific technologies, and in specific cases the economic results after some years of exercise are not always positive or anyway a they need to be deeply analyzed.

The performance of ORC systems is strongly influenced by heat exchangers design, strongly dependant on external thermophysical parameters (reservoir potential, ambient temperature variation). In practical applications, the space limitation in which the heat exchanger will be installed is also an important constraint. Therefore, the heat exchanger size must also be more compact to satisfy this condition by an enhancement in its thermal design.

Since there are competing effects of increased heat transfer with increased hot-side and cold-side frictional pressure drop, this makes the optimization design of heat exchanger a complex and difficult task. Two efforts are then significant: the evaluation of the optimal trade-off between fluid friction and heat transfer irreversibility and the analysis of augmentation techniques to decrease frontal area at a specified thermal capacity, without causing a damaging increase in pumping power.

It is found that the capital cost of heat exchanger accounts for a large proportion of ORC, and even reaches about 80% in case of air cooled condenser [6]. Therefore, the selection of heat exchanger in ORC is extremely important.

Considering this, the objective of the present paper is to define a kind of balance about the recent diffusion of ORC power plants and the perspectives of future development of geothermal ORCs, analyzing the different perspectives coming from industrial and academic research, mainly in connection with modular small power solutions (often in the range between 200 kW and 10 MW).

When power is recovered from low temperature heat sources, due to the quite low value of the First Law efficiency, large heat exchanger areas and circulation mass flow rates (geofluid, working fluid and cooling media) are necessary. This impacts in particular on heat transfer devices such as condenser, preheater and evaporator, and the cost of those largely contributes to the total power plant cost increase. The importance of the correct sizing of the heat exchangers is well recognized.

This is mainly true for the condenser surface in case of dry type cooling towers. They usually consume much electricity than wet cooling towers, regarding auxiliary power consumption, leading to a reduced power output of the plant. Moreover, due to the need of higher heat exchange surface and of a large volume of air that has to be moved through them, the condenser section can cost about 5-10 times as much as a wet type one, depending on the condensing temperature of the turbine, in particular when low condensing temperatures are considered (as minimum approach against environmental temperature decreases). So a suitable selection of operating conditions (condenser temperature and ambient temperature) of the plant and an optimum thermal design of the heat transfer surfaces is necessary in order to solve the relative optimum design problem. Scaling and solid deposition have a strong impact on the heat exchangers operation too.

Considering the previous elements, the aim of this paper is to analyze the status of the development of heat exchangers with special attention to the development of ORC for geothermal energy use, analysing the negative influence determined on the operation of the plant of an incorrect sizing of the heat exchangers. Also a focus on the design methods is reported including the idea that heat exchangers are not only components of the systems but important elements for an optimum design.

2. ORC Power Plants: state of the art

The binary power plants have a small environmental impact due to the "confinement" of the geofluid respect to atmosphere and soil. In a binary cycle power plant the heat of the geothermal water is transferred to a secondary working fluid, usually an organic fluid that has a low boiling point and high vapor pressure when compared to water at a given temperature. The cooled geothermal water is then returned to the ground by the re-injection well to recharge the reservoir [2].

As clearly evidenced in Fig. 1, the plant comprises two separate sections: the geothermal fluid loop and the power cycle. Geothermal fluid loop and power cycle are completely separated. The connection point is represented by recovery heat exchanger (RHE).

In this way "Nearly zero emission" plants (mainly for all-liquid geofluid) can be conceived. Moreover, this kind of plant is suitable for integration with other energy sources (solar, biomass, or other waste heat recovery).

Binary technology allows the use of medium to low temperature water dominant reservoirs and makes geothermal power production feasible even in countries where high enthalpy resources at shallow depth are not available. Moreover, geothermal resources at temperatures below 150 °C can be available resources at lower depths than high enthalpy geothermal fields. The availability of the resource at a quite low deep (e.g. 1000-1500 m below the ground level instead of 3000-4000 m) permits interesting reduction of the costs connected to exploration and the drilling phase.

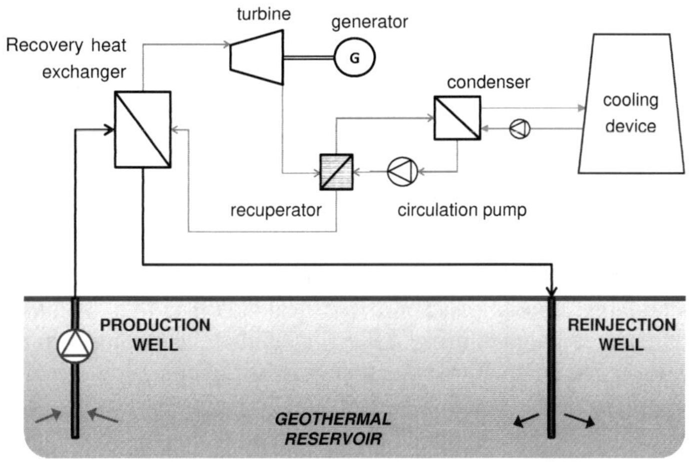

Figure 1. ORC power plant, typical configuration

The geothermal binary plants currently in operation can be divided according to different classifications: the first and more meaningful is surely the dichotomy between "stand-alone" or "bottoming cycles". Another difference is related to the power output. Binary plants can also be classified basing on the cooling system used: plants with a wet cooling system (the working fluid is condensed by cooling water) and plants with a dry cooling system (heat is rejected directly to air). Relevant differences can be evidenced in the operating fluid and in the thermodynamic variables. [7]

The worldwide installed geothermal power plants reach approximately 13 GW and ORC plants (or steam/ORC combined cycles) are about 1800 MW, namely 280 power plants about 14% of the total. The diffusion of such kind of plants has increased in a meaningful way in the last ten years, as clearly discussed in the next section. It is possible to estimate that about 1050 MW have been installed in the last ten years, mainly under the impulse of some manufacturers, [8].

Mainly in the last ten years the perspective of installation of new binary cycle power plants has driven in a significant way the geothermal energy branch. More than 1050 MW of binary plants, corresponding to about 78 new power plants have been installed between 2005 and 2015 increasing the installed capacity from 738.9 MW up to 1790 MW (Table 1).

Table 1. Binary cycle power plants installed.

Period	Plants installed (n)	Cumulative power (MW)
Before 2005	~200	738.9
2005-2009	27	465.8
2010-2014	51	585.3
Total	~280	1790.0

Even if geothermal power plants have the characteristics of operating for a lot of time (up to 8500 hours in a year), the total amount of installed power generation considering geothermal energy is well below the level of the other renewables. Considering the same period, a less remarkable installed capacity of flash-steam plants has been observed and few hundreds additional MW of dry steam power plants have been installed.

In the industrial practice, the power production capability is impacted by practical issues and by constraints individuated by Institutions, customers, and all of these are case specific. These constraints are apparently contrasting with the necessity of standardization (e.g. size, equipment selection, environmental issues). In the technical field of ORC technology, the recent evolution of the research, slightly evidenced in this paper, mainly deal with a lot of topics, summarized in the following lines:

- Working fluid chemical studies and stability;
- Plants layout and compact configuration;
- Combined plants optimization;
- Thermal cycle definition;
- Supercritical cycles configuration;
- Machinery optimization;
- Exergetic analysis;
- Thermoeconomic and Exergonomic analysis;
- Optimization engineering;
- Dynamic analysis;
- Plant equipment design.

The interest of the researchers in the field of the binary cycle power plants and ORC technology is clearly identified by the large amount of scientific contributions in this field. About 1300 papers can be found in the most used research journal collector (www.sciencedirect.com), and about 200 are produced in the first months of 2017. While the papers dealing with geothermal heat exchangers, both for power conversion and direct use, are about 7400, and about 600 are dated 2017.

3. ORC Plant Technology: technical solutions, design problems and role of heat exchangers

One key point to understand ORCs development along history of geothermal branch is that since the beginning and for a quite long time, each installation has been designed for the conditions at a given location and the different systems have been tailored to specific geothermal fluid characteristics. Fig. 1 provides a typical scheme of the most common schematization of a ORC power systems in which the role of the different exchangers is well evident: the recovery heat exchanger (RHE), the Recuperator, in which the temperature of the fluid at the outlet from the turbine is used for preheating the same fluid before entering in the RHE, and the cooling system (condenser).

Mainly in the last years, some attempts have been made by manufacturers to design "standard machinery". Some manufacturers have proposed the use of standard sizes and systems, and this approach could be a key driving factor for a larger diffusion of small size geothermal plants. Unfortunately, it is very difficult to pursue this standardization in practical design purpose, for different reasons dealing with the nature of the geothermal resources and the environmental conditions (that have importance for the definition of the condensation temperature).

The temperature of the condenser in geothermal power plants may have different values depending on its location in the world. Indeed, the climate around a geothermal power plant dictates the outdoor temperature, which then impacts on the nominal temperature in the condenser. This value fixes the turbine discharge pressure, due to the environmental air design temperature, leading to cut the possibility of exploiting a wider useful enthalpy range in the expansion machinery.

3.1. Working fluids and thermodynamic optimization of the heat recovery cycle

The thermodynamic properties of the working fluids are considered as key parameters. The selection of suitable fluids for use in binary cycle plant is a quite complex problem and cannot be disjointed by the heat recovery cycle selected. In general, the working fluids deserve the following characteristics: low critical temperature and pressure, small specific volume, low viscosity and surface tension, quite high thermal conductivity, suitable thermal stability and environmental acceptability. A desirable thermodynamic property of the fluid is a near vertical saturated liquid and vapor line for the possibility of "carnotizing the cycle" but this is a quite complex question. The saturation vapor curve is the most crucial characteristic of a working fluid in an ORC. This characteristic affects the fluid applicability, cycle efficiency, and arrangement of associated equipment in a power-generation system.

Working fluid selection for geothermal ORCs has been largely addressed in scientific literature, also by one of the authors in [5] and it deals with thermodynamic issues respect to more practical heat transfer issues and feasibility and industrial questions.

There are two types of vapor saturation curves in the temperature-entropy (T-s) diagram: "dry fluids" with positive slopes (dT/ds), "wet fluids" with negative slopes. In the second case, since the vapor expands along a quite vertical line on the T-S diagram, vapor saturated at the turbine inlet will remain vapor phase throughout the turbine exhaust duct, then it seems not particularly convenient to resort to Rankine cycle with superheater. In some cases the superheating of "dry fluids" (with positive slope of the vapor saturation curve) is done, but this to increase the specific enthalpy content of the expanding stream and to make sure that the vapor is dry also in transient conditions (e.g. start-up).

The minimization of the exergy losses must be performed to perform the optimization of a certain ORC arrangement. It is mainly related to the recovery heat exchanger (RHE), this can be obtained theoretically by increasing the number of pressure levels (PL) of the thermodynamic cycle, in order to keep close the temperature profiles of the two exchanging fluids. A complete analysis of the system involves the evaluation of exergy losses in the working fluid turbine as well as the residual exergy of the auxiliary working fluid at the end of expansion. The residual exergy of the geothermal brine at reinjection temperature are important element too, but it is usually kept higher than a controlling value depending on the scaling threshold due to the chemical composition of the geofluid.

Different types of ORCs exist in the literature and in commercial application. The investigated cycles can be of the simple or recuperated type, be subcritical or supercritical and can have one or two pressure levels. Different examples are illustrated in Fig. 2, in which the scheme of single-pressure with or without superheat ORC, a double-pressure ORC and supercritical ORC, using two different operating fluids, are shown. A great number of heat exchangers (economizer, evaporator, superheater, desuperheater, condenser and recuperator) are necessary.

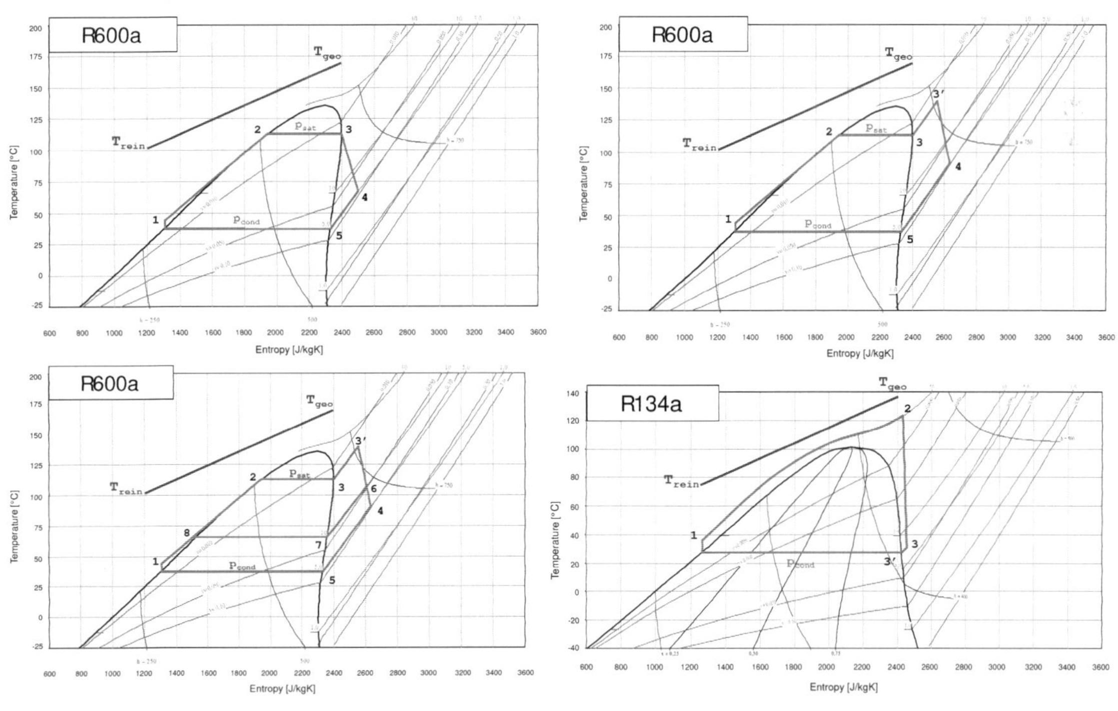

Figure 2. Thermodynamic cycles for binary plant

A critical point in the design of the binary cycle power plants is the design of the Recovery Heat Exchanger (RHE). Anyway, it is possible to state that a good match between thermodynamic cycle and working fluid can be obtained for each kind of combination, so that the influence of the working fluid could be considered marginal for the optimization of ORC performances. For this reason, as it will be discussed in the further section, in some cases the working fluid choice is driven by economic or supply-oriented evaluations, mainly driven by end-users.

3.2. Turbine (expander) efficiency and new turbine concept

Increase in the turbine efficiency (through continuous improvements in thermo-fluid-dynamic, finite element and vibration analysis) is object of the research too. Titanium blades have been developed by some leading manufacturers in the industry, displaying improved corrosion resistance. However, these have not gained widespread acceptance due to concerns about their performance in relation to erosion resistance. At the opposite end of the spectrum, there is reputedly intriguing ongoing work to develop small, 'high speed' turbine wheels that rotate at tens of thousands of RPM a concentrated power extraction. Clearly the benefits here would be reductions in materials usage, although questions remain about use of such devices with wet steam that would cause erosion, and also about gearbox coupling and efficiency. Recently, within the indirect geothermal market (i.e. non-flash) such turbo-expanders are commonly used with binary fluids.

3.3. Heat exchangers and auxiliary systems

The heat exchangers technology for power production, although well-known, also deserves some attention, particularly in light of equipment sizing and selection and integration into ORC systems. Heat exchangers represent a major share of the total cost. Their design should therefore be carefully performed. Key characteristics regarding heat exchangers design are the efficiency (or pinch point) and pressure drops. Different types of heat exchangers can be used, the most common are: shell & tube; and plate (or plate and shell) heat exchangers (mainly in small-scale systems, due to their compactness), [13]. Some basic configurations of the heat exchangers are provided in Fig. 3

Concerning the various components of ORC plants, a critical heat exchanger is surely the one installed directly on the heat source, with the purpose or energy recovery form geothermal fluid.

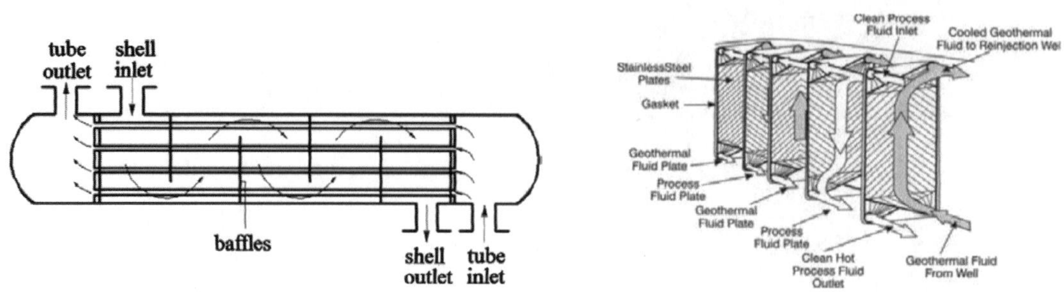

(a) shell and tube heat exchanger (b) plate heat exchanger

Figure 3. Typical heat exchangers for geothermal energy systems application

This particular heat exchanger (transferring heat from the geofluid to the working fluid) is the critical equipment of the whole plant, as it is affected both by strict pinch point and from chemical aggressiveness of the geofluid. Depending on its nature, this heat exchanger operates with a fluid at a quite high temperature and with a not simply defined chemical composition.

Preliminary evaluation of the global heat transfer coefficient (U) is important in order to perform cycle calculation: this is not always a simple task due to the fact that it depends on the following terms: inside and outside convective heat transfer coefficients, wall thermal resistance and fouling factors. The presence of an organic working fluid made sometimes difficult the estimation of the heat transfer coefficient mainly with respect to operating conditions different from the nominal values.

RHE can be often subject to fouling and corrosion, due to the presence of acids in the geothermal fluid, and/or reactants. Scaling and solid deposition can reduce the efficiency of this equipment, leading to critical area reduction or catastrophic rupture, if not prevented or mitigated with chemical cleaning and inhibitors injection. The cost of these devices largely contributes to the total power plant cost increase. This is particularly true for the condenser surface in case of dry type cooling towers, as they consume much more electricity than wet cooling towers (reducing the global power output of the plant). Moreover, due to the need of higher heat exchange surface and of a large volume of air flow, the condenser section costs about 5-10 times as much as a wet type one, depending on the condensing temperature of the turbine, mainly if low condensing temperatures are considered (as minimum approach against environmental temperature decreases). A suitable selection of operating conditions (condenser temperature and ambient temperature) of the plant and an optimum thermal design of the heat transfer surfaces is necessary for optimum design problem.

4. Heat exchangers design and development with respect to geothermal ORC industry

As it has been underlined in the previous section, the heat exchangers are the most critical components of a geothermal ORC power plant. A lot of issues and technical problems (thermodynamic, corrosion, scaling, heat transfer, huge mass flow rates) are concentrated into these exchangers because both geofluid and working fluid are processed. The design of a heat exchanger is really linked to the analysis of the whole system. The preliminary sizing and the optimum design of the heat exchangers used in the ORC plants is a problem with a great number of variables

The three main heat exchangers can be treated. The most important role is surely played by the Evaporator or Regenerative Heat Exchanger (RHE). The Recuperator (REC), necessary in case of regenerative cycle, is less impacted by geofluid side issues, but it can be critical for the whole plant efficiency and in transient operative conditions. The largest heat exchanger area used in a geothermal ORC is then the condenser surface, mainly if dry cooling must be performed. Not always it is possible to use a liquid heat source (that would give significant benefit on the condenser equipment sizing and auto consumption). The typical ORC location does not allow reaching water bodies for condensation.

Air cooled condenser could be also critical when the plant facility is located close to inhabited areas, as the fans release noise. Low-noise fans can also be proposed but it increases the cost of the equipment. Considering the recent studies about ORC systems, the real innovation element is the proposal of including the heat exchangers and the connected variable in the optimum design process of ORC. The optimum design of ORC system can arranged considering two steps, as represented in Fig. 4. A system optimization of an ORC is taken by including the heat exchangers in the optimum design. The heat exchangers can be modelled with the methods available in the literature that can be used for single-phase flow, condensing and evaporation.

4.1. Heat exchanger configurations

The type of heat exchangers that can be used in ORC plants are basically two: plate heat exchangers and shell-and-tube heat exchangers. The use of different heat exchangers and heat exchanger configuration is argument of some interesting papers, like [14-16]. In general, considering the two main types of heat exchangers, the following dimensions can be considered.

Table 2. Typical values of the variables for the shell and tube heat exchangers in ORC application.

Variable	Lower boundary	Upper boundary
Shell diameter (D)	0.3 m	2 m
Tube outside diameter (d)	5 mm	50 mm
Relative tube pitch (p/d)	1.2	2.5
Ratio length/shell diameter (L/D)	6	10
Ratio of tube diameter to shell diameter	0.001	0.1

IOP Conf. Series: Journal of Physics: Conf. Series **923** (2017) 012044 doi:10.1088/1742-6596/923/1/012044

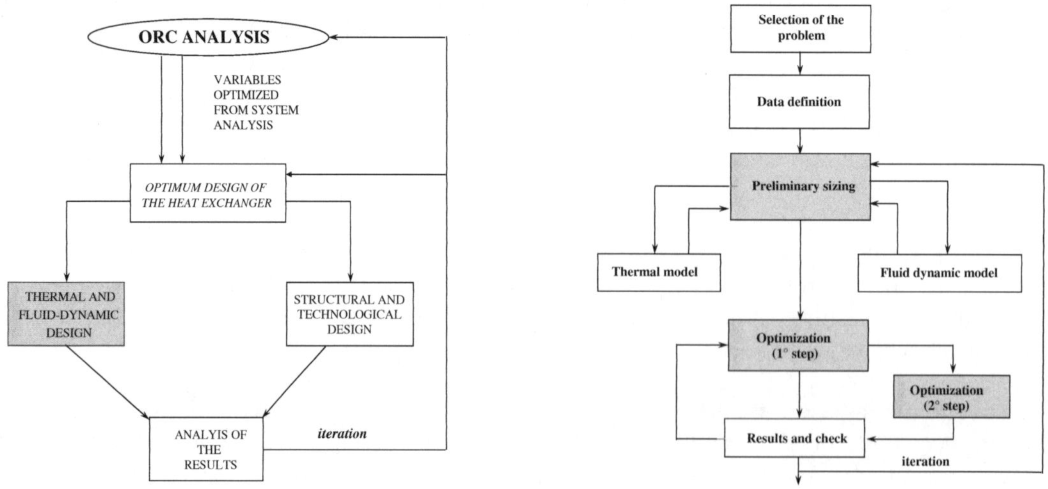

(a) First step: main heat exchanger variables (b) Second step: detailed sizing of heat exchanger

Figure 4. Optimum design of heat exchanger for ORC plant

4.2. Recovery Heat Exchanger (evaporator): critical elements and perspectives

The evaporation of the working fluid (and/or superheating, in case of both "wet fluids" and "dry fluids") is possible because of the heat given by the hot geofluid. Shell and tube heat exchanger type can be used achieving standard cost structure. The source fluid outlet temperature is critical as it is linked to scaling phenomena. This temperature must be kept sufficiently high to avoid scaling on the source fluid side of the exchanger. Cleaning of the source fluid side may be necessary, so the vaporizer design must take this into account. Any geothermal fluid may be corrosive, so an appropriate material must be identified (e.g. also special steels) to be used for the vaporizer.

As the evaporation of a low-boiling point fluid (respect to water) is to be performed, two types of heat exchanger can be proposed: kettle type boilers and once-through heat exchangers. The first type is more traditional, belonging to the more experienced practice of combined cycle plants (see Fig. 5). The second type (see Fig. 3-a) leads to a different path of flow arrangement (particularly for the working fluid), as a tidy liquid-vapor interface is not identified or observed along the heat exchanger. This second typology often implies longer heat exchangers (with geofluid passing through the tube side), with reduced shell diameter.

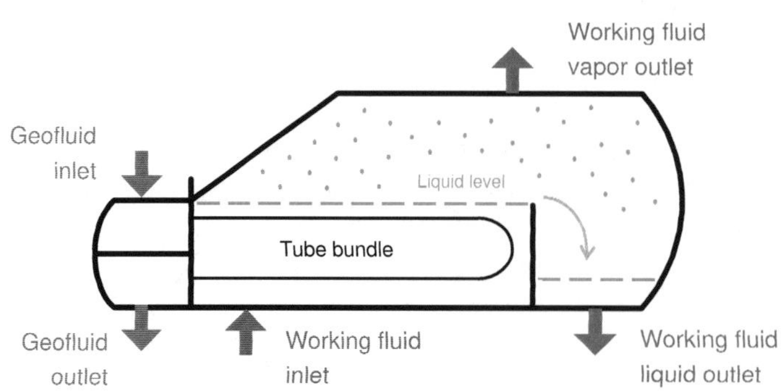

Figure 5. Kettle type evaporator (shell and tube)

One possible perspective for standardization and development from manufacturer is truly given by these aspects, trying to match clearly identified and modular typology of heat exchanger, mainly for the RHE. As for machinery manufacturing, for which standardized sizes of turbines and expanders should be a perspective, also in case of geothermal heat exchangers modular equipment could be an enhancement. Reinjection temperature (usually in the range between 70 °C and 100 °C) is often a driven factor, as the scaling phenomena. Another operating problem is surely represented by non-condensable gases (NCG). When vapor phase geofluid is used instead of liquid phase, then NCG must be removed by the geofluid side of the equipment. This operation impacts on the efficiency of the RHE and a careful design of the NCG removal system must be performed. A small compressor could be used for this purpose, increasing the auxiliary consumption of the plant and also the noise level emission.

4.3. Recuperator design
The regenerative heat exchanger, as clearly illustrated in Fig. 1, is the heat exchanger between the hot exhaust vapor from the turbine and the liquid being pumped from the condenser. The objective of this component is the heat transfer from the exhaust vapor exiting the turbine to the cold liquid from the condenser. The fluid behavior is usually close to linear, so it is normally not necessary to divide the regenerator into sections. The Recuperator is not impacted by geofluid issues, but it contributes to the whole energy recovery arrangement of the cycle. Its sizing is then also important, as it could lead to reduce the size of the RHE.

4.4. ORC condenser design
The condenser may be either water or air cooled. Three different schemes of condensers are available as illustrated in Fig. 6. The calculations for the condenser are roughly the same in both cases, as the cooling fluid (air or water) is very close to have a thermally linear behavior. In a typical ORC plant the condenser is a heat exchanger between the hot vapor from the regenerator/turbine and a cooling medium. The temperature of the hot fluid is higher than the one of the cold fluid throughout the condenser, but this difference is not so marked. As well it must be kept in mind that the relation between the enthalpy and the temperature is non-linear, requiring that the vaporizer is divided into appropriate sections for the calculation. This is especially valid for Kalina cycles, in the ORC cycle there is only a property change at the dew point, where de-superheat ends and condensation begins.

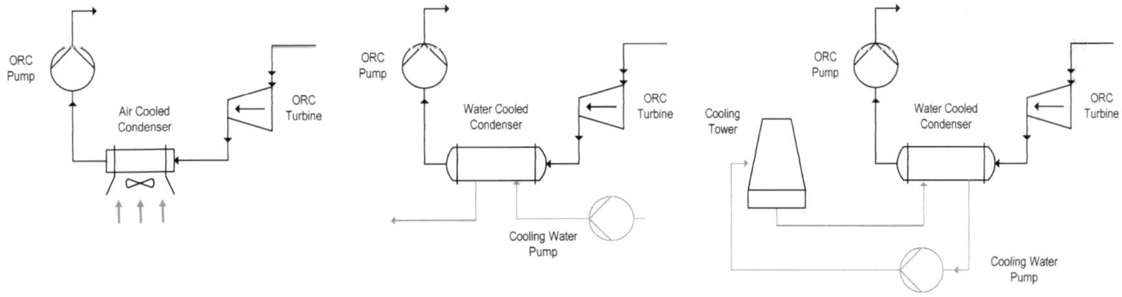

Figure 6. Schematic of three possible condenser design for ORC plants

4.5. Scaling and material selection
As it has been addressed from research and technical applications, scaling is one the more impacting problems in the design of binary cycle power plants. Scaling (usually silica or other typology of deposition) is directly connected with the chemical composition of the fluid. Its peculiarity is that it can only be analyzed basing on manufacturer experience and geothermal exploration (it depends on the chemical composition of the geofluid extracted), which are lacking when facing a new developing reservoir, with no drilling historical data. Its real impact can be measured only when the wells are drilled and a dedicated analysis is performed. Scaling and deposition problem are currently being faced through experience and background of operators and it is one of the sectors in this field less involved in technology transfer. Material selection is an obvious criticality, being strongly influenced by the geofluid chemical composition and aggressiveness.

IOP Conf. Series: Journal of Physics: Conf. Series **923** (2017) 012044 doi:10.1088/1742-6596/923/1/012044

The RHE must resist to both aggressiveness and potential scaling, and a periodic chemical cleaning should be planned. For small application also particular plate type heat exchanger could be used, however the correct design and chemical issues must be treated in a more severe way, as the small channels could be impacted in a more critical way.

5. Conclusions

The design of ORC systems is a challenging task in energy systems engineering. Many design alternatives are available at each level of design, from working fluid selection via cycle optimization and control through to the efficient integration of the various components of the system.

An important role is covered by heat exchangers and heat transfer sections optimum design. The problem concerns the thermal and fluid dynamic elements as well as the materials.

Significant possible approaches moving from systematic strategies to detailed design issue have emerged over the past years with the aim at supporting the designer in making optimal choices at each level of ORC development.

The paper has analyzed a state-of-the-art of the application of the heat exchangers in the ORC plants considering both conventional and emerging studies with a particular emphasis on the optimum design of the system that have to consider the heat exchangers.

References

[1] Stefansson V, 2005. Proc. of the World Geothermal Congress, Antalya, Turkey.
[2] DiPippo R, 2008. Butterworth-Heinemann, New York, NY, USA.
[3] Gnutek Z and Bryszewska-Mazurek A, 2001. Energy, **26**, 1075-1082.
[4] Madhawa Hettiarachchi HD, Golubovic M, Worek WM and Ikegami Y, 2007. Energy **32**, 1698–1706.
[5] Franco A and Villani M, 2009. Geothermics **38**, 379–391.
[6] Walraven D, Laenen B and D'haeseleer W, 2015. Energy **80**, 104-113.
[7] Zarrouk SJ and Moon H, 2014. Geothermics **51**, 142–153.
[8] Bertani R, 2015. Proc. of the World Geothermal Congress, Melbourne, Australia.
[9] Drescher U and Brüggemann D, 2007. Appl Therm Eng **27**, 223–228.
[10] Cataldo F, Mastrullo R, Mauro AW and Vanoli GP, 2014. Energy **72**, 159-167.
[11] Chagnon-Lessard N, Mathieu-Potvin F and Gosselin L, 2016. Geothermics **64**, 111–124.
[12] Invernizzi CM, Iora P, Manzolini G and Lasala S, 2017. Appl Therm Eng **121**, 172–179.
[13] Kuppan T, 2000. Marcel Dekker, Inc., New York.
[14] Lecompte S, Van den Broek M and De Paepe M, 2014. Proc. of the 15th International Heat Transfer Conference, IHTC-15 August 10-15, Kyoto, Japan.
[15] Walraven D, Laenen B and D'haeseleer W, 2014. Energ Convers Manage **83**, 177-187.
[16] Walraven D, Laenen B and D'haeseleer W, 2014. Energ Convers Manage **87**, 227-237.

IOP Conf. Series: Journal of Physics: Conf. Series **923** (2017) 012045 doi:10.1088/1742-6596/923/1/012045

Analytical solutions for tomato peeling with combined heat flux and convective boundary conditions

G Cuccurullo, L Giordano and A Metallo
Department of Industrial Engineering, University of Salerno,
Via Giovanni Paolo II 132, 84084 Fisciano (SA), Italy

E-mail: gcuccurullo@unisa.it

Abstract. Peeling of tomatoes by radiative heating is a valid alternative to steam or lye, which are expensive and pollutant methods. Suitable energy densities are required in order to realize short time operations, thus involving only a thin layer under the tomato surface. This paper aims to predict the temperature field in rotating tomatoes exposed to the source irradiation. Therefore, a 1D unsteady analytical model is presented, which involves a semi-infinite slab subjected to time dependent heating while convective heat transfer takes place on the exposed surface. In order to account for the tomato rotation, the heat source is described as the positive half-wave of a sinusoidal function. The problem being linear, the solution is derived following the Laplace Transform Method. In addition, an easy-to-handle solution for the problem at hand is presented, which assumes a differentiable function for approximating the source while neglecting convective cooling, the latter contribution turning out to be negligible for the context at hand. A satisfying agreement between the two analytical solutions is found, therefore, an easy procedure for a proper design of the dry heating system can be set up avoiding the use of numerical simulations.

1. Introduction

Food safety reasons induce to remove the skin from vegetables, where a lot of bacteria, yeasts and moulds are housed which are difficult to remove. Since a suitable warming up ensures the skins to come off easily, radiative heating appears to be a sustainable means for peel removal as opposed to the ones traditionally used for peeling [1-6]. In facts, peeling is usually realized by using steam or lye, both techniques being recognized to be expensive and/or polluting. Consider that one of the most important advantages of the radiative peeling is that it is mostly waterless. The technique could not only cut the cost of bringing water into the cannery, but may also reduce the expense of recycling or properly disposing of it. Moreover, it is claimed that infrared peeling helps to reduce the wasteful "overpeeling" because the technique primarily affects only the peel and few thin layers beneath it. Reduced overpeeling improves sensorial quality as well, because deeper layers are typically paler than surface deep-red upper layers.

A typical scheme for realizing industrial plants based on radiative heating adopts a plane matrix of infrared emitters placed alongside the conveyor belts. These “illuminate” tomatoes traveling on belts which ensure their rotation in order to improve heating uniformity all over the surface, [8]. Predicting the thermal response of the tomato during the radiative dry peeling process can lead to good peeling

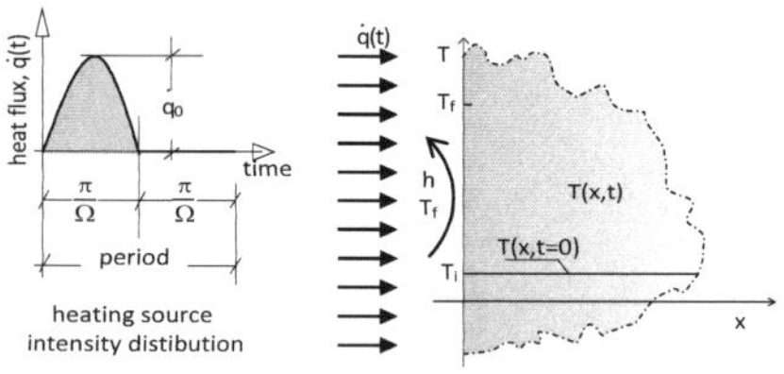

Figure 1. A schematic representation of the problem

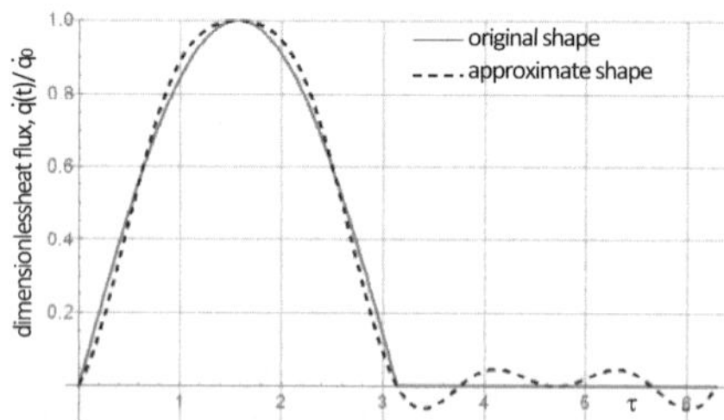

Figure 2. The original and approximate dimensionless shapes for heat flux

performance and, in addition, it can be of help in designing a suitable radiative heating system, ensuring high quality of peeled tomatoes and energy saving. To this purpose, numerical simulations are usually adopted [1-4;7] or models are selected which are based on semi-empirical exponentially decaying curves with a finite penetration depth [9-11;12-13].

In this context, the analytical solutions herein presented aim to be easy to handle yet accurate models in order to predict the tomato thermal response when subjected to both radiative and convective heat transfer at the surface. Both the analytical models consider a semi-infinite slab, since warming up involves only a few thin surface layers. With reference to the boundary conditions, tomato rotation exposes a point on the surface to a time variable heat flux even considering a radiative source featured by a constant and uniform heat flux; such effect is taken into account by introducing a suitable time varying source at boundary, which radiates oscillating sinusoidally in the first half-period, while is switched off in the following half. Of course, the source angular velocity corresponds to the tomato rotating speed. Most previous works considered pulsed heat flux or sudden temperature changes at the exposed surface [14-15]; however, as far as authors know, none are found in the literature of such kind as that above described. The solution is presented and discussed as depending on the main dimensionless parameters. Its poor sensitivity to the Biot numbers in the range of interest for tomato peeling allows to introduce a simplified yet accurate set of equations for predicting the onset of the peeling conditions.

2. Basic equations

In order to realize successful peeling, surface temperatures are to be raised to values as high as 100°C in a very short warming up period, typically no more than 60 s. Assuming water thermophysical properties apply for tomato, it can be easily checked that in such a short period, the thermal penetration depth is confined to really few millimetres under the tomato skin thus enabling to employ the semi-infinite body model [16]. At the boundary, a cycling radiative heating extended to the first half period of a sinusoidal source is allowed while convective heat transfer takes place as well.

2.1 Analytical solution with combined second and third kind boundary conditions

The present investigation concerns a semi-infinite medium with combined second and third kind boundary conditions, accounting for both heating by a periodic heat source and convective heat transfer at the solid-fluid interface, Figure 1. One-dimensional heat conduction and constant thermal properties are considered. The medium is assumed initially in equilibrium with the ambient at temperature T_i.

In order to describe the heating felt by the rotating tomatoes, the surface at $x = 0$ is assumed to be exposed to a periodic on/off heat flux with ½ duty cycle; when the source is on, i.e. in the first half period, the heating intensity is assumed to vary sinusoidally with time, while it is zero in the second half, Figure 2. Then, a periodic non-homogenous boundary condition is established, which is featured by the semi-amplitude $\dot{q}_0$ and characteristic angular velocity Ω related to the tomato rotation speed around its main axis. The semi-infinite slab model is justified because small heating times are required by the peeling process; it is easy to check that the temperature penetration depth is confined to the very first layers of the tomato skin [16]. By considering constant properties and internal conduction heat transfer for the tomato, while neglecting water evaporation effect, the dimensionless energy balance equation and the related boundary conditions lead to a non-homogeneous linear problem:

$$\begin{cases} \dfrac{\partial^2\theta}{\partial\xi^2} = 2\dfrac{\partial\theta}{\partial\tau} \\ -\left.\dfrac{\partial\theta}{\partial\xi}\right|_{\xi=0,\tau} = \hat{q}(\tau) - Bi\cdot(\theta-\theta_f) \\ \theta(\xi\to\infty,\tau)=0 \\ \theta(\xi,\tau=0)=0 \end{cases} \tag{1}$$

where the following dimensionless parameters have been introduced:, $\theta=(T-T_i)/\Delta T_{ref}$ is the temperature; the group $\Delta T_{ref} = \dot{q}_0 x_{ref}/k$ is a reference temperature difference related to the maximum wall heat flux, $\dot{q}_0$, and to a reference length $x_{ref} = (2\alpha\ t_{ref})^{1/2}$, k and α being the tomato thermal conductivity and diffusivity; the reference time, $t_{ref}=\Omega^1$, was chosen such as the dimensionless time resulted $\tau=\Omega t$, Ω being the angular velocity of the source; the dimensionless space variable was defined such as $\xi=x/x_{ref}$ and, finally, $Bi = h\ x_{ref}/k$ is the Biot number, $\hat{q}(\tau)=\sin(\tau)\cdot\left[1+\mathrm{sign}(\sin(\tau))\right]/2$ is the normalized wall heat flux.

Since the problem is linear, the solution can be sought as the sum of two partial solutions, θ_1 and θ_2, each one affected by a single non-homogeneity arising from the boundary condition at the wall; the former is due to the heating source and the latter by the convective heat transfer to the surroundings. Then, the two partial solutions have to satisfy two distinct problems derived from the basic one:

$$\begin{cases} \dfrac{\partial^2\theta_1}{\partial\xi^2} = 2\dfrac{\partial\theta_1}{\partial\tau} \\ \left.\dfrac{\partial\theta_1}{\partial\xi}\right|_{\xi=0,\tau} = -\hat{q}(\tau) \\ \theta_1(\xi\to\infty,\tau)=0 \\ \theta_1(\xi,\tau=0)=0 \end{cases} \qquad \begin{cases} \dfrac{\partial^2\theta_2}{\partial\xi^2} = 2\dfrac{\partial\theta_2}{\partial\tau} \\ \left.\dfrac{\partial\theta_2}{\partial\xi}\right|_{\xi=0,\tau} = Bi\cdot(\theta_1+\theta_2-\theta_f) \\ \theta_2(\xi\to\infty,\tau)=0 \\ \theta_2(\xi,\tau=0)=0 \end{cases} \tag{2}$$

The analytical solution of both the above systems for θ_1 and θ_2 is obtained by the application of the Laplace transform technique; the transformation eliminates the time variable and reduces both the partial differential equations into ODEs in the space variable ξ:

$$\begin{cases} \dfrac{\partial^2\bar{\theta}_1}{\partial\xi^2} = 2s\cdot\bar{\theta}_1 - 2\theta_1(\xi,\tau=0) \\ \left.\dfrac{\partial\bar{\theta}_1}{\partial\xi}\right|_{\xi=0,\tau} = \dfrac{1}{(e^{-\pi\cdot s}-1)\cdot(1+s^2)} \\ \bar{\theta}_1(\xi\to\infty)=0 \end{cases} \qquad \begin{cases} \dfrac{\partial^2\bar{\theta}_2}{\partial\xi^2} = 2s\cdot\bar{\theta}_2 - 2\theta_1(\xi,\tau=0) \\ \left.\dfrac{\partial\bar{\theta}_2}{\partial\xi}\right|_{\xi=0,\tau} = Bi\cdot\left(\bar{\theta}_1(\xi=0)+\bar{\theta}_2(\xi=0)-\dfrac{\theta_f}{s}\right) \\ \bar{\theta}_2(\xi\to\infty)=0 \end{cases} \tag{3}$$

Here s is the transformed variable, while $\theta_1(\xi, \tau=0)$ and $\theta_2(\xi, \tau=0)$ turn out to be zero for the respective initial conditions. The θ_2-problem is coupled to the θ_1-problem via the boundary condition at $\xi = 0$, that requiring the latter system to be formerly solved.
The complete solution in the transformed plane is:

$$\bar{\theta}(\xi) = \frac{Bi \cdot \theta_f}{s\left(Bi + \sqrt{2s}\right)} \cdot \mathrm{e}^{-\xi\sqrt{2s}} - \frac{\mathrm{e}^{-\xi\sqrt{2s}}}{\left(Bi + \sqrt{2s}\right)} \frac{1}{\left(e^{-\pi \cdot s} - 1\right) \cdot \left(1 + s^2\right)} \tag{4}$$

The former term of the above solution is easily inverted, turning out to be

$$\theta_h(\xi, \tau) = L^{-1}\left[\frac{Bi \cdot \theta_f}{s\left(Bi + \sqrt{2s}\right)} \cdot \mathrm{e}^{-\xi\sqrt{2s}}\right] = \theta_f\left(\mathrm{erfc}\left[\frac{\xi}{\sqrt{2\tau}}\right] - e^{Bi \cdot \sqrt{2}\xi + \tau \cdot Bi^2} \mathrm{erfc}\left[\frac{\xi}{\sqrt{2\tau}} + Bi\sqrt{\tau}\right]\right) \tag{5}$$

It represents the thermal response due to the convective heat transfer driven by the initial temperature excess $(T_i - T_f)$.
The latter term of eq. (4) is inverted by noting that it can be regarded as a product of two s-functions for which the inverse Laplace transform is well known:

$$F(\xi, \tau) = L^{-1}\left[\frac{e^{-\xi\sqrt{2s}}}{Bi + \sqrt{2s}}\right] = \left(\frac{1}{\pi\tau}\right)^{1/2} \cdot e^{-\frac{\xi^2}{2\tau}} - Bi \cdot e^{\left(Bi \cdot \sqrt{2}\xi + \tau \cdot Bi^2\right)} \mathrm{erfc}\left[\frac{\xi}{\sqrt{2\tau}} + Bi\sqrt{\tau}\right] \tag{6}$$

$$G(\tau) = L^{-1}\left[-\frac{1}{\left(e^{-\pi \cdot s} - 1\right) \cdot \left(1 + s^2\right)}\right] = \hat{q}(\tau) \tag{7}$$

Therefore, a convolution integral may be built:

$$\int_0^\tau F(\xi, \tau - \tilde{\tau}) \cdot G(\tilde{\tau}) d\tilde{\tau} = L^{-1}\left[-\frac{\mathrm{e}^{-\xi\sqrt{2s}}}{Bi + \sqrt{2s}} \frac{1}{\left(e^{-\pi \cdot s} - 1\right) \cdot \left(1 + s^2\right)}\right] \tag{8}$$

In order to perform the integration of the piecewise function $F(\xi, \tau - \tilde{\tau}) \cdot G(\tilde{\tau})$, considering a fixed value for τ, care is taken to break up the integral into n intervals of amplitude π over which $\hat{q}(\tau)$ remains continuous:

$$\int_0^\tau F(\xi, \tau - \tilde{\tau}) \cdot G(\tilde{\tau}) d\tilde{\tau} = \left(1 + (-1)^n\right) \cdot \tilde{\theta}_\tau(\xi, \tau) - 2 \cdot \sum_{i=0}^{n} (-1)^i \, \tilde{\theta}_{i\pi}(\xi, \tau) \tag{9}$$

where

$$\tilde{\theta}_{\tilde{\tau}}(\xi, \tau) = \int\left[\frac{e^{-\frac{\xi^2}{2(\tau - \tilde{\tau})}}}{\sqrt{4\pi\,(\tau - \tilde{\tau})}} - \frac{\mathrm{Bi}}{2} \exp\left[Bi\,\sqrt{2}\xi + Bi^2\,(\tau - \tilde{\tau})\right]\,\mathrm{erfc}\left[\frac{\xi}{\sqrt{2(\tau - \tilde{\tau})}} + Bi\sqrt{\tau - \tilde{\tau}}\right]\right] \sin(\tilde{\tau})\;\mathrm{d}\tilde{\tau} \tag{10}$$

The above integral is expressed in explicit form in the Appendices.
Finally, the full solution turns out to be:

$$\theta(\xi, \tau) = \theta_h(\xi, \tau) + \left(1 + (-1)^n\right) \cdot \tilde{\theta}_\tau(\xi, \tau) - 2 \cdot \sum_{i=0}^{n} (-1)^i \, \tilde{\theta}_{i\pi}(\xi, \tau) \tag{11}$$

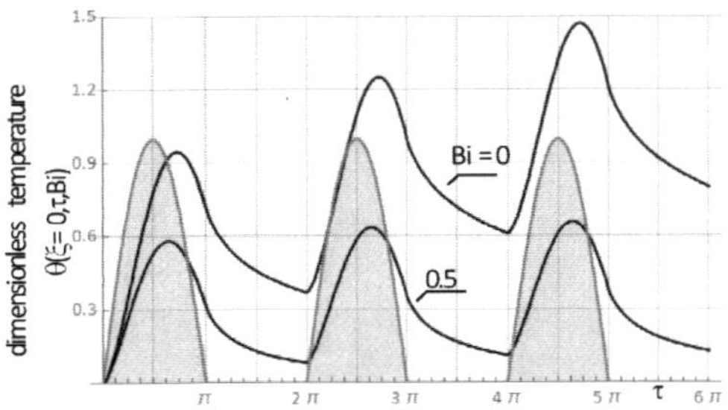

Figure 3. The dimensionless surface temperature @ $\xi = 0$

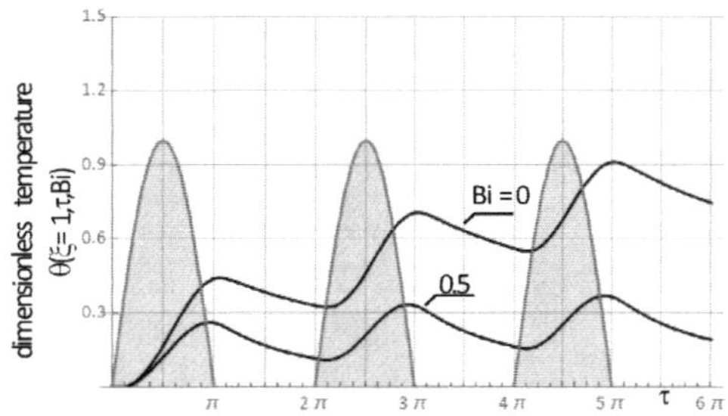

Figure 4. The dimensionless inner temperature @ $\xi = 1$

As expected, the solution depends not only on the values at time τ, it also depends on the heating previously experienced at each revolution, i.e. at each $\tilde{\tau} = i\,\pi$.

2.2 A simpler analytical solution

The solution obtained in previous paragraph appears rather involved, thus a further approximate approach to the same problem is attempted in order to obtain an easy-to-handle solution, which takes into account only radiative heating.

In fact, considering that tomato diameters are typically lower than 3 cm and anticipating that the rotational speed of the tomatoes has to be at least as high as 10 rpm for suitably reducing surface temperature fluctuations, the Biot numbers always turn out to be lower than 0.01. Therefore, the convective cooling contribution turns out to be negligible, as witnessed by the boundary condition @ $\xi = 0$ in the problem (1).

The simplification stems from considering a time-differentiable function for approximating the heat source distribution:

$$\dot{q}_{\text{approx}}(\tau) = \frac{1}{2}\left(\sin(\tau) + \sin(\tau)^2 + \frac{4-\pi}{\pi}\sin(2\tau)^2\right) \tag{12}$$

The latter shape exhibits the same average value of the original source over a period, meanwhile it is such as the relative variation with respect to the original shape is always lower than 5%; its structure is clearly simpler than the previous one being expressed by a single, differentiable, term. Then, by making use of the Laplace Transform once again, the solution to the problem is readily obtained:

$$\theta_{\text{approx}}(\xi,\tau) = \int_0^\tau \frac{e^{-\frac{\xi^2}{2\sqrt{\tau-\tilde{\tau}}}}}{\sqrt{4\pi\,(\tau-\tilde{\tau})}}\left(\sin(\tilde{\tau}) + \sin(\tilde{\tau})^2 + \frac{4-\pi}{\pi}\sin(2\tilde{\tau})^2\right) d\tilde{\tau} \tag{13}$$

The explicit integral is reported in the Appendices.

It can be shown that the magnitude of temperature oscillations given by the previous equation differs from that given by the corresponding exact ones, eq. (11), no more than 2.6%.

It will be shown in the next paragraph that it is of interest to evaluate the average temperature on the surface during the last revolution at a given time τ; it can be calculated as

$$\theta_{\text{ave,approx}}(\tau) = \int_\tau^{\tau+2\pi} \vartheta_{\text{approx}}(0,z)\, dz / 2\pi \tag{14}$$

The explicit integral is given in the Appendices again. Such a function approximates the corresponding exact one by a maximum relative error less than 2%.

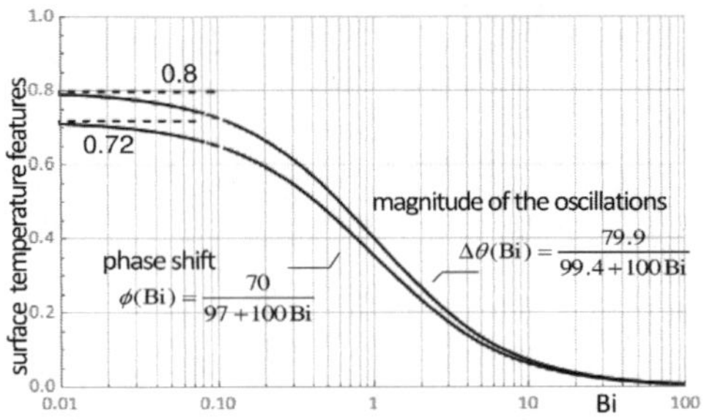

Figure 5. Surface temperature features

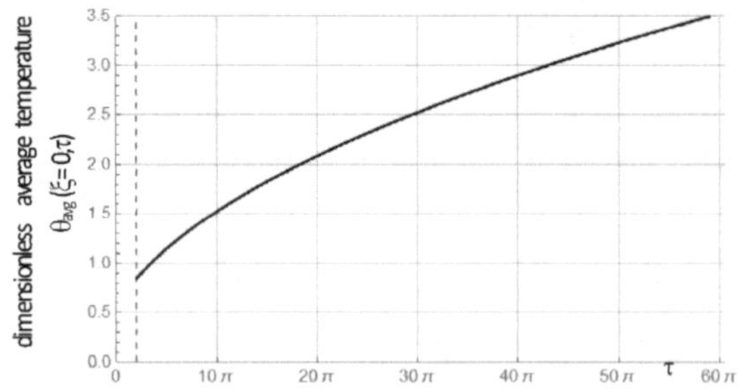

Figure 6. The dimensionless average temperature

3. Results and discussion

The solution given by eq. (11) is readily validated by considering the limiting case for $\tau < \pi$ when the initial temperature equals the fluid temperature, i.e. $\theta_f = 0$: for such conditions, the full temperature distribution should reduce to that well-known resulting from a simple sinusoidal heating source. That is the case at hand since it can be shown that the first term in the summation, i.e. $-2\,\tilde{\theta}_0(\xi,\tau)$, fully recovers expected solution.

The unsteady temperature field turns out to depend on the spatial and time coordinates and on the Biot number. The oscillating temperature field behaviour is represented in Figures 3 and 4 by taking temperature profiles on the surface and at a distance $\xi=1$ from it and for different values of the Biot number. The plots clearly show different thermal responses to different cooling rates: the greater the Biot number, the more both temperature levels and the amplitude of the oscillations decrease at each location. It can be further observed that, the phase shift with respect to the driving heating flux decreases with the Biot number. Attenuation and phase shift behaviour as a function of depth ξ are qualitatively the same discussed before for the Biot number, however thermal diffusion deeper under the surface determines a smoothing effect with increasing ξ so that the time-discontinuous shape of the source is scarcely resembled. Profiles still keep their asymmetric shapes since different heating and cooling rates are realized.

As noted before, it is of primary interest for proper dry tomato peeling design to characterize the surface temperature. For the present context, reduced machining times are to be considered, meanwhile $Bi \to 0$: therefore, a periodic sustained solution is not useful and the complete solution should be handled which requires high computational efforts. In the light of this consideration the magnitude of the oscillations and the phase shift are evaluated numerically at the surface after a reference time is elapsed while the Biot number varies. The reference time is chosen to correspond to the completion of 30 revolutions which, for usually accepted rotating rates, requires no more than a minute and is compliant with tomato peeling. Such results are then fitted with Hyperbolae equations, which are shown in Figure 5; hyperbolae fine adapt to the computed data, so that the corresponding coefficients of determination differ from one only in the fourth decimal place, both for the magnitude of the oscillations and the phase shift. Inspection of figures shows that the solution is basically insensitive to Biot numbers up to 0.01. Such a behaviour is of help wishing to answer the question of the frequency at which the tomato has to rotate so that the source appears to it almost as a continuous source. In fact, provided small Biot numbers are realized, the amplitude of the oscillations @ the surface, ΔT_0, can be estimated as:

$$\Delta T_0 = 0.8\,\dot{q}_0 \frac{\sqrt{2\alpha/\Omega}}{k} \tag{15}$$

from which the tomato angular speed can be obtained. The knowledge of Ω allows the subsequent evaluation of the time at which the heating process is complete, t_{proc}; the latter event is triggered when a target value, usually fixed @ $T_{\text{targ}} = 90°\text{C}$, for the surface average temperature is attained. The above purpose is suitably fulfilled by invoking the approximate solution; then, making use of eq. (14) or equivalently of Figure 6, one can write:

$$T_{\text{targ}} = T_i + \theta_{\text{ave,approx}}\left(\Omega\, t_{\text{proc}}\right)\, \dot{q}_0 \frac{\sqrt{2\alpha/\Omega}}{k} \tag{16}$$

in order to derive t_{proc}. A plot of the average temperature could be of help as well, see Figure 6.

4. Conclusions

Having in mind to study temperature patterns through tomatoes subjected to radiative heating during dry peeling process, an analytic solution of the Fourier heat conduction in a semi-infinite medium is obtained by making use of the Laplace transform technique. The thermal response fully recovers the temperature field due to a sinusoidal varying heat source as a limiting case. Simplified expressions for evaluating the surface phase shift and magnitude of temperature oscillations are given. Moreover, in order to facilitate the average temperature evaluation, a simplified model is introduced which properly recovers the average surface temperature. Since peeling takes place when the surface temperature attains a threshold value with contained magnitude of temperature fluctuations, an easy way to help the design of a proper radiative heating system is proposed, thus avoiding the use of numerical simulations.

5. Appendices

Performing the integration of eq. (10) gives:

$$\begin{aligned}
\tilde{\theta}_{\tilde{\tau}}(\xi,\tau) = {} & \frac{1}{8\sqrt{2}\left(1+\text{Bi}^4\right)}\Big(\left(\sqrt{2}+2\text{Bi}+\sqrt{2}\text{Bi}^2\right)\,\left(e^{-\xi}\cos(\xi-\tau)-e^{\xi}\cos(\xi+\tau)\right)+ \\
& +4\text{Bi}\, e^{\left(\xi\sqrt{2}+\text{Bi}(\tau-\tilde{\tau})\right)} erfc\left(\frac{\xi\sqrt{2}+2\text{Bi}\,(\tau-\tilde{\tau})}{2\sqrt{\tau-\tilde{\tau}}}\right)\,\left(\cos(\tilde{\tau})+\text{Bi}^2\sin(\tilde{\tau})\right)+ \\
& +\left(\sqrt{2}-\sqrt{2}\text{Bi}^2-2\text{Bi}^3\right)\, e^{-\xi}\left(\sin(\xi-\tau)+e^{2\xi}\sin(\xi+\tau)\right)+e^{-\xi}\Big(\left(2\text{Bi}\text{-}\sqrt{2}-\sqrt{2}\text{Bi}^2\right)\big(\cos(\xi-\tau)a_1(\xi,\tau,\tilde{\tau})+ \\
& +\sin(\xi-\tau)\big)b_1(\xi,\tau-\tilde{\tau})+\left(-\sqrt{2}+\sqrt{2}\text{Bi}^2-2\text{Bi}^3\right)\,\left(\text{-}\cos(\xi-\tau)b_1(\xi,\tau-\tilde{\tau})+\sin(\xi-\tau)a_1(\xi,\tau-\tilde{\tau})\right)\Big)+ \\
& +e^{\xi}\Big(\Big(\left(\sqrt{2}+2\text{Bi}\right)b_2(\xi,\tau,\tilde{\tau})+\sqrt{2}\,a_2(\xi,\tau-\tilde{\tau})\Big)\,\left(\text{Bi}^2\cos(\xi+\tau)-\sin(\xi+\tau)\right)+\Big(-\sqrt{2}b_3(\xi,\tau-\tilde{\tau})+ \\
& +\left(\sqrt{2}+2\text{Bi}\right)a_3(\xi,\tau-\tilde{\tau})\Big)\,\left(\cos(\xi-\tau)+\text{Bi}^2\sin(\xi+\tau)\right)\Big)\Big)
\end{aligned}$$

where $a_k(\xi,z)$ and $b_k(\xi,z)$ are the real and the imaginary coefficients of the following error functions:

$$\text{erf}_k(\xi,z) = \text{erf}\left(\frac{\xi\sqrt{2}+(-1)^k\sqrt{2z}}{2\sqrt{z}}+(-1)^k i\frac{\sqrt{z}}{\sqrt{2}}\right),\ k=1,2$$

Performing the integration of eq. (13) gives:

$$\vartheta_{\text{approx}}(\xi,\tau)=2\left(\sqrt{\tau}/\left(\sqrt{2}\pi^{3/2}\right)\right)e^{-\xi^2/2\tau}-\xi/\pi+\left(1/\left(16\sqrt{2}\right)\right)\left(2\sqrt{2}\left(-e^{-\xi}\left(\cos(\xi-\tau)+\sin(\xi-\tau)\right)+\right.\right.$$
$$+e^{\xi}\left(\cos(\xi+\tau)-\sin(\xi+\tau)\right)\Big)+e^{-\xi\sqrt{2}}\left(-\cos\left(\xi\sqrt{2}-2\tau\right)+\sin\left(\xi\sqrt{2}-2\tau\right)\right)+$$
$$+((4-\pi)/(\pi\sqrt{2}))\left(e^{-2\xi}\left(-\cos(2\xi-4\tau)+\sin(2\xi-4\tau)\right)+e^{2\xi}\left(\cos(2\xi+4\tau)+\sin(2\xi+4\tau)\right)\right)+$$
$$+e^{\xi\sqrt{2}}\left(\cos\left(\xi\sqrt{2}+2\tau\right)+\sin\left(\xi\sqrt{2}+2\tau\right)\right)+(16/\pi)\sqrt{2}\xi\ \text{erf}\left(\xi/\sqrt{2\tau}\right)+$$
$$+e^{-\xi\sqrt{2}}\left(\cos\left(\xi\sqrt{2}-2\tau\right)\left(c_1(\xi,\tau)-d_1(\xi,\tau)\right)-\sin\left(\xi\sqrt{2}-2\tau\right)\left(c_1(\xi,\tau)+d_1(\xi,\tau)\right)\right)+$$
$$2\sqrt{2}\left(e^{-\xi}\left(-\cos(\xi-\tau)\left(n_1(\xi,\tau)-m_1(\xi,\tau)\right)+\sin(\xi-\tau)\left(n_1(\xi,\tau)+m_1(\xi,\tau)\right)\right)+\right.$$
$$+e^{\xi}\left(\cos(\xi+\tau)\left(n_2(\xi,\tau)-m_2(\xi,\tau)\right)+\sin(\xi+\tau)\left(n_2(\xi,\tau)+m_2(\xi,\tau)\right)\right)\Big)+$$
$$-e^{\xi\sqrt{2}}\left(\cos\left(\xi\sqrt{2}+2\tau\right)\left(c_2(\xi,\tau)-d_2(\xi,\tau)\right)+\sin\left(\xi\sqrt{2}+2\tau\right)\left(c_2(\xi,\tau)+d_2(\xi,\tau)\right)\right)+$$
$$+((4-\pi)/(\pi\sqrt{2}))\left(e^{-2\xi}\left(\cos(2\xi-4\tau)\left(p_1(\xi,\tau)+q_1(\xi,\tau)\right)-\sin(2\xi-4\tau)\left(p_1(\xi,\tau)-q_1(\xi,\tau)\right)\right)+\right.$$
$$\left.\left.-e^{2\xi}\left(\cos(2\xi+4\tau)\left(p_2(\xi,\tau)+q_2(\xi,\tau)\right)+\sin(2\xi+4\tau)\left(p_2(\xi,\tau)-q_2(\xi,\tau)\right)\right)\right)\right)$$

where:

$c_k(\xi,\tau)$ and $d_k(\xi,\tau)$ are the real and the imaginary coefficients of the following error functions:

$$\text{erf}_k(\xi,\tau)=\text{erf}\left(\frac{\xi\sqrt{2}+(-1)^k 2\tau}{2\sqrt{\tau}}-(-1)^k i\sqrt{\tau}\right),\ k=1,2$$

$m_k(\xi,\tau)$ and $n_k(\xi,\tau)$ are the real and the imaginary coefficients of the following error functions:

$$\text{erf}_k(\xi,\tau)=\text{erf}\left(\frac{\xi\sqrt{2}+(-1)^k\tau\sqrt{2}}{2\sqrt{\tau}}+(-1)^k i\frac{\sqrt{2\tau}}{2}\right),\ k=1,2$$

$p_k(\xi,\tau)$ and $q_k(\xi,\tau)$ are the real and the imaginary coefficients of the following error functions:

$$\text{erf}_k(\xi,\tau)=\text{erf}\left(\frac{\xi\sqrt{2}+(-1)^k 2\sqrt{2}\tau}{2\sqrt{\tau}}+(-1)^k i\sqrt{2\tau}\right),\ k=1,2$$

Performing the integration of eq (14) the approximate average temperature on the surface over a period is given by

$$\theta_{\text{ave,approx}}(\tau)=\frac{-3\pi\sqrt{\tau}-4\tau^{3/2}+11\pi\sqrt{2\pi+\tau}+4\tau\sqrt{2\pi+\tau}}{6\sqrt{2}\,\pi^{5/2}}$$

6. References

[1] Li X, Pan Z, Upadhyaya S K, Atungulu G G and Delwiche M 2011 Three-dimensional geometric modeling of processing tomatoes, *Transactions of the ASABE* **54** 2287–2296.

[2] Li X 2012 *A study of infrared heating technology for tomato peeling: Process characterization and modeling*, Ph.D. Dissertation (Department of Biological and Agricultural Engineering, University of California at Davis, Davis, CA, USA).

[3] Li X, Pan Z 2014 Dry peeling of Tomato by Infrared Radiative Heating: Part I. Model Development, *Food Bioprocess Technol* **7** 1996–2004.

[4] Li X, Pan Z 2014 Dry peeling of tomato by infrared radiative heating: Part II. Model validation and sensitivity analysis, *Food and Bioprocess Technology* **7** 2005–2013.

[5] Rock C, Yang W, Goodrich-Schneider R & Feng H 2011 Conventional and alternative methods for tomato peeling, *Food Engineering Reviews*, **4** 1–15.

[6] Tanaka F, Verboven P, Scheerlinck N, Morita K, Iwasaki K, Nicola B 2007 Investigation of far infrared radiation heating as an alternative technique for surface decontamination of strawberry, *Journal of Food Engineering*, **79** 445–452.

[7] Cuccurullo G, Giordano L and Viccione G 2016 Numerical and experimental modeling for thermal developing pipe flow with microwave heating, *International Journal of Mechanics*, **10** 68–15.

[8] Pan Z, Li X, Khir R, El-Mashad H M, Atungulu G G, McHugh T H., Delwiche M 2015 A pilot scale electrical infrared dry-peeling system for tomatoes: Design and performance evaluation, *Biosystems engineering* **137**, 1-8.

[9] Datta A, Ni H 2002 Infrared and hot-air-assisted microwave heating of foods for control of surface moisture, *Journal of Food Engineering*, **51** 355–364.

[10] Prakash B 2012, *Mathematical modeling of moisture movement within a rice kernel during convective and infrared drying*, Ph.D. Dissertation (Department of Biological and Agricultural Engineering, University of California at Davis, Davis, CA, USA).

[11] Trivittayasil V, Tanaka F, Uchino T 2011 Investigation of deactivation of mold conidia by infrared heating in a model-based approach, *Journal of Food Engineering*, **104** 565–570.

[12] Cuccurullo G, Pierro V 2004, A procedure to measure electromagnetic skin depth in microwave heating, *Infrared Physics and Technology*, **46** 49-55

[13] Cuccurullo G, Giordano L, Viccione G 2013, An Analytical approximation for continuous flow microwave heating of liquids, *Advances in Mechanical Engineering*, **2013**

[14] Ylbas B S 1997 Analytical solution for time unsteady laser pulse heating of semi-infinite solid, *Int. J. of Mech. Sci.*, **39** 671-682

[15] Bishri Abdel-Hamid 1999 Modelling non-Fourier heat conduction with periodic thermal oscillation using the finite integral transform, *Applied Mathematical Modelling* **23** 899–914

[16] Cuccurullo G, Giordano L 2017 Temperature field for radiative tomato peeling, *Journal of Physics Conference Series* **796**

How accelerated biological aging can affect solar reflective polymeric based building materials

C Ferrari[1], G Santunione[1], A Libbra[1], A Muscio[1] and E Sgarbi[2]

[1] EELab – Dip. di Ingegneria "Enzo Ferrari" – Univ. di Modena e Reggio E., Italy
[2] Dip. di Scienze della Vita – Univ. di Modena e Reggio E., Italy

E-mail: alberto.muscio@unimore.it

Abstract. Among the main issues concerning building materials, in particular outdoor ones, one can identify the colonization by microorganisms referred to as biological aggression. This can affect not only the aesthetical aspect but also the thermal performance of solar reflective materials. In order to improve the reliability of tests aimed to assess the resistance to biological aggression and contextually reduce the test duration, an accelerated test method has been developed. It is based on a lab reproducible setup where specific and controlled environmental and boundary conditions are imposed to accelerate as much as possible biological growth on building materials. Due to their widespread use, polymeric materials have been selected for the present analysis, in the aim of reaching an advanced bio-aged level in a relatively short time (8 weeks or less) and at the same time comparatively evaluate different materials under a given set of ageing conditions. Surface properties before, during and after ageing have been investigated by surface, microstructural and chemical analyses, as well as by examination of time progressive images to assess bacterial and algal growth rate.

1. Introduction

Building materials can be significantly affected and damaged by the deleterious effects of weathering processes, which involve both chemical and physical deterioration, mechanical failure and biological deterioration. While chemical or physical deterioration and mechanical failure are generally countered by optimization of the manufacturing process, the effects of debris and ambient dust deposition, which can also provide an opportunity for colonization by biological organisms [1], are not so easy to address. Generally speaking, studying the biological deterioration, or bio-fouling, is becoming more and more interesting for the community of civil engineers and architects because of its consequences on building structures and roof aesthetical and energy-related properties.

With regard to biological growth, it can cause changes in the aesthetic appearance of materials and also reduce their durability due to deterioration and corrosion induced by the microorganisms [2]. Another aspect to investigate, in combination with physical and chemical degradation caused by weathering and soiling, is that a worsening of the degradation process of outdoor materials can be led by bio-deterioration [3]. This phenomenon is caused by a multitude of species such as algae, cyanobacteria, heterotrophic bacteria, fungi, lichens, protozoa, and a variety of small animals (arthropods) and plants (bryophyte) [4] that often create diffuse dark colored stains. Such biofilm formation origins metabolites (acids and polysaccharides, phospholipid fatty acids) that eventually contribute to the deterioration and corrosion processes [1,5,6] in combination with climate, local conditions, building design and roof or façade materials [7].

All building materials are affected by bio-deterioration in terms of aesthetical, physical and chemical damages caused by microorganisms. A lot of material types are involved in the bio-deterioration phenomenon and the most important factor promoting microbial growth on roofs and façades is the formation of moisture, from rain and also from condensation in the period of overcooling. Reducing the time of condensation (i.e. using materials with latent heat effect) could avoid such problem [8]. The type of support, together with moisture, light conditions, temperature, roughness, and porosity can create the condition to promote biological development [9]. In this work, particular attention has been paid to polymeric roofing materials, with a focus on their properties relevant to the thermal behavior of buildings. More specifically, efficiency of solar reflective roofing solutions, the so-called cool roofs, is reduced due to weathering, those lose their initial solar reflectance and thus see their cooling energy saving potential compromised [10,11]. If natural exposure is analyzed [6,10], data collected on field show that loss of solar reflectance is cyclic with the onset of seasons having more rainfall.

2. Aim of the work

The best solution to understand biological aggression on building materials is represented by natural exposure experiments. Several studies have been carried out on natural weathering in order to analyze and understand the energy impact of soiling on buildings [4, 6, 10-13]. Analyzing aged samples, high bacterial and fungal growth is documented, if compared with fresh materials [11]. However, despite the theoretical faithfulness of this approach, the time required by outdoor exposure does not match with the needs of the industry for research and development. Another critical point is represented by the low repeatability of natural ageing due to the random variability of the environmental conditions and their strong dependence on the geographical location (*i.e.* the climate) of the exposure field.

The most effective way to reduce the time of exposure is to develop a laboratory approach that can provide a reliable and reproducible test method to assess the long-term performance of building materials, especially with regards to changes in solar reflectance over time [13]. With artificial ageing, in fact, natural ageing processes can be compressed in a very short time by stimulating the growth of microorganisms and also strictly controlling the environmental conditions [3]. This work is aimed at a deep investigation of the biological colonization processes that occur on building surfaces, and especially on polymeric based ones, through the setup of a highly automatized test chamber for accelerated bio-ageing. In order to overcome the high variability of biological growth in dependence of external factors, the developed device is aimed to make the growth conditions as much reproducible, repeatable, and of course fast as possible.

This study is aimed to be a preliminary survey on the behavior of eight different polymeric roofing materials under forced biological growth, to understand how the biofilm growing on the surface alters not only the physical properties but also energy-related ones. The final goal is to assess the long term performance of different surfaces.

3. Materials and methods

Biofilm growth dynamic on exposed surfaces is experiencing a growing interest due to its impact on energy efficiency of building. Previously, accelerated ageing setups have been studied in order to understand the bio-deterioration phenomena after a short period of exposure [9, 13-16]. Analyses performed by Kultur et al. [13] treated the samples by using xenon lamps simulating the solar spectrum in the range from 300 to 800 nm, excluding all other environmental parameters that may induce biological growth (moisture, wind, pollutants, chemical and physical degradation). A further step was made by Barberousse et al. [15], Dirkx et al. [9] and D'Orazio et al. [16], providing reproducible protocols to accelerate micro-organisms growth by means of standard light/dark and watering cycles on the samples. Künzel et al. [14] considered biological inoculation by means of hand brushing of the sample surface, which represents a heterogeneous way to spread biological colonies.

Most of the above mentioned studies, however, don't seem adequately focused on the effects of bio-fouling in terms of loss of solar reflective performance of building surfaces.

3.1. Set of tested materials and energy-related performance

Different roofing materials and membranes currently employed in the building sector have been tested. Being this an exploratory study, materials with different physical and chemical features were chosen. In details, polymeric based coatings or membranes were analyzed. Figure 1 shows the visible details of each sample before the aging period and the measured initial value of solar reflectance ρ_{sol}, that is the ratio of reflected and incident solar radiation. Details on the measurement method are given in the following sections. Building materials with high solar reflectance, usually with white or very light visible color (see samples A-E), are very effective solutions to avoid overheating of buildings and urban areas, but retaining a high reflectance value along time is very difficult due to weathering, pollution, and biological growth of course.

A Painted polyurea ρ_{sol}=0.745	B Structured membrane ρ_{sol}=0.855	C Double layer membrane ρ_{sol}=0.852	D Field applied coating ρ_{sol}=0.763	E White single ply membrane ρ_{sol}=0.828	F Green single ply membrane ρ_{sol}=0.095	G Grey single ply membrane ρ_{sol}=0.335

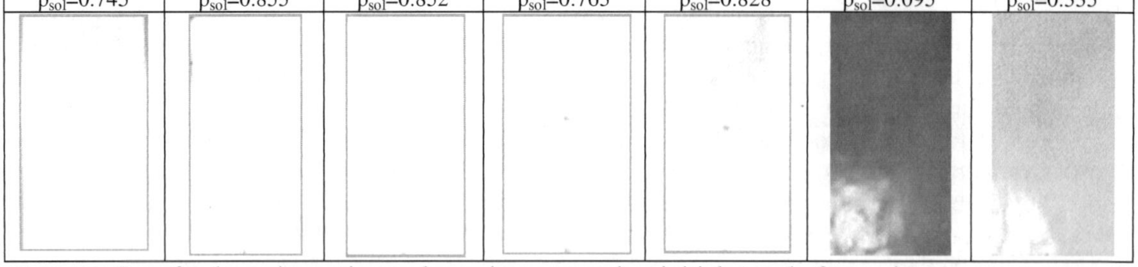

Figure 1. Set of polymeric coating and membrane samples: initial state before aging.

The effect of solar reflectance on the thermal performance can be expressed through the 'solar reflectance index' (SRI) [17], a parameter contemplated by voluntary rating systems such as LEED [18] and calculated as

$$SRI = 100 \times \frac{T_{sb} - T_{se}}{T_{sb} - T_{sw}} \tag{1}$$

where T_{se} (K) is the temperature that the analyzed surface would steadily reach when irradiated by a reference solar flux $I_{sol,max}$ = 1000 W/m^2 with atmospheric air temperature T_{air} = 37°C ≡ 310 K, sky temperature T_{sky} = 27°C ≡ 300 K, and convection heat transfer coefficient h_{ce} to which three different values can be assigned, equal to 5, 12, and 30 W/(m^{2}K) for, respectively, low (v_{wind} < 2 m/s), intermediate (2 m/s < v_{wind} < 6 m/s), and high (6 m/s < v_{wind} < 10 m/s) wind speed. T_{sb} (K) and T_{sw} (K) are the temperatures that would be reached by two reference surfaces, respectively a black one ($\rho_{sol,b}$ = 0.05) and a white one ($\rho_{sol,w}$ = 0.80), both ones having high thermal emittance (ε = 0.90). SRI represents the decrement of surface temperature that the analyzed surface would allow with respect to the reference black one in the reference conditions, divided by the analogous decrement allowed by the reference white surface and eventually given in percentage terms.

The surface temperature T_{se} (as well as T_{sb} and T_{sw}) is determined by iteratively solving the following thermal balance, based on the hypothesis of adiabatic roof or wall external surface:

$$(1 - \rho_{sol}) I_{sol} = \varepsilon\sigma_0 \left(T_{se}^4 - T_{sky}^4\right) + h_{ce} \left(T_{se} - T_{air}\right) \tag{2}$$

The SRI works very well even with non-adiabatic surfaces since the heat flow rate conducted inside can be lower by one or two orders of magnitude than solar irradiance. Values of SRI and T_{se} for the considered materials will be shown in Figure 5 for intermediate wind speed and both initial or aged solar reflectance. The value $\varepsilon \cong 0.90$ is generally assumed for the thermal emittance, that is the ratio of energy re-emitted in the far infrared range and maximum theoretical emission at the same temperature, since this is typical of non-metallic surfaces. The higher is SRI, which under its definition can exceed 100% for materials with very high solar reflectance, the lower is the temperature reached by the insolated surface.

3.2. Accelerated bio-aging set up

The methodology of investigation presented here is focused on the acceleration of microbiological growth. In order to develop the aging procedure, it is essential to define a specific set up with reproducible study parameters.

Among the organisms involved in the bio-deterioration, the majority is represented by phototrophic organisms such as cyanobacteria and algae [3]. There are several microbiological species involved in natural bio-aging, considering the climate conditions and the microclimate of a particular site. It has therefore been necessary to find and chose the most representative species responsible of biological growth on building surfaces. By means of an extensive literature review [3], some ubiquitous green algae and cyanobacteria have been selected for the testing activities. More specifically, some significant studies [9,19-20] permitted the selection of the following algae species: Chlorella mirabilis, Chlorella vulgaris, Klebsormidium flaccidum, Chlorhormidium spp. (spp. is addressed to several species of the same genus), Stichococcus bacillaris; the strains were isolated from the Algal collection of the University Federico II (ACUF) [21]. A Cyanobacteria specie has also been selected, Gloeocapsa sp. (sp. is addressed to one specie only, not need to be specified), isolated from the Culture collection of algae and protozoa, UK [22]. A batch culture of each strain was grown in Bold's Basal Medium (BBM) [23], in a chamber with a light/dark alternate photoperiod of 12h/12h, and a temperature cycle 23°C/15°C.

The experimental setup consist of a 120 cm x 40 cm x 55 cm glass chamber (Figure 2) where the samples are placed on a 45° sloped stainless steel support. The slope angle was selected to optimize the water runoff on the sample surface and then the probability for algae to be deposited and root [24]. The weathering room was filled with 50 L of BBM, the same above mentioned medium used in the batch cultures, enriched with 100 ml of each algae and cyanobacteria species.

Figure 2. Accelerated bio-ageing set up: 1) lamps, 2) sprinkling rails, 3) samples, 4) samples holder, 5) pump, 6) growing medium which is sprinkled over the samples.

The system has been inspired by the device developed by Barberousse et al. [15]. The glass chamber has been equipped with two sprinkling rails made of stainless steel tubes with 1 mm diameter holes drilled every centimeter. Every rail is connected by a rubber tube to a pump submersed in the liquid solution, allowing a 50 L/h flow rate. The solution, containing different types of organisms, is driven by the pump through the sprinkling rails, where it drops onto the specimens and runs on the surface, allowing algal and cyanobacterial cells to adhere and proliferate. The solution is eventually collected at the bottom of the glass chamber and here recycled by the pump and sprayed again. The chamber is equipped with a lighting system working with a cycle of 12/24 light. Lamps were specifically selected for the biological growth, otherwise the darkness is maintained. Two 39 W neon lamps are placed in parallel to the rails, on top of the chamber. During the daylight period, 3 sprinkling cycles lasting 90 minutes each are set, each one starting 4 hours after start of the previous one. During the darkness period no sprinkling cycle is applied. Daylight is the most favorable time for biological growth and this is the reason why the sprinkling cycles are turned on only during this period. The chamber is covered with a non-hermetic lid. Humidity and temperature are measured by a data logger (Lascar Electronics) that records data at least every hour. The temperature range inside the chamber is as low as 15°C during the dark period and it gradually increases up to 23°C during the daylight period. The relative humidity is 90 ± 5% during the dark period and 70 ± 5% during the day.

3.3. Characterization techniques

Different techniques were used to analyze the samples in order to evaluate the biological growth. Solar reflectance, Hunter parameters (L*, a*, b*) [25] and surface roughness have been measured before and after the exposition. Solar reflectance is measured according to ASTM E903 standard test method [26] by a UV-VIS-NIR Spectrophotometer Jasco V-670 with a 150mm integrating sphere. Solar reflectance ρ_{sol} is calculated averaging the spectrum of the surface reflectivity ρ_λ measured in the range from 300 nm to 2500 nm weighted by the solar spectral irradiance $I_{sol,\lambda}$ ($W/(m^2nm)$). The irradiance spectrum used to calculate solar reflectance is AM1GH (air mass 1, global irradiance on a flat horizontal surface) [27,28]:

$$\rho_{sol} = \frac{\int_{300}^{2500} \rho_\lambda I_{sol,\lambda}\, d\lambda}{\int_{300}^{2500} I_{sol,\lambda}\, d\lambda} \tag{3}$$

After 8 weeks of exposure, both non-weathered and weathered samples have been measured to calculate L*, a*, b* values by means of a colorimeter X-Rite SP60. The total color difference ΔE between initial and aged samples has also been calculated.

$$\Delta E = \sqrt{\Delta L^{*2} + \Delta a^{*2} + \Delta b^{*2}} \tag{4}$$

Roughness has been measured using an Alicona IF-Portable profilometer. The measured area is 3.5 mm x 2.6 mm wide, and different profiles have been extracted every 0.25 mm both horizontally and vertically on the selected area, then the average and the standard deviation have been calculated.

Microstructural characterization by an environmental scanning electron microscope (ESEM FEI QUANTA 200) operating in low vacuum mode has been performed to analyze the biological growth on the samples. Moreover, semi-quantitative chemical analyses have been performed on non-weathered samples by using the same device in high vacuum mode and using EDS technique. Pictures of the tested samples have also been taken on a weekly basis and then elaborated trough a code written in Python, in order to obtain a quantitative representation of the progress of biological colonization.

4. Results

On each sample, the different characterizations have been performed before and after the campaign, which lasted 8 weeks. Most of the analyzed materials showed a considerable biological growth on their surfaces. Since only one sample was used for each material, in this preliminary study the only characterization performed on a weekly basis has been image analysis.

4.1. Solar reflectance

As expected, a decrease of solar reflectance generally occurs after the 8 weeks of the campaign. Results are shown in Figure 3. There is just one exception: sample A shows a significant increase in solar reflectance, from 0.745 to 0.822 (see also Figure 4). All other samples show a decrease in solar reflectance which is small in single ply membrane, in particular the green one shows a decrease of just

1%, whereas the highest decrease was shown by sample D with a reduction from 0.763 to 0.459. SRI and T_s change consequently, as shown in Fig. 5.

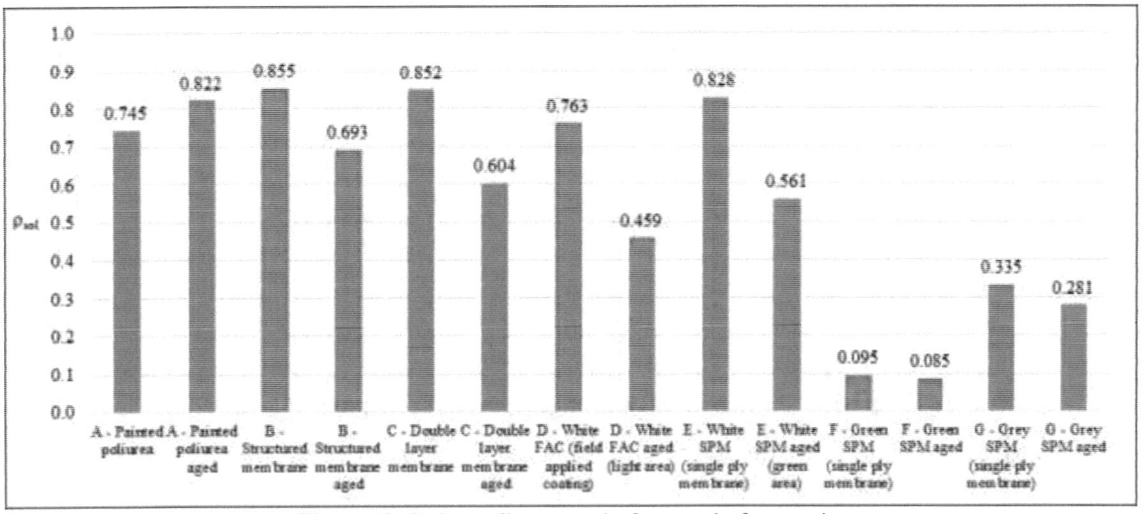

Figure 3. Solar reflectance before and after ageing
(ASTM 903 test method with AM1GH solar spectrum).

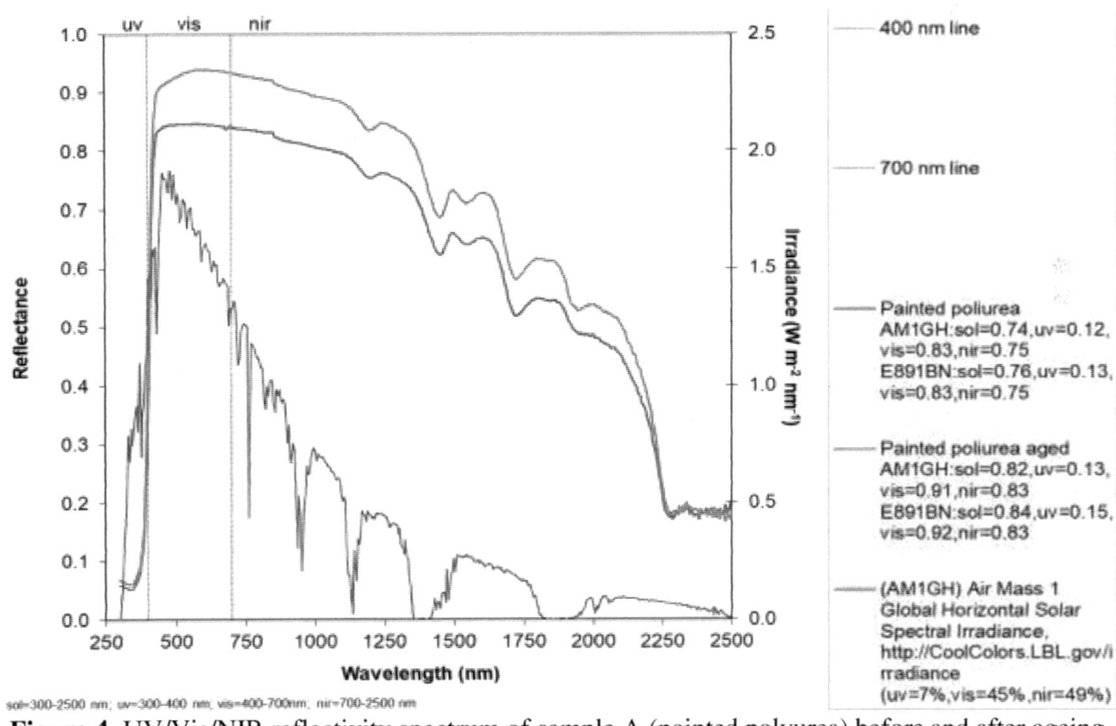

Figure 4. UV/Vis/NIR reflectivity spectrum of sample A (painted polyurea) before and after ageing.

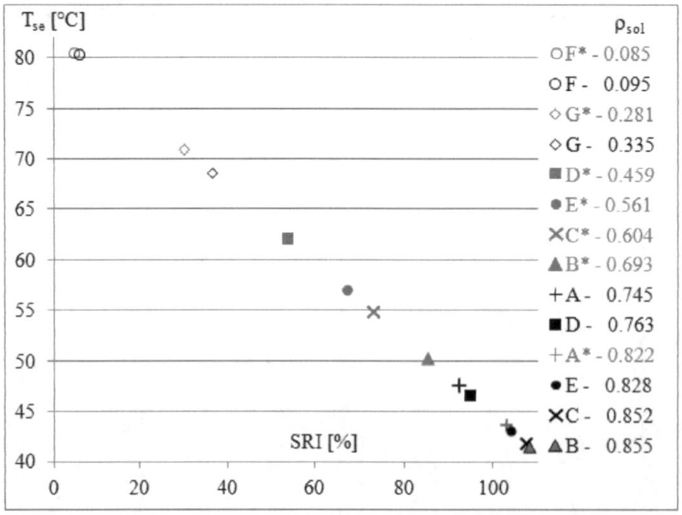

Figure 5. Solar reflectance index (SRI) and surface temperature T_{se} of the new and aged test samples calculated from the measured solar reflectance and typical thermal emittance ($\varepsilon \cong 0.90$) for intermediate wind conditions (6 m/s < v_{wind} < 10 m/s): the same symbol is used for a given material, data in red fonts and marked with an asterisk are those for aged samples.

4.2. Image analysis

To estimate the amount of colonized area, image analysis was used since it provides a nondestructive testing approach and permits subsequent measurement of algae development on the same surface in the different weeks of the test. Photos of each sample have been taken before the aging period and thereafter every week in order to collect images adequate to assess the evolution of the biological film versus time (Figures 6 and 7).

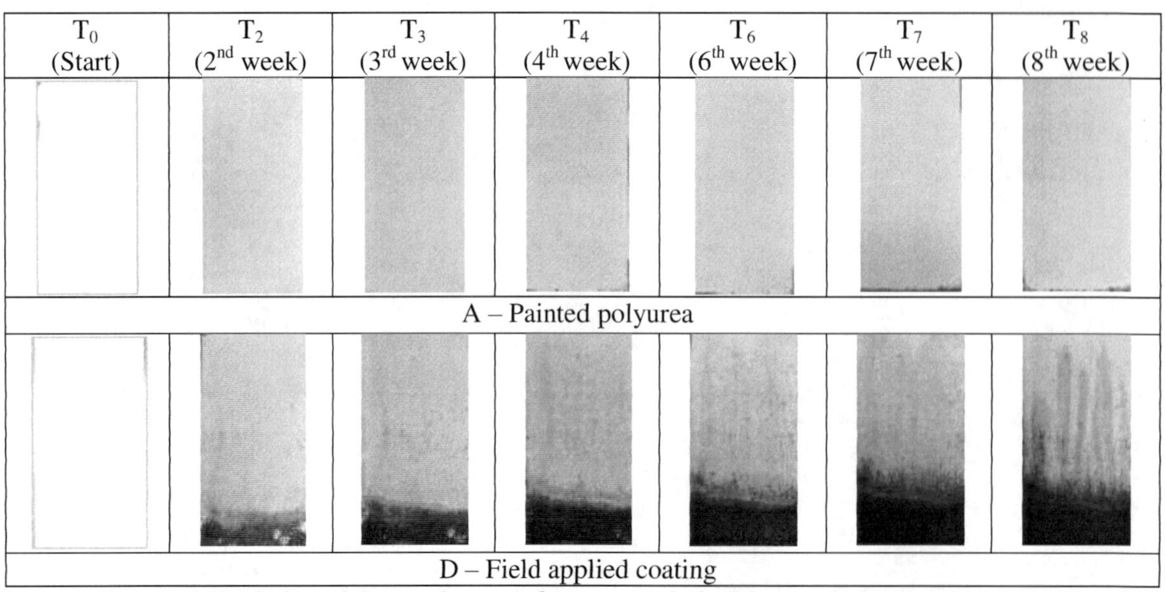

Figure 6. Evolution of the specimen surface aspect during the accelerated testing period for two samples (A and D) with opposite behavior.

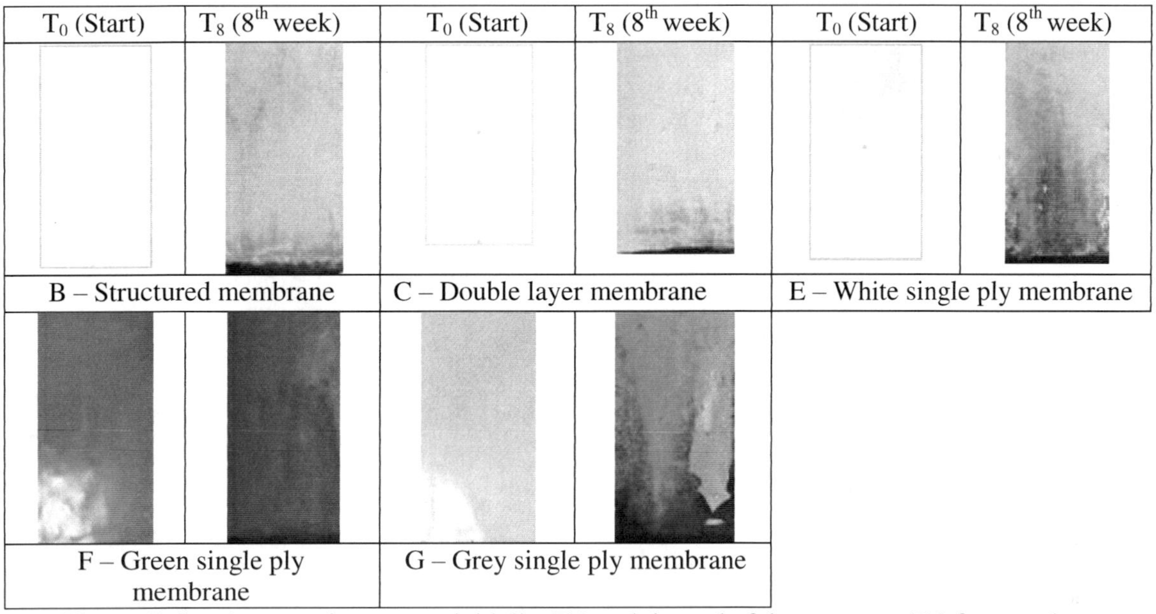

Figure 7. Specimen surface aspect initially (T_0) and the end of the exposure (T_8) for samples B,C,E,F,G.

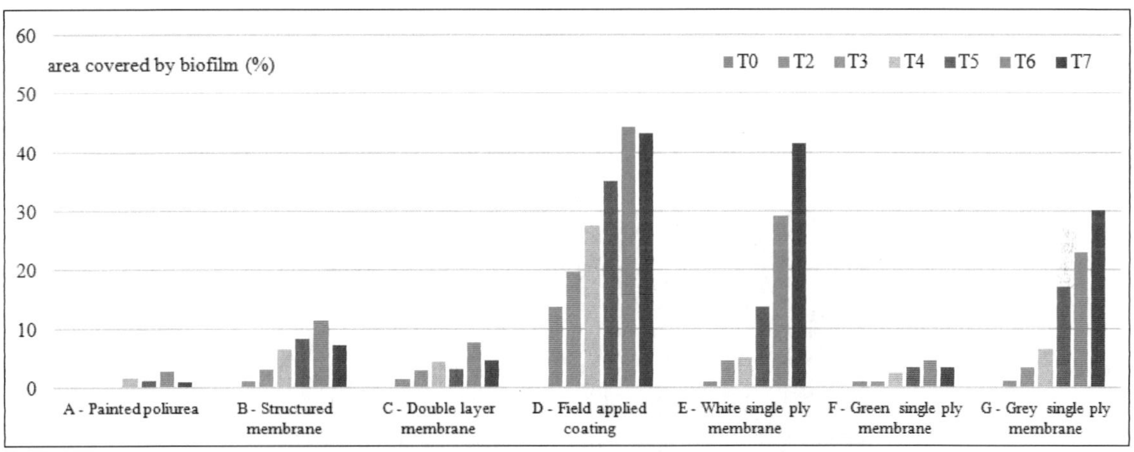

Figure 8. Percentage of surface area covered by biofilm versus time: the condition T_0 is that at start of the ageing period, when no bio growth is visible, whereas for T_2,T_3,T_4,T_5,T_6,T_7 the percentage of covered area is shown week by week, starting from the end of the second week.

Weekly pictures (Figure 6) have been compared in order to monitor the progressive change of the surface aspect and identify a color threshold associated to the onset of biological growth. All pictures have been cropped to exclude every portion but the sample surface and then digitally elaborated. Figure 8 reports the percentage of surface area covered by algae and bacteria week by week during the test period. It is interesting to see how sample A (painted polyuria) was colonized for a percentage of area lower than 5% while sample D (field applied coating) is the most colonized one. It is also worth to notice that week 1 (*i.e.* T_1) was not considered since no growth was visible on the surface. Moreover, except for samples E and G the colonization seems to reach its maximum at T_7 with a small decrease in colonization in the last week of exposure. This can be related to the micro-organisms life cycle.

IOP Conf. Series: Journal of Physics: Conf. Series **923** (2017) 012046 doi:10.1088/1742-6596/923/1/012046

*4.3. L*a*b* and ΔE values.*

To evaluate quantitatively colors and color change [29], five different measurements have been performed on each sample. From these values, mean and standard deviation have been calculated. On aged samples, due to heterogeneity of the surfaces, the number of measured spot was increased to analyze all the shades of colonization. The difference between non-weathered and weathered samples was calculated by means of the total color difference ΔE as defined by Eq. (4). Initial and aged values of L* a* b* are reported in Table 1 together with ΔE values. To calculate ΔE, the average of all aged measurements was calculated. Since in this case only the visible part of the solar spectrum has been considered, the darker is the surface of the aged sample, the higher is ΔE, as expected.

Table 1. L*, a*, b* and ΔE values of analyzed samples

	Initial			Aged			
	L*	**a***	**b***	**L***	**a***	**b***	**ΔE**
A	96.64±0.51	-0.74±0.05	1.73±0.11	89.52±1.43	0.43±2.96	-6.05±6.79	10.61
B	97.45±0.04	-0.95±0.03	2.23±0.1	72.96±5.91	-20.36±4.87	26.40±6.24	39.51
C	97.44±2.27	-0.80±0.06	1.92±0.13	80.89±4.31	-3.42±1.24	27.36±8.05	30.46
D	93.93±0.24	-1.32±0.08	4.88±0.15	66.40±7.36	-24.75±2.37	39.01±4.06	49.71
E	94.16±0.08	-0.75±0.02	2.86±0.04	51.42±6.75	-32.40±5.53	42.51±4.02	66.33
F	30.68±0.18	-20.09±0.15	6.43±0.12	42.82±1.61	-30.35±0.84	14.48±2.38	17.82
G	67.44±0.04	-1.25±0.03	-2.42±0.04	51.4±8.06	-26.11±3.01	41.13±5.5	52.65

4.4. Roughness

Roughness values have been measured by means of a portable profilometer that can also produce, in addition to usual values of R_a, 3D images of the analyzed sample (Figure 9). Because of the technical specifications of the device combined with the high roughness of sample B, it was not possible measure roughness of this materials.

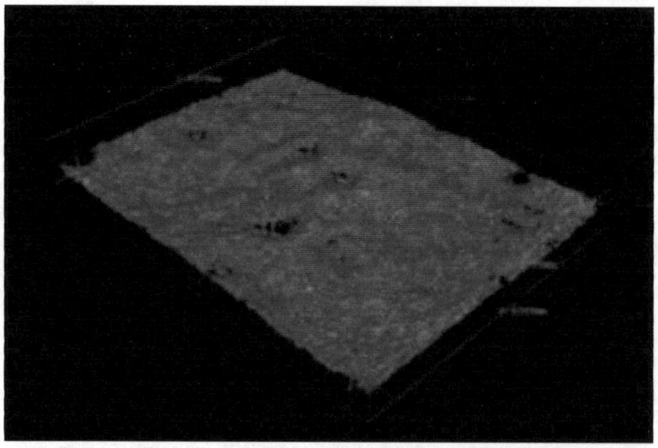

Figure 9. Roughness profile of sample A (Painted polyurea).

As proposed by other authors [30] a correlation between roughness and surface colonization can be made. The only exception was sample A which, despite relevant roughness values, was colonized only for 1% of its area, and F, a green single ply membrane where the image analysis can be affected by the original color of the sample.

Table 2. Sample roughness compared with percentage of biofilm covered area.

	R_a (μm)	St. Dev.(μm)	Covered area (%)
A	1.9	0.54	1.0
C	1.2	0.37	4.6
D	1.4	0.32	43.3
E	1.4	0.58	41.6
F	2.1	0.63	3.4
G	1.4	0.39	30.2

4.5. Chemical analysis

Chemical analysis has been performed on samples before soiling by means of SEM – EDS. In Table 3 the results related only to white samples are reported together with the percentage of covered area. Except for sample B, a lower amount of covered area is associated to samples with a higher amount of TiO_2 pigment such as A and C. The particular texture of sample B (structured membrane) could be the reason why the percentage of colonized area remains lower than 10% since the water runoff probably was helped by the structure of the surface.

Table 3. Results of EDS analysis on white samples correlated with % of covered area

	% Covered area	C	Al	Si	Cl	Ca	Ti	O
A	1.025	21.04	1.98			1.39	10.32	65.26
B	7.280	22.71		0.32	13.10	0.19	1.64	62.04
C	4.588	17.04	0.52	0.61			21.14	60.68
D	43.334	21.61	0.41	3.60	0.17	1.31	6.21	66.70
E	41.647	22.35	0.90		13.70	0.69	1.04	61.31
G	30.160	23.19		0.08	10.71	1.47	1.26	63.29

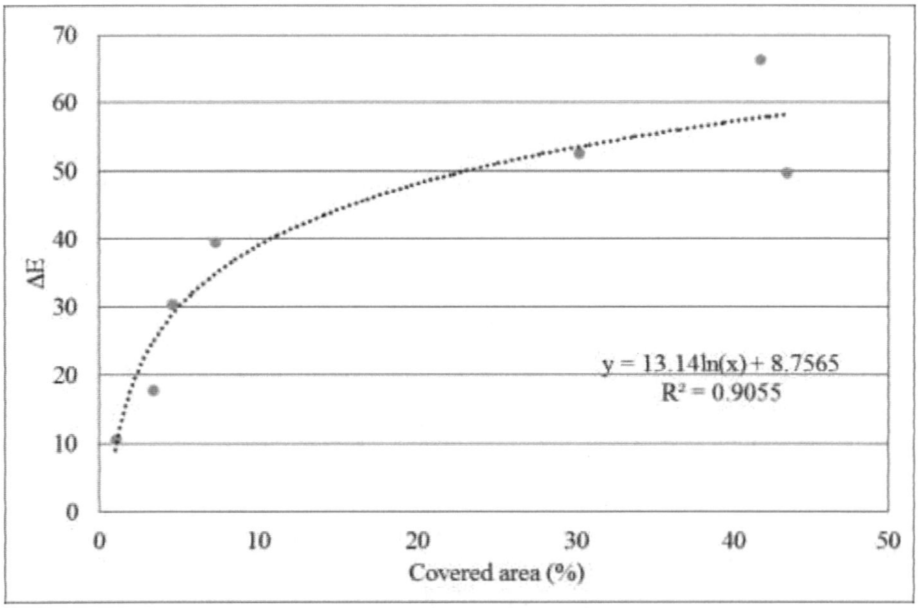

Figure 10. Relations between % of area covered by biofilm and ΔE.

5. Discussion

Biological soiling of building materials influences not just the aesthetic properties but also the thermal performance, with specific regard to the decrease of solar reflectance and its impact on cooling energy. Such decrease can generally be correlated to the percentage of surface area covered by biological growth and denote a remarkable loss such that reported for samples B-G. Sample A represents the only exception in this dataset since it seems not being affected by biofilm, likely thanks to its chemical composition (high amount of TiO_2), thus maintaining initial color and surface characteristics and presenting even a slight increase of solar reflectance after ageing. The percentage of covered area can also be correlated to the total color difference ΔE with a logarithmic function which can be drawn with a $R^2=0.9055$ (Figure 10).

Other tests helped to investigate factors such as roughness, color and composition and their correlation with biological film growth. Generally speaking, low roughness seems to be correlated with a low degree of biological growth. This however presents exceptions which let to think that not just one parameter alone but a synergy between roughness, color, reflectance and chemical composition can influence the growth of biofilm.

6. Conclusions

By this study we found that the test method developed for accelerated ageing can be useful to mimic and accelerate the growth of biological films on different types of building materials which, under the same ageing conditions, behave in deeply different ways in terms of solar reflectance and color, depending on several parameters among which chemical composition, physical properties and roughness can be numbered. The progressive biofilm growth, in fact, can cause a strong decrease in solar reflectance and thus result in summer overheating of built surfaces.

This first experimental campaign was designed to provide an overview of the bio-ageing process on a panel of seven different polymeric based building materials. For each material just one specimen was selected allowing a wider material selection. Just one specimen for each sample could be aged, thus preventing destructive analyses on the surfaces along the ageing time. Further studies will be organized by reducing the range of analyzed materials and thus include at least one specimen to be extracted per week, in order to allow a complete weekly characterization. Analyzing the growing patterns, further changes will be made on the bio-ageing setup such as using lamps with emitting spectra that more effectively stimulate algal growth, and replacing sprinkling rails with nebulizer to guarantee a more homogeneous spraying. Moreover, in order to integrate this method with ASTM D7897-15 for soiling due to simple fouling [31], attempts will be done to use the D7897 soiling mixture for the algae growing medium. In addition to that, natural exposure should be organized in order to find a correlation between laboratory method and what really happens in nature – even if bio-ageing is subject to such a large number of variables that a precise correlation is likely to not exist and the devised approach should be intended more as a comparative benchmark of the long-term resistance to biological ageing rather than the accurate representation of an actual natural process.

References

[1] Berdahl P, Akbari H, Levinson R and Miller W 2008 *Constr. Build. Mater.* **22** 423
[2] Maury-Ramirez A, De Muynck W, Stevens R, Demeestere K and De Belie N 2013 *Cement Concrete Comp.* **36** 93
[3] Ferrari C, Santunione G, Libbra A, Muscio A, Sgarbi E, Siligardi C and Barozzi GS 2015 *Int. J. Design Nature Ecodynamics* **10** 21
[4] Gaylarde CC and Gaylarde PM 2005 *Int. Biodeter. Biodegr.* **55** 131
[5] Tomaselli L, Lamenti G, Bosco M and Tiano P 2000 *Int. Biodeter. Biodegr.* **46** 251
[6] Cheng MD, Pfiffner SM, Miller WA and Berdahl P 2011 *Build. Environ.* **46** 999
[7] Tran TH, Govin A, Guyonnet R, Grosseau P, Lors C, Garcia-Diaz E, Damidot D, Devès O and Ruot B 2014 *Int. Biodeter. Biodegr.* **86** 334
[8] Muscio A 2016 *Int. J. Environ. Sci. Sust. Dev.* **1**, 1

[9] Drirkx I and Grégoire Y 2012 *4th APFAC Congress.*
[10] Cheng MD, Miller W, New J and Berdahl P 2012 *Constr. Build. Mater.* **26**, 516
[11] Mastrapostoli E, Santamouris M, Kolokotsa D, Vassilis P, Venieri D and Gompakis K 2016 *Energ. Buildings* **114** 191
[12] Pisello AL, Castaldo VL, Pignatta G, Cotana F and Santamouris M 2016 *Energ. Buildings* **114** 180
[13] Kultur S and Turkeri N 2012 *Build. Environ.* **48** 164
[14] Kunzel HM, Krus M, Fitz C, Hofbauer W, Scherer C and Breuer K 2011 *XII Int. Conf. on Durability of Bulding Materials & Components*
[15] Barberousse H, Rupt B, Yepremian C and Boulon G 2007 *Build. Environ.* **42** 2555
[16] D'Orazio M and Cursio G 2015 *Costruire in Laterizio* **161** 50
[17] ASTM Int. 2011 *ASTM E1980-11*
[18] USGBC 2015 *LEED v4 for Building Design and Construction* (updated July 1, 2015)
[19] Escadeillas G, Bertron A, Blanc P and Dubosc A 2007 *Mater. Struct.* **40** 1061
[20] Dubosc A, Escadeillas G and Blanc PJ 2001 *Cement Concrete Res.* **31** 1613
[21] *Algal collection Univ. Federico II* 2017 website: www.acuf.net
[22] CCAP 2017 *Culture Collection of Algae & Protozoa* website: www.ccap.ac.uk
[23] CCAP 2017 *Bold's Basal Medium* website: www.ccap.ac.uk/media/documents/BB.pdf
[24] Knight TL and Hammett M 1993 *Masonry Int.* **7** 9
[25] HunterLab 2008 *Application Notes* **8** (9)
[26] ASTM Int. 2012 *ASTM E903-12*
[27] Ferrari C, Muscio A, Libbra A and Siligardi C 2013 *Advances Building Energ. Res.* **7** 244
[28] Levinson R, Akbari H and Berdahl P 2010 *Sol. Energy* **84** 1717
[29] Comm. Int. de l'Eclairage 2008 *CIE S 014-4/E:2007 – ISO 11664-4:2008* (joint standard)
[30] Coutinho ML, Miller AZ and Macedo MF 2015 *J. Cult. Herit.* **16** 759
[31] ASTM Int. 2015 *ASTM D7897-15*

A Solar Chimney for renewable energy production: thermo-fluid dynamic optimization by CFD analyses

S Montelpare[1], V D'Alessandro[2], A Zoppi[2] and E Costanzo[2]

[1] University "G. d'Annunzio" of Chieti-Pescara, Engineering and Geology Department (INGEO)
Viale Pindaro 42, 65127 Pescara (Italy)
[2] Marche Polytechnic University, Industrial Engineering and Mathematical Sciences Department (DIISM)
Via Brecce Bianche 1, 60131 Ancona (Italy)

E-mail: s.montelpare@unich.it

Abstract. This paper analyzes the performance of a solar tower designed for renewable energy production. The Solar Chimney Power Plant (SCPP) involves technology that converts solar energy by means of three basic components: a large circular solar collector, a high tower in the center of the collector and a turbine generator inside the chimney. SCPPs are characterized by long term operational life, low maintenance costs, zero use of fuels, no use of water and no emissions of greenhouse gases. The main problem of this technology is the low energy global conversion coefficient due to the presence of four conversions: solar radiation > thermal energy > kinetic energy > mechanical energy > electric energy. This paper defines its starting point from the well known power plant of Manzanares in order to calibrate a numerical model based on finite volumes. Following that, a solar tower with reduced dimensions was designed and an analysis on various geometric parameters was conducted: on the inlet section, on the collector slope, and on the fillet radius among the SUPP sections. Once the optimal solution was identified, a curved deflectors able to induce a flow swirl along the vertical tower axis was designed.

1. Introduction

The demand for energy has significantly increased in the last decades and, as a direct consequence, the use of fossil fuels, deforestation and increased air pollution are contributing to the exponential growth of greenhouse gases. These are the main causes of global warming and of the recent climatic changes. In this scenario there is a growing need for new technologies able to produce electricity with renewable sources. Solar energy is one of these and many people consider it the main resource for the future.

The Solar Chimney Power Plant (SCPP) or Solar Updraft Power Plant (SUPP) is a technology that combines solar and wind energies in order to produce electricity by means of a turbine. The power plant is composed of three main components: a very large collector operating as a greenhouse, a high tower allowing the updraft movement of the warmer air and a wind turbine placed at the base of the tower and elaborating the ascending flow. These systems are characterized by a very long operational life, poor maintenance costs, the absence of fossil

IOP Conf. Series: Journal of Physics: Conf. Series **923** (2017) 012047 doi:10.1088/1742-6596/923/1/012047

fuels, the absence of water consumption and no greenhouse gases emissions. Thanks to this easy technology this kind of power plant is normally realized with economic and locally accessible materials. The main disadvantage of the SUPP is its very low efficiency due to the high number of energy conversions with multiple η: solar radiation $\rightarrow$ air heating $\rightarrow$ air movement $\rightarrow$ turbine rotation $\rightarrow$ electric production.

The first experimental prototype was realized in 1982 by Prof. Schlaich with his colleagues in Manzanares (Spain), as a result of a joint venture between the German government and a Spanish utility. The pilot prototype had a 194.6 $[m]$ high tower with a diameter of 10 $[m]$ and a collector with a radius of 122 $[m]$. The collector roof was mainly realized in plastic with the inner glass part connecting to the tower. The collector height was about 1.85 $[m]$ above ground. The air temperature increase in the collector was measured up to 17 $[^{\circ}C]$ and the air velocity upcoming to the turbine was 12 $[m/s]$. The turbine had four blades and was arranged in a vertical axis configuration at the base of the tower [1, 2]. The prototype operated with a peak power of about 50 $[kW]$ for seven years from 1983 to 1989 [3].

The Manzanares project launched the start of a series of research works aimed at evaluating the energy potential of SUPP systems all around the world. A full literature review is reported by Zhou et al. [4, 5], by Dhahri and Omri [6], and by Hussain H. Al-Kayiem et al. [7]. Pasumarthi and Sherif [8, 9] studied the performance of a small-scale physical prototype and concluded that an intermediate absorber in the collector had the potential to boost mass flow rate. Gannon and von Backström [10] introduced a thermodynamic cycle approach with an isobaric collector model and analyzed the energy conversion processes with a temperature-entropy diagram of the air standard cycle also considering system losses (Fig.2). Bernardes et al. [11] and Pretorius and Kröger [12] developed more comprehensive analytic collector steady-state models using the Boussinesq approximation to calculate the density. Von Backström and Gannon analyzed turbine performance in a SUPP with radial inflow through inlet guide vanes at the base of the chimney [13]. Schlaich et al., [14] studied a water-filled system placed on the ground under the collector roof as an additional cheap and effective heat storage system. Cottam [15] analyzed the role of the exponential canopy profile shape in power output, and concluded that an exponential canopy with a collector constant cross-sectional flow area produces the greatest power output. Sangi et al. [16] described an analytic model for the SUPP starting from Navier-Stokes equations and validated their findings with respect to a CFD model by also considering the ground heat storage. Fasel [17] numerically investigated the SUPP with unsteady RANS calculations by using ANSYS Fluent and an in-house developed Computational Fluid Dynamics (CFD) code, confirming that the output power follows the cubic scaling predicted by theory. Koonsrisuk and Chitsomboon [18] used CFD technology to investigate the changes in flow properties caused by the variations of flow area.

This paper shows the results of analyses carried out on a SUPP having a reduced dimension with respect to Manzanares (Fig.1); i.e. a tower height of 60 $[m]$ with a diameter of 4 $[m]$ and a collector with a 60 $[m]$ radius. The choice to analyze a different model is related to the advantages of using small dimensions both for numerical and experimental approaches.

2. Thermodynamic Cycle

As shown by Gannon [10], it is possible to analyze the SUPP as an ideal air standard cycle similar to a gas turbine one (Fig.2). Assuming that all components are ideal and that all processes are loss free, the aim was to find a relationship between the plant performance and variables such as chimney height and collector temperature increase.

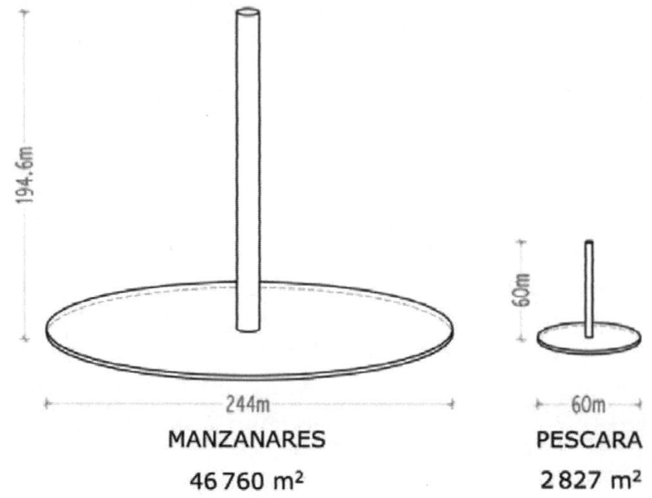

Figure 1: Scaled model.

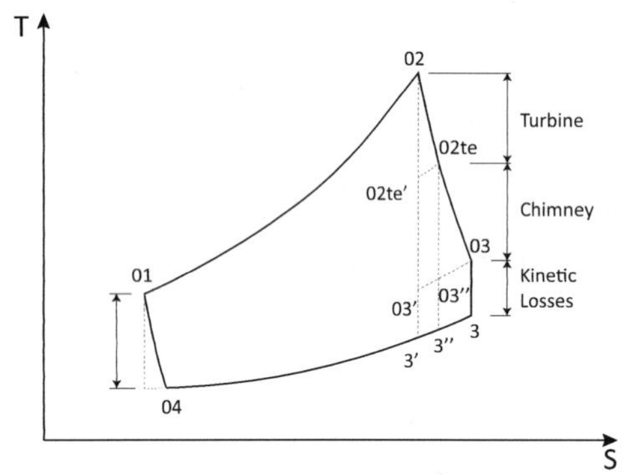

Figure 2: SUPP Thermodynamic air cycle.

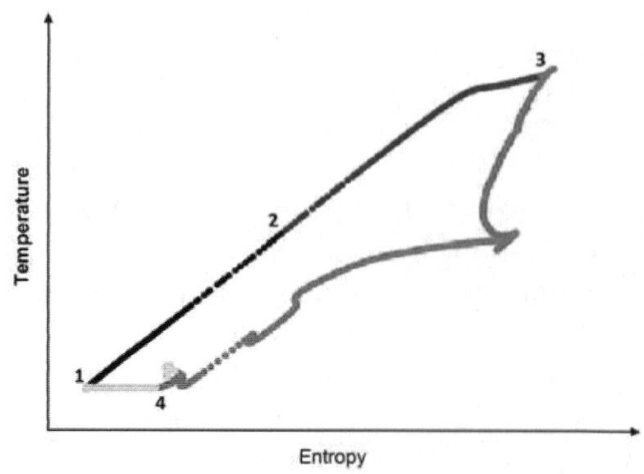

Figure 3: CFD SUPP Thermodynamic cycle.

The process fluid (air) gains heat and so does enthalpy inside the collector (phase $1 \rightarrow 2$) due to the convective exchanges with the roof's internal surface and the absorber upper surface. Kinematic losses related to the collector inlet flow and to the collector-tower junction are normally disregarded in the literature thermodynamic analyses, but actually they significantly modify the available energy of flow that should be converted downstream in turbine and that is instead converted in entropy. Point 2 corresponds to the vertical axis turbine inlet at the base of the chimney tower; the power extracted in the section $02 \rightarrow 02_{te}$ corresponds only to pressure and temperature variations, because of a constant cross sectional area corresponds to same velocities upstream and downstream the turbine. For this reason some authors do not agree with the use of the Betz theory, that is instead largely used in the SUPP analyses; in fact, this paper does not consider the Betz limit suitable. The section $02_{te} \rightarrow 3$ corresponds to the negative gravitational work $g/\Delta z$ plus the friction losses up to the chimney exit and to the kinetic losses of the flow leaving the SUPP. The $3 \rightarrow 4$ and $4 \rightarrow 1$ phases correspond to the fluid loop external to the SUPP.

IOP Conf. Series: Journal of Physics: Conf. Series **923** (2017) 012047 doi:10.1088/1742-6596/923/1/012047

The cycle efficiency may be expressed as:

$$\eta = \frac{\text{ShaftPowerOut}}{\text{SolarPowerIn}} \tag{1}$$

The energy exchanges in the SUPP collector can be obtained by using the thermodynamic first law for flowing fluid:

$$\dot{Q} - \underbrace{\dot{W}'}_{=0} = \dot{m}\left[(h_{02} - h_{01}) + \underbrace{\frac{v_{02}^2 - v_{01}^2}{2}}_{\ll} + \underbrace{g\,(z_{02} - z_{01})}_{\simeq 0}\right] \tag{2}$$

by assuming a constant specific heat value:

$$\dot{Q}_{solar} = \dot{m} \cdot c_p \cdot (T_{01} - T_{02}) \tag{3}$$

The power extracted by the turbine is:

$$\underbrace{\dot{Q}}_{\simeq 0} - \dot{W}_{turb} = \dot{m}\left[(h_{02,\,te} - h_{02}) + \underbrace{\frac{v_{02,\,te}^2 - v_{02}^2}{2}}_{=0} + \underbrace{g\,(z_{02,\,te} - z_{02})}_{\simeq 0}\right] \tag{4}$$

$$\dot{W}_{turb} \cong \dot{m} \cdot c_p \cdot \eta_{is,\,turb} \cdot (T_{02,\,te'} - T_{02}) \tag{5}$$

On the basis of the reported equations, the main solutions for increasing the system efficiency are:

- an increase of the turbine enthalpy drop, with a maximum limit due to a near zero mass flow [10];
- a more detailed design of the collector section in order to reduce energy losses and also to increase the turbine inlet enthalpy.

This paper focuses its attention on solutions able to reduce the kinetic losses at the inlet and outlet of the collector and on the use of a swirl flow to modify the turbine inlet velocity profile. Figure 3 shows the thermodynamic cycle along a streamline of the flow pattern; it is possible to observe a different behavior from point 3 to point 4 because the CFD model does not consider the presence of the turbine. The CFD model exhibits a simultaneous temperature and entropy increase due to the heat transfer with the tower internal surface and to the friction losses; instead, the wind turbine, that is present in the real system, extracts the mechanical work from the flow's enthalpy by inducing a temperature decrease and an entropy increase.

3. Analytic Model

The SUPP collector is the part of the system where the main heat exchanges take place and where the flow enthalpy increase occurs. Several authors have proposed more or less simplified models able to solve the temperature and pressure distributions with an iterative approach.
Figure 5 shows the equivalent electric scheme for a collector section of the analyzed model; the solar radiation **S** directly induces a temperature increase in the absorber surface and in the collector roof by way of a radiative heat exchange, while the air flow exhibits a temperature increase mainly due to the convective heat exchange with the collector internal surfaces.
For the collector roof:

$$\begin{aligned} S_R + h_{r,\,abs\to roof}\,(T_{abs} - T_{roof}) &= h_{c,\,roof\to amb}\,(T_{roof} - T_{amb}) + ... \\ &...h_{r,\,roof\to sky}\,(T_{roof} - T_{sky}) + h_{c,\,roof\to air}\,(T_{roof} - T_{air}) \end{aligned} \tag{6}$$

For the absorber surface:

$$S_P = h_{r,\,abs \to roof}\,(T_{abs} - T_{roof}) + h_{k,\,abs \to ground}\,(T_{abs} - T_{ground}) + \ldots$$
$$\ldots + h_{c,\,abs \to air}\,(T_{abs} - T_{air}) \tag{7}$$

For the air flow:

$$\frac{\dot{m} \cdot c_p}{2\pi r} \cdot \frac{\Delta T_f}{\Delta r} = h_{c,\,roof \to air}\,(T_{roof} - T_{air}) + h_{c,\,abs \to air}\,(T_{abs} - T_{air}) \tag{8}$$

The radiative heat fluxes with the roof collector and with the absorber surface which are related to the solar radiation q_{solar} and to the glass transmission coefficient of the roof τ:

$$S_R = (1 - \tau) \cdot \dot{q}_{solar} \tag{9}$$

$$S_P = \tau \cdot \dot{q}_{solar} \tag{10}$$

The radiative heat exchange coefficient may be evaluated with the subsequent relations [19]:

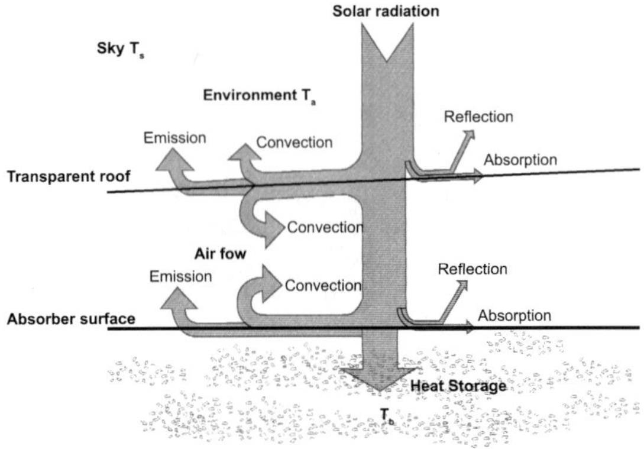

Figure 4: Collector heat fluxes.

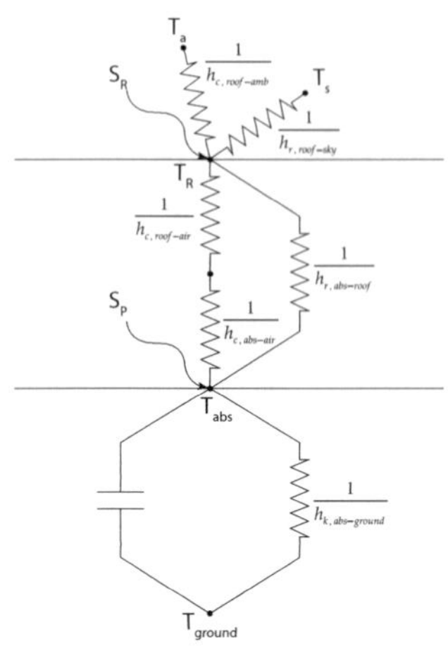

Figure 5: Collector thermal resistance network.

$$h_{r,\,roof \to sky} = \varepsilon_{roof} \cdot \sigma \cdot \left(T_{roof}^2 + T_{sky}^2\right) \cdot (T_{roof} + T_{sky}) \tag{11}$$

$$h_{r,\,abs \to roof} = \frac{1}{\dfrac{1}{\varepsilon_{abs}} + \dfrac{1}{\varepsilon_{roof}} - 1} \cdot \sigma \cdot \left(T_{abs}^2 + T_{roof}^2\right) \cdot (T_{abs} + T_{roof}) \tag{12}$$

The radiative heat exchange coefficient may be evaluated with the subsequent relations [5]:

$$h_{c,\,roof \to amb} = 5.7 + 3.8 \cdot v_{wind} \tag{13}$$

$$h_{c,\,roof \to air} = \frac{\left(f/8\right)(\mathrm{Re} - 1000)\,\mathrm{Pr}}{1 + 12.7\left(f/8\right)^{1/2}\left(\mathrm{Pr}^{2/3} - 1\right)}\left(\frac{k}{d_h}\right) \tag{14}$$

The pressure equation of the system needs to be used in order to evaluate the unknown mass flow rate, which is necessary to iteratively solve the analytic model is finally reported [16]:

$$p = H_{ch} \cdot \rho_a \beta_a g \left(T_f \left(r = r_i\right) - T_a\right) + \frac{\dot{m}^2 \rho_a \left(r_o^{-2} - r_i^{-2}\right)}{8 H_c^2 \pi^2 \rho_f^2} - \frac{\dot{m}^2}{A_c^2 \rho_{f,c}} \frac{c_F}{H_c} \left(r_o - r_i\right) + \ldots$$
$$\ldots - 2 \frac{\dot{m}^2}{A_{ch}^2 \rho_{f,ch}} \frac{H_{ch} c_F}{d_{ch}} - \frac{8}{27} \frac{\dot{m}^2}{A_T^2 \rho_{ch}} \tag{15}$$

The solution to this equations system allows reveals the temperature of SUPP components by varying the solar radiation. As reported by Sangi [16], for a solar flux of $1000[W/m^2]$, the absorber temperature shows an average temperature of $70[°C]$, while the internal roof surface increases up to about $50[°C]$(Fig.6). These results were used in this paper as boundary conditions

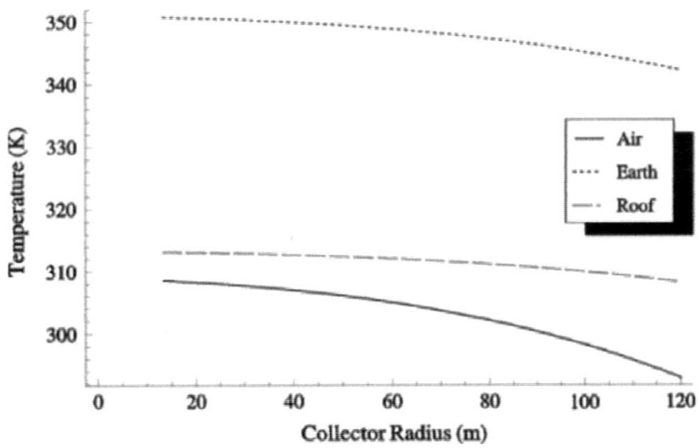

Figure 6: SUPP temperature trends @ q=1000 $[W/m^2]$.

for the numerical analysis, by setting fixed temperatures for the absorber surface and for the internal roof surface. The power of the SUPP is instead evaluated by scaling the available pressure obtained in the numerical simulation in absence of the turbine [5]:

$$x = \frac{\Delta p_{turb}}{\Delta p_{poten}} \tag{16}$$

$$P = \eta_{turb} \left(x \Delta p_{poten,\, no\, turb}\right) \left(\sqrt{1-x} \cdot v_{inlet\, turb,\, no\, turb}\right) \tag{17}$$

The output power estimated using this scaling method, however, was recently found to be significantly lower than that calculated under a turbine load condition, based on the assumed x value of 0.8 [20].

4. Numerical Analysis

Numerical simulations were carried out by using the STAR-CCM+ commercial software with an Unsteady Reynolds-Averaged Navier-Stokes (URANS) approach. A summary of the selected options:

- A turbulent flow; i.e. the Rayleigh number for the collector dimensions and the temperature assumptions is largely greater than 10^8.
- A realizable $k - \epsilon$ turbulence model.

- Implicit with a first order temporal discretization.
- Air as ideal gas with compressible flow and a gravity model that allows the inclusion of the body force due to gravity in the momentum equations. In the full buoyancy model, the buoyancy force is evaluated directly from the difference between air density and a constant reference density though the compressibility factor for air is considered. Other authors [21] observed that neglecting air compressibility inside or outside the SUPP results in inaccuracy in the calculated results.
- Segregated approach for $p - u$ decoupling.

A 2D approach was selected for all simulations, thanks to the axial symmetry of the problem, except for the curved guide vanes analysis where a flow swirl is present and also a three-dimensional behavior is observed. By using a 2D model the overall number of mesh elements was reduced and so the computational effort. Polyhedral mesh was used for all domain extent (Fig.8) except

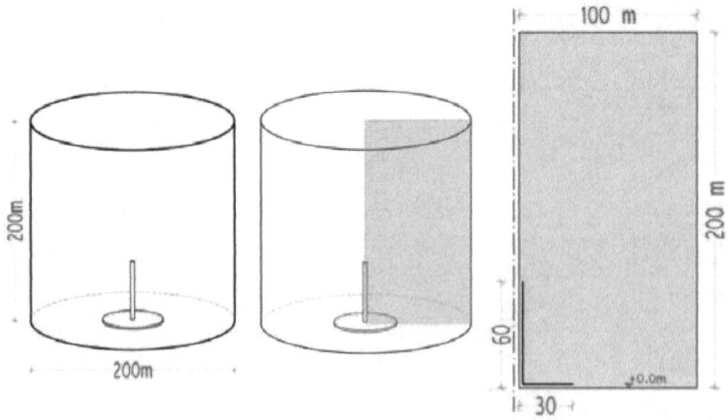

Figure 7: Computational domain.

near walls where 5 prism layers were selected with a stretch of 1.2 and a prism layer thickness of 0.3$[m]$; in this way and by selecting the two layer all y^+ wall treatment model, it was possible to obtain the values of the y^+ between 70 and 160 inside the SUPP. The CFD domain extends 3

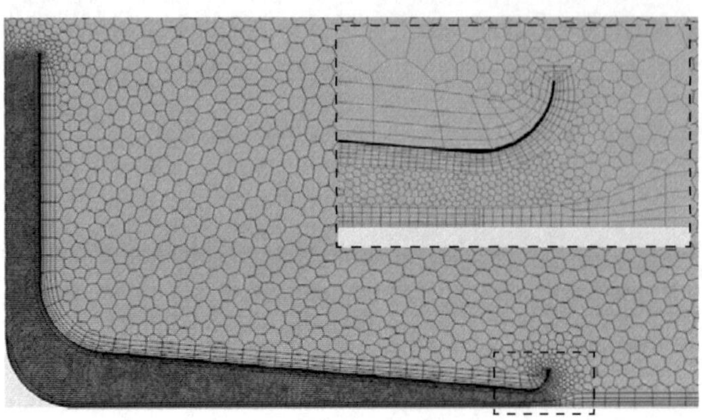

Figure 8: Adopted mesh.

times the SUPP characteristic length both in the vertical and horizontal directions; this distance between the external boundaries and the solar tower is large enough to ensure the development of an undisturbed flow pattern. The choice to include the outside domain by avoiding a boundary condition at the inlet and outlet of the SUPP derives from the results of previous analyses

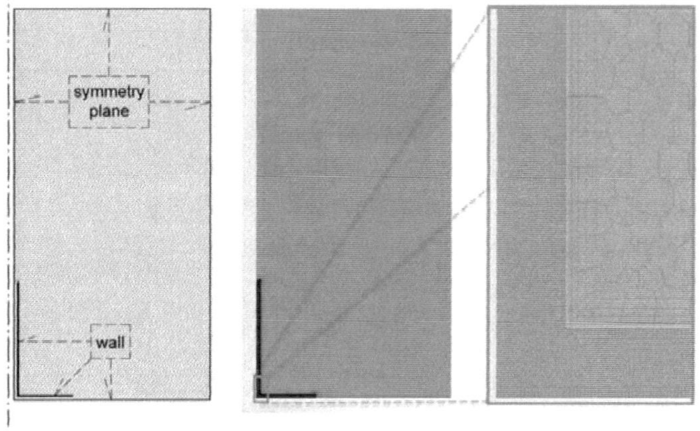

Figure 9: Bounday conditions.

of other authors that concluded it to be more realistic solar chimney inlet and outlet conditions [22].
As reported in table 1 several CFD analyses were carried out: i.e. six 2D models and one 3D model. Model 0 simulates the full scale Manzanares SUPP, while the others refer to the 1:2 scaled model prototype (Fig.10). Model 1 is a scaled version of the Manzanares plant so it has a flat roof and no corner fillet. Model 2 exhibits no round corners but a sloped roof, as suggested by several authors [15, 5]. Model 3 introduces aerodynamic optimization so as to reduce kinematic losses at the collector inlet, by using a bellmouth profile, and at the chimney section change, by rounding both the corner with the collector roof and the floor profile in the proximity of the plant central axis. Model 4 exhibits an additional change with a sloped floor that ensures a quasi-constant flow passage area and a small divergence to contrast the wall boundary layer development. Model 5 is designed as model 3, but with a lower chimney; i.e. $20[m]$ vs. $60[m]$. Finally, model 6 introduces, when compared to model 4, curved guide vanes so to induce a swirl motion. The numerical simulation for the model corresponding to the Manzanares prototype

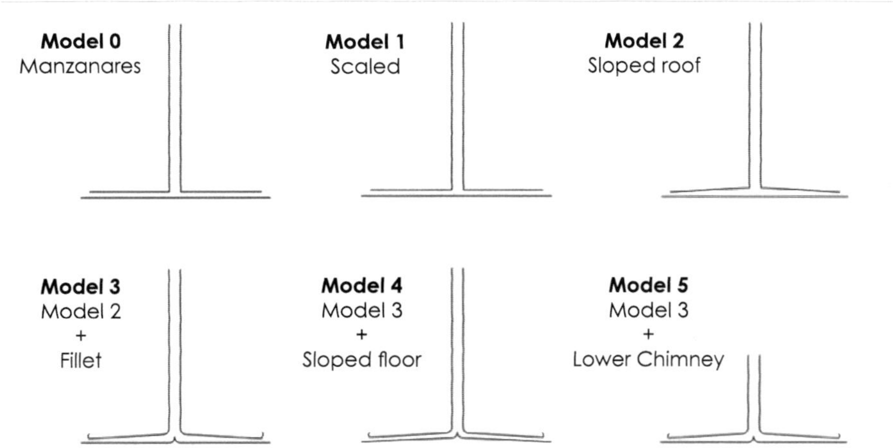

Figure 10: CFD Models.

(without turbine) shows a temperature increase of about 15 $[°C]$ and an updraft velocity of about 10 $[m/s]$ (Fig.12) with respect to real values (with turbine) of 7–9 $[m/s]$ (Fig.11) and a maximum collector temperature rise of 15–18 $[°C]$. It is so possible to affirm that there is a

Table 1: CFD models.

Name	Description	Used approach
Model 0	Manzanares SUPP	2D
Model 1	Scaled SUPP	2D
Model 2	Model 1 with roof sloped	2D
Model 3	Model 2 with shape optimization	2D
Model 4	Model 3 with floor sloped	2D
Model 5	Model 3 with lower chimney	2D
Model 6	Model 4 with curved vanes	3D

good correspondence for the CFD model because it was observed [17] that the turbine presence induces lower updraft velocities in the order of 2-3 $[m/s]$ and higher temperatures due to the lower flow velocity and longer exchange time in the collector. The comparison shows a good agreement also in terms of available power: Manzanares experimental data exhibit values ranging from 26.3 to 38.6$[kW]$ for a solar irradiation of 1000$[Wm^{-2}]$, while numerical CFD analyses show a potential value, estimated on the updraft velocity, of 26.2$[kW]$.

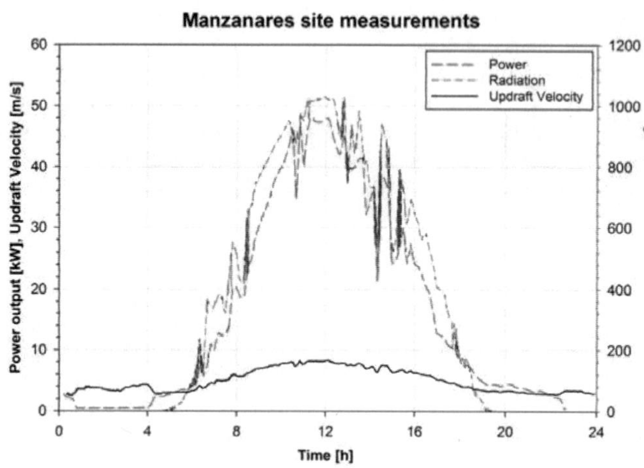

Figure 11: Manzanares measurements.

On this basis subsequent analyses were carried out with the same boundary conditions.

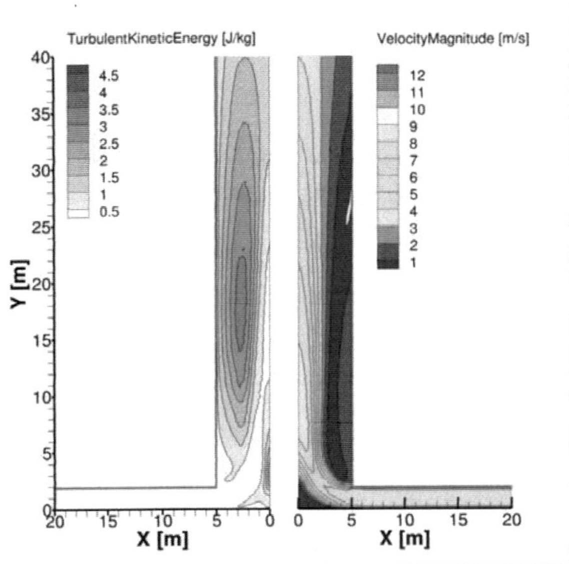

Figure 12: Model-0 Turbulent Kinetic Energy and Velocity Magnitude.

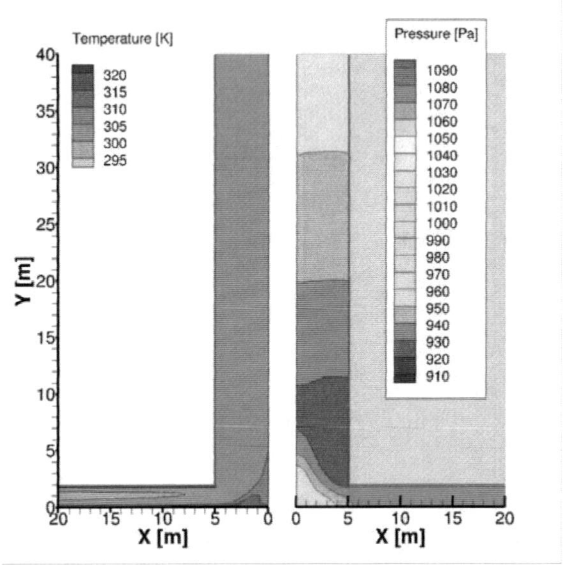

Figure 13: Model-0 Temperature and Relative pressure.

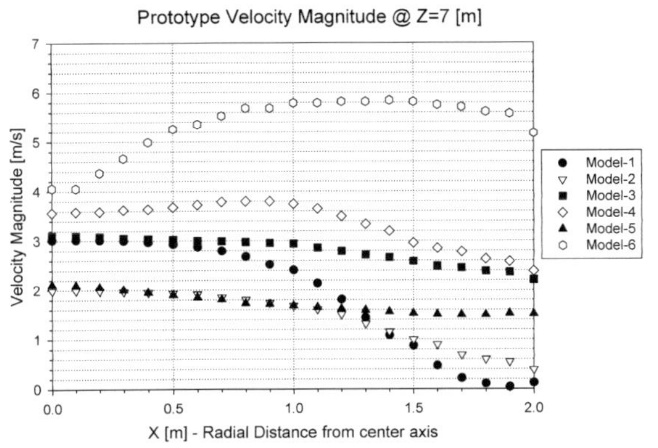

Figure 14: Velocity comparison @ Z=7 [m].

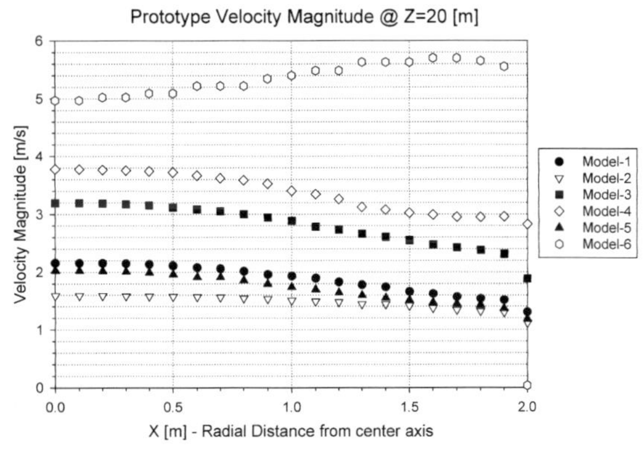

Figure 15: Velocity comparison @ Z=20 [m].

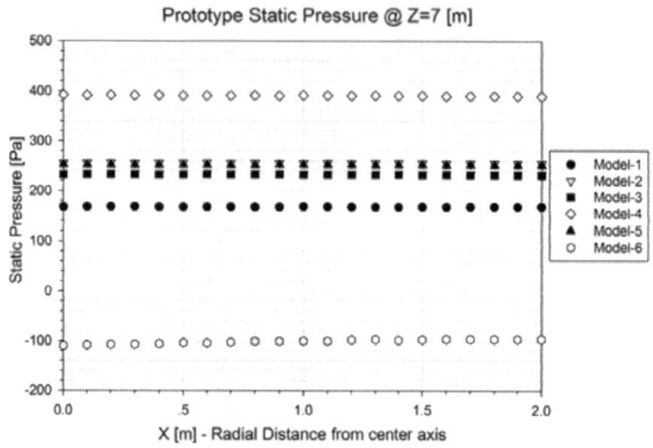

Figure 16: Static pressure comparison @ Z=7 [m].

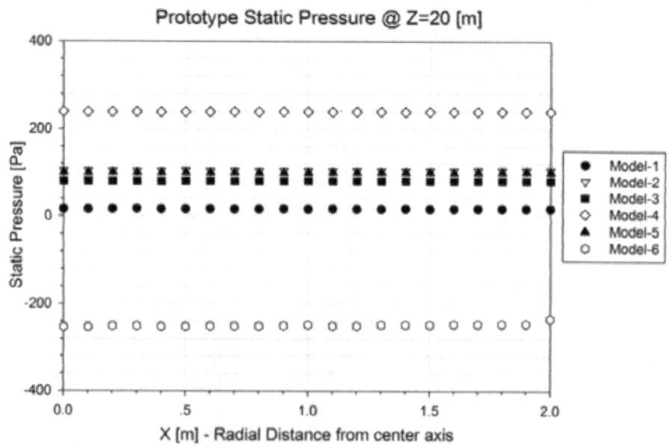

Figure 17: Static pressure comparison @ Z=20 [m].

Model 1 shows, as expected, a drastic reduction of SUPP performance parameters, with a maximum updraft velocity of about 3 $[m/s]$, a very low temperature increase and also lower pressure decrease to be used in the turbine energy production. A strong flow separation occurs around the corner between the collector roof and the chimney and a stagnation zone is instead present over the floor near the plant central axis.
Model 2 shows the worst behavior among the tested models, with lower velocities than model 1 and with a maximum updraft value of about 2$[m/s]$; it however exhibits an higher temperature increase and also lower pressure decrease with respect to the model 1. The lower velocities can be associated to the flow expansion in the final section of the collector.
Model 3 shows better behavior than model 1, with a maximum updraft value of the same order (about 3 $[m/s]$), but with a more uniform velocity profile along the radius. It also shows higher temperatures and also lower pressure.
Model 4 has very good performance with maximum velocities of 3.8 $[m/s]$ and an almost uniform velocity distribution. This better performance is associated to the constant flow passage area that ensures a reduction of kinetic losses.
Model 5 has very poor performance that is comparable to model 2; this behavior is due to the lower chimney that does not promote the updraft natural convection.
Finally model 6 exhibits the best performance with velocities up to 5.8 $[m/s]$ and a radial distribution that shifts higher velocities to the outer radius. This behavior is related to the swirl motion and could be more useful for the turbine energy conversion developing higher torques. Also the figures 16 and 17 underline the observations reported by analyzing the radial distribution of the updraft velocities: the suggested adjustments show an improvement of the available pressure starting from the model 1 up to the model 4, with the exception of model 5 that behaves worse due to the Chimney's lower height. A completely different behavior is showed by the model 6 due to the pressure depression induced by the swirling flow, that increases the elaborated flow and substantially alters the SUPP's behavior.

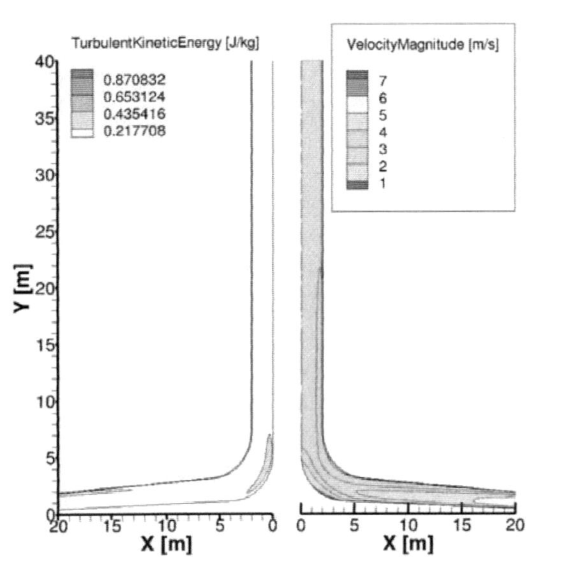

Figure 18: Model-4 Turbulent Kinetic Energy and Velocity Magnitude.

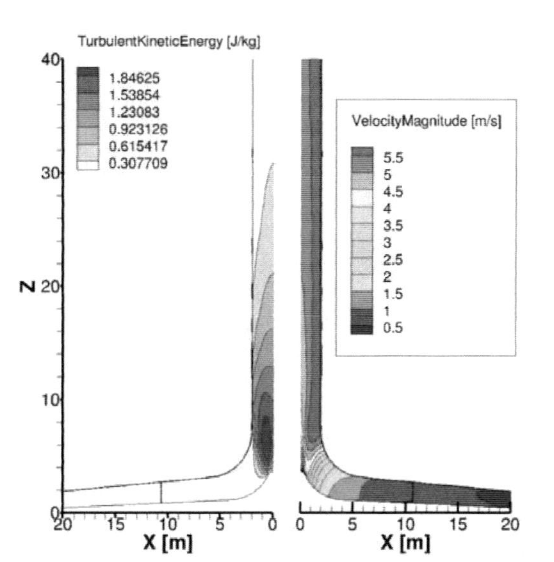

Figure 19: Model-6 Turbulent Kinetic Energy and Velocity Magnitude.

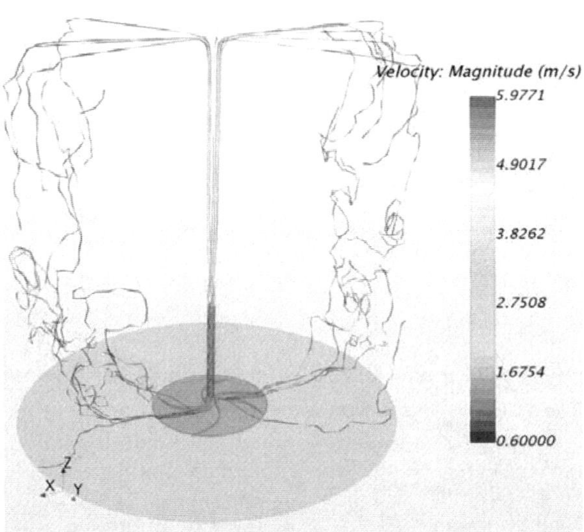

Figure 20: Model-6 Flow pattern.

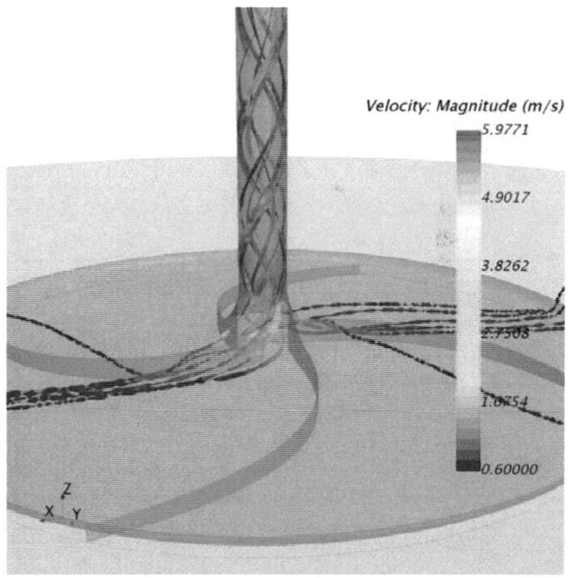

Figure 21: Model-6 Flow swirl.

5. Conclusions

This paper is aimed to numerically analyze the behavior of a SUPP scaled prototype. Once analyzed, the analytic model defined the boundary conditions for the subsequent numerical simulations. Several models were tested: initially a model 0 corresponding to the Manzanares experimental setup was computed in order to verify the correctness of the used approach. Following that, other configurations were tested so as to identify the best solutions; these correspond to sloped roof and floor, to a bellmouth profile at the collector inlet, and to the use of curved vanes able to induce swirl.

6. Nomenclature

Pr	Prandtl number
Ra	Rayleigh number
Re	Reynolds number
η	Efficiency
β	Volumetric expansion coefficient
ρ	Density $[\frac{kg}{m^3}]$
σ	Stephan-Boltzmann constant $[\frac{W}{m^2K^4}]$
τ	Radiative trasmissivity coefficient
ε	Radiative emissivity
a	Ambient
f	Flow
i	Inner
o	Outer
ch	Chimney
is	Isoentropic
$\dot{m}$	Mass flow rate $[\frac{kg}{s}]$
$\dot{Q}$	Heat flux $[W]$
$\dot{W}'$	Mechanical Power $[W]$
A_c	Collector cross sectional area $[m^2]$
A_ch	Chimney cross sectional area $[m^2]$
A_T	Turbine cross sectional area $[m^2]$
g	Gravitational acceleration $[\frac{m}{s^2}]$
h	Enthalpy $[\frac{J}{kg}]$
H_{ch}	Chimney height $[m]$
p	Pressure $[Pa]$
T	Temperature $[°C]$
v	Velocity $[\frac{m}{s}]$
z	Geodetical height $[m]$
c_p	Specfic Heat $[\frac{J}{kgK}]$
f	Friction coefficient
h_c	Convective heat transfer coefficient $[\frac{W}{m^2K}]$
h_k	Conductive heat transfer coefficient $[\frac{W}{m^2K}]$
h_r	Radiative heat transfer coefficient $[\frac{W}{m^2K}]$
r	Collector radius $[m]$
S_P	Floor absorbed solar radiation $[\frac{W}{m^2}]$
S_R	Roof absorbed solar radiation $[\frac{W}{m^2}]$
$\dot{q}_{solar}$	Solar irradiation $[\frac{W}{m^2}]$

References

[1] Haaf W, Friedrich K, Mayr G and Schlaich J 1983 *Int. J. Solar Energy* **2** 3–20
[2] Haaf W 1984 *Int. J. Solar Energy* **2** 141–161 ISSN 0142-5919
[3] Schlaich J 1995 *The Solar Chimney: Electricity From the Sun.* (Edition Axel Menges)
[4] Zhou X, Wang F and Ochieng R M 2010 *Renewable Sustainable Energy Rev.* **14** 2315–2338 ISSN 13640321
[5] Zhou X and Xu Y 2016 *Sol. Energy* **128** 95–125 ISSN 0038092X
[6] Dhahri A and Omri A 2013 *Int. J. Eng. Adv. Technol.* **2** 2249–8958
[7] Al-Kayiem H H and Aja O C 2016 *Renewable Sustainable Energy Rev.* **58** 1269–1292 ISSN 13640321
[8] PASUMARTHI N and SHERIF S A 1998 *Int. J. Energy Res.* **22** 443—461
[9] PASUMARTHI N and SHERIF S A 1998 *Int. J. Energy Res.* **22** 277—288
[10] Gannon A J and von Backström T W 2000 *J. Sol. Energy Eng.* **122** 133 ISSN 01996231
[11] dos S Bernardes M, Voß A and Weinrebe G 2003 *Sol. Energy* **75** 511–524 ISSN 0038092X
[12] Pretorius J and Kröger D 2006 *Sol. Energy* **80** 535–544 ISSN 0038092X
[13] von Backström T and Gannon A 2004 *Solar Energy* **76** 235–241 ISSN 0038092X
[14] Schlaich J, Bergermann R, Schiel W and Weinrebe G 2005 *J. Sol. Energy Eng.* **127** 117–124
[15] Cottam P, Duffour P, Lindstrand P and Fromme P 2016 *Sol. Energy* **129** 286–296 ISSN 0038092X
[16] Sangi R, Amidpour M and Hosseinizadeh B 2011 *Sol. Energy* **85** 829–838 ISSN 0038092X
[17] Fasel H F, Meng F, Shams E and Gross A 2013 *Sol. Energy* **98** 12–22 ISSN 0038092X
[18] Koonsrisuk A and Chitsomboon T 2013 *Energy* **51** 400–406 ISSN 03605442
[19] Choi Y J, Kam D H, Park Y W and Jeong Y H 2016 *Energy* **112** 200–207 ISSN 03605442

[20] Zhou X, Yuan S and Bernardes M A d S 2013 *Int. J. Heat Mass Transfer* **66** 798–807 ISSN 00179310
[21] Zhou X, Yang J, Xiao B, Hou G and Xing F 2009 *Appl. Therm. Eng.* **29** 178–185 ISSN 13594311
[22] Shams E, Gross A and Fasel H 2011 *Performance analysis of solar chimneys of different physical scales using CFD* parts a, b, and c ed ISBN 9780791854686

IOP Conf. Series: Journal of Physics: Conf. Series **923** (2017) 012048 doi:10.1088/1742-6596/923/1/012048

Experimental characterization of a solar cooker with thermal energy storage based on solar salt

G Coccia[1], G Di Nicola[1], S Tomassetti[1], G Gabrielli[1], M Chieruzzi[2] and M Pierantozzi[3]

[1]Università Politecnica delle Marche, Dipartimento di Ingegneria Industriale e Scienze Matematiche, Via Brecce Bianche 12, 60131, Ancona (Italy)

[2]Università di Perugia, Dipartimento di Ingegneria Civile e Ambientale, UdR INSTM, Strada di Pentima 4, 05100, Terni (Italy)

[3]Università degli Studi di Camerino, Scuola di Ateneo Architettura e Design, Lungo Castellano Sisto V 36, 63100, Ascoli Piceno (Italy)

E-mail: g.coccia@univpm.it

Abstract. High temperature solar cooking allows to cook food fast and with good efficiency. An unavoidable drawback of this technology is that it requires nearly clear-sky conditions. In addition, evening cooking is difficult to be accomplished, particularly on the winter season during which solar radiation availability is limited to a few hours in the afternoon in most of countries. These restrictions could be overcome using a cooker thermal storage unit (TSU). In this work, a TSU based on solar salt was studied. The unit consists of two metal concentric cylindrical vessels, connected together to form a double-walled vessel. The volume between walls was filled with a certain amount of nitrate based phase change material (solar salt). In order to characterize the TSU, a test bench used to assess solar cooker performance was adopted. Experimental load tests with the TSU were carried out to evaluate the cooker performance. The obtained preliminary results show that the adoption of the solar salt TSU seems to allow both the opportunity of evening cooking and the possibility to better stabilize the cooker temperature when sky conditions are variable.

1. Introduction

Solar cooking is considered as one of the simplest and attractive ways of the utilization of solar thermal energy [1]. Energy for cooking is one of the fundamental uses in developing countries, where wood is still the primary energy source. In most of rural areas of Africa, the energy demand for cooking is supplied by non-commercial fuels (e.g., firewood, agricultural waste, cow dung, kerosene); in India, the energy required for cooking accounts for 36% of total primary energy consumption and 90% of rural households depend on biomass fuels [2]. In addition to the environmental and economic issues, the firewood use also causes some serious health problems such as burns, eye disorders, and lung diseases.

Solar cookers represent a possibility to meet the energy demand in the domestic sector of developing countries where, generally, there is abundance of solar radiation; e.g., a mean daily solar radiation of 5-7 kWh/m^2 and more than 275 sunny days in a year have been estimated [3]. However, the large-scale dissemination of solar cookers still remains limited as most of these devices are only used for research purposes [4]. The main obstacles to the dissemination of the technology are the resistance to acceptance as it is a new technology, variable nature of solar radiation, limited space availability in urban areas, and higher initial costs [5].

In the last years, solar box cookers showed considerable developments in terms of performance and design [1]. A solar box cooker usually includes an insulated box with a transparent glass cover and mirrors to reflect direct solar radiation into the box. The inner part of the box is painted in black in order to maximize the absorption of solar energy. Box cookers are simple to manufacture, can be employed with minimal attendance during the cooking process, and can keep food warm for a long period of time.

A solar box cooker prototype with a high concentration ratio (10.78) was designed, realized, and tested by some of the authors in a previous work [6], based on their experience on parabolic trough collectors [7, 8, 9, 10]. Experimental tests with and without load were carried out, and showed that the prototype is able to cook food fast and at high temperature. Load tests were conducted with both water and peanut oil; 1 kg of the former fluid could be boiled in about 11 minutes, while the latter fluid was able to reach temperatures higher than 200 °C in around 41 minutes.

However, an unavoidable drawback of high temperature cooking obtained through elevated solar concentration is that only direct solar radiation can be exploited. This condition limits the cooker usability to clear-sky days and makes evening cooking difficult. These restrictions could be removed by adopting a cooker thermal storage unit (TSU). In particular, a TSU based on phase change materials (PCMs) allows both the opportunity of evening cooking and the possibility to better stabilize the cooker temperature when sky conditions are variable. Although a number of PCM-based TSUs was designed for solar cooker applications [11], none of those adopted solar salts to be used for temperatures up to about 150 °C. Therefore, a solar-salt-based TSU was designed, realized, and tested in DIISM (Department of Industrial Engineering and Mathematical Sciences) to be used with our solar cooker prototype.

2. Materials and Methods

The PCM-based thermal storage unit (TSU) described in this paper was designed and manufactured to store solar energy during sunshine periods. The stored thermal energy can be utilized to cook food in the evening or when solar radiation is low. A mixture composed by KNO_3, $NaNO_2$, and $NaNO_3$ (also commercially known as HITEC [12]) was used as latent heat storage material.

The following subsections describe the solar box cooker, the TSU, and the PCM mixture, respectively.

2.1. Solar Box Cooker

A schematic of the considered solar box cooker prototype is shown in Figure 1. The cooker consists of a wooden large box containing a zinc-coated steel frame with the function of cooking chamber (also referred to as absorber). The chamber was painted with a special selective coating (SOLKOTE HI/SORB-II) that, respect to a common black paint, has an absorptance in the solar spectrum ranging from 0.88 to 0.94, and an emissivity from 0.20 to 0.49, depending on the dry film thickness and substrate. In the chamber, there is a vessel support able to rotate of 360°, in order to maintain the pots in balance.

A double transparent cover on the top of the box allows the solar radiation to be transmitted to the cooking chamber. Each cover has a diameter of 460 mm and is made of tempered glass, which is high resistant and suitable for solar thermal applications (the declared transmittance in the solar spectrum is about 0.90 and the reflection coefficient is 0.08).

The higher part of the box is circumscribed by a double row of booster mirrors (Figure 2). The mirrors were manufactured with special aluminum reflective foils (MIRO-SUN Weatherproof Reflective 90) glued on phenolic compound elements. Respect to traditional aluminum foils, these

IOP Conf. Series: Journal of Physics: Conf. Series **923** (2017) 012048 doi:10.1088/1742-6596/923/1/012048

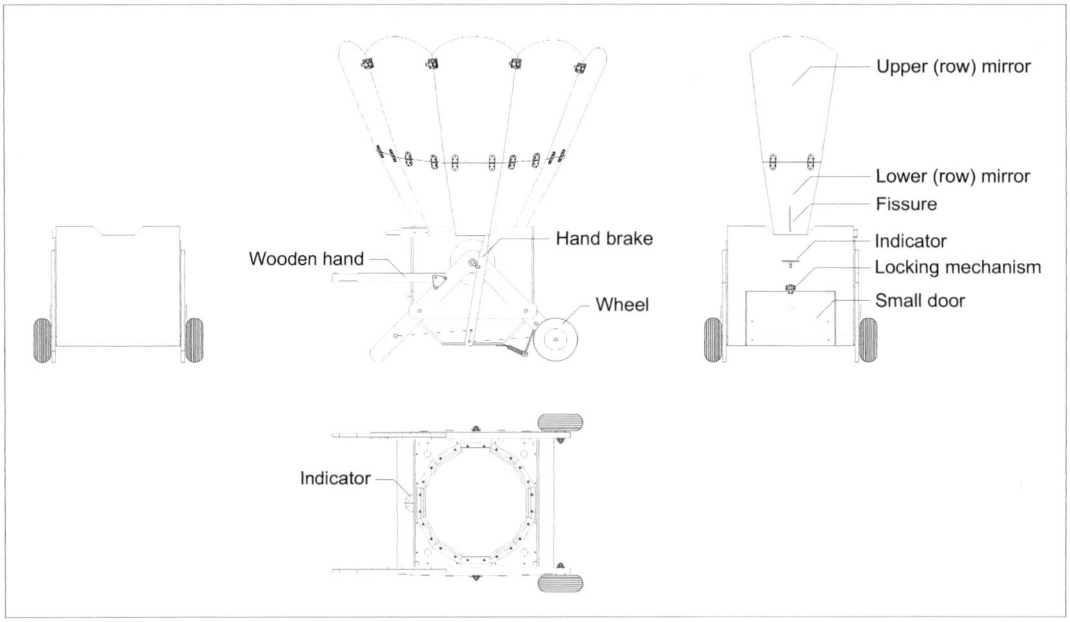

Figure 1. Solar box cooker prototype.

mirrors are able to withstand atmospheric agents and guarantee an overall reflection of about 0.94 in the solar spectrum, with a negligible dependence on the solar incident angle. Each row consists of 12 mirrors and has a different inclination angle respect to the horizontal plane. The upper row has an inclination angle of 70.45° and each mirror, which is 630 mm long, can be reclined to reduce the space occupied by the cooker when it is not working. Instead, the mirrors in the lower row (inclination angle of 63.80°) are trapezoidal and 616 mm long. One of the lower mirrors includes a fissure which allows solar radiation to be projected on an indicator, used to evaluate the correct cooker alignment respect to the sun. The overall aperture area is $1.789\,\mathrm{m}^2$.

The prototype has two border wooden hands and two wheels that allow both its movement and its azimuthal orientation. In addition, a zenithal orientation is possible as the cooker is able to rotate around the horizontal axis.

2.2. Thermal Storage Unit

A thermal storage unit used to hold the cooking vessel was designed and realized as shown in Figure 3. The TSU has two hollow concentric stainless steel cylinders of diameter 260 mm and 220 mm; the internal pot is 150 mm deep, while the external pot is black in order to enhance solar energy absorption. The annular space was filled with PCM, while the internal pot was used to load the testing fluid. The two pots were connected through four stainless steel bolts. Two small stainless steel tubes were welded to two opposite bolts in order to host two K-type thermocouples, used to measure the PCM temperature in two different points ($T_{\mathrm{PCM},1}$ and $T_{\mathrm{PCM},2}$). Another K-type thermocouple, fixed at the center of the inner pot, was used to measure the testing fluid temperature.

2.3. Molten Salt Based PCM

The saline nitrates used in this work were $NaNO_2$, $NaNO_3$, and KNO_3 (purchased from Sigma-Aldrich and Campoverde Srl). Starting from these base components, the PCM was prepared according to the recommendations reported by [12]. The anhydrous nitrates were obtained by heating in two stages: at 140 °C for 4 hours, and at 250 °C for 8 hours. Once this thermal treatment was accomplished, the salts were carefully handled to avoid water absorption.

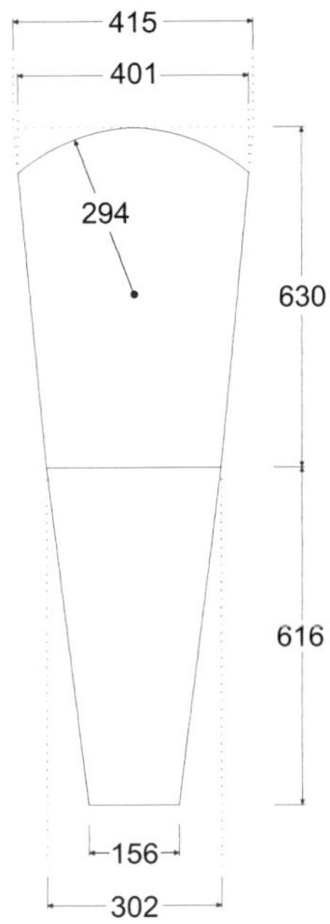

Figure 2. Solar box cooker mirrors.

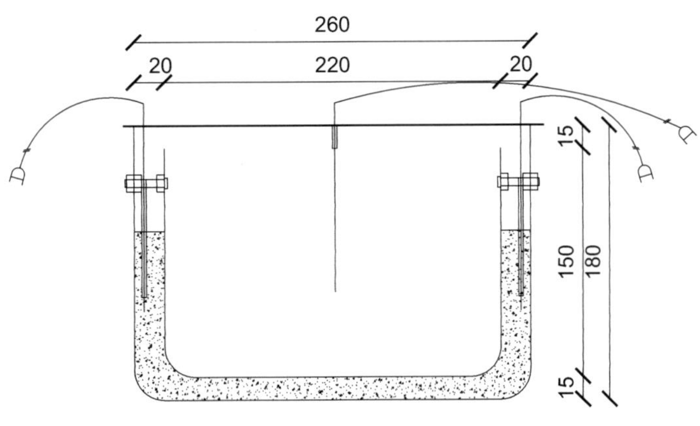

Figure 3. Thermal storage unit.

Afterwards, the mixed formulation (53 wt% KNO_3, 40 wt% $NaNO_2$, and 7 wt% $NaNO_3$ for an overall mass of about 4 kg) was inserted in the annular volume of the thermal storage unit. Finally, the TSU including the salts was inserted in an electric furnace and heated at 350 °C for about 2 hours. In this way, the PCM was definitely realized and the TSU was ready to be used for testing.

The ternary mixture is reported to have a melting point of about 142 °C and a fusion enthalpy of 82 kJ/kg [12]. In order to verify the phase transitions, the thermal storage unit with the PCM was put in an electric furnace to be heated up to melting. Then, the furnace was turned off to let the PCM cooling and solidifying. In this way, it was possible to determine the PCM temperature curve, as shown in Figure 4 and 5.

As can be noted from Figure 4, a first transition occurs at about 94 °C due to water evaporation. Instead, the actual PCM melting point can be observed at about 143 °C, confirming

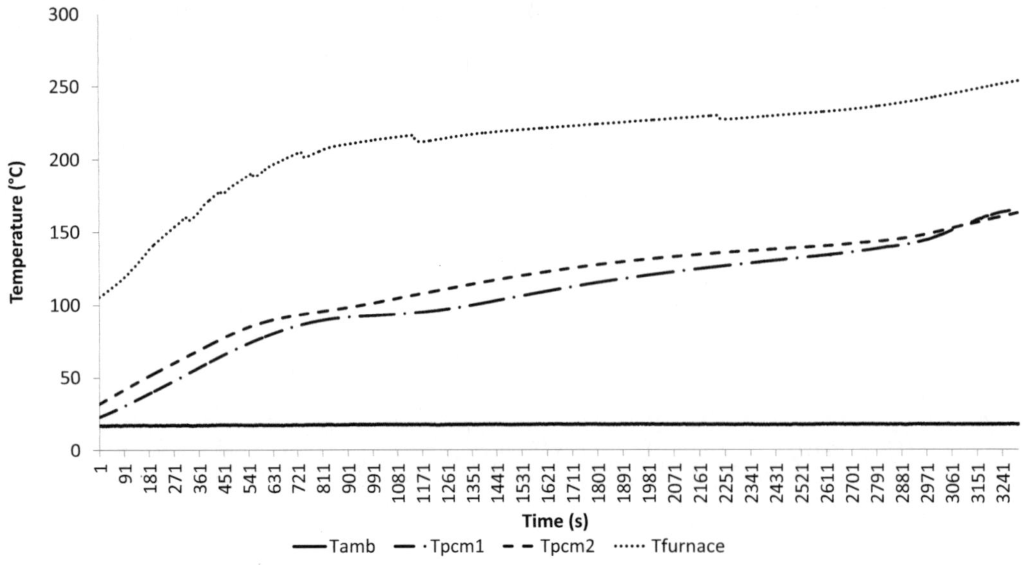

Figure 4. Solar salt temperature curve: melting.

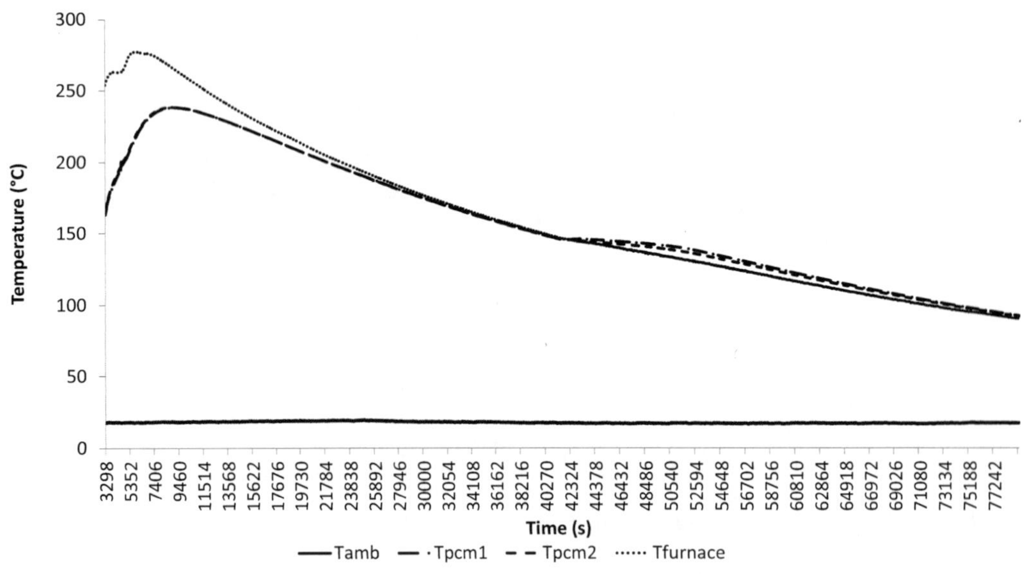

Figure 5. Solar salt temperature curve: solidification.

the results obtained by [12]. The furnace was turned off when the solar salt reached a temperature of approximately 240 °C (Figure 5). The cooling curve shows that solidification took place in the same melting temperature range.

3. Experimental Tests and Results

Two outdoor tests were carried out in March 2017 on the DIISM roof (latitude 43.5867 N, longitude 13.5150 E) to characterize the PCM-based solar box cooker. Silicone oil (Rhodorsil Oil 47 V 100) was used as testing fluid to find the prototype most relevant thermodynamical parameters. This fluid was chosen for its high stability and good thermal conductivity.

IOP Conf. Series: Journal of Physics: Conf. Series **923** (2017) 012048 doi:10.1088/1742-6596/923/1/012048

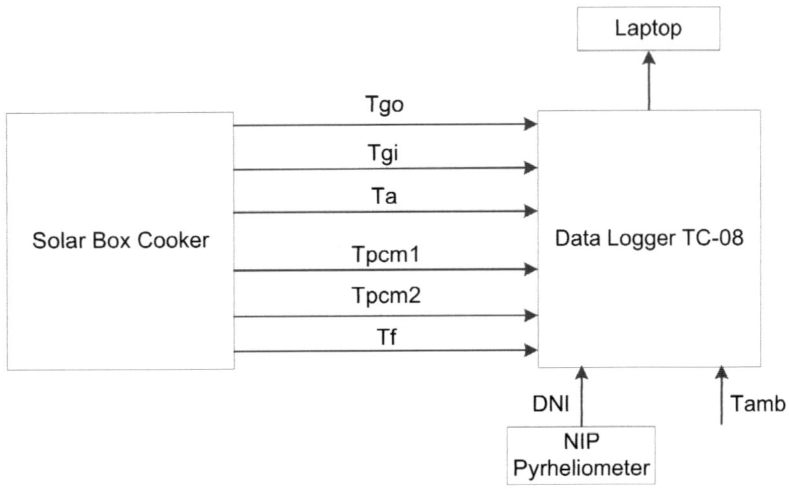

Figure 6. Solar box cooker test bench.

3.1. Test Bench

The solar box cooker was characterized adopting the test bench reported in Figure 6.

K-type thermocouples were used to measure temperature of ambient air (T_{amb}) and in different points of the cooker: outer (T_{go}) and inner glass cover (T_{gi}), absorber (T_{a}), PCM (two opposite points, $T_{\text{PCM},1}$ and $T_{\text{PCM},2}$), testing fluid (T_{f}). The thermocouple measuring T_{go} was positioned on the external side of the outer glass, near the wooden frame, and it was shielded by solar radiation. Instead, the sensor measuring T_{gi} was attached on the internal side of the inner glass, near the wooden frame (the thermocouple was not allowed to "see" solar radiation). The absorber temperature (T_{a}) was measured with a thermocouple fixed on the bottom of the cooking chamber.

The direct normal irradiance, *DNI*, was measured through a first-class normal-incidence pyrheliometer (NIP) mounted on a solar tracker. The pyrheliometer has a 1 second time response, a temperature dependence of $\pm 1\%$ in the range from -20 to $40\,^{\circ}\text{C}$, and shows a linear relationship of $\pm 0.5\%$ in the range 0–$1400\,\text{W/m}^2$. Only direct solar radiation was measured since the solar box prototype considered in this work has a high concentration ratio and hence it is not able to exploit diffuse solar energy [9].

The signals provided by the K-type thermocouples and the pyrheliometer were acquired by a Pico Technology TC-08 data logger.

3.2. Load Tests

The first load test, carried out on March 14, 2017, started at 10:32 and ended at 17:14. The average direct normal irradiance was $462.48\,\text{W/m}^2$ and the average ambient temperature was $12.73\,^{\circ}\text{C}$. The internal cylinder of the TSU was filled with 2.9 kg of silicone oil.

When the PCM was completely melted (at about 14:15), the solar cooker was darkened to simulate absence of solar radiation and was left cooling down. Figure 7 shows the temperatures and the solar radiation detected during the test. It is possible to note that solar radiation was intermittent during the test, especially in the first hours; however, the presence of the thermal storage unit facilitated the temperature stability of the testing fluid respect to other cooker parts (note the 12:00–13:00 time period). In addition, it is evident that both the oil and the PCM temperature decrease slowly when solar radiation is no longer available, in contrast to the other cooker elements.

The second load test was conducted on March 16, 2017. It began at 09:46 and ended at 16:49.

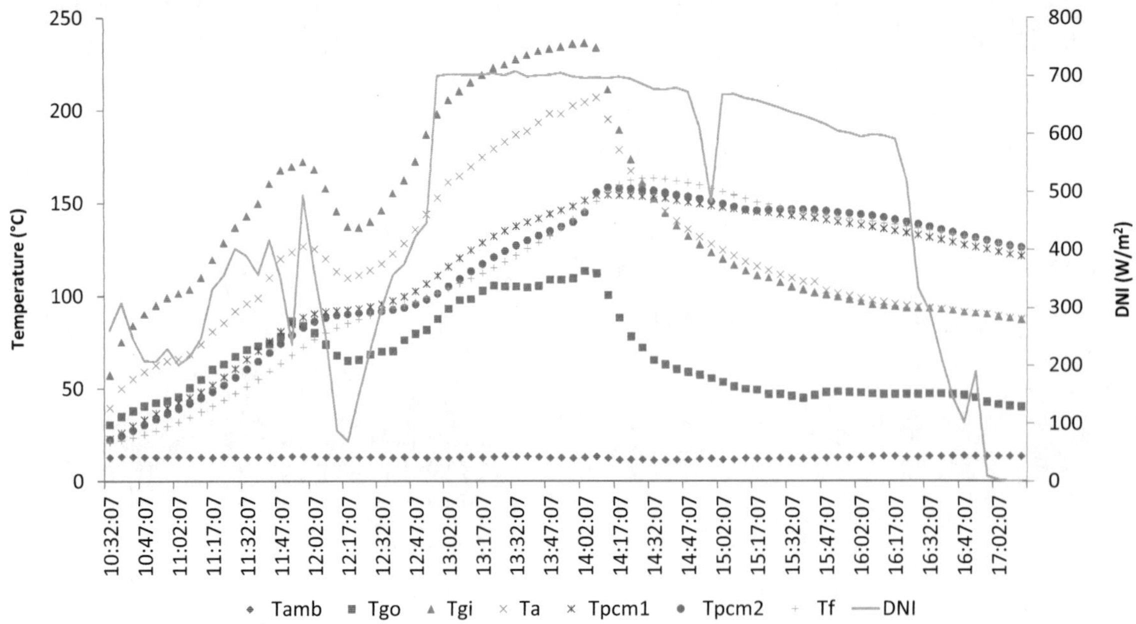

Figure 7. Test 1 (14/03/2017).

The average direct normal irradiance was 656.75 W/m^2, while the average ambient temperature was 17.67 °C. Again, the TSU was filled with 2.9 kg of silicone oil. The test results are reported in Figure 8. Thanks to a greater and constant solar radiation, in this second test the silicone oil was able to reach a temperature higher than the one obtained in the first test. In the same fashion of the previous experiment, the solar cooker prototype was obscured when the PCM reached melting (12:00); results show that the oil and the PCM temperatures are effectively stabilized during the cooling down period, respect to the other cooker elements.

In order to characterize the cooker performance, several parameters were calculated based on the obtained experimental data. The first parameter was the time required to take the oil temperature from 40 °C to the PCM melting, Δt_1 (s). This parameter was used to derive the specific boiling time (min m^2/kg), defined in this work as the time required for a cooker of aperture A_a to take 1 kg of oil from 40 °C to the PCM melting temperature [13]:

$$t_\mathrm{s} = \frac{\Delta t_1 A_\mathrm{a}}{m} \tag{1}$$

where A_a (m^2) is the cooker aperture area and m (kg) is the mass of testing fluid. Another parameter is the characteristic boiling time (min m^2/kg), generally used as a parameter for making comparisons between various solar cooker designs under different solar radiation levels [13]:

$$t_\mathrm{c} = t_\mathrm{s} \frac{DNI_\mathrm{av}}{DNI_\mathrm{ref}} \tag{2}$$

where DNI_av (W/m^2) is the average direct normal irradiance during the time interval Δt_1, while DNI_ref is a reference direct normal irradiance equal to 900 W/m^2.

The average overall solar cooker thermal efficiency is [13]:

$$\eta_\mathrm{av} = \frac{m\, c\, \Delta T}{DNI_\mathrm{av} A_\mathrm{a} \Delta t_1} \tag{3}$$

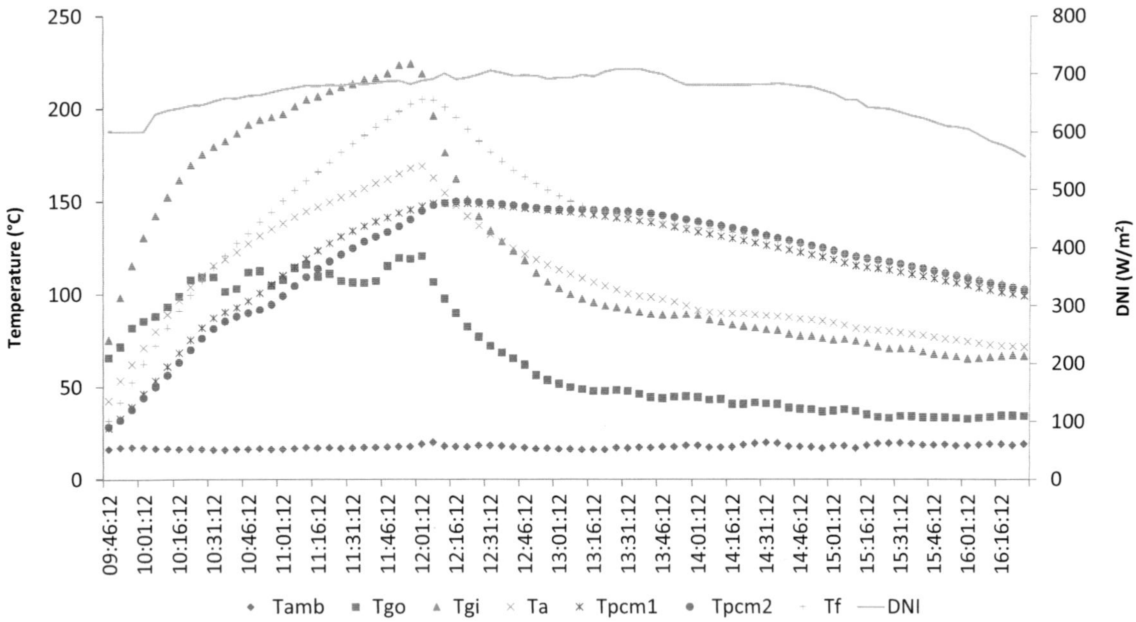

Figure 8. Test 2 (16/03/2017).

where c (J/(kg K)) is the oil specific heat and ΔT (°C) is the temperature difference between the maximum cooking fluid temperature and the ambient temperature.

In addition, [14] introduced the first figure of merit, F_1 (°C/(W/m^2)), which is defined as:

$$F_1 = \frac{T_{\text{a,max}} - T_{\text{amb}}}{DNI} \tag{4}$$

where $T_{\text{a,max}}$ (°C) is the maximum temperature reached by the absorber, while T_{amb} (°C) and DNI (W/m^2) are, respectively, the corresponding ambient temperature and direct normal irradiance measured when the stagnation temperature is reached. For the solar cooker under study, F_1 was determined to be equal to 0.39 °C/(W/m^2) [6].

A second figure of merit, F_2, was introduced to involve the temperature increase measurement with time of a known amount of fluid placed in the cooker. It is defined as [14]:

$$F_2 = \frac{F_1 m c}{A_{\text{a}} \Delta t_1} \ln \left[\frac{1 - \frac{1}{F_1}(T_1 - T_{\text{amb,av}})/DNI_{\text{av}}}{1 - \frac{1}{F_1}(T_2 - T_{\text{amb,av}})/DNI_{\text{av}}} \right] \tag{5}$$

where Δt_1 is the time interval during which the fluid temperature rises from T_1 to T_2, while DNI_{av} and $T_{\text{amb,av}}$ are, respectively, the average direct normal irradiance and the average ambient temperature over the time interval Δt_1.

Table 1 reports a summary of the aforementioned parameters calculated during the oil heating period. It can be noted that, due to higher solar radiation and ambient temperature, test 2 allowed to reach better results respect to test 1. In particular, the results obtained during test 2 are only slightly worse than those obtained by the authors with a comparable test [6], where the testing fluid was peanut oil and the environmental conditions were more favorable. This seems to confirm that the solar salt thermal conductivity is good and that the chosen thermal storage mass does not penalize too much the fluid heating phase.

A summary of the cooling phase is reported in Table 2. The time required by the oil to drop its temperature from T_2 to 130 °C, Δt_2 (s), is similar for the two tests; however, it should be

IOP Conf. Series: Journal of Physics: Conf. Series **923** (2017) 012048 doi:10.1088/1742-6596/923/1/012048

Table 1. Heating test summary.

Quantity	Test 1	Test 2
Date	14/03/2017	16/03/2017
Start	11:19	09:53
End	14:34	12:03
A_a (m^2)	1.789	1.789
T_1 (°C)	40	40
T_2 (°C)	163.59	205.73
Δt_1 (s)	11 706	7782
$T_{amb,av}$ (°C)	12.77	16.79
DNI_{av} (W/m^2)	511.11	662.04
m (kg)	2.9	2.9
t_s (min m^2/kg)	120.36	80.01
t_c (min m^2/kg)	68.35	58.86
η_{av}	0.05	0.09
F_1 (°C/(W/m^2))	0.39	0.39
F_2	0.11	0.17

Table 2. Cooling test summary.

Quantity	Test 1	Test 2
Date	14/03/2017	16/03/2017
Start	14:34	12:03
End	16:58	14:30
$T_{amb,av}$ (°C)	12.50	17.59
Δt_2 (s)	8664	8788
$\Delta T_f/\Delta t_2$ (°C/min)	0.23	0.52

noted that T_2 was equal to 163.59 °C during the former test, while it was 205.73 °C during the latter. In other words, the silicone oil cooling down ratio, $\Delta T_f/\Delta t_2$ (°C/min), was worse for test 2. In this test, the temperature difference between the silicone oil and the PCM is due to the fact that no lid was used. In fact, thanks to a higher solar radiation, the oil was able to heat faster than the PCM. This behavior could be avoided if a lid was adopted, but in this case the heating phase would have been slower. Anyway, from Figure 8 it can be observed that, once the testing fluid and the PCM reach the same temperature, the TSU is able to stabilize the fluid temperature with good results.

4. Conclusions

This study concerns the thermodynamic characterization of a solar box cooker equipped with a thermal storage unit (TSU) based on a solar salt mixture PCM (53 wt% KNO_3, 40 wt% $NaNO_2$, 7 wt% $NaNO_3$). The mixture, having a melting point of about 142 °C, allowed to keep the testing fluid temperature stabilized in a near temperature range. Starting from the experimental results, the most relevant thermodynamic parameters were determined. It was found that the cooker performance during the heating phase is not significantly penalized. On the other hand, the presence of the TSU proved to stabilize and extend the utilization of a solar box cooker when solar radiation is absent or intermittent. Even though the reported results are promising, further tests are necessary to determine other cooker and TSU thermodynamic characteristics, e.g. their behavior when a lid is used.

References

[1] Cuce E and Cuce P 2013 *Appl. Energy* **102** 1399–1421
[2] Pohekar S, Kumar D and Ramachandran M 2005 *Renew. Sustain. Energy Rev.* **9** 379–393
[3] Nahar N 2003 *Energy Convers. Manage.* **44** 1323–1331
[4] Yettou F, Azoui B, Malek A, Gama A and Panwar N 2014 *Renew. Sustain. Energy Rev.* **37** 288–306
[5] Muthusivagami R, Velraj R and Sethumadhavan R 2010 *Renew. Sustain. Energy Rev.* **14** 691–701
[6] Coccia G, Di Nicola G, Pierantozzi M, Tomassetti S and Aquilanti A 2017 *Sol. Energy* **155** 781–792
[7] Coccia G, Latini G and Sotte M 2012 *J. Renew. Sustain. Energy* **4** 023110
[8] Coccia G, Di Nicola G and Sotte M 2015 *Renew. Energy* **74** 727–736
[9] Coccia G, Di Nicola G and Hidalgo A 2016 *Parabolic Trough Collector Prototypes for Low-Temperature Process Heat* (Springer)
[10] Coccia G, Di Nicola G, Colla L, Fedele L and Scattolini M 2016 *Energy Convers. Manage.* **118** 306–319
[11] Sharma A, Chen C, Murty V and Shukla A 2009 *Renew. Sustain. Energy Rev.* **13** 1599–1605
[12] Fernández A, Galleguillos H, Fuentealba E and Pérez F 2015 *J. Therm. Anal. Calorim.* **122** 3–9
[13] Khalifa A, Taha M and Akyurt M 1985 *Energy* **10** 819–829
[14] Mullick S, Kandpal T and Saxena A 1987 *Sol. Energy* **39** 353–360

Numerical investigation on natural convection in horizontal channel partially filled with aluminium foam and heated from above

B Buonomo[1], A Diana[2], O Manca[1] and S Nardini[1]

[1] Dipartimento di Ingegneria Industriale e dell'Informazione, Università degli Studi della Campania "Luigi Vanvitelli", Aversa (CE), Italy
[2] Dipartimento di Ingegneria Meccanica, Energetica, Gestionale e dei Trasporti, Università degli Studi di Genova, Genova (GE), Italy

Corresponding author's e-mail address: dialex.ad@gmail.com

Abstract. Natural convection gets a great attention for its importance in many thermal engineering applications, such as cooling of electronic components and devices, chemical vapor deposition systems and solar energy systems. In this work, a numerical investigation on steady state natural convection in a horizontal channel partially filled with a porous medium and heated at uniform heat flux from above is carried out. A three-dimensional model is realized and solved by means of the ANSYS-FLUENT code. The computational domain is made up of the principal channel and two lateral extended reservoirs at the open vertical sections. Furthermore, a porous plate is considered near the upper heated plate and the aluminium foam has different values of PPI. The numerical simulations are performed with working fluid air. Different values of assigned wall heat flux at top surface are considered and the configuration of the channel partially filled with metal foam is compared to the configuration without foam. Results are presented in terms of velocity and temperature fields, and both temperature and velocity profiles at different significant sections are shown. Results show that the use of metal foams, with low values of PPI, promotes the cooling of the heated wall and it causes a reduction of Nusselt Number values with high values of PPI.

1. Introduction

Natural convection has a great importance in several practical applications, such as cooling of electronic components and devices, chemical vapor deposition systems and solar energy systems, solar energy system, drying process, fire safety research and safety in nuclear reactors.
In this work, a numerical investigation on steady state natural convection in a horizontal channel partially filled with a porous medium and heated at uniform heat flux from above is carried out. The flow motion is strongly affected by the location of the heated surface, positioned on the upper or the lower wall of the horizontal channel [1-8]. Secondary flows are induced by buoyancy force due to the heating of the cavity walls, hence the local heat transfer increases [2-5, 7]. The onset point of the secondary flows delineates the region after which the two-dimensional laminar flow becomes three-dimensional and a transition motion of the flow from laminar to turbulent is observed [6-7]. The main flow shows a "C" loop behaviour very close to the flow inside open-ended channels and open cavities [6, 9]. The main flaw in natural convection is caused by the low heat exchanged between flow and

walls, and one of the technique to enhance the heat transfer is expanding the exchange surfaces. Porous media provided with high thermal conductivity, like metal foams, are an adequate method of heat transfer enhancement due to their large surface area to volume ratio and to intense mixing of the flow [10]. In [8, 11, 12], natural convection in high porosity metal foams heated from below is studied numerically and experimentally. Aluminum foam samples of different pore sizes and porosities were used to show the effects of metal foam geometry on heat transfer. The use of metal foams entails significant enhancements in heat transfer and metal foam disk is considered in [3, 12] to evaluate the effect of natural convection on the effective thermal conductivity. In [13], the design of aluminum foam was experimentally built and the calibration was done comparing the results of a flat plate with literature data and the agreement resulted excellent. The investigated foams had a pore density of 10 and 20 PPI and the bonding of the foam was performed employing a single epoxy or via brazing. A metal foam heat sink with rectangular section was placed in a large Plexiglas housing and a two dimensional model was considered for the experimental tests and for the numerical simulations [14]. A numerical investigation on natural convection for the idealized regular metal blocks is realized in [15] to reveal the detailed flow and the thermal behaviors at the pore level using the lattice Boltzmann method. Results show that the heat transfer enhances increasing the pore density and it weakens increasing the porosity. An experimental investigation on natural convection heat transfer in superposed metal foams with internal heat sources is studied in [16-18]. Numerical results of laminar fully developed natural convection in an inclined channel partially filled with metal foam are studied in [19]. An experimental investigation of air natural convection in horizontally-positioned copper metallic foams with open cells ia performed in [20]: results show that the porosity influence on the heat transfer performance is more remarkable when the pore density is higher, and natural convection in the copper foam weakens its thermal resistance and enhances its heat transfer performance. The natural convection on metallic foam-sintered plate at different inclination angles is experimentally studied in [21]. The sintered foam surface with lower porosity and pore density results more suitable for heat transfer enhancement. Studies on partially opened cavities filled with porous media are reported in [17-20] and investigations on partially filled horizontal channels with porous media are reported in [22, 25].

In this work, a numerical investigation on steady state natural convection in a horizontal channel partially filled with a porous medium and heated from above is carried out. A three-dimensional model is realized and solved by means of the ANSYS-FLUENT code. The computational domain is made up of the principal channel and two lateral extended reservoirs at the open vertical sections. Furthermore, a porous plate is considered near the upper heated plate and it fills the channel partially. The numerical simulations are performed with working fluid air. Different values of assigned wall heat flux at top surface are considered and the configuration of the channel partially filled with metal foam is compared to the configuration without foam.

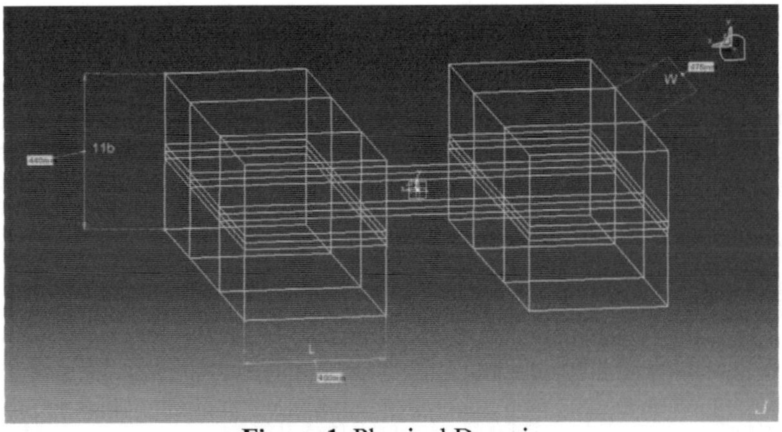

Figure 1. Physical Domain.

2. Mathematical Model

The analyzed system is composed by two horizontal parallel plates, with the upper plate heated at uniform heat flux, $\dot{q}$, and the lower insulated plate. As shown in Figure 1, the distance between the two horizontal plates is *b*, the plate length is *L*, as the dimension of the external reservoir, and the plate width is *W*. The dimension values of the model are reported in Table 1.

Table 1. Geometric values of the model.

b	40 mm
b/2	20 mm
L	400 mm
W	475 mm

The natural convective flow between the two horizontal plates is considered to be laminar, incompressible, and two-dimensional. Under steady-state condition, the governing equations, with constant thermophysical properties and with the Boussinesq approximation, are in conservative form [22].

$$\frac{\partial u}{\partial x}+\frac{\partial v}{\partial y}=0 \tag{1}$$

$$\frac{\partial (u^2)}{\partial x}+\frac{\partial (uv)}{\partial y}=-\frac{1}{\rho}\frac{\partial p}{\partial x}+\upsilon\nabla^2 u \tag{2}$$

$$\frac{\partial (uv)}{\partial x}+\frac{\partial (v^2)}{\partial y}=-\frac{1}{\rho}\frac{\partial p}{\partial y}+\upsilon\nabla^2 v+g\beta(T-T_\infty) \tag{3}$$

$$\frac{\partial (uT)}{\partial x}+\frac{\partial (vT)}{\partial y}=a\nabla^2 T \tag{4}$$

In Equation (4), the terms of dissipation and the total derivative of pressure are neglected, since this is usually done in natural convection phenomena when temperature gradients are relatively weak [22]. These equations are made nondimensional by introducing the following variables.

$$X=\frac{x}{b},\ Y=\frac{y}{b},\ U=\frac{ub}{\upsilon},\ V=\frac{vb}{\upsilon},\ P=\frac{(p-p_\infty)b^2}{\rho\upsilon^2}$$
$$\theta=\frac{(T-T_\infty)}{\dot{q}b/k},\ Gr=\frac{g\beta\dot{q}b^4}{k\upsilon^2},\ Pr=\frac{\upsilon}{a},\ Ra=GrPr \tag{5}$$

An enlarged computational domain of finite extension has been employed in this investigation to simulate the free-stream conditions of the flow far away from the region of the thermal disturbance induced by the heated plate, as suggested in reference [1, 4, 13]. The local convective heat transfer over the heated plate is described by the local Nusselt number, defined as:

$$Nu(x)=\frac{h(x)b}{k}=\frac{1}{\theta_{w,upper}(x)} \tag{6}$$

The corresponding average value can be written as

$$Nu = \frac{b}{L}\int_0^{L/b} Nu(X)dX \quad (7)$$

3. The numerical solution

The numerical solutions of the governing equations, reminding that they represent a set of coupled, non-linear, partial differential equations, are carried out using the commercial code Ansys-Fluent 12.2 [26].

The operating conditions are set as follow: the operating temperature is equal to 300 K and the pressure is equal to 1 bar. In the analysis a uniform heat flux is applied on the top wall of the channel and it is equal to 220 W/m2 and to 350 W/m2. Furthermore, a porous plate is considered near the upper heated plate and it has a thickness equal to 20 mm whereas the length and the width are the same of the channel. The aluminum foam has 10, 20 and 30 PPI and its porosity is equal to 93%.

Different structured mesh distributions were tested to accomplish the grid-independence analysis, paying attention, in particular, to regions characterized to unexpected temperature and velocity gradients. The analysis is carried out in steady regime and considering the metal foam with 10 PPI. The evaluated parameters are the Nusselt number and the average temperature on the heated plate. Three simulation are performed to analyze the sensitivity of the meshes employed. Richardson's extrapolation equation allows to estimate a reference value of a generic quantity, comparing two successive configurations related to the considered meshes. The percentage error is evaluated and it is chosen to compare the results. In Table 2, the values are reported.

The choice of the mesh, used for the subsequent simulations, was performed looking for the right compromise between the accuracy of the simulation and calculation time. Based on these two parameters, the Mesh 2 is chosen. It presents a slightly higher error than the Mesh 3 but, with respect to the latter, presents a lower calculation times.

Table 2. Percentage error.

mesh	$\bar{T}$w(plate) [K]	Nu_{avg}(plate)	%Δ$\bar{T}$w(plate)	%ΔNuavg(plate)
mesh1	462.55	1.90	0.56	2.78
mesh2	460.62	1.91	0.14	2.00
mesh3	459.97	1.95	0	0

4. Results and discussions

The objective of this study is to analyze natural convection in a horizontal channel partially filled with a porous medium and heated at uniform heat flux from above. Different values of assigned wall heat flux at top surface are considered and the configuration of the channel partially filled with metal foam is compared to the configuration without foam.

(a) Section 1, in the plane z = 0 mm. (b) Section 2, in the plane y = 20 mm.

Figure 2. Analyzed sections.

Results are presented in terms of velocity and temperature fields, referring to the planes shown in Figure 2. Furthermore, both temperature and velocity profiles are shown at different three significant sections (Figure 3), to obtain a description of the natural convection inside the open-ended cavity.

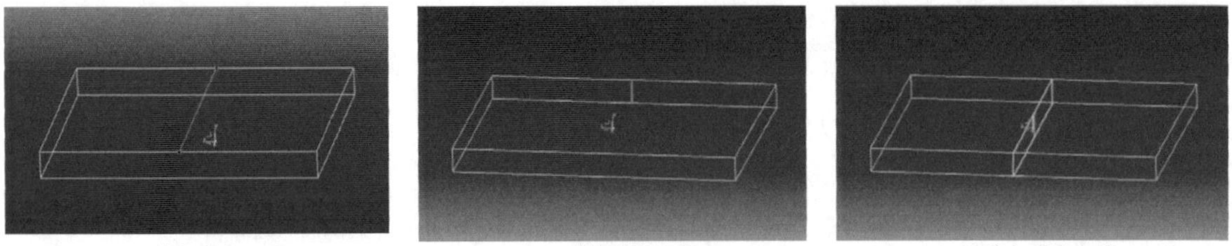

(a) Line 1. (b) Line 2. (c) Line 3.

Figure 3. Analyzed sections.

Figures 4 and 5 show temperature fields in sections 1 and 2 with $\dot{q}$ = 220 W/m^2, comparing the configurations with and without foam. In Figure 1, temperature values are higher in the metal foam, where the convective heat transfer is more difficult. On the heated plate, temperature values are higher in the central zone and they decrease in proximity of the edges in all configurations, as shown in Figure 5. Temperature values increase for higher PPI values.

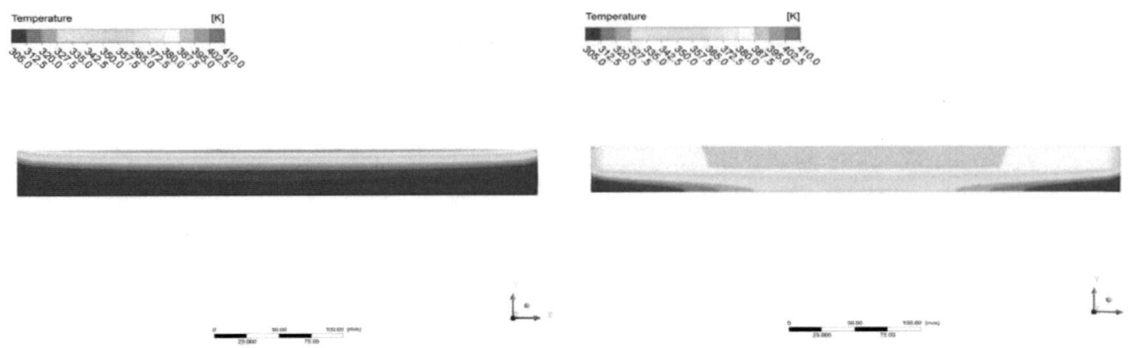

(a) Clean configuration. (b) With foam, 10 PPI.

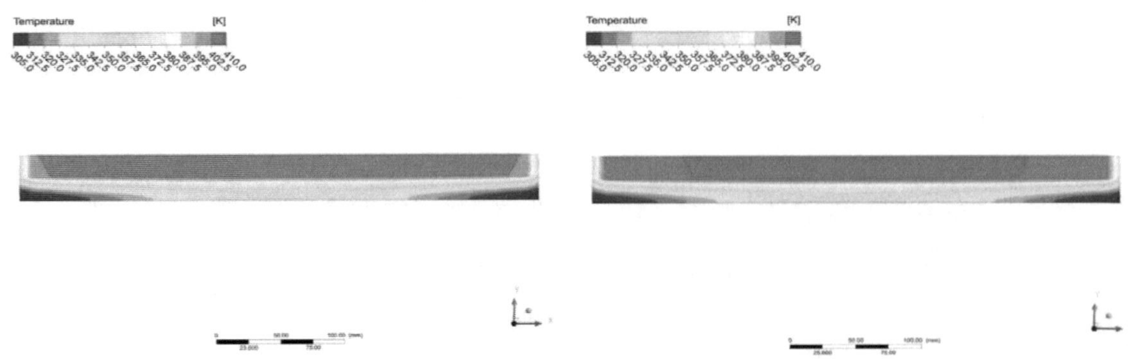

(c) With foam, 20 PPI. (d) With foam, 30 PPI.

Figure 4. Temperature field, Section 1, $\dot{q}$ = 220 W/m^2.

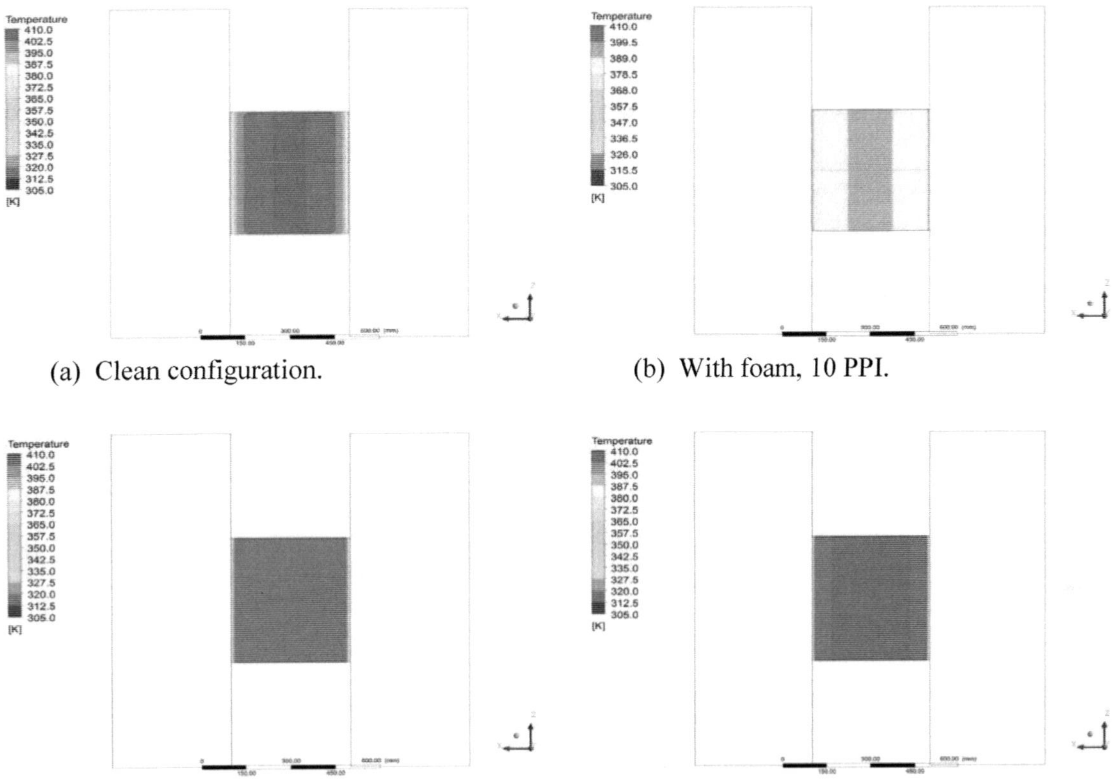

(a) Clean configuration. (b) With foam, 10 PPI.

(c) With foam, 20 PPI. (d) With foam, 30 PPI.

Figure 5. Temperature field, Section 2, $\dot{q} = 220$ W/m^2.

Figures 6, 7 and 8 show temperature fields in Sections 1 and 2, considering different values of PPI foam, and varying the heat flux value.

Figure 6 shows the effect of the foam with 10 PPI: temperature values are lower at the exterior sections of the foam, then they rise in the centre part of it. In the lower part of the channel, far from the heated plate and without foam, temperature values are close to ambient temperature value.

As shown in Figures 7 and 8, the qualitative trend of temperature values is the same with a 20 PPI or 30 PPI foam, but values are higher, thanks to the foam effect. In each configuration, values are higher when the heat flux applied on the top wall of the channel $\dot{q}$ is equal to 350 W/m^2.

The maximum temperature value is observed in Figure 8, when heat flux and PPI foam are higher.

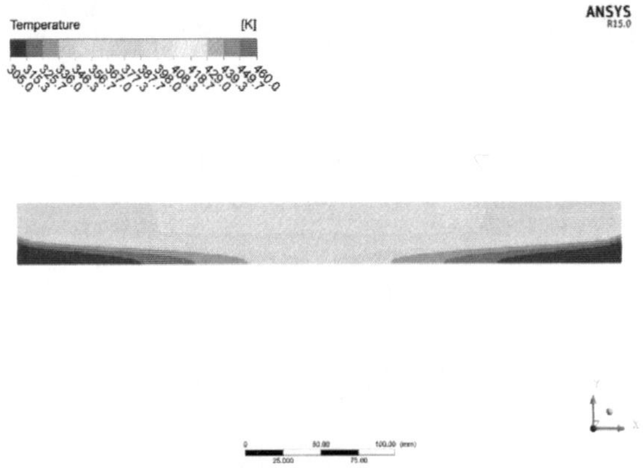

(a) Section 1, $\dot{q} = 220$ W/m^2.

(b) Section 2, $\dot{q} = 220$ W/m^2.

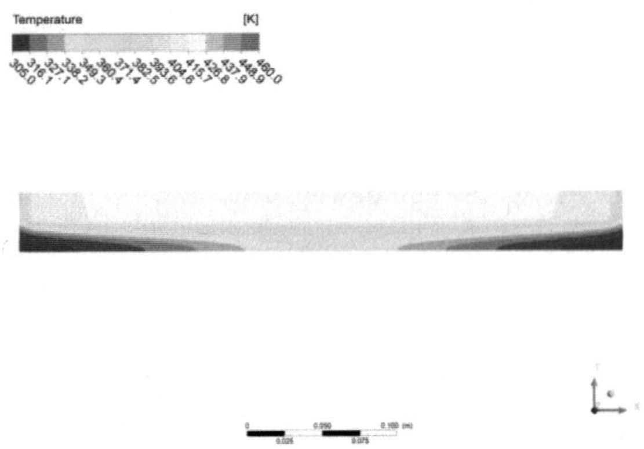

(c) Section 1, $\dot{q} = 350$ W/m^2.

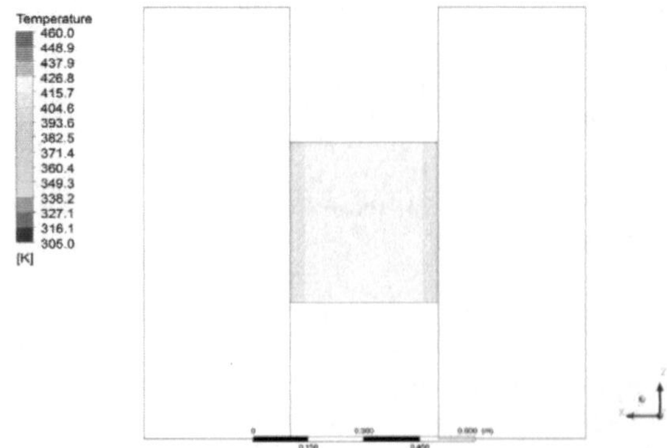

(d) Section 2, $\dot{q} = 350$ W/m^2.

Figure 6. Temperature field, Section 1 and 2, with foam 10 PPI.

7

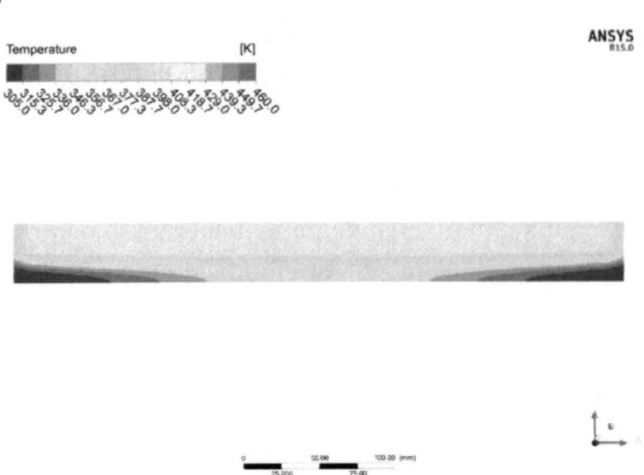

(a) Section 1, $\dot{q} = 220$ W/m^2.

(b) Section 2, $\dot{q} = 220$ W/m^2.

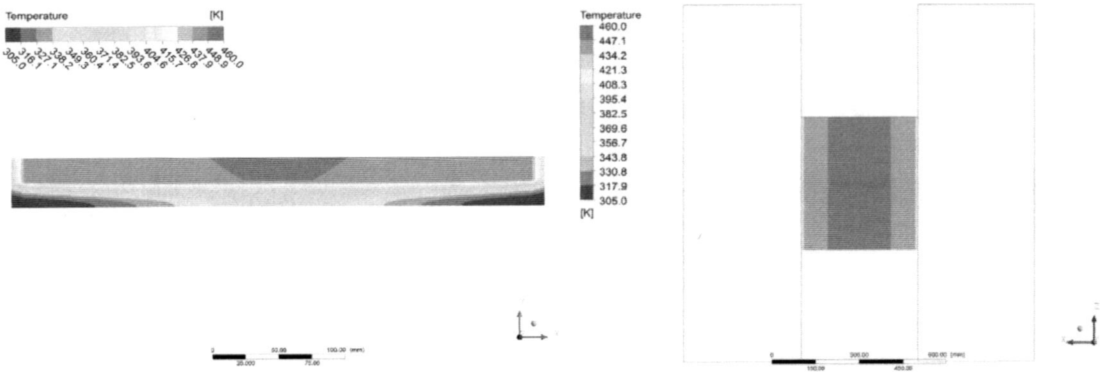

(c) Section 1, $\dot{q}$ = 350 W/m^2. (d) Section 2, $\dot{q}$ = 350 W/m^2.

Figure 7. Temperature field, Section 1 and 2, with foam 20 PPI.

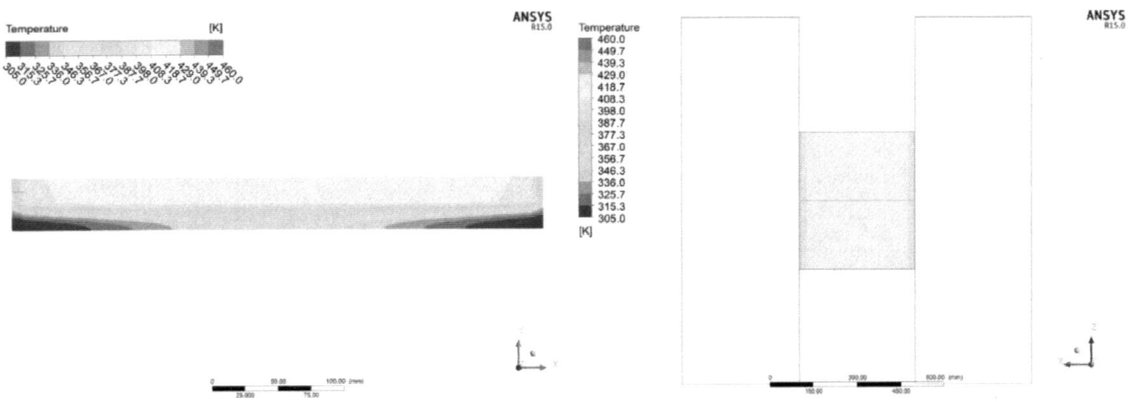

(a) Section 1, $\dot{q}$ = 220 W/m^2. (b) Section 2, $\dot{q}$ = 220 W/m^2.

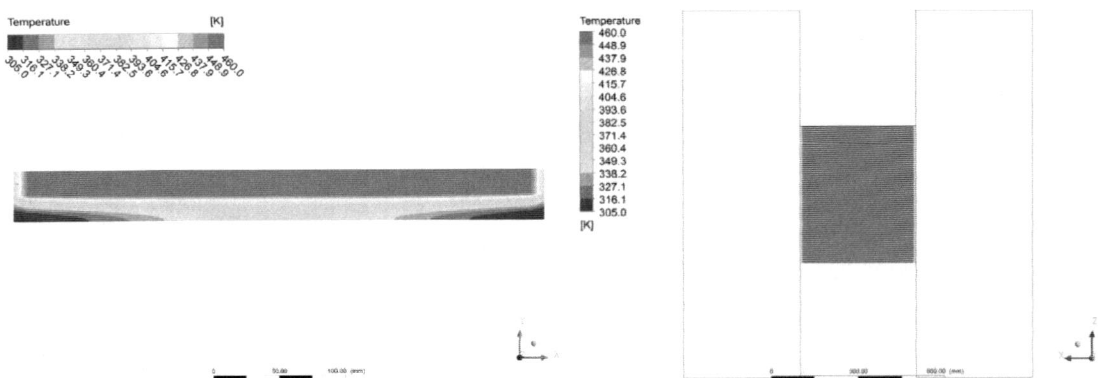

(c) Section 1, $\dot{q}$ = 350 W/m^2. (d) Section 2, $\dot{q}$ = 350 W/m^2.

Figure 8. Temperature field, Section 1 and 2, with foam 30 PPI.

IOP Conf. Series: Journal of Physics: Conf. Series **923** (2017) 012049 doi:10.1088/1742-6596/923/1/012049

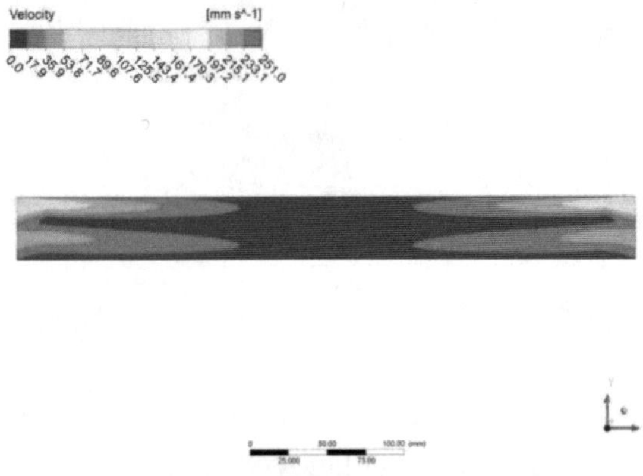

(a) Clean configuration.

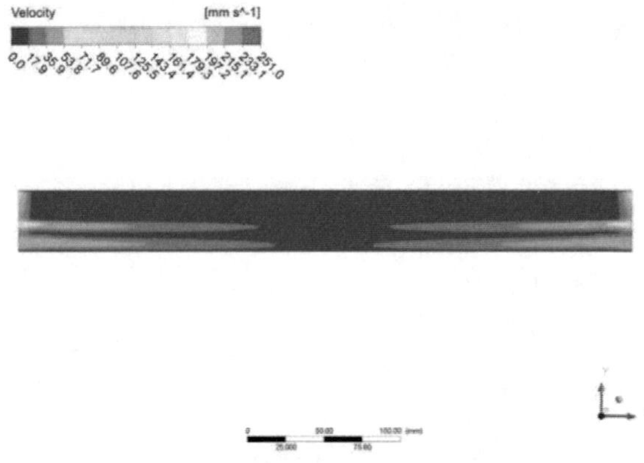

(b) With foam, 10 PPI.

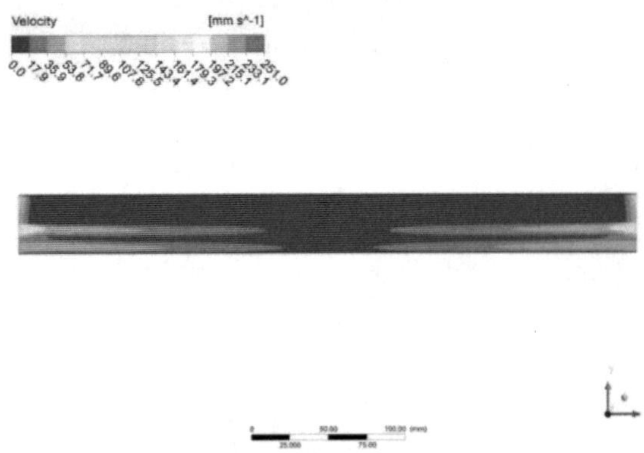

(c) With foam, 20 PPI.

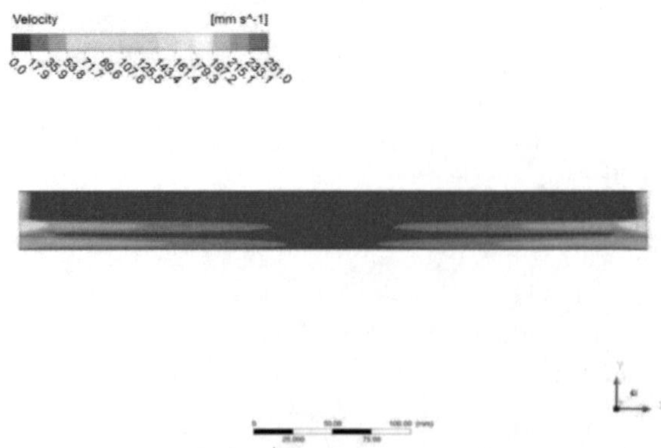

(d) With foam, 30 PPI.

Figure 9. Velocity field, Section 1, $\dot{q}$ = 220 W/m^2.

Figure 9 shows velocity fields in sections 1 and 2 with $\dot{q}$ = 220 W/m^2, comparing the configurations with and without foam. In clean configuration, convective motions are not obstructed. Due to the foam, air gets slower and velocity values decrease.

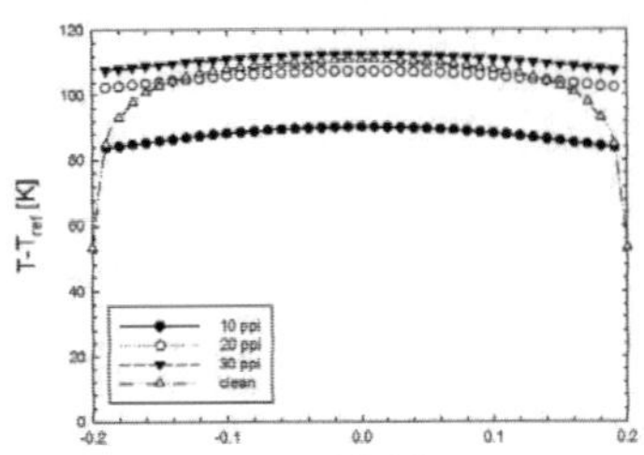

(a) Line 1.

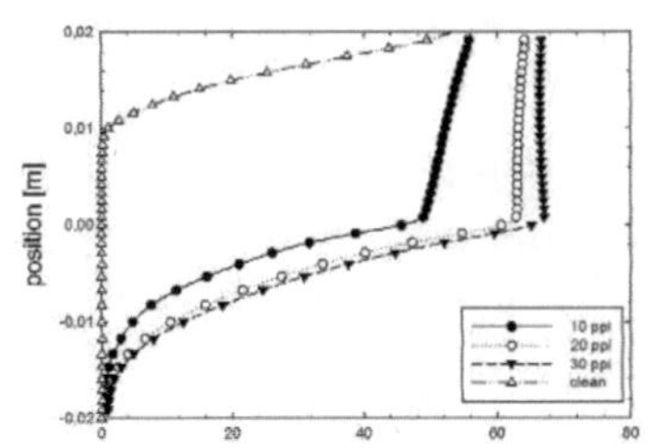

(b) Line 2.

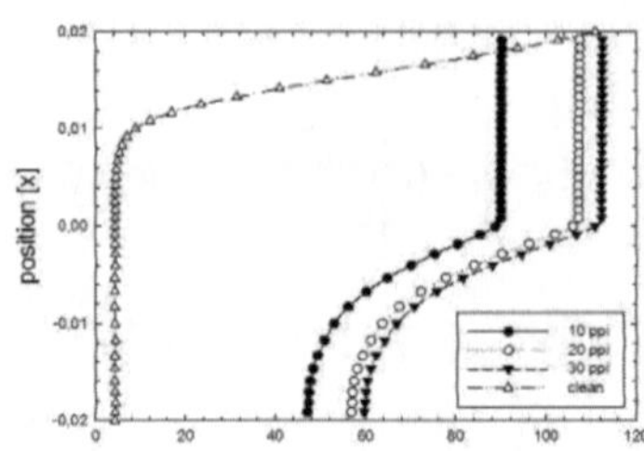

(c) Line 3.

Figure 10. Temperature profile, $\dot{q}$ = 220 W/m^2.

Figure 10 shows temperature profiles along Lines 1, 2 and 3, considering the heat flux equal to 220 W/m^2, and comparing the configurations with and without foam. In Figure 10(a), the maximum value is reached in the middle of Line 1 and temperature values are lower at the far end, due to the contact with a colder air. Considering foams with 20 and 30 PPI, temperature values are higher; the heat transfer improves with foam 10 PPI. In Figure 10(b), when the channel is clean, temperature values are close to

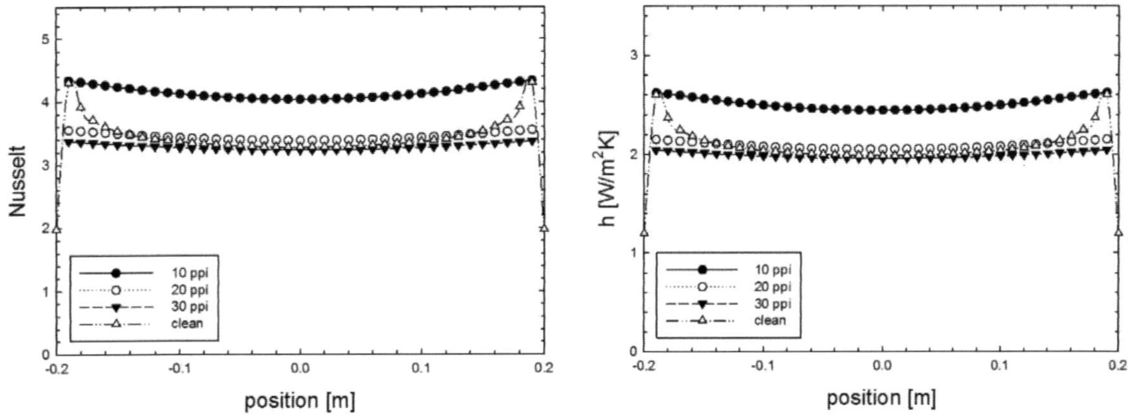

(a) Nusselt number. (b) Heat transfer coefficient.

Figure 11. Profiles of Nusselt number and heat transfer coefficient along Line 1, $\dot{q}$ = 220 W/m^2.

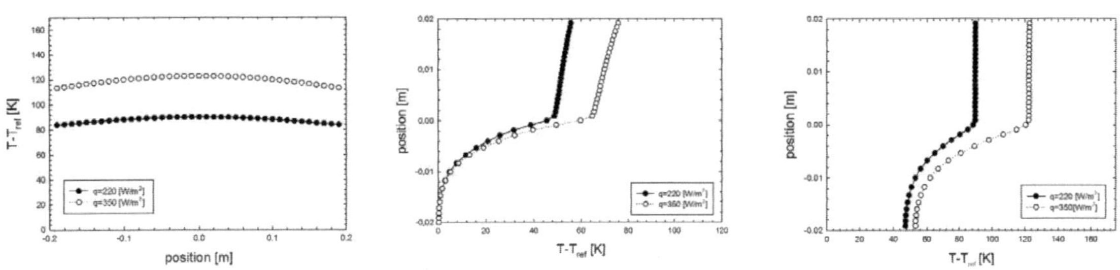

(a) Line 1. (b) Line 2. (c) Line 3.

Figure 12. Temperature profile, foam 10 PPI.

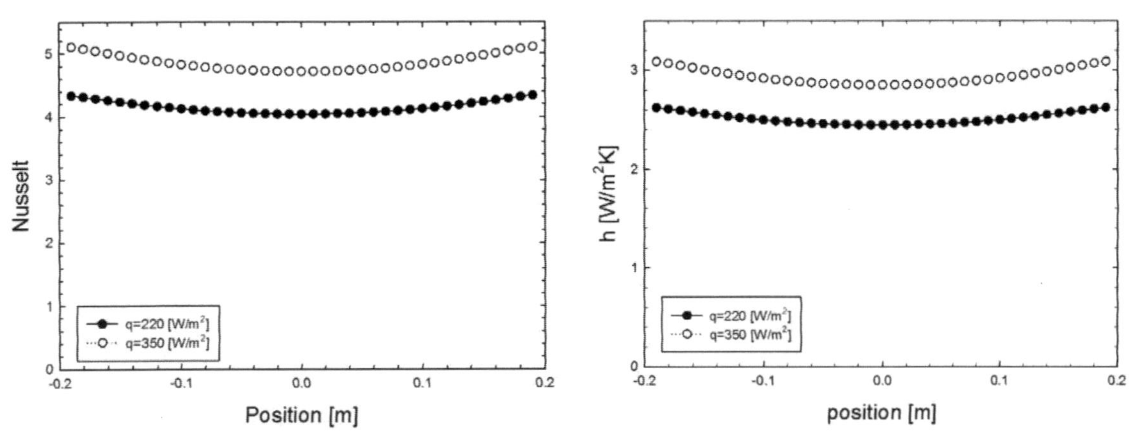

(a) Nusselt number. (b) Heat transfer coefficient.

Figure 13. Profiles of Nusselt number and heat transfer coefficient along Line 1, foam 10 PPI.

ambient temperature, then they increase in proximity of the heated plate. Temperature values are higher in presence of the foam. In Figure 10(c), in the clean configuration, temperature values are constant at the first, then they grow rapidly. When the foam partially fill the channel, temperature values grows rapidly at the fist, then they become quite constant.
Profiles of Nusselt number and heat transfer coefficient present the same qualitative trend, as shown in Figure 11. Values are lower in the middle of the Line 1 in all configurations.
Figure 12 shows temperature profile along Lines 1, 2 and 3 varying the heat flux and considering the foam with 10 PPI. In Figure 12(a), temperature values grow with heat flux and are higher in the middle of the Line 1. In Figures 12(b) and 12(c), temperature values increase in the lower part of the channel, then they are quite constant in the metal foam.
In Figure 13, the values of Nusselt number and heat transfer coefficient are higher when $\dot{q}$ is equal to 350 W/m^2 and they decrease in the middle of Line 1.

5. Conclusions

In this work, a numerical investigation on steady state natural convection in a horizontal channel partially filled with a porous medium and heated at uniform heat flux from above is carried out. A three-dimensional model is realized and a porous plate is considered near the upper heated plate. Different values of assigned wall heat flux at top surface are considered and the configuration of the channel partially filled with metal foam is compared to the configuration without foam. Results are presented in terms of velocity and temperature fields, and both temperature and velocity profiles at different significant sections are shown, to obtain a description of the natural convection inside the open-ended cavity. Finally, Average Nusselt number values are evaluated. Results show that the use of metal foams with low values of PPI promotes the cooling of the heated wall. The presence of the metal foam causes a reduction of Nusselt Number values and it is more evident with high values of PPI. The use of metal foam deteriorates the heat transfer for higher values of PPI, because of the reduction of the volume in which convective motion develops. A foam with 10 PPI increases the heat transfer and temperature values on the heated wall are lower than in clean configuration.

6. Nomenclature

a	thermal diffusivity, m^2/s
b	horizontal channel gap, m
g	acceleration due to the gravity, m/s^2
Gr	Grashof number
h(x)	local convective coefficient, W/m^{2}K
k	thermal conductivity, W/mK
L	plate length, m
L_x	length of the reservoir, m
L_y	height of the reservoir, m
L_z	width of the reservoir, m
Nu(x)	local Nusselt number
Nu	average Nusselt number
P	pressure, Pa
P	dimensionless pressure
Pr	Prandtl number
$\dot{q}$	heat flux, W/m^2

Ra	Rayleigh number
Ra*	modified Rayleigh number, Ra(b=L)
T	temperature, K
u, v, w	velocity components along x, y and z, m=s
U, V, W	dimensionless velocity components
x, y, z	Cartesian coordinates, m
X, Y, Z	dimensionless Cartesian coordinates,
B	volumetric coefficient of expansion, 1/K
ɵ	dimensionless temperature
ʋ	kinematic viscosity, m^2/s
ρ	density, kg/m^3

Subscripts

∞	free stream condition
lower	lower plate
max	maximum value
upper	upper plate
w	wall

References

[1] Manca O. and Nardini S., 2009, Experimental Investigation of Radiation Effects on Natural Convection in Horizontal Channels Heated From Above, *J. Heat Transfer*, **131**, art. no. 062503.
[2] Fu W.-S., Wang W.-H. and Huang S.-H., 2012, An Investigation of Natural Convection of Three Dimensional Horizontal Parallel Plates from a Steady to an Unsteady Situation by a CUDA Computation Platform, *Int. J. Heat Mass Transfer*, **55**, pp. 4638-4650.
[3] Mouhtadi D., Amahmid A.,Hasnaoui M.,Bennacer R., 2012, Natural convection in a horizontal channel provided with heat generating blocks: Discussion of the isothermal blocks validity, *Energy Conversion and Management*, **53**, pp. 45-54.
[4] Manca O., Morrone B, and Nardini S., 2000, Experimental analysis of thermal instability in natural convection between horizontal parallel plates uniformly heated, *J. Heat Transfer*, **122**, pp. 50–57.
[5] Yang H., Zhu Z. and Gilleard J, 2001, Numerical Simulation of Thermal Fluid Instability between Two Horizontal Parallel Plates, *Int. J. Heat Mass Transfer*, **44**, pp 1485–1493.
[6] Manca O. and Nardini S., 2007 Experimental investigation on natural convection in horizontal channels with the upper wall at uniform heat flux, *Int. J. Heat Mass Transfer*, **50**, pp. 1075–1086.
[7] Andreozzi A., Jaluria Y. and Manca O., 2007, Numerical Investigation of Transient Natural Convection in a Horizontal Channel Heated from the Upper Wall, *Numer. Heat Transfer A*, **51**, pp. 815–842.
[8] Buonomo B., Diana A., Manca O. and Nardini S., 2016, Experimental and numerical investigation on natural convection in horizontal channels partially filled with aluminium foam and heated from below, *Proceedings of the ASME 2016 Heat Transfer Summer Conference* vol. 1 (July 10-14, Washington, DC, USA).
[9] Vafai K. and Ettefagh J., 1990, The Effects of Sharp Corners on Buoyancy-Driven Flows with Particular Emphasis on Outer Boundaries, *Int. J. Heat Mass Transfer*, **33**, pp. 2311–2328.
[10] Zhao C.Y., Review on Thermal Transport in High Porosity Cellular Metal Foams with Open Cells, 2012 *Int. J. Heat Mass Transfer*, **55**, pp. 3618–3632.
[11] Cheng K., Li J.-W. and Pao H., Numerical simulation of natural convection with metal foams, 2008, *Journal of Engineering Thermophysics*, **29**, pp. 1722-1724.

[12] Zhao C.Y., Lu T.J. and H.P. Hodson, Natural Convection in Metal Foams With Open Cells, 2005, *Int. J. Heat Mass Transfer*, **48**, pp. 2452–2463.
[13] De Jaeger P., Reynders R., De Schampheleire S., Joen C.-T., Huisseune H., Ameel B., De Paepe M., An experimental study of natural convection in open-cell aluminum foam, 2012, *Journal of Phisics: Conference* Series 395.
[14] Phanikumar M.S. and Mahajan R.L., Non-Darcy Natural Convection in High Porosity Metal Foams, 2002, *Int. J. Heat Mass Transfer*, **45**, pp. 3781–3793.
[15] Zhao C.Y., Dai L.N., Tang G.H. and Qu Z.G., Numerical Study of Natural Convection in Porous Media (Metals) Using Lattice Boltzmann Method (LBM), 2010, *Int. J. Heat Fluid Flow*, **31**, pp. 925–934.
[16] Kathare V., Kulacki F.A and Davidson J.H., Buoyant Convection in Superposed Metal Foam and Water Layers, 2010, *J. Heat Transfer*, **132**, art. no. 014503.
[17] Hetsroni G., Gurevich M. and Rozenblit R., 2008, Natural Convection in Metal Foam Strips with Internal Heat Generation, *Exp. Therm. Fluid Sci.*, **32**, pp. 1740–1747.
[18] Khezzar L., Siginer D. and Vinogradov I., 2012, Natural convection in inclined two dimensional rectangular cavities, *Heat Mass Transfer*, **48**, pp. 227-239.
[19] Piller M. and Stalio E, 2012, Numerical Investigation of Natural Convection in Inclined Parallel-Plate Channels Partly Filled with Metal Foams, *Int. J. Heat Mass Transfer*, **55**, pp. 6506–6513.
[20] Xu Z.G., Qu Z.G. and Zhao C.Y., Experimental Study of Natural Convection in Horizontally-Positioned Open- Celled Metal Foams, 2011, *Int. Conf. Materials. Renew. Energy Environ.*, **1**, pp. 923 - 928.
[21] Ettefagh J., Vafai K. and Kim S.J., Non-Darcian Effects in Open-Ended Cavities Filled with a Porous Medium, 1991, *ASME J. Heat Transfer*, **113**, pp. 747–756.
[22] Haghshenas A., Rafati Nasr M. and Rahimian M.H., Numerical Simulation of Natural Convection in an Open-Ended Square Cavity Filled with Porous Medium by Lattice Boltzmann Method, 2010, *Int. Comm. Heat Mass Transfer*, **37**, pp. 1513–1519.
[23] Oztop H.F., Al-Salem K., Varol Y. and Pop I., Natural Convection Heat Transfer in a Partially Opened Cavity Filled with Porous Media, 2011, *Int. J. Heat Mass Transfer*, **54**, pp. 2253–2261.
[24] Oztop H.F., Al-Salem K., Varol Y., Pop I. and Firat M., Effects of inclination angle on natural convection in an inclined open porous cavity with non-isothermally heated wall, 2012, *Int. J. Num. Meth. Heat Fluid Flow*, **22**, pp. 1053-1072.
[25] Carr M. and Straughan B., Penetrative Convection in a Fluid Overlying a Porous Layer, 2003, *Adv. Water Res.*, **26**, pp. 263–276.
[26] Moffat R. J., Describing the Uncertainties in Experimental Results, 1988, *Experimental Thermal and Fluid Science*, **1**, pp. 3-17.
[27] Ansys-Fluent Incorporated, Fluent 6.3.26, User Manual. 2011.
[28] Buonomo B., D'Ambrosio S., Manca O. and Nardini S., Numerical investigation on mixed convention in a horizontal channel heated from below with upper heat losses, 2010, *Atti 65° Congresso Nazionale ATI* (Chia Laguna Resort Domus de Maria Cagliari, IT 13 - 17 Settembre 2010) 11-191.

Analytical models of Ohmic heating and conventional heating in food processing

A Serventi[1], F Bozzoli[1,2] and S Rainieri[1,2]

[1] Department of Industrial Engineering, University of Parma,
Parco Area delle Scienze 181/A I-43124 Parma, Italy
[2] SITEIA.PARMA Interdepartmental Centre, University of Parma,
Parco Area delle Scienze 181/A, I-43124 Parma, Italy

E-mail: fabio.bozzoli@unipr.it

Abstract. Ohmic heating is a food processing operation in which an electric current is passed through a food and the electrical resistance of the food causes the electric power to be transformed directly into heat. The heat is not delivered through a surface as in conventional heat exchangers but it is internally generated by Joule effect. Therefore, no temperature gradient is required and it origins quicker and more uniform heating within the food. On the other hand, it is associated with high energy costs and its use is limited to a particular range of food products with an appropriate electrical conductivity. Sterilization of foods by Ohmic heating has gained growing interest in the last few years. The aim of this study is to evaluate the benefits of Ohmic heating with respect to conventional heat exchangers under uniform wall temperature, a condition that is often present in industrial plants. This comparison is carried out by means of analytical models. The two different heating conditions are simulated under typical circumstances for the food industry. Particular attention is paid to the uniformity of the heat treatment and to the heating section length required in the two different conditions.

1. Introduction

Ohmic heating is an innovative thermal processing method for the food industry wherein an alternating current is passed through a food with the specific purpose of increasing its temperature. Heating occurs in the form of internal energy generation and offers significant advantages over conventional food processing methods [1]. In fact, during conventional thermal processing, heat transfer occurs from a heated surface to the inner product by means of conduction and convection phenomena. Traditional heat exchangers are therefore related with problems such as fouling and burning of the product directly in contact with the heat transfer surfaces. On the other hand, in Ohmic heating the electrical energy is dissipated into heat and results in more rapid and uniform heating of the entire mass of the food material. However, it is associated with high energy costs and its use is limited to a particular range of products with an appropriate electrical conductivity.

Excellent investigation on theoretical models for predicting the temperature field distribution, for both conventional heat exchangers and Ohmic heaters, may be found in literature [2-6].

Keys and Crawford [2] report the solution of the thermal entry length problem for a Newtonian fluid flowing in a circular tube with prescribed uniform surface temperature. Sparrow and Siegel [3] proposed an analytical solution for the thermal entrance region in circular ducts with adiabatic wall and uniform internal heat generation. This particular case was further investigated experimentally by Inman [4], who found good agreement with the theoretical model.

The thermal entrance region for a non-Newtonian fluid in a circular duct with uniform wall temperature and internal heat generation source term that depends on temperature was studied by Faraboschi and Di Federico [5]. Pesso and Piva [6] studied a similar problem, but for Newtonian fluids. They proposed analytical solutions for the case of boundary condition of the third kind and temperature dependant heat generation term.

However no Author compared the two different heating conditions in order to understand the real theoretical advantages of Ohmic heating. In the present work comparisons between performances of the two technologies, simulated under typical circumstances for the food industry, are drawn by means of analytical approach.

2. Analytical Models

In order to compare the quality of the product treatment by Ohmic heating with conventional heat exchangers two models are here presented. These models are based on the analytical solution of the convection transfer equations for the case of Newtonian fluid with constant properties in the laminar flow regime. For the case of conventional heat transfer device, the fixed temperature boundary condition is considered. This is a typical situation that may be encountered in many industrial applications, such as condensers. Finally a model for Ohmic heating with uniform internal heat generation and adiabatic wall is presented.

2.1 Constant surface temperature heat exchanger

The convection problem considered is schematically shown in figure 1. The fluid enters the tube at uniform temperature $T(r,0)=T_0$, that is lower than the constant surface temperature T_s.

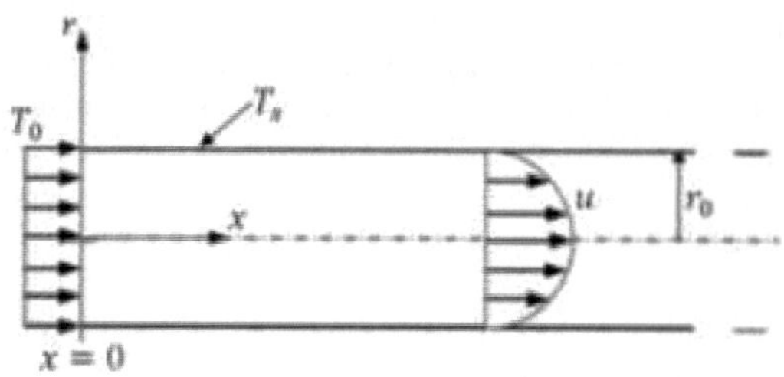

Figure 1. Convection problem: geometry and boundary conditions.

Assuming a steady fully developed flow and neglecting viscous dissipation, axial conduction and internal thermal energy sources, the energy equation for a constant properties fluid reads as follows:

$$\frac{u}{\alpha}\frac{\partial T}{\partial x} = \frac{\partial^2 T}{\partial r^2} + \frac{1}{r}\frac{\partial T}{\partial r} \tag{1}$$

The considered boundary conditions are:

$$T = T_0 \qquad \text{at } x = 0$$

$$\left.\frac{\partial T}{\partial r}\right|_{r=0} = 0 \qquad \text{for } x > 0 \tag{2}$$

$$T(x, r_0) = T_s$$

To solve the problem it is convenient to rewrite equations (1) and (2) in the dimensionless form substituting the following dimensionless variables:

$$x^* = \frac{x}{Pe\, r_0} \qquad \xi = \frac{r}{r_0} \qquad \Theta(x^*,\xi) = \frac{T_s - T(x,r)}{T_s - T_0} \qquad u^* = \frac{u}{u_m} \tag{3}$$

Where *Pe* is the Peclet number defined as $Pe=2u_m r_0/\alpha$. The energy equation in the dimensionless form becomes:

$$\frac{\partial^2\Theta}{\partial\xi^2} + \frac{1}{\xi}\frac{\partial\Theta}{\partial\xi} = \frac{u^*}{2}\frac{\partial\Theta}{\partial x^*} \tag{4}$$

with the following dimensionless boundary conditions

$$\begin{aligned} \Theta(0,\xi) &= 1 \\ \Theta(x^*,1) &= 0 \\ \frac{\partial\Theta}{\partial\xi}(x^*,0) &= 0 \end{aligned} \tag{5}$$

Equation (4) is a linear homogeneous partial differential equation, which may be solved using the method of separation of variables [2]. Therefore, the solution can be expressed by the product:

$$\Theta(x^*,\xi) = R(\xi)\cdot X(x^*) \tag{6}$$

The method of separation of variables yields an infinite series solution for the temperature field [2]:

$$\Theta(x^*,\xi) = \sum_{n=0}^{\infty} C_n R_n(\xi) \exp\left(-\lambda_n{}^2 x^*\right) \tag{7}$$

In the above solution, λ_n^2 are the so called eigenvalues and R_n are the corresponding eigenfunctions of the following Sturm Liouville equation:

$$\frac{d^2R}{d\xi^2} + \frac{1}{\xi}\frac{dR}{d\xi} + \lambda^2(1-\xi^2)R = 0 \tag{8}$$

Eigenvalues can be obtained from the following transcendental equation:

$${}_1F_1\left(\frac{1}{2} - \frac{\lambda}{4}, 1, \lambda\right) = 0 \tag{9}$$

Where ${}_1F_1$ is the confluent Hypergeometric function. The expression of the constant C_n in the series can be obtained exploiting the orthogonality property of the eigenfunctions:

$$C_n = \frac{\int_0^1 R_n(\xi)\,\xi\left(1-\xi^2\right)d\xi}{\int_0^1 R_n{}^2(\xi)\,\xi\left(1-\xi^2\right)d\xi} \tag{10}$$

The bulk temperature can be calculated from the following expression:

$$\Theta_b(x^*) = 8\sum_{n=0}^{\infty} \frac{G_n}{\lambda_n{}^2} \exp\left(-\lambda_n{}^2 x^*\right) \tag{11}$$

Where constants $G_n = -\frac{C_n}{2}\frac{dR_n(1)}{d\xi}$ are introduced to shorten the above expression.

2.2 Ohmic Heating with uniform internal heat generation

The problem considered is schematically shown in figure 2:

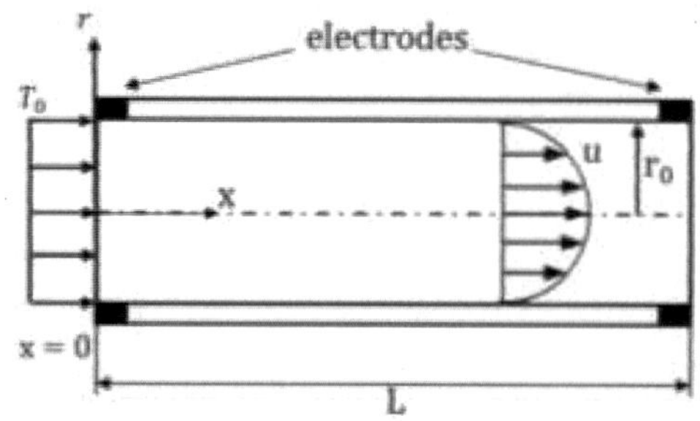

Figure 2. Ohmic heating problem: geometry and boundary conditions.

The geometry consists of a collinear cylindrical Ohmic heater with two electrodes at the pipe ends. Considering a steady fully developed flow and neglecting viscous dissipation, axial conduction, the energy equation becomes:

$$\rho c_p u \frac{\partial T(x,r)}{\partial x} = \frac{k}{r}\frac{\partial}{\partial r}\left(r\frac{\partial T(x,r)}{\partial r}\right) + q_{gen} \tag{12}$$

To simplify the problem, the wall of the Ohmic heater is considered adiabatic and the internal heat generation is assumed uniform. Therefore, the boundary conditions related to equation (12) read:

$$\begin{aligned} &\text{at } x = 0 && T = T_0 \\ &\text{for } x > 0 && \left.\frac{\partial T}{\partial r}\right|_{r=0} = 0 \\ & && \left.\frac{\partial T}{\partial r}\right|_{r=r_0} = 0 \end{aligned} \tag{13}$$

Equations (12) and (13) are then expressed in a dimensionless form, defining the dimensionless temperature as follows:

$$\Theta(x^*, \xi) = -\frac{T(x,r)}{T_0} \tag{14}$$

While the axial and the radial dimensionless coordinates as the same used in section 2.1. Substituting the velocity profile and the dimensionless terms the energy equation (12) becomes [6]:

$$(1-\xi^2)\frac{\partial \Theta}{\partial x^*} = \frac{1}{\xi}\frac{\partial}{\partial \xi}\left(\xi\frac{\partial \Theta}{\partial \xi}\right) + S_0 \tag{15}$$

Where $S_0 = (q_{gen} \cdot r_0^2)/(k \cdot (T_a - T_0))$. The boundary conditions are expressed as follows:

$$\begin{aligned} &\Theta = \Theta(\xi, 0) = -1 && \text{at } x^* = 0 \\ &\left.\frac{\partial \Theta}{\partial \xi}\right|_{\xi=0} = 0 && \text{for } x^* > 0 \\ &\left.\frac{\partial \Theta}{\partial \xi}\right|_{\xi=1} = 0 && \end{aligned} \tag{16}$$

Equation (15) may be solved using the method of separation of variables [6]. This yields to the following dimensionless temperature expression:

$$\Theta(x^*,\xi) = \sum_{i=1}^{\infty} A_i \exp(-\lambda_i x^*) f_i(\xi) + 2S_0 x^* - 1 + \frac{S_0}{4}\left(\xi^2 - \frac{1}{2}\xi^4 - \frac{1}{4}\right) \tag{17}$$

Where $f_i(\xi)$ are the Poiseuille functions defined as follows:

$$f_i(\xi) = pe(\xi,\mu_i) = \exp\left(-\mu_i \frac{\xi^2}{2}\right) {}_1F_1(\alpha_i, 1, \mu_i \xi^2) \tag{18}$$

in which:

$$\mu_i = \lambda_i^{\ 1/2} \qquad \alpha_i = \frac{1}{2}\left(1 - \frac{\mu_i}{2}\right) \tag{19}$$

The admissible values of the constant of separation λ_i are the root of the transcendental equation:

$$\text{at } \xi = 1 \qquad \frac{\partial f_i}{\partial \xi} = 0 \tag{20}$$

The expression of the constant A_i can be obtained from the orthogonality property of the functions f_i:

$$A_i = \frac{\int_0^1 \left(-\frac{S_0}{4}\left(\xi^2 - \frac{1}{2}\xi^4 - \frac{1}{4}\right)\right) f_i(\xi)(1-\xi^2)\xi \, d\xi}{\int_0^1 f_i^{\ 2}(\xi)(1-\xi^2)\xi \, d\xi} \tag{21}$$

Finally the dimensionless bulk temperature can be calculated from the expression:

$$\Theta_b(x^*) = 2S_0 x^* - 1 \tag{22}$$

3 Models Implementation

3.1 Constant surface temperature heat exchanger

Finding the solutions of the transcendental equation (9) is a fundamental step for the implementation of the analytical model. In Matlab® environment, by *"fzero"* function and *Symbolic Math Toolbox*, eigenvalues can be efficiently calculated noting that their periodicity is $\lambda_n \approx \lambda_{n-1}+4$. In the available literature, Kays and Crawford [2] report only the first five values but in many cases they are not enough for the implementation of accurate models. Table (1) reports the values of the first twenty eigenvalues λ_n and of the corresponding constants C_n to provide to the reader a powerful tools for further investigations.

Table 1. Calculated eigenvalues λ_n and constants C_n.

n	λ_n	C_n	n	λ_n	C_n
1	2.70436	1.47643	11	42.6677	0.23322
2	6.67903	-0.80612	12	46.6676	-0.21969
3	10.6734	0.58876	13	50.6675	0.20796
4	14.6711	-0.47585	14	54.6674	-0.19768
5	18.6699	0.40502	15	58.6674	0.188586
6	22.6691	-0.35575	16	62.6673	-0.18046
7	26.6687	0.31916	17	66.6673	0.17317
8	30.6683	-0.29073	18	70.6672	-0.16657
9	34.6681	0.26789	19	74.6672	0.16056
10	38.6679	-0.24906	20	78.66714	-0.15507

Finally, for the sake of completeness, in figure 3 the first five eigenfunctions, calculated again with the help of the *Symbolic Math Toolbox*, are represented graphically from $\xi=0$ *to 1*.

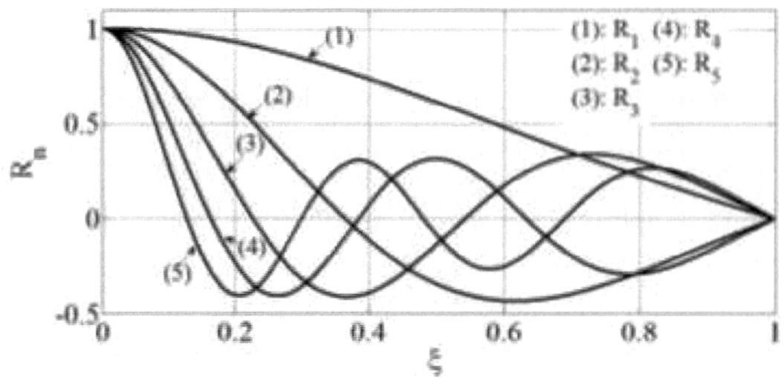

Figure 3. Calculated eigenfunctions R_n for $n=1$ *to 5*.

With the calculated eigenvalues λ_n, constants C_n, and eigenfunctions R_n it is finally possible to obtain the temperature distribution by equation (7).

3.2 Ohmic Heating with uniform internal heat generation

Finding the solutions of the transcendental equation (20) is a key point for the implementation of the analytical model for Ohmic heating. In Matlab® environment, by "*fzero*" function and *Symbolic Math Toolbox*, the eigenvalues are efficiently calculated noting that their periodicity is $\lambda_i \approx (\lambda_{i-1}^{1/2}+4)^2$. Twenty values of the eigenvalues λ_i and constants A_i are reported in Table 2.

Table 2. Calculated eigenvalues λ_i and constants A_i.

i	λ_i	A_i	i	λ_i	A_i
1	25.679612	-7.25210	11	2049.8425	-0.09934
2	83.861755	2.30424	12	2428.3660	0.08414
3	174.16674	-1.129115	13	2838.8968	-0.07188
4	296.53630	0.67020	14	3281.4341	0.06245
5	450.94719	-0.443768	15	3755.9772	-0.05440
6	637.38733	0.315644	16	4262.5254	0.04820
7	855.84949	-0.23589	17	4801.0783	-0.04258
8	1106.3289	0.18315	18	5371.6354	0.03833
9	1388.8224	-0.14613	19	5974.1963	-0.03421

10	1703.3275	0.11953	20	6608.7608	0.03123

In figure first five Poiseuille functions f_i , calculated with the *Symbolic Math Toolbox*, are shown graphically from $\xi=0$ to 1.

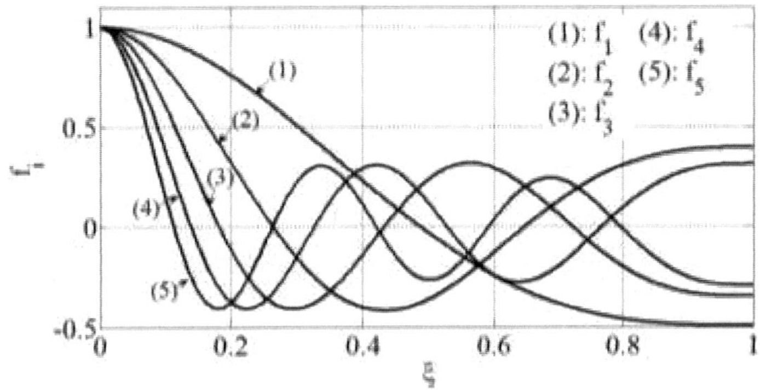

Figure 4. Calculated f_i function for $i=1$ to 5.

With the calculated eigenvalues λ_i, constants A_i, and functions f_i it is finally possible to obtain the temperature distribution by equation (17).

4 Models Comparison

In order to understand the advantages of Ohmic heating over conventional technologies, the two models presented in section 2 have to be compared under representative circumstances for the food industry. Attention is paid particular on the uniformity of the heat treatment and on the length of the heating section required in the two different conditions.

Firstly, simulations are performed on the Ohmic model (*ohm*). Flow rates, product properties, inlet and outlet product bulk temperatures are fixed. In this case, the volumetric flow rate is fixed at $1.4 \cdot 10^{-4}\ m^3/s$, while the diameter and the length of the Ohmic heater are $D=63.5\ mm$ and $L_{ohm}=3m$. The inlet and outlet product bulk temperature are fixed at $T_{b,in}=60°C$ and $T_{b,out}=100°C$. Those values are chosen in order to simulate a typical situation that may be encountered in industrial plants. Temperature profiles at different axial coordinates of the heater are therefore obtained. For each of these sections, the attention is focused on the minimum and maximum temperature values, in order to evaluate the product stratification and its evolution through the heater. Some representative temperature profiles are presented in figure 5. The maximum temperature value is always close to the tube wall while the minimum in the center of the tube section.

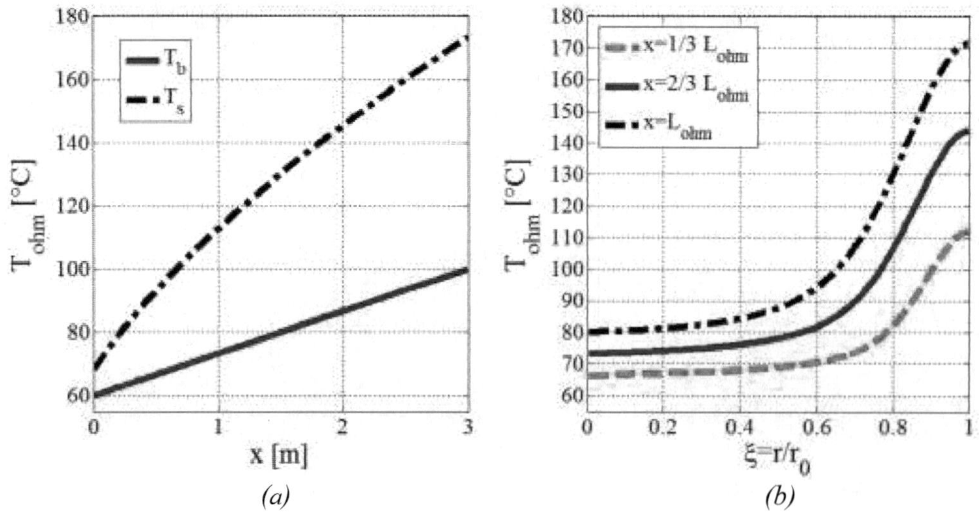

Figure 5. *(a)*Bulk and *(b)* radial temperature profiles at three selected axial positions for the Ohmic heater.

Then simulations are performed on the conventional heat exchanger model (*che*) with prescribed surface temperature with the same volumetric flow and the same diameter as in the Ohmic case. The basic idea of the comparison is to have, for the two different models, an equivalent thermal treatment which means having the same bulk temperature at the outlet section and the same maximum temperature reached by the product. From the practical point of view, the surface temperature of the conventional heat exchanger is taken equal to the maximum product temperature of the product in the Ohmic model. The required length of the equivalent conventional heat exchanger is obtained by imposing the same inlet and outlet product bulk temperature. In figure 6 the bulk, surface and temperature profiles for the conventional heat exchanger are reported. The required heat exchanger length for having an equivalent thermal treatment is, in this case, $L_{che}=20.9\ m$.

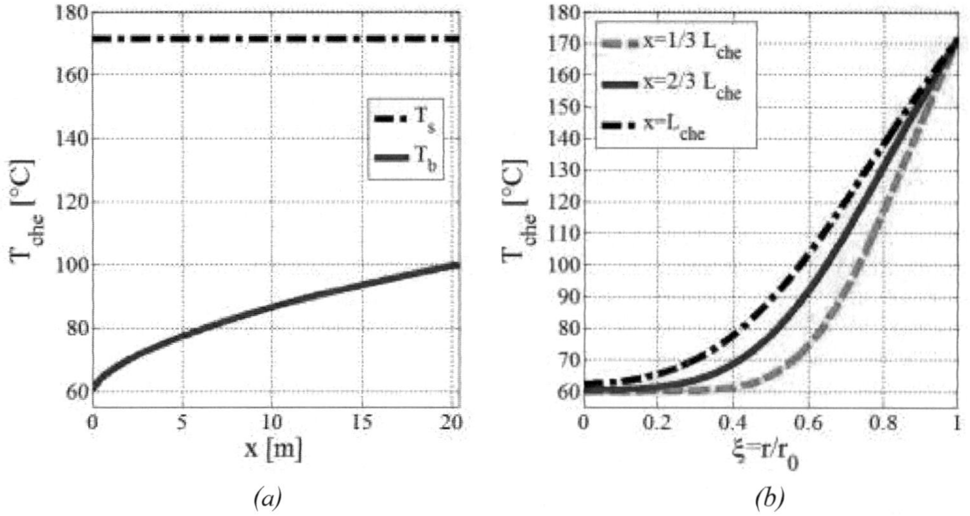

Figure 6. *(a)* Bulk and *(b)* radial temperature profiles at three selected axial positions for the constant surface temperature heat exchanger.

Simulations are then made for different flow rates varying from $1 \cdot 10^{-4}\ m^3/s$ to $7 \cdot 10^{-4} m^3/s$. The thermal process was always the same (i.e., same maximum product temperature, same inlet and outlet temperature). In figure 7 it is reported the ratio between the length of the conventional heat exchanger and of the Ohmic heater as a function of the volumetric product flow rate. It is possible to say that the Ohmic heating is always better in terms of heating section lengths; in particular as the flow rate increase this advantage is further amplified.

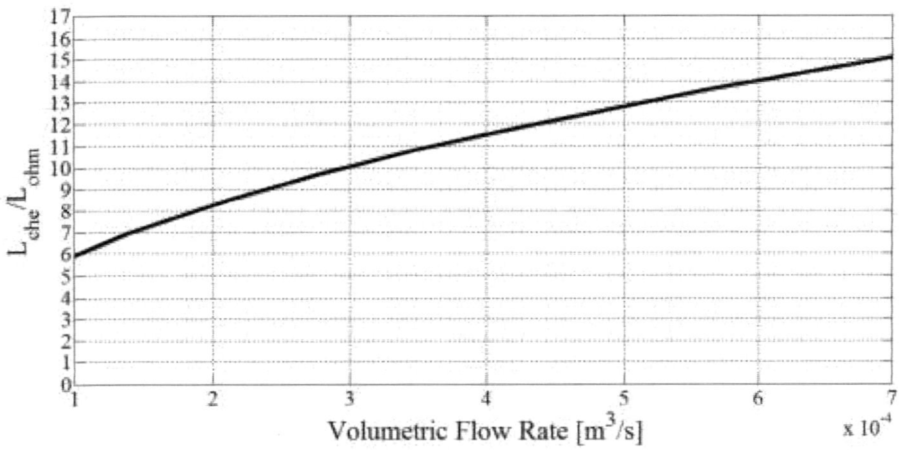

Figure 7. Heating section lengths ratio

To evaluate the product stratification through the heater in the two different heating processes it is introduced a dimensionless stratification factor *SF* defined as:

$$SF = \frac{T_{max} - T_{min}}{T_b} \quad (22)$$

Where T_{max}, T_{min} are the maximum and minimum temperature found at the surface and center of the tube respectively, while T_b is the corresponding bulk temperature for a particular section of the heater. SF is then obtained at different sections of the heater. In figure 8 stratification factors are reported as a function of dimensionless axial coordinate of the two heaters.

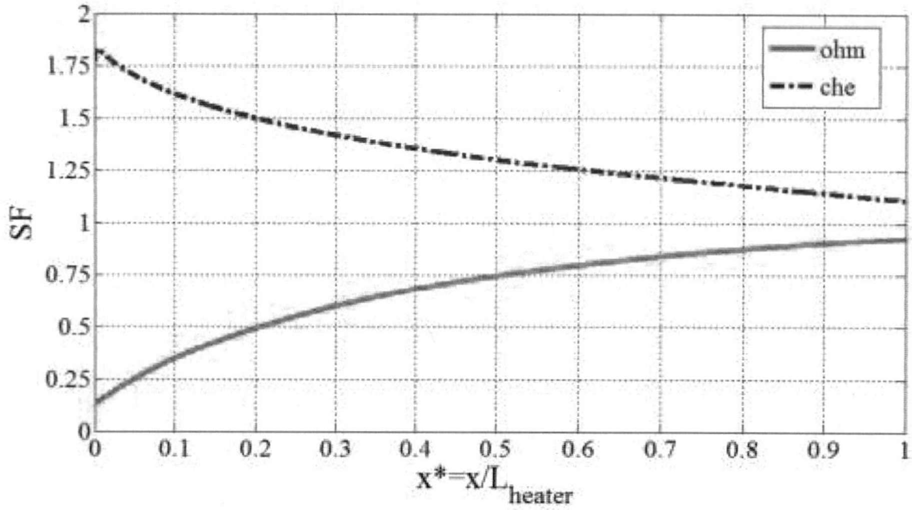

Figure 8. Stratification factor

Simulations again are repeated varying the volumetric flow rate. In figure 9 it is reported the maximum value of the stratification factor as a function of the volumetric flow rate for both the Ohmic and the traditional heating condition.

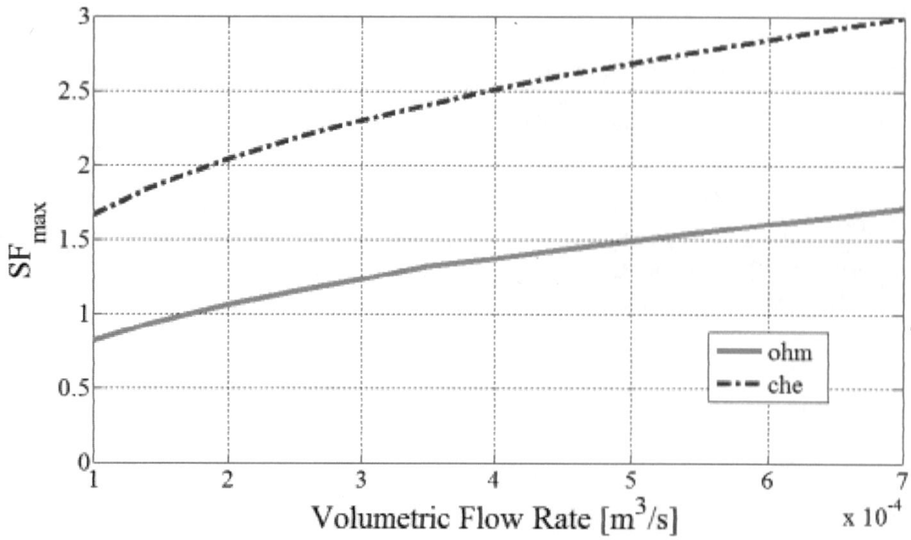

Figure 9. *Maximum stratification factor*

The stratification factor is always lower in the Ohmic heating than in the conventional heat exchanger. Therefore, as expected, it assures a more uniform thermal treatment of the product.

Finally, it should be stressed that while in Ohmic heaters there is a direct use of electrical energy, in conventional heat exchangers the constant surface temperature is generally obtained by condensing steam in the external jacket. Therefore, a comparison in term of exergy consumption between these two technologies is undoubtedly very interesting but is not easy because it strongly depends on the nature of the energy available [7]. Further efforts are required to investigate this aspect.

IOP Conf. Series: Journal of Physics: Conf. Series **923** (2017) 012050 doi:10.1088/1742-6596/923/1/012050

5 Conclusions

Analytical models for predicting the temperature field for fully developed laminar circular-tube flow under two different heating condition have been presented. The first model described a typical heat exchanger with constant wall temperature. The second one describes an Ohmic heater with uniform internal heat generation and adiabatic walls. The eigenvalues and eigenfunctions embedded in the two analytical models were reported.

Temperature profiles at different axial position were given for the two heat exchangers by considering the same operating conditions. The stratification of the product and the heating section lengths required by the two different heating conditions were studied and discussed.

For the condition here investigated, which is representative for the food industry, Ohmic heating always showed much more efficient and uniform heating than traditional heat exchangers.

Finally, it has to be highlighted than the analytical models here presented represent a powerful tool which could be useful in the design of this kind of complex devices, with very limited computational costs.

Acknowledgments

This work was partially supported by the Emilia-Romagna Region (POR-FESR 2014-2020). JBT SpA (Parma, Italy) is gratefully acknowledged for the support.

References

[1] Hosahalli S R, Marcotte M, Sastry S and Abdelrahim K 2014 *Ohmic Heating in Food Processing* (New York:CRC Press-Taylor & Francis Group)

[2] Kays W M and Crawford M E 1993 *Convective Heat and Mass Transfer* (New York: McGraw Hill)

[3] Sparrow E M and Siegel R 1958 *Nucl Sci Eng* **4** 239-254

[4] Inman R M 1962 *Int J Heat Mass Tran* **5** 1053-1058,

[5] Faraboschi F P and Di Federico I 1964 *Int J Heat Mass Tran* **7** 315-325

[6] Pesso T and Piva S 2015 *J Appl Fluid Mech* **8** 614-650

[7] Sakr M and Liu S 2014 *Renew. Sustainable Energy Rev.* **39** 262-269

On the effect of the aspect ratio on the mixed convection in a vertical cylindrical cavity with rotating inner wall

A Fichera[1], M Marcoux[2] and A Pagano[1,3]

[1] Università degli Studi di Catania, Dipartimento di Ingegneria Elettrica, Elettronica ed Informatica, viale A. Doria 6, 95125 Catania, Italy

[2] Institut de Mécanique des Fluides de Toulouse, 2 Allée du Professeur Camille Soula, Toulouse, France

[3] Corresponding author email: arturo.pagano@dieei.unict.it

Abstract. The thermal fluidynamic behaviour of a variety of fluids filling either vertical cylinders or vertical annuli under external conditions giving rise to natural convective motions has been thoroughly investigated, both experimentally and numerically, because of the several possible engineering applications that can be well approximated and described by these kinds of configurations. This study aims at investing through a numerical analysis the mixed convection arising inside cavities formed between two horizontal adiabatic disks and two isothermally and differentially heated vertical cylindrical walls, one of which is rotating. Previous studies reported and discussed the role played by both the Rayleigh number the Reynolds number for the case of the rotation of either the inner or the external wall and for a specified geometry of the cavity. The focus of the present study is to analyse the role played by the aspect ratio, $A=H/(R_e-R_i)$, for the case of rotating inner cylinder. It is reported that this role is particular relevant with respect to the flow structures established within the cavity and, as a consequence, also with respect to the Nusselt number.

1. Introduction

The natural convection of a fluid in an annulus between two vertical rotating concentric cylinders is a topic of major interest in many fields of applications, such as electrical machineries, thermal energy storage systems and chemical mixing equipment. In literature, some interest has been focused on the analysis of "buoyancy dominated flows", i.e. natural convective flows that are generated in static systems. Example of this are the studies of Vahl Davis and Thomos [1] and Kumar and Kalman [2], where correlations between the number of Nusselt and physical parameters have been determined. Other works deal with the dynamical contribution occurring in vertical annular cavity in which one (or both) of the walls, either of the cylinder or of the annulus, rotates. The numerical study of free convection in a rotating annulus was conducted by Williams [3]. De Vahl Davis et alii. [4] analyzed a vertical annular cavity with rotating inner wall and top and bottom surfaces. G. Urquiza et alii. [5] proposed a numerical study to evaluate the influence of the Prandtl number on the flow and the heat transfer rate on a rotating cylindrical annulus.

A specific configuration is the annular cavity between two concentric vertical cylinders filled with a fluid and closed by two horizontal disks. The heat exchange occurs between the fluid and the isothermally inner heated and outer cooled surfaces of the cylinder, whilst both the bottom and top horizontal disks are assumed to be adiabatic. Moreover, one or both of the cylindrical walls are considered to be rotating with constant angular velocity. As consequence, mixed convection arises in the system due to the contribution of natural convection, originated by the differential heating imposed at the vertical walls, and the forced convection established by the rotation of one of the cylinders.
Such a system is studied by M. Venkatachalappa [6] for different parameters as the speed rotation of both the inner and outer surfaces delimiting the vertical annular cavity, the buoyancy as expressed by the Rayleigh number, and the type of fluid as expressed by specified values of the Prandtl number (Pr=0.01, 0.1, 1). Therefore, the results reported in [6] hold for liquid metals and for gases.
With the aim of characterising the behaviour of liquids of moderate Pr, the same configuration, schematized in Fig. 1, was analyzed by the authors for a fluid with $Pr=5$ in the case of rotation of either the inner [7] or the external wall [8], for an enclosure specified by an aspect ratio $A=H/(R_e-R_i)=1$ and a radius ratio $RK=R_e/R_i=5$. In particular, the thermal and the flow fields, the profiles of the temperature and of the velocity components and the Nusselt number were described for the case of the cavity differentially heated inside and cooled outside, under a variety of values of the Rayleigh number and of the Reynolds number, within the limit of mixed convection.
In the present study the attention is focused on the relative influence of the forced component on both the flow structures and heat transfer between the vertical walls by varying the aspect ratio of the enclosure, for Pr=5.

2. Mathematical Model

The system considered in the present study is an enclosure, as shown in Fig. 1, formed by two differentially and isothermally heated vertical coaxial cylindrical walls of height H and inner and outer radius R_i and R_e, respectively, and two horizontal adiabatic disks.

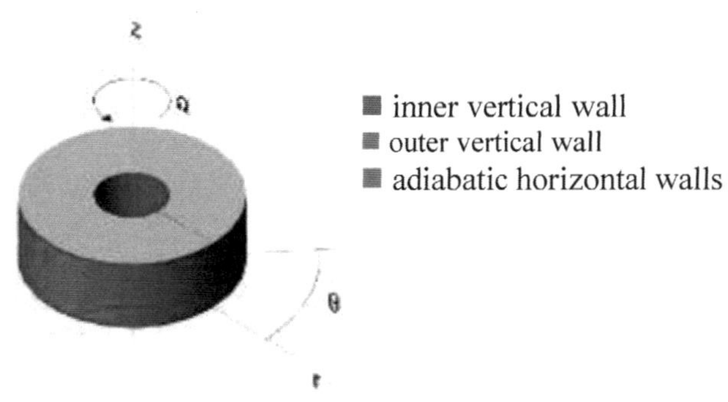

Fig. 1 Scheme of the system geometry

The reference coordinate system is cylindrical (r, θ, z) and three velocity components u, w, v, can be consequently identified along the three coordinates. The system symmetry with respect to the vertical axis ensures the flow independence with respect to the angular position, θ; therefore, the system can be modelled in 2D just with respect to r and z. Fixed coordinates are considered with the corresponding choice of boundary conditions at the walls. In line with Boussinesq's approximation, the fluid density is assumed to be constant in all terms, except for the buoyancy term that linearly depends on the temperature. The model was derived in terms of primitive variables u, v, w, p and T by expressing the mass, energy and momentum along the directions r, z and θ, as it follows:

$$\frac{1}{r}\frac{\partial ru}{\partial r} + \frac{\partial v}{\partial z} = 0 \tag{1}$$

$$\frac{\partial T}{\partial t} + u\frac{\partial T}{\partial r} + v\frac{\partial T}{\partial z} = a\nabla^2 T \tag{2}$$

$$\frac{\partial u}{\partial t} + u\frac{\partial u}{\partial r} + v\frac{\partial u}{\partial z} = -\frac{1}{\rho_0}\frac{\partial p}{\partial r} + \frac{\mu}{\rho_0}\left(\nabla^2 u - \frac{u}{r^2}\right) + \frac{\rho}{\rho_0}\cdot\left(\frac{w^2}{r}\right) \tag{3}$$

$$\frac{\partial v}{\partial t} + u\frac{\partial v}{\partial r} + v\frac{\partial v}{\partial z} = -\frac{1}{\rho_0}\frac{\partial p}{\partial z} + \frac{\mu}{\rho_0}\nabla^2 v + g\beta(T - T_0) \tag{4}$$

$$\frac{\partial w}{\partial t} + u\frac{\partial w}{\partial r} + v\frac{\partial w}{\partial z} = \frac{\mu}{\rho_0}\left(\nabla^2 w - \frac{w}{r^2}\right) - \frac{\rho}{\rho_0}\cdot\left(\frac{uw}{r}\right) \tag{5}$$

The model was rewritten in non-dimensional form considering the following scaling factors:

- Vertical coordinate: $H\,[m]$
- Radial coordinate: $\Delta r/\eta\,[m]$
- Reference velocity: $V_0 = \Omega \cdot R_i$ [m/s]
- Reference time: H/V_0 [s]
- Reference pressure: $2\rho_0 V_0$ [Pa].

which corresponds to the following choice of the non-dimensional system variables:

$$A = \frac{H}{\Delta R};\quad RK = \frac{R_e}{R_i};\quad \eta = RK - 1;$$

$$z' = \frac{z}{H};\quad r' = \frac{r}{R_i} = \eta\frac{r}{\Delta R};\quad t' = t\frac{V_0}{H};$$

$$u' = \frac{u}{V_0};\quad w' = \frac{w}{V_0};\quad v' = \frac{v}{V_0};$$

$$p' = \frac{p + \rho_0 gH}{\rho_0 V_0^2};\quad T' = \frac{2(T - T_0)}{\Delta T};\quad \mathrm{Re} = \frac{2\Omega\Delta R^2}{\nu};$$

$$Ra = \frac{\beta \cdot g\,\Delta T \Delta R^3}{\nu a};\quad Pr = \frac{c_p\mu}{k} = \frac{\nu}{a};\quad Fr = \frac{\Omega^2 \Delta R}{g}. \tag{6}$$

Under the previous choices the non-dimensional model was therefore reformulated as:

$$(\eta A)\frac{\partial ru}{\partial r} + \frac{\partial v}{\partial z} = 0 \tag{7}$$

$$\frac{\partial T}{\partial t}+(\eta A)\cdot u\frac{\partial T}{\partial r}+v\frac{\partial T}{\partial z}=\frac{2}{Pr\ Re}\left[(\eta^3 A)\cdot\left(\frac{\partial^2 T}{\partial r^2}+\frac{1}{r}\frac{\partial T}{\partial r}\right)+\left(\frac{\eta}{A}\right)\frac{\partial^2 T}{\partial z^2}\right] \tag{8}$$

$$\frac{\partial u}{\partial t}+(\eta A)\cdot u\frac{\partial u}{\partial r}+v\frac{\partial u}{\partial z}=(\eta A)\cdot\left(1-2\frac{RaFr}{\Pr\ \mathrm{Re}^2}T\right)\cdot\left(\frac{w^2}{r}\right)+$$
$$+\frac{2}{\mathrm{Re}}\left[(\eta^3 A)\cdot\left(\frac{\partial^2 u}{\partial r^2}+\frac{1}{r}\frac{\partial u}{\partial r}\right)+\left(\frac{\eta}{A}\right)\frac{\partial^2 u}{\partial z^2}\right]-\frac{\partial p}{\partial r} \tag{9}$$

$$\frac{\partial v}{\partial t}+(\eta A)\cdot u\frac{\partial v}{\partial r}+v\frac{\partial v}{\partial z}=-(\eta A)\cdot\frac{1}{Fr}\left(1-2\frac{RaFr}{\Pr\ \mathrm{Re}^2}T\right)+$$
$$+\frac{2}{\mathrm{Re}}\left[(\eta^3 A)\cdot\left(\frac{\partial^2 v}{\partial r^2}+\frac{1}{r}\frac{\partial v}{\partial r}\right)+\left(\frac{\eta}{A}\right)\frac{\partial^2 v}{\partial z^2}\right]-\frac{\partial p}{\partial z} \tag{10}$$

$$\frac{\partial w}{\partial t}+(\eta A)\cdot u\frac{\partial w}{\partial r}+v\frac{\partial w}{\partial z}=(\eta A)\cdot\left(1-2\frac{RaFr}{\Pr\ \mathrm{Re}^2}T\right)\cdot\left(\frac{uw}{r}\right)+$$
$$+\frac{2}{\mathrm{Re}}\left[(\eta^3 A)\cdot\left(\frac{\partial^2 w}{\partial r^2}+\frac{1}{r}\frac{\partial w}{\partial r}\right)+\left(\frac{\eta}{A}\right)\frac{\partial^2 w}{\partial z^2}\right] \tag{11}$$

where the prime symbol has been dropped. Moreover, though not used for calculations, for the analysis of the simulated flow field it was convenient to refer to the stream function ψ [9]:

$$u=\frac{1}{r}\frac{\partial\psi}{\partial z}\quad ;\quad v=-\frac{1}{r}\frac{\partial\psi}{\partial r} \tag{12}$$

The model solution allows to evaluate the temperature profile and, on this basis, it is possible to calculate the Nusselt number at either of the two vertical walls. In the present study, reported results refer to the Nusselt number evaluated at the inner wall, defined as follows:

$$Nu=Nu_{r=0}=\int_0^H Q(r,y)\cdot dy=\int_0^H\frac{\partial T}{\partial r}\cdot dy \tag{13}$$

where $Q(r,y)$ is the local heat flux in the radial direction.
A finite volume scheme, assessed and validated in previous studies [10], was adopted to solve the model Equations (7)–(11), defining a two-dimensional grid and differentiating the equations for each control volume. The pressure-velocity coupling was solved adopting SIMPLER algorithm, performing nodal variable calculation on the basis of *CDS* algorithm. The code inputs were the parameters characterizing the operating condition (*Re*, *Ra*, *Pr*, *Fr*), the cavity shape (*A* and *RK*), the grid characteristics (number of cells and their distribution along *r* and *z*) and the number of iterations. The output of the code was a set of matrices, one for each of the primitive variables plus an additional one for the stream function. Moreover, the code produced as outputs also the value at each iteration of the Nusselt number, of the primitive variables and of their residues, and of the number of iterations of the pressure-velocity algorithm. In this way it was possible to check the iteration process and to verify the simulation reliability. With this respect, errors below 10^{-4} were assumed satisfactory for all primitive variables, with zero iteration of the pressure algorithm at the end of the calculation. It is just mentioned that the model defined in the present study was satisfactorily validated through the

comparison with benchmark results obtained by Khellaf [11] and Ho and Tu [12] for a vertical cylindrical cavity formed by two coaxial cylinders with a rotating inner wall.

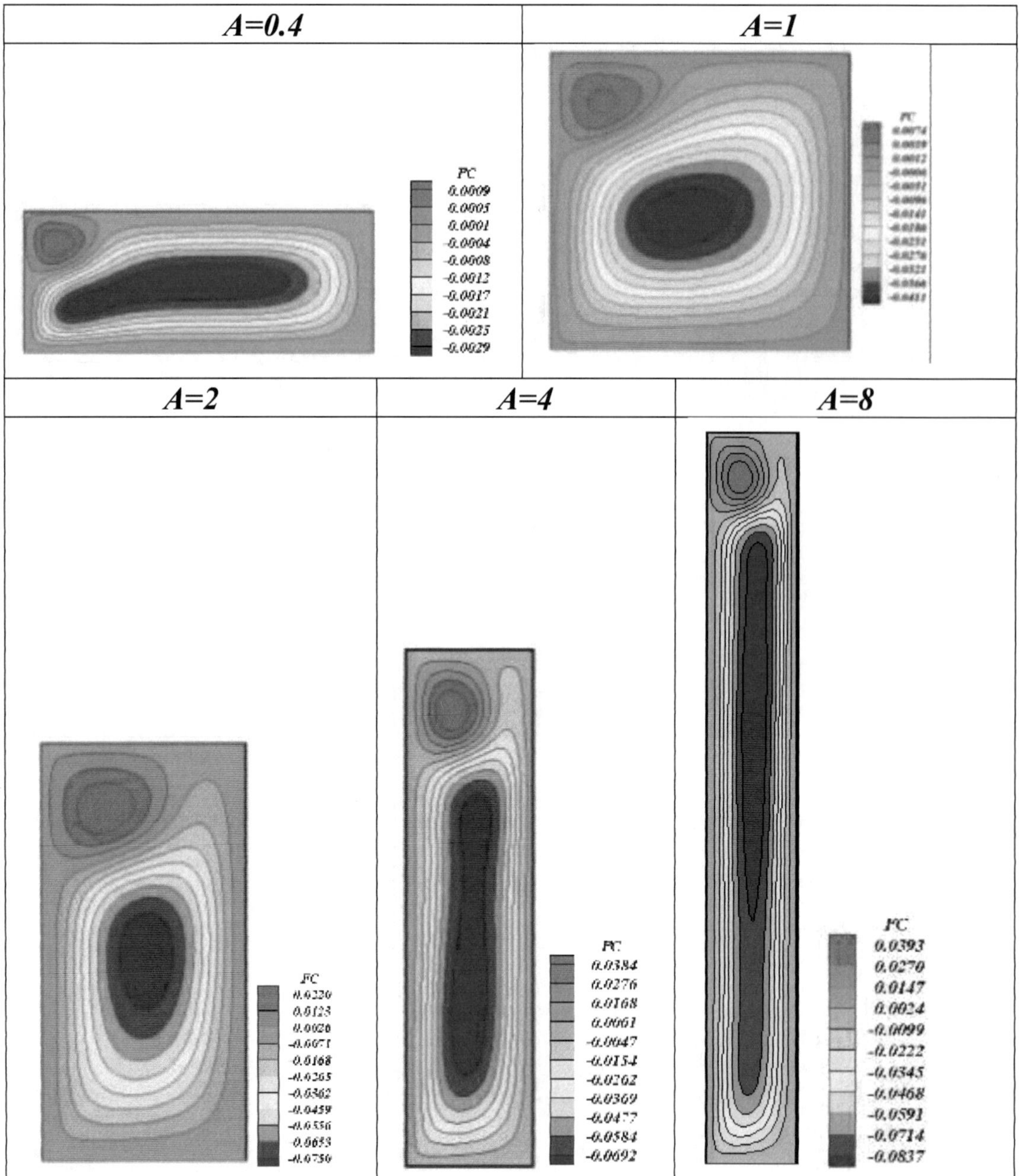

Fig. 2 Stream function for $Ra=10^4$ and A=[0.4, 1, 2, 4, 8]

3. Results

The mixed convection of a fluid of $Pr=5$ within a vertical annular enclosure with isothermally heated and cooled vertical walls was simulated for the case of rotation of the inner wall. As the interest was focused on analysing the influence of the aspect ratio, A, both the radius ratio and the Reynolds number were specified at, respectively, $RK=5$ (i.e. $\eta=4$), $Re=100$. The flow structures reported in the following refers to aspect ratio values A=[0.4, 1, 2, 4, 8] and are evaluated by varying the Rayleigh number within the range $Ra \subset [10^4:10^5]$. It is worth observing that the calculations were performed within ranges of Ra and Re for which it is possible to neglect both viscous dissipation and the influence of density variations on centrifugal and Coriolis forces.

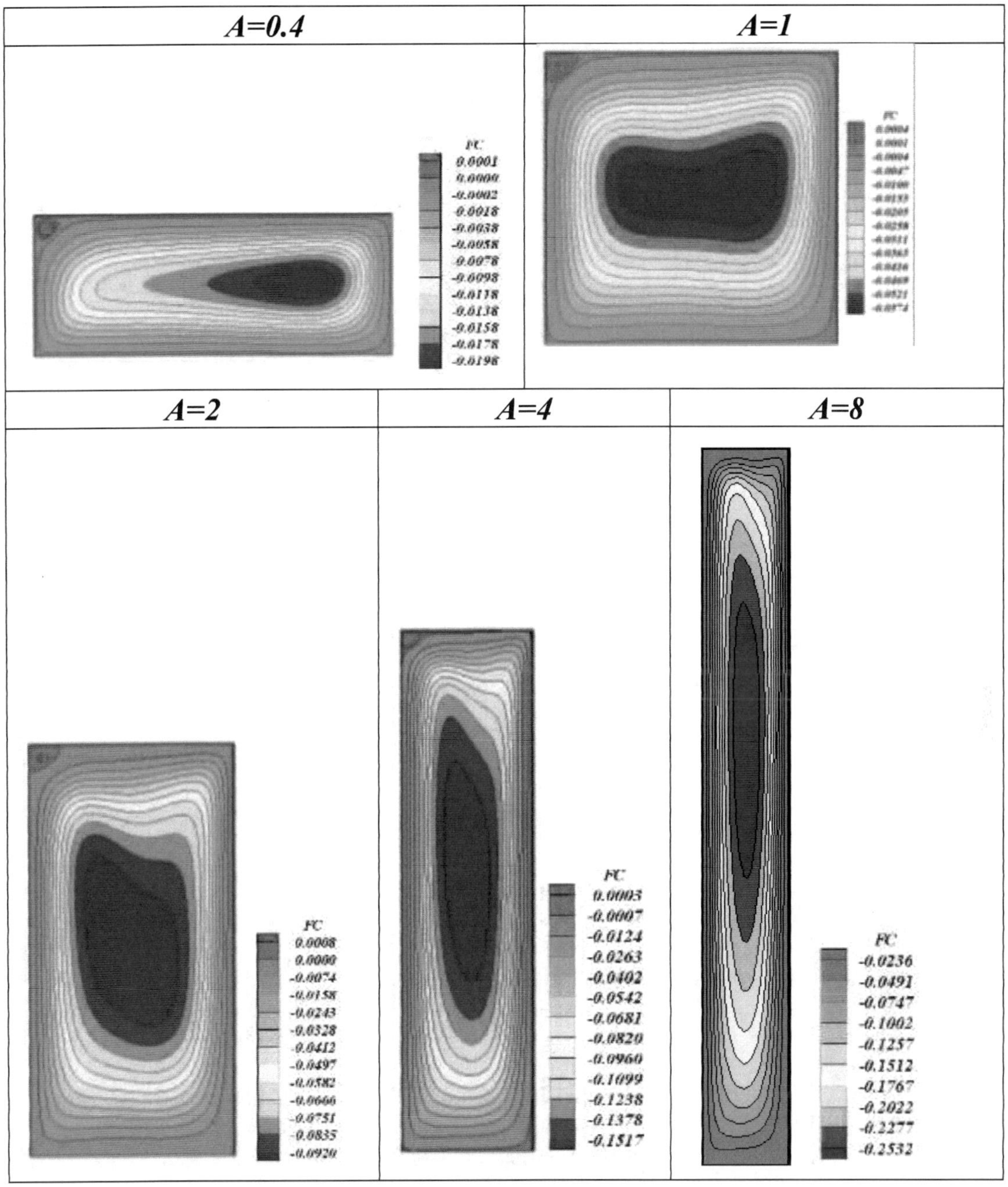

Fig. 3 Stream function for $Ra=10^5$ and A=[0.4, 1, 2, 4, 8]

The dynamics of the fluid within the cavity is obtained by solving the numerical model. Fig. 2 and 3 report the stream functions describing the distributions of the flow structures within the right-hand side cross section of the enclosure corresponding respectively to $Ra=10^4$ and $Ra=10^5$ at the simulated values of the aspect ratio A.

It is possible to observe that the structure of the flow is generally bi-cellular, due to the contribution of both natural convection, caused by the radial thermal gradient, and forced convection, caused by the rotation of the inner wall. In particular, the forced convection contribution to the structure of the flow is the birth of the counter-rotating cell at the inner wall that occupies the upper part of the cavity. The natural convection, instead, is responsible of the cell that occupies the main part of the cavity.

From the analysis of Fig. 2 it is possible to observe that, considering $Ra=10^4$, the bi-cellular flow pattern structure is maintained for all of the values of A. Fig. 3 allows to observe that the growth of the Rayleigh number to $Ra=10^5$ reinforces the overall relevance of the buoyancy whilst the forced convection is dumped. This is reflected in the reduction of the size of the counter-rotating cell at the inner rotating wall that can be observed for growing aspect ratio, until A=4, and its disappearance for A=8, where the buoyancy-generates cell occupies the entire enclosure.

In order to evaluate the influence of the aspect ratio on the heat exchange, i.e. the number of Nusselt, simulations for different values of Ra and A were conducted. Fig. 4 shows the effect that the increase of A has on Nu for $Ra=10^4$ and $Ra=10^5$ and for $Re=100$. In particular, the same general dependence of Nu on A can be observed. This consists in a steep increase of Nu for $A< 1$, after which a maximum is reached around $A=1$, followed by a monotonic decrease. Moreover, the increase of Ra causes the growth of the Nu-values calculated by the numerical model with a slight decrease of the value of A corresponding to the maximum reached by Nu. Therefore, it is possible to conclude that cavities with either square or only lightly stretched cross sections show the highest values of Nu.

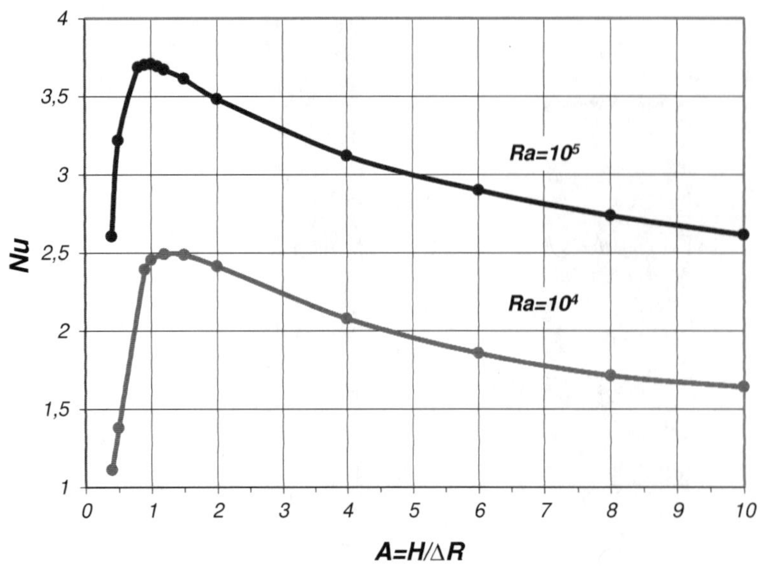

Fig. 4 Dependence of the Nusselt number on A for Ra=10^4 and Ra=10^5.

4. Conclusions

A numerical model was defined and analysed for the description of mixed convection in a vertical annular cavity differentially heated inside and cooled outside and with horizontal adiabatic disk. The case described in this study refers to influence of the aspect ratio, A, on the mixed convection arising

IOP Conf. Series: Journal of Physics: Conf. Series **923** (2017) 012051 doi:10.1088/1742-6596/923/1/012051

as a consequence of the buoyancy, driven by the temperature difference between the vertical walls, and of the rotation at constant angular velocity of the inner. Results have been reported for $Re=100$ and for $Ra=10^4$ and $Ra=10^5$ and allow to observe that a bi-cellular structure of the flow is generally observed which is lost in slender enclosures when the Rayleigh number is increased. The maximum of the Nusselt number can be observed to correspond to cross sections around the squared geometry; in fact, for such configurations it was observed that the relative importance of the force convection-driven cell is the highest, and, hence, the forced convection contribution determines the maximum increase to the Nusselt number.

References

[1] Williams G P 1967 *J. Atmos. Sc.* **24** 144.
[2] Williams G P 1967 *J. Atmos. Sci.* **24** 162
[3] Samuels M R 1968 *Phys. Fluids* **11** 1889
[4] De Vahl Davis G, Leonardi E, Reizes J A 1982 *Pr. XIV ICHMT-Symposium* (Dubrovnik) 131
[5] Urquiza G, Castro L, García J, Basurto M, Bogarin E 2013 *Adv. Mech. Eng.* **2013**
[6] Venkatachalappa M, Sankar M, Natarajan A A 2001 *Acta Mechanica* **147** 173
[7] Desrayaud G, Fichera A, Marcoux M, Pagano A 2006 *Progr. Comp. Fluid Dyn.* **6**
[8] Desrayaud G, Fichera A, Marcoux M, Pagano A, 2009 *Pr. XXVII UIT* (Reggio Emilia)
[9] Marcoux M, Charrier-Mojtabi M C, Azaiez M 1999 *Int. J. Heat Mass Transfer* **42(13)** 2313
[10] Desrayaud G, Fichera A 2002 *Int. J. Heat and Fluid Flow* **23** 519
[11] Khellaf K 1994 *Thèse de Doctorat* Conservatoire national des arts et des métiers, Paris.
[12] Ho C J, Tu F J 1992 *ASME J Heat Transfer* **114(2)** 418

New data processing of local heat transfer coefficient inside a rectangular channel

P Gramazio, L Vitali, D Fustinoni, and A Niro

Politecnico di Milano, Department of Energy,
Campus Bovisa, Via Lambruschini 4, 20156 Milano, Italy

E-mail: alfonso.niro@polimi.it

Abstract. In this paper, we critically reconsider and discuss the models used in one of our previous work to calculate the local convective heat flux for forced air-flows inside narrow rectangular channel, in order to evaluate the limitations which may be inherent with them. To this end, several numerical FEM models have been developed in COMSOL™ and used to analyze in depth previous data processing procedures; furthermore, experimental tests are made on the channel materials, to determine their thermal conductivities and radiative properties. Here, the results of this analysis are presented and discussed.

1 Introduction

Forced air-flows inside narrow rectangular channels are frequently encountered in many "compact" heat transfer devices. Because of small passages and relatively low air velocity, however, heat transfer is rather poor, and hence a number of specially configured extended-surfaces are used to overcome this limit. Ribs are an efficient and cost-effective way to enhance heat transfer without overmuch penalizing pressure drops. For this reason, a long-term experimental campaign has been carried out at the ThermALab of Energy Department of Politecnico di Milano on heat transfer characteristics in forced convection through rectangular channels with one or two rib-roughened surfaces in a very large variety of configurations.

The present work originates from our previous endeavors in measuring the local convective heat transfer coefficient in transitioning flow regimes inside a channel with only the bottom surface heated and ribbed [1]. Indeed, their spanwise-averaged values, for both ribbed and flat reference configurations, were in good agreement with the corresponding average results directly measured in previous works [2] and, for the flat reference channel, validated with the literature results [3]. However, the local heat transfer coefficient, as shown in figure 1, displayed an unexpected increasing trend while approaching the channel side walls. Although it could drop to zero within a thickness too small to be seen, we conjecture it to be an effect of the adopted processing technique. More precisely, to derive the convective heat flux from the heating rate measurement, we have to take into account and evaluate all the other contributions to heat transfer. Thus, accurate conductive and radiative models are needed to retrieve the information on heat fluxes from local temperature measurements; on other hand, the thermophysical properties of materials should be known with the greater possible accuracy.

In this paper, a critical assessment of the models already used to calculate the local convective heat flux has been carried out, in order to evaluate the limitations which may be inherent with them. Furthermore, experimental tests are made on the channel materials, to determine their thermal conductivities and radiative properties. The final objective is to understand and assess how each calculated heat flux affects the evaluation of the local heat transfer coefficient, and to improve the models to reduce the estimated error. To this end, several models have been developed in COMSOL™ FEM and are used as tools to better understand the limitations of the previously utilized models.

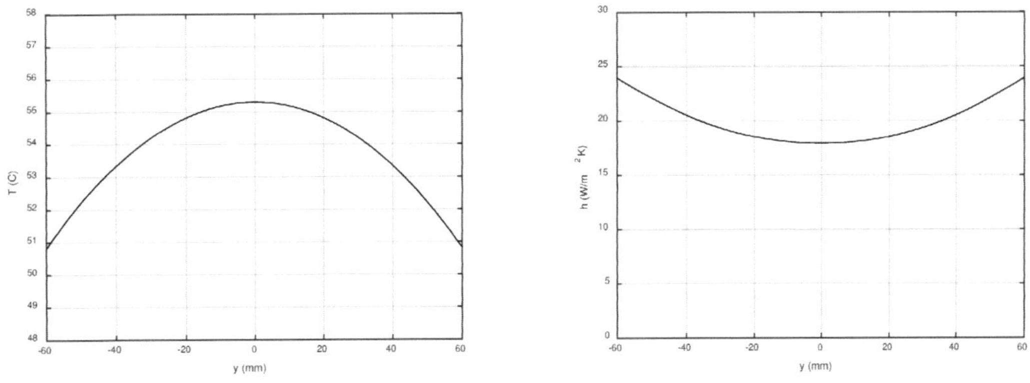

Figure 1. Temperature distribution (a) and local heat transfer coefficient (b), Re = 5400, presented in [1], non-ribbed bottom surface.

2 Materials and methods

2.1 Open-loop wind channel

The experimental setup has been thoroughly presented in [1]; here, just a brief description is reported in order to better focus the problem. It consists of an insulated open-loop wind tunnel operated in suction, schematically shown in figure 2, with a 120 mm wide, 12 mm high test section, whose main features are summarized in table 1. The channel is heated with two independently powered and controlled printed circuit boards (PCB), and the air flow is regulated by means of a valve system. A Hendress Hauser T-Mass 65F15 flow meter is used for automatic mass-flow readings. A double-glazing 130 mm x 130 mm germanium window grants the optical access to a FLIR T650sc IR-camera, which measures the temperature field of the breadboard surface due to the calibration procedure also described in [1]. The channel is also equipped with 10 temperature sensors, 4 RTDs and 6 thermocouples, to measure the average temperatures of the inlet and outlet flow, of the internal germanium window, and of the channel wall beneath the heater.

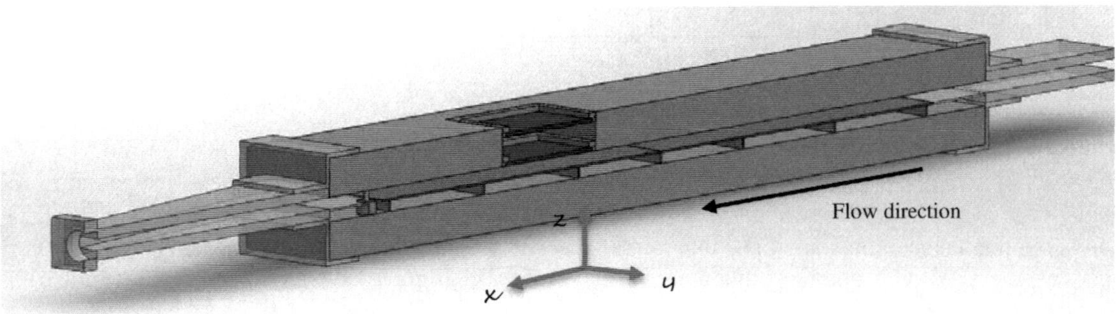

Figure 2. Open-loop wind tunnel design, vertical cross-section

Table 1. main experimental features

H	Channel height	12	mm
L	Channel length	840	mm
w	Channel width	120	mm
D*h*	Hydraulic diameter	21.82	mm
H/w	Height to width ratio	0.10	
s	Rib side dimension	4	mm
d	Distance rib-to-rib	80	mm
Re	Reynolds number	700 ÷ 8000	

2.2 Radiative properties measurements

For an opaque surface, the hemispherical spectral emissivity ε_λ equals the hemispherical spectral absorptance α_λ which in turn, for diffuse radiation, is the complement to 1 of the hemispherical spectral reflectance ρ_λ. On other hand, in term of directional-hemispherical spectral reflectance $\rho'_\lambda(\lambda,\theta_i)$, i.e., the energy reflected into all directions due to intensity incoming from a single direction, the hemispherical spectral reflectance ρ_λ is

$$\rho_\lambda(\lambda)=\frac{\int_{2\pi}\rho'_\lambda(\lambda,\vartheta_i)\,I_\lambda(\lambda,\vartheta_i)\cos\vartheta_i\,d\Omega_i}{\int_{2\pi}I_\lambda(\lambda,\vartheta_i)\cos\vartheta_i\,d\Omega_i} \tag{1}$$

If the incident intensity is independent of direction (diffuse radiation), the previous equation becomes

$$\rho_\lambda(\lambda)=\frac{1}{\pi}\int_{2\pi[sr]}\rho'_\lambda(\lambda,\vartheta_i)\cos\vartheta_i\,d\Omega_i=2\int_0^{\pi/2[rad]}\rho'_\lambda(\lambda,\vartheta_i)\cos\vartheta_i\sin\vartheta_i\,d\vartheta_i \tag{2}$$

For optically smooth non-conductors and for unpolarized radiation incident from a gas, the directional-hemispherical spectral reflectance $\rho'_\lambda(\lambda,\theta_i)$ can be evaluated by the following expression derived from the electromagnetic wave theory [4]

$$\rho'_\lambda(\lambda,\vartheta_i)=\frac{1}{2}\left[\left(\frac{n^2\cos\vartheta_i-\sqrt{n^2-\sin^2\vartheta_i}}{n^2\cos\vartheta_i+\sqrt{n^2-\sin^2\vartheta_i}}\right)^2+\left(\frac{\cos\vartheta_i-\sqrt{n^2-\sin^2\vartheta_i}}{\cos\vartheta_i+\sqrt{n^2-\sin^2\vartheta_i}}\right)^2\right] \tag{3}$$

where n the refractive index of the medium. By means of the Perkin Elmer Frontier FT-IR spectrophotometer equipped with a 3-inch integrating gold sphere, $\rho'_\lambda(\lambda,\theta_i)$ has been measured in the Mid-IR range and for a near normal incidence direction, i.e., $\theta_i = 12°$, for the three main kinds of surfaces enclosing the test-section sensed portion, namely, panels in extruded polystyrene foam (XPS), parts in acrylic glass (PMMA), and the printed circuit board (PCB). Figure 3 displays these values plotted vs wavelength. By solving equation 3 with respect n for $\theta_i = 12°$ and using the measured value of $\rho'_\lambda(\lambda,\theta_i)$ of each surface, the corresponding refractive index has been evaluated. Finally, by substituting equation 3 into equation 2 and then numerically integrating, the values of ρ_λ and hence ε_λ were found.

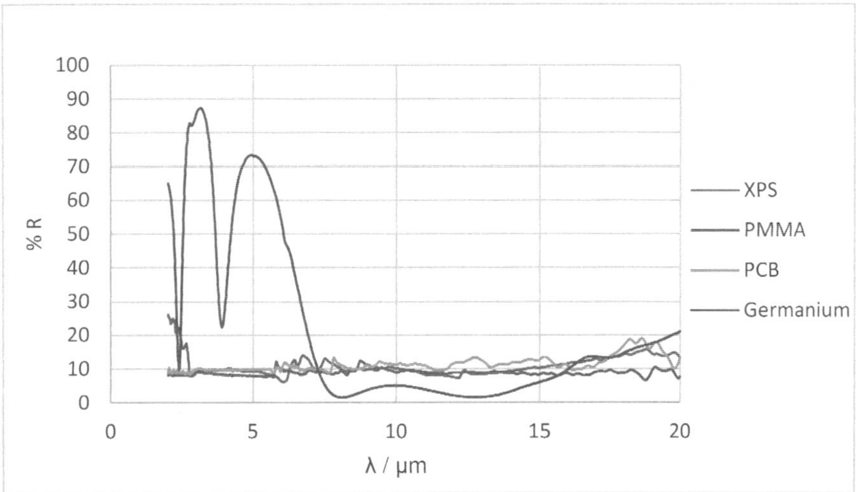

Figure 3. Directional-hemispherical spectral reflectance, Mid-IR wavelength range

2.3 Thermal conductivity measurements

Thermal conductivity for the same three materials, i.e., XPS, PCB and PMMA, has been measured by means of the C-Therm TCi™ instrument by C-Therm Technologies, whose accuracy and precision are better than 5% and 1%, respectively, as tested with reference materials. Table 2 lists the measured values of thermal conductivity.

Table 2. Thermal conductivities of the three main materials in the test-section

Material	Conductivity [W/mK]
XPS	0.02
PMMA	0.30
PCB	0.30

3 Modeling of the experiment

3.1 Local convective heat flux evaluation

By a steady-state energy balance for a unit-volume of the heater of infinitesimal base area $dx \cdot dy$ and height $s = (s_{cu} + s_{pcb})$, i.e., equal to the sum of copper track and breadboard thicknesses, we obtain

$$q_c(x,y) = q_{el} - q_r(x,y) - q_{kz}(x,y) + \int_{z=0}^{z=s_{Cu}+s_{PCB}} k(z)\left(\frac{\partial^2 T(x,y,z)}{\partial x^2} + \frac{\partial^2 T(x,y,z)}{\partial y^2}\right) dz \tag{4}$$

where q_{el} the heat generated by Joule effect in the PCB copper tracks, which is supposed to be uniform, $q_r(x,y)$ the heat loss by radiation from the PCB upward surface, and $q_{kz}(x,y)$ the heat loss from the PCB backside. The last term in equation 4 is the divergence of the conductive heat flux through the PCB plane integrated along its thickness which, for the sake of brevity, we refer to as the pseudo-conductive heat flux q_{kxy}. Each term needs a dedicated model in order to be evaluated from the experimental measurements. In the following subsections, we focus on the discussion of these models. Once all these terms are evaluated, the local convective heat transfer coefficient can be retrieved as

$$h(x,y) = \frac{q_c(x,y)}{T_W(x,y) - T_b(x)} \tag{5}$$

The experiments reported in our previous work produced the estimate of the average relative contribution of the heat flux terms to the overall energy balance, as shown in figure 4. The radiative heat flux could seemingly be the most relevant term, whereas the pseudo-conductive heat flux and backside one should lose importance with increasing Reynolds numbers. However, it is noteworthy to keep in mind that local relative contribution can be significantly different from the average one, especially in proximity of the side walls.

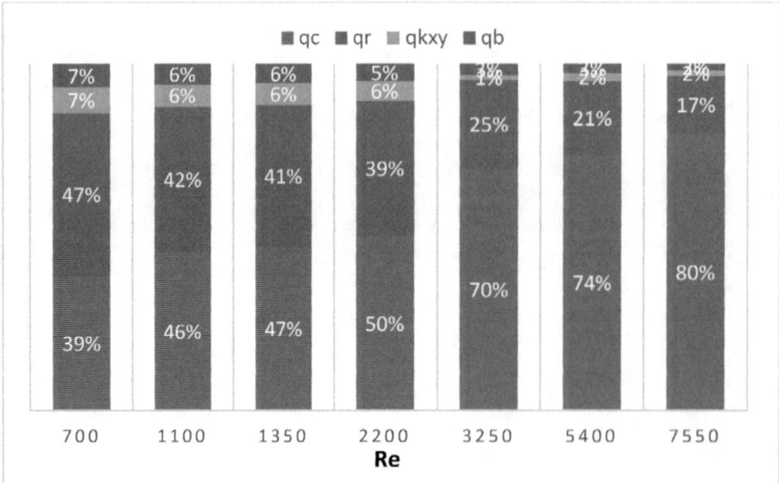

Figure 4. Relative heat flux distribution, as a function of Reynolds number

3.2 Numerical FEM model of the section

A detailed 2D numerical FEM heat transfer model of a *y-z* transverse section of the test-section has been developed in COMSOL™, featuring all its components as shown in figure 5, namely, PCB, PMMA frames, germanium double window, insulating walls, and the layer of still air underlying the PCB. The convective heat transfer coefficient and the bulk flow temperature are imposed as boundary conditions in the only non-modelled region, i.e., the internal channel where air flows through. Finally, the radiative heat exchange between the channel inside walls is solved too by COMSOL™ but with a model for opaque surfaces. Convective and radiative heat transfer from the test-section outside walls to laboratory environment is imposed on them as boundary conditions. The simulations yield the temperature field in the computational domain, and are used as benchmark to test the heat flux modes.

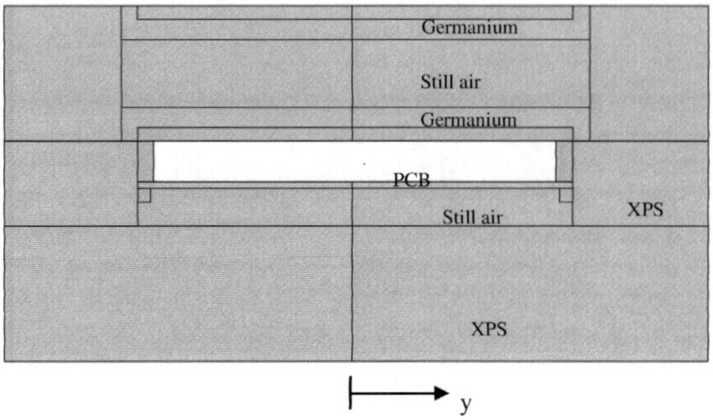

Figure 5. Geometry of the COMSOL™ FEM model of the complete test section

3.3 Electric heat flux

The electric heater is made of thin copper tracks that lay on the board substrate; they cover 97.5% of the surface exposed to the air flow. In our calculation, a spatially uniform generated heat flux is assumed. The correctness of this assumption has been specifically verified by measuring the heater surface temperature by means of IR thermography while being operated in open air and, hence, subject only to natural convection. The acquired temperature field has been compared with the results of a CFD non stationary model of natural convection for the same geometry, with uniform heat generation. The local temperature difference between model and experiment was less than 5%, therefore proving the hypothesis to be adequate, to our best knowledge.

3.4 Radiative heat flux

A radiative heat transfer model is needed to precisely evaluate the local radiative heat flux $q_r(x,y)$, from the thermographic measurement of the local temperature on the PCB surface and the temperature measurements by means of thermocouples on the germanium window. To this end, a first Matlab® model was developed in [1], for the ease of interfacing it with the acquired thermograms, based on Modest's model [4] for opaque, gray and diffuse enclosures, with an added term that takes into account the estimated radiative losses, due to the partial IR-transparency of the germanium windows. The model solves the radiative heat transfer problem in a 2D cavity, which is a *y-z* slice of the test section, by discretizing the bottom surface, i.e. the heater, with N_y elements, while the remaining three surfaces were modelled with one element each, as their temperature was supposed to be position-independent. The view factors were calculated by means of the Hottel's method. This model showed some deviation from the expected results, especially near the edges, when tested with low emissivity top surfaces.

This model has been improved, using a radiative COMSOL™ model as benchmark to evaluate the accuracy in specific test cases for opaque surfaces. The equation set is now based on a slight modified version of the Siegel's model [5], which describes the radiative heat transfer for cavities composed by semi-transparent, diffuse and selective surfaces, as shown in equations 6 – 10. In addition, all surfaces are now discretized, with the same number of elements for opposite surfaces, i.e. N_y for the top and bottom surfaces, and N_z for each side, being N_y the pixel number in *y*-direction of a thermogram, equal to 433. Furthermore, the spectral behaviour was introduced, by considering three wavelength regions with constant optical properties, i.e., between 0 and 7 μm, 7 and 13 μm, and from 13 μm up.

Moreover, whereas the top temperature is considered to be constant because of the high conductivity of the Germanium window, several numerical tests have also been carried out to evaluate the best estimate of the side walls temperature $T_s(z)$, which is not experimentally determined, on the basis of the top and bottom-side measurements. The linear approximation showed a good agreement in the resulting heat flux trend, while a constant approximation showed a better matching in a broad central region; so the latter was eventually adopted.

Finally, the better geometrical discretization led to an improvement of the accuracy of the view factors, especially near the edges, influencing the trend of the radiative heat flux, while the experimental measurement of the radiative properties affected its estimated value, which now results 20% larger than that previously modelled, as it can be seen in figure 6.

$$\sum_{j=1}^{N} \frac{1}{E_{w,j}^{(k)}}\left(\delta_{kj} - R_{w,j}^{(k)}F_{kj}\right)\left[q_j^{(k)} - E_{w,j}^{(k)}\sigma T_j^4 F_{0\lambda}^{(k)} + \left(1 - R_{w,j}^{(k)}\right) q_{e,j}^{(k)}\right] = \sum_{j=1}^{N}\left(\sigma T_j^4 F_{0\lambda}^{(k)} + \frac{T_{w,j}^{(k)}}{E_{w,j}^{(k)}} q_{e,j}^{(k)}\right)\left[\delta_{kj} - \left(1 - T_{w,j}^{(k)}\right) F_{kj}\right] \qquad \begin{matrix} 1 \le j \le N \\ k: \lambda_1 \le \lambda \le \lambda_2 \end{matrix} \qquad (6)$$

$$q_j = \sum_{k=1}^{N_\lambda} q_j^{(k)} \tag{7}$$

$$R_{w,j}^{(k)} = \rho^{(k)} \left[1 + \frac{\left(1-\rho^{(k)}\right)^2 \tau^{(k)^2}}{1-\rho^{(k)^2}\tau^{(k)^2}} \right] \tag{8}$$

$$T_{w,j}^{(k)} = \tau^{(k)} \frac{1-\rho^{(k)}}{1+\rho^{(k)}} \frac{1-\rho^{(k)^2}}{1-\rho^{(k)^2}\tau^{(k)^2}} \tag{9}$$

$$E_{w,j}^{(k)} = A_{w,j}^{(k)} = \frac{\left(1-\rho^{(k)}\right)\left(1-\tau^{(k)}\right)}{1-\rho^{(k)}\tau^{(k)}} \tag{10}$$

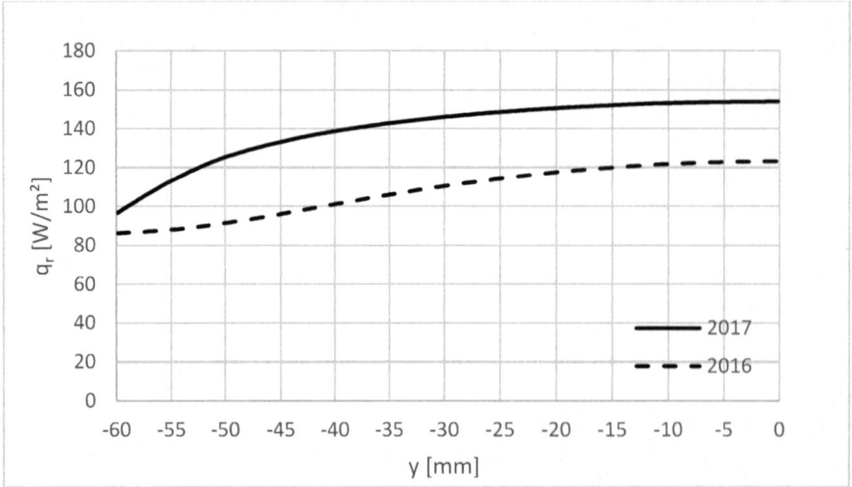

Figure 6. Radiative heat flux, comparison between ref. [1] (2016) and present work (2017).

3.5 Breadboard conductive heat flux

In our previous work, the pseudo-conductive heat flux q_{kxy} in the breadboard plane was calculated with equation 11, under the hypothesis of the uniformity of the surface temperature over the copper tracks thickness. The model can be extended by using a combined copper and PCB conductivity, as shown in equation 12, yet considering temperature to be uniform over the entire PCB thickness. The effectiveness of the latter assumption, has been tested by using the numerical FEM model presented in subsection 3.2, which allowed to compare the span-wise distribution of the pseudo-conductive heat flux evaluated with equation 12, i.e., using only the surface-temperature field numerically obtained, with that evaluated by means of equation 13 taking advantage of fully knowing the *y-z* temperature field inside the PCB.

$$q_{kxy.1}(x,y) = k_{Cu} s_{Cu} \left(\frac{\partial^2 T_w(x,y)}{\partial x^2} + \frac{\partial^2 T_w(x,y)}{\partial y^2} \right) \tag{11}$$

$$q_{kxy.2}(x,y) = (k_{Cu} s_{Cu} + k_{PCB} s_{PCB}) \left(\frac{\partial^2 T_w(x,y)}{\partial x^2} + \frac{\partial^2 T_w(x,y)}{\partial y^2} \right) \tag{12}$$

$$q_{kxy}(x,y) = k_{Cu}\int_0^{z_{Cu}}\left(\frac{\partial^2 T(x,y,z)}{\partial x^2}+\frac{\partial^2 T(x,y,z)}{\partial y^2}\right)dz + k_{PCB}\int_0^{z_{PCB}}\left(\frac{\partial^2 T(x,y,z)}{\partial x^2}+\frac{\partial^2 T(x,y,z}{\partial y^2}\right)dz \quad (13)$$

The results are compared in figure 7, which highlights that the two pseudo-conductive heat flux distributions differ each other by a mere 1% in the central region, i.e., for *y* ranging from -50 to 50 mm, but they are completely different, even changing their concavity, while moving toward the edges. Since experimentally it is possible only measuring the surface-temperature distribution, we are obliged to evaluate q_{kxy} by means of equation 12, but we have to keep well in mind however that it adequately predicts the pseudo-conductive heat flux in a quite wide region but not in nearness of the side walls where error becomes huge.

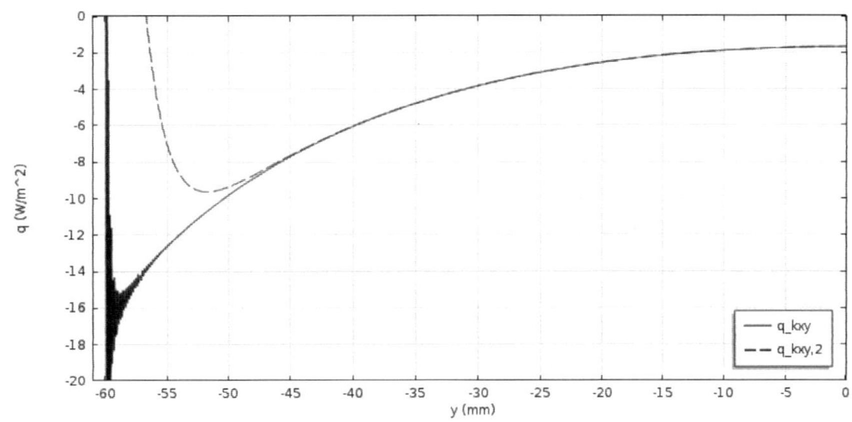

Figure 7. Pseudo-conductive heat flux, numerical results, model based on surface temperature and average over the thickness, Re = 5400.

3.6 Backside conductive heat flux

The heat flux dispersed from the PCB backside was evaluated under the assumption of using a simple 1D conductive model, as shown in equation 14, applied on each thermogram pixel

$$q_{kz.1}(x,y) = \frac{k_{air}}{s_{air}}[T_w(x,y) - T_{int}] \quad (14)$$

where $T_w(x,y)$ the PCB upper surface temperature, and T_{int} the temperature on the bottom side of the air layer. It can be seen that only the still air insulating layer was taken into account, and that an average internal temperature was used for the calculation. This model can be improved by considering the conductive resistance of the PCB in addition to that of air layer (equation 15), but further considerations are needed because of the results of the numerical FEM model shown in figure 8. Indeed, the spanwise trend of q_{kz} numerically evaluated shows a very different behavior with respect to that of the 1D model.

$$q_{kz.2}(x,y) = \left(\frac{k_{air}}{s_{air}}+\frac{k_{PCB}}{s_{PCB}}\right)[T_w(x,y) - T_{int}] \quad (15)$$

IOP Conf. Series: Journal of Physics: Conf. Series **923** (2017) 012052 doi:10.1088/1742-6596/923/1/012052

The adopted solution is therefore to use instead of the local temperature $T_w(x,y)$ its spanwise-averaged value $T_{w.av}(x)$, which yields our best estimate as it reduces the error by about 50% in the most part of the spanwise y-direction

$$q_{kz.3}(x,y) = \frac{k_{air}}{s_{air}}\left[T_{w.av}(x) - T_{int}\right] \tag{16}$$

The accuracy of the q_{kz} estimate can be improved only by considering a more complex numerical 2D model, which by now is beyond the scope of this work but that may be integrated in the future.
A summary of the main improvements and modifications is reported in table 3.

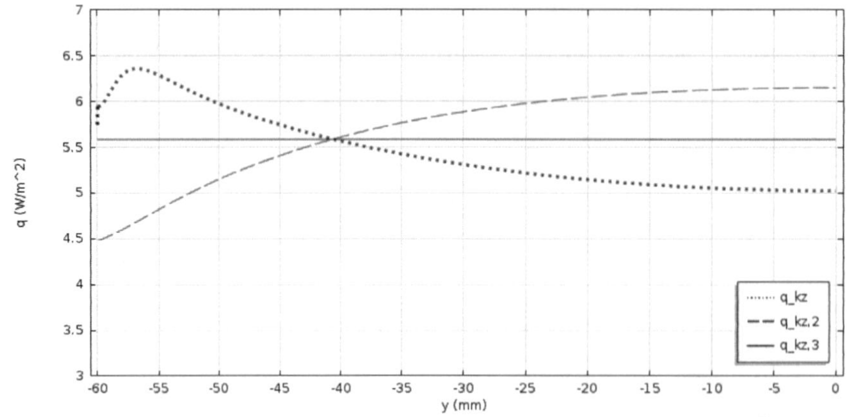

Figure 8. Comparison between three ways of calculating the backside conductive heat flux, and the numerical trend, Re = 5400.

Table 3. Model comparison

Model	Previous version	New version
Radiative (q_r)	Modest's model	Modified Siegel's model
	One element top and sides	N_y elements top/bottom, N_z elements each side
	No spectral behavior	Constant properties in N_λ custom bands
	Linear T_s	Constant T_s
Conductive (q_{kxy})	Copper-only conductivity	Combined copper-PCB conductivity $q_{kxy.2}$
Conductive back	1D, $q_{kz.1}$	1D $q_{kz.3}$

4 Results

4.1 Experimental local convective coefficient

By using the experimental data for the flat channel reported in [1], and calculating the convective heat flux as

$$q_c'(x,y) = q_{el} - q_r(x,y) + q_{kxy.2}(x,y) - q_{kz.3}(x,y) \tag{17}$$

equation 5 yields the local heat transfer coefficient h', where the prime for both q_c and h remarks they are approximations. In figure 9, h' is plotted versus the y direction. As it can be seen, the new processing procedure reduces slightly, and almost uniformly, the value of h' mainly because of a larger value of the radiative heat flux as estimated with the new model, but it still displays the increasing trend towards the edges. This outcome may appear a quite poor result of the revision process but, before drawing any conclusion, it is worthwhile discussing about that.

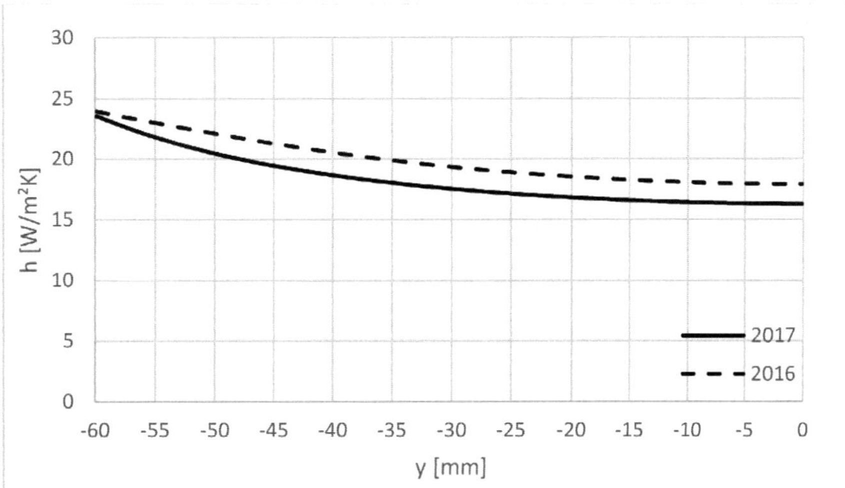

Figure 9. Ref. [1] (2016) vs present work (2017) local heat transfer estimate from the same experimental data, Re=5400

4.2 Numerical assessment of the model

An estimate of the overall error due to modeling issues can be carried out by comparing the local heat transfer coefficient based on q_c' from equation 17, to the ones evaluated by using for q_{kz} and q_{kxy} the values calculated by means of the numerical 2D FEM model; more precisely, we define

$$err = \frac{q_c'}{q_c} - 1 \equiv \frac{h'}{h} - 1 \tag{18}$$

It is important to notice that this estimated error is based solely on the numerical model, therefore eventual experimental issues are not taken into account. As it can be seen in figure 10, errors lie beneath 4% up to 40 mm from the centerline, and tend to diverge towards the edges. An increase of the Reynolds number reduces local errors, while the trend is unchanged. Figure 11 shows in detail the extent of the region where error is equal to or larger than 5% as a function of Reynolds numbers. As it can be see this region extends from the wall sides for a length ranging from 8 to 12 mm while Reynolds spacing from 8000 to 700.

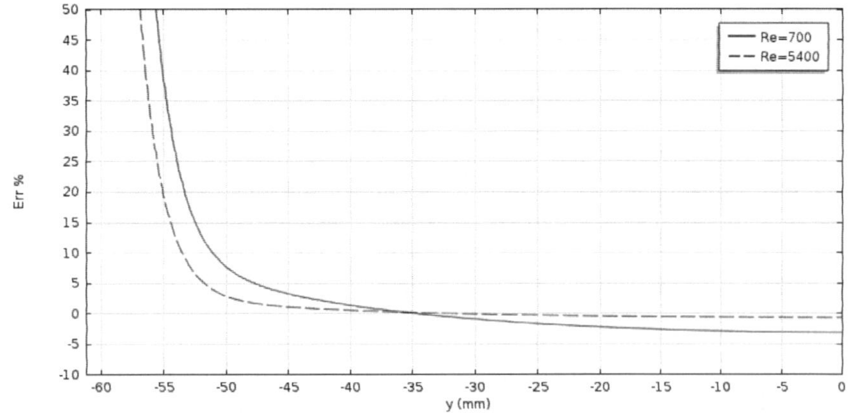

Figure 10. Error maps of h, for Re=700 and for Re=5400

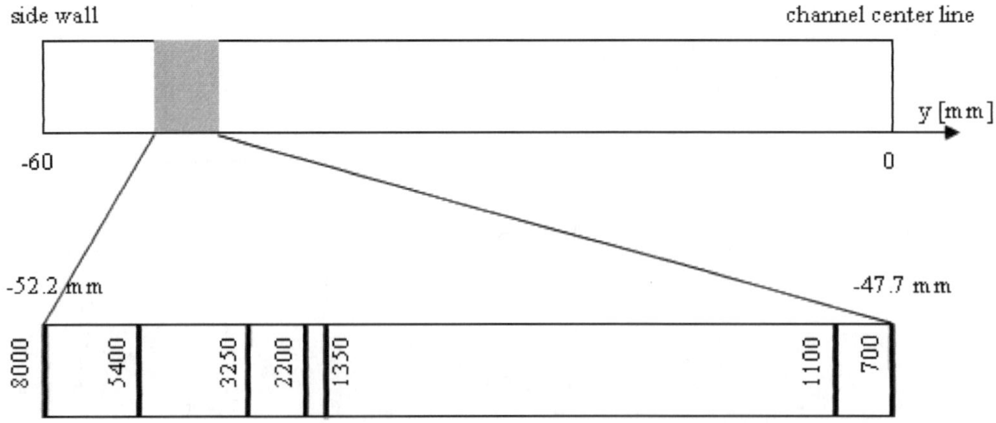

Figure 11. Spanwise positions of 5% error, as function of Reynolds number

Finally, a new estimate of the local convective coefficient can be made by correcting the approximation *h'* with the calculated error, as in the following:

$$h'_{mod} = \frac{h'}{err - 1} \tag{19}$$

The results are shown in figure 12 and are much closer to the expected values. In particular, h'_{mod} shows a flatter behavior in the central zone while tending to zero towards the edges. The conductive contribution near the edges has been proved to be the main source of error while estimating the spanwise distribution of *h*. Further issues could be linked to other aspects, which are already taken into account to our best knowledge but that could be more deeply investigated, like the influence of directional radiative properties on the radiative heat flux.

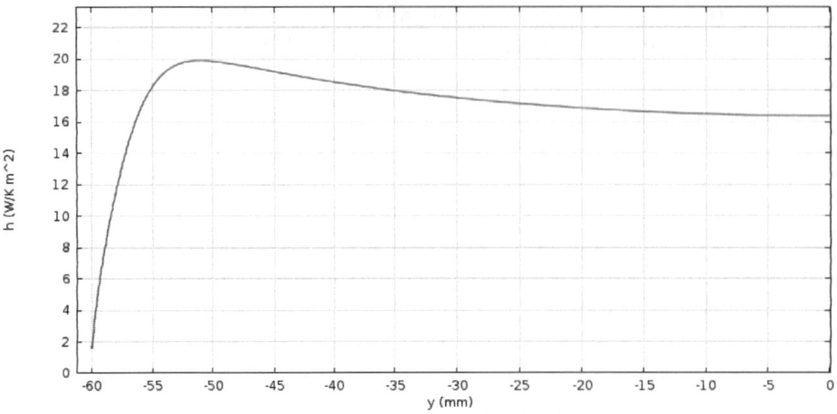

Figure 12. Estimate of local heat corrected by equation 19, Re=5400

5 Conclusions

This work consisted in a critical assessment of the processing techniques adopted to retrieve from experimental thermographic measurements the distribution of local convective heat flux, and therefore

IOP Conf. Series: Journal of Physics: Conf. Series **923** (2017) 012052 doi:10.1088/1742-6596/923/1/012052

of local convective heat transfer coefficient, in a forced air flow through a rectangular channel. The radiative model has been improved, and now it takes into account semi-transparent surface, with spectral-dependant properties. The contribution of the conductive terms, both inside PCB and towards its backside, has been deeply evaluated with the aid of numerical FEM models. The results show that the models based only on the surface and backside temperatures can approximate well the pseudo-conductive heat flux in the central part, but fail while moving towards the edges. Then, the estimate of the error has been used to provide a better evaluation of the convective heat transfer coefficient, which is actually closer to the expected behaviour in the non-ribbed channel. This kind of correction will be applied to ribbed geometry, while future work by means of DNS simulations will also provide a further support for the results.

References

[1] D Fustinoni, P Gramazio, L Vitali and A Niro 2017 *J. of Phys.: Conf. Series* **796** 012015
[2] D Fustinoni, P Gramazio, L P M Colombo and A Niro 2012 *J. Phys.: Conf. Ser.* **395** 012042
[3] S Kakaç, R K Shah, W Aung, Handbook of single-phase convective heat transfer, Wiley-Interscience, 1987
[4] M F Modest, Thermal radiation heat transfer, Elsevier Inc, 2013
[5] R Siegel 1973 Net Radiation Method for Enclosure Systems Involving Partially Transparent Walls, NASA technical note D-7384

IOP Conf. Series: Journal of Physics: Conf. Series **923** (2017) 012053 doi:10.1088/1742-6596/923/1/012053

Influence of the boundary conditions on heat and mass transfer in spacer-filled channels

M Ciofalo, M F La Cerva, M Di Liberto and A Tamburini

Dipartimento dell'Innovazione Industriale e Digitale, Università di Palermo, Italy
E-mail michele.ciofalo@unipa.it

Abstract. The purpose of this study is to discuss some problems which arise in heat or mass transfer in complex channels, with special reference to the spacer-filled channels adopted in membrane processes. Among the issues addressed are the consistent definition of local and mean heat or mass transfer coefficients; the influence of the wall boundary conditions; the influence of one-side versus two-side heat/mass transfer. Most of the results discussed were obtained by finite volume CFD simulations concerning heat transfer in Membrane Distillation or mass transfer in Electrodialysis and Reverse Electrodialysis, but many of the conclusions apply also to different processes involving geometrically complex channels.

1. Heat and mass transfer in spacer-filled channels

Membrane processes offer examples of heat or mass transfer in plane channels bearing mixing promoters of more or less tortuous geometry, aimed at reducing temperature or concentration polarization phenomena thus improving the process performance [1, 2]. Promoters usually consist of polymeric filaments and play also the role of spacers, keeping a fixed distance between the opposite channel walls. For example, the top row in Figure 1 shows real spacers made of overlapped (a) or woven (b) filaments. The bottom row shows corresponding unit cells (computational domains).

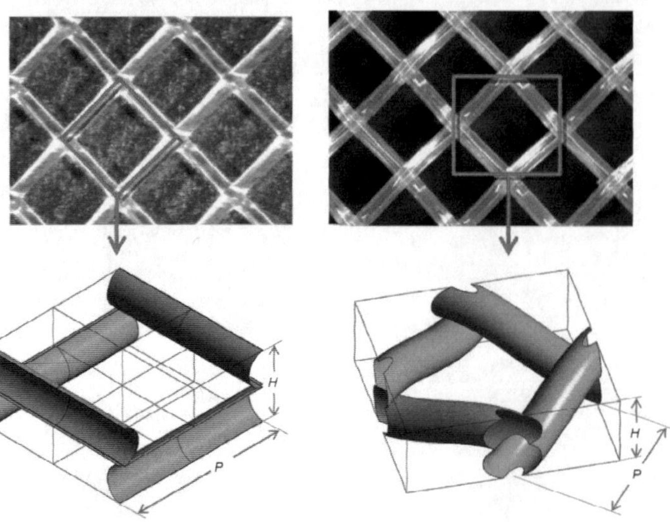

(a) (b)

Figure 1. Lattice (top row) and unit cells (bottom row) of spacer-filled channels. (a) overlapped; (b) woven.

CFD simulations of spacer-filled channels for membrane processes started to appear in the literature since the year 2001 and have been presented at an increasing rate since then. Hitsov *et al.* [3], Fimbres-Wehis and Wiley [4], Karabelas *et al.* [5] and La Cerva *et al.* [6] have presented reviews addressing a number of issues in the modeling of such channels.

The present paper aims to complement the above reviews, by discussing a number of aspects not sufficiently clarified so far. The main issues addressed here are:

- the consistent definition of local and mean heat or mass transfer coefficients;
- the influence of the wall boundary conditions;
- the influence of one-side versus two-side heat/mass transfer.

Most of the results that follow make reference to the spacer geometry Diamond_90 in Figure 1(a), consisting of identical overlapping cylindrical filaments with an intrinsic angle α=90° and different flow attack angles γ and Reynolds numbers Re.

The problem is described by the continuity, Navier-Stokes and scalar transport equations for a constant-property fluid. For simplicity, in the following the transported scalar will be assumed to be the specific enthalpy c_pT; obvious changes apply in the case of mass transport.

In order to simulate fully developed flow and thermal fields by simply applying periodicity conditions at the opposite faces, "true" pressure p^* and temperature T^* are replaced by their periodic components p, T such that

$$p^* = p + p_s s \tag{1}$$

$$T^* = T + T_s s \tag{2}$$

in which s is the distance along the streamwise (main flow) direction while p_s and T_s are the large-scale gradients of pressure and temperature, respectively, in the same direction. Note that p_s is always negative while T_s is positive for a heated fluid and negative for a cooled fluid.

In the simulations, the pressure gradient p_s is imposed while the flow rate and thus the bulk flow Reynolds number Re follow as results of the simulation.

As regards the large-scale temperature gradient, a heat balance yields $T_s=q/(\rho c_pUV)$ in which U is the mean "void channel" velocity in the direction s, V is the volume of the unit cell and q is the thermal power entering the fluid from the walls of the same cell. Both q and U are evaluated as parts of the solution during the course of the simulation.

Substituting eqs. (1)-(2) for p^* and T^* in the standard steady-state momentum and energy equations, after some manipulation one obtains the following governing equations (written in Cartesian tensor notation):

$$\frac{\partial \rho u_j}{\partial x_j} = 0 \tag{3}$$

$$\frac{\partial \rho u_i u_j}{\partial x_j} = -\frac{\partial p}{\partial x_i} + \frac{\partial}{\partial x_j}\mu\frac{\partial u_i}{\partial x_j} - p_s\frac{\partial x_i}{\partial s} \tag{4}$$

$$\frac{\partial \rho c_p u_j T}{\partial x_j} = \frac{\partial}{\partial x_j}\lambda\frac{\partial T}{\partial x_j} - \frac{q}{V}\frac{u_s}{U} \tag{5}$$

in which u_s is the component of the local velocity along the direction s.

Translational periodic boundary conditions were now adopted for all variables at the opposite faces of the computational domain. No slip conditions (u_i=0) were imposed at the top and bottom walls and on the filaments' surface. As regards the thermal boundary conditions, the filaments were assumed to be adiabatic (q''=0), while at either one or both the walls different thermal boundary conditions were imposed, as will be discussed below.

All simulations were conducted by the ANSYS-CFX® code. The code uses a finite volume approach and a co-located (non-staggered) grid layout, such that the control volumes are identical for all transport equations. A coupled solver is adopted, in which the hydrodynamic equations for the three velocity components and pressure are treated as a single large system. A Multi-grid (MG) accelerated Incomplete Lower Upper (ILU) factorization technique is adopted for solving the discrete system of linearized equations.

In the present simulations, the "high resolution" (higher order upwind) interpolation scheme was used for the advection terms. Further numerical methods and computational details have been discussed elsewhere [7, 8].

2 Definition of local or averaged heat / mass transfer coefficients

2.1 Local coefficients

Making reference, for the purpose of clarity, to heat transfer, the most common definition of the local heat transfer coefficient is:

$$h = \frac{q''_w}{T_b - T_w} \tag{6}$$

In Eq. (6) q''_w is the local wall heat flux (assumed to be positive for a cooled fluid), T_w is the local wall temperature and T_b is the bulk temperature, usually defined as:

$$T_b = \int_A \rho c_p T u_s \mathrm{d}A \;\Big/\; \int_A \rho c_p u_s \mathrm{d}A \tag{7}$$

A being the cross sectional area of the channel. Often ρ and c_p can be treated as constant properties. Note that T_b remains a function of s. When the computational domain is a single repetitive unit of a large lattice (unit cell approach) it may be preferable to define the bulk temperature by replacing the area integrals in Eq. (7) with volume integrals taken over a whole unit cell. Yet another alternative is to replace the bulk temperature in the definition (6) of h with a different reference value, e.g. the fluid's temperature at the channel midplane [9] or the fluid's inlet temperature [10].

In the case of mass transfer, temperature is replaced by concentration and h is replaced by a mass transfer coefficient k (in m s^{-1}).

The problem with the above definition of h or k is that, in complex geometries involving flow recirculation, the iso-surface $T=T_b$ (or $C=C_b$) may intersect the channel walls; in this case, h (k) diverges on the line $T_b=T_w$ ($C_b=C_w$) and becomes negative inside this line. An example is reported in Figure 2, obtained by post-processing the results of a three-dimensional CFD simulation. It shows the iso-surface $T=T_b$ (bulk temperature) for an overlapped spacer with $P/H=4$, $\gamma=0°$, Re≈42 and two-side heat transfer with uniform imposed wall heat flux. Even in the absence of singularities, the simple fact that the iso-surface $T=T_b$ (or $C=C_b$) approaches the wall closely may give rise to abnormal and unphysically large values of h or k. This kind of distribution is rather common, especially for imposed flux (2nd type) or mixed (3rd type) boundary conditions, two-side heat / mass transfer, high Reynolds numbers and high Prandtl / Schmidt numbers. The only condition in which singularities cannot occur is that of imposed wall temperature or concentration (1st type, or Dirichlet).

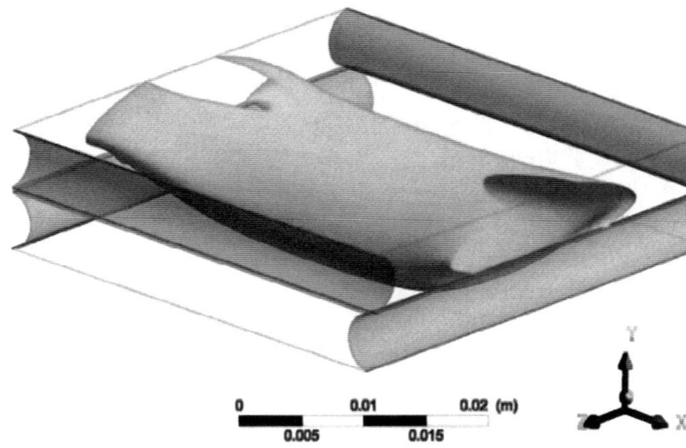

Figure 2. Iso-surface $T=T_b$ (bulk temperature) for an overlapped spacer with $P/H=4$, $\gamma=0°$, Re≈42 and two-side heat transfer with uniform imposed wall heat flux.

2.2 Average coefficients

In most papers dealing with heat or mass transfer in spacer-filled channels [9, 11, 12, 13], the average transfer coefficient is simply defined as the area average

$$h^{(1)} = \langle h \rangle = \frac{1}{S}\int_S h \, \mathrm{d}S = \frac{1}{S}\int_S \frac{q''_w}{T_b - T_w} \mathrm{d}S \quad (8)$$

where S may be the whole active transfer surface or a sub-region of it (e.g., one of two active channel walls). Obvious modifications apply if mass transfer is considered.

The problem with this definition is that, if the local h (or K) is singular or attains unphysical values (for the reasons discussed just above), also its average $\langle h\rangle$ (or $\langle K\rangle$) will be affected by this behavior. A preferable definition of the averages, which does not suffer from this problem, is [7, 8]

$$h^{(2)} = \frac{\langle q''_w \rangle}{T_b - \langle T_w \rangle} \quad (9)$$

in which $\langle q''_w\rangle$ and $\langle T_w\rangle$ are surface averages of q''_w or T_w over the surface of interest (e.g., one or both walls). Obvious modifications apply in the case of mass transfer. Dimensionless numbers $\mathrm{Nu}^{(1)}$, $\mathrm{Nu}^{(2)}$ or $\mathrm{Sh}^{(1)}$, $\mathrm{Sh}^{(2)}$ can be obtained from the above definitions by choosing a suitable length scale.

$\mathrm{Nu}^{(1)}$ and $\mathrm{Nu}^{(2)}$ coincide when a uniform wall temperature is imposed, while they differ when a uniform wall heat flux or a wall thermal resistance are imposed. In fact, one can write:

$$\mathrm{Nu}^{(1)} = \mathrm{constant} \times \left\langle \frac{q''_w}{\Delta T_{bw}} \right\rangle, \quad \mathrm{Nu}^{(2)} = \mathrm{constant} \times \frac{\langle q''_w \rangle}{\langle \Delta T_{bw} \rangle} \quad (10)$$

in which $\Delta T_{bw} = T_b - T_w$. When T_w is imposed, ΔT_{bw} is a constant (say, ΔT_{bw}^0) and one has

$$\mathrm{Nu}^{(1)} = \mathrm{Nu}^{(2)} = \mathrm{constant} \times \frac{\langle q''_w \rangle}{\Delta T_{bw}^0} \quad (11)$$

On the contrary, when a uniform heat flux is imposed, q''_w is a constant (say, q''^0_w) and one has

$$\mathrm{Nu}^{(1)} = \mathrm{constant} \times q''^0_w \left\langle \frac{1}{\Delta T_{bw}} \right\rangle \neq \mathrm{Nu}^{(2)} = \mathrm{constant} \times q''^0_w \frac{1}{\langle \Delta T_{bw} \rangle} \quad (12)$$

since the average of the reciprocal is *not* the reciprocal of the average.

The difference between the two definitions is illustrated in Figure 3, which reports average Nusselt and Sherwood numbers, computed by using definitions (8) and (9) for the average heat (mass) transfer coefficients and $2H$ as the length scale. They are reported as functions of the Reynolds number for an overlapped spacer with P/H=4, γ=45° and two-side heat / mass transfer. The values 4 and 600 were assumed for the Prandtl and Schmidt numbers Pr and Sc, respectively, and 3rd type boundary conditions were imposed at the walls. It can be observed that $Nu^{(2)}$ and $Sh^{(2)}$ are always lower than $Nu^{(1)}$ and $Sh^{(1)}$, and that the difference increases with Re. Note also that the difference between $Sh^{(1)}$ and $Sh^{(2)}$ (Sc=600) is larger than that between $Nu^{(1)}$ and $Nu^{(2)}$ (Pr=4).

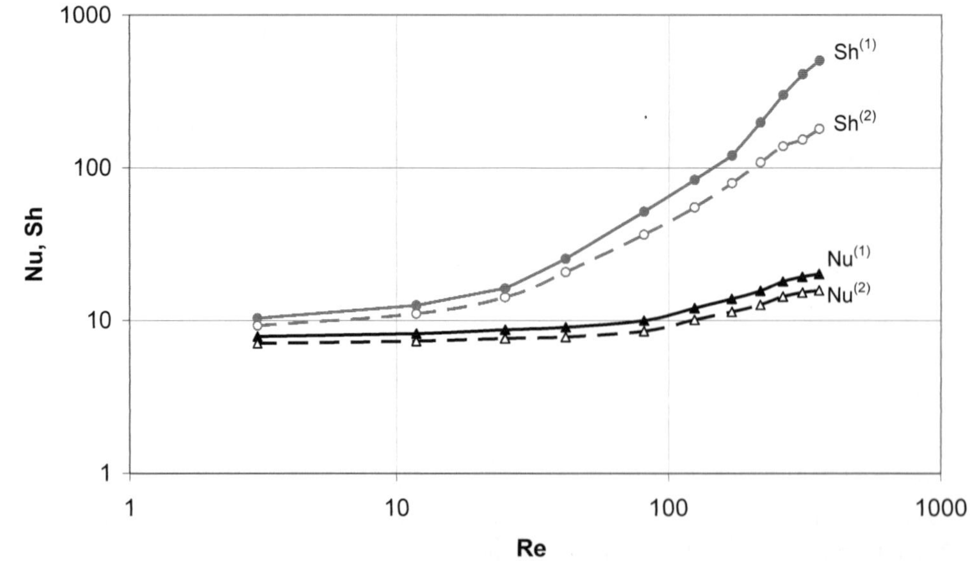

Figure 3. Average Nusselt and Sherwood numbers, computed by definitions (1) and (2), as functions of the Reynolds number for an overlapped spacer with l/H=4, flow attack angle γ=45°, two-side heat / mass transfer and 3rd type boundary conditions (Prandtl number Pr=4, Schmidt number Sc=600).

3 Influence of the boundary conditions at the walls

The different boundary conditions that can be imposed at a wall for a scalar transport equation are schematically illustrated in Figure 4. For the purpose of clarity, reference is made to temperature and heat transfer; obvious modifications apply if concentration and mass transfer are considered instead.

The different BC's can be regarded as particular cases of the most general 3rd type condition :

$$T_w - T_{ext} = r_{ext}\, q''_w \tag{13}$$

in which T_w is the wall temperature; T_{ext} is the temperature of some external medium; r_{ext} is the interposed thermal resistance per unit wall area, inclusive of the conductive resistance of the membrane and (if appropriate) of the convective resistance on the opposite fluid flow; and q''_w is the wall heat flux. This is made clear by writing

$$q''_w = \frac{T_b - T_{ext}}{r_{ext} + 1/h} \tag{14}$$

in which h is the convective heat transfer coefficient. Case (a) is obtained for $r_{ext} \ll 1/h$ and approaches uniform wall temperature (1st type, or Dirichlet) conditions; case (b) applies to a generic condition; case (c) is obtained for $r_{ext} \gg 1/h$, and approaches uniform heat flux (2nd type, or Neumann) conditions.

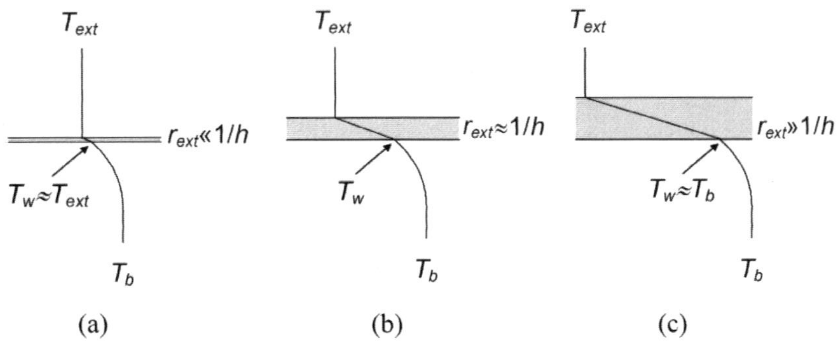

Figure 4. Thermal boundary conditions at a wall obtained as particular cases of the general 3rd type condition for different values of the ratio between the wall thermal resistance r_{ext} and the convective heat transfer resistance $1/h$. (a) Case r_{ext}«$1/h$, approaching uniform wall temperature (1st type, or Dirichlet) conditions; (b) generic case ($r_{ext} \approx 1/h$); (c) case r_{ext}»$1/h$, approaching uniform heat flux (2nd type, or Neumann) conditions.

The thermal resistance r_{ext} can be made dimensionless as $R=r_{ext}\lambda_f/\delta$, in which λ_f is the thermal conductivity of the fluid and δ is the channel's half-height (other choices of the reference length are, of course, possible). Accurate solutions for the Nusselt number under any prescribed value of R can be obtained for the case of a (void, or spacerless) plane channel, as described in detail in the Appendix. Figure A.1 therein reports the Nusselt number (based on the hydraulic diameter $2H$ of the void channel) as a function of R both for two-side and one-side heat transfer. The limiting conditions $R\to 0$ (uniform wall temperature) and $R\to\infty$ (uniform wall heat flux) are also reported. It can be observed that thermal boundary conditions affect the Nusselt number to a significant extent, and that Nu is always increasing with R both for one-side and for two-side transfer.

In a real spacer-filled channel for membrane processes, the actual boundary conditions will be, in general, of some intermediate type, with an external resistance r_{ext} and an external temperature T_{ext} that will depend on the specific process. For example, in Direct Contact Membrane Distillation r_{ext} may represent the thermal resistance of the membrane itself in series with the condensation thermal resistance on the permeate side, and T_{ext} will be the condensing temperature. In numerical simulations, most authors imposed at the walls either a uniform temperature / concentration [11] or a uniform heat / mass flux [14, 15, 16]. Tamburini *et al.* [7] simulated their own experiments on Membrane Distillation, conducted under one-sided heat transfer (see also following Section), by imposing the mixed (3rd type) boundary condition of Eq. (7) with r_{ext}=0.00625 m^2K/W (corresponding to R=0.75 for the geometry investigated) at one of the walls (active, top wall), the opposite wall being adiabatic. Only Shakaib *et al.* [10] explicitly simulated both the hot and the cold fluid, thermally coupled through a membrane of prescribed thickness and thermal conductivity.

In the present work, in order to study how the Nusselt number in a spacer-filled channel is affected by the thermal resistance, simulations were repeated for different geometric configurations and Reynolds numbers letting R vary between 0 and 100. Figure 5 shows the behavior of the averages $Nu^{(1)}$ and $Nu^{(2)}$ as functions of R for one-side heat transfer, γ=45° and P/H=4, Re≈42 (graph a) or P/H=2, Re≈126 (graph b).

The more reliable of the two averages, i.e., $Nu^{(2)}$, decreases monotonically with R in both cases; the highest values of $Nu^{(2)}$ are attained for R=0 (isothermal wall conditions) and the lowest for $R\to\infty$. Note that this behavior is opposite to that reported in the Appendix for the case of a spacerless channel, where Nu *increases* monotonically with R and thus is larger for an imposed uniform wall heat flux than for an imposed uniform wall temperature. This is not surprising since spacer-filled channels exhibit a strong and complex non-uniformity of heat transfer rates over the active wall surface of each unit cell, whereas a spacerless plane channel exhibits only a large-scale longitudinal variation of wall temperature and/or heat flux.

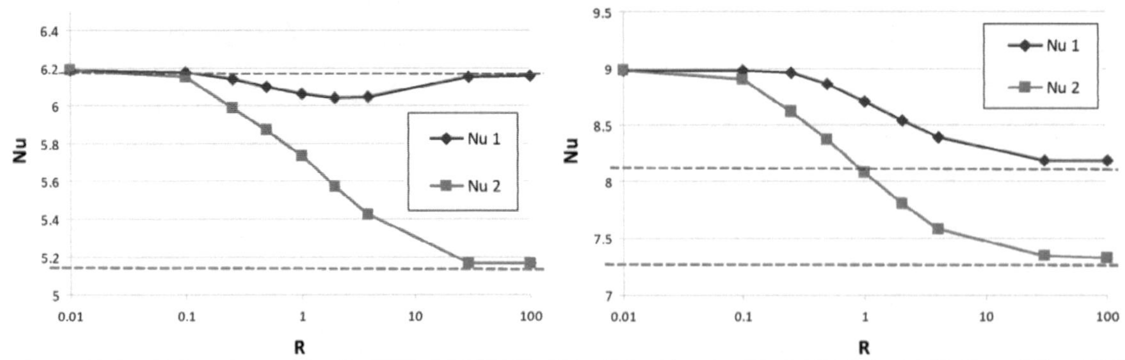

Figure 5. Dependence of the average Nusselt numbers $Nu^{(1)}$ and $Nu^{(2)}$ on the dimensionless wall resistance R for one-side heat transfer in a spacer-filled channel (overlapped, γ=45°). (a) P/H=4, Re≈42; (b) P/H=2, Re≈126. Broken lines are the theoretical values of Nu in a plane (spacerless) channel for $R\to0$ (red) and $R\to\infty$ (blue).

For P/H=4, γ=0°, Re≈42 and one-side heat transfer, Figure 6 shows the influence of thermal boundary conditions on the local distributions of wall temperature (top row), local Nusselt number (middle row) and heat flux (bottom row). The left column is for R=0 (uniform wall temperature); the central column is for R=0.75 (mixed, or 3rd type boundary condition); finally, the right column is for $R\to\infty$ (uniform wall heat flux).

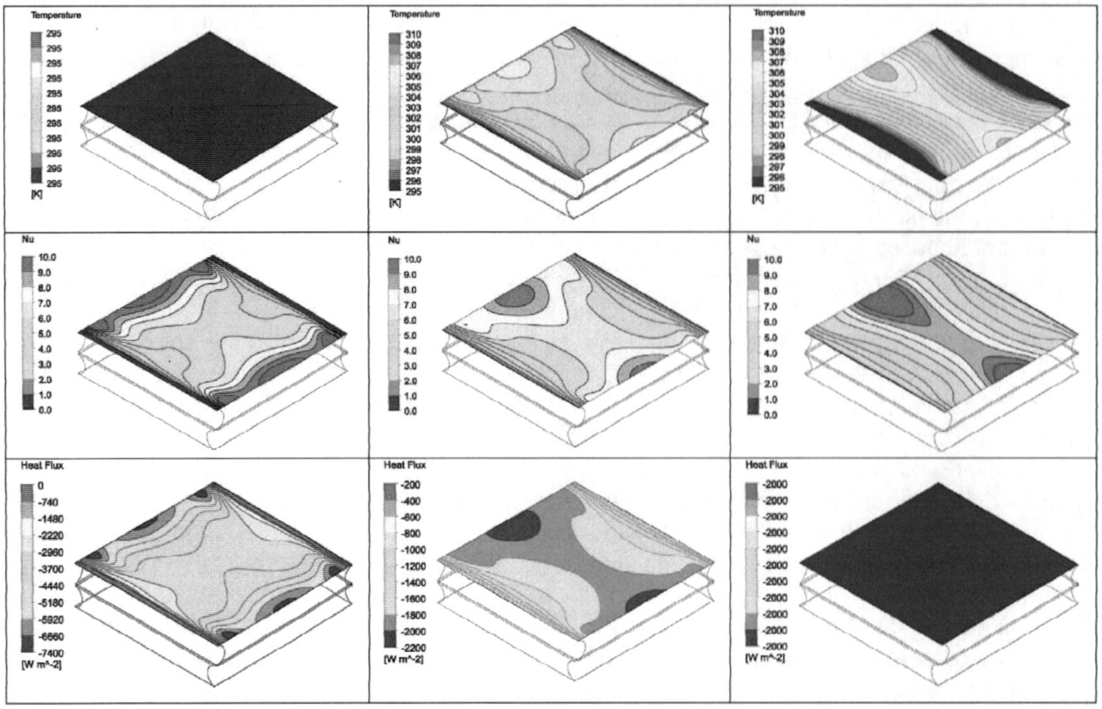

Figure 6. Computed top wall distributions for P/H=4, γ=0°, Re≈42, one-side heat transfer and different thermal boundary conditions: R=0, or uniform wall temperature (left), R=0.75, or general 3rd type boundary conditions (centre), and $R\to\infty$, or uniform wall heat flux (right). Top row: temperature; middle row: heat transfer coefficient; bottom row: heat flux.

Figure 6 shows that the choice of the thermal boundary conditions at the wall affects the local distributions of thermal quantities even more than it affects average quantities such as $Nu^{(2)}$. Therefore, in studies discussing the wall distributions of thermal quantities, and even more so in studies comparing experimental and predicted results for the same quantities, a realistic choice of the thermal boundary conditions appears to be crucial.

Needless to say, the above considerations can be adapted to *mass* transfer by a suitable change of notation, provided realistic values are adopted for the membrane resistance to mass transfer.

4 One-side against two-side heat / mass transfer

A fluid-filled channel can be subjected to different types of heat or mass transfer on the two sides, as sketched in Figure 7 for the case of heat transfer. Case (a) is heat transfer from one-side with the opposite side wall adiabatic. Case (b) is two-side asymmetric heat transfer with heat inflow from one side and outflow from the opposite side. Finally, case (c) is two-side symmetric heat transfer with heat inflow or outflow from both sides. Indicative bulk and wall temperatures are also shown in the figures; of course, in all cases the sign of T_w-T_b is arbitrary.

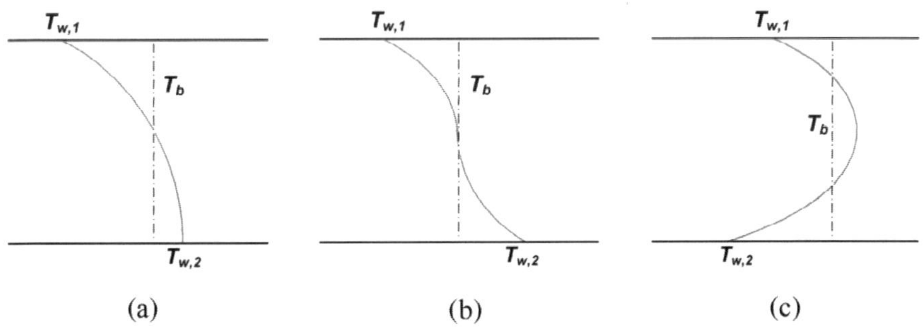

Figure 7. Possible heat transfer arrangements in membrane processes. (a) one-side heat transfer with one adiabatic wall; (b) two-side asymmetric heat transfer with heat inflow from one side and outflow from the opposite side; (c) two-side symmetric heat transfer with heat inflow or outflow from both sides. Representative temperature profiles and bulk and wall values of T are shown.

In the case of Membrane Distillation (MD), case (a) (one-side heat transfer) is usually realized in laboratory-scale experimental rigs [7, 17, 18] or small-scale, plate-and-frame, MD equipment [19, 20] because it is simpler to build and allows an easier experimental characterization of temperatures and heat fluxes.

Case (b) (heat inflow from one side of the channel, outflow from the opposite side) is found, for example, in the Memsys™ modules adopted in Vacuum Multi Effect Membrane Distillation (V-MEMD) [21], in which the generic stage is powered by the condensation heat from a previous stage, thus saving the thermal energy that otherwise should be provided from the outside. One side of each channel is provided with a microporous membrane allowing the passage of vapor, while the other side is a thin polypropylene sheet, on the outside of which the vapor from the previous stage condenses.

Most commercial MD plant usually exhibit outwards (though not necessarily symmetric) heat and mass transfer from both sides of the feed channels, case (c). An example are the spiral wound modules developed at the Fraunhofer Institute [22], in which the hot (feed) channel is placed between two membranes followed by two condensate channels, two conductive walls and two cooling channels, and the whole arrangement is then spirally wound to achieve compactness and reduce heat losses. The feed channel is usually provided with a spacer (not shown).

Among computational studies of membrane processes, both the one-side configuration (a) [7, 10, 14, 23, 24] and the two-side, symmetric configuration (c) [11, 9, 13] have been considered.

For the case of a spacerless plane channel subjected to hydrodynamically and thermally fully developed flow, the difference in the Nusselt numbers for one-side and two-side heat transfer is illustrated in Figure A.1. It can be observed that two-side heat transfer yields Nusselt numbers higher by a factor ~1.5 with respect to one-side transfer for any given dimensionless thermal resistance *R*. Therefore, switching from one-side to two-side heat transfer causes an increase of the total heat exchanged by about three times.

The physical interpretation of the increase in the Nusselt number on a given wall (say, the top wall) caused by the occurrence of symmetric heat transfer also on the *opposite* wall (say, the bottom wall) rests on the observation that switching from one-side to two-side heat transfer moves the adiabatic condition $\partial T/\partial y=0$ at the bottom wall to a formally identical symmetry condition $\partial T/\partial y=0$ at the midplane, thus halving the actual distance between bulk fluid and wall. Since velocity profiles are not affected by thermal boundary conditions, the increase in Nu is not twofold but ~1.5 times.

For spacer-filled channels, the behavior of the Nusselt number is more complex but follows the same general trend. For example, Figure 8 compares $Nu^{(2)}$ on both walls in two-side heat transfer with $Nu^{(2)}$ on the single active (top) wall in one-side heat transfer for an overlapped spacer having $P/H=2$ and $\gamma=0°$ (a) or 45° (b). Note that the second definition of the average Nusselt number was used in order to obtain more regular and meaningful results. Note also that for $\gamma=45°$, graph (b), the values of $Nu^{(2)}$ on the two walls are identical for symmetry reasons and, of course, coincide also with their average. It can be observed that, for $\gamma=0°$ (main flow aligned with the spacer filaments adjacent to the top wall), the Nusselt number in the case of heat transfer from the top wall only exhibits a completely different behavior than either the top- or the bottom-wall Nu in two-side heat transfer; in particular, it exhibits a large *plateau* which shrinks considerably in two-side transfer.

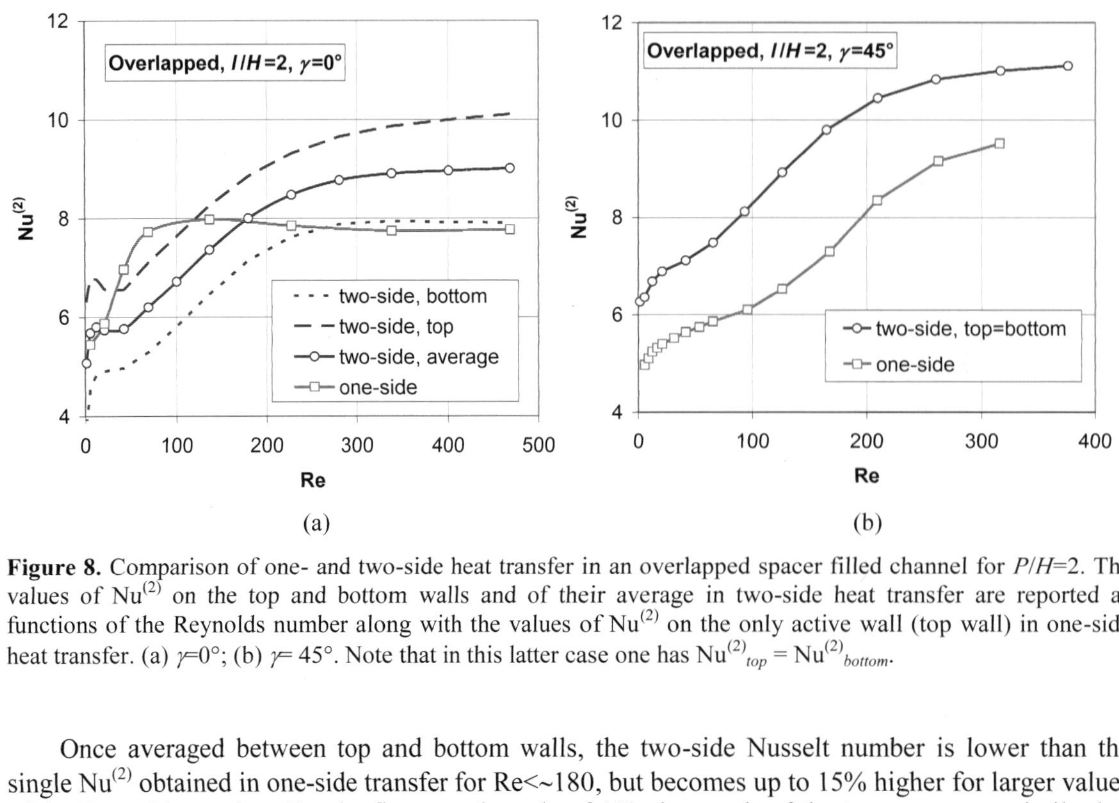

Figure 8. Comparison of one- and two-side heat transfer in an overlapped spacer filled channel for $P/H=2$. The values of $Nu^{(2)}$ on the top and bottom walls and of their average in two-side heat transfer are reported as functions of the Reynolds number along with the values of $Nu^{(2)}$ on the only active wall (top wall) in one-side heat transfer. (a) $\gamma=0°$; (b) $\gamma=45°$. Note that in this latter case one has $Nu^{(2)}_{top} = Nu^{(2)}_{bottom}$.

Once averaged between top and bottom walls, the two-side Nusselt number is lower than the single $Nu^{(2)}$ obtained in one-side transfer for Re<~180, but becomes up to 15% higher for larger values of the Reynolds number. For the flow attack angle of 45°, the trends of the two curves are similar but the two-side Nusselt number is 1.15~1.35 times higher than the one-side Nusselt number at all Re, so that the overall thermal power exchanged for any bulk to wall temperature difference is from 2.3 to 2.7 times larger (of course, the heat transferred *per unit membrane area* is only 1.15~1.35 times larger).

From these results it can be concluded that:

- in Membrane Distillation, two-side heat and mass transfer should be preferred as it generally provides larger heat transfer rates per unit membrane surface area at the Reynolds numbers (>200) expected in practical applications;
- experimental and computational results obtained for one-side transfer cannot be extended to two-side transfer, because not only the absolute values of the average Nusselt number, but also its trend with Re and other parameters, may differ significantly, and the relative performance ranking of the various configurations examined may be altered.

5 Conclusions

The main objective of the present paper was to clarify a number of issues arising in the study of fluid flow and heat/mass transfer in membrane processes involving flat, spacer-filled, channels (either plane or spirally wound). Other designs, including dead-end arrangements and tubular or hollow-fiber membranes, raise different problems which were not discussed here. The conclusions reached were based on many years' experience with Membrane Distillation and Reverse Electrodialysis, but many of them can be extended to other membrane processes as well. The results of the study can be summarized as follows:

- In regard to the definition of *local* heat or mass transfer coefficients, the definition based on the local heat / mass flux and on the difference between the local temperature or concentration at the wall and the corresponding bulk values, i.e. $h=q''_w/(T_b-T_w)$ or $k=j_w/(C_b-C_w)$, remains the most commonly adopted, but we have shown that these quantities may become singular or negative at some points of the wall in complex geometries. For this and other reasons, the definition of the average coefficients as surface averages $h^{(1)}=\langle h\rangle$, $k^{(1)}=\langle k\rangle$ (definition 1) should be avoided, and the alternative averages $h^{(2)}=\langle q''_w\rangle/(T_b-\langle T_w\rangle)$, $k^{(2)}=\langle j_w\rangle/(C_b-\langle C_w\rangle)$ (definition 2), which do not suffer from singularities, are preferable both in computational and in experimental work.
- In regard to the choice of temperature or concentration boundary conditions at the walls representing the fluid-membrane interfaces in CFD simulations, we have shown that it may have a significant impact on the computed Nusselt or Sherwood numbers, especially in laminar flow. Therefore, experimental data ought to be compared only with CFD results based on a realistic description of the actual boundary conditions, rather than with idealized cases based on uniform temperature / concentration (Dirichlet) or uniform heat / mass flux (Neumann) conditions. For design purposes, it is preferable to use convection coefficients based on generalized (3rd type) boundary conditions with some intermediate value of the wall heat / mass transfer resistance.
- In regard to the comparison between one-side and two-side heat or mass transfer, this choice has a dramatic impact on the corresponding coefficients: these may be up to 30% higher in a two-side configuration, yielding correspondingly higher heat or mass fluxes per unit membrane area. Experimental or computational results obtained for one-side transfer should not be extended to two-side transfer, because not only the values of the average Nusselt / Sherwood numbers, but also their dependence on Re and other parameters, may differ significantly, and the relative performance ranking of different configurations may be altered.

References

[1] P. Onsekizoglu, Membrane Distillation: principle, advances, limitations and future prospects in food industry, in *Distillation-Advances from Modeling to Applications*, Zereshki S. (Ed.), InTech (2012).

[2] L. Gurreri, A. Tamburini, A. Cipollina, G. Micale, M. Ciofalo, Flow and mass transfer in spacer-filled channels for Reverse Electrodialysis: a CFD parametrical study, *J. Membr. Sci.* **497** (2016) 300–317.

[3] I. Hitsov, T. Maere, K. DeSitter, C. Dotremont, I. Nopens, Modelling approaches in Membrane Distillation: a critical review, *Sep. Purif. Technol.* **142** (2015) 48–64.

[4] G. A. Fimbres-Weihs, D. E. Wiley, Review of 3D CFD modeling of flow and mass transfer in narrow spacer-filled channels in membrane modules, *Chem. Eng. Proc.* **49** (2010) 759–781.

[5] A. J. Karabelas, M. Kostoglou, C. P. Koutsou, Modeling of spiral wound membrane desalination modules and plants: review and research priorities, *Desalination* **356** (2015) 165–186.

[6] M. La Cerva, M. Ciofalo, L. Gurreri, A. Tamburini, A. Cipollina, G. Micale, On some issues in the computational modelling of spacer-filled channels for membrane distillation, *Desalination* **411** (2017) 101-111.

[7] A. Tamburini, M. Renda, A. Cipollina, G. Micale, M. Ciofalo, Investigation of heat transfer in spacer-filled channels by experiments and direct numerical simulations, *Int. J. Heat Mass Transf.* **93** (2016) 1190-1205.

[8] M. Ciofalo, F. Ponzio, A. Tamburini, A. Cipollina, G. Micale, Unsteadiness and transition to turbulence in woven-spacer filled channels for Membrane Distillation, *Journal of Physics*, in press (paper presented at the *34th UIT Heat Transfer Conf.*, Ferrara, Italy, 2-6 July 2016).

[9] C. P. Koutsou, S. G. Yiantsios, A. J. Karabelas, A numerical and experimental study of mass transfer in spacer-filled channels: effects of spacer geometrical characteristics and Schmidt number, *J. Membr. Sci.* **326** (2009) 234–251.

[10] M. Shakaib, S. M. F. Hasani, I. Ahmed, R. M. Yunus, A CFD study on the effect of spacer orientation on temperature polarization in Membrane Distillation modules, *Desalination* **284** (2012) 332–340.

[11] F. Li, W. Meindersma, A. B. de Haan, T. Reith, Optimization of commercial net spacers in spiral wound membrane modules, *J. Membr. Sci.* **208** (2002) 289–302.

[12] C. P. Koutsou, A. J. Karabelas, A novel retentate spacer geometry for improved spiral wound membrane (SWM) module performance, *J. Membr. Sci.* **488** (2015) 129–142.

[13] A. Saeed, R. Vuthaluru, H. B. Vuthaluru, Investigations into the effects of mass transport and flow dynamics of spacer filled membrane modules using CFD, *Chem. Eng. Res. and Des.* **93** (2015) 79–99.

[14] A. Cipollina, G. Micale, L. Rizzuti, Membrane Distillation heat transfer enhancement by CFD analysis of internal module geometry, *Des. and Water Treat.* **25** (2011) 195–209.

[15] M. Qureshi, M. Shakaib, CFD study for temperature and concentration profiles in membrane channels, *Proceedings Int. Conf. Energy and Sustainability*, NED University, Karachi, Pakistan (2013).

[16] S. Al-Sharif, M. Albeirutty, A. Cipollina, G. Micale, Modelling flow and heat transfer in spacer-filled Membrane Distillation channels using open source CFD code, *Desalination* **311** (2013) 103–112.

[17] A. Ali, F. Macedonio, E. Drioli, S. Aljlil, O. A. Alharbi, Experimental and theoretical evaluation of temperature polarization phenomenon in Direct Contact Membrane Distillation, *Chem. Eng. Res. and Des.* **91** (2013) 1966-1977.

[18] M. Khayet, C. Cojocaru, A. Baroudi, Modeling and optimization of sweeping gas Membrane Distillation, *Desalination* **287** (2012) 159–166.

[19] H. Chang, C.-L. Chang, C.-Y. Hung, T.-W. Cheng, C.-D. Ho, Optimization study of small-scale solar Membrane Distillation desalination systems (s-SMDDS), *Int. J. Environ. Res. Public Health* **11** (2014), 12064-12087.

[20] L. Francis, N. Ghaffour, A. S. Alsaadi, S. P. Nunes, G. L. Amy, Performance evaluation of the DCMD desalination process under bench scale and large scale module operating conditions, *J. Membr. Sci.* **455** (2014) 103–112.

[21] W. Heinzl, S. Büttner, G. Lange, Industrialized modules for MED Desalination with polymer surfaces, *Des. and Water Treat.* **42** (2012) 177–180.

[22] D. Winter, J. Koschikowski, M. Wieghaus, Desalination using Membrane Distillation: Experimental studies on full scale spiral wound modules, *J. Membr. Sci.* **375** (2011) 104–112.

[23] E. K. Summers, H. A. Arafat, J. H. Lienhard, Energy efficiency comparison of single-stage Membrane Distillation (MD) desalination cycles in different configurations, *Desalination* **290** (2012) 54–66.

[24] S. Wardeh, H. P. Morvan, CFD simulations of flow and concentration polarization in spacer-filled channels for application to water desalination, *Chem. Eng. Res. and Des.* **86** (2008) 1107–1116.

Appendix: Nusselt number in parallel channel flow for different thermal boundary conditions

Although, in the present paper, the attention is focused on complex, spacer-filled channels (which can only be studied by means of fully three-dimensional numerical simulations), in order to appreciate the influence of thermal boundary conditions on heat transfer it is instructive to consider a simpler, essentially one-dimensional, problem (plane Poiseuille flow) since it leads to ordinary differential equations which can be solved to a very high accuracy. The mathematical treatment and the results obtained can easily be adapted also to mass transfer with obvious changes.

Consider the steady, hydrodynamically and thermally fully developed flow of a constant-property fluid in a plane channel of half-thickness δ. Let x be the streamwise direction and y the cross stream one, with y=0 at the midplane, and let $u(y)$ be the velocity along x, U its average, $T(x, y)$ the temperature and $\alpha=\lambda/(\rho c_p)$ the thermal diffusivity.

Consider first the case of symmetric two-side cooling, which can be characterized by a symmetry condition at the midplane and a general (3rd type) boundary conditions at one wall:

$$\left[\frac{\partial T}{\partial y}\right]_0 = 0; \quad -\lambda\left[\frac{\partial T}{\partial y}\right]_\delta = \frac{1}{r}\left[T(x,\delta) - T_\infty\right] \tag{A.1}$$

T_∞ being a uniform external temperature and r an interposed thermal resistance. Let ϑ=T-T_∞ and

$$\vartheta_b = \frac{1}{2\delta\rho U}\int_{-\delta}^{\delta} \vartheta\, u\, dy \tag{A.2}$$

its bulk value. The following dimensionless variables can be introduced

$$\eta \equiv \frac{y}{\delta}; \quad \xi \equiv \frac{x}{\delta}; \quad \tilde{u} \equiv \frac{u}{U}; \quad \mathrm{Pe}_\delta \equiv \frac{U\delta}{\alpha} \tag{A.3}$$

The Poiseuille velocity distribution and the energy balance equation can be written

$$\tilde{u} = \frac{3}{2}\left(1-\eta^2\right) \tag{A.4}$$

$$\frac{\partial^2\vartheta}{\partial\eta^2} = Pe_\delta\,\tilde{u}\,\frac{\partial\vartheta}{\partial\xi} \tag{A.5}$$

The assumption of thermally fully developed flow implies

$$\vartheta = \vartheta_b(x)\,\varphi(y) \tag{A.6}$$

so that Eq. (A.5) can be written

$$\frac{1}{\varphi\tilde{u}}\frac{\partial^2\varphi}{\partial\eta^2} = Pe_\delta\,\frac{\partial\vartheta_b}{\partial\xi} \tag{A.7}$$

Following the method of separation of variables, Eq. (A.7) splits into the two distinct equations

$$\frac{1}{\varphi\tilde{u}}\frac{\partial^2\varphi}{\partial\eta^2} = -B^2 \tag{A.8}$$

$$\mathrm{Pe}_\delta\,\frac{1}{\vartheta_b}\frac{\mathrm{d}\vartheta_b}{\mathrm{d}\xi} = -B^2 \tag{A.9}$$

in which the constant $-B^2$ must be negative because, for example, ϑ_b>0 implies $d\vartheta_b/d\xi < 0$ (cooled fluid). By substituting the velocity profile of Eq. (A.4) into Eq. (A.8) one obtains

$$\frac{\mathrm{d}^2\varphi}{\mathrm{d}\eta^2} + \frac{3}{2}B^2\left(1-\eta^2\right)\varphi = 0 \tag{A.10}$$

This is a non-integrable ODE with non-constant coefficients which can easily be solved numerically. The boundary conditions for φ are derived from eqs. (A.4) and (A.6):

$$\left[\frac{\partial \varphi}{\partial \eta}\right]_0 = 0; \quad \varphi(1) = B^2 R \tag{A.11}$$

in which $R=r\lambda/\delta$ (dimensionless thermal resistance). The following condition must also be satisfied:

$$\left[\frac{\partial \varphi}{\partial \eta}\right]_1 = -B^2 \tag{A.12}$$

which is derived from the enthalpy balance:

$$-\lambda\left[\frac{\partial \vartheta}{\partial y}\right]_\delta = -\delta \rho c_p U \frac{d\vartheta_b}{dx} \tag{A.13}$$

Conceptually, Eq. (A.10) for $\varphi(\eta)$, with B.C.'s (A.11), must be solved numerically for each generic B^2, and B^2 has to be varied so as to satisfy Eq. (A.12). A simple iterative procedure was implemented in Fortran to perform this calculation. After convergence, the heat transfer coefficient is

$$h = \frac{q''_w}{\vartheta_b} = -\frac{\lambda}{\delta \vartheta_b}\left[\frac{d\vartheta}{d\eta}\right]_\delta \tag{A.14}$$

which can be made dimensionless as a Nusselt number $\mathrm{Nu}=h(4\delta/\lambda)$ (based on the channel hydraulic diameter 4δ). From the above equations and definitions it follows that

$$\mathrm{Nu} = \frac{4}{R}\frac{\varphi(1)}{1-\varphi(1)} \tag{A.15}$$

Note that a different value of Nu and a different profile $\varphi(\eta)$ are computed for each choice of the dimensionless thermal resistance R. Note also that the Dirichlet and Neumann conditions are obtained as limiting cases for R=0 and $R\rightarrow\infty$, respectively. The function Nu(R) is represented in Figure A.1 by the "Two-side" line.

Consider now the case of one-side cooling, i.e. cooling from one wall with the opposite wall adiabatic. The thermal boundary conditions (A.1) are replaced by

$$\left[\frac{\partial T}{\partial y}\right]_{-\delta} = 0; \quad -\lambda\left[\frac{\partial T}{\partial y}\right]_\delta = \frac{1}{r}[T(x,\delta) - T_\infty] \tag{A.16}$$

and their dimensionless counterparts (A.11) are replaced by

$$\left[\frac{\partial \varphi}{\partial \eta}\right]_{-1} = 0; \quad \varphi(1) = 2B^2 R \tag{A.17}$$

while all other equations and definitions remain unchanged. The solution for Nu is reported in Figure A.1 as the "One-side" line. Note the lower levels of Nu with respect to the "Two-side" case.

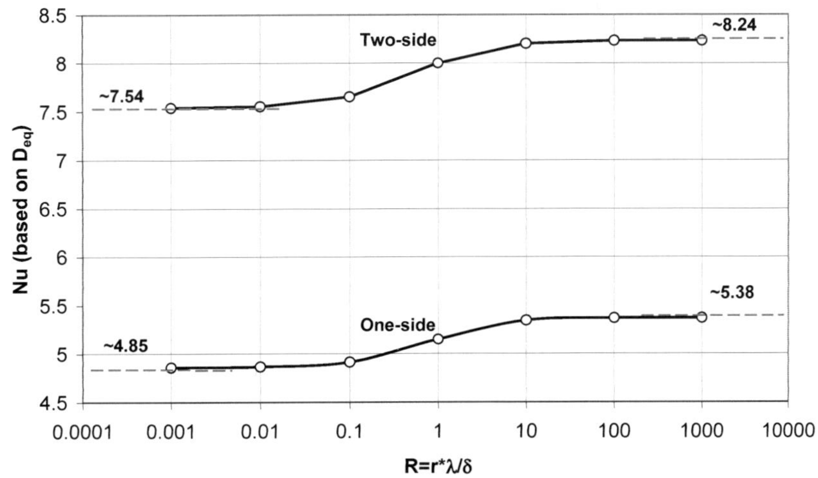

Figure A.1 Nusselt number in a plane channel as a function of the dimensionless thermal resistance *R*.

IOP Conf. Series: Journal of Physics: Conf. Series **923** (2017) 012054 doi:10.1088/1742-6596/923/1/012054

Artificial Neural Network Modeling of Liquid Thermal Conductivity for alkanes, ketones and silanes

G Latini[1], G Di Nicola[1], M Pierantozzi[2], G Coccia[1] and S Tomassetti[1]

[1] DIISM, Università Politecnica delle Marche, Ancona, Italy
[2] SAAD, Università di Camerino, Ascoli Piceno, Italy

E-mail: mariano.pierantozzi@unicam.it

Abstract. The values of thermal conductivity λ at different temperatures for organic and inorganic compounds in the liquid phase is essential in the study of numerous processes, but experimental data are frequently not available with acceptable reliability or not available at all, since rigorous theoretical or semi-theoretical models of the liquid state are usually of poor practical use for engineering purposes. The Artificial Neural Network (ANN) approach is a very powerful tool and it can be a good indicator of the lowest limit achievable with a selected database and with a selected set of inputs. This study investigates the applicability of the ANN as an efficient tool for the prediction of pure organic compounds' thermal conductivity of three important families such as alkanes, ketones and silanes, for a wide range of temperatures. The families of n-alkanes, ketones and silanes were chosen to verify the general reliability of the proposed method when used in large temperature ranges for very different organic families, above all the silanes (compounds containing silicon), whose liquid thermal conductivity is experimentally investigated in very few cases. This method appears to be successful: in all reduced temperature range, along or near the saturation line, the average absolute deviations between calculated and experimental thermal conductivity data are 0.19% and the maximum absolute ones 2.44%

1. Introduction

Recently our research group focused its attention on the study of some important thermo-physical properties [1–17]. One of the most important of them is thermal conductivity. The values of the thermal conductivity of liquids are required in several phenomena, but its experimental values at different temperatures are often not available with acceptable reliability or sometimes not available at all; theoretical or semi-theoretical models of the liquid state ("gas-like" or "solid-like" models) are useless or of poor practical use for engineering purposes, either for the errors they finally lead to, or for the excessive mathematical difficulties.

On a microscopic scale the liquids are obviously characterized by a potential function determining the energy transfer mechanisms, but the different models appeared in the scientific literature generally do not allow to find reliable values for thermal conductivity, and empirical or semi-empirical correlations are usually necessary for many practical or engineering purposes. In using these correlations, a reasonable equilibrium has to be reached between the simplicity of the method and the accuracy of the values of the estimated thermal conductivity.

The thermal conductivity of liquids usually decreases with the increase of temperature, with the exception of some compounds, for example water and some aqueous solutions. In the range from the normal melting point to the normal boiling point and over, the thermal conductivity dependence on the

IOP Conf. Series: Journal of Physics: Conf. Series **923** (2017) 012054 doi:10.1088/1742-6596/923/1/012054

temperature is usually almost linear, but this behavior near the critical point dramatically ceases and the critical point represents a singularity.

From a theoretical point of view [18, 19], the thermal conductivity of liquids depends upon temperature and density as sum of three contributions: the "dilute gas contribution", the "excess contribution" and the "critical enhancement contribution"; the density dependence is usually substituted by pressure dependence because the pressure variations are negligible up to 3-4 MPa, and in this range it is acceptable to express thermal conductivity as a function of temperature only. In this work a suitable equation for the liquid phase is proposed linking liquid thermal conductivity directly with temperature. In a recent past [4], among the different organic families, n-alkanes, ketones and silanes were chosen as a test of the proposed estimation method. In this paper, the method is extended to three very different families: n-alkanes, ketones and silanes. The results are compared with nine of the best empirical or semi-empirical equations proposed in the scientific and technical literature.

In order to develop an acceptable comparison between estimated and experimental thermal conductivity data, the thermal conductivity values (experimental, smoothed and predicted) available from DIPPR801 [20] database for the n-alkanes, ketones and silanes are analyzed. During the data collection, a fluid by fluid analysis was performed and only experimental or experimental&predicted data with claimed accuracy less than 10% were taken into account for the check of the proposed method and for the comparison with the other investigated methods.

2. The database used for the test of Artificial Neural Network

Since reliable experimental thermal conductivity data of liquids available in literature are due to different authors who use different apparatuses at different temperatures (and, obviously, with different accuracies), a valid check would be developed by disposing of a unique collection of experimental thermal conductivity data. A reliable and updated source of thermal conductivity values can be found in the DIPPR801 database [20], where experimental, predicted and smoothed data are reported. The DIPPR 801 database collects data from a wide range of sources and considers them critically, giving quality codes for all data points. The DIPPR801 database was chosen as the source of the thermal conductivity data because the database satisfies the following requirements: the thermal conductivity data are clearly indicated as experimental, predicted or experimental&predicted; the accuracy of the thermal conductivity data is clearly indicated; if experimental thermal conductivity data are not available, the used estimation method is clearly indicated with the claimed accuracy.

The data are collected for the three selected families of n-alkanes, ketones and silanes.

The choices of the authors, as extensively exposed in a previous work [4] concerning the families considered, were as follows:

- only experimental or experimental&predicted data for each fluid with claimed accuracy better than 10% were used in the check of the model, thus the predicted thermal conductivity data presented in DIPPR801 were not taken into account;
- the results obtained through the Artificial Neural Network (ANN) were compared with the results of other ten estimation methods: seven methods are proposed in DIPPR801 [21–29], while three methods were taken from the recent literature. [3, 5, 30]

The compounds taken into account were the following ones:

- n-alkanes: 23 investigated compounds;
- ketones: 14 investigated compounds;
- silanes: 18 investigated compound.

The check was developed exploring the reduced temperature range 0.30 to 0.80.

3. An artificial Neural Network

For the modelling of thermal conductivity of n-alkanes, ketones and silanes, the Wolfram Mathematica artificial neural network toolbox was employed. A neural network is a structure involving weighted interconnections among neurons, or units, which are most often nonlinear scalar transformations, but they can also be linear. A neuron is structured to process multiple inputs, including the unity bias, in a nonlinear way, producing a single output. Specifically, all inputs to a neuron are first augmented by multiplicative weights. These weighted inputs are summed and then transformed via an activation function (also called neuron function).

The training goal is to find values of the parameters so that, for any input x, the network output $\hat{y}$ is a good approximation of the desired output y. Training is carried out via suitable algorithms that tune the parameters so that input training data map well the corresponding desired outputs. These algorithms are iterative in nature, starting at some initial value for the parameter vector θ, and incrementally updating it to improve the performance of the network.

The network is generally divided into layers. The input layer consists of just the inputs to the network. Then follows a hidden layer, which consists of any number of neurons, or hidden units, placed in parallel. The network output is formed by another weighted summation of the outputs of the neurons in the hidden layer. This summation on the output is called the output layer.

The required number of training data points and hidden layer neurons are the main goals of ANN modelling. They can be determined by the constructive approach [31]. A network that has only one hidden layer is able to approximate almost any type of nonlinear mapping [32].

Each neuron performs a weighted summation of the inputs, which then passes through a nonlinear activation function s. Mathematically, the functionality of a hidden neuron is described by:

$$s(a) = s\left(\sum_{j=1}^{n} w_j x_j + b\right) \tag{1}$$

where a is the overall weighted input of the neuron and the weights w_j, b are symbolized with the arrows feeding into the neuron itself. For this feedforward network, a sigmoid function was employed as the activation function:

$$s(a) = \frac{1}{1+e^{-a}} \tag{2}$$

The number of output neurons equals the number of outputs of the approximation problem. The output of this network is given by:

$$\hat{y} = g(\theta, x) = g\left(\sum_{i=1}^{nh} w_i s\left(\sum_{j=1}^{n} w_{i,j} x_j + b_i\right) + b_{\text{out}}\right) \tag{3}$$

where n is the number of inputs to the ith neuron, nh is the number of neurons in the hidden layer, and b_{out} is the output bias. In this work, the output activation function g was chosen to be linear as this is a convenient and sufficient choice for interpolation problems.

The parameters of the network model are represented collectively by the parameter vector θ. In general, the neural network model will be represented by the compact notation $g(\theta, x)$. The size of the input and output layers are defined by the number of inputs and outputs of the network and, therefore, only the number of hidden neurons has to be specified when the network is defined.

Keeping the number of hidden layers fixed, it is possible to modulate the network by changing the number of hidden neurons in each layer. In this way, we tested each time the network with a different number of neurons trying to limit the error produced by the model. Too few neurons in the hidden layer prevent the network from better fitting the experimental values and approximate points maintaining a low deviation. On the other hand, too many neurons lead to a network that has a very good response when it is tested with the training points, but the structure of this network is too rigid with the provided values.

Often it is more convenient to use the *RMSE* when evaluating the quality of a model during and after training, because it can be compared with the output signal directly. Different neural networks were compared adopting the Levenberg-Marquardt algorithm [33] using their Root Mean Square Errors (*RMSEs*), defined as follows:

$$RMSE = \sqrt{\frac{1}{N}\sum_{i=1}^{N}\left(y_i - g(\theta, x_i)\right)^2} \tag{4}$$

where N is the number of data points.

When the network is trained with new data there is the risk of committing unacceptably large errors. These problems, occurring during the neural network training, are referred to as overfitting. To avoid and overcome this overfitting problem, it is possible to divide the data points into training, validation and test data.

Therefore, before the trained network is accepted, it should be validated. Roughly, this means running a number of tests to determine whether the network model meets certain requirements. In this case, it was decided to split the database into three parts: one that comprises 70% of the data, which were used for the "training", a second which employs 20% of the data, which were used to perform the "validation" test, and a third which employs 10% of the data which were used for the ''test'' carried out to investigate the prediction capability and validity of the obtained model. In all cases, data were randomly chosen within the database.

Figure 1 illustrates a diagram of a one-hidden-layer feedforward network adopting the following inputs: experimental reduced temperature, T_r, the molecular weight M, and the acentric factor, ω. In particular, the acentric factor is an important characterization of substances that was introduced by Pitzer in 1955 [34]. Pitzer came up with this factor by analyzing the vapor pressure curves of various pure substances [35]. Table 1 contains the molecular weight and the acentric factor for all the fluids taken into account.

The output is thermal conductivity, λ. Each arrow in the figure symbolises a parameter in the network. In summary, the selected ANN architecture includes three input parameters (Tr, M, ω), one hidden layers with 37 neurons, and a output neuron giving the thermal conductivity value for each fluid and each temperature. The values for the weights and bias for each neuron are given in Table 2 and 3.

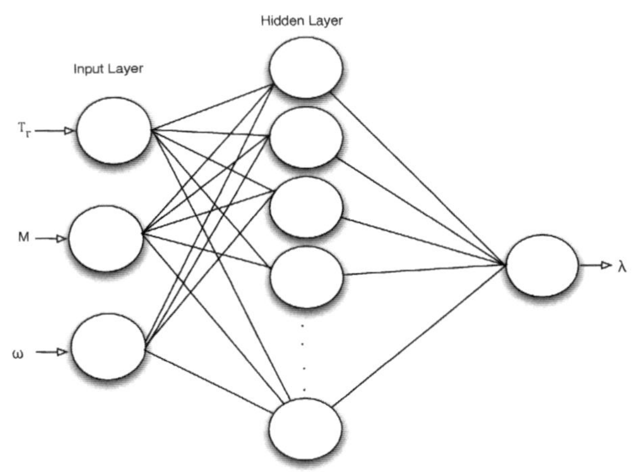

Figure1. Schematic diagram of the ANN model. Inputs: T_r, reduced temperature, M molecular weight, ω acentric factor. Output: λ thermal conductivity

In order to compare the equations, deviations were calculated as follows:

$$AAD = \left|\frac{\lambda_{exp}-\lambda_{calc}}{\lambda_{exp}}\right| \cdot 100 \qquad (5)$$

Figure 2 illustrates the *AAD* for the training, validation and test, and the overall set of data obtained for various neural network configurations with one hidden layer, achieved after 1000 iterations during the modeling of the thermal conductivity of n-alkanes, ketones and silanes. From the figure it is evident that, as expected, *AAD* decreases with the increase of the number of neurons and that a good configuration is reached when the number of neurons in the hidden layer is 37. This configuration was selected as the best network architecture.

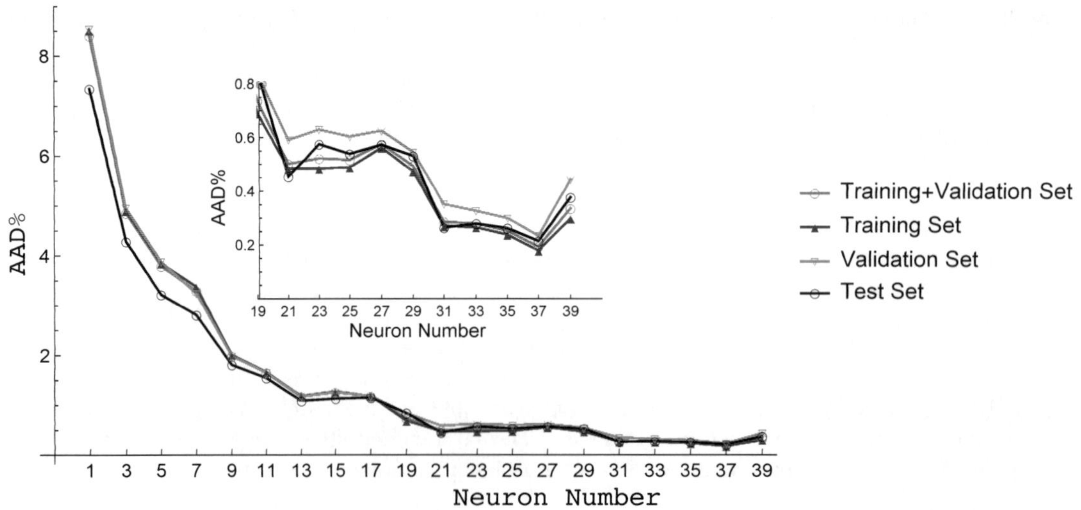

Figure2. Absolute average percent deviation between data collected and ANN results vs number of neurons in hidden layer

The *AADs* for the proposed ANN are reported in Table 1.

4. Comparison between the results obtained through the proposed method (9) and the results obtained through different equations appeared in literature

Together with the methods from the literature and those proposed in DIPPR810 [5,8,9,17-24], two additional methods are taken into account [25,26]. All the methods, even if already appeared in [37], are shortly illustrated below for a convenient comparison.

Tables 1 shows the results of the comparison between the ANN model proposed by the authors and the correlations proposed in the literature by the authors indicated in the same Table for the n-alkanes, ketones and silanes, respectively.

The blank spaces indicate that the estimation method is not acceptable for the corresponding compound.

Fig. 3 shows the calculated thermal conductivity values versus the ones in the database. As can be seen, there are not significant disagreements in the data considered.

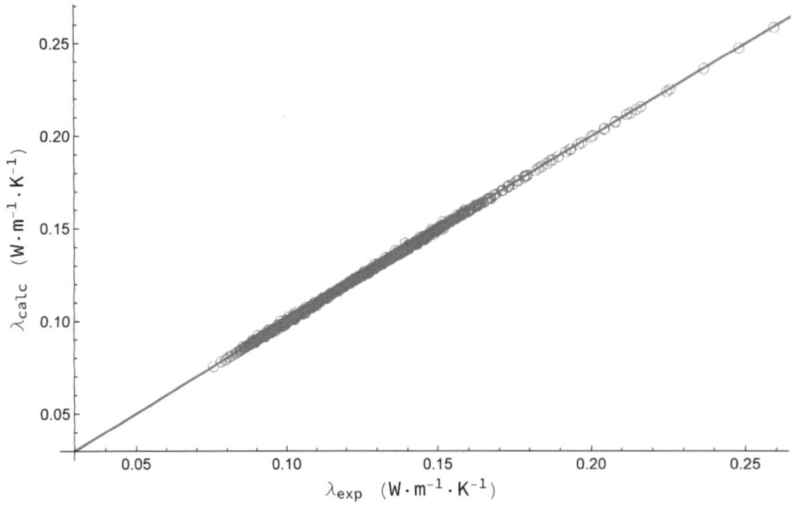

Figure 3. Calculated thermal conductivity values versus all the values in the datasets when the selected ANN architecture is used.

In order to clearly show the improvement obtained with the use of the ANN with respect to the best corresponding-states model, Fig. 4 shows the calculated and dataset values for the proposed ANN and for the Latini et al. model [3]. As can be seen, the corresponding-state model cannot adequately reproduce a significant number of data. Some excessive over-predictions and under-predictions are clearly obtained.

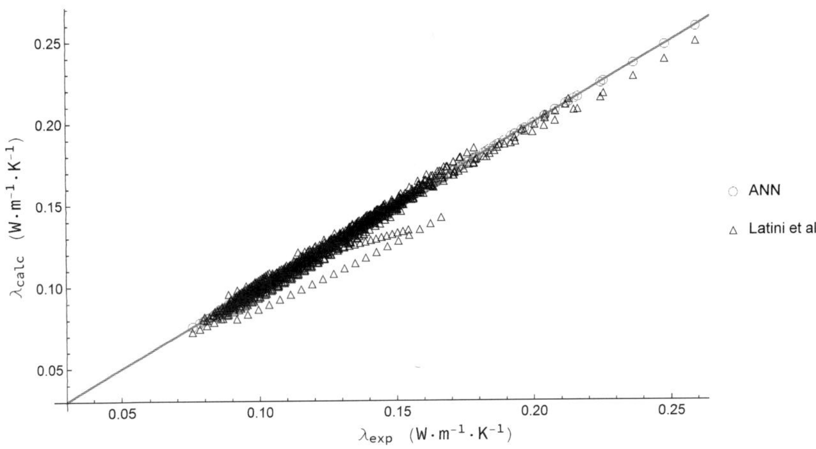

Figure 4. Calculated thermal conductivity values versus values in the datasets when the selected ANN model and the Latini et al. [3] corresponding-state model are used

Fig. 5 shows the deviations values versus temperature. As can be seen, only a very few number of data give deviations values higher than 2%, in fact except very few points, residuals are well within ±2.0 W·(m·K)$^{-1}$.

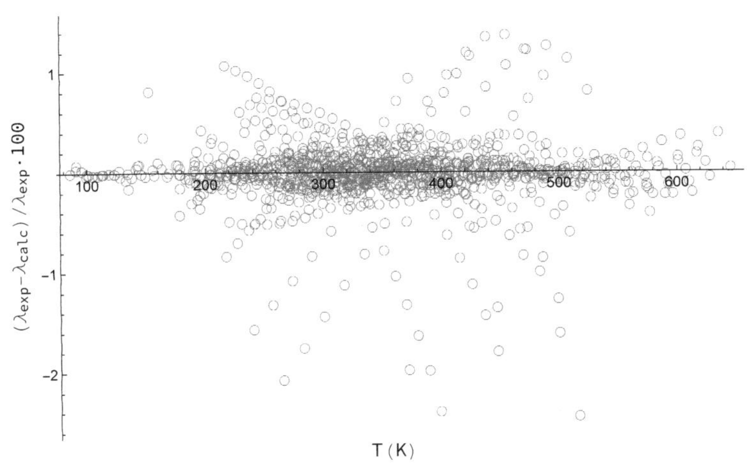

Figure 5. Deviations values for each dataset obtained by using the selected ANN architecture

5. **Conlcusions**

In this paper a multilayer perceptron neural network are proposed to thermal conductivity of three important chemical families: ketones, alkanes and silanes. After checking different architectures and

different input variables, the ANN model giving the better results includes three input properties for each fluid (Tr, M, ω), one hidden layers with 37 neurons, and a output neuron.
The model was validated, trained and tested for a wide set of data, showing that the accuracy of the neural network model is higher than the accuracy of the methods proposed in the literature. This selected model can reproduce the data with an overall AAD value of 0.19%, and a maximum AAD value of 2.44%
For the sake of comparison, we have shown that the presently available corresponding-states models cannot reproduce these data with AAD below 1%. The proposed ANN model can be considered then as a very accurate tool to reproduce the currently available thermal conductivity data for the considered families and even for prediction purposes. Needed input properties, bias, and weights values are given in Table 2 and 3.

References

[1] Di Nicola G, Coccia G, Malvagi L, Pierantozzi M 2017 *J. Thermophys. Heat Transf.* 1–9.
[2] Khosharay S, Khosharay K, Di Nicola G, Pierantozzi M 2017 *Phys. Chem. Liq.* 1–17.
[3] Latini G, Di Nicola G, Pierantozzi M 2017 *Phys. Chem. Liq.* 1–19.
[4] Latini G, Di Nicola G, Pierantozzi M 2016 *Fluid Phase Equilib.* **427** 488–497.
[5] Di Nicola G, Pierantozzi M, Petrucci G, Stryjek R 2016 *J. Thermophys. Heat Transf.* **30** 651–660.
[6] Di Nicola G, Coccia G, Pierantozzi M 2016 *Fluid Phase Equilib.* **417** 229–236.
[7] Di Nicola G, Coccia G, Pierantozzi M, Falone M 2016 *Int. J. Refrig.* **68** 242–251.
[8] Di Nicola G, Coccia G, Pierantozzi M 2016 *Fluid Phase Equilib.* **418** 88–93.
[9] Di Nicola G, Pierantozzi M 2015 *Fluid Phase Equilib.* **389** 16–27.
[10] Latini G, Di Nicola G, Pierantozzi M 2014 *Energy Procedia* **45** 616–625.
[11] Di Nicola G, Pierantozzi M 2014 *J. Therm. Anal. Calorim.* **116** 129–134.
[12] Di Nicola G, Falone M, Pierantozzi M, Stryjek R 2014 *Ind. Eng. Chem. Res.* **53** 13804–13809.
[13] Di Nicola G, Ciarrocchi E, Coccia G, Pierantozzi M 2014 *Int. J. Refrig.* **45** 168–176.
[14] Di Nicola G, Ciarrocchi E, Pierantozzi M, Stryjek R 2014 *J. Therm. Anal. Calorim.* **116** 135–140.
[15] Di Nicola G, Pierantozzi M 2013 *Int. J. Refrig.* **36** 562–566.
[16] Mulero Á, Pierantozzi M, Cachadiña I, Di Nicola G 2017 *Fluid Phase Equilib.* **449** 28–40.
[17] Pierantozzi M, Nicola G Di, Latini G, Coccia G *Phys. Chem. Liq.* **0** 1–18.
[18] Assael MJ, Trusler JPM, Tsolakis TF 1996 *Thermophysical Properties of Fluids*.
[19] Assael MJ, Charitidou E, Wakeham WA 1989 *Int. J. Thermophys.* **10** 779–791.
[20] The DIPPR Information and Data Evaluation Manager for the Design Institute for Physical Properties 2016.
[21] Sastri SRS, Rao KK 1999 *Chem. Eng. J.* **74** 161–169.
[22] Missenard A 1965 *Conductivité thermique des solides, liquides, gaz et de leurs mélanges*.
[23] Riedel L 1951 *Chemie Ing. Tech.* **23** 321–324.
[24] Reid RC, Prausnitz JM, Poling BE 1987 *The Properties of Gases and Liquids* (McGraw-Hill)
[25] Baroncini C, Di Filippo P, Latini G, Pacetti M 1981 *Int. J. Thermophys.* **2** 21–38.
[26] Poling BE., Prausnitz JM, O apos Connell JP 2007 *Exp. Therm. Fluid Sci.* **1** 1–803.
[27] Nagvekar M, Daubert TE 1987 *Ind. Eng. Chem. Res.* **26** 1362–1365.
[28] Robbins L, Kingrea C 1962 *Hydrocarb. Process. Pet. Refinerining* **41** 133–136.
[29] Pachaiyappan V, Ibrahim S, Kuloor N 1967 *Chem. Eng.* **74** 140–144.
[30] Gharagheizi F, et al. 2013 *AIChE J.* **59** 1702–1708.
[31] Haykin SS 1999 *Neural Networks: A Comprehensive Foundation* (Prentice Hall)
[32] Cybenko G 1989 *Math. Control. Signals, Syst.* **2** 303–314.
[33] Marquardt DW 1963 *J. Soc. Ind. Appl. Math.* **11** 431–441.
[34] Pitzer KS, Lippmann DZ, Curl RF, Huggins CM, Petersen DE 1955 *J. Am. Chem. Soc.* **77** 3433–3440.
[35] Pitzer KS 1977 Origin of the Acentric Factor, pp 1–10.
[36] Missenard A 1965 *Conductivité thermique des solides, liquides, gaz et de leurs mélanges* (Eyrolles, Paris).
[37] Latini G, Di Nicola G, Pierantozzi M 2016 *Fluid Phase Equilib.* **427** 488–497.

Table1. Dataset of the fluids considered in the present work and absolute relative deviations (*AAD*) for each fluids for the ANN and the comparison with literature models taken into account.

Fluid	Molecular Weight [kg/kmol]	Acentric Factor	Present work	Latini et al. [3]	Sastri-Rao [20]	Missenard [21]	Sato-Riedel [25]	Latini et al [24]	Nagvekar-Daubert [26]	Robbins-Kingrea [27]	Pachaiyappan et al [28]	Gharagheizi et al [29]	Di Nicola et al [5]
methane	16.04	0.01	0.11	2.3	28.5	26.3	48.8	55.8	-	2.6	58.3	13.5	7.6
ethane	30.07	0.1	0.04	2.8	11.1	56.5	17.9	42.1	29.8	18.2	36.9	11.8	12.8
propane	44.1	0.15	0.05	3.8	34	14.6	25.3	23.8	9.2	17.6	17.2	13.4	5.9
n-butane	58.12	0.2	0.07	3.2	46.1	5.5	19.4	15.9	3.8	9.4	9.8	10.4	5.1
n-pentane	72.15	0.25	0.18	0.7	56.2	4.5	16	8.8	2.3	11.6	11.9	9.2	5
n-hexane	86.18	0.3	0.17	1.9	61.5	3.1	11.4	5.1	3.7	9.1	13.7	6.2	3.1
n-heptane	100.2	0.35	0.31	1.2	63.2	1.5	6.4	3.2	4.6	10.1	16.1	4.8	1.4
n-octane	114.23	0.4	0.87	0.5	69.5	1.9	3.6	1.7	5.4	5.8	22.9	4.3	1.7
n-nonane	128.26	0.44	0.46	1.1	72.4	1	0.9	1.1	5.4	7	25.9	4	0.9
n-decane	142.28	0.49	0.64	1.2	77.4	1.2	1.8	2.7	5.9	4.5	30.5	3.4	1.7
n-undecane	156.31	0.53	0.23	1.2	79.7	0.9	4.4	3.4	5.7	5.1	32.7	3.7	1.3
n-dodecane	170.33	0.57	0.26	1	83.2	1	6.2	4.3	5.7	3.3	35.9	3.6	1.8
n-tridecane	184.36	0.62	0.13	1.1	85.3	0.6	8.2	4.5	5.2	6.5	37	3.9	1.8
n-tetradecane	198.39	0.66	0.06	1.1	87.9	0.6	9.9	5.2	5.3	6.8	39.8	3.7	1.6
n-pentadecane	212.41	0.7	0.08	1.9	87.9	1.4	12.4	4.3	3.9	6.8	39.8	4.8	1.1
n-hexadecane	226.44	0.73	0.11	1.5	91.3	1.2	13.2	5.2	4.5	8.6	42.7	4.1	2
n-heptadecane	240.47	0.77	0.31	1.4	94.9	1.4	13.7	6.2	5.1	9.7	45.3	3.1	4.5
n-octadecane	254.49	0.8	0.17	1.1	95.2	2.2	15.6	5	4.1	12.1	45.3	4	4.3
n-nonadecane	268.52	0.84	0.12	0.8	97.8	3.7	16.3	5.3	4.4	12.7	47.3	3.5	5.8
n-eicosane	282.55	0.91	0.05	2.5	95.7	7.6	18.8	3.3	3.7	15	44.4	5.3	5.5
n-heneicosane	296.57	0.94	0.11	4	95.2	7.7	20.3	2.6	3.6	15.7	44.1	5.9	5.1
n-docosane	310.6	0.97	0.04	4.1	94.1	8.7	22.2	2	3.8	16.9	43	7.1	4
n-tricosane	324.63	1.03	0.13	2.7	94.6	9.5	23.3	1.8	4.1	19.2	43	7.4	5.6
n-tetracosane	338.65	1.07	0.06	0.7	93.7	10.7	24.9	2.5	4.8	21.5	41.8	8.4	5.8
acetone	58.08	0.31	0.05	1.8	31.5	5.2	2.3	9.7	7.2	12.1	4.9	8.8	13.8
methylethylketone	72.11	0.32	0.02	1.5	46.4	7.2	4.1	2.4	13.4	26.4	13.5	3.3	4.6
2-pentanone	86.13	0.34	0.17	2.6	59.9	3.4	1.6	3.9	17.8	19.9	15.7	5	4.6
3-pentanone	86.13	0.34	0.13	0.9	58.4	6.1	1.6	2	21.7	15.9	16.1	6	6.2
2-hexanone	100.16	0.38	0.32	0.4	69.5	4.2	0.9	7.8	22.5	14.9	19	8.8	5.6
2-heptanone	114.19	0.42	0.36	1	78.8	5.4	1.4	11.5	27	12.2	22.4	3	1.1
3-heptanone	86.13	0.34	0.15	0.5	79.4	5.6	3.1	9.3	32.1	11	25.7	3.2	6.7
4-heptanone	114.19	0.41	0.89	3.2	75.6	4.6	5.5	5.8	29.1	10.5	22.8	5.7	4.7
2-octanone	128.21	0.46	0.31	2.7	86.4	4.9	4.1	12.9	29.5	6.7	25.8	3.2	1.1
2-nonanone	142.24	0.49	0.96	3.1	93.1	6.7	5.2	15.3	32.3	8.4	28	1.8	2.3
5-nonanone	142.24	0.51	0.52	0.6	92.3	5.4	5.6	12.5	36.6	7.9	28.1	3.4	2.2
cyclopentanone	84.12	0.29	0.11	3.9	57.2	8.5	3.4	-	19.8	18.6	30.9	1.7	6.9
cyclohexanone	98.14	0.4	0.18	1.7	70.9	3.9	3.8	-	28.5	40.3	38.2	3.8	4.9
acetophenone	120.15	0.4	0.1	1.8	76.7	5.4	9.4	2.2	57.2	21	28.7	1.3	6.9

Methyl trichlorosilane	149.48	0.24	0.11	3.1	-	*26.6*	*29.9*	-	-	-	-	*35.8*	*29.8*
Dichlorodiethylsilane	157.11	0.32	0.11	0.8	-	19.6	19	-	5.6	-	-	19.5	17.9
Vinyltrichlorosilane	161.49	0.28	0.14	14.1	-	*23.4*	*22.7*	-	*78.9*	-	-	*26.4*	*23.2*
Phenyltrichlorosilane	211.55	0.4	0.26	4.9	-	16.6	16.9	-	2.7	-	*97.9*	7.8	9.2
Diphenyldichlorosilane	253.2	0.43	0.11	5	-	5.7	7.7	-	*141.8*	-	*107.5-*	14.6	4.4
Hexamethyl-cyclotrisiloxane	222.46	0.5	0.06	4.7	-	2.6	15.1	-	-	-	-	16.3	10.1
Octamethyl-cyclotetrasiloxane	296.62	0.55	0.12	0.2	-	2.7	19.5	-	-	-	-	15.3	10.7
Decamethyl-cyclopentasiloxane	370.77	0.64	0.21	1.4	-	14.1	*31.1*	-	-	-	-	*22*	*21.7*
Dodecamethyl-cyclohexasiloxane	444.92	0.71	0.08	5.3	-	*24.1*	*37.1*	-	-	-	-	*24.6*	*26*
Hexadecamethyl-cyclooctasiloxane	593.23	0.88	0.04	1.4	-	*31*	*43.1*	-	-	-	-	*25.6*	*27.7*
Hexamethyldi-siloxane	162.38	0.41	0.17	2.5	-	9.8	8.5	-	-	-	-	6.8	9.6
Octamethyltri-siloxane	236.53	0.54	0.3	1.4	-	6.7	7.8	-	-	-	-	8.2	9.1
Decamethyltetra-siloxane	310.69	0.66	0.06	3.2	-	7.1	*15*	-	-	-	-	8.9	8.2
Dodecamethylpenta-siloxane	384.84	0.68	0.07	2.1	-	*16.3*	*23.1*	-	-	-	-	12	9.6
Tetradecamethyl hexa-siloxane	458.99	0.8	0.08	6.4	-	*38.3*	*45.9*	-	-	-	-	15.6	12.2
Hexadecamethyl hepta-siloxane	533.15	0.89	0.07	0.6	-	*29.8*	*35.2*	-	-	-	-	*17.7*	*15*
Methyl silicate	152.22	0.44	0.05	0.8	-	*28.4*	*29.9*	-	*85.5*	-	-	*33*	*26*
Tetraethoxysilane	208.33	0.63	0.12	2.9	-	14.2	20.8		*65.6*	-	-	20.6	7.7

Table 2. Hidden layer weights

Hidden neurons Neurons	bias	Tr	Mass	Acentric factor
1	0.32	-4.41	5.65	-4.41
2	-8.61	-0.02	34.52	21.49
3	19.66	-0.76	-20.12	-26.47
4	-12.56	0.17	22.76	43.27
5	15.31	1.05	-14.75	-0.98
6	-5.66	0.19	43.06	3.62
7	10.11	-0.07	-52.58	-2.51
8	4.09	-0.27	42.95	-43.39
9	4.16	-0.64	37.79	-37.19
10	18.87	0.33	-19.43	-1.74
11	16.64	0.08	6.91	-43.33
12	8.55	-0.24	49.42	-59.57
13	26.90	-0.21	-26.36	-7.64
14	3.17	-0.01	-41.58	9.08
15	-19.83	1.04	11.51	38.43
16	-5.25	-0.14	-52.33	38.98
17	14.87	-21.78	-1.62	8.33
18	14.12	-0.31	-25.21	-2.69

19	8.69	0.30	-29.17	-18.43
20	-7.62	0.49	20.56	-4.63
21	4.27	0.10	1.27	-6.27
22	-6.87	-0.62	14.96	34.81
23	4.27	1.92	-8.64	-1.12
24	-11.40	-0.22	11.12	22.57
25	-8.30	-0.36	3.80	24.48
26	-3.76	-0.60	4.71	1.86
27	-9.59	0.36	35.03	13.27
28	-11.50	0.32	-45.52	63.01
29	-15.39	-0.98	13.96	39.93
30	2.64	-6.43	-1.66	-6.79
31	14.41	-0.32	-16.43	-8.09
32	2.22	0.07	43.18	-28.50
33	9.21	-0.18	-12.57	-1.08
34	0.10	0.39	-5.91	-20.84
35	19.03	-0.53	51.78	-86.00
36	18.59	-22.21	-0.81	7.41
37	0.10	0.02	58.92	-33.71

Table 3. Output layer weights

Output of Neurons	
1	-31.12
2	-0.22
3	-0.56
4	-0.44
5	-0.42
6	-0.26
7	-27.46
8	-0.89
bias	29.77

On the temperature prediction in a fire escape passage

G Casano and S Piva

DE Department of Engineering, Università di Ferrara, via Saragat 1, 44122 Ferrara

E-mail: stefano.piva@unife.it

Abstract. Fire safety engineering requires a detailed understanding of fire behaviour and of its effects on structures and people. Many factors may condition the fire scenario, as for example, heat transfer between the flame and the boundary structures. Currently advanced numerical codes for the prediction of the fire behaviour are available. However, these solutions often require heavy calculations and long times. In this context analytical solutions can be useful for a fast analysis of simplified schematizations. After that, it is more effective the final utilization of the advanced fire codes. In this contribution, the temperature in a fire escape passage, separated with a thermally resistant wall from a fire room, is analysed. The escape space is included in a building where the neighbouring rooms are at a constant undisturbed temperature. The presence of the neighbouring rooms is considered with an equivalent heat transfer coefficient, in a boundary condition of the third type. An analytical solution is used to predict the temperature distribution during the fire. It allows to obtain useful information on the temperature reached in the escape area in contact with a burning room; it is useful also for a fast choice of the thermal characteristics of a firewall.

1. Introduction

To enable the occupants to move to a place of safety before the environment in any evacuation route becomes untenable from the effects of a fire, is the main objective of "fire hazard management" in buildings [1]. In general terms, in an event of fire inside a building the conditions of tenability in an evacuation route must be maintained for the period of time that occupants need to evacuate the building. These conditions have been synthesized in [2] as:
"In the event of a fire in a building the conditions in any evacuation route must be maintained for the period of time occupants take to evacuate the part of the building so that:

(i) the temperature will not endanger human life; and
(ii) the level of visibility will enable the evacuation route to be determined; and
(iii) the level of toxicity will not endanger human life".

In the present paper we concentrate our attention on the *tenability criterion for temperature*. Purser [3, 4] gives a comprehensive review of the threshold times to temperature exposure for the human body, beyond which incapacitation, serious injury or death occurs. This tolerance time depends on the temperature (higher temperature / shorter tolerance time) up to a lower limiting value that is estimated around 50 °C. Based on available experimental data, Purser [3] was able to derive expressions for these tolerance times, intended as maximum exposure time, to incapacitation, serious injury / severe incapacitation, fatal exposure conditions with extensive third-degree burns. This information is particularly useful in a performance based design approach.

Differently, in a prescription based design approach is the "fire resistance" of building components and constructions to be commonly used. The building elements are classified in different fire resistance classes, or combinations of those, by specifying different performance criteria. Normally the fire resistance classification is expressed as a time limit in minutes (15, 30, 45, 60, 90, 120, 180, 240, or 360) which shows the time the performance criteria is fulfilled during a standardized fire test. Specific requirements for the classes are: R Load bearing capacity, E Integrity, I Insulation.

By following the Eurocode standard EN 1992-1-2 [5], "where compartmentation is required, the elements forming the boundaries of the fire compartment, including joints, shall be designed and constructed in such a way that they maintain their separating function during the relevant fire exposure. This shall ensure, where relevant, that:

- integrity failure does not occur
- insulation failure does not occur
- thermal radiation from the unexposed side is limited."

By following EN 1992-1-2 [5], "criterion "I" may be assumed to be satisfied where the average temperature rise over the whole of the non-exposed surface is limited to 140 K, and the maximum temperature rise at any point of that surface does not exceed 180 K."

In a previous paper [6] we concentrated our attention on the *insulation, I*, that is the ability of a separating element of a building, when exposed to fire on one side, to restrict the temperature rise of the unexposed face below specified levels. In the present paper we now concentrate our attention to the *tenability criterion for temperature*, expressed in terms of tolerance time.

The predictions of the time-space temperature distribution in the separating wall and in the evacuation route are obtained by means of the analytical solution already discussed in [6]. The limits of this approach were also discussed in [6] and are not further analysed here. Now, the mathematical model is completed to take into account the presence of rooms surrounding the escape passage. This is a configuration often encountered in special buildings, like hospitals and multi-residence housing. To the best of our knowledge, there are no bibliographic references dealing with this subject.

The temperature trend in the escape route is obtained in a simplified manner with a suitable choice of the boundary condition placed on the face of the wall not exposed to the fire. This boundary condition is obtained with an energy balance between the escape passage and the neighbouring rooms. In this configuration, dimensions and thermal characteristics of the walls (floor, ceiling and other unexposed walls) become influent for the evolution of the temperature. Numerical simulations obtained with specialized software, like FDS [7] or Smartfire [8], are certainly a right approach for the prediction of the temperature in complex escape routes, characterized by complex geometries and stratified wall. Nevertheless, our simplifyed analytical approach, with simplified geometry and walls characterization, can be certainly useful to individuate the driving parameters of the problem and to give them the right weight in a further detailed analysis.

The present work investigates the thermal performance of a fire resistant wall in an escape passage when exposed to a fire, in order to predict the temperature trend in the passage itself. The goal is the estimation of the tolerance time to temperature, it means the time needed to reach a value of temperature after that pains or hyperthermia occurs for a human body.

2. Formulation and solution

The problem is schematically shown in Fig. 1. In [6] we examined a fire resistant wall in the typical configuration used for the assessment of its fire resistance (Fig. 1.a). In the present paper the wall is now part of a fire-isolated passageway (Figs. 1.b and 1.c). This is a corridor of fire-resisting construction, "which provides egress to or from a fire-isolated stairway or fire-isolated ramp or to a road or open space" [2]. The air temperature in the fire-isolated passageway is free to change and follows the evolution of the fire. The remaining walls of the fire-isolated passageway face rooms unaffected by the fire.

A sketch of the geometry considered for the mathematical formulation of the problem, is shown in Fig. 2. A single layer wall (thickness L) is exposed to a fire on one face (x = 0). The temperature in the fire room changes stepwise, with a final value typical of a post-flashover condition. The boundary condition on the surface exposed to fire is of the third type. As already demonstrated in [6], with a proper choice of the heat transfer coefficient, the temperature on this wall is that due to the standard

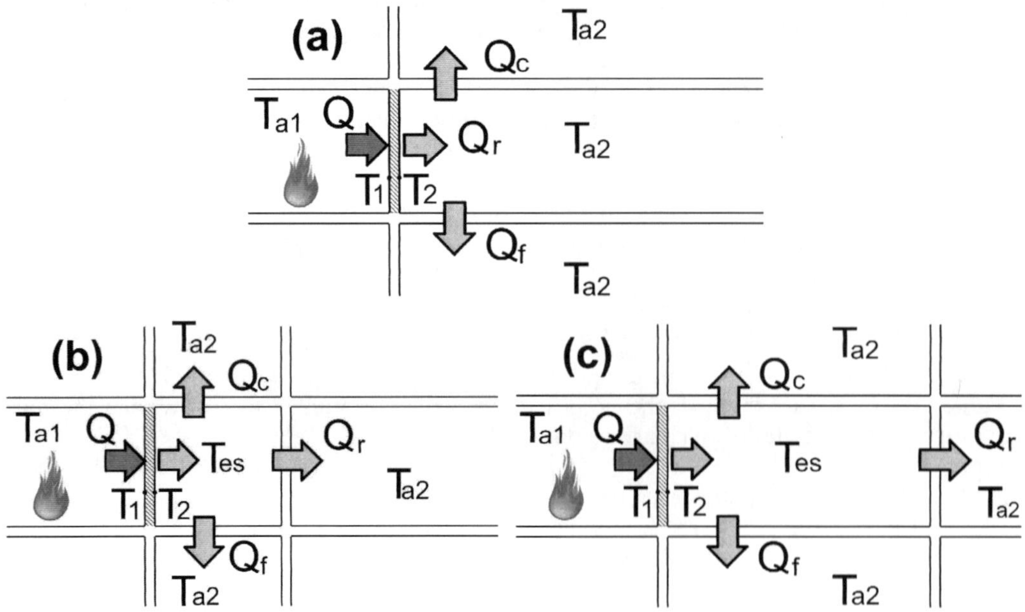

Figure 1. Schematic of the problem.

temperature-time curve suggested by EN 1991-1-2 [9]. The thickness of the wall is small against its height and width, so that the wall can be considered of unlimited extension and its heat flow is one-dimensional in the direction of the thickness. The properties of the fire resistant wall are uniform, constant and evaluated at a mean value of its temperature, significant for the solution of the problem. The prediction of the temperature in the wall is obtained with the procedure discussed in [6].

In the present application the thermal effect due to the presence of the neighbouring rooms, where the temperature is assumed to be maintained at a constant value in time, is simulated through an equivalent heat transfer coefficient. This approach is based on two assumptions:

- constant temperature in the neighbouring rooms;
- negligible thermal inertia of the walls separating the escape passage from the neighbouring rooms.

The first assumption is justified by considering that the time commonly involved in the evacuation of a building is relatively short and at the beginning of the fire. In [6] we discussed the cases described

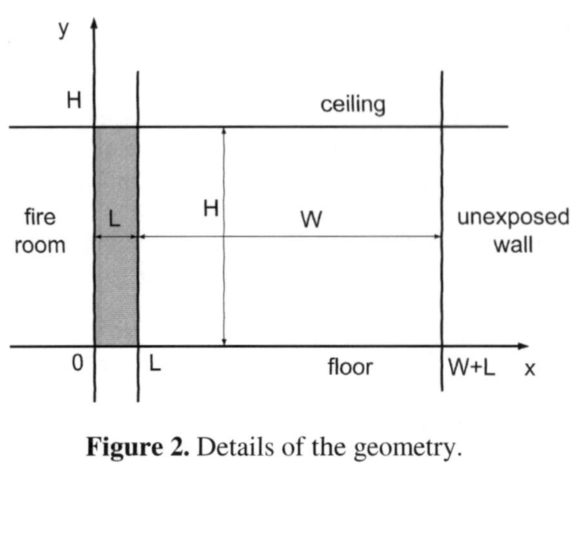

Figure 2. Details of the geometry.

by Fig. 1.a for applications similar to that analyzed here. In that cases the heat flux exchanged through the fire resistant wall is that highest possible. Nevertheless, during the initial phase of the fire, the heat flux transmitted by the wall is small for a time comparable to that requested for the evacuation. The thermal inertia of the fire resistant wall is large enough to accumulate the heat before transmitting it in the protected room. It follows that the neighbouring rooms are affected by the fire only after a significant time and that in the intermediate time their temperature can be assumed constant in time.

The second assumption is justified by the slow evolution of the temperature in the escape passage during the time necessary for the evacuation of the building, that is a direct consequence of the former assumption. It follows that during this time the heat transfer process between the escape passage and the neighbouring rooms can be modelled as quasi-steady.

Furthermore, a 2D effect (something like a fin effect) could be considered where floor and ceiling intersect the fire resistant wall. However, the low thermal conductivity of their materials and the significant thickness of these walls ensure a negligible contribution.

The problem is stated mathematically as follows:

Conservation of energy:
$$\frac{\partial T}{\partial t} = \alpha \frac{\partial^2 T}{\partial x^2} \tag{1}$$

Boundary conditions:

$x = 0$
$$h_1 \left[T_{a1} - T(x=0,t) \right] = -\lambda \frac{\partial T}{\partial x}(x=0,t) \tag{2}$$

$x = L$
$$-\lambda \frac{\partial T}{\partial x}(x=L,t) = h_2 \left[T(x=L,t) - T_{a2} \right] \tag{3}$$

Initial condition:

$t = 0$
$$T(x,t=0) = T_{a2} \tag{4}$$

Equations (1-4) are now expressed as dimensionless, according to the following dimensionless variables:

$$\Theta = \frac{T(x,t) - T_{a2}}{T_{a1} - T_{a2}} \qquad \xi = \frac{x}{L} \qquad \tau = \frac{t}{t_0}$$

Equation (1) and related boundary and initial conditions, Eqs. (2-4), become:

$$\frac{\partial \Theta}{\partial \tau} = \mathrm{T} \frac{\partial^2 \Theta}{\partial \xi^2} \tag{5}$$

for $\xi = 0$
$$\left. \frac{\partial \Theta}{\partial \xi} \right|_{0,\tau} = Bi_1 \left(\Theta(0,\tau) - 1 \right) \tag{6}$$

for $\xi = 1$
$$\left. \frac{\partial \Theta}{\partial \xi} \right|_{1,\tau} = -Bi_2 \Theta(1,\tau) \tag{7}$$

for $\tau = 0$
$$\Theta(\xi, 0) = 0 \tag{8}$$

The dimensionless parameters Bi_1, Bi_2 and T, are given by:

$$Bi_1 = \frac{h_1 L}{\lambda} \qquad Bi_2 = \frac{h_2 L}{\lambda} \qquad \mathrm{T} = \frac{t_0}{L^2/\alpha}$$

The solution of Eqs. (5-8) can be decomposed as follows:

$$\Theta(\xi,\tau) = U(\xi) + V(\xi,\tau) \tag{9}$$

When substituting Eq. (9) in Eqs. (5), it follows that the problem is now given by the solution of the following ordinary differential equations:

$$0=\frac{d^2U}{d\xi^2} \tag{10}$$

$$\frac{\partial V}{\partial \tau}=\mathrm{T}\frac{\partial^2 V}{\partial \xi^2} \tag{11}$$

where Eq. (10) deals with the Steady state component of the solution and Eq. (11) with its Time dependent component.

The Steady state component of the solution is coupled to the following boundary conditions:

$\xi=0$
$$\left.\frac{dU}{d\xi}\right|_0 = Bi_1\left(U(0)-1\right) \tag{12}$$

$\xi=1$
$$\left.\frac{dU}{d\xi}\right|_1 = -Bi_2 U(1) \tag{13}$$

The solution of Eqs. (10) and (12-13) is given by:

$$U(\xi)=A+B\xi \tag{14}$$

where:

$$A=\frac{Bi_1(1+Bi_2)}{Bi_1+Bi_2+Bi_1Bi_2} \tag{15}$$

$$B=-\frac{Bi_1Bi_2}{Bi_1+Bi_2+Bi_1Bi_2} \tag{16}$$

The time dependent term of the dimensionless solution (Eq. (9)) comes from Eq. (11), to be solved with its dimensionless boundary and initial conditions:

for $\xi=0$
$$\left.\frac{\partial V}{\partial \xi}\right|_{\xi=0,\tau} = Bi_1 V(\xi=0,\tau) \tag{17}$$

for $\xi=1$
$$\left.\frac{\partial V}{\partial \xi}\right|_{\xi=1,\tau} = -Bi_2 V(\xi=1,\tau) \tag{18}$$

for $\tau=0$
$$V(\xi,\tau=0)=-U(\xi) \tag{19}$$

The solution of the differential problem given by Eq. (11) and Eqs. (17-19) is obtained through the well known method of separation of the variables [10-14]. The details are given in [6]; here only the final result is reported:

$$V(\xi,\tau)=2\sum_{n=1}^{+\infty}\frac{\left(Bi_2^2+\beta_n^2\right)\left[\frac{ABi_1-B}{\beta_n}+\left(A+B+\frac{Bi_1 B}{\beta_n^2}\right)\sin(\beta_n)+\frac{B-(A+B)Bi_1}{\beta_n}\cos(\beta_n)\right]}{\left\{\left(Bi_1^2+\beta_n^2\right)\left[Bi_2+\left(Bi_2^2+\beta_n^2\right)\right]+Bi_1\left(Bi_2^2+\beta_n^2\right)\right\}}\cdot \left[Bi_1\sin(\beta_n\xi)+\beta_n\cos(\beta_n\xi)\right]e^{\left(-\mathrm{T}\beta_n^2\tau\right)} \tag{20}$$

The admissible values of the constant of separation β_n are the roots of the transcendental equation:

$$\frac{\sin(\beta_n)}{\cos(\beta_n)}=\frac{\beta_n(Bi_1+Bi_2)}{\beta_n^2-Bi_2Bi_1} \tag{21}$$

The search for these roots can be easily executed by means of a numerical procedure.

Finally, the temperature distribution is given by:

$$T(x,t)=T_{a2}+(T_{a1}-T_{a2})(U(\xi)+V(\xi,\tau)) \tag{22}$$

In the boundary condition given by Eq. (3), an equivalent overall heat transfer coefficient representing the overall heat transfer resistance between the wall surface facing the fire passageway and the neighbouring rooms not affected by the fire is used.

An energy balance written for the escape passage gives:

$$h_{2b}H(T(L,t)-T_{es}(t))=K_fW(T_{es}(t)-T_{a2})+K_cW(T_{es}(t)-T_{a2})+K_{un}H(T_{es}(t)-T_{a2}) \tag{23}$$

In terms of overall heat transfer, this can be written as:

$$h_{2b}H(T(L,t)-T_{es}(t))=h_2H(T(L,t)-T_{a2}) \tag{24}$$

The equivalent overall heat transfer coefficient, is then given by:

$$h_2=\frac{1}{H\left[\frac{1}{h_{2b}H}+\frac{1}{K_fW+K_cW+K_{un}H}\right]} \tag{25}$$

and in dimensionless form,

$$Bi_2=\frac{1}{\frac{1}{Bi_{2b}}+\frac{\xi_H}{\mathrm{K}_f\xi_W+\mathrm{K}_c\xi_W+\mathrm{K}_{un}\xi_H}} \tag{26}$$

The dimensionless parameter Bi_2 is now expressed as a function of a new set of dimensionless parameters:

$$Bi_{2b}=\frac{h_{2b}L}{\lambda}\quad \xi_H=\frac{H}{L}\quad \xi_W=\frac{W}{L}\quad \mathrm{K}_f=\frac{K_fL}{\lambda}\quad \mathrm{K}_c=\frac{K_cL}{\lambda}\quad \mathrm{K}_{un}=\frac{K_{un}L}{\lambda}$$

Finally, the temperature in the fire isolated passageway is given by:

$$T_{es}(t)=T(x=L,t)-\frac{Q(t)}{h_{2b}H} \tag{27}$$

where:

$$Q(t)=h_2H(T(x=L,t)-T_{a2}) \tag{28}$$

In dimensionless form Eqs. (27-28) become:

$$\Theta_{es}(\tau)=\left(1-\frac{Bi_2}{Bi_{2b}}\right)\Theta(1,\tau) \tag{29}$$

$$\Gamma(\tau)=(Bi_2/\mathrm{K}_w)\Theta(1,\tau) \tag{30}$$

Finally, the dimensionless temperature in the fire isolated passageway depends on the following dimensionless parameters:

$$Bi_1\ ,\ \mathrm{T}\ ,\ Bi_{2b}\ ,\ \xi_H\ ,\ \xi_W\ ,\ \mathrm{K}_f\ ,\ \mathrm{K}_c\ ,\ \mathrm{K}_{un}$$

3. Results and Discussion

For the calculation of the tolerance time to be used as the tenability limit for design purposes, we refer to the correlation under midhumidity conditions given by Purser [3]:

$$t_{tol}=2\cdot 10^{31}T^{-16.963}+4\cdot 10^{8}T^{-3.7561} \tag{31}$$

where T is the ambient temperature (°C) and t_{tol} is the tolerance time (min).

IOP Conf. Series: Journal of Physics: Conf. Series **923** (2017) 012055 doi:10.1088/1742-6596/923/1/012055

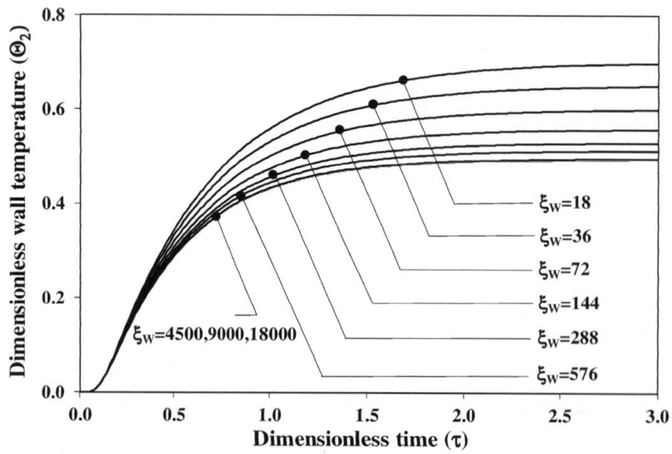

Figure 3. Wall temperature as a function of time for different values of ξ_W and Bi_2.

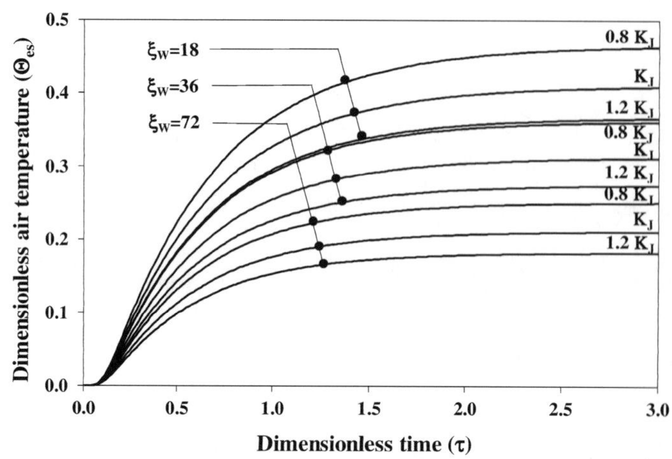

Figure 4. Temperature in the escape room as a function of time for different values of ξ_W and Bi_2.

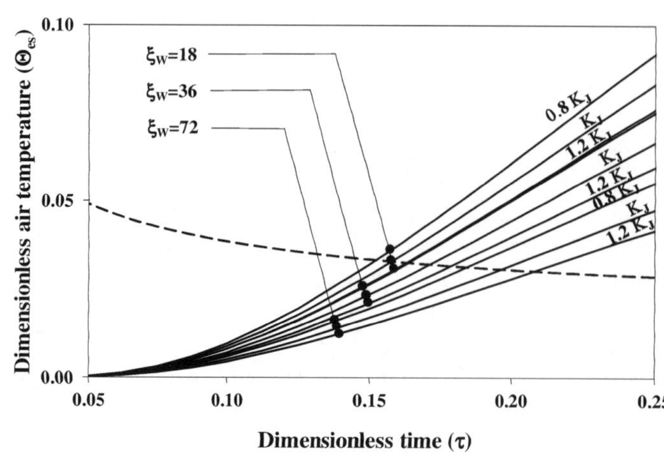

Figure 5. Temperature in the escape passage (continuous line) and tolerance time (dashed line) as a function of time.

Due to the large number of dimensionless parameters characterizing the problem, the calculations are done for assigned values of a part of these. In particular, Bi_1, T, Bi_{2b}, ξ_H, are assumed constant. In particular these dimensionless values are: T = 1, $Bi_1 = 2.857$, $Bi_{2b} = 0.756$ and $\xi_H = 30$.

The free dimensionless numbers are ξ_W, K_f, K_c, K_{un} ; with these parameters it is possible to discuss the effect on the temperature trend in the escape passage, of the thermal characteristics of the different walls of the corridor (K_f, K_c, K_{un}) and of their heat transfer surface (ξ_W). In the reference configuration the conductance of ceiling, floor and unexposed wall are the same of the wall exposed to the fire. Due to the different total (convection and radiation) heat transfer coefficients (vertical, horizontal with upward heat flow, horizontal with downward heat flow) the trasmittance of these walls are different (K_c, K_f, K_{un}). In the discussion of the results these parameters are incremented / decremented, at the same time, of the same amount.

In Figure 3 the dimensionless temperature of a wall heated by a fire is shown as a function of the dimensionless time. In particular, the dimensionless temperature refers to the unexposed face of the wall ($\xi = 1$), and is given by two contribution, a steady one (Eq. (14)) and a time dependent one (Eq. (20)). The time dependent component is given by a summation of terms tending to zero for τ tending to infinity. Necessarily, the final temperature tends to the asymptotic steady value given by Eq. (14). This asymptotic steady value depends on Bi_1, Bi_2 and T. Since Bi_2 depends on a set of parameters through Eq. (26), for the same values of Bi_{2b}, ξ_H, K_f , K_c, K_{un}, the wall temperature dependens only on ξ_W. This behaviour is evident in Fig. 3 where, for constant values of Bi_1, T, Bi_{2b}, ξ_H, K_f, K_c, K_{un}, for increasing values of ξ_W a decrease of the asymptotic wall temperature can be observed. This is due to the better heat transfer conditions resumed in a higher Bi_2 and due to the larger heat transfer surface. Furthermore, for very large values of ξ_W, the increase tends to reduce and the dimensionless wall temperature tends to a constant distribution, independent of ξ_W. This is the case of Fig. 1.a, corresponding to the condition examined in [6] and used to determinate the fire resistance of a wall in terms of *Criterion "I"*, insulation of the wall.

The dimensionless temperature of the air in the escape room is shown in Fig. 4. For three values of ξ_W, three curves are shown, calculated for $0.8K_j, K_j$ and $1.2K_j$ (the subscript j means: f (floor), c (ceiling) and un (unexposed), respectively). For increasing values of K_j, the asymptotic dimensionless air temperature in the fire protected room decreases. A thermal insulation of the walls is the source of an increase of the temperature in the escape room. At the same time an increase of ξ_W produces a decrease of the asymptotic dimensionless air temperature in the fire protected room. This is clearly due to the combined effect of heat transfer surface and wall transmittance. Large corridors and slightly insulated walls are positive elements if we want to keep low the temperature in an escape, fire protected, corridor.

In Figure 5 the intersection of the temperature in a corridor with the tolerance time curve, Eq. (31), is detailed. The tolerance time, intended as the maximum exposure time before incapacitation, changes with the characteristics of the fire resistant wall, but also with that of the remaining walls of an escape passage. The results are shown in Tab. 1, where the dimensionless tolerance times are detailed for the same fixed data used to produce Fig. 6. For the examined wall exposed to fire, the tolerance time

Table 1. Dimensionless tolerance time as a function of the free dimensionless parameters ξ_W and K_j.

ξ_W	K_C	K_f	K_{un}	τ_{tol}
18	0.220	0.161	0.201	0.153
18	0.274	0.201	0.252	0.158
18	0.329	0.242	0.302	0.162
36	0.220	0.161	0.201	0.163
36	0.274	0.201	0.252	0.171
36	0.329	0.242	0.302	0.178
72	0.220	0.161	0.201	0.184
72	0.274	0.201	0.252	0.196
72	0.329	0.242	0.302	0.208

varies from $\tau_{tol} = 0.15$ for the most insulated walls and the smallest width of the corridor, up to $\tau_{tol} = 0.21$ for the less insulated walls and the largest width of the corridor.

In dimensional terms, for a wall characterized by $L = 0.10$ m, $\lambda = 1.19$ Wm^{-1}K^{-1}, $c = 1.3$ kJkg^{-1}K^{-1}, $\rho = 2200$ kgm^{-3}, $h_1 = 34$ Wm^{-2}K^{-1}, $h_{2b} = 9$ Wm^{-2}K^{-1}, and neigbouring walls with $U_f = 2.4$ Wm^{-2}K^{-1}, $U_c = 3.3$ Wm^{-2}K^{-1}, $U_{un} = 3.0$ Wm^{-2}K^{-1}, the characteristic time of the wall is $L^2/\alpha = 24034$ s. Taking this characteristic time as the reference t_0, for the values of τ_{tol} reported in Tab. 1, the dimensional values of tolerance time are calculated. For the largest corridor (W = 7.2 m) and the less insulated walls (1.2 K_j) the tolerance time is $t_{tol} = 83.3$ min; for the smallest corridor (W = 1.8 m) and the most insulated walls (0.8 K_j) the tolerance time is $t_{tol} = 61.3$ min. The difference amounts to 22 min. In terms of time useful for the evacuation of a building, the difference is significant.

4. Concluding Remarks

The analytical solution proposed for the study of the performance of a wall that separates a room with a fire from an escape space has provided useful results. With this solution we are able to easily calculate the air temperature in the escape route. This temperature depends on the boundary conditions, in particular on the width of the corridor and on the thermal transmittance of the walls separating the corridor from the neighbouring ambients.

The air temperature in the escape route varies significantly with geometry and boundary conditions. All this reflects into a tolerance time that significantly shorten if the walls are thermally insulated and the corridor is narrow.

Finally, the reference conditions used to determine the fire resistance class are those giving the longest tolerance time, while in the practical situations the dimensions the corridor and the thermal characteristics of the walls can reduce significantly this tolerance time.

5. Nomenclature

A	constant coefficient	
B	constant coefficient	
Bi	Biot number	
h	heat transfer coefficient	(Wm^{-2}K^{-1})
H	height of the wall	(m)
K	overall wall heat transfer coefficient	(Wm^{-2}K^{-1})
L	thickness of the wall	(m)
t	time	(s)
Q	heat flow	(W/m)
T	temperature	(°C)
U	dimensionless steady state term of the solution	
V	dimensionless time dependent term of the solution	
W	width of the passage	(m)
x	horizontal coordinate	
y	vertical coordinate	

Greek symbols

α	thermal diffusivity	(m^2/s)
β	eigenvalue	
Γ	dimensionless heat flux, Eq. (30)	
K	dimensionless heat transfer coefficient	
λ	thermal conductivity	(W/m°C)
Θ	dimensionless temperature	
τ	dimensionless time	
T	dimensionless reference time	
ξ	dimensionless coordinate	

Subscripts

$a1$	air in the fire room

$a2$ air in the neighbouring rooms
b boundary
c ceiling
es escape space
f floor
H height of the escape space
tol tolerance
un unexposed wall of the corridor
w wall
W width of the escape space
1 upstream of the compartment wall
2 downstream of the compartment wall
0 reference

References

[1] Poh W 2011 *Ecolibrium*, AIRAH, 32 – 37.
[2] ABCB 2016, *National construction code 2016* Vol. One (Camberra - Building code of Australia Board)
[3] Purser D A 2016 In *The SFPE Handbook of Fire Protection Engineering* 5th ed. (NY Springer).
[4] Purser D A 1989 *Fire Safety Science* **2** 391-400.
[5] EN 1992-1-2:2004 Eurocode 2: Design of concrete structures - Part 1-2: General rules - Structural fire design.
[6] Casano G, Piva S 2017 *IOP Conf. Series: Journal of Physics: Conf. Series* **796** 012036.
[7] NIST 2016 *Fire Dynamics Simulator User's Guide*.
[8] FSEG 2013 *Smartfire v.4.3* University of Greenwich.
[9] EN 1991-1-2:2009 Eurocode 1: Actions on structures - Part 1-2: General actions - Actions on structures exposed to fire.
[10] Carslaw H S Jaeger J C 1959 *Conduction of Heat in Solids* 2nd ed. (Oxford Science Publ.)
[11] Arpaci V S 1966 *Conduction heat transfer* (Reading MA Addison-Wesley Publ. Co.)
[12] Hahn D W, Necati Özisik M 2012 *Heat Conduction* 3rd Ed. (NY John Wiley & Sons Inc.).
[13] Kakac S, Yener Y 2008 *Heat Conduction* 4th Ed. (NY Taylor & Francis Groop).
[14] Necati Ozisik M 1968 *Boundary Value Problems of Heat Conduction* (Scranton PA International Textbook Corp.).

A brief compendium of correlations and analytical formulae for the thermal field generated by a heat source embedded in porous and purely-conductive media

P Conti, D Testi and W Grassi

Department of Energy, Systems, Territory and Constructions Engineering (DESTEC), University of Pisa, 56122 Pisa, Italy

E-mail: paolo.conti@unipi.it

Abstract. This work reviews and compares suitable models for the thermal analysis of forced convection over a heat source in a porous medium. The set of available models refers to an infinite medium in which a fluid moves over different three heat source geometries: i.e. the moving infinite line source, the moving finite line source, and the moving infinite cylindrical source. In this perspective, the present work presents a plain and handy compendium of the above-mentioned models for forced external convection in porous media; besides, we propose a dimensionless analysis to figure out the reciprocal deviation among available models, helping the selection of the most suitable one in the specific case of interest. Under specific conditions, the advection term becomes ineffective in terms of heat transfer performances, allowing the use of purely-conductive models. For that reason, available analytical and numerical solutions for purely-conductive media are also reviewed and compared, again, by dimensionless criteria. Therefore, one can choose the simplest solution, with significant benefits in terms of computational effort and interpretation of the results. The main outcomes presented in the paper are: the conditions under which the system can be considered subject to a Darcy flow, the minimal distance beyond which the finite dimension of the heat source does not affect the thermal field, and the critical fluid velocity needed to have a significant contribution of the advection term in the overall heat transfer process.

1. Introduction

The heat transfer over a heat source embedded in a saturated porous medium represents the reference model for many engineering problems in different contexts, e.g., geothermal energy, ground-source heat pump applications, nuclear engineering, storage of nuclear waste material, groundwater flows, solar power collectors, compact heat exchangers, food industries and many others [1–3].

A proper design of the heat source (e.g., the dimension of the heat exchanger apparatus) is strictly related to a sufficiently-accurate modeling of the thermal field evolution in the surrounding medium. Many works have been published on this topic, including both numerical and analytical methods [4–8]. Despite the recent development of computer science, we believe that numerical methods are particularly appropriate for a detailed design and the analysis of a specific case or device, but they do not seem able to provide fast and general indications on the physical phenomena involved in the heat transfer process. In this perspective, analytical models represent a very suitable tool for a more general analysis of the problem, preliminary sizing, and feasibility analysis, because of their short computational time and flexibility in parametric designs.

This notwithstanding, analytical models are always affected by some simplifying hypotheses to make the problem solvable. In the context of heat sources in porous media, typical assumptions refer to:

- the relevance of the fluid movement in terms of heat transfer performance, i.e. purely-conductive medium vs. saturated porous medium;
- the radial dimension of the heat source, i.e. line vs. cylindrical heat source models;
- temperature gradients in the axial direction of the heat source, i.e. infinite vs. finite axial extension.

We classify the solutions proposed in literature as shown in Table 1, in accordance with the above-mentioned hypotheses.

Table 1. Classification scheme and acronyms for analytical solutions of heat source in purely-conductive and porous media models.

Geometry	Purely-conductive media		Porous media[a]	
Infinite axial extension	ILS Infinite line source	ICS Infinite cylindrical source	MILS Moving infinite line source	MICS Moving infinite cylindrical source
Finite axial extension	FLS Finite line source	FCS Finite cylindrical source	MFLS Moving finite line source	MFCS Moving finite cylindrical source (not yet developed)

[a] Generally, the porous medium is considered saturated.

Purely-conductive models consist of the limit solution of porous media when the fluid velocity is sufficiently low. Similarly, line models represent the limit of cylindrical solutions when the radial dimension of the heat source is negligible with respect to the space scales of the problem. Finally, the deviation between finite and infinite axial models is mainly related to the time scales of interest, resulting practically equivalent at short times [9,10].

These preliminary considerations suggest that specific conditions exist, in terms of time and space scales, at which the mentioned models result practically equivalent to each other. Therefore, one should be able to readily choose the most appropriate model for the specific analysis, in order to achieve sound and interpretable results with the least amount of computational effort.

To this aim, we started a systematic work aimed at developing a handy compendium of formulae for the analysis of the thermal field generated by a heat source embedded in purely-conductive and porous media. Our goal consists of developing analytical correlations and maps of the solutions, together with some quantitative, general and dimensionless criteria to assess when one model is practically equivalent to another one. Purely-conductive models have already been reviewed and discussed in [9]. In this work, we introduce the discussion on porous media, addressing the main features of the two analytical models currently available in literature: the MILS and the MFLS. A numerical model for the MICS has been recently developed by the authors in [11], while the analysis of the MFCS model is still ongoing.

In section 2, we will present the geometries and the constitutive equations of the discussed problems, together with a dimensional analysis aimed at identifying the dimensionless groups governing the solutions. In section 3, we will present the compendium of formulae and maps for the above-mentioned porous models. Finally, in section 4, we briefly recall purely-conductive models that can be used as reference when the advection effects become negligible (see, for instance, Fig. 1).

2. Momentum and energy equations

The discussed models (see Table 1) differ from each other in terms of geometry and boundary conditions. However, the mathematical structure and the constitutive equations of the problem are always the same. The momentum equation consists of the classical Darcy's formulation for a porous medium, namely:

$$\mathbf{u} = -\frac{K}{\mu}\nabla p \tag{1}$$

We recall that Eq. 1 is valid when the so-called *local pore Reynolds number,* Re_K , is lower than one [12].

The formulation of the energy balance (Eq. 2) is based upon the *local thermal equilibrium* hypothesis, i.e. both solid and fluid phases have the same local temperature, θ [12].

$$\frac{\partial \theta}{\partial t} + \mathbf{u}_m \cdot \nabla \theta = \alpha_m \nabla^2 \theta \tag{2}$$

where:

$$\alpha_m = \lambda_m / (\rho c)_m \qquad \mathbf{u}_m = \frac{(\rho c)_f}{(\rho c)_m}\mathbf{u}$$

$$\lambda_m = (1-\varepsilon_p)\lambda_s + \varepsilon_p \lambda_f \qquad (\rho c)_m = (1-\varepsilon_p)(\rho c)_s + \varepsilon_p (\rho c)_f$$

and $\mathbf{u}$ is evaluated through Eq. 1. We can make the following remarks:

- Eq. 2 is equivalent to Fourier's law when $\mathbf{u}$ is zero;
- infinite-axial-extension models (e.g., MILS) neglect the temperature gradient over the z-axis, therefore the problem becomes a 2D problem in the $x-y$ (or $r-\phi$) plane;
- the boundary conditions needed to set the thermal problem depend on the specific geometry; Table 2 shows the full set of equations for all the models included in Table 1;
- the porosity ε_p is the main parameter affecting the effective thermo-physical properties of the porous medium; all the parameters are volume-average quantities depending on ε_p .

Table 2. Boundary and initial conditions for the thermal problem.

Geometry	Line models	Cylindrical models
Infinite axial extension	B.C.: $\begin{cases} \theta(r\to\infty,\phi,t)=\theta_\infty \\ -(2\pi r)\lambda_m \left.\frac{\partial\theta}{\partial r}\right\|_r = \dot{q}_b \quad r\to 0 \end{cases}$ I.C.: $\theta(r,\phi,t=0)=\theta_\infty$	B.C.: $\begin{cases} \theta(r\to\infty,\phi,t)=\theta_\infty \\ -(2\pi r_b)\lambda_m \left.\frac{\partial\theta}{\partial r}\right\|_{r_b} = \dot{q}_b \end{cases}$ I.C.: $\theta(r,\phi,t=0)=\theta_\infty$
Finite axial extension	B.C.:	B.C.:

$$\begin{cases} \theta(r \to \infty, \phi, z, t) = \theta_\infty \\ -(2\pi r)\lambda_m \left.\dfrac{\partial \theta}{\partial r}\right|_r = \dot{q}_b \quad r \to 0 \quad 0 < z < H \end{cases} \qquad \begin{cases} \theta(r \to \infty, \phi, z, t) = \theta_\infty \\ -(2\pi r_b)\lambda_m \left.\dfrac{\partial \theta}{\partial r}\right|_{r_b} = \dot{q}_b \quad 0 < z < H \end{cases}$$

I.C.: $\theta(r, \phi, z, t = 0) = \theta_\infty$ I.C.: $\theta(r, \phi, z, t = 0) = \theta_\infty$

2.1 Dimensional analysis

According to the Buckingham theorem [13], the thermal problem can be rewritten as a function of 6 dimensionless parameters given by the combination of 10 physical variables: the temperature variation, $\theta - \theta_\infty$, the time, t, the radial coordinate, r, the angular coordinate, ϕ, the effective fluid velocity, $\mathbf{u}_m$, the effective thermal conductivity, λ_m, the effective thermal diffusivity, α_m, the source radius, r_b, the source depth, H, and the source magnitude, $\dot{q}_b$.

The resulting dimensionless quantities are: the dimensionless temperature, Θ, the Fourier number, Fo the Péclet number, Pe, the dimensionless radial coordinate, R, the angular coordinate, ϕ, the dimensionless axial coordinate, Z. The specific length to define the just-mentioned groups depends on the specific reference model, as well as the actual number of groups needed to evaluate the dimensionless temperature Θ.

According to [14], it is possible to define the overall characteristic length and time of the advection-diffusion problems and the corresponding dimensionless coordinate and time as:

$$L = \frac{\alpha}{u_m} \qquad \mathrm{T} = \frac{L}{u_m}$$

$$X = \frac{x}{L} = Pe_x \qquad \tau = \frac{t}{T} = Fo_x Pe_x^2$$

However, as we will show in Table 3, sometimes it is more convenient to refer to the Péclet and Fourier numbers distinctly, especially when we want to compare porous models with purely-conductive ones.

3. Compendium of analytical solutions for a heat source in Darcy flow

Table 3 shows the analytical solutions of the three models already available in literature, i.e. the MILS, the MFLS, and the MICS. The MFCS model will be presented in next works. Table 3 also shows some dimensionless criteria to evaluate when a given model is practically equivalent to another one. In this work, the term "practically equivalent" refers to a relative deviation of 5% in terms of Θ.

Table 3. Available solutions of heat sources in porous media.

Model	References	Description	Solution
MILS	[6,14,15]	Infinite line heat source subject to a Darcy crossflow. The source is located in $(0,0)$.	$\Theta(\phi, Pe_r, Fo_r) = \frac{1}{4\pi}\exp\left(\frac{Pe_r}{2}\cos\phi\right)\Gamma\left(0, \frac{1}{4Fo_r}, \frac{Pe_r^2}{16}\right)$ $\Theta = \frac{(\theta - \theta_\infty)\lambda_m}{\dot{q}_b} \quad Pe_r = \frac{u_m r}{\alpha_m} \quad Fo_r = \frac{\alpha_m t}{r^2}$ $\Gamma(a, x, b) = \int_x^\infty \beta^{a-1}\exp\left(-\beta - \frac{b}{\beta}\right)d\beta$

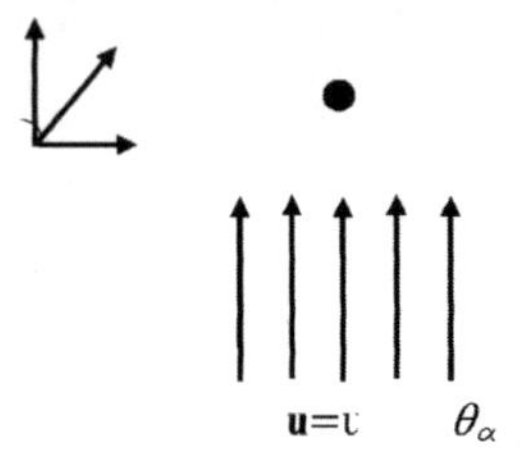

Steady-state solution (i.e. $\tau > \tau_s$)

$$\tau_s = 6.14 Pe_r^{0.24}, \quad 10^{-3} \le Pe_r \le 10$$

$$\Theta_s(\phi, Pe_r) = \frac{1}{2\pi} \exp\left(\frac{Pe_r}{2}\cos\phi\right) K_0\left(\frac{Pe_r}{2}\right)$$

Equivalence with other models

- MILS is practically equivalent to the ILS when: $Fo_r < Fo_{r,c}$ (see Fig. 1)

MFLS [16]

Finite line heat source subject to a Darcy crossflow. The source is located in $(0,0)$.

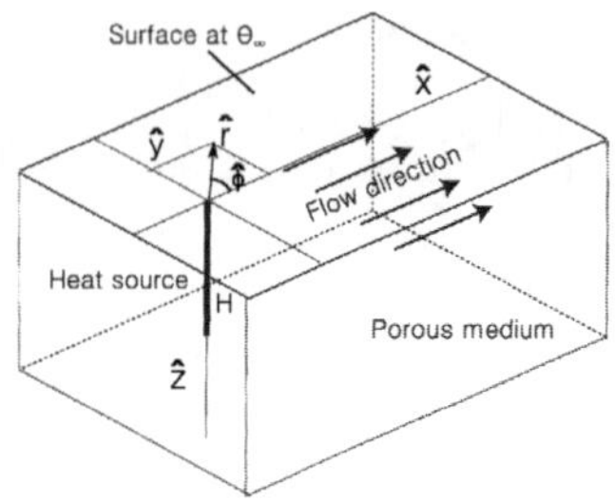

$$\Theta(R, Z, \phi, Fo_H, Pe_H) = \frac{1}{2\pi} \exp\left(\frac{Pe_H}{2} R\cos\phi\right) \times$$

$$\left[\int_0^1 f(R, Z, Z', Fo_H, Pe_H)\, dZ' - \int_{-1}^0 f(R, Z, Z', Fo_H, Pe_H)\, dZ'\right]$$

$$f(R, Z, Z', Fo_H, Pe_H) = \frac{1}{4\tilde{R}}\left[\exp\left(-\frac{Pe}{2}\tilde{R}\right) erfc\left(\frac{\tilde{R} - Pe_H Fo_H}{2\sqrt{Fo_H}}\right) + \right.$$

$$\left. + \exp\left(\frac{Pe}{2}\tilde{R}\right) erfc\left(\frac{\tilde{R} + Pe_H Fo_H}{2\sqrt{Fo_H}}\right)\right]$$

$$\Theta = \frac{(\theta - \theta_\infty)\lambda_m}{\dot{q}_b} \quad Pe_H = \frac{u_m H}{\alpha_m} \quad Fo_H = \frac{\alpha_m t}{H^2} \quad R = \frac{r}{H}$$

$$\tilde{R} = \sqrt{R^2 + (Z - Z')^2}$$

Steady-state solution (i.e. $\tau > \tau_s$)

τ_s expression is under development

$$\Theta_s(R, Z, \phi, Pe_H) = \frac{1}{4\pi} \exp\left(\frac{Pe_H R}{2}\cos\phi\right) \times$$

$$\left[\int_0^1 \frac{1}{\tilde{R}} \exp\left(-\frac{Pe_H}{2}\tilde{R}\right) dZ' - \int_{-1}^0 \frac{1}{\tilde{R}} \exp\left(-\frac{Pe_H}{2}\tilde{R}\right) dZ'\right]$$

MICS [11]

Infinite cylindrical heat source subject to a Darcy crossflow. The centre of the source is located in $(0,0)$.

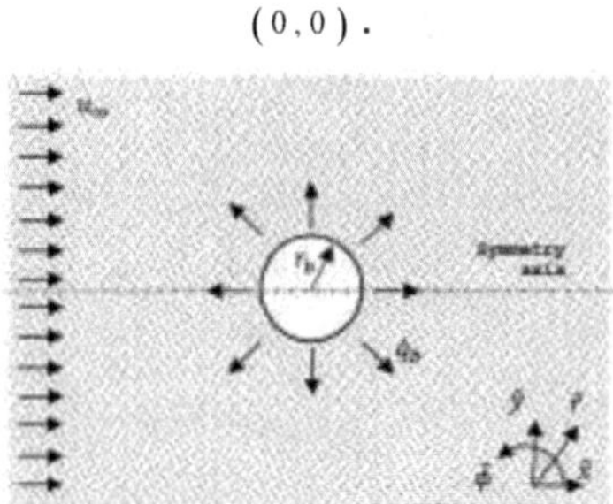

$$\overline{\Theta}_b(Pe_{r_b}, Fo_{r_b}) = 10 \wedge \left\{\left[p_1\tilde{\tau}^2 + p_2\tilde{\tau} + p_3\right] / \left[\tilde{\tau}^2 + q_1\tilde{\tau} + q_2\right]\right\}$$

$$\tilde{\tau} = \log_{10}(\tau)$$

(ϕ-averaged temperature of the heat source, $r = r_b$)

The values of the coefficients p and q are shown in [11]

$$\overline{\Theta}_b = \frac{(\overline{\theta}_b - \theta_\infty)\lambda_m}{\dot{q}_b} \quad Pe_{r_b} = \frac{u_m r_b}{\alpha_m} \quad Fo_{r_b} = \frac{\alpha_m t}{r_b^2} \quad \tau = Fo_{r_b} Pe_{r_b}^2$$

Steady-state solution (i.e. $\tau > \tau_s$)

$$\tau_s = 1.99 Pe_{r_b} + 3.56, \quad 5.0 \times 10^{-3} \le Pe_{r_b} \le 10^2$$

$$\overline{\Theta}_{b,s}(Pe_{r_b}) = \begin{cases} 0.291 Pe_{r_b}^{-0.22}, & Pe_{r_b} \le 0.1 \\ 0.1745 Pe_{r_b}^{-0.48}, & Pe_{r_b} > 0.1 \end{cases}$$

Equivalence with other models

- MICS is practically equivalent to the MILS at:

$$r/r_b > \begin{cases} 10, & Pe_{r_b} < 1, \forall t \\ 5\ , & Pe_{r_b} > 1, \forall t \end{cases}$$

- MICS is practically equivalent to the ICS in the range $\tau_{dev,1} \le \tau \le \tau_{dev,2}$, where:

$$\tau_{dev,1} = 3.6 \times 10^{-3} Pe_{r_b}^{2.02}$$

$$\tau_{dev,2} = \begin{cases} 0.69 Pe_{r_b}^{-0.26}, & Pe_{r_b} < 0.5 \\ 0.72 Pe_{r_b} + 0.88, & Pe_{r_b} \ge 0.5 \end{cases}$$

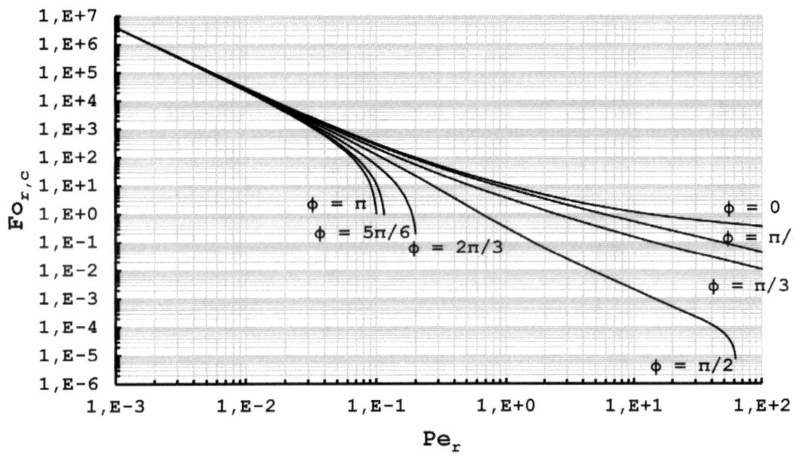

Figure 1. MILS model: $Fo_{r,c}$ as a function of Péclet number, Pe_r, and angular coordinate, ϕ .

4. Purely-conductive models

In this section, we briefly recall the analytical expressions of the ILS, ICS, FLS, and FCS solutions. The reader can refer to [9] for a comprehensive analysis of the just-mentioned models.

Table 4. Purely-conductive models

Model	References	Description	Solution
ILS	[6,9]	Infinite line heat source in a semi-infinite medium. The source is located in $(0,0)$. 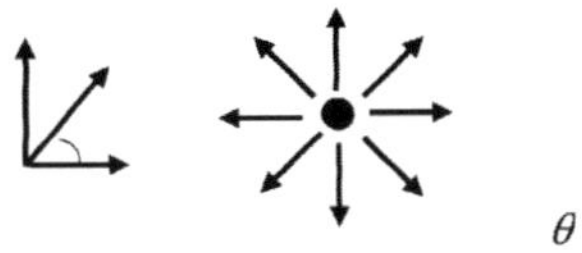	$\Theta(Fo_r) = \frac{1}{2\pi}\int_{r/2\sqrt{\alpha t}}^{\infty} \frac{\exp(-\beta^2)}{\beta} d\beta = \frac{1}{4\pi} Ei\left(\frac{1}{4 Fo_r}\right)$ $Ei(x)$ is the *exponential integral* function. **Equivalence with other models** - ILS is practically equivalent to the MILS when: $Fo_r < Fo_{r,c}$ (see Fig. 1) - ILS is practically equivalent to the ICS when: $Fo_{r_b} \ge 3.1(r/r_b) + 7.1$ - ILS is practically equivalent to the FLS when: $Fo_H \le 3.0 \times 10^{-4} R^{-0.67}$

<table>
<tr>
<td>ICS</td>
<td>[9,10,17]</td>
<td>Infinite cylindrical heat source in a semi-infinite medium. The centre of the source is located in $(0,0)$.
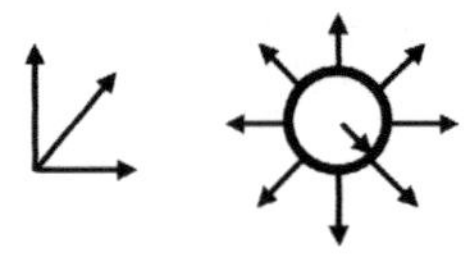</td>
<td>

$$\Theta\left(Fo_{r_b},\frac{r}{r_b}\right)=\frac{1}{\pi^2}\int_0^{\infty}\frac{\exp\left(-\beta^2 Fo_{r_b}\right)-1}{J_1^2(\beta)+Y_1^2(\beta)}\times\left[J_0\left(\frac{r}{r_b}\beta\right)Y(\beta)+J_1(\beta)Y_0\left(\frac{r}{r_b}\beta\right)\right]\frac{d\beta}{\beta^2}$$

Equivalence with other models
- ICS is practically equivalent to the ILS when: $Fo_{r_b}\geq 3.1(r/r_b)+7.1$
- ICS is practically equivalent to the FCS when: $Fo_H\geq 10^{-7}$
</td>
</tr>
<tr>
<td>FLS</td>
<td>[6,9,18]</td>
<td>Finite line heat source $0<z<H$ in a semi-infinite medium. The source is located in $(0,0)$.
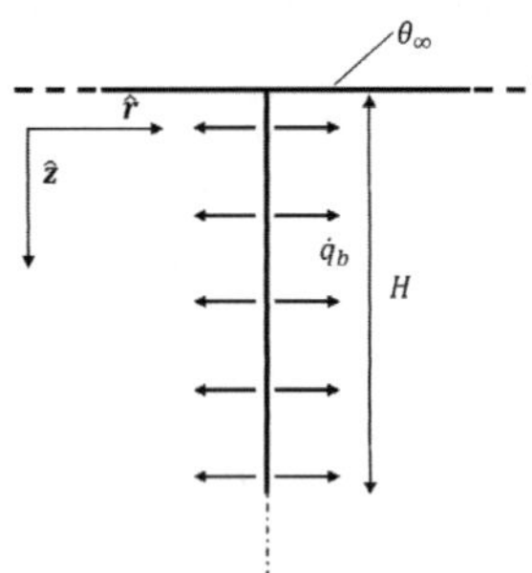
</td>
<td>

$$\Theta(Fo_H,R,Z)=\frac{1}{4\pi}\int_0^1\left[\frac{1}{d/H}erfc\left(\frac{d/H}{2\sqrt{Fo_H}}\right)-\frac{1}{d'/H}erfc\left(\frac{d'/H}{2\sqrt{Fo_H}}\right)\right]dZ'$$

where:

$$d/H=\sqrt{R^2+(Z-Z')^2}\quad d/H=\sqrt{R^2+(Z+Z')^2}$$

Axial-average temperature

$$\bar{\Theta}(Fo_H,R)=\int_0^1\Theta(Fo_H,R,Z)\,dZ=\frac{1}{4\pi}\int_{1/\sqrt{4Fo_H}}^{\infty}\exp\left(-R^2\beta^2\right)\frac{I_{ls}(\beta,0)}{\beta^2}d\beta$$

$$I_{ls}(x,y)=2\cdot ierf(x)+2\cdot ierf(x+2y)-ierf(2x+2y)-ierf(2y)$$

$$ierf(x)=\int_0^x erf(\beta)\,d\beta=x\cdot erf(x)-\frac{1}{\sqrt{\pi}}\left[1-\exp\left(-x^2\right)\right]$$

Steady-state solution (i.e. $Fo_H>Fo_{H,s}$)

$$\left.\begin{aligned}Fo_{H,s}&=-7.83R+3.84R+10\\ \bar{\Theta}_s(R)&=-0.36\log_{10}R-0.13\end{aligned}\right\}\quad 10^{-4}\leq R\leq 10^{-1}$$

$$\Theta_s(R,Z)=\frac{1}{4\pi}\int_0^1\left[\frac{1}{d/H}-\frac{1}{d'/H}\right]dZ'$$

Equivalence with other models
- FLS is practically equivalent to the ILS when: $Fo_H\leq 3.0\times 10^{-4}R^{-0.67},\quad 10^{-4}\leq R\leq 10^{-1}$
- FLS is practically equivalent to the FCS when $Fo_{r_b}\geq 6.3\times 10^{-2}(r/r_b)^2+5.4\times 10^{-2}(r/r_b)+17.8,\quad 1\leq r/r_b\leq 100$
</td>
</tr>
<tr>
<td>FCS</td>
<td>[9,19]</td>
<td>Finite cylindrical heat source $0<z<H$ in a semi-infinite medium. The axis of the source is located in $r=0$.</td>
<td>Analytical solution is not available. The reader can refer to the dimensionless maps in *[9]*.
Steady-state solution (i.e. $Fo_H>Fo_{H,s}$)</td>
</tr>
</table>

IOP Conf. Series: Journal of Physics: Conf. Series **923** (2017) 012056 doi:10.1088/1742-6596/923/1/012056

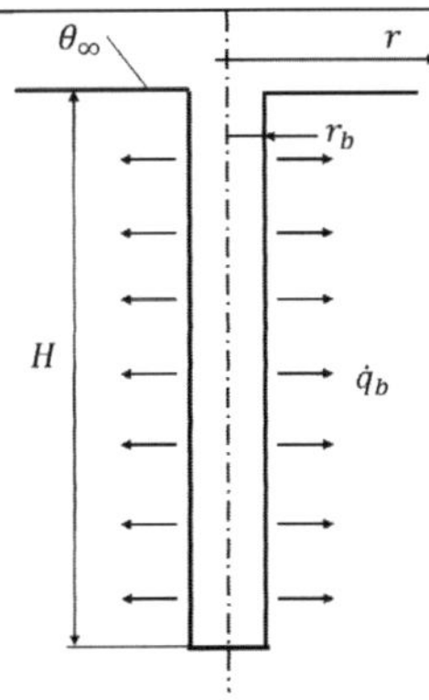

The same of FLS model

Equivalence with other models

- FCS is practically equivalent to the ICS when:

  $Fo_H \geq 10^{-7}$

- FCS is practically equivalent to the FLS when:

  $Fo_{r_b} \geq 6.3 \times 10^{-2} \left(r/r_b\right)^2 + 5.4 \times 10^{-2} \left(r/r_b\right) + 17.8,$

  $1 \leq r/r_b \leq 100$

5. Conclusions

In this work, we reviewed the available analytical solutions for the problem of a heat source embedded in a porous medium, i.e. MILS, MFLS and MICS models. Specifically, we provided a handy compendium of formulae to evaluate the thermal field evolution around a constant heat source at any space and time. The assumptions and the characteristics of each model were presented and discussed. The main references were selected for each model, to provide a fast bibliographic tool to the reader. Purely-conductive models were also reviewed and presented for the same bibliographic purposes.

Together with the general solution of the thermal field, we presented some dimensionless expressions to evaluate the dimensionless time needed to reach the steady-state condition and the corresponding temperature values. We investigated the conditions under which the fluid movement does not produce significant effects in terms of heat transfer. In other words, we presented some dimensionless criteria to evaluate the threshold Péclet numbers at which MILS and MICS are practically equivalent to the ILS and ICS, respectively.

We suggest using the MICS solution to evaluate the temperature of the heat source in the presence of a significant fluid movement. In fact, the MILS model is not able to account for the actual velocity field in proximity of the source, thus it overestimates the heat exchange between the source and the fluid (the reader can refer to [11] for further details). On the other hand, the MILS model can be used to evaluate the temperature at a sufficient distance from the source, namely when the actual radial dimension of the source does not affect the velocity field.

We believe that the present compendium, with possible additional developments on the MFCS model, represents a fast and useful tool for researchers and professional operators, helping the design and the performance analysis of such heat transfer systems, especially when coupled to models of engineering devices – such as ground-coupled borehole heat exchangers [20], energy piles [21], and heat pump units [22] – in lifetime simulations of the overall system [23].

Nomenclature

H	heat source depth [m]
K	permeability [m^{-2}]
$L = \alpha_m / u_m$	characteristic length of advection-diffusion problems [m]
$T = L / u_m$	characteristic time of advection-diffusion problems [s]
c	specific heat [J kg^{-1} K^{-1}]
p	pressure [Pa]
$\dot{q}_b$	linear heat source strength [J s^{-1} m^{-1}]
r	radial coordinate [m]

r_b	heat source radius [m]
t	time [s]
$\mathbf{u}$	Darcy velocity or seepage velocity vector [m s^{-1}]
u	magnitude of the Darcy velocity or seepage velocity [m s^{-1}]
x	x coordinate [m]
y	y coordinate [m]

Greek letters

Γ	generalized incomplete gamma function
ϕ	angular coordinate [rad]
α	thermal diffusivity [m^2 s^{-1}]
β	auxiliary variable [-]
ε_p	porosity [-]
θ	temperature [K]
λ	thermal conductivity [W m^{-1} K^{-1}]
μ	dynamic viscosity [Pa s]
ρ	density [kg m^3]

Dimensionless numbers and groups

$\Theta = \dfrac{(\theta - \theta_\infty)\lambda}{\dot{q}_b}$	dimensionless temperature
$\tau = t/T = Fo_x Pe_x^2$	dimensionless time
τ_s	dimensionless time needed to reach the steady-state condition
$Fo_x = \dfrac{\alpha t}{x^2}$	generic Fourier number referred to the length x.
$Pe_x = \dfrac{x}{L} = \dfrac{ux}{\alpha}$	generic Péclet number referred to the length x
$R = \dfrac{r}{H}$	dimensionless radial coordinate
$\mathrm{Re}_K = \dfrac{\rho_f u K^{0.5}}{\mu}$	local pore Reynolds number
$Z = \dfrac{z}{H}$	dimensionless axial coordinate

Subscripts

∞	undisturbed condition
b	heat source
c	threshold value from a purely-conductive to a Darcy's regime
f	liquid phase
m	volume-averaged/effective quantity
s	solid phase or steady-state quantity

Acronyms

ILS	Infinite line source model
ICS	Infinite cylindrical source model

FLS	Finite line source model
FCS	Finite cylindrical source model
MILS	Moving infinite line source model
MICS	Moving infinite cylindrical source model
MFLS	Moving finite line source model
MFCS	Moving finite cylindrical source model

References

[1] Nield D A and Bejan A 2006 *Convection in Porous Media* (New York, NY: Springer Science+Business Media, Inc)
[2] Thevenin J 1995 *Int. Commun. Heat Mass Transf.* **22** 507–16
[3] Grassi W, Conti P, Schito E and Testi D 2015 *J. Phys. Conf. Ser.* **655** 12003
[4] Layeghi M and Nouri-Borujerdi A 2004 *J. Porous Media* **7** 239–47
[5] Zeng Z, Brown J M B and Vardy A E 1997 *Heat Mass Transf.* **33** 41–9
[6] Carslaw H S and Jeager J C 1959 *Conduction of heat in solids* (Oxford, UK: Clarendon Press)
[7] Layeghi M and Nouri-Borujerdi A 2006 *Int. J. Comput. Methods Eng. Sci. Mech.* **7** 323–9
[8] Kimura S 1990 *J. Geotherm. Res. Soc. Japan* **12** 79-90
[9] Conti P 2016 *Energies* **9** 890
[10] Philippe M, Bernier M and Marchio D 2009 *Geothermics* **38** 407–13
[11] Conti P, Testi D and Grassi W 2017 *Int. J. Heat Mass Tranfer* **117C** 154–66
[12] Bejan A 2003 Porous Media *Heat Transfer Handbook* ed A Bejan and A D Kraus (Hoboken, NJ: John Wiley & Sons Incorporated) chapter 15 pp 1131–80
[13] Buckingham E 1915 *Nature* **96** 396–7
[14] Diao N, Li Q and Fang Z 2004 *Int. J. Therm. Sci.* **43** 1203–11
[15] Sutton M G, Nutter D W and Couvillion R J 2003 *J. Energy Resour. Technol.* **125** 183–9
[16] Molina-Giraldo N, Blum P, Zhu K, Bayer P and Fang Z 2011 *Int. J. Therm. Sci.* **50** 2506–13
[17] Ingersoll L R, Zobel O J and Ingersoll A C 1954 *Heat conduction with engineering, geological and other applications* (New York, NY: McGraw-Hill)
[18] Claesson J and Javed S 2011 *ASHRAE Trans.* **117** 279–88
[19] Priarone A and Fossa M 2016 *Appl. Therm. Eng.* **103** 934–44
[20] Conti P, Testi D and Grassi W 2016 *Appl. Therm. Eng.* **106** 1257–67
[21] Batini N, Rotta Loria A F, Conti P, Testi D, Grassi W and Laloui L 2015 *Appl. Therm. Eng.* **86** 199–213
[22] Casarosa C, Conti P, Franco A, Grassi W and Testi D 2014 *J. Phys. Conf. Ser.* **547** 12006
[23] Conti P 2015 *Proc. 5th Int. Youth Conf. on Energy* (Pisa, IT) (IEEE) pp 8

Dry deposition models for radionuclides dispersed in air: a new approach for deposition velocity evaluation schema

M. Giardina[1], P. Buffa[1], A. Cervone[2], F. De Rosa[2], C. Lombardo[2], M. Casamirra[3]

[1] Department of Energy, Information Engineering and Mathematical Models (DEIM), University of Palermo, Viale delle Scienze, Edificio 6, 90128 Palermo, Italy
[2] Italian Agency for New Technologies, Energy and Sustainable Economic Development (ENEA), Via Martiri di Monte Sole, 4, 40129 Bologna, Italy
[3] Servizio di Radioprotezione di Ateneo, University of Palermo, Viale delle Scienze, Edificio 6, 90128 Palermo, Italy

E-mail: mariarosa.giardina@unipa.it

Abstract. In the framework of a National Research Program funded by the Italian Minister of Economic Development, the Department of Energy, Information Engineering and Mathematical Models (DEIM) of Palermo University and ENEA Research Centre of Bologna, Italy are performing several research activities to study physical models and mathematical approaches aimed at investigating dry deposition mechanisms of radioactive pollutants. On the basis of such studies, a new approach to evaluate the dry deposition velocity for particles is proposed. Comparisons with some literature experimental data show that the proposed dry deposition scheme can capture the main phenomena involved in the dry deposition process successfully.

1. Introduction

As highlighted in ATMES (Atmospheric Transport Model Evaluation Study) report [1], the greatest number of uncertainties in numerical evaluations of pollutant transport and dispersion in air are introduced by the parameterizations both of the source term and of dry or wet deposition processes.

In the nuclear field, in case of severe accident that causes a release of radionuclides in the atmosphere, key challenges are the characterization of the specific type of release, the analysis of the dispersion and deposition phenomena and, consequently, the definition of appropriate corrective actions for the protection of the population in case of these extreme emergency conditions.

Recently, ENEA research center at Bologna, Italy, performed some research activities concerning "Studies in support to emergency preparedness" in case of accident in a nuclear power plant near the border of Italy, in collaboration with the Department of Energy, Information Engineering and Models Mathematicians (DEIM) of the University of Palermo. These activities included the study of physical-mathematical models able to properly simulate the dry deposition flux of radioactive pollutants.

Until now, due to limited understanding of some key aspects of dry deposition processes, there isn't a single accepted theoretical description of the involved phenomena because of the multiplicity and complexity of the fluid-dynamic processes which can influence the deposition rate.

Various experimental campaigns have been performed in international laboratories to evaluate the deposition velocity for different types of pollutants. Nevertheless, it is difficult to obtain a global generalization since the velocity values differ by four orders of magnitude for gases and three orders for particles [2]. These issues limit the possibility to study the dry deposition process with a single modelling approach.

In literature there are different semi-empirical correlations characterized by well-defined criteria for application, and, in this field, parameterizations of the dry deposition velocity cover only some general conditions which may be of interest in nuclear applications. There are some models that can be applied only to one type of surface or canopy configuration, others that can work with different types of surface. The comparison between these two groups however highlights large differences between their predictions.

In this paper a new schema for parameterization of dry deposition velocity for particles is described. The work has involved comparisons with the experimental data obtained in different international laboratories. The results shows that the proposed approach allows to catch, with good agreement, some aspects of the phenomena involved in the dry deposition process for the examined environmental conditions.

2. Description of dry deposition processes

Dry deposition process refers to all phenomena of meteorological, chemical and biological nature that can influence a pollutant flux interacting with a ground surface, without involving the water into the atmosphere. The process is very complex and involves both gaseous pollutants and particulate of various granulometry and density, even if general phenomenological factors involving the two pollutant categories can be very different. The knowledge on particle dry deposition is far from being completely understood due to the complex dependences of deposition on particle size, density, terrain, vegetation, meteorological conditions and chemical species.

The main phenomena that are considered to influence the deposition process, are described as follows:

- transport due to atmospheric turbulence in the lower layer of the Planetary Boundary Layer (PBL) very near to the ground, called Surface Layer (SL). This process is independent of the physical and chemical nature of the pollutant and it depends only on the turbulence level;
- diffusion in the thin layer of air which overlooks the air-ground interface (named quasi-laminar sublayer resistance), where the dominant component becomes molecular diffusion for gasses, Brownian motion for particles and gravity for heavier particles;
- transfer to the ground that exhibits a pronounced dependence on surface type with which the pollutant interacts (i.e. urban context, grass, forest, etc.).

In the case of gas, containment of a pollutant by a surface depends on the surface chemical property, which absorbs, dissolves or involves pollutant in chemical reactions; in case of particles the efficiency of pollutant capture is connected to resuspension and redeposition phenomena that depend both on the surface type and wind velocity. Moreover, the deposition process changes quite a lot over the year, for example due to the seasonal variation of vegetation (with or without leaf) or over the day in connection with meteorological conditions (e.g. influence of temperature on leaf stoma).

3. Dry deposition models

A key concept to study the dry deposition process is the deposition velocity v_d [m/s] (i.e. deposition velocity at a given height z) that links the pollutant vertical flux F_d [g/(m^{2}s)] to the concentration C(z) [g/m^3] measured at quota z [m] above the ground reference level:

$$v_d = \frac{F_d}{C(z)} \quad (1)$$

Considering that the reciprocal of v_d is the overall resistance to mass transfer, the influence of the various phenomena on deposition velocity can be expressed in terms of an electrical analogy. Resistance to the deposition is configured as resistance in parallel and series combinations to describe transfer factor between air and surface.

For gaseous pollutant collection, it is possible to schematize the deposition process as resistances in series, so we can write the relationship:

$$v_d = (r_a + r_b + r_s)^{-1} \quad (2)$$

where r_a is the aerodynamic resistance considering the turbulence phenomenon in SL; r_b is the quasi-laminar sublayer resistance related to the diffusion phenomenon for gas and collisions due to the Brownian motion for particles; and r_s is the surface resistance, which depends on the nature of the receptor ground.

The evaluation of parameters r_a and r_b for gas pollution is reported in [3-7]. The calculation of gas surface resistance r_s is complicated and depends on the primary pathways for uptake such as diffusion through leaf stomata. A revised parameterization that includes a realistic treatment of cuticle and ground resistance in winter (low temperature and snow-covered surfaces) as well as the handling of seasonally-dependent input parameters is reported in [8].

For particle pollutant, in SL region the turbulence acts on particle motion exactly like on gas, however the process is influenced also by gravity. Into the quasi-laminar sublayer, the particle diffusion is influenced especially by Brownian motion. In fact, dusts are deposited also because of collisions with less dense but more numerous particles, such as gasses.

In this case the aerodynamic resistance r_a, the resistance to travel across the quasi-laminar surface layer r_b and the resistance to surface uptake r_s are considered in parallel to a second pathway: the gravitational settling defined as the reciprocal of settling velocity, v_s [9-11].

Seinfeld and Pandis in [11] derived a dry deposition flux relationship based on the assumption that r_s=0, and by equating the vertical fluxes in two layers over a surface to the total resistance.

The velocity v_d is obtained by using the equation reported below:

$$v_d = v_s + (r_a + r_b + r_a r_b v_s)^{-1} \tag{3}$$

where the product r_a r_b v_s represents a virtual resistance.

The settling velocity v_s grows in proportion to the square of particle diameter, d_p, according to the law of Stokes valid for particles with a diameter of up to 50 μm:

$$v_s = \frac{d_p^2 g(\rho_p - \rho) C_c}{18\nu} \tag{4}$$

where g is the acceleration due to gravity, ρ_p the particle density, ρ the air density, ν the air kinematic viscosity and c_c the correction factor evaluated as follows:

$$c_c = 1 + \frac{\lambda_{air}}{d_p}\left(2{,}514 + 0{,}8\, e^{\frac{-0{,}55 d_p}{\lambda_{air}}}\right) \tag{5}$$

with λ_{air} the mean free path of air.

3.1. Estimation of aerodynamic resistance

In [11-14] the parametrization of the aerodynamic resistance is as follows:

$$r_a = \frac{1}{k u_*}\left[\ln\frac{z}{z_0} - \Psi_h\right] \tag{6}$$

with z_0 the height above the displacement plane at which the mean wind becomes zero downward through the surface layer, u_* the shear velocity, k the von Karman constant (generally equal to 0.4).

Brandt et al. in [15] suggest the following relationship for calculating the parameter Ψ_h:

$$\Psi_h = -5\frac{z}{L} \qquad \text{with } \frac{z}{L} > 0 \tag{7}$$

$$\Psi_h = e^{\left\{0{,}598 + 0{,}390 \ln\left(-\frac{z}{L}\right) - 0{,}09\left[\ln\left(-\frac{z}{L}\right)\right]^2\right\}} \qquad \text{with } \frac{z}{L} < 0 \tag{8}$$

with L the Monin-Obukhov length that characterizes the stability of PBL layer in the lower part of atmosphere, computed as follows:

$$L = \frac{u_*^3 c_p \rho \overline{T}}{kgH} \quad (9)$$

where c_p is the specific heat at constant pressure, $\overline{T}$ is the average temperature in SL, and H is the sensible heat.

It is worth to note that Eq. (6) is a general expression that may be used for every pollutant and the parameter r_a depends only on mechanical and convective turbulence levels that are typical of the examined condition.

3.2. Estimation of the quasi-laminar sublayer resistance for particles

Once in the quasi-laminar sublayer, most particle dry deposition models treat surface uptake in terms of inertial impaction by inertial forces, interception, Brownian diffusion and gravitational settling.

The resistance to transfer across the quasi-laminar surface layer, that includes the effects of brownian motion (Schmidt number, Sc) as well as the influence of inertial impaction (Stokes number, St), can be evaluated by using the following relationship [9, 13, 16-18]

$$r_b = \frac{1}{(Sc^{-2/3} + 10^{-3/St})u_*} \quad (10)$$

where Sc is defined as Sc=D/ν, and St is evaluated as:

$$S_t = \frac{v_S}{g}\frac{u_*^2}{\nu} \quad (11)$$

with v_s sedimentation velocity from Eq. (4)

4. Influence of captation processes on particle dry deposition parameterization

A bibliographical analysis of the determination of dry deposition velocities for vegetative canopy (experiment in situ) or derived by wind tunnel measurements demonstrates a substantial number of variations that are more pronounced for forest canopies or in the accumulation range. Comparisons of several models reveal that they differ from each other greatly and the largest uncertainty is for the accumulation mode particles (around 0.1–1.0 micron diameter range).

Although Brownian diffusion and gravitational settling are well-known to dominate small and large particle deposition, the dominant deposition mechanism for particles of intermediate d_p (in the range about of 0.1–1 μm) depends strongly on atmospheric conditions, surface characteristics and particle size.

Depending on the value of the roughness length z_0, the eddy turbulence effect is a major contribution to the total dry deposition velocity for particles in the size range from 0.01 μm to approximately a few micrometers. Above this range, the dry deposition velocity is dominated by the inertial impaction term, and the Brownian diffusion and the eddy turbulence terms can be neglected.

Therefore the deposition of aerosol requires to describe Brownian diffusion phenomena and sedimentation process together with the following processes:

- Interception. This occurs when the particles, that follow the streamlines of the mean flow field, have small inertia. In the vicinity of an obstacle, they are held back because the distance between the particle center and the surface is smaller than half the particle diameter [19].
- Inertial impaction. If the particle inertia is too large, the particle, which is transported by the flow towards an obstacle, cannot follow the flow deviation in the vicinity of the obstacle. Thus it collides with the obstacle and remains on the surface.
- Turbulent impaction. By increasing the particle diameter, the inertial effect influences the deposition

process. However, in this mechanism, the particle has a sufficiently high velocity that turbulent eddies give some of them a transverse "free fight velocity", so that some particles possess sufficient momentum to reach the wall [20-22]. When gravity is in the direction of the flow, the lift force enhances the particle deposition rate. For upward flow, however, the lift force moves the particle away from the wall and reduces the deposition rate [23].

- Rebound. It is thought to influence coarse particle deposition (size typically larger than 5 μm) [24]. From a mechanistic point of view, rebound is related to the kinetic energy of the incident particle, obtained from the particle velocity component normal to the deposit surface, and to the nature of the impact [25-27].

As above said, a review of the existing mechanicistic models emphasizes the wide variety of ways that captation can take place, however, a comparison of two similar scenarios provides large discrepancies, generating values with different order of magnitude for the same particle diameter.

In the framework of the work performed to revise the various model as reported in literature, the comparison with the models from Slinn [16], Zhang et al. [28], and Wesely et al. [29] indicates that the Slinn model underestimates the deposition velocities by a factor of 40 to 100, while the Zhang and Wesely models better represent the experimental measurements, with the measurement/model differences generally lower than 30. In particular, the model of Zhang et al. [28] shows rather better agreement with the observations but tends to overestimate v_d for d_p under approximately 1.0 μm.

4.1. A modified approach for dry deposition as inferred from various parameterization schemes

To evaluate the effects of turbulent impaction processes on dry deposition phenomena of particles the following considerations can be performed. The new schematization for parametrization of particle deposition process, based on the electrical analogy, is depicted in Figure 1.

The resistance r_a is connected in series to two resistances in parallel, that is the resistance r_{bd} which represents the Brownian diffusion and the resistance r_i which allows to treat impaction processes. This last resistance is evaluated by considering two resistances in series: resistance r_{ii} that takes into account the inertial impact condition, and resistance r_{ti} that considers the effects resulting from turbulent impaction. This assumption allows to take into consideration effects on particle concentration coming from both the inertial and turbulent impaction (i.e. reciprocal influence of the two impact processes on dry deposition efficiency). It is worth to note that all the resistances regard processes which influence the flux over the canopy.

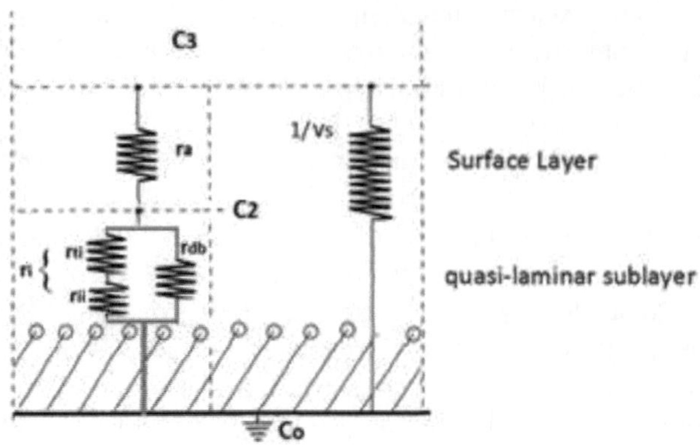

Figure 1. New schematization for parametrization of particle deposition process.

On the basis of the above described hypothesis, it is possible to write the following relationship:

$$v_d = v_s + (r_a + r_i)^{-1} \tag{12}$$

where the resistance r_a is calculated by using Eq.s (6)-(9) and the resistance r_i as follows:

$$\frac{1}{r_i} = \frac{1}{r_{bd}} + \frac{1}{r_{ii}+r_{ti}} \tag{13}$$

where $r_{bd} = 1/(u_* Sc^{-2/3})$, $r_{ii} = 1/(u_* 10^{-3/St})$.

The transport of particles by Brownian diffusion represented as function of $Sc^{-2/3}$ is recommended in [12, 25, 30, 31]. Moreover some Authors [9, 16, 32, 33] have parameterized the inertial impaction by using the term of $10^{-3/St}$. This was also confirmed by the observations conducted by Aluko and Noll [34] for large airborne particles (diameter greater than about 10 μm).

For the calculation of resistance r_{ti} in Eq. (13) the following considerations are taken into account.

As well known, empirical relations of turbulent deposition are typically presented in terms of the dimensionless particle relaxation time τ_+ [35] that, for a spherical particle, is defined as:

$$\tau_+ = \tau \frac{u_*^2}{\nu} \tag{14}$$

where τ is the particle relaxation time that, for a spherical particle, is defined as:

$$\tau = \frac{d_p^2 \rho_p C_c}{18\mu} \tag{15}$$

where μ is the dynamic viscosity of the air.

For turbulent deposition in pipe flows, the following regimes have been observed [32,33]:

- Regime with about $\tau_+ < 0.1$ (very small particles); Brownian diffusion becomes significant and deposition is affected by a combination of Brownian and eddy diffusion. For the Brownian regime the deposition velocity is proportional to $Sc^{-2/3}$;
- Regime with about $0.1 < \tau_+ < 30$; particle motion is strongly dependent on turbulent fluctuation in the fluid flow and the deposition velocity increases with τ_+. For larger particles the deposition velocity can be considered proportional to τ_+ squared.
- Regime with about $\tau_+ > 30$; the particles have large inertia and the effect of turbulence on particle motion is significantly reduced; deposition velocity decreases with increasing particle size.

On the basis of these considerations, the parameter r_{ti} is assumed to be an exponential function of τ_+ as reported below:

$$r_{ti} = \frac{1}{u_*(a\tau_+^b)} \tag{16}$$

with a and b constants.

In Eq. (16) the constants m and n have been evaluated by fitting Eq.s (4), (12), (13) and (16) to the experimental data reported in [36, 37] where the Authors investigated the deposition of aerosol particles in a turbulent vertical pipe flow. The results reported in [36] show that the dimensionless deposition velocity increases linearly with the square of the dimensionless relaxation time for $\tau_+<10$. Furthermore, they found that the particle deposition velocity increases until about $\tau_+ = 30$ and then slightly decreases beyond this value.

The fitting procedure provides for the constants a and b the values of 2 and 1, respectively.

5. Comparison with experimental data

The new scheme for computing the deposition velocity v_d is validated by comparison with experimental data reported in literature for several typology of soil and meteorological conditions. Some of the obtained results are shown in Figures 2-6.

In Figure 2, comparisons with the experimental data reported by Liu and Agarwal in [36], valid for grass surfaces, and Chamberlain in [37], for sticky grass surfaces, are reported. As expected, a good agreement is obtained, comparable in trend and values.

A further comparison has been carried out by using experimental measurements reported in [38], for several surface typologies. The obtained results are reported in Figures 3-6. It is worth to note that Authors in [38] reveal uncertainties on experimental data especially significant for d_p values less than 10 μm.

From the comparison we can say that in case of "sticky wood" surfaces (Figure 3), "sand" (Figure 4) and "plant" (Figure 5) prediction of v_d is good, both in trend and values, for u_* with high values, whereas it is underestimated for u_* with the lowest value within range, especially for "plant" surface (Figure 5).

Note that the experimental data reported in [34, 39, 40] are also shown in Figures 3 and 5. These data refer to experimental activities performed in similar surface conditions and various range of velocity u*.

The comparison between v_d prediction and the experimental data obtained by using very close values of u_* (i.e. $u_* = 0.52$ m/s) shows a good agreement.

By comparison of prediction of v_d with experimental data related to "water" surface (Figure 6), we can see that the model underestimates the experimental data for d_p below 5 μm. In this Figure the experimental data reported in [41] for water surface are also shown. It is worth to note that as a whole the prediction of v_d allows to capture the trend of the experimental data also in case of low values of d_p.

6. Conclusion

A scheme to evaluate particles deposition velocity is proposed. The validation work, carried out by using a significant number of experimental data collected from literature, shows that the proposed model reproduces with accuracy the data. It is worth to note that the correlation obtained with the new approach are able to catch the main phenomenology involved in the deposition process and partially to reduce those singularities that characterize the different models reported in literature that limit the applicability of the model to the entire canopy. However, further research activities are under-way to improve the model with the aim of take into account the rebound phenomenon that influences the coarse particle deposition.

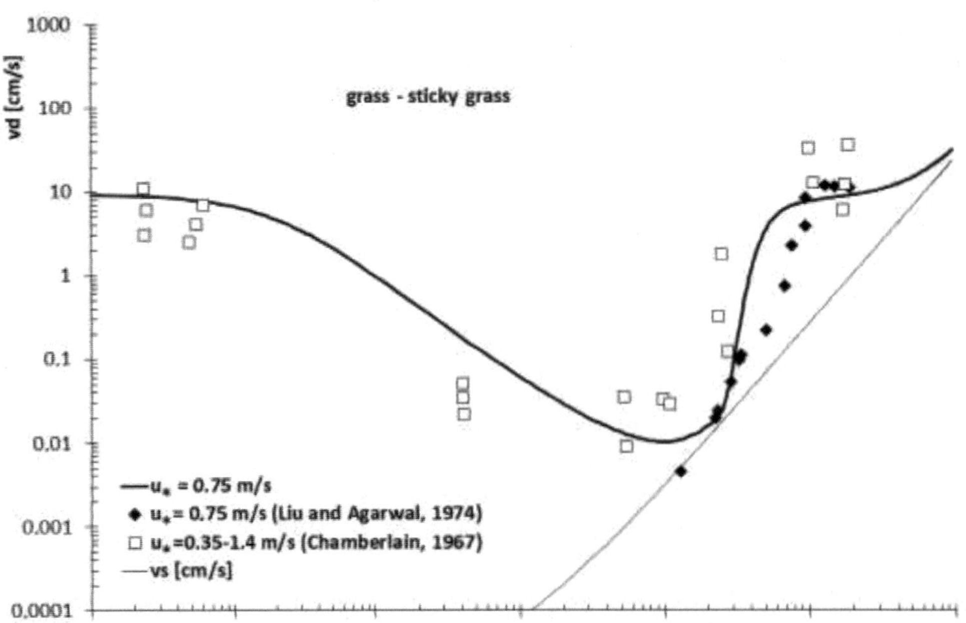

Figure 2. Comparison between the calculated deposition velocity and the experimental data reported in [36, 37] for grass surfaces and sticky artificial grass.

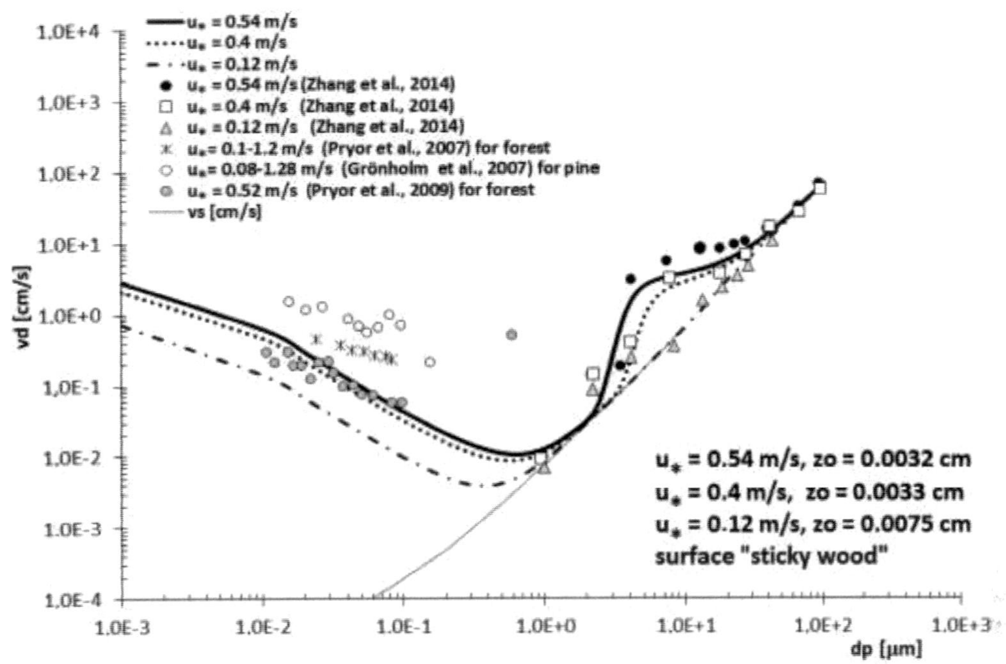

Figure 3. Comparison between the calculated deposition velocity and the measurements reported in [34, 38-40] for "sticky wood" surfaces, "forest" and "pine".

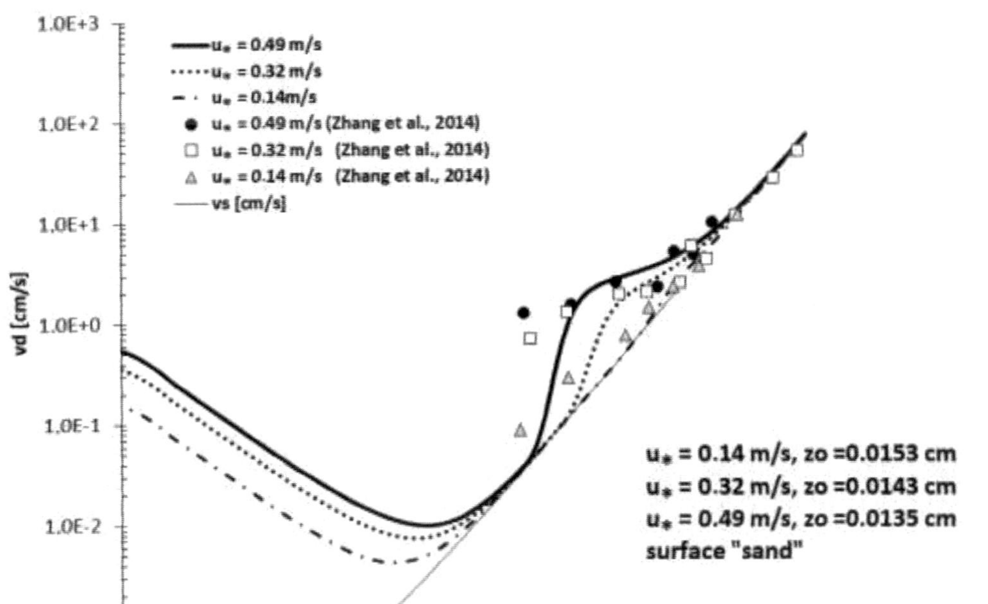

Figure 4. Comparison between the calculated deposition velocity and the experimental data reported in [38] for "sand" surface.

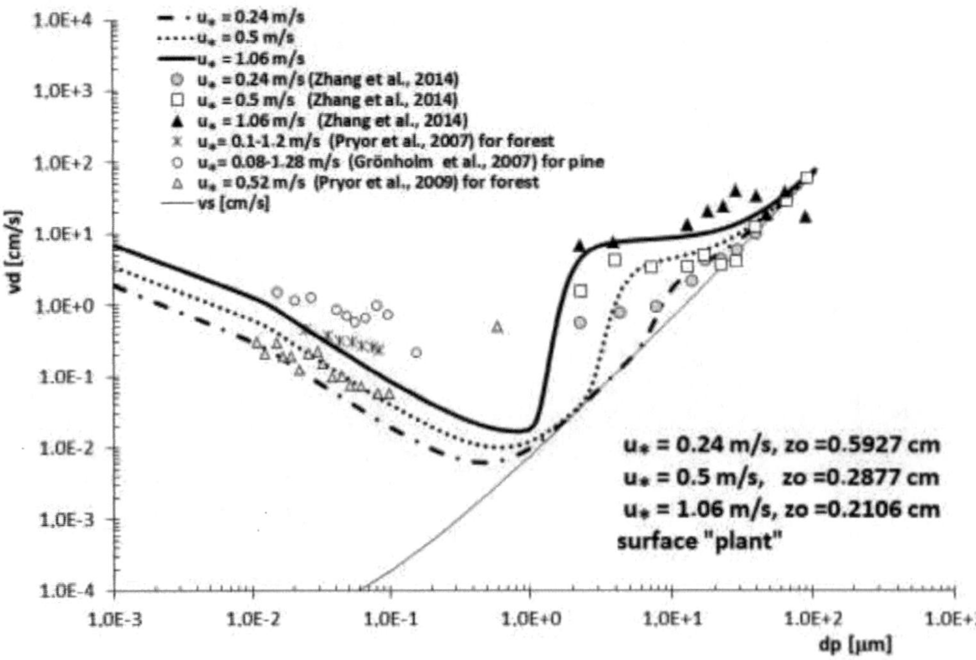

Figure 5. Comparison between the calculated deposition velocity and the experimental data reported in [34, 38-40] for "plant" surface, "forest" and "pine".

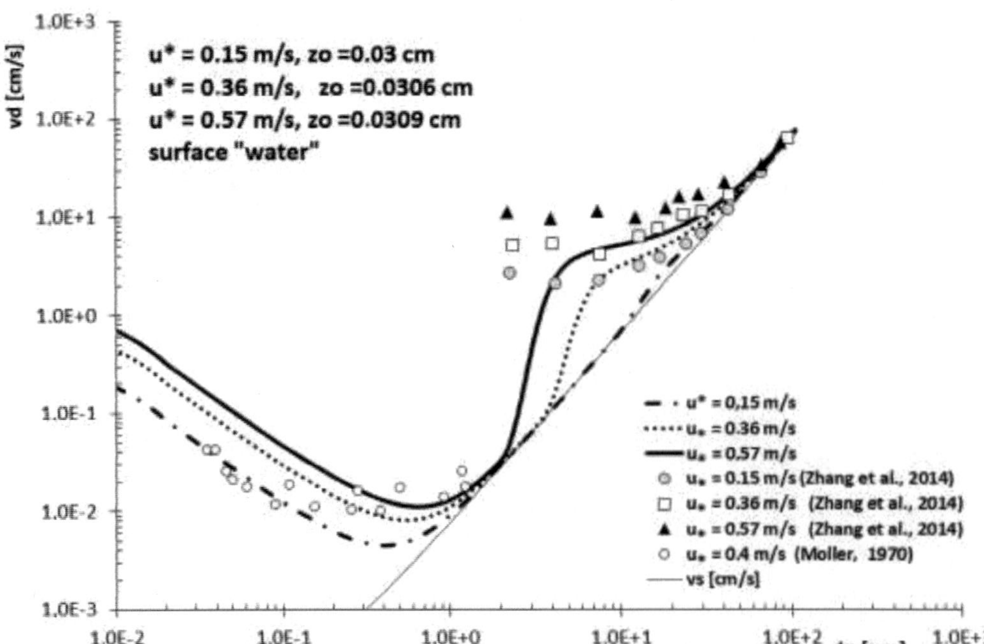

Figure 6. Comparison between the calculated deposition velocity and experimental data reported in [38, 41] for "water" surface.

IOP Conf. Series: Journal of Physics: Conf. Series **923** (2017) 012057 doi:10.1088/1742-6596/923/1/012057

Acknowledgments
The activity has been funded by Italian Minister of Economic Development.

References

[1] Klug W, Graziani G, Grippa G, Pierce D and Tassone C 1992 *The ATMES Report (*London and Klug W, Graziani G, Grippa G, Pierce D and Tassone C 1992 *The ATMES Report (*London and New York: Elsevier Applied Science*)* **1992**.
[2] Sehmel G A 1980 *Atmos. Environ.* **14** 983
[3] Padro J, den Hartog G and Neumann H H 1991 *Atmos. Environ.* **25** 1689
[4] Erisman J W, Van Pul A and Wyers G P *Atmos. Environ.* **28** 2595
[5] Massman W J, Pederson J, Delany A, Grantz D, den Hartog G, Neumann H H, Oncley SP, Pearson R and Shaw R H 1994 *J. Geophys. Res.* **99** 8281
[6] Padro J 1996 *Atmos. Environ.* **30** 2363
[7] Wesely M L, Doskey PV and Shannon J D 2001 *Deposition Parameterizations for the Industrial Source Complex (ISC3) Model, report to U.S. EPA* (USA: Argonne National Laboratory)
[8] Zhang L, Brook J R and Vet R 2003 *Atmos. Chem. Phys.* **3** 2067
[9] Slinn S A and Slinn W G N 1980 *Atmos. Environ.* **14** 1013
[10] Hicks B B Baldocchi D D, Hosker R P, Hutchison B A, Matt D R, McMillen R T and Satterfield LC 1985 *On the Use of Monitores Air Concentrations to Infer Dry Deposition* (NOAA Technical Memorandum ERLO ARLZ141)
[11] Seinfeld J H and Pandis S N 1998 *Atmospheric chemistry and Physics* (New York: John Wiley&Sons)
[12] Wesely M L and Hicks B B 1977 *J. Air Polka. Control Ass.* **27** 1110
[13] Hicks B B 1982 *Critical assessment document on acid deposition, ATDL Contributory file 81/24.* (Oak Ridge Tennessee: Atmospheric turbulence and diffusion laboratory, NOAA)
[14] Slinn W G N, Hasse L, Hicks B B, Hogan A W, Lal D, Liss P S, Munnich K O, Sehmel G A and Vittorri O 1978 *Atmos. Environ.* **12** 2055
[15] Brandt J, Christensen J H and Frohn L M 2002 *Atmos. Chem. Phys.* **2** 397
[16] Slinn W G N 1982 *Atmos. Environ.* **16** 1785
[17] Pleim J, Venkatram A and Yamartino R 1984 *ADOM/TADAM model development program. ERT P-b980-520* (OME AES of Canada and the Umweltbundesamt, West Germany)
[18] Binkowski F S and Shankar U 1995 *J. Geophys. Res.* **100**, 26191
[19] Fuchs N A 1964 *The Mechanics of Aerosols* (New York: Pergamon Press)
[20] Epstein N 1997 *Exp. Therm. Fluid Sci.* **14** (4) 323
[21] Kor P and Kharrat R 2016 *Modeling of asphaltene particle deposition from turbulent oil flow in tubing: Model validation and a parametric study. Petroleum.* **2** (4) 393
[22] Almohammed N and Breuer M 2016 *Int. J. of Multiphase Flow* **85** 142
[23] Chen Q and Ahmadi G 1997 *J. Aerosol Sci.* **28** (5) 789
[24] Clough W S 1975 *Atmos. Environ.* **9** 1113
[25] Paw U K T 1983 J. of Colloid Interface Sci. **93** 442
[26] Paw U K T and Braaten D A 1992 *Aerosol Sci. and Tech.* **17** 278
[27] Giorgi F 1986 *J. Geophjx Res.* **91** 9794
[28] Zhang L, Gong S, Padro J and Barrie L 2001 *Atmos. Environ,* **35** 549
[29] Wesely M L, Cook, D R, Hart R L, Speer R E 1985 *J. of Geophysical Res.* **90** 2131
[30] Hicks B B, Baldocchi D D, Meyers T P, Hosker R P and Matt D R 1987 *Water Air Soil Pollut.* **36** 311
[31] Pryor S C , Barthelmie R J, Spaulding A M, Larsen S E and Petroff A 2009 J. of Geophysical Res. **114** D18212
[32] Guha A 1997 J. of Aerosol Sci. **28**(8) 1517
[33] Guha A 2008 *Annual Review of Fluid Mechanics* **40** 311
[34] Aluko O and Noll K E 2006 *Aerosol Sci. Technol.* **40** 503

[35] Hinds W C 1999 *Aerosol technology - properties, behavior, and measurement of airborne particles. Hoboken* (NJ: Wiley-Interscience)
[36] Liu B Y H and Agarwal J K 1974 *J. Aerosol Sci.* **5** 145
[37] Chamberlain A C. 1967 *Proc. Roy. Soc. Lnd.* **A 296** 45
[38] Zhang J, Shao Y and Huang N *Atmos. Chem. Phys.* **14** 8869
[39] Pryor S C, Gallagher M, Sievering H, Larsen S E, Barthelmie R J, Birsan F, Nemitz E, Rinne J, Kulmala M, Gro Nholm T, Taipale R and Vesala T 2007 *Review of measurement and modelling results of particle atmosphere–surface exchange* **60** (1) 42
[40] Grönholm T, Aalto P P, Hiltunen V, Rannik Ü, Rinne J, Laakso L, Hyvönen S, Vesala T and Kulmala M 2007 *Measurements of aerosol particle dry deposition velocity using the relaxed eddy accumulation technique* Tellus B **59** 381
[41] Möller U, Schumann G 1970 *Mechanisms of transport from the atmosphere to the Earth's surface. Oceans and Atmospheres* **75** 3013